FINITE MATHEMATICS
& ITS
APPLICATIONS

NINTH EDITION

Larry J. Goldstein
Goldstein Educational Technologies

David I. Schneider
University of Maryland

Martha J. Siegel
Towson University

PEARSON

Prentice
Hall

Upper Saddle River, NJ 07458

Library of Congress Cataloging-in-Publication Data

Goldstein, Larry Joel.
  Finite mathematics & its applications/Larry J. Goldstein, David I. Schneider, Martha J.
Siegel—9th ed.
    p.   cm.
  ISBN 0-13-187364-4
    1. Mathematics. I. Title: Finite mathematics & its applications. II. Schneider, David I.
III. Siegel, Martha J. IV. Title

QA39.2.G643 2007
510--dc22   2005056570

Executive Aquisitions Editor: *Petra Recter*
Editor in Chief: *Sally Yagan*
Project Manager: *Michael Bell*
Production Editor: *Lynn Savino Wendel*
Editorial Assistant/Supplements Editor: *Joanne Wendelken*
Senior Managing Editor: *Linda Mihatov Behrens*
Executive Managing Editor: *Kathleen Schiaparelli*
Manufacturing Buyer: *Maura Zaldivar*
Manufacturing Manager: *Alexis Heydt-Long*
Director of Marketing: *Patrice Jones*
Marketing Assistant: *Jennifer De Leeuwerk*
Art Director: *John Christiana*
Interior Designer: *Kenny Beck*
Cover Designer: *John Christiana*
Art Editor: *Thomas Benfatti*
Creative Director: *Juan R. López*
Director of Creative Services: *Paul Belfanti*
Cover Art: *Origami by Ilene Kahn*
Art Studio: *Interactive Composition Corporation/Laserwords*
Compositor: *Dennis Kletzing*

Printed in the United States of America
6   5   4   3   2

ISBN 0-13-187364-4

Pearson Education LTD., *London*
Pearson Education Australia PTY, Limited, *Sydney*
Pearson Education Singapore, Pte. Ltd
Pearson Education North Asia Ltd, *Hong Kong*
Pearson Education Canada, Ltd., *Toronto*
Pearson Educación de Mexico, S.A. de C.V.
Pearson Education—Japan, *Tokyo*
Pearson Education Malaysia, Pte. Ltd

Contents

Preface

THIS work is the ninth edition of our text for the traditional finite mathematics course taught to first- and second-year college students, especially those majoring in business and the social and biological sciences. Finite mathematics courses exhibit tremendous diversity with respect to both content and approach. Therefore, in developing this book, we incorporated a wide range of topics from which an instructor may design a curriculum, as well as a high degree of flexibility in the order in which the topics may be presented. For the mathematics of finance, we even allow for flexibility in the approach of the presentation.

In this edition, we attempt to maintain our popular student-oriented approach throughout and, in particular, by the following features:

Applications

We provide realistic applications that illustrate the uses of finite mathematics in other disciplines. The reader may survey the variety of applications by referring to the Index of Applications. Wherever possible, we attempt to use applications to motivate the mathematics. For example, the concept of linear programming is introduced in Chapter 3 via a discussion of production options for a factory with labor limitations.

Examples

We include 350 worked examples, many more than is customary. Furthermore, we include computational details to enhance comprehension by students whose basic skills are weak.

Exercises

The 2,860 exercises comprise about one-quarter of the book, the most important part of the text in our opinion. The exercises at the ends of the sections are usually arranged in the order in which the text proceeds, so that homework assignments may be made easily after only part of a section is discussed. Interesting applications and more challenging problems tend to be located near the ends of the exercise sets. Supplementary exercises at the end of each chapter amplify the other exercise sets and provide cumulative exercises that require skills acquired from earlier chapters. Answers to the odd-numbered exercises are included at the back of the book.

Practice Problems

The practice problems are a popular and useful feature of the book. They are carefully selected exercises located at the end of each section, just before the exercise set. Complete solutions follow the exercise set. The practice problems often focus on points that are potentially confusing or are likely to be overlooked. We recommend that the reader seriously attempt to do the practice problems and study their solutions before moving on to the exercises.

Use of Technology

Although the use of technology is optional for this text, many of the topics can be enhanced with graphing calculators, electronic spreadsheets, and mathematical software. Whenever relevant, we explicitly show the student how to use graphing calculators and electronic spreadsheets effectively to assist in understanding the fundamental concepts of the course. In addition, the text contains appendices on the fundamentals of using graphing calculators and electronic spreadsheets. The powerful mathematical software Explorations in Finite Mathematics is packaged with each Student Solutions Manual. Many sections contain specially designed technology exercises intended to be solved with one of these technologies.

In our discussions of graphing calculators, we specifically refer to the TI-83 family of calculators and the TI-89 since these are the most popular graphing calculators. Therefore, most students will have a book customized to their calculator. Students with other graphing calculators can consult their guidebooks to learn how to make adjustments. Had the calculator material been written generically, *every* student would have to make adjustments. For the same reasons, the discussions of electronic spreadsheets refer to Microsoft Excel.

Examples from Professional Exams

We have included questions similar to those found on CPA and GMAT exams to further illustrate the relevance of the material in the course. These multiple-choice questions are identified with the notation PE. Examples of this can be found in Chapter One, pages 38–39.

Review of Fundamental Concepts

Near the end of each chapter is a set of questions that help the student recall the key ideas of the chapter and focus on the relevance of these concepts.

Chapter Summaries

Each chapter contains a detailed summary of the important definitions and results from the chapter.

Chapter Tests

Each chapter has a sample test that can be used by the student to help determine if he or she has mastered the important concepts of the chapter.

Chapter Projects

Most chapters have extended projects that can be used as in-class or out-of-class group projects, or special assignments. These projects develop interesting applications or enhance key concepts of the chapters. The Chapter two project, entitled "Population Dynamics," focuses on how logging directly affects the spotted owl population. This environmental application is one that will hold relevance for students.

New in This Edition

Among the changes in this edition, the following are the most significant.

1. *Approximately 15% of the exercises are new or have been revised.* We have added many new exercises and have updated the real-world data appearing in exercises and examples.

2. *Now Try Exercises.* Most of the examples in the book are followed with a parallel odd-numbered exercise. This allows students to immediately strengthen their understanding of the concepts presented in the examples.

3. *Circled Exercises.* When the number for an exercise is circled, the complete solution of the exercise is given in the back of the book. Circled exercises are usually a little different than the other exercises and the inclusion of full solutions allows students to quickly confirm their understanding.

4. *Key Formulas.* A new section has been added near the end of each chapter presenting the formulas developed in the chapter. These lists of formulas will be helpful to students as they study for tests.

5. *Captions.* Every example and applied exercise has been labeled with a caption to make it more easily identifiable.

6. *Conceptual Exercises.* These challenging exercises probe fundamental concepts, often involve verbalization of ideas, and usually do not have a numerical answer. An example of this can be found in Chapter 6, page 340.

7. *Calculator Upgrade.* Although graphing calculators are optional, many students have calculators from high school. The discussion of the use of calculators has been upgraded by including the TI-89 along with the previous discussion on the use of the TI-83 family of calculators. The material on the TI-82 was removed.

8. *Logic Circuits.* A section on logic circuits has been added to the logic chapter (Chapter 12). This is one of the most exciting applications of pure mathematics.

9. *Graph Theory.* The chapter on graph theory (Chapter 13 of the 8th edition) has been dropped, but will be available on the Web.

Minimal Prerequisites

Because of great variation in student preparation, we keep formal prerequisites to a minimum. We assume only a first year of high school algebra. Furthermore, we review, as needed, those topics that are typically weak spots for students.

Topics Included

This edition has more material than can be covered in most one-semester courses. Therefore, the instructor can structure the course to the students' needs and interests. The book divides naturally into four parts. The first part consists of linear mathematics: linear equations, matrices, and linear programming (Chapters 1-4); the second part is devoted to probability and statistics (Chapters 5-7); the third part covers topics utilizing the ideas of the other parts (Chapters 8-10); and the fourth part explores key topics from discrete mathematics that are sometimes included in the modern finite mathematics curriculum (Chapters 11 and 12). We prefer to begin with linear mathematics since it makes for a smooth transition from high school mathematics and leads quickly to interesting applications, especially linear programming. Our preference notwithstanding, the instructor may begin this book with Chapter 5 (Sets and Counting) and then do either the linear mathematics or the probability and statistics.

Acknowledgments

While writing this book, we have received assistance from many persons. And our heartfelt thanks go out to them all. Especially, we should like to thank the following reviewers, who took the time and energy to share their ideas, preferences, and often their enthusiasm, with us.

Reviewers

ALABAMA Stephen H. Brown, Auburn University   **ARIZONA** Joan McCarter, Arizona State University; Donald E. Myers, University of Arizona, Tempe **CALIFORNIA** Sam Councilman, California State University, Long Beach; Ralph James, California State University at Stanislaus; Charles J. Miller, Foothill Community College; Hugh Walters, Contra Costa College; Peter Williams, California State University at San Bernardino   **COLORADO** William Ramaley, Fort Lewis College   **CONNECTICUT** James F. Hurley, University of Connecticut **FLORIDA** Manoug Manougian, University of South Florida **GEORGIA** Cynthia Wilson, Life College   **ILLINOIS** William D. Blair, Northern Illinois University; Paul Boisvert, Robert Morris College; Hiram Paley, Univeristy of Illinois, Urbana-Champaign; Linda Schultz, McHenry County College; Martin C. Tangora, University of Illinois, Chicago   **INDIANA** Peter Majumdar, Indiana-Purdue University at Fort Wayne; Michael McLane, Purdue University; Juan Migliore, University of Notre Dame; Joseph Stampfli, Indiana University; Robin G. Symonds, Indiana University at Kokomo **IOWA** Juan Gatica, University of Iowa; Philip Kutzko, University of Iowa **KENTUCKY** Bart Braden, Northern Kentucky University; Rebecca Wells, Henderson Community College **MARYLAND** Brenda Bloomgarden, Chesapeake College; James P. Coughlin, Towson University; Gary W. Frick, The College of Southern Maryland; Eric Heinz, Catonsville Community College; Russell Hendel, Towson University; Elizabeth Teles, Montgomery College **MAS-SACHUSETTS** Richard Porter, Northeastern University; Arthur Rosenthal, Salem State University; Tan Vovan, Suffolk University   **MICHIGAN** D.R. Dunninger, Michigan State University; Dennis Pence, Western Michigan University; Randy Schwartz, Schoolcraft College   **MISSOURI** Robert Carmignani, University of Missouri; Ruthanne Harre, Missouri Valley College; James C. Thorpe, University of Missouri, St. Louis   **NEW HAMPSHIRE** Larry G. Blaine, Plymouth State College; Jeanette L. McGillicuddy, River College **NEW YORK** Karen J. Edwards, Paul Smith's College; Theodore Faticoni, Fordham University; Henry J. Ricardo, Medgar Evers College of the City University of New York; David C. Vella, Skidmore College; Chungkuang Wang, Fashion Institute of Technology   **NORTH CAROLINA** Carl D. Meyer, Jr., North Carolina State University   **OKLAHOMA** Thomas J. Hill, University of Oklahoma, Norman   **OREGON** Donald A. Jones, Oregon State University **PENNSYLVANIA** Josephine Battaglia, Pennsylvania State University; Frank Warner, University of Pennsylvania   **SOUTH CAROLINA** Robert Gusalnick, University of South Carolina   **TEXAS** Roger Osborn, University of Texas, Austin; Mary Pearce, Austin Community College   **VERMONT** Karla Karstens, University of Vermont   **VIRGINIA** Eric Chandler, Randolph-Macon Womens College; Richard Pellerin, Northern Virginia Community College **WISCONSIN** Karl A. Beres, Ripon College; Richard O'Malley, University of Wisconsin at Milwaukee; Phil Steitz, Beloit College; Elisabeth Schuster, DeVry Institute of Technology; Barry Cipra

We also thank the many people at Prentice Hall who have contributed to the success of this book. We appreciate the efforts of the production, art, manufacturing, marketing, and sales departments. We are grateful to William Radulovich and Hossein Hamadani for their careful creation and thorough checking of solutions to exercises. Our sincere thanks to production editor Lynn Savino Wendel for her most able work. We would also like to thank Erica O'Leary and Amanda Lubell for their assistance throughout the revision of the book. The authors wish to extend special thanks to our editor Petra Recter and our typesetter Dennis Kletzing. Petra's help in planning and executing this new edition was greatly appreciated. Dennis' considerable skills and congenial manner made for an uncomplicated and pleasant production process.

If you have comments or suggestions, we would like to hear from you. We hope that you enjoy using this book as much as we have enjoyed writing it.

Larry J. Goldstein
larrygoldstein@comcast.net

David I. Schneider
dis@math.umd.edu

Martha J. Siegel
msiegel@towson.edu

Student and Instructor Resources

Student Resources

Printed Resources

Student Study-Pack

- Text and Student Study Pack Bundle ISBN: 0-13-175064-X
- Student Study Pack Stand-alone ISBN: 0-13-223200-6

Everything a student needs to succeed in one place. Provided at no charge when packaged with the book or available for purchase stand-alone. The **Student Study-Pack** contains:

➢ **Student Solutions Manual**
Fully worked solutions to odd-numbered exercises

➢ **CD Lecture Videos**
A comprehensive set of topical videos, in which examples for each topic are worked out by an instructor. The videos provide excellent support for students who require additional assistance, for distance learning and self-paced programs, or for students who missed class.

➢ **Pearson Tutor Center**
Tutors provide one-on-one tutoring for any problem with an answer at the back of the book. Students access the Tutor Center via toll-free phone, fax, or e-mail.

➢ **Explorations in Finite Mathematics**
Built from the ground up by David Schneider (University of Maryland), *Explorations in Finite Mathematics* enhances the understanding of many of the topics covered in the text through the visualization of said topics. Part one of this software covers finite mathematics topics involving linear equations, linear inequalities, and matrices. Students will be able to perform routines such as Gaussian elimination, matrix operations and inversions, the simplex method, and Markov chains. Part two contains routines for Venn diagrams, computation of combinations, permutations and factorials, statistical analysis of data, financial calculations and much more. Find *Explorations in Finite Mathematics* in MyMathLab and on the CD Lecture Videos.

Internet Resources

MyMathLab

- Text and MyMathLab Bundle ISBN: 0-13-147894-X
- MyMathLab Stand-alone ISBN: 0-13-223233-2

MyMathLab is a text-specific, online course that provides unlimited tutorial exercises correlated to the exercises in the text, a multimedia textbook with links to learning aids, such as videos and student supplements in electronic format—the student solutions manual, PowerPoint presentations and much more! Since 2001, over one million students have done better in mathematics with **MyMathLab**'s dependable and easy-to-use online homework, guided solutions, multimedia, tests, and e-books.

MathXL®

- 12-month MathXL and Text Bundle ISBN: 0-13-238720-4
- 24-month MathXL and Text Bundle ISBN: 0-13-238721-2
- 12-month Stand-alone ISBN: 0-13-147892-3
- 24-month Stand-alone ISBN: 0-13-147897-4

MathXL offers unlimited self-paced tutorials, practice-guided mathematical instructions, and multimedia learning aids. **MathXL** allows students to learn and practice the concepts so they will succeed. **MathXL** also creates personalized study plans based on the student's test results and offers unlimited (algorithmically generated) practice exercises correlated directly to the exercises in the textbook.

MathXL CD (0-13-187391-1)
Everything found in **MathXL** but on CD! **MathXL** offers unlimited self-paced tutorials, guided mathematical instructions, and multimedia learning aids. **MathXL** allows students to learn and practice the concepts so they will succeed. **MathXL** also creates personalized study plans based on the student's test results and offers unlimited (algorithmically generated) practice exercises correlated directly to the exercises in the textbook.

Instructor Resources

Instructor Resource Center
All instructor resources can be downloaded from **www.prenhall.com** (then search for Goldstein, *Finite Math*, and click on "Instructor Resources" in left column). This is a password protected site that requires instructors to set up an account or, alternatively, instructor resources can be ordered from your Prentice Hall sales representative.

Instructor's Solutions Manual (ISBN: 0-13-187385-7)
Fully worked solutions to every textbook exercise.

TestGen (ISBN: 0-13-187387-3)
Test-generating software—create tests from textbook section objectives. Questions are algorithmically generated, allowing for unlimited test versions. Instructors may also edit problems or create their own.

PowerPoint Lecture Slides
Fully editable and printable slides that follow the textbook. Use during lecture or post to a Web site in an online course. For download from the Instructor Resource Center and available within MyMathLab.

MyMathLab
Powered by CourseCompass, the course management system, and **MathXL**, the online homework, tutorial and assessment system, **MyMathLab** is a text-specific, customizable course that allows you to deliver all or a portion of your course online. You can create and assign online homework and tests that are automatically graded and tightly correlated to the text. Manage your students' results in the online gradebook—designed specifically for mathematics—which automatically tracks students' homework and test results and gives you control over how to calculate

final grades. You can customize your course to meet your course objectives. In addition to the e-book, you gain access to all instructor and student supplements electronically—PowerPoint Presentations, solutions manuals, videos, software, etc. For added flexibility, instructors can import TestGen tests. Request an access code from your local PH Sales Representative.

MathXL®

Offered as a standalone product, on CD or as a component of **MyMathLab** (online), **MathXL** is the powerful online homework, tutorial and assessment system designed to make teaching your course easier and help your students succeed! With **MathXL**, instructors can create, edit, and assign online homework and tests and track all student work in the online class gradebook.

Explorations in Finite Mathematics

Built from the ground up by David Schneider (University of Maryland), *Explorations in Finite Mathematics* enhances the understanding of many of the topics covered in the text through the visualization of said topics. Part one of this software covers finite mathematics topics involving linear equations, linear inequalities, and matrices. Students will be able to perform routines such as Gaussian elimination, matrix operations and inversions, the simplex method, and Markov chains. Part two contains routines for Venn diagrams, computation of combinations, permutations and factorials, statistical analysis of data, financial calculations and much more. Find *Explorations in Finite Mathematics* in **MyMathLab** and the **CD Lecture Videos**.

Linear Equations and Straight Lines

1.1 Coordinate Systems and Graphs
1.2 Linear Inequalities
1.3 The Intersection Point of a Pair of Lines
1.4 The Slope of a Straight Line
1.5 The Method of Least Squares

ANY applications considered later in this text involve linear equations and their geometric counterparts—straight lines. So let us begin by studying the basic facts about these two important notions.

1.1 Coordinate Systems and Graphs

Often we can display numerical data by using a *Cartesian coordinate system* on either a line or a plane. We construct a Cartesian coordinate system on a line by choosing an arbitrary point O (the *origin*) on the line and a unit of distance along the line. We then assign to each point on the line a number that reflects its directed distance from the origin. Positive numbers refer to points on the right of the origin, negative numbers to points on the left. In Fig. 1 we have drawn a Cartesian coordinate system on the line and have labeled a number of points with their corresponding numbers. Each point on the line corresponds to a number (positive, negative, or zero). Conversely, every number corresponds to a point on the line.

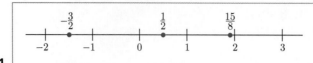

Figure 1

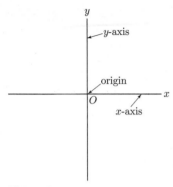

Figure 2

A Cartesian coordinate system may be used to numerically describe points on a line. In a similar fashion, we can construct a Cartesian coordinate system to numerically locate points on a plane. Such a system consists of two perpendicular lines called the *coordinate axes*. These lines are usually drawn so that one is horizontal and one is vertical. The horizontal line is called the *x-axis*, the vertical line the *y-axis*. Their point of intersection is called the *origin* (Fig. 2). Each point of the plane is identified by a pair of numbers (a, b). The first number, a, tells the number of units from the point to the y-axis (Fig. 3). When a is positive, the point is to the right of the y-axis; when a is negative, the point is to the left of the y-axis. The second number, b, gives the number of units from the point to the x-axis (Fig. 4). When b is positive, the point is above the x-axis; when b is negative, the point is below. The numbers a and b are called, respectively, the x- and *y-coordinates of the point*. In order to plot the point (a, b), begin at the origin and move a units in the x-direction and then b units in the y-direction.

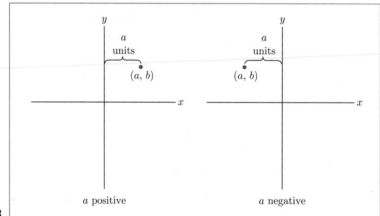

Figure 3

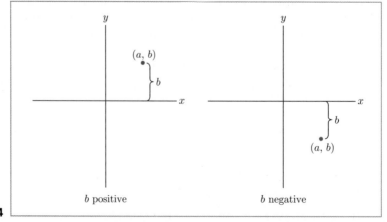

Figure 4

EXAMPLE 1	**Plotting points** Plot the following points.

(a) $(2,1)$ (b) $(-1,3)$ (c) $(-2,-1)$ (d) $(0,-3)$

Solution (a)

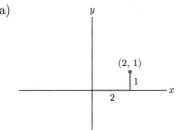

(b)

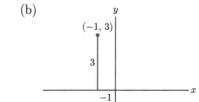

(c)

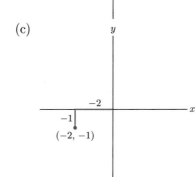

(d)

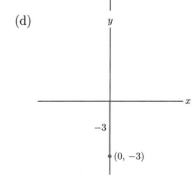

Now Try Exercise 1

In many applications one encounters equations expressing relationships between variables x and y. Some typical equations are

$$y = 2x - 1$$
$$5x^2 + 3y^3 = 7$$
$$y = 2x^2 - 5x + 8.$$

To any equation in x and y one can associate a certain collection of points in the plane. Namely, the point (a, b) belongs to the collection provided that the equation is satisfied when we substitute a for each occurrence of x and substitute b for each occurrence of y. This collection of points is usually a curve of some sort and is called the *graph of the equation*.

EXAMPLE 2	**Solution of an equation** Are the following points on the graph of the equation $8x - 4y = 4$?

(a) $(3,5)$ (b) $(5,17)$

Solution (a) Substitute 3 for each occurrence of x and 5 for each occurrence of y in the equation.

$$8x - 4y = 4$$
$$8 \cdot 3 - 4 \cdot 5 = 4$$
$$24 - 20 = 4$$
$$4 = 4$$

Since the equation is satisfied, the point $(3,5)$ is on the graph of the equation.

(b) Replace x by 5 and y by 17 in the equation.

$$8x - 4y = 4$$
$$8 \cdot 5 - 4 \cdot 17 = 4$$
$$40 - 68 = 4$$
$$-28 = 4$$

The equation is clearly not satisfied, so the point $(5, 17)$ is *not* on the graph of the equation.

Now Try Exercises 11 and 13   ■

EXAMPLE 3   **Graph of an equation** Sketch the graph of the equation $y = 2x - 1$.

Solution First find some points on the graph by choosing various values for x and determining the corresponding values for y:

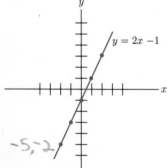

x	y
-2	$2 \cdot (-2) - 1 = -5$
-1	$2 \cdot (-1) - 1 = -3$
0	$2 \cdot 0 - 1 = -1$
1	$2 \cdot 1 - 1 = 1$
2	$2 \cdot 2 - 1 = 3$

Figure 5

Thus the points $(-2, -5)$, $(-1, -3)$, $(0, -1)$, $(1, 1)$, and $(2, 3)$ are all on the graph. Plot these points (Fig. 5). It appears that the points lie on a straight line. By taking more values for x and plotting the corresponding points, it is easy to become convinced that the graph of $y = 2x - 1$ is indeed a straight line.

Now Try Exercise 27   ■

EXAMPLE 4   **Graph of an equation** Sketch the graph of the equation $x = 3$.

Solution It is clear that the x-coordinate of any point on the graph must be 3. The y-coordinate can be anything. So some points on the graph are $(3, 0)$, $(3, 5)$, $(3, -4)$, $(3, -2)$. Again the graph is a straight line (Fig. 6).

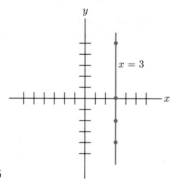

Figure 6

Now Try Exercise 29   ■

EXAMPLE 5   **Graph of an equation**   Sketch the graph of the equation $x = a$, where a is any given number.

Solution   The x-coordinate of any point on the graph must be a. Reasoning as in Example 4, the graph is a vertical line a units away from the y-axis (Fig. 7). (Of course, if a is negative, then the line lies to the left of the y-axis.)

Now Try Exercise 33

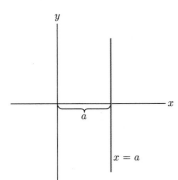

Figure 7

The equations in Examples 3 to 5 are all special cases of the general equation

$$cx + dy = e,$$

corresponding to particular choices of the constants c, d, and e. Any such equation is called a *linear equation* (in the two variables x and y).

The *standard form* of a linear equation is obtained by solving for y if y appears and for x if y does not appear. In the former case the standard form looks like

$$y = mx + b \qquad (m, b \text{ constants}),$$

whereas in the latter it looks like

$$x = a \qquad (a \text{ constant}).$$

EXAMPLE 6   **Standard form of a linear equation**   Find the standard form of the following equations.

(a) $8x - 4y = 4$ (b) $2x + 3y = 3$ (c) $2x = 6$

Solution   (a) Since y appears, we obtain the standard form by solving for y in terms of x:

$$8x - 4y = 4$$
$$-4y = -8x + 4$$
$$y = 2x - 1.$$

Thus the standard form of $8x - 4y = 4$ is $y = 2x - 1$—that is, $y = mx + b$ with $m = 2$, $b = -1$.

(b) Again y occurs, so we solve for y:

$$2x + 3y = 3$$
$$3y = -2x + 3$$
$$y = -\tfrac{2}{3}x + 1.$$

So the standard form is $y = -(2/3)x + 1$. Here $m = -2/3$ and $b = 1$.

(c) Here y does not occur, so we solve for x:

$$2x = 6$$
$$x = 3.$$

Thus the standard form of $2x = 6$ is $x = 3$—that is, $x = a$, where a is 3.

Now Try Exercise 19

We have seen that any linear equation has one of the two standard forms $y = mx + b$ and $x = a$. From Example 5, the graph of $x = a$ is a vertical line, a units from the y-axis. What can be said about the graph of $y = mx + b$? In Example 3, we saw that the graph is a straight line in the special case $y = 2x - 1$. Actually, the graph of $y = mx + b$ is always a straight line. To sketch the graph, we need only locate two points. Two convenient points to locate are the *intercepts*, the points where the line crosses the x- and y-axes. When x is 0, $y = m \cdot 0 + b = b$. Thus $(0, b)$ is on the graph of $y = mx + b$ and is the y-intercept of the line. The x-intercept is found as follows: A point on the x-axis has y-coordinate 0. So the x-coordinate of the x-intercept can be found by setting $y = 0$—that is, $mx + b = 0$—and solving this equation for x.

EXAMPLE 7 **Graph of a linear equation** Sketch the graph of the equation $y = 2x - 1$.

Solution Here $m = 2$, $b = -1$. The y-intercept is $(0, b) = (0, -1)$. To find the x-intercept, we must solve $2x - 1 = 0$. But then $2x = 1$ and $x = 1/2$. So the x-intercept is $(1/2, 0)$. Plot the two points $(0, -1)$ and $(1/2, 0)$, and draw the straight line through them (Fig. 8).

Now Try Exercise 23

◼

▶ **Note** The line in Example 7 had different x- and y-intercepts. Two other circumstances can occur, however. First, the two intercepts may be the same, as in $y = 3x$. Second, there may be no x-intercept, as in $y = 1$. In both of these circumstances, we must plot some point other than an intercept in order to graph the line. ◀

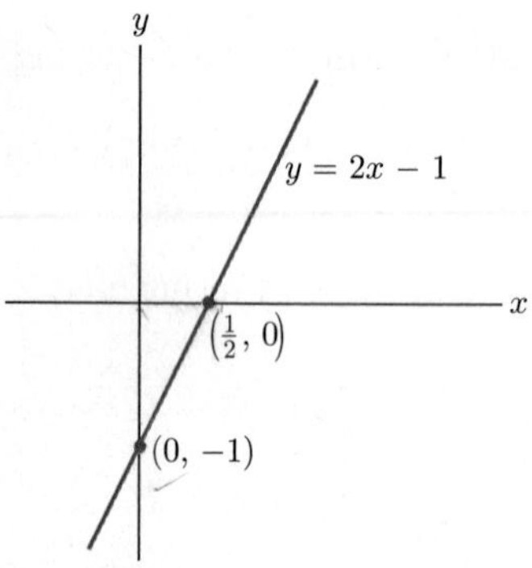

To summarize,

To graph the equation $y = mx + b$,

1. Plot the y-intercept $(0, b)$.

2. Plot some other point. [The most convenient choice is often the x-intercept $(x, 0)$, where x is determined by solving $mx + b = 0$.]

3. Draw a line through the two points.

Figure 8

The next example gives an application of linear equations.

EXAMPLE 8 **Linear depreciation** For tax purposes, businesses must keep track of the current values of each of their assets. A common mathematical model is to assume that the current value y is related to the age x of the asset by a linear equation. A moving company buys a 40-foot van with a useful lifetime of 5 years. After x months of use, the value y of the van is estimated by the linear equation

$$y = 25{,}000 - 400x.$$

(a) Sketch the graph of this linear equation.

(b) What is the value of the van after 5 years?

(c) What economic interpretation can be given to the y-intercept of the graph?

Solution (a) The y-intercept is $(0, 25{,}000)$. The x-intercept is found from the equation

$$25{,}000 - 400x = 0$$

$$x = \frac{125}{2},$$

so the x-intercept is $(125/2, 0)$. The graph of the linear equation is sketched in Fig. 9. Note how the value decreases as the age of the truck increases. The value of the truck reaches 0 after $125/2$ months. Note also that we have sketched only the portion of the graph that has physical meaning, namely the portion for x between 0 and $125/2$.

(b) After 5 years (or 60 months), the value of the van is

$$y = 25{,}000 - 400(60) = 25{,}000 - 24{,}000 = 1000.$$

Since the useful life of the van is 5 years, this value represents the *salvage value* of the van.

(c) The y-intercept corresponds to the value of the truck at $x = 0$ months, that is, the initial value of the truck, $\$25{,}000$.

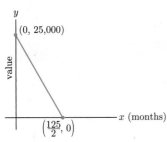

Figure 9

Now Try Exercise 41   ■

GC[1]   Throughout this text we discuss specifics of using the TI-83[2] and TI-89 graphing calculators. Additional details are provided in Appendices B and D. For the specific details of using other calculators, consult the guidebook that comes with the calculator.

Accurate graphs of straight lines can be obtained with a *graphing utility*; that is, a graphing calculator or a computer with graphing software. The equation is usually entered in standard form. Figures 10, 11, and 12 show how the equation from Example 7 is entered and graphed on a TI-83 graphing calculator. The window setting in Fig. 11 is referred to as $[-2, 4]$ *by* $[-2, 2]$. The graphing utility also can be used to find the coordinates of points on the line (see Fig. 13) and to determine the intercepts (see Figs. 14 and 15). With the TI-83, vertical lines can be drawn with the VERTICAL command from the DRAW menu. For instance, to draw the vertical line through the cursor, go to the home screen and press $\boxed{\text{2nd}}$ [DRAW] **4**. The TI-89 uses screens similar to those in Figs. 10 and 11 to specify functions and viewing windows; and can produce graphs and special points similar to those in Figs. 12–15. With the TI-89, the vertical line $x = k$ can be drawn by entering **LineVert k** into the home screen. The word **LineVert** either can be typed into the entry line or can be selected from the CATALOG menu.

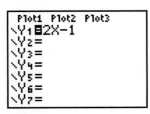

Figure 10

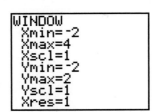

Figure 11

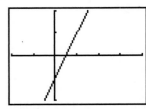

Figure 12

---

[1]GC is an abbreviation for "Graphing Calculator."

[2]The designation TI-83 refers to the TI-83, TI-83+, and TI-84+ family of calculators. When necessary, we will distinguish between the different members of the family.

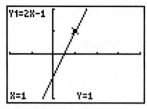
Figure 13

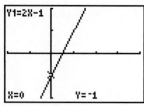
Figure 14

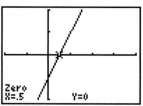
Figure 15

PRACTICE PROBLEMS 1.1

1. Plot the point $(500, 200)$.

2. Is the point $(4, -7)$ on the graph of the linear equation $2x - 3y = 1$? Is the point $(5, 3)$?

3. Graph the linear equation $5x + y = 10$.

4. Graph the straight line $y = 3x$.

EXERCISES 1.1

In Exercises 1–8, plot the given point.

1. $(2, 3)$

2. $(-1, 4)$

3. $(0, -2)$

4. $(2, 0)$

5. $(-2, 1)$

6. $\left(-1, -\frac{5}{2}\right)$

7. $(-20, 40)$

8. $(25, 30)$

9. $(\mathbf{PE})^3$ In Fig. 16, the point P has coordinates
   (a) $(5, 3)$ **(b)** $(-3, 5)$ **(c)** $(-5, 3)$
   (d) $(3, -5)$ **(e)** $(5, -3)$

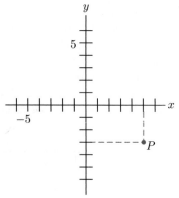
Figure 16

10. **(PE)** In Fig. 17, the point P has coordinates
   (a) $(-4, 5)$ **(b)** $(5, -4)$ **(c)** $(10, -8)$
   (d) $(-8, 10)$ **(e)** $(4, 5)$

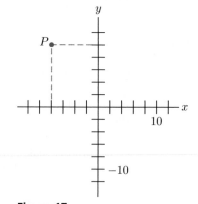
Figure 17

In Exercises 11–14, determine if the point is on the graph of the equation $-2x + \frac{1}{3}y = -1$.

11. $(1, 3)$ **12.** $(2, 6)$ **13.** $\left(\frac{1}{2}, 3\right)$ **14.** $\left(\frac{1}{3}, -1\right)$

In Exercises 15–18, each linear equation is in the standard form $y = mx + b$. Identify m and b.

15. $y = 5x + 8$ **16.** $y = -2x - 6$

17. $y = 3$ **18.** $y = \frac{2}{3}x$

In Exercises 19–22, put the linear equations into standard form.

19. $14x + 7y = 21$ **20.** $x - y = 3$

21. $3x = 5$ **22.** $-\frac{1}{2}x + \frac{2}{3}y = 10$

In Exercises 23–26, find the x-intercept and the y-intercept of each line.

23. $y = -4x + 8$ **24.** $y = 5$

25. $x = 7$ **26.** $y = -8x$

---

[3]Exercises denoted PE are similar to questions appearing in professional exams such as GMAT and CPA exams.

In Exercises 27–34, graph the given linear equation.

27. $y = \frac{1}{3}x - 1$ **28.** $y = 2x$ **29.** $y = \frac{5}{2}$

30. $x = 0$ **31.** $3x + 4y = 24$ **32.** $x + y = 3$

33. $x = -\frac{5}{2}$ **34.** $\frac{1}{2}x - \frac{1}{3}y = -1$

35. Which of the following equations describe the same line as the equation $2x + 3y = 6$?

(a) $4x + 6y = 12$ (b) $y = -\frac{2}{3}x + 2$

(c) $x = 3 - \frac{3}{2}y$ (d) $6 - 2x - y = 0$

(e) $y = 2 - \frac{2}{3}x$ (f) $x + y = 1$

36. Which of the following equations describe the same line as the equation $\frac{1}{2}x - 5y = 1$?

(a) $2x - \frac{1}{5}y = 1$ (b) $x = 5y + 2$

(c) $2 - 5x + 10y = 0$ (d) $y = .1(x - 2)$

(e) $10y - x = -2$ (f) $1 + .5x = 2 + 5y$

37. Each of the lines L_1, L_2, and L_3 in Fig. 18 is the graph of one of the equations (a), (b), and (c). Match each of the equations with its corresponding line.

(a) $x + y = 3$ (b) $2x - y = -2$ (c) $x = 3y + 3$

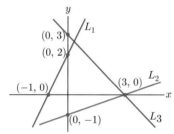

Figure 18

38. Which of the following equations is graphed in Fig. 19?

(a) $x + y = 3$ (b) $y = x - 1$ (c) $2y = x + 3$

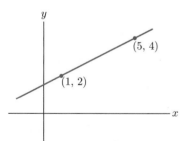

Figure 19

39. (*Heating Water*) The temperature of water in a heating tea kettle rises according to the equation $y = 30x + 72$, where y is the temperature (in degrees Fahrenheit) x minutes after the kettle was put on the burner.

(a) After how many minutes will the water boil?

(b) What physical interpretation can be given to the y-intercept of the graph?

(c) Does the x-intercept have a meaningful physical interpretation?

40. (*Deforestation*) The amount of tropical rainforest area in Central America has been decreasing steadily in recent years. The amount y (in thousands of square miles) x years after 1969 is estimated by the linear equation

$$y = \left(-\frac{25}{8}\right)x + 130.$$

(a) Sketch the graph of this linear equation.

(b) What interpretation can be given to the y-intercept of the graph?

(c) When were there 80,000 square miles of tropical rain forests?

(d) If this trend continues, how large will the rain forest be in the year 2007?

41. (*Cigarette Consumption*) The worldwide consumption of cigarettes has been increasing steadily in recent years. The number of trillions of cigarettes, y, purchased x years after 1960, is estimated by the linear equation

$$y = .075x + 2.5.$$

(a) Sketch the graph of this linear equation.

(b) What interpretation can be given to the y-intercept of the graph?

(c) When were there 4 trillion cigarettes sold?

(d) If this trend continues, how many cigarettes will be sold in the year 2020?

42. (*Ecotourism Income*) In a certain developing country, ecotourism income has been increasing in recent years. The income y (in thousands of dollars) x years after 2000 can be modeled by $y = 1.15x + 14$.

(a) Sketch the graph of this linear equation.

(b) What interpretation can be given to the y-intercept of this graph?

(c) When was there $20,000 in ecotourism income?

(d) If this trend continues, how much ecotourism income will there be in 2016?

43. (*Insurance Rates*) Yearly car insurance rates have risen in the last few years. The rate y (in dollars) for a small car x years after 1997 can be modeled by $y = 60x + 678$.

(a) Sketch the graph of this linear equation.

(b) What interpretation can be given to the y-intercept of this graph?

(c) What was the yearly rate in 2000?

(d) If this trend continues, when will the yearly rate be $1300?

44. Find an equation of the line having y-intercept $(0, 5)$ and x-intercept $(4, 0)$.

45. What is the equation of the x-axis?

46. Can a line have more than one x-intercept?

47. What is the general form of the equation of a line that is parallel to the x-axis?

48. Give the x- and y-intercepts of the line $\frac{x}{a} + \frac{y}{b} = 1$.

*Every straight line has an equation of the form $cx + dy = e$, where not both c and d are 0. Such an equation is said to be a **general form** of the equation of the line. In Exercises 49–52, find a general form of the given equation.*

49. $y = 2x + 3$ **50.** $x = 5$

51. $x = -3$ **52.** $y = 3x - 4$

53. $y = -\frac{2}{3}x - 5$ **54.** $y = 4x - \frac{5}{6}$

55. Show that the straight line with x-intercept $(a, 0)$ and y-intercept $(0, b)$, where a and b are not both zero, has $bx + ay = ab$ as a general form of its equation.

56. Use the result of Exercise 55 to find a general form of the equation of the line having x-intercept $(5, 0)$ and y-intercept $(0, 6)$.

Exercises 57–62 require the use of a graphing calculator.

In Exercises 57–60, (a) graph the line in the standard window $[-10, 10]$ by $[-10, 10]$, (b) use the calculator to find the y-coordinate of the point on the line with x-coordinate 2, (c) use the calculator to determine the two intercepts.

57. $y = -3x + 6$ **58.** $y = .25x - 2$

59. $3y - 2x = 9$ **60.** $2y + 5x = 8$

61. Determine an appropriate window and graph the line $2y + x = 100$.

62. Determine an appropriate window and graph the line $x - 3y = 60$.

SOLUTIONS TO PRACTICE PROBLEMS 1.1

1. Since the numbers are large, make each hatchmark correspond to 100. Then the point $(500, 200)$ is found by starting at the origin, moving 500 units to the right and then 200 units up (Fig. 20).

2. Replace x by 4 and y by -7 in the equation $2x - 3y = 1$.

$$2(4) - 3(-7) = 1$$
$$8 + 21 = 1$$
$$29 = 1$$

The equation is clearly not satisfied, so $(4, -7)$ is not on the graph. Similarly, if we substitute $(5, 3)$ into the equation, we find that

$$2 \cdot 5 - 3 \cdot 3 = 1$$
$$10 - 9 = 1$$
$$1 = 1.$$

So the equation is satisfied and $(5, 3)$ is on the graph.

3. The standard form is obtained by solving for y:

$$y = -5x + 10.$$

Thus $m = -5$ and $b = 10$. The y-intercept is $(0, 10)$. To find the x-intercept, set $y = 0$:

$$0 = -5x + 10$$
$$5x = 10$$
$$x = 2.$$

So the x-intercept is $(2, 0)$ (Fig. 21).

4. To find the y-intercept, set $x = 0$; then $y = 3 \cdot 0 = 0$. To find the x-intercept, set $y = 0$; then $3x = 0$ or $x = 0$. The two intercepts are the same point $(0, 0)$. We must therefore plot some other point. Setting $x = 1$, we obtain $y = 3 \cdot 1 = 3$, so another point on the line is $(1, 3)$ (Fig. 22).

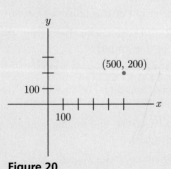

Figure 20

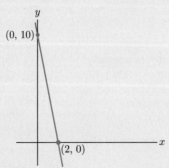

Figure 21

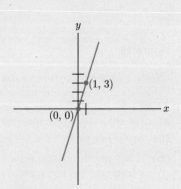

Figure 22

1.2 Linear Inequalities

In this section we study the properties of inequalities. Start with a Cartesian coordinate system on the line (Fig. 1). Recall that it is possible to associate with each point of the line a number; and conversely, with each number (positive, negative, or zero) it is possible to associate a point on the line.

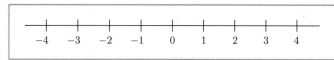

Figure 1

Let a and b be any numbers. We say that a *is less than* b if a lies to the left of b on the line. When a is less than b, we write $a < b$ (read: a is less than b). Thus, for example, $2 < 3$, $-1 < 2$, $-3 < -1$, and $0 < 4$. When a is less than b, we also say that b *is greater than* a ($b > a$). Thus, for example, $3 > 2$, $2 > -1$, $-1 > -3$, and $4 > 0$. Sometimes it is convenient to have a notation that means that a is no larger than b. The notation used for this is $a \le b$ (read: a is less than or equal to b). Similarly, the notation $a \ge b$ (read: a is greater than or equal to b) means that a is no smaller than b. The symbols $>$, $<$, $\ge$, and $\le$ are called *inequality signs*. It is easiest to remember the meaning of these various symbols by noting that the symbols $>$ and $<$ always *point toward the smaller* of a and b.

An inequality expresses a relationship between the quantities on both of its sides. This relationship is very similar to the relationship expressed by an equation. And just as some problems require solving equations, others require solving inequalities. Our next task will be to state and illustrate the arithmetic operations permissible in dealing with inequalities. These permissible operations form the basis of a technique for solving the inequalities occurring in applications.

Inequality Property 1   Suppose that $a < b$ and that c is any number. Then $a + c < b + c$ and $a - c < b - c$. In other words, the same number can be added or subtracted from both sides of the inequality.

For example, start with the inequality $2 < 3$ and add 4 to both sides to get

$$2 + 4 < 3 + 4.$$

That is,

$$6 < 7,$$

a correct inequality.

EXAMPLE 1   **Solving a linear inequality**   Solve the inequality $x + 4 > 3$. That is, determine all values of x for which the inequality holds.

Solution   We proceed as we would in dealing with an equation. Isolate x on the left by subtracting 4 from both sides, which is permissible by Inequality Property 1:

$$x + 4 > 3$$
$$(x + 4) - 4 > 3 - 4$$
$$x > -1.$$

That is, the values of x for which the inequality holds are exactly those x greater than -1.   ∎

In dealing with an equation, both sides may be multiplied or divided by a number. However, multiplying or dividing an inequality by a number requires some care. The result depends on whether the number is positive or negative. More precisely,

---

Inequality Property 2

2A. If $a < b$ and c is positive, then $ac < bc$.

2B. If $a < b$ and c is negative, then $ac > bc$.

In other words, an inequality may be multiplied by a positive number, just as in the case of equations. But to multiply an inequality by a negative number, it is necessary to reverse the inequality sign. For example, the inequality $-1 < 2$ can be multiplied by 4 to get $-4 < 8$, a correct statement. But if we were to multiply by -4, it would be necessary to reverse the inequality sign, since -4 is negative. In this latter case we would get $4 > -8$, a correct statement.

▶ *Note* Inequality Properties 1 and 2 are stated using only $<$. However, exactly the same properties hold if $<$ is replaced by $>$, $\leq$, or $\geq$. ◀

EXAMPLE 2 **Solving a linear inequality** Solve the inequality $-3x + 2 \geq -1$.

Solution Treat the inequality as if it were an equation. The goal is to isolate x on one side. To this end, first subtract 2 from both sides (Property 1). This gives

$$(-3x + 2) - 2 \geq -1 - 2$$
$$-3x \geq -3.$$

Next, we multiply by $-\frac{1}{3}$. (This gives x on the left.) But $-\frac{1}{3}$ is negative, so by Property 2B we must reverse the inequality sign. Thus,

$$-\tfrac{1}{3}(-3x) \leq -\tfrac{1}{3}(-3)$$
$$x \leq 1.$$

Therefore, the values of x satisfying the inequality are precisely those values that are ≤ 1.

Now Try Exercise 7 ■

The inequalities of greatest interest to us[1] are those in two variables, x and y, and having the form $cx + dy \leq e$ or $cx + dy \geq e$, where c, d, and e are given numbers, with c and d not both 0. We will call such inequalities *linear inequalities*. When $d \neq 0$ (that is, when y actually appears), the inequality can be put into one of the *standard forms* $y \leq mx + b$ or $y \geq mx + b$. When $d = 0$, the inequality can be put into one of the *standard forms* $x \leq a$ or $x \geq a$. The procedure for putting a linear inequality into standard form is analogous to that for putting a linear equation into standard form.

---

[1] Such inequalities arise in our discussion of linear programming in Chapter 3.

EXAMPLE 3 **Standard form of a linear inequality** Put the linear inequality $2x - 3y \geq -9$ into standard form.

Solution Since we want the y-term on the left and all other terms on the right, begin by subtracting $2x$ from both sides:

$$-3y \geq -2x - 9.$$

Next, multiply by $-\frac{1}{3}$, remembering to change the inequality sign, since $-\frac{1}{3}$ is negative:

$$y \leq -\tfrac{1}{3}(-2x - 9)$$
$$y \leq \tfrac{2}{3}x + 3.$$

The last inequality is in standard form.

Now Try Exercise 9   ∎

EXAMPLE 4 **Standard form of a linear inequality** Find the standard form of the inequality $\frac{1}{2}x \geq 4$.

Solution Note that y does not appear in the inequality. Just as was the case in finding the standard form of a linear equation when y does not appear, solve for x. To do this, multiply by 2 to get

$$x \geq 8,$$

the standard form.

Now Try Exercise 13   ∎

Graphing Linear Inequalities Associated with every linear inequality is a set of points of the plane, the set of all those points that satisfy the inequality. This set of points is called the *graph* of the inequality.

EXAMPLE 5 **Solution of a linear inequality** Determine whether or not the given point satisfies the inequality $y \geq -\frac{2}{3}x + 4$.

(a) $(3, 4)$ (b) $(0, 0)$

Solution Substitute the x-coordinate of the point for x and the y-coordinate for y, and determine if the resulting inequality is correct or not.

(a) $4 \geq -\frac{2}{3}(3) + 4$ (b) $0 \geq -\frac{2}{3}(0) + 4$

   $4 \geq -2 + 4$ $0 \geq 0 + 4$

   $4 \geq 2$ (correct) $0 \geq 4$ (not correct)

Therefore, the point $(3, 4)$ satisfies the inequality and the point $(0, 0)$ does not.

Now Try Exercises 15 and 17   ∎

It is easiest to determine the graph of a given inequality after it has been written in standard form. Therefore, let us describe the graphs of each of the standard forms. The easiest to handle are the forms $x \geq a$ and $x \leq a$.

A point satisfies the inequality $x \geq a$ if and only if its x-coordinate is greater than or equal to a. The y-coordinate can be anything. Therefore, the graph of $x \geq a$ consists of all points to the right of and on the vertical line $x = a$. We will display the graph by crossing out the portion of the plane to the left of the line (see Fig. 2). Similarly, the graph of $x \leq a$ consists of the points to the left of and on the line $x = a$. This graph is shown in Fig. 3.

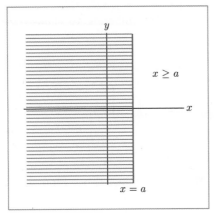

Figure 2

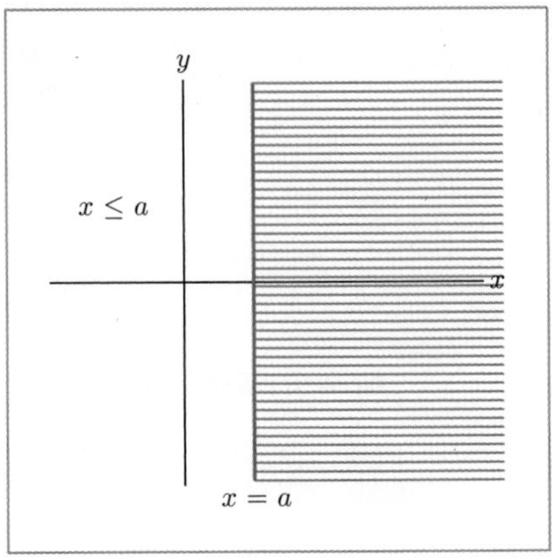

Figure 3

Here is a simple procedure for graphing the other two standard forms.

To graph $y \geq mx + b$ or $y \leq mx + b$,

1. Draw the graph of $y = mx + b$.

2. Throw away, that is, "cross out," the portion of the plane not satisfying the inequality. The graph of $y \geq mx + b$ consists of all points above or on the line. The graph of $y \leq mx + b$ consists of all points below or on the line.

The graphs of the inequalities $y \geq mx + b$ and $y \leq mx + b$ are shown in Fig. 4. Some simple reasoning suffices to show why these graphs are correct. Draw the line $y = mx + b$, as in Fig. 5, and pick any point P above the line. Suppose that P has coordinates (x, y). Let Q be the point on the line that lies directly below P. Then Q has the same first coordinate as P, that is, x. Since Q lies on the line, the second

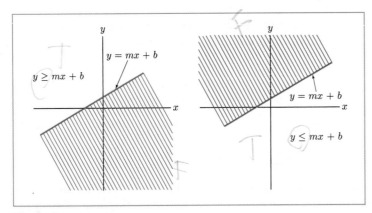

Figure 4

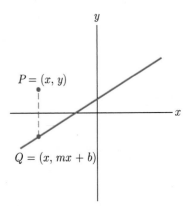

Figure 5

coordinate of Q is $mx + b$. Since the second coordinate of P is clearly larger than the second coordinate of Q, we must have $y \geq mx + b$. Similarly, any point below the line satisfies $y \leq mx + b$. Thus the two graphs are as given in Fig. 4.

EXAMPLE 6 **Graphing a linear inequality** Graph the inequality $2x + 3y \geq 15$.

Solution In order to apply the aforementioned procedure, the inequality must first be put into standard form:

$$2x + 3y \geq 15$$
$$3y \geq -2x + 15$$
$$y \geq -\tfrac{2}{3}x + 5.$$

The last inequality is in standard form. Next, we graph the line $y = -\tfrac{2}{3}x + 5$. Its intercepts are $(0, 5)$ and $\left(\tfrac{15}{2}, 0\right)$. Since the inequality is "$y \geq$" we cross out the region below the line and label the region above with the inequality. The graph consists of all points above or on the line (Fig. 6).

Now Try Exercise 31

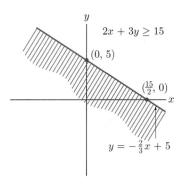

Figure 6

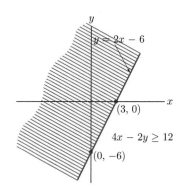

Figure 7

EXAMPLE 7 **Graphing a linear inequality** Graph the inequality $4x - 2y \geq 12$.

Solution First, put the inequality in standard form:

$$4x - 2y \geq 12$$
$$-2y \geq -4x + 12$$
$$y \leq 2x - 6$$

(note the change in the inequality sign!). Next, graph $y = 2x - 6$. The intercepts are $(0, -6)$ and $(3, 0)$. Since the inequality is "$y \leq$" the graph consists of all points below or on the line (Fig. 7).

Now Try Exercise 27

So far, we have been concerned only with graphing single inequalities. The next example concerns graphing a system of inequalities. That is, it asks us to determine all points of the plane that *simultaneously* satisfy all inequalities of a system.

EXAMPLE 8   **Graphing a system of linear inequalities**   Graph the system of inequalities

$$\begin{cases} 2x + 3y \geq 15 \\ 4x - 2y \geq 12 \\ y \geq 0. \end{cases}$$

Solution   The first two inequalities have already been graphed in Examples 6 and 7. The graph of $y \geq 0$ consists of all points above or on the x-axis. In Fig. 8, any point that is crossed out is *not* on the graph of at least one inequality. So the points that simultaneously satisfy all three inequalities are those in the remaining clear region and its border.

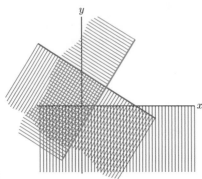

Figure 8

Now Try Exercise 41   ■

▶ *Remark*   At first, our convention of crossing out those points *not* on the graph of an inequality (instead of shading the points *on* the graph) may have seemed odd. However, the real advantage of this convention becomes apparent when graphing a system of inequalities. Imagine trying to find the graph of the system of Example 8 if the points *on* the graph of each inequality had been shaded. It would have been necessary to locate the points that had been shaded three times. This is hard to do.   ◀

The graph of a system of inequalities is called a *feasible set*. The feasible set associated to the system of Example 8 is a three-sided, unbounded region.

Given a specific point, we should be able to decide whether or not the point lies in the feasible set. The next example shows how this is done.

EXAMPLE 9   **Feasible set of a system of linear inequalities**   Determine whether the points $(5, 3)$ and $(4, 2)$ are in the feasible set of the system of inequalities of Example 8.

Solution   If we had very accurate measuring devices, we could plot the points in the graph of Fig. 8 and determine whether or not they lie in the feasible set. However, there is a simpler and more reliable algebraic method. Just substitute the coordinates of the points into each of the inequalities of the system and see whether or not *all* of the inequalities are satisfied. So doing, we find that $(5, 3)$ is in the feasible set and

$(4, 2)$ is not.

yes (5, 3) $\begin{cases} 2(5) + 3(3) \geq 15 \\ 4(5) - 2(3) \geq 12 \\ (3) \geq 0 \end{cases}$ $\begin{cases} 19 \geq 15 \quad T \\ 14 \geq 12 \quad T \\ 3 \geq 0 \quad T \end{cases}$

No (4, 2) $\begin{cases} 2(4) + 3(2) \geq 15 \\ 4(4) - 2(2) \geq 12 \\ (2) \geq 0 \end{cases}$ $\begin{cases} 14 \geq 15 \quad F \\ 12 \geq 12 \quad F \\ 2 \geq 0 \quad T \end{cases}$

Now Try Exercises 45 and 47 ■

GC Graphing calculators will display feasible sets of inequalities. With the TI-83 and TI-89, the **Shade** command (invoked on the TI-83 with [2nd] [DRAW] **7** from the home screen, and typed directly into the TI-89 entry line or selected from the CATALOG menu) can be used to shade one side of the straight line $\mathbf{Y_1}$. Suppose that the window is $[a, b]$ *by* $[c, d]$. The region above the line can be thought of as lying between the lower function $\mathbf{Y_1}$ and the upper function $y = d + 1$. The region below the line can be thought of as lying between the lower function $y = c - 1$ and the upper function $\mathbf{Y_1}$. (With the TI-89, we can use **y=ymax** as the upper function and **y=ymin** as the lower function.) These regions can be shaded with the commands

$$\textbf{Shade}(lowerfunction, upperfunction, a, b, 2, 3) \quad \text{[TI-83]}$$

$$\textbf{Shade} \, lowerfunction, upperfunction \quad \text{[TI-89]}$$

See Figs. 9 and 10, which were drawn with the window $[-5, 13]$ *by* $[-6, 6]$.

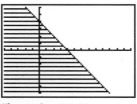

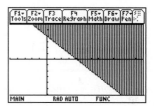

Figure 9. TI-83
Shade$(-7, -\mathbf{X} + 4, -5, 13, 2, 3)$

Figure 10. TI-89
Shade $- \mathbf{x} + 4, \mathbf{ymax}$

The regions to the left and right, respectively, of the vertical line $x = k$ can be shaded with the commands

$$\textbf{Shade(c-1,d+1,a,k,2,3)} \quad \text{and} \quad \textbf{Shade(c-1,d+1,k,b,2,3)} \quad \text{[TI-83]}$$

$$\textbf{Shade ymin,ymax,xmin,k} \quad \text{and} \quad \textbf{Shade ymin,ymax,k,xmax} \quad \text{[TI-89]}$$

See Figs. 11 and 12, which were drawn with the window $[-5, 13]$ *by* $[-6, 6]$.

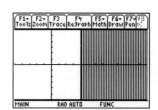

Figure 11. TI-83
Shade$(-7, 7, -5, 4, 2, 3)$

Figure 12. TI-89
Shade $\mathbf{ymin, ymax, 4, xmax}$

With the TI-83, there is a style icon (initially a line of four dots) to the left of each function in the **Y=** editor. To change the style icon of a function, move the cursor to the icon and press the ENTER key. Of the seven different icons, the icons ◣ and ◤ specify that the line be graphed and the points above and below the line, respectively, be crossed out. With the TI-89, a style can be set for each function in the **Y=** editor. To set the style so that the points above (below) a function are shaded when the function is graphed, highlight the function definition in the **Y=** editor, press 2nd [F6], and then select **Above (Below)**. Figure 13 shows the **Y=** editor and the feasible set for Example 8 with the window setting $[-10, 20]$ *by* $[-10, 10]$.

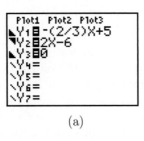

(a)

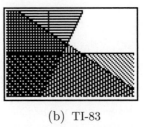

(b) TI-83

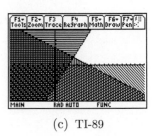

(c) TI-89

Figure 13

PRACTICE PROBLEMS 1.2

1. Graph the inequality $3x - y \geq 3$.

2. Graph the feasible set for the system of inequalities

$$\begin{cases} x \geq 0, \ y \geq 0 \\ x + 2y \leq \ \ 4 \\ 4x - 4y \geq -4. \end{cases}$$

EXERCISES 1.2

In Exercises 1–4, state whether the inequality is true or false.

1. $2 \leq -3$ **2.** $-2 \leq 0$ **3.** $7 \leq 7$ **4.** $0 \geq \frac{1}{2}$

In Exercises 5–7, solve for x.

5. $2x - 5 \geq 3$ **6.** $3x - 7 \leq 2$

7. $-5x + 13 \leq -2$

8. Which of the following results from solving $-x + 1 \leq 3$ for x?

(a) $x \leq 4$ (b) $x \leq 2$

(c) $x \geq -4$ (d) $x \geq -2$

In Exercises 9–14, put the linear inequality into standard form.

9. $2x + y \leq 5$ **10.** $-3x + y \geq 1$

11. $5x - \frac{1}{3}y \leq 6$ **12.** $\frac{1}{2}x - y \leq -1$

13. $4x \geq -3$ **14.** $-2x \leq 4$

In Exercises 15–22, determine whether or not the given point satisfies the given inequality.

15. $3x + 5y \leq 12$, $(2, 1)$ **16.** $-2x + y \geq 9$, $(3, 15)$

17. $y \geq -2x + 7$, $(3, 0)$ **18.** $y \leq \frac{1}{2}x + 3$, $(4, 6)$

19. $y \leq 3x - 4$, $(3, 5)$ **20.** $y \geq x$, $(-3, -2)$

21. $x \geq 5$, $(7, -2)$ **22.** $x \leq 7$, $(0, 0)$

In Exercises 23–26, graph the given inequality by crossing out (i.e., discarding) the points not satisfying the inequality.

23. $y \leq \frac{1}{3}x + 1$

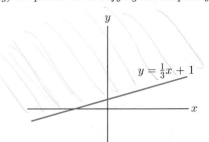

24. $y \geq -x + 1$

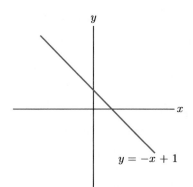

$y = -x + 1$

25. $x \geq 4$

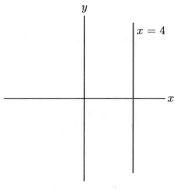

$x = 4$

26. $y \leq 2$

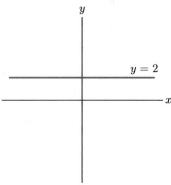

$y = 2$

In Exercises 27–38, graph the given inequality.

27. $y \leq 2x + 1$ **28.** $y \geq -3x + 6$

29. $x \geq 2$ **30.** $x \geq 0$

31. $x + 4y \geq 12$ **32.** $4x - 4y \geq 8$

33. $4x - 5y + 25 \geq 0$ **34.** $.2y - x \geq .4$

35. $\frac{1}{2}x - \frac{1}{3}y \leq 1$ **36.** $3y - 3x \leq 2y + x + 1$

37. $.5x + .4y \leq 2$ **38.** $y - 4x \geq \frac{1}{2}y + 2$

In Exercises 39–44, graph the feasible set for the system of inequalities.

39. $\begin{cases} y \leq 2x - 4 \\ y \geq 0 \end{cases}$ **40.** $\begin{cases} y \geq -\frac{1}{3}x + 1 \\ x \geq 0 \end{cases}$

41. $\begin{cases} x + 2y \geq 2 \\ 3x - y \geq 3 \end{cases}$ **42.** $\begin{cases} 3x + 6y \geq 24 \\ 3x + y \geq 6 \end{cases}$

43. $\begin{cases} x + 5y \leq 10 \\ x + y \leq 3 \\ x \geq 0, \ y \geq 0 \end{cases}$ **44.** $\begin{cases} x + 2y \geq 6 \\ x + y \geq 5 \\ x \geq 1 \end{cases}$

In Exercises 45–48, determine whether the given point is in the feasible set of this system of inequalities:

$$\begin{cases} 6x + 3y \leq 96 \\ x + y \leq 18 \\ 2x + 6y \leq 72 \\ x \geq 0, \ y \geq 0. \end{cases}$$

45. $(8, 7)$ **46.** $(14, 3)$ **47.** $(9, 10)$ **48.** $(16, 0)$

In Exercises 49–52, determine whether the given point is above or below the given line.

49. $y = 2x + 5, \ (3, 9)$ **50.** $3x - y = 4, \ (2, 3)$

51. $7 - 4x + 5y = 0, \ (0, 0)$ **52.** $x = 2y + 5, \ (6, 1)$

53. Give a system of inequalities for which the graph is the region between the pair of lines $8x - 4y - 4 = 0$ and $8x - 4y = 0$.

54. (PE) The shaded region in Fig. 14 is bounded by four straight lines. Which of the following is NOT an equation of one of the boundary lines?

(a) $y = 0$ (b) $y = 2$ (c) $x = 0$

(d) $2x + 3y = 12$ (e) $3x + 2y = 12$

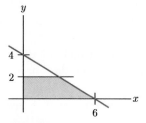

Figure 14

55. (PE) The shaded region in Fig. 15 is bounded by four straight lines. Which of the following is NOT an equation of one of the boundary lines?

(a) $x = 0$ (b) $y = x$ (c) $y = 5$

(d) $y = 0$ (e) $x + 2y = 6$

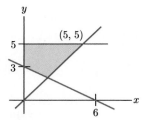

Figure 15

56. (PE) Which quadrant in Fig. 16 contains no points that satisfy the inequality $x + 2y \geq 6$?

    (a) none   **(b)** I   **(c)** II   **(d)** III   **(e)** IV

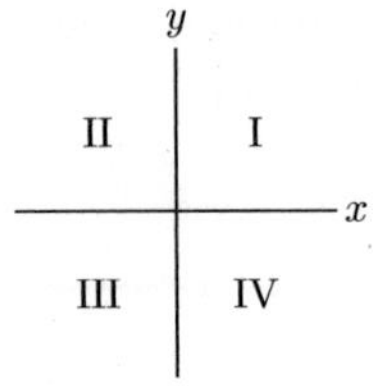

Figure 16

57. (PE) Which quadrant in Fig. 16 contains no points that satisfy the inequality $3y - 2x \geq 6$?

    (a) none   **(b)** I   **(c)** II   **(d)** III   **(e)** IV

Use a graphing calculator to solve Exercises 58–61.

58. Graph the line $4x - 2y = 7$.

    (a) Locate the point on the line with x-coordinate 3.6.

    (b) Does the point $(3.6, 3.5)$ lie above or below the line? Explain.

59. Graph the line $x + 2y = 11$.

    (a) Locate the point on the line with x-coordinate 6.

    (b) Does the point $(6, 2.6)$ lie above or below the line? Explain.

60. If your graphing calculator can shade feasible sets of inequalities, display the feasible set in Exercise 39.

61. If your graphing calculator can shade feasible sets of inequalities, display the feasible set in Exercise 42.

SOLUTIONS TO PRACTICE PROBLEMS 1.2

1. Linear inequalities are easiest to graph if they are first put into standard form. Subtract $3x$ from both sides and multiply by -1:

$$3x - y \geq 3$$
$$-y \geq -3x + 3$$
$$y \leq 3x - 3.$$

Now graph the line $y = 3x - 3$ (Fig. 17). The graph of the inequality is the portion of the plane below and on the line ("$\leq$" corresponds to below), so throw away (that is, cross out) the portion above the line.

2. Begin by putting the linear inequalities into standard form and then graphing them all on the same coordinate system.

$$\begin{cases} x \geq 0, \ y \geq 0 \\ x + 2y \leq 4 \\ 4x - 4y \geq -4 \end{cases}$$

has standard form

$$\begin{cases} x \geq 0, \ y \geq 0 \\ y \leq -\tfrac{1}{2}x + 2 \\ y \leq x + 1 \end{cases}$$

A good procedure to follow is to graph all of the linear equations and then cross out the regions to be thrown away one at a time (Fig. 18). The inequalities $x \geq 0$ and $y \geq 0$ arise frequently in applications. The first has the form $x \geq a$, where $a = 0$, and the second has the form $y \geq mx + b$, where $m = 0$ and $b = 0$. To graph them, just cross out all points to the left of the y-axis and all points below the x-axis, respectively.

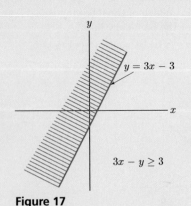

Figure 17

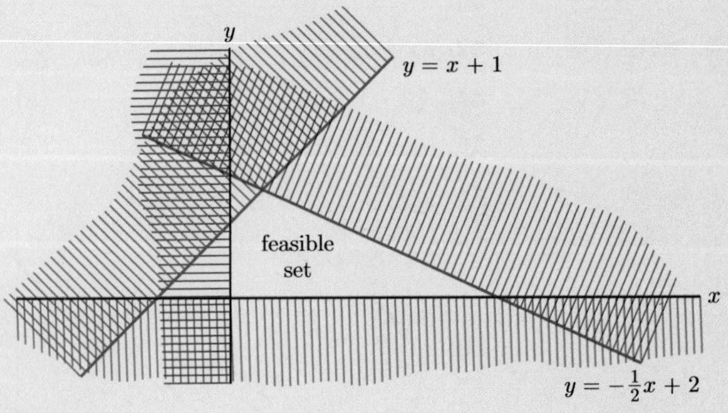

Figure 18

1.3 The Intersection Point of a Pair of Lines

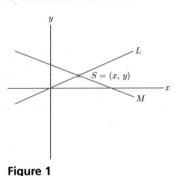

Figure 1

Suppose that we are given a pair of intersecting straight lines L and M. Let us consider the problem of determining the coordinates of the point of intersection $S = (x, y)$ (see Fig. 1). We may as well assume that the equations of L and M are given in standard form. First, let us assume that both lines are in the first standard form — that is, that the equations are

$$L:\ y = mx + b, \qquad M:\ y = nx + c.$$

Since the point S is on both lines, its coordinates satisfy both equations. In particular, we have two expressions for its y-coordinate:

$$y = mx + b = nx + c.$$

The last equality gives an equation from which x can easily be determined. Then the value of y can be determined as $mx + b$ (or $nx + c$). Let us see how this works in a particular example.

EXAMPLE 1 **Finding the point of intersection** Find the point of intersection of the two lines $y = 2x - 3$ and $y = x + 1$.

$5 = 2(4) - 3 \qquad 5 = 4 + 1$

Solution To find the x-coordinate of the point of intersection, equate the two expressions for y and solve for x:

$$2x - 3 = x + 1$$
$$2x - x = 1 + 3$$
$$x = 4.$$

To find the value of y, set $x = 4$ in either equation, say the first. Then

$$y = 2 \cdot 4 - 3 = 5.$$

So the point of intersection is $(4, 5)$.

Now Try Exercise 1   ∎

EXAMPLE 2 **Finding the point of intersection** Find the point of intersection of the two lines $x + 2y = 6$ and $5x + 2y = 18$.

Solution To use the method described, the equations must be in standard form. Solving both equations for y, we get the standard forms

$$y = -\tfrac{1}{2}x + 3$$
$$y = -\tfrac{5}{2}x + 9.$$

Equating the expressions for y gives

$$-\tfrac{1}{2}x + 3 = -\tfrac{5}{2}x + 9$$
$$\tfrac{5}{2}x - \tfrac{1}{2}x = 9 - 3$$
$$2x = 6$$
$$x = 3.$$

Setting $x = 3$ in the first equation gives

$$y = -\tfrac{1}{2}(3) + 3 = \tfrac{3}{2}.$$

So the intersection point is $\left(3, \tfrac{3}{2}\right)$.

Now Try Exercise 3   ■

The preceding method works when both equations have the first standard form $(y = mx + b)$. In case one equation has the standard form $x = a$, things are much simpler. The value of x is then given directly without any work, namely $x = a$. The value of y can be found by substituting a for x in the other equation.

EXAMPLE 3   **Finding the point of intersection**   Find the point of intersection of the lines $y = 2x - 1$ and $x = 2$.

Solution   The x-coordinate of the intersection point is 2, and the y-coordinate is $y = 2 \cdot 2 - 1 = 3$. Therefore, the intersection point is $(2, 3)$.

Now Try Exercise 5   ■

The method just introduced may be used to solve systems of two equations in two variables.

EXAMPLE 4   **Solving a system of equations**   Solve the following system of linear equations:

$$\begin{cases} 2x + 3y = 7 \\ 4x - 2y = 9. \end{cases}$$

Solution   First convert the equations to standard form:

$$2x + 3y = 7 \qquad\qquad 4x - 2y = 9$$
$$3y = -2x + 7 \qquad\qquad -2y = -4x + 9$$
$$y = -\tfrac{2}{3}x + \tfrac{7}{3}; \qquad\qquad y = 2x - \tfrac{9}{2}.$$

Now equate the two expressions for y:

$$2x - \tfrac{9}{2} = -\tfrac{2}{3}x + \tfrac{7}{3}$$
$$\tfrac{8}{3}x = \tfrac{7}{3} + \tfrac{9}{2} = \tfrac{14}{6} + \tfrac{27}{6} = \tfrac{41}{6}$$
$$x = \tfrac{3}{8} \cdot \tfrac{41}{6} = \tfrac{41}{16}$$
$$y = 2x - \tfrac{9}{2} = 2\left(\tfrac{41}{16}\right) - \tfrac{9}{2} = \tfrac{5}{8}.$$

So the solution of the given system is $x = \tfrac{41}{16}$, $y = \tfrac{5}{8}$.

Now Try Exercise 9   ■

Supply and Demand Curves   The price p that a commodity sells for is related to the quantity q available. Economists study two sorts of graphs that express relationships between q and p. To describe these graphs, let us plot *quantity* along the horizontal axis and *price* along the vertical axis. The first graph relating q and p is called a *supply curve* (Fig. 2) and expresses the relationship between q and p from a manufacturer's point of view. For every quantity q, the supply curve specifies the price p for which the manufacturer is willing to produce the quantity q. The greater the quantity to be supplied, the higher the price must be. So supply curves rise when viewed from left to right.

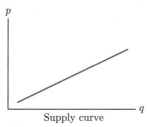

Supply curve

Figure 2

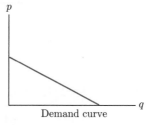

Demand curve

Figure 3

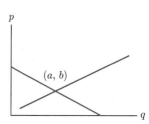

(a, b)

Figure 4

The second curve relating q and p is called a *demand curve* (Fig. 3) and expresses the relationship between q and p from the consumer's viewpoint. For each quantity q, the demand curve gives the price p that must be charged in order for q units of the commodity to be sold. The greater the quantity that must be sold, the lower the price that consumers must be asked to pay. So demand curves fall when viewed from left to right.

Suppose the supply and demand curves for a commodity are drawn on a single coordinate system (Fig. 4). The intersection point (a, b) of the two curves has an economic significance: The quantity produced will stabilize at a units and the price will be b dollars per unit.

EXAMPLE 5   **Applying the law of supply and demand**   Suppose the supply curve for a certain commodity is the straight line whose equation is $p = .0002q + 2$ (p in dollars). Suppose the demand curve for the same commodity is the straight line whose equation is $p = -.0005q + 5.5$. Determine both the quantity of the commodity that will be produced and the price at which it will sell.

Solution   We must solve the system of linear equations

$$\begin{cases} p = .0002q + 2 \\ p = -.0005q + 5.5 \end{cases}$$

$$.0002q + 2 = -.0005q + 5.5$$

$$.0007q = 3.5$$

$$q = \frac{3.5}{.0007} = 5000$$

$$p = .0002(5000) + 2 = 1 + 2 = 3.$$

Thus 5000 units of the commodity will be produced and it will sell for $3 per unit.

Now Try Exercise 27   ■

GC   Graphing utilities have commands that find the intersection point of a pair of lines. Figure 5 shows the result of solving Example 4 on the TI-83 with the intersect command of the CALC menu. Since the x-coordinate of the intersection point is assigned to ANS, the x-intercept can be converted to a fraction by pressing MATH 1 ENTER from the home screen. Figure 6 shows the result of solving Example 4 on the TI-89 with the Intersection routine in the menu resulting from pressing F5.

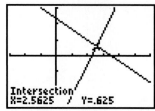

Figure 5. $[-3, 6]$ by $[-3, 3]$

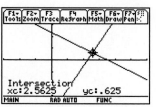

Figure 6. $[-3, 6]$ by $[-3, 3]$

PRACTICE PROBLEMS 1.3

Figure 7 shows the feasible set of a system of linear inequalities; its four vertices are labeled A, B, C, and D.

1. Use the method of this section to find the coordinates of the point C.

2. Determine the coordinates of the points A and B by inspection.

3. Find the coordinates of the point D.

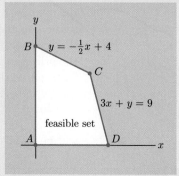

Figure 7

EXERCISES 1.3

In Exercises 1–6, find the point of intersection of the given pair of straight lines.

1. $\begin{cases} y = 4x - 5 \\ y = -2x + 7 \end{cases}$ (2,3)

2. $\begin{cases} y = 3x - 15 \\ y = -2x + 10 \end{cases}$ (2,1)

3. $\begin{cases} x - 4y = -2 \\ x + 2y = 4 \end{cases}$

4. $\begin{cases} 2x - 3y = 3 \\ \quad\quad y = 3 \end{cases}$

5. $\begin{cases} y = \frac{1}{3}x - 1 \\ x = 12 \end{cases}$

6. $\begin{cases} 2x - 3y = 3 \\ x = 6 \end{cases}$

7. Does $(6, 4)$ satisfy the following system of linear equations?
$$\begin{cases} x - 3y = -6 \\ 3x - 2y = 10 \end{cases}$$

8. Does $(12, 4)$ satisfy the following system of linear equations?
$$\begin{cases} y = \frac{1}{3}x - 1 \\ x = 12 \end{cases}$$

In Exercises 9–12, solve the systems of linear equations.

9. $\begin{cases} 2x + y = 7 \\ x - y = 3 \end{cases}$

10. $\begin{cases} x + 2y = 4 \\ \frac{1}{2}x + \frac{1}{2}y = 3 \end{cases}$

11. $\begin{cases} 5x - 2y = 1 \\ 2x + y = -4 \end{cases}$

12. $\begin{cases} x + 2y = 6 \\ x - \frac{1}{3}y = 4 \end{cases}$

In Exercises 13–16, find the coordinates of the vertices of the feasible sets.

13.

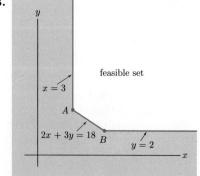

14.

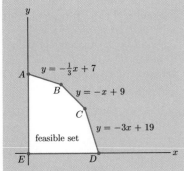

15.

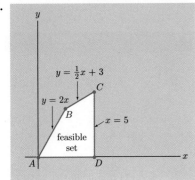

16.

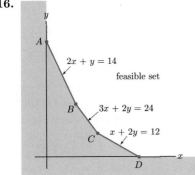

In Exercises 17–22, graph the feasible set for the system of inequalities and find the coordinates of the vertices.

17. $\begin{cases} 2y - x \le 6 \\ x + 2y \ge 10 \\ x \le 6 \end{cases}$

18. $\begin{cases} 2x + y \ge 10 \\ x \ge 2 \\ y \ge 2 \end{cases}$

19. $\begin{cases} x + 3y \le 18 \\ 2x + y \le 16 \\ x \ge 0,\ y \ge 0 \end{cases}$

20. $\begin{cases} 5x + 2y \ge 14 \\ x + 3y \ge 8 \\ x \ge 0,\ y \ge 0 \end{cases}$

21. $\begin{cases} 4x + y \ge 8 \\ x + y \ge 5 \\ x + 3y \ge 9 \\ x \ge 0,\ y \ge 0 \end{cases}$

22. $\begin{cases} x + 4y \le 28 \\ x + y \le 10 \\ 3x + y \le 24 \\ x \ge 0,\ y \ge 0 \end{cases}$

23. (*Supply Curve*) The supply curve for a certain commodity is $p = .0001q + .05$.

(a) What price must be offered in order for 19,500 units of the commodity to be supplied?

(b) What prices result in no units of the commodity being supplied?

24. (*Demand Curve*) The demand curve for a certain commodity is $p = -.001q + 32.5$.

(a) At what price can 31,500 units of the commodity be sold?

(b) What quantities are so large that that many units of the commodity cannot possibly all be sold no matter how low the price?

25. (*Supply and Demand*) Suppose that supply and demand for a certain commodity are described by the supply and demand curves of Exercises 23 and 24. Determine the quantity of the commodity that will be produced and the selling price.

26. The x- and y-intercepts of line L_1 are 1 unit to the left and 2 units below the x- and y-intercepts of line L_2, respectively. The equation for L_1 is $y = 3x + 6$. Find the point of intersection of L_1 and L_2.

27. (*Supply and Demand*) A discount book seller has determined that the supply curve for a certain author's newest paperback book is $p = \frac{1}{300}q + 13$. The demand curve for this book is $p = -.03q + 19$. What quantity of sales would result in supply exactly meeting demand, and for what price should the book be sold?

28. (*Time Apportionment*) A plant supervisor must apportion her 40-hour workweek between hours working on the assembly line and hours supervising the work of others. She is paid $12 per hour for working and $15 per hour for supervising. If her earnings for a certain week are $504, how much time does she spend on each task?

29. (*Calling Card Options*) A calling card offers two methods of paying for a phone call. Method A charges 1 cent per minute but has a 45-cent connection fee. Method B charges 3.5 cents per minute but has no connection fee. Write the equations that show the total cost, y, of a call of x minutes for methods A and B. Graph the two equations and determine their intersection point. What does the intersection point represent?

30. (*Weight Determination*) (**PE**) In a wrestling competition, the total weight of the two contestants is 700 pounds. If twice the weight of the first contestant is 275 pounds more than the weight of the second contestant, what is the weight (in pounds) of the first contestant?
(a) 275 **(b)** 300 **(c)** 325 **(d)** 350 **(e)** 375

31. (*Sales Determination*) (**PE**) An appliance store sells a 13″ TV for $90 and a 19″ TV of the same brand for $120. During a one-week period the store sold 5 more 19″ TVs than 13″ TVs and collected $4800. What was the total number of TV sets sold?
(a) 42   **(b)** 43   **(c)** 44   **(d)** 45   **(e)** 46

Exercises 32–37 require the use of a graphing calculator.

In Exercises 32–35, graph the lines and use the INTERSECT command to estimate the point of intersection to two decimal places.

32. $\begin{cases} y = .25x + 1.3 \\ y = -.5x + 4.1 \end{cases}$ **33.** $\begin{cases} y = -\frac{2}{3}x + 4.5 \\ y = 2x \end{cases}$

34. $\begin{cases} x - 4y = -5 \\ 3x - 2y = 4.2 \end{cases}$ **35.** $\begin{cases} 2x + 3y = 5 \\ -4x + 5y = 1 \end{cases}$

In Exercises 36 and 37, (a) draw the straight lines corresponding to the inequalities (use a window such as ZDecimal or ZoomDec that gives nice values of the coordinates of points); (b) find the coordinates of the vertices of the feasible set; (c) locate the given point with the cursor; (d) determine whether or not the given point lies in the feasible set.

36. $\begin{cases} -x + 3y \geq 3 \\ .4x + y \geq 3.2 \end{cases}$ $(3.2, 2)$

37. $\begin{cases} 2x + y \geq 5 \\ x - 2y \leq 0 \end{cases}$ $(2.2, 1.4)$

SOLUTIONS TO PRACTICE PROBLEMS 1.3

1. Point C is the point of intersection of the lines with equations $y = -\frac{1}{2}x + 4$ and $3x + y = 9$. To use the method of this section, the second equation must be put into its standard form $y = -3x + 9$. Now equate the two expressions for y and solve.

$$-\tfrac{1}{2}x + 4 = -3x + 9$$
$$\tfrac{5}{2}x = 5$$
$$x = \tfrac{2}{5} \cdot 5 = 2$$
$$y = -\tfrac{1}{2}(2) + 4 = 3$$

Therefore, $C = (2, 3)$.

2. $A = (0, 0)$, since the point A is the origin. $B = (0, 4)$, since it is the y-intercept of the line with equation $y = -\frac{1}{2}x + 4$.

3. D is the x-intercept of the line $3x + y = 9$. Its first coordinate is found by setting $y = 0$ and solving for x.

$$3x + (0) = 9$$
$$x = 3$$

Therefore, $D = (3, 0)$.

1.4 The Slope of a Straight Line

As we have seen, any linear equation can be put into one of the two standard forms, $y = mx + b$ or $x = a$. In this section, let us exclude linear equations whose standard form is of the latter type. *Geometrically, this means that we will consider only nonvertical lines.*

Suppose that we are given a nonvertical line L whose equation is $y = mx + b$. The number m is called the *slope of L*. That is, the slope is the coefficient of x in the standard form of the equation of the line.

EXAMPLE 1   **Finding the slope of a line**   Find the slopes of the lines having the following equations.

(a) $y = 2x + 1$ (b) $y = -\frac{3}{4}x + 2$ (c) $y = 3$ (d) $-8x + 2y = 4$

Solution   (a) $m = 2$.

(b) $m = -\frac{3}{4}$.

(c) When we write the equation in the form $y = 0 \cdot x + 3$, we see that $m = 0$.

(d) First, we put the equation in standard form:

$$-8x + 2y = 4$$
$$2y = 8x + 4$$
$$y = 4x + 2.$$

Thus $m = 4$.

Now Try Exercise 1   ■

The definition of the slope is given in terms of the standard form of the equation of the line. Let us give an alternative definition.

> **Geometric Definition Of Slope**   Let L be a line passing through the points (x_1, y_1) and (x_2, y_2), where $x_1 \neq x_2$. Then the slope of L is given by the formula
> $$m = \frac{y_2 - y_1}{x_2 - x_1}. \tag{1}$$

That is, the slope is the difference in the y-coordinates divided by the difference in the x-coordinates, with both differences formed in the same order.

Before proving this definition equivalent to the first one given, let us show how it can be used.

EXAMPLE 2   **Finding the slope of a line**   Find the slope of the line passing through the points $(1, 3)$ and $(4, 6)$.

Solution   We have
$$m = \frac{[\text{difference in } y\text{-coordinates}]}{[\text{difference in } x\text{-coordinates}]} = \frac{6 - 3}{4 - 1} = \frac{3}{3} = 1.$$

Thus $m = 1$. [Note that if we reverse the order of the points and use formula (1) to compute the slope, then we get
$$m = \frac{3 - 6}{1 - 4} = \frac{-3}{-3} = 1,$$

which is the same answer. The order of the points is immaterial. The important concern is to make sure that the differences in the x- and y-coordinates are formed in the same order.]

Now Try Exercise 5   ■

The slope of a line does not depend on which pair of points is chosen as (x_1, y_1) and (x_2, y_2). Consider the line $y = 4x - 3$ and two points $(1, 1)$ and $(3, 9)$, which are on the line. Using these two points, we calculate the slope to be
$$m = \frac{9 - 1}{3 - 1} = \frac{8}{2} = 4.$$

Now let's choose two other points on the line, say $(2, 5)$ and $(-1, -7)$, and use these points to determine m. We obtain
$$m = \frac{-7 - 5}{-1 - 2} = \frac{-12}{-3} = 4.$$

The two pairs of points give the same slope.

Justification of Formula (1)   Since (x_1, y_1) and (x_2, y_2) are both on the line, both points satisfy the equation of the line, which has the form $y = mx + b$. Thus
$$y_2 = mx_2 + b$$
$$y_1 = mx_1 + b.$$

Subtracting these two equations gives

$$y_2 - y_1 = mx_2 - mx_1 = m(x_2 - x_1).$$

Dividing by $x_2 - x_1$, we have

$$m = \frac{y_2 - y_1}{x_2 - x_1},$$

which is formula (1). So the two definitions of slope lead to the same number. ■

Let us now study four of the most important properties of the slope of a straight line. We begin with the *steepness property*, since it provides us with a geometric interpretation for the number m.

Steepness Property   Let the line L have slope m. If we start at any point on the line and move 1 unit to the right, then we must move m units vertically in order to return to the line (Fig. 1). (Of course, if m is positive, then we move up; and if m is negative, we move down.)

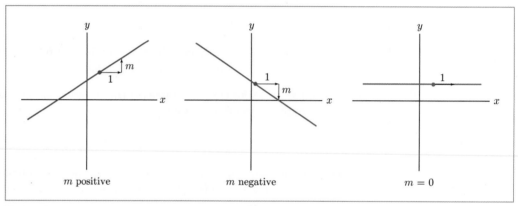

m positive m negative $m = 0$

Figure 1

EXAMPLE 3   **Steepness property of a line**   Illustrate the steepness property for each of the lines.

(a) $y = 2x + 1$ (b) $y = -\frac{3}{4}x + 2$ (c) $y = 3$

Solution   (a) Here $m = 2$. So starting from any point on the line, proceeding 1 unit to the right, we must go 2 units up to return to the line (Fig. 2).

(b) Here $m = -\frac{3}{4}$. So starting from any point on the line, proceeding 1 unit to the right, we must go $\frac{3}{4}$ unit down to return to the line (Fig. 3).

(c) Here $m = 0$. So going 1 unit to the right requires going 0 units vertically to return to the line (Fig. 4).

Now Try Exercise 51   ■

In the next example we introduce a new method for graphing a linear equation. This method relies on the steepness property and is much more efficient than finding two points on the line (e.g., the two intercepts).

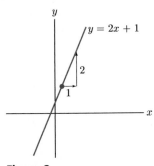

Figure 2

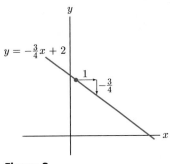

Figure 3

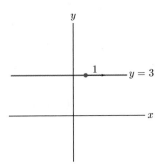

Figure 4

EXAMPLE 4 **Using the steepness property to graph a line** Use the steepness property to draw the graph of $y = \frac{1}{2}x + \frac{3}{2}$.

Solution The y-intercept is $\left(0, \frac{3}{2}\right)$, as we read from the equation. We can find another point on the line using the steepness property. Start at $\left(0, \frac{3}{2}\right)$. Go 1 unit to the right. Since the slope is $\frac{1}{2}$, we must move vertically $\frac{1}{2}$ unit to return to the line. But this locates a second point on the line. So we draw the line through the two points. The entire procedure is illustrated in Fig. 5.

Now Try Exercise 11 ■

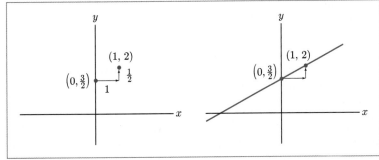

Figure 5

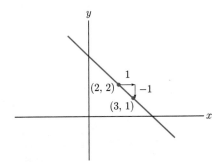

Figure 6

Actually, to use the steepness property to graph an equation, all that is needed is the slope plus *any* point (not necessarily the y-intercept).

EXAMPLE 5 **Using the steepness property to graph a line** Graph the line of slope -1 which passes through the point $(2, 2)$.

Solution Start at $(2, 2)$, move 1 unit to the right and then -1 unit vertically, that is, 1 unit down. The line through $(2, 2)$ and the resulting point is the desired line (see Fig. 6).

Now Try Exercise 25 ■

Slope measures the *steepness* of a line. Namely, the slope of a line tells whether it is rising or falling, and how fast. Specifically, lines of positive slope rise as we move from left to right. Lines of negative slope fall, and lines of zero slope stay level. The larger the magnitude of the slope, the steeper the ascent or descent. These facts are directly implied by the steepness property (see Fig. 7).

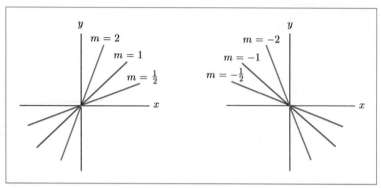

Figure 7

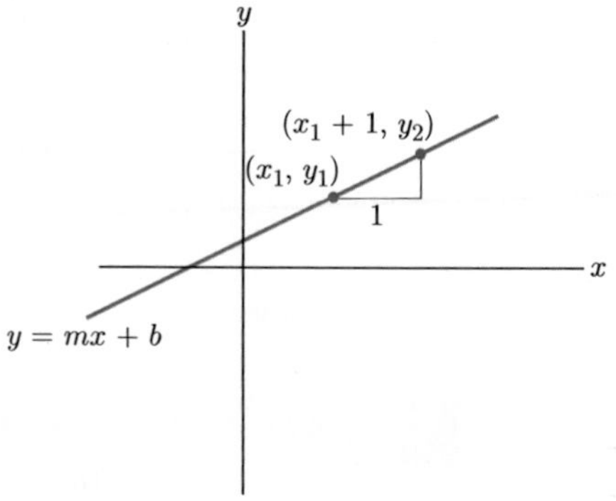

Figure 8

Justification of the Steepness Property Consider a line with equation $y = mx + b$, and let (x_1, y_1) be any point on the line. If we start from this point and move 1 unit to the right, the first coordinate of the new point is $x_1 + 1$ since the x-coordinate is increased by 1. Now go far enough vertically to return to the line. Denote the y-coordinate of this new point by y_2 (see Fig. 8). We must show that to get y_2, we add m to y_1. That is, $y_2 = y_1 + m$. By equation (1), we can compute m as

$$m = \frac{[\text{difference in } y\text{-coordinates}]}{[\text{difference in } x\text{-coordinates}]} = \frac{y_2 - y_1}{1} = y_2 - y_1.$$

In other words, $y_2 = y_1 + m$, which is what we desired to show. ∎

Often the slopes of the straight lines that occur in applications have interesting and significant interpretations. An application in the field of economics is illustrated in the next example.

EXAMPLE 6 **Cost line in economics** Suppose a manufacturer finds that the cost y of producing x units of a certain commodity is given by the formula $y = 2x + 5000$. What interpretation can be given to the slope of the graph of this equation?

Solution Suppose that the firm is producing at a certain level and increases production by 1 unit. That is, x is increased by 1 unit. By the steepness property, the value of y then increases by 2, which is the slope of the line whose equation is $y = 2x + 5000$. Thus each additional unit of production costs \$2. (The graph of $y = 2x + 5000$ is called a *cost curve*. It relates the size of production to total cost. The graph is a straight line, and economists call its slope the *marginal cost of production*. The y-coordinate of the y-intercept is called the *fixed cost*. In this case the fixed cost is \$5000, and it includes costs such as rent and insurance which are incurred even if no units are produced.)

Now Try Exercise 27 ∎

In applied problems having time as a variable, the letter t is often used in place of the letter x. If so, straight lines have equations of the form $y = mt + b$ and are graphed on a ty-coordinate system.

EXAMPLE 7   **Straight-line depreciation**   The federal government allows an income tax deduction for the decrease in value (or *depreciation*) of capital assets (such as buildings and equipment). One method of calculating the depreciation is to take equal amounts over the expected lifetime of the asset. This method is called *straight-line depreciation*. For tax purposes the value V of an asset t years after purchase is figured according to the equation $V = -100{,}000t + 700{,}000$. The expected life of the asset is 5 years.

(a) How much did the asset originally cost?

(b) What is the annual deduction for depreciation?

(c) What is the salvage value of the asset? (That is, what is the value of the asset after 5 years?)

Solution (a) The original cost is the value of V at $t = 0$, namely

$$V = -100{,}000(0) + 700{,}000 = 700{,}000.$$

That is, the asset originally cost $700,000.

(b) By the steepness property, each increase of 1 in t causes a decrease in V of 100,000. That is, the value is decreasing by $100,000 per year. So the depreciation deduction is $100,000 each year.

(c) After 5 years, the value of V is given by

$$V = -100{,}000(5) + 700{,}000 = 200{,}000.$$

The salvage value is $200,000.   ■

We have seen in Example 5 how to sketch a straight line when given its slope and one point on it. Let us now see how to find the equation of the line from this data.

> **Point-Slope Formula**   The equation of the straight line passing through (x_1, y_1) and having slope m is given by $y - y_1 = m(x - x_1)$.

EXAMPLE 8   **Finding the equation of a line**   Find the equation of the line that passes through $(2, 3)$ and has slope $\frac{1}{2}$.

Solution Here $x_1 = 2$, $y_1 = 3$, and $m = \frac{1}{2}$. So the equation is

$$y - 3 = \tfrac{1}{2}(x - 2)$$
$$y - 3 = \tfrac{1}{2}x - 1$$
$$y = \tfrac{1}{2}x + 2.$$

Now Try Exercise 41   ■

EXAMPLE 9   **Finding the equation of a line**   Find the equation of the line through the points $(3, 1)$ and $(6, 0)$.

Solution We can compute the slope from equation (1):

$$[\text{slope}] = \frac{[\text{difference in } y\text{-coordinates}]}{[\text{difference in } x\text{-coordinates}]} = \frac{1 - 0}{3 - 6} = -\frac{1}{3}.$$

Now we can determine the equation from the point-slope formula with $(x_1, y_1) = (3, 1)$ and $m = -\frac{1}{3}$:

$$y - 1 = -\tfrac{1}{3}(x - 3)$$
$$y = -\tfrac{1}{3}x + 2.$$

[*Question:* What would the equation be if we had chosen $(x_1, y_1) = (6, 0)$?]

Now Try Exercise 47   ∎

EXAMPLE 10 **Finding the revenue generated by advertising** For each dollar in monthly advertising expenditure, a store experiences a 6-dollar increase in sales revenue. Even without advertising, the store has $30,000 in sales revenue per month. Let x be the number of dollars of advertising expenditures per month and let y be the number of dollars in sales revenue per month.

(a) Find the equation of the line that expresses the relationship between x and y.

(b) If the store spends $10,000 in advertising, what will be the sales revenue for the month?

(c) How much would the store have to spend on advertising to attain $150,000 in sales revenue for the month?

Solution (a) The steepness property tells us that the line has slope $m = 6$. Since $x = 0$ (no advertising expenditures) yields $y = \$30,000$, the y-intercept of the line is (0, 30,000). Therefore, the standard form of the equation of the line is

$$y = 6x + 30,000.$$

(b) If $x = 10,000$, then $y = 6(10,000) + 30,000$. Therefore, the sales revenue for the month will be $90,000.

(c) We are given that $y = 150,000$, and we must find the value of x for which

$$150,000 = 6x + 30,000.$$

Solving for x, we obtain $6x = 120,000$, and hence $x = \$20,000$. To attain $150,000 in sales revenue, the store should invest $20,000 in advertising.

Now Try Exercise 37   ∎

Verification of the Point-Slope Formula Let (x, y) be any point on the line passing through the point (x_1, y_1) and having slope m. Then, by equation (1), we have

$$m = \frac{y - y_1}{x - x_1}.$$

Multiplying through by $x - x_1$ gives

$$y - y_1 = m(x - x_1). \tag{2}$$

Thus every point (x, y) on the line satisfies equation (2). So (2) gives the equation of the line through (x_1, y_1) and having slope m.   ∎

The next property of slope relates the slopes of two perpendicular lines.

Perpendicular Property When two lines are perpendicular, their slopes are negative reciprocals of one another. That is, if two lines with slopes m and n are perpendicular to one another, then[1]

$$m = -\frac{1}{n}.$$

Conversely, if two lines have slopes that are negative reciprocals of one another, they are perpendicular.

A proof of the perpendicular property is outlined in Exercise 75. Let us show how it can be used to help find equations of lines.

EXAMPLE 11 **Perpendicular lines** Find the equation of the line perpendicular to the graph of $y = 2x - 5$ and passing through $(1, 2)$.

Solution The slope of the graph of $y = 2x - 5$ is 2. By the perpendicular property, the slope of a line perpendicular to it is $-\frac{1}{2}$. If a line has slope $-\frac{1}{2}$ and passes through $(1, 2)$, it has the equation

$$y - 2 = -\tfrac{1}{2}(x - 1) \quad \text{or} \quad y = -\tfrac{1}{2}x + \tfrac{5}{2}$$

(by the point-slope formula).

Now Try Exercise 19   ■

The final property of slopes gives the relationship between slopes of parallel lines. A proof is outlined in Exercise 74.

Parallel Property Parallel lines have the same slope. Conversely, if two lines have the same slope, they are parallel.

EXAMPLE 12 **Parallel lines** Find the equation of the line through $(2, 0)$ and parallel to the line whose equation is $y = \tfrac{1}{3}x - 11$.

Solution The slope of the line having equation $y = \tfrac{1}{3}x - 11$ is $\tfrac{1}{3}$. Therefore, any line parallel to it also has slope $\tfrac{1}{3}$. Thus the desired line passes through $(2, 0)$ and has slope $\tfrac{1}{3}$, so its equation is

$$y - 0 = \tfrac{1}{3}(x - 2) \quad \text{or} \quad y = \tfrac{1}{3}x - \tfrac{2}{3}.$$

Now Try Exercise 21   ■

GC Figure 9 shows the perpendicular lines $y = x$ and $y = -x$ drawn on a TI calculator with the windows **ZStandard** and **ZSquare**. With the **ZSquare** window, the lines visually bisect the quadrants and appear perpendicular. Also, with the **ZSquare** window, one unit on the x-axis has the same length as one unit on the y-axis. We say that the **ZSquare** window has "true aspect." On the TI-83 calculator, a window will approximately exhibit true aspect when (**Ymax** − **Ymin**) ≈

---

[1] If $n = 0$, this formula does not say anything, since $1/0$ is undefined. However, in this case, one line is horizontal and one is vertical, the vertical one having an undefined slope.

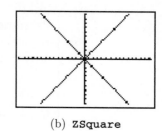

(a) **ZStandard**   (b) **ZSquare**

Figure 9

$\frac{2}{3}$(**Xmax** − **Xmin**). On the TI-89, a window will approximately exhibit true aspect when (**Ymax** − **Ymin**) ≈ $\frac{12}{25}$(**Xmax** − **Xmin**). Only with true aspect do two lines whose slopes are negative reciprocals of each other actually look perpendicular.

PRACTICE PROBLEMS 1.4

Suppose that the revenue y from selling x units of a certain commodity is given by the formula y = 4x. (Revenue is the amount of money received from the sale of the commodity.)

1. What interpretation can be given to the slope of the

graph of this equation?

2. (See Example 6.) Find the coordinates of the point of intersection of $y = 4x$ and $y = 2x + 5000$.

3. What interpretation can be given to the value of the x-coordinate of the point found in Problem 2?

EXERCISES 1.4

In Exercises 1–4, find the slope of the line having the given equation.

1. $y = \frac{2}{3}x + 7$ **2.** $y = -4$

3. $y - 3 = 5(x + 4)$ **4.** $7x + 5y = 10$

In Exercises 5–8, plot each pair of points, draw the straight line between them, and find the slope.

5. $(3, 4)$, $(7, 9)$ **6.** $(-2, 1)$, $(3, -3)$

7. $(0, 0)$, $(5, 4)$ **8.** $(4, 17)$, $(-2, 17)$

9. What is the slope of any line parallel to the y-axis?

10. Why doesn't it make sense to talk about the slope of the line between the two points $(2, 3)$ and $(2, -1)$?

In Exercises 11–14, graph the given linear equation by beginning at the y-intercept, moving 1 unit to the right and m units in the y-direction.

11. $y = -2x + 1$ **12.** $y = 4x - 2$

13. $y = 3x$ **14.** $y = -2$

In Exercises 15–22, find the equation of line L.

15.

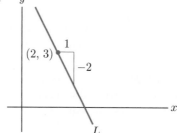

16.

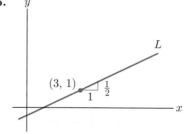

17.

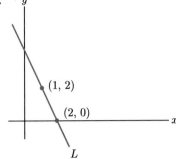

18.

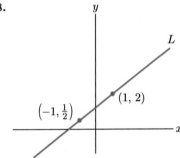

19.

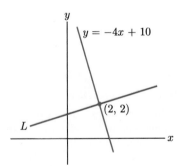

L perpendicular to $y = -4x + 10$

20.

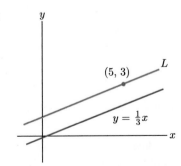

L parallel to $y = \frac{1}{3}x$

21.

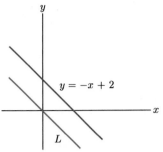

L parallel to $y = -x + 2$

22.

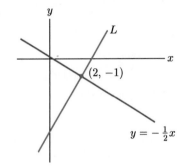

L perpendicular to $y = -\frac{1}{2}x$

23. Find the equation of the line passing through the point $(2, 3)$ and parallel to the x-axis.

24. Find the equation of the line passing through $(0, 0)$ and having slope 1.5.

25. Find the y-intercept of the line passing through the point $(5, 6)$ and having slope $\frac{3}{5}$.

26. Find the slope of the line passing through the point $(1, 4)$ and having y-intercept $(0, 4)$.

27. (*Weekly Pay*) A salesperson's weekly pay depends on the volume of sales. If she sells x units of goods, then her pay is $y = 5x + 60$ dollars. Give an interpretation to the slope and the y-intercept of this straight line.

28. (*Cost Curve*) A manufacturer has fixed costs (such as rent and insurance) of $2000 per month. The cost of producing each unit of goods is $4. Give the linear equation for the cost of producing x units per month.

29. (*Demand Curve*) Suppose the price that must be set in order to sell q items is given by the equation $p = -3q + 1200$.

   (a) Find and interpret the p-intercept of the graph of the equation.

   (b) Find and interpret the q-intercept of the graph of the equation.

   (c) Find and interpret the slope of the graph of the equation.

   (d) What price must be set in order to sell 350 items?

   (e) What quantity will be sold if the price is $300?

   (f) Sketch the graph of the equation.

30. (*Cricket Chirps*) Biologists have found that the number of chirps that crickets of a certain species make per minute is related to the temperature. The relationship is very close to linear. At 68°F those crickets chirp about 124 times a minute. At 80°F they chirp about 172 times a minute.

   (a) Find the linear equation relating Fahrenheit temperature F and the number of chirps c.

   (b) If you only count chirps for 15 seconds, how can you quickly estimate the temperature?

31. (*Cost Equation*) Suppose that the cost of making 20 radios is $6800 and the cost of making 50 radios is $9500.

   (a) Find the cost equation.

   (b) What is the fixed cost?

   (c) What is the marginal cost of production?

   (d) Sketch the graph of the equation.

Exercises 32–34 are related.

32. (*Cost Equation*) Suppose that the total cost y of making x coats is given by the formula $y = 40x + 2400$.

   (a) What is the cost of making 100 coats?

   (b) How many coats can be made for $3600?

   (c) Find and interpret the y-intercept of the graph of the equation.

   (d) Find and interpret the slope of the graph of the equation.

33. (*Revenue Equation*) Suppose that the total revenue y from the sale of x coats is given by the formula $y = 100x$.

   (a) What is the revenue if 300 coats are sold?

   (b) How many coats must be sold to have a revenue of $6000?

   (c) Find and interpret the y-intercept of the graph of the equation.

   (d) Find and interpret the slope of the graph of the equation.

34. (*Profit Equation*) Consider a coat factory with the cost and revenue equations given in Exercises 32 and 33.

   (a) Find the equation giving the profit y resulting from making and selling x coats.

   (b) Find and interpret the y-intercept of the graph of the equation.

   (c) Find and interpret the x-intercept of the graph of the equation.

   (d) Find and interpret the slope of the graph of the equation.

   (e) How much profit will be made if 80 coats are sold?

   (f) How many coats must be sold to have a profit of $6000?

   (g) Sketch the graph of the equation found in part (a).

35. (*Heating Oil*) An apartment complex has a storage tank to hold its heating oil. The tank was filled on January 1, but no more deliveries of oil will be made until sometime in March. Let t denote the number of days after January 1 and let y denote the number of gallons of fuel oil in the tank. Current records show that y and t will be related by the equation $y = 30{,}000 - 400t$.

   (a) Graph the equation $y = 30{,}000 - 400t$.

   (b) How much oil will be in the tank on February 1?

   (c) How much oil will be in the tank on February 15?

   (d) Determine the y-intercept of the graph. Explain its significance.

   (e) Determine the t-intercept of the graph. Explain its significance.

36. (*Cash Reserves*) A corporation receives payment for a large contract on July 1, bringing its cash reserves to $2.3 million. Let y denote its cash reserves (in millions) t days after July 1. The corporation's accountants estimate that y and t will be related by the equation $y = 2.3 - .15t$.

   (a) Graph the equation $y = 2.3 - .15t$.

   (b) How much cash does the corporation have on the morning of July 16?

   (c) Determine the y-intercept of the graph. Explain its significance.

   (d) Determine the t-intercept of the graph. Explain its significance.

   (e) Determine the cash reserves on July 4.

   (f) When will the cash reserves be $.8 million?

37. (*Weekly Pay*) A furniture salesperson earns $160 a week plus 10% commission on her sales. Let x denote her sales and y her income for a week.

   (a) Express y in terms of x.

   (b) Determine her week's income if she sells $1000 in merchandise that week.

   (c) How much must she sell in a week in order to earn $500?

Find the equations of the following lines.

38. Slope is $-\frac{1}{2}$; y-intercept is $(0, 0)$

39. Slope is 3; y-intercept is $(0, -1)$

40. Slope is $-\frac{1}{3}$; $(6, -2)$ on line

41. Slope is 1; $(1, 2)$ on line

42. Slope is $\frac{1}{2}$; $(2, -3)$ on line

43. Slope is -7; $(5, 0)$ on line

44. Slope is $-\frac{2}{5}$; $(0, 5)$ on line

45. Slope is 0; $(7, 4)$ on line

46. $(5, -3)$ and $(-1, 3)$ on line

47. $(2,1)$ and $(4,2)$ on line

48. $(2,-1)$ and $(3,-1)$ on line

49. $(0,0)$ and $(1,-2)$ on line

50. $(3,-1)$ and $(3,1)$ on line

In each of Exercises 51–54 we specify a line by giving the slope and one point on the line. We give the first coordinate of some points on the line. Without deriving the equation of the line, find the second coordinate of each of the points.

51. Slope is 2, $(1,3)$ on line; $(2,\ )$; $(0,\ )$; $(-1,\ )$

52. Slope is -3, $(2,2)$ on line; $(3,\ )$; $(4,\ )$; $(1,\ )$

53. Slope is $-\frac{1}{4}$, $(-1,-1)$ on line; $(0,\ )$; $(1,\ )$; $(-2,\ )$

54. Slope is $\frac{1}{3}$, $(-5,2)$ on line; $(-4,\ )$; $(-3,\ )$; $(-2,\ )$

55. Each of the lines (A), (B), (C), and (D) in Fig. 10 is the graph of one of the linear equations (a), (b), (c), and (d). Match each line with its equation.

(A)

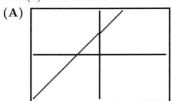

(B)

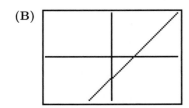

(C)

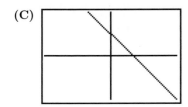

(D)

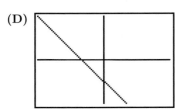

Figure 10

(a) $x+y=1$ **(b)** $x-y=1$

(c) $x+y=-1$ **(d)** $x-y=-1$

56. The following table gives several points on the line $\mathbf{Y}_1 = mx + b$. Find m and b.

X	Y₁
4.8	3.6
4.9	4.8
5	6
5.1	7.2
5.2	8.4
5.3	9.6
5.4	10.8

Y₁=10.8

57. (*Temperature Conversion*) Celsius and Fahrenheit temperatures are related by a linear equation. Use the fact that $0°\text{C} = 32°\text{F}$ and $100°\text{C} = 212°\text{F}$ to find an equation.

58. (*Dating of Artifacts*) An archaeologist dates a bone fragment discovered at a depth of 4 feet as approximately 1500 B.C. and dates a pottery shard at a depth of 8 feet as approximately 2100 B.C. Assuming there is a linear relationship between depths and dates at this archeological site, find the equation that relates depth to date. How deep should the archaeologist dig to look for relics from 3000 B.C.?

59. (*College Tuition*) The average college tuition and fees at four-year public colleges increased from \$2035 in 1990 to \$5132 in 2004. Assuming that average tuition and fees increased linearly with respect to time, find the equation that relates the average tuition and fees, y, to the number of years after 1990, x. What were the average tuition and fees in 1994?

60. (*College Enrollments*) Two-year college enrollments increased from 4.5 million in 1980 to 5.8 million in 2000. Assuming that enrollments increased linearly with respect to time, find the equation that relates the enrollment, y, to the number of years after 1980, x. When was the enrollment 6 million?

61. (*Gas Mileage*) A certain car gets 25 miles per gallon when the tires are properly inflated. For every pound of pressure that the tires are underinflated, the gas mileage decreases by $\frac{1}{2}$ miles per gallon. Find the equation that relates miles per gallon, y, to the amount the tires are underinflated, x. Use the equation to calculate the gas mileage when the tires are underinflated by 8 pounds of pressure.

62. (*Home Health Aid Jobs*) According to the U.S. Department of Labor, home health aid jobs are expected to increase from 580,000 in 2000 to 859,000 in 2012. Assuming that the number of home health aid jobs increases linearly during that time, find the equation that relates the number of jobs, y, to the number of years after 2000, x. Use the equation to predict the number of home health aid jobs in 2008.

63. (*Bachelor Degrees in Business*) According to the U.S. National Center of Educational Statistics, 184,867 bachelor degrees in business were awarded in 1980 and 281,330 were awarded in 2002. If the number of bachelor degrees in business continues to grow linearly, how many bachelor degrees in business will be awarded in 2007?

64. (*Pizza Stores*) According to the Pizza Marketing Quarterly, the number of Domino's Pizza stores grew from 4818 in 2001 to 4904 in 2003. If the number of stores continues to grow linearly, when will there be 6000 stores?

65. Consider a linear equation of the form $y = mx + b$. When the value of b remains fixed and the value of m changes, the graph rotates about the point $(0, b)$. As the value of m increases, will the graph rotate in a clockwise or counterclockwise direction?

66. Write an inequality whose graph consists of the points on or below the straight line passing through the two points $(2, 5)$ and $(4, 9)$.

67. Write an inequality whose graph consists of the points on or above the line with slope 4 and y-intercept $(0, 3)$.

68. Find a system of inequalities having the feasible set in Fig. 11.

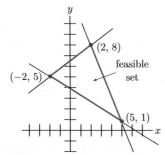

Figure 11

69. Find a system of inequalities having the feasible set in Fig. 12.

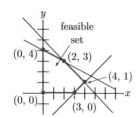

Figure 12

70. Show that the points $(1, 3)$, $(2, 4)$, and $(3, -1)$ are *not* on the same line.

71. For what value of k will the three points $(1, 5)$, $(2, 7)$, and $(3, k)$ be on the same line?

72. Find the value of a for which the line through the points $(a, 1)$ and $(2, -3.1)$ is parallel to the line through the points $(-1, 0)$ and $(3.8, 2.4)$.

73. Rework Exercise 72, where the word "parallel" is replaced by the word "perpendicular."

74. Prove the parallel property. [*Hint*: If $y = mx + b$ and $y = m'x + b'$ are the equations of two lines, then the two lines have a point in common if and only if the equation $mx + b = m'x + b'$ has a solution x.]

75. Prove the perpendicular property. [*Hint*: Without loss of generality, assume that both lines pass through the origin. Use the point-slope formula, the Pythagorean theorem, and Fig. 13.]

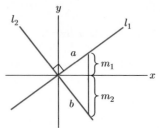

Figure 13

76. (*Temperature Conversion*) (**PE**) Figure 14 gives the conversion of temperatures from Centigrade to Fahrenheit. What is the Fahrenheit equivalent of $30°C$?

(a) $85°F$ (b) $86°F$ (c) $87°F$

(d) $88°F$ (e) $89°F$

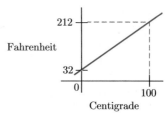

Figure 14

77. (*Shipping Costs*) (**PE**) Figure 15 gives the cost of shipping a package from coast to coast. What is the cost of shipping a 20-pound package?

(a) $15.00   (b) $15.50   (c) $16.00

(d) $16.50   (e) $17.00

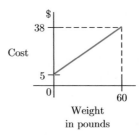

Figure 15

78. (*Costs and Revenue*) (**PE**) A T-shirt company has fixed costs of $25,000 per year. Each T-shirt costs $8.00 to produce and sells for $12.50. How many T-shirts must the company produce and sell each year in order to make a profit of $65,000?

(a) 9200   (b) 11,250   (c) 14,375

(d) 20,000   (e) 25,556

79. (*Costs and Revenue*) (**PE**) A company produces a single product for which variable costs are $100 per unit and annual fixed costs are $1,000,000. If the product sells for $130 per unit, how many units must the company produce and sell in order to attain an annual profit of $2,000,000?

    (a) 10,000   (b) 15,385   (c) 20,000

    (d) 66,667   (e) 100,000

80. (*Revenue Equation*) (**PE**) Suppose the quantity q of a certain brand of mountain bike sold each week depends on price according to the equation $q = 800 - 4p$. What is the total weekly revenue if a bike sells for $150?

    (a) $200   (b) $600   (c) $2000

    (d) $30,000   (e) $80,000

81. (*Revenue Equation*) (**PE**) Suppose the number n of single-use cameras sold each month varies with the price according to the equation $n = 2200 - 25p$. What is the monthly revenue if the price of each camera is $8?

    (a) $200   (b) $2,000   (c) $16,000

    (d) $17,600   (e) $18,000

82. (*Setting a Price*) (**PE**) During 2005 a manufacturer produced 50,000 items that sold for $100 each. The manufacturer had fixed costs of $600,000 and made a profit before income taxes of $400,000. In 2006, rent and insurance combined increased by $200,000. Assuming that the quantity produced and all other costs were unchanged, what should the 2006 price be if the manufacturer is to make the same $400,000 profit before income taxes?

    (a) $96   (b) $100   (c) $102   (d) $104   (e) $106

83. (*Setting a Price*) (**PE**) Rework Exercise 82 with a 2005 fixed cost of $800,000 and a profit before income taxes of $300,000.

    (a) $96   (b) $100   (c) $102   (d) $104   (e) $106

Exercises 84–88 require the use of a graphing calculator.

84. Graph the three lines $y = 2x - 3$, $y = 2x$, and $y = 2x + 3$ together and then identify each line without using TRACE.

85. Graph the two lines $y = .5x + 1$ and $y = -2x + 9$ in the standard window $[-10, 10]$ by $[-10, 10]$. Do they appear perpendicular? If not, use **ZSquare** or **ZoomSqr** to obtain true aspect, and look at the graphs.

86. Graph the line $y = -.5x + 2$ with the window **ZDecimal** or **ZoomDec**. Without pressing TRACE, move the cursor to a point on the line. Then move the cursor one unit to the right and down $\frac{1}{2}$ unit to return to the line. If you start at a point on the line and move 2 units to the right, how many units down will you have to move the cursor to return to the line? Test your answer.

87. Graph the three lines $y = 2x + 1$, $y = x + 1$, and $y = .5x + 1$ together and then identify each line without using TRACE.

88. Repeat Exercise 86 for the line $y = .75x - 2$, using "up" instead of "down" and $\frac{3}{4}$ instead of $\frac{1}{2}$.

SOLUTIONS TO PRACTICE PROBLEMS 1.4

1. By the steepness property, whenever x is increased by 1 unit, the value of y is increased by 4 units. Therefore, each additional unit of production brings in $4 of revenue. (The graph of $y = 4x$ is called a *revenue curve* and its slope is called the *marginal revenue of production*.)

2. $\begin{cases} y = 4x \\ y = 2x + 5000 \end{cases}$

Equate expressions for y:

$$4x = 2x + 5000$$
$$2x = 5000$$
$$x = 2500$$
$$y = 4(2500) = 10,000.$$

3. When producing 2500 units, the revenue equals the cost. This value of x is called the *break-even point*. Since profit = (revenue) − (cost), the company will make a profit only if its level of production is greater than the break-even point (Fig. 16).

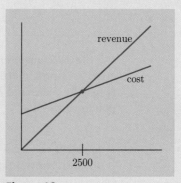

Figure 16

1.5 The Method of Least Squares

Modern people compile graphs of literally thousands of different quantities: the purchasing value of the dollar as a function of time; the pressure of a fixed volume of air as a function of temperature; the average income of people as a function of their years of formal education; or the incidence of strokes as a function of blood pressure. The observed points on such graphs tend to be irregularly distributed due to the complicated nature of the phenomena underlying them as well as to errors made in observation (for example, a given procedure for measuring average income may not count certain groups). In spite of the imperfect nature of the data, we are often faced with the problem of making assessments and predictions based on them. Roughly speaking, this problem amounts to filtering the sources of errors in the data and isolating the basic underlying trend. Frequently, on the basis of a suspicion or a working hypothesis, we may suspect that the underlying trend is linear—that is, the data should lie on a straight line. But which straight line? This is the problem that the *method of least squares* attempts to answer. To be more specific, let us consider the following problem:

> **Problem**   Given observed data points $(x_1, y_1), (x_2, y_2), \ldots, (x_N, y_N)$ in the plane, find the straight line that "best" fits these points.

In order to completely understand the statement of the problem being considered, we must define what it means for a line to "best" fit a set of points. If (x_i, y_i) is one of our observed points, then we will measure how far it is from a given line $y = ax + b$ by the vertical distance, E_i, from the point to the line. (See Fig. 1.)

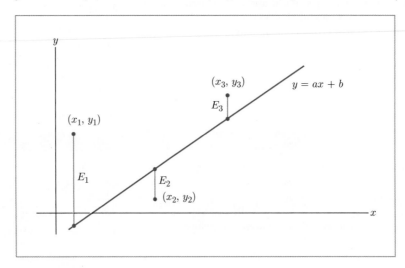

Figure 1. Fitting a line to data points.

Statisticians prefer to work with the square of the vertical distance E_i. The total error in approximating the data points $(x_1, y_1), \ldots, (x_N, y_N)$ by the line $y = ax + b$ is usually measured by the sum E of the squares of the vertical distances from the points to the line,

$$E = E_1^2 + E_2^2 + \cdots + E_N^2.$$

E is called the *least-squares error* of the observed points with respect to the line. If all the observed points lie on the line $y = ax + b$, then all the E_i are zero and the error E is zero. If a given observed point is far away from the line, the corresponding E_i^2 is large and hence makes a large contribution to the error E.

EXAMPLE 1 **Finding the least-squares error** Determine the least-squares error when the line $y = 1.5x + 3$ is used to approximate the data points $(1, 6)$, $(4, 5)$, and $(6, 14)$.

Solution Figure 2 shows the line, the points, and the vertical distances. The vertical distance of a point from the line is determined by finding the second coordinate of the point on the line having the same x-coordinate as the point. For instance, for the data point $(1, 6)$, the point on the line with x-coordinate 1 has y-coordinate $y = 1.5(1) + 3 = 4.5$ and therefore vertical distance $6 - 4.5 = 1.5$. Table 1 summarizes the vertical distances. The table shows that the least-squares error is $1.5^2 + 4^2 + 2^2 = 2.25 + 16 + 4 = 22.25$.

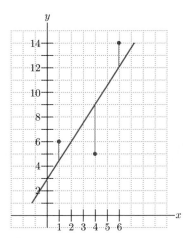

Figure 2

TABLE 1	Vertical distances from the line $y = 1.5x + 3$	
Data Point	**Point on Line**	**Vertical Distance**
$(1, 6)$	$(1, 4.5)$	1.5
$(4, 5)$	$(4, 9)$	4
$(6, 14)$	$(6, 12)$	2

Now Try Exercise 1 ■

In general, we cannot expect to find a line $y = mx + b$ that fits the observed points so well that the error E is zero. Actually, this situation will occur only if the observed points lie on a straight line. However, we can rephrase our original problem as follows:

Problem Given observed data points $(x_1, y_1), (x_2, y_2), \dots, (x_N, y_N)$ in the plane, find the straight line $y = mx + b$ for which the error E is as small as possible.

This line, called the *least-squares line* or the *regression line*, can be found from the data points with the formulas

$$m = \frac{N \cdot \sum xy - \sum x \cdot \sum y}{N \cdot \sum x^2 - \left(\sum x\right)^2}$$

$$b = \frac{\sum y - m \cdot \sum x}{N},$$

where

$$\sum x = \text{sum of the } x\text{-coordinates of the data points}$$

$$\sum y = \text{sum of the } y\text{-coordinates of the data points}$$

$$\sum xy = \text{sum of the products of the coordinates of the data points}$$

$$\sum x^2 = \text{sum of the squares of the } x\text{-coordinates of the data points}$$

$$N = \text{number of data points.}$$

That is,

$$\sum x = x_1 + x_2 + \cdots + x_N$$

$$\sum y = y_1 + y_2 + \cdots + y_N$$

$$\sum xy = x_1 \cdot y_1 + x_2 \cdot y_2 + \cdots + x_N \cdot y_N$$

$$\sum x^2 = x_1^2 + x_2^2 + \cdots + x_N^2.$$

EXAMPLE 2 **Finding the least-squares line** Find the least-squares line for the data points of Example 1.

Solution The sums are calculated in Table 2 and then used to determine the values of m and b.

TABLE 2

x	y	xy	x^2
1	6	6	1
4	5	20	16
6	14	84	36
$\sum x = 11$	$\sum y = 25$	$\sum xy = 110$	$\sum x^2 = 53$

$$m = \frac{3 \cdot 110 - 11 \cdot 25}{3 \cdot 53 - 11^2} = \frac{55}{38} \approx 1.45$$

$$b = \frac{25 - \frac{55}{38} \cdot 11}{3} = \frac{38 \cdot 25 - 55 \cdot 11}{38 \cdot 3} = \frac{345}{114} = \frac{115}{38} \approx 3.03$$

Therefore, the equation of the least-squares line is $y = \frac{55}{38}x + \frac{115}{38}$. With this line, the least-squares error can be shown to be about 22.13.

Now Try Exercise 7 ■

Obtaining the Least-Squares Line with Technology Least-squares lines can be obtained with graphing calculators, spreadsheets, or mathematics software. For instance, on the TI-83 graphing calculator screens in Fig. 3, the data points are entered into lists, the least-squares line is calculated with the item **LinReg(ax + b)** of the STAT/CALC menu, and the data points and line are plotted with [STAT PLOT] and GRAPH. On the TI-89, the least-squares line is calculated with the **LinReg** command found in the MATH/Statistics/Regressions menu. The end of this section contains the details for obtaining least-squares lines with TI-83 and TI-89 graphing calculators. For other graphing calculators, see the guidebook for that particular calculator.

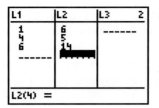 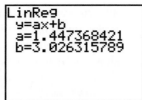 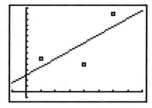

Figure 3. Obtaining a least-squares line with a TI-83.

Spreadsheet programs, such as Excel, have special functions that calculate the slope and y-intercept of the least-squares line for a collection of data points. In Fig. 4 the least-squares line of Example 2 is calculated and graphed in Excel. The end of this section shows how to obtain the graph in Fig. 4.

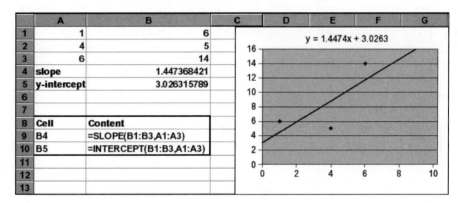

Figure 4. Obtaining a least-squares line with Excel.

The next example obtains a least-squares line and uses the line to make projections.

EXAMPLE 3   **Finding and using the least-squares line**   Table 3 gives the U.S. per capita health care expenditures[1] for several years.

TABLE 3	U.S per capita health care expenditures				
Year	1998	1999	2000	2001	2002
Dollars	4147	4377	4637	5039	5427

(a) Find the least-squares line for this data.

(b) Use the least-squares line to estimate the per capita health care expenditures for the year 2005.

(c) Use the least-squares line to estimate when the per capita health care expenditures will reach $7000.

Solution   (a) The numbers can be made more manageable by counting years beginning with 1998 and measuring dollars in thousands. See Table 4.

TABLE 4	U.S per capita health care expenditures				
Year	0	1	2	3	4
Dollars	4.147	4.377	4.637	5.039	5.427

---

[1]Centers for Medicare and Medicaid Services, Office of the Actuary.

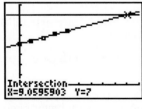

The slope and y-intercept of the least-squares line can be found with the formulas involving sums or with technology. The sums are calculated in Table 5 and then used to determine the values of m and b. The screens in the margin show the results of calculating m and b with a graphing calculator.

TABLE 5

x	y	xy	x^2
0	4.147	0	0
1	4.377	4.377	1
2	4.637	9.274	4
3	5.039	15.117	9
4	5.427	21.708	16
$\sum x = 10$	$\sum y = 23.627$	$\sum xy = 50.476$	$\sum x^2 = 30$

$$m = \frac{5 \cdot 50.476 - 10 \cdot 23.627}{5 \cdot 30 - 10^2} = \frac{16.11}{50} = .3222$$

$$b = \frac{23.627 - .3222 \cdot 10}{5} = 4.081$$

Therefore, the least-squares line is $y = .3222x + 4.081$.

(b) The year 2005 corresponds to $x = 7$. The value of y is

$$y = .3222(7) + 4.081 = 6.3364.$$

Therefore, an estimate of per capita health care expenditures in the year 2005 is \$6336.40.

(c) Set the value of y equal to 7 and solve for x.

$$7 = .3222x + 4.081$$

$$x = \frac{7 - 4.081}{.3222} \approx 9.06$$

Therefore, expenditures are projected to reach \$7000 in 9.06 years after 1998; that is, shortly after the beginning of the year 2007.

Now Try Exercise 13

■

Obtaining the Least-Squares Line with a TI-83

The following steps find the straight line that minimizes the least-squares error for the points $(1, 4)$, $(2, 5)$, and $(3, 8)$.

1. Press $\boxed{\text{STAT}}$ **1** to obtain a table used for entering the data.
2. If there are no data in columns labeled $\mathbf{L}_1$ and $\mathbf{L}_2$, proceed to step 4.
3. Move the cursor up to $\mathbf{L}_1$ and press $\boxed{\text{CLEAR}}$ $\boxed{\text{ENTER}}$ to delete all data in $\mathbf{L}_1$'s column. Move the cursor right and up to $\mathbf{L}_2$ and press $\boxed{\text{CLEAR}}$ $\boxed{\text{ENTER}}$ to delete all data in $\mathbf{L}_2$'s column.
4. If necessary, move the cursor left to the first blank row of the $\mathbf{L}_1$ column. Press **1** $\boxed{\text{ENTER}}$ **2** $\boxed{\text{ENTER}}$ **3** $\boxed{\text{ENTER}}$ to place the x-coordinates of the three points into the $\mathbf{L}_1$ column.
5. Move the cursor right to the $\mathbf{L}_2$ column and press **4** $\boxed{\text{ENTER}}$ **5** $\boxed{\text{ENTER}}$ **8** $\boxed{\text{ENTER}}$ to place the y-coordinates of the three points into the $\mathbf{L}_2$ column. The screen should now appear as in Fig. 5.

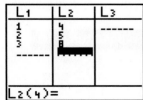

Figure 5

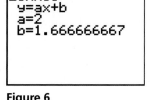

Figure 6

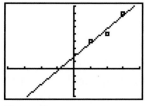

Figure 7

6. Press $\boxed{\text{STAT}}$ $\boxed{\blacktriangleright}$ and press the number for **LinReg(ax + b)**.

7. Press $\boxed{\text{ENTER}}$. The TI-83 screen should now appear as in Fig. 6. The least-squares line is $y = 2x + \frac{5}{3}$. (*Note*: $\frac{5}{3} \approx 1.666666667$.)

8. If desired, the linear function can be assigned to $\mathbf{Y}_1$ with the following steps:

   (a) Press $\boxed{\text{Y=}}$ $\boxed{\text{CLEAR}}$ to erase the current expression in $\mathbf{Y}_1$.

   (b) Press $\boxed{\text{VARS}}$ **5** $\boxed{\blacktriangleright}$ $\boxed{\blacktriangleright}$ and the number for **RegEQ** to assign the linear function to $\mathbf{Y}_1$.

9. The original points can be easily plotted along with the least-squares line. Assume that the linear function has been assigned to $\mathbf{Y}_1$, that all other functions have been cleared or deselected, and that the window has been set to $[-4, 4]$ *by* $[-4, 9]$. Press $\boxed{\text{2nd}}$ [STAT PLOT] $\boxed{\text{ENTER}}$ $\boxed{\text{ENTER}}$ $\boxed{\text{GRAPH}}$ to see the display in Fig. 7. (*Note 1*: To turn off the point-plotting feature, press $\boxed{\text{2nd}}$ [STAT PLOT] $\boxed{\text{ENTER}}$ $\boxed{\blacktriangleright}$ $\boxed{\text{ENTER}}$. *Note 2*: The point-plotting feature can be toggled from the **Y=** editor by moving the cursor to the word **"Plot1"** on the top line and pressing $\boxed{\text{ENTER}}$.)

Obtaining the Least-Squares Line with a TI-89 The following steps find the straight line that minimizes the least-squares error for the points $(1, 4)$, $(2, 5)$, and $(3, 8)$.

1. Use $\boxed{\text{APPS}}$ to display the Data/Matrix editor.

2. Press **3** to display a New dialog box.

3. Press $\boxed{\blacktriangledown}$ $\boxed{\blacktriangledown}$ and give a name to the variable, such as **dat**.

4. Press $\boxed{\text{ENTER}}$ twice to display a spreadsheet.

5. Enter the three x-values (1, 2, and 3) into the first column.

6. Enter the three y-values (4, 5, and 8) into the second column.

7. Press $\boxed{\text{F5}}$ to invoke a Calculate dialog box.

8. Set **Calculation Type** to **LinReg**; that is press **5**.

9. Set x to **c1**, and set y to **c2**.

10. Set **Store RegEQ** to **y1(x)**.

11. Press $\boxed{\text{ENTER}}$ to invoke a STAT VARS window displaying $\mathbf{y = a \cdot x + b}$, $\mathbf{a = 2}$, $\mathbf{b = 1.666667}$, **corr** $= .960769$, $\mathbf{R^2} = .923077$. (*Note*: $\frac{5}{3} \approx 1.666667$. The number *corr* is called the correlation coefficient and is a measure of the degree of linear association between the two variables. If the absolute value of *corr* is close to 1, then there is a high degree of linear association. The number $\mathbf{R^2}$ is the square of *corr*. Also, $\mathbf{2 \cdot x + 1.6666666666667}$ will be assigned to $\mathbf{y1(x)}$ in the **Y=** editor.)

12. Press $\boxed{\text{ENTER}}$ or $\boxed{\text{ESC}}$ to return to the spreadsheet.

13. Press $\boxed{\text{F2}}$ to invoke the Plot window.

14. Press $\boxed{\text{F1}}$ to invoke the dialog box for setting Plot1.

15. Set **Plot Type** to **Scatter**, set **Mark** to **Box**, set **x** to **c1**, set **y** to **c2**, and then press $\boxed{\text{ENTER}}$.

16. Press $\boxed{\blacklozenge}$ $\boxed{\text{F2}}$ and set the window to $[-4, 4]$ *by* $[-4, 9]$.

17. Press $\boxed{\blacklozenge}$ $\boxed{\text{F3}}$ to produce a graph similar to the one in Figure 7. (*Note*: The point plotting feature can be suspended by going to the **Y=** window, moving the cursor to Plot1 and pressing $\boxed{\text{F4}}$. Pressing $\boxed{\text{F5}}$ instead, clears the value assigned to Plot1.)

Obtaining the Least-Squares Line with an Excel Spreadsheet The following steps obtain the graph in Fig. 4.

1. Enter the six numbers in the range A1:B3, and then select the range.
2. Click on the Chart Wizard icon in the Standard toolbar.
3. Click on XY (Scatter) in the "Chart type" list.
4. Click on the Finish button in the lower right corner of the window.
5. Right click on one of the diamond-shaped points in the graph and then click on the Add Trendline.
6. Click on the Options tab.
7. Check the box in front of "Display equation on chart," and click OK.
8. Click on the equation and then drag it to the top of the graph.
9. Right click on the legend located to the right of the graph and click on Clear.

PRACTICE PROBLEMS 1.5

1. Can a vertical distance be negative?

2. Under what condition will a vertical distance be zero?

EXERCISES 1.5

1. Suppose the line $y = 3x + 1$ is used to fit the four data points in Table 6. Complete the table and determine the least-squares error E.

TABLE 6

Data Point	Point on Line	Vertical Distance
$(1, 3)$		
$(2, 6)$		
$(3, 11)$		
$(4, 12)$		

2. Suppose the line $y = -2x + 12$ is used to fit the four data points in Table 7. Complete the table and determine the least-squares error E.

TABLE 7

Data Point	Point on Line	Vertical Distance
$(1, 11)$		
$(2, 7)$		
$(3, 5)$		
$(4, 5)$		

3. Find the least-squares error E for the least-squares line fit to the four points in Fig. 8.

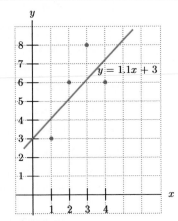

Figure 8

4. Find the least-squares error E for the least-squares line fit to the five points in Fig. 9.

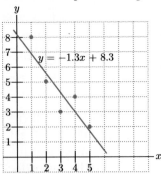

Figure 9

5. Complete Table 8 and find the values of m and b for the straight line that provides the best least-squares fit to the data.

TABLE 8

x	y	xy	x^2
1	7		
2	6		
3	4		
4	3		
$\sum x =$	$\sum y =$	$\sum xy =$	$\sum x^2 =$

6. Complete Table 9 and find the values of m and b for the straight line that provides the best least-squares fit to the data.

TABLE 9

x	y	xy	x^2
1	2		
2	4		
3	7		
4	9		
5	12		
$\sum x =$	$\sum y =$	$\sum xy =$	$\sum x^2 =$

7. Consider the data points $(1, 2)$, $(2, 5)$, and $(3, 11)$. Find the straight line that provides the best least-squares fit to these data.

8. Consider the data points $(1, 8)$, $(2, 4)$, and $(4, 3)$. Find the straight line that provides the best least-squares fit to these data.

9. Consider the data points $(1, 9)$, $(2, 8)$, $(3, 6)$, and $(4, 3)$. Find the straight line that provides the best least-squares fit to these data.

10. Consider the data points $(1, 5)$, $(2, 7)$, $(3, 6)$, and $(4, 10)$. Find the straight line that provides the best least-squares fit to these data.

Since Exercises 11–16 use real data, they are best answered with a graphing calculator or spreadsheet.

11. (*Lung Cancer and Smoking*) The following table[2] gives the crude male death rate for lung cancer in 1950 and the per capita consumption of cigarettes in 1930 in various countries.

Country	Cigarette Consumption (per Capita)	Lung Cancer Deaths (per Million Males)
Norway	250	95
Sweden	300	120
Denmark	350	165
Australia	470	170

(a) Use the method of least-squares to obtain the straight line that best fits these data.

(b) In 1930 the per capita cigarette consumption in Finland was 1100. Use the straight line found in part (a) to estimate the male lung cancer death rate in Finland in 1950.

12. (*Cost and Usage of Fuel*) The accompanying table shows the 1999 price of a gallon (in U.S. dollars) of fuel and the average miles driven per automobile for several countries.[3]

Country	Price per Gallon	Average Miles per Automobile
France	$4.01	9090
Germany	$3.78	7752
Japan	$3.65	6009
Sweden	$4.22	9339
United Kingdom	$5.13	10,680

(a) Find the straight line that provides the best least-squares fit to these data.

(b) In 1999, the price of gas in Canada was $2.04 per gallon. Use the straight line of part (a) to estimate the average number of miles automobiles were driven in Canada.

(c) In 1999 the average miles driven in the United States was 11,868. Use the straight line of part (a) to estimate the 1999 price of a gallon of gasoline in the United States.

13. (*College Graduates*) The following table gives the percent of persons 25 years and over who have completed four or more years of college.[4]

(a) Use the method of least squares to obtain the straight line that best fits these data.

(b) Estimate the percent for the year 1993.

(c) If the trend determined by the straight line in part (a) continues, when will the percent reach 28?

---

[2]These data were obtained from *Smoking and Health*, Report of the Advisory Committee to the Surgeon General of the Public Health Service, U.S. Department of Health, Education, and Welfare, Washington, D.C., Public Health Service Publication No. 1103, p. 176.

[3]U.S. Department of Transportation, Federal Highway Administration, *Highway Statistics*, 2000.

[4]U.S. Bureau of the Census.

Year	Percent
1975	13.9
1980	17.0
1985	19.4
1990	21.3
1995	23.0
2000	25.6

14. (*College Enrollments*) The following table gives enrollment (in millions) in public colleges in the United States[5] for certain years.

Year	Enrollment
1985	9.5
1990	10.8
1995	11.1
1997	11.2
1998	11.2
1999	11.3
2000	11.8
2001	12.2

(a) Use the method of least squares to obtain the straight line that best fits these data.

(b) Estimate the enrollment in 1988.

(c) If the trend determined by the straight line in part (a) continues, when will the enrollment reach 13 million?

15. (*Life Expectancy*) The following table is an abbreviated life expectancy table for U.S. males.

TABLE 10	Life Expectancy for U.S. Males
Current Age	**Life Expectancy**
0	74
20	75
40	77
60	80
80	87

(a) Find the straight line that provides the best least-squares fit to these data.

(b) Use the straight line of part (a) to estimate the life expectancy of a 30-year-old U.S. male. (*Note*: The actual life expectancy is 76 years.)

(c) Use the straight line of part (a) to estimate the life expectancy of a 50-year-old U.S. male. (*Note*: The actual life expectancy is 78 years.)

(d) Use the straight line of part (a) to estimate the life expectancy of a 90-year-old U.S. male. (*Note*: The actual life expectancy is 94 years.)

16. (*Banking*) Two Harvard economists studied countries' relationships between the independence of banks and inflation rates from 1955 to 1990.[6] The independence of banks was rated on a scale of −1.5 to 2.5, with −1.5, 0, and 2.5 corresponding to least, average, and most independence, respectively. The following table gives the values for various countries.

Country	Independence Rating	Inflation Rate (%)
New Zealand	−1.4	7.6
Italy	−.75	7.2
Belgium	.3	4.0
France	.4	6.0
Canada	.9	4.5
United States	1.6	4.0
Switzerland	2.2	3.1

(a) Use the method of least squares to obtain the straight line that best fits these data.

(b) What relationship between independence of banks and inflation is indicated by the least-squares line?

(c) Japan has a .6 independence rating. Use the least-squares line to estimate Japan's inflation rate.

(d) The inflation rate for Britain is 6.8. Use the least-squares line to estimate Britain's independence rating.

17. (*Consumer Price Index*) The following table gives the average price of a pound of potato chips in January of the given years. (*Source*: U.S. Bureau of Labor Statistics, Consumer Price Index.)

Year	Price
1987	$2.75
1993	$2.83
1996	$3.00
2000	$3.39
2004	$3.43

(a) Use the method of least squares to obtain the straight line that best fits these data.

(b) Estimate the average price of a pound of potato chips in January 1999.

---

[5]U.S. Dept. of Education, National Center for Education Statistics, *Digest of Education Statistics, 2003*.

[6]J. Bradford DeLong (Harvard) and L. H. Summers (World Bank).

(c) If this trend continues, when will the average price of a pound of potato chips be $3.63?

18. Find the best least-squares fit to the points $(5, 4)$ and $(7, 3)$. Show that it is the straight line through the two points.

SOLUTIONS TO PRACTICE PROBLEMS 1.5

1. No. The word "distance" implies a nonnegative number. It is the absolute value of the difference between the y-coordinate of the data point and the y-coordinate of the point on the line.

2. The vertical distance will be zero when the point actually lies on the least-squares line.

CHAPTER 1 SUMMARY

1. Cartesian coordinate systems associate a number with each point of a line and associate a pair of numbers with each point of a plane.

2. The collection of points in the plane that satisfy the linear equation $ax + by = c$ (where a and b are not both zero) lies on a straight line. After this equation is put into one of the standard forms $y = mx + b$ or $x = a$, the graph can be easily drawn.

3. The direction of the inequality sign in an inequality is unchanged when a number is added to or subtracted from both sides of the inequality, or when both sides of the inequality are multiplied by the same positive number. The direction of the inequality sign is reversed when both sides of the inequality are multiplied by the same negative number.

4. The collection of points in the plane that satisfy the linear inequality $ax + by \leq c$ or $ax + by \geq c$ consists of all points on and to one side of the graph of the corresponding linear equation. After this inequality is put into standard form, the graph can be easily pictured by crossing out the half-plane consisting of the points that do not satisfy the inequality.

5. The feasible set of a system of linear inequalities (that is, the collection of points that satisfy all the inequalities) is best obtained by crossing out the points not satisfied by each inequality.

6. The point of intersection of a pair of lines can be obtained by first converting the equations to standard form and then either equating the two expressions for y or substituting the value of x from the form $x = a$ into the other equation.

7. The slope of the line $y = mx + b$ is the number m. It is also the ratio of the difference between the y-coordinates and the difference between the x-coordinates of any pair of points on the line.

8. The steepness property states that if we start at any point on a line of slope m and move 1 unit to the right, then we must move m units vertically to return to the line.

9. The point-slope formula states that the line of slope m passing through the point (x_1, y_1) has the equation $y - y_1 = m(x - x_1)$.

10. Two lines are parallel if and only if they have the same slope. Two lines are perpendicular if and only if the product of their slopes is -1.

11. The method of least squares finds the straight line that gives the best fit to a collection of points in the sense that the sum of the squares of the vertical distances from the points to the line is as small as possible. The slope and y-intercept of the least-squares line are usually found with formulas involving sums of coordinates or by using technology.

REVIEW OF FUNDAMENTAL CONCEPTS

1. How do you determine the coordinates of a point in the plane?

2. What is meant by the graph of an equation in x and y?

3. What is the general form of a linear equation in x and y?

4. What is the standard form of a linear equation in x and y?

5. What is the y-intercept of a line? How do you find the y-intercept from an equation of a line?

6. What is the x-intercept of a line? How do you find the x-intercept from an equation of a line?

7. Give a method for graphing the equation $y = mx + b$.

8. State the inequality properties for addition, subtraction, and multiplication.

9. What are the general and standard forms of a linear inequality in x and y?

10. Explain how to obtain the graph of a linear inequality.

11. What is meant by the *feasible set* of a system of linear inequalities?

12. Describe how to obtain the point of intersection of two lines.

13. Define the slope of a line, and give a physical description.

14. Suppose you know the slope and the coordinates of a point on a line. How could you draw the line without first finding its equation?

15. What is the point-slope form of the equation of a line?

16. Describe how to find the equation for a line when you know the coordinates of two points on the line.

17. What can you say about the slopes of perpendicular lines?

18. What can you say about the slopes of parallel lines?

19. What is the least-squares line approximation to a set of data points?

KEY FORMULAS

Equation of line with slope m and y-intercept $(0, b)$:

$$y = mx + b$$

Equation of line with slope m and passing through (x_1, y_1):

$$y - y_1 = m(x - x_1)$$

Slope of line passing through (x_1, y_1) and (x_2, y_2):

$$m = \frac{y_2 - y_1}{x_2 - x_1}$$

Let m and n be slopes of parallel lines:

$$m = n$$

Let m and n be slopes of perpendicular lines:

$$m = -\frac{1}{n}$$

Let $y = mx + b$ be the least-squares line for the points $(x_1, y_1), (x_2, y_2), \ldots, (x_N, y_N)$:

$$m = \frac{N \cdot \sum xy - \sum x \cdot \sum y}{N \cdot \sum x^2 - \left(\sum x\right)^2}$$

$$b = \frac{\sum y - m \cdot \sum x}{N},$$

where

$$\sum x = x_1 + x_2 + \ldots + x_N$$

$$\sum y = y_1 + y_2 + \ldots + y_N$$

$$\sum xy = x_1 \cdot y_1 + x_2 \cdot y_2 + \cdots + x_N y_N$$

$$\sum x^2 = x_1^2 + x_2^2 + \cdots + x_N^2.$$

SUPPLEMENTARY EXERCISES

1. What is the equation of the y-axis?

2. Graph the linear equation $y = -\frac{1}{2}x$.

3. Find the point of intersection of the pair of straight lines $x - 5y = 6$ and $3x = 6$.

4. Find the slope of the line having the equation $3x - 4y = 8$.

5. Find the equation of the line having y-intercept $(0, 5)$ and x-intercept $(10, 0)$.

6. Graph the linear inequality $x - 3y \geq 12$.

7. Does the point $(1, 2)$ satisfy the linear inequality $3x + 4y \geq 11$?

8. Find the point of intersection of the pair of straight lines $2x - y = 1$ and $x + 2y = 13$.

9. Find the equation of the straight line passing through the point $(15, 16)$ and parallel to the line $2x - 10y = 7$.

10. Find the y-coordinate of the point having x-coordinate 1 and lying on the line $y = 3x + 7$.

11. Find the x-intercept of the straight line with equation $x = 5$.

12. Graph the linear inequality $y \leq 6$.

13. Solve the following system of linear equations:

$$\begin{cases} 3x - 2y = 1 \\ 2x + y = 24. \end{cases}$$

14. Graph the feasible set for the following system of inequalities:

$$\begin{cases} 2y + 7x \geq 30 \\ 4y - x \geq 0 \\ y \leq 8. \end{cases}$$

15. Find the y-intercept of the line passing through the point $(4, 9)$ and having slope $\frac{1}{2}$.

16. (*Cost of Moving*) The fee charged by a local moving company depends on the amount of time required for the move. If t hours are required, then the fee is $y = 35t + 20$ dollars. Give an interpretation of the slope and y-intercept of this line.

17. Are the points $(1, 2)$, $(2, 0)$, and $(3, 1)$ on the same line?

18. Write an equation of the line with x-intercept $(3, 0)$ and y-intercept $(0, -2)$.

19. (**PE**) If $x + 7y = 30$ and $x = -2y$, then y equals
(**a**) -12 (**b**) 12 (**c**) $30/9$ (**d**) 6 (**e**) -6

20. Write the inequality whose graph is the half-plane below the line with slope $\frac{2}{3}$ and y-intercept $\left(0, \frac{3}{2}\right)$.

21. Write the inequality whose graph is the half-plane above the line through $(2, -1)$ and $(6, 8.6)$.

22. Solve the system of linear equations

$$\begin{cases} 1.2x + 2.4y = .6 \\ 4.8y - 1.6x = 2.4. \end{cases}$$

23. Find the equation of the line through $(1, 1)$ and the intersection point of the lines $y = -x + 1$ and $y = 2x + 3$.

24. Find all numbers x such that $2x + 3(x - 2) \geq 0$.

25. Graph the equation $x + \frac{1}{2}y = 4$, and give the slope and both intercepts.

26. Do the three graphs of the linear equations $2x - 3y = 1$, $5x + 2y = 0$, and $x + y = 1$ contain a common point?

27. Show that the lines with equations $2x - 3y = 1$ and $3x + 2y = 4$ are perpendicular.

28. Each of the half-planes (A), (B), (C), and (D) is the graph of one of the linear inequalities (a), (b), (c), and (d). Match each half-plane with its inequality.

(A)

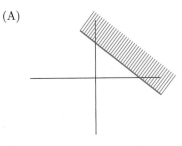

(B)

(C)

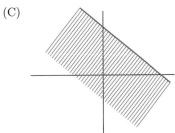

(D)

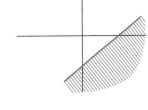

(**a**) $x + y \geq 1$ (**b**) $x + y \leq 1$
(**c**) $x - y \leq 1$ (**d**) $y - x \leq -1$

29. Each of the lines L_1, L_2, and L_3 in Fig. 1 is the graph of one of the equations (a), (b), and (c). Match each of the following equations with its corresponding line.

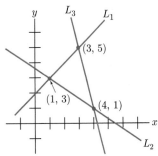

Figure 1

(**a**) $4x + y = 17$ (**b**) $y = x + 2$
(**c**) $2x + 3y = 11$

30. Find a system of inequalities having the feasible set in Fig. 2 and find the coordinates of the unspecified vertex. [*Note*: There is a right angle at the vertex $\left(4, \frac{3}{2}\right)$.]

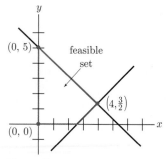

Figure 2

31. Consider the following four equations:

$$p = .01q - 5$$
$$p = .005q + .5$$
$$p = -.01q - 10$$
$$p = -.01q + 5.$$

One is the equation of a supply curve and another is the equation of a demand curve. Identify the two equations and then find the intersection point of those two curves.

32. Find the vertices of the following feasible set.

$$\begin{cases} x \geq 0 \\ y \geq 0 \\ 5x + y \leq 50 \\ 2x + 3y \leq 33 \\ x - 2y \geq -8 \end{cases}$$

33. It is not possible to draw a straight line through all three of the points $(2, 4)$, $(5, 8)$, and $(7, 9)$. However, it *is* possible for a straight line to *miss* all three points by the same amount; that is, there is a line that makes the vertical distances d_1, d_2, and d_3 in Fig. 3 equal. Find the equation of this line. [*Hint:* Let the equation of the line be $y = mx + b$. Then the point P has coordinates $(2, 2m + b)$.]

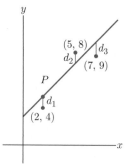

Figure 3

34. (*Profit Equation*) For a certain manufacturer, the production and sale of each additional unit yields an additional profit of $10. The sale of 1000 units yields a profit of $4000.

(a) Find the profit equation.

(b) Find the y- and x-intercepts of the graph of the profit equation.

(c) Sketch the graph of the profit equation.

35. (*Car Rentals*) One-day car rentals cost $50 plus 10 cents per mile from company A and $40 plus 20 cents per mile from company B.

(a) For each company, give the linear equations for the cost, y, when x miles are driven.

(b) Which company offers the best value when the car is driven for 80 miles?

(c) Which company offers the best value when the car is driven for 160 miles?

(d) For what mileage do the two companies offer the same value?

36. (*Inflation*) In 1980, the average loaf of white bread cost $0.51. In 2004, the average loaf of white bread cost $0.97. (*Source:* U.S. Bureau of Labor Statistics, Consumer Price Index)

(a) Assuming a linear increase in the price per loaf of bread, find the equation that relates the cost, y, to the number of years after 1980, x.

(b) When was the average cost of a loaf of white bread $0.89?

37. Graph the linear inequality $x \leq 3y + 2$.

38. (*Sales Commission*) A furniture store offers its new employees a weekly salary of $200 plus a 3% commission on sales. After one year, employees receive $100 per week plus a 5% sales commission. For what weekly sales level will the two scales produce the same salary?

39. Find a system of linear inequalities having the feasible set of Fig. 4.

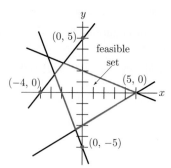

Figure 4

40. Find a system of linear inequalities having the feasible set of Fig. 5.

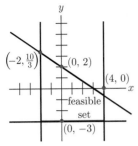

Figure 5

41. (*Medical Assistant Jobs*) According to the U.S. Department of Labor, medical assistant jobs are expected to increase from 365,000 in 2000 to 579,000 in 2012. Assuming that the number of medical assistant jobs increases linearly during that time, find the equation that relates the number of jobs, y, to the number of years after 2000, x. Use the equation to predict the number of medical assistant jobs in 2008.

42. (*Bachelor Degrees in Education*) According to the U.S. National Center of Educational Staistics, 118,038 bachelor degrees in education were awarded in 1980 and 106,383 were awarded in 2002. If the number of bachelor degrees in education continues to decline linearly, how many bachelor degrees in education will be awarded in 2010?

43. (*Soft Drinks*) According to *Beverage Digest*, Coke Classic's percentage of the soft drink market declined from 20.4 in 2000 to 18.6 in 2003. If the percentage declined linearly during that time, estimate Coke Classic's percentage of the soft drink market in 2002.

44. (*Life Expectancy*) The following table gives the 2005 life expectancy at birth for several countries[7]

	Male	**Female**
Finland	74.82	82.02
United States	74.89	80.67
New Zealand	75.67	81.78
Japan	77.86	84.61

(a) Use the method of least squares to obtain the straight line that best fits these data.

(b) In Switzerland the life expectancy of men is 77.58 years. Use the least-squares line from (a) to estimate the life expectancy for women.

(c) In Denmark the life expectancy for women is 80.03 years. Use the least-squares line from (a) to estimate the life expectancy for men.

45. (*Prospective Nurses*) Students entering college in 2004 expressed the greatest interest in nursing careers in many years. The following table gives the percent of college freshmen in 2000 through 2004 who said that their probable career choice would be in nursing.[8]

(a) Use the method of least squares to obtain the straight line that best fits these data.

(b) Estimate the percentage of students who entered college in 2006 whose probable career was in nursing.

(c) If the trend continues, when will 5% of freshmen have nursing as their probable career choice?

Year	Percent
2000	2.1
2001	2.4
2002	2.7
2003	3.5
2004	3.9

46. (*Cancer and Diet*) The following table[9] gives the (age-adjusted) death rate per 100,000 women from breast cancer and the daily dietary fat intake (in grams per day) for various countries.

Country	Fat Intake	Death Rate
Japan	41	4
Poland	90	10
Finland	118	13
United States	148	21

(a) Use the method of least squares to obtain the straight line that best fits these data.

(b) In Denmark women consume an average of 160 grams of fat per day. Use the least-squares line to estimate the breast cancer death rate.

(c) In New Zealand the breast cancer death rate is 22 women per 100,000. Use the least-squares line to estimate the daily fat intake in New Zealand.

Conceptual Exercises

47. Consider an equation of the form $y = mx + b$. When the value of m remains fixed and the value of b changes, the graph is translated vertically. As the value of b increases, does the graph move up or down?

48. Does every line have an x-intercept? y-intercept?

49. When is the x-intercept of a line the same as the y-intercept?

50. What is the difference between a line having no slope and zero slope?

---

[7] U.S. Bureau of the Census, International Database.

[8] *The American Freshman: National Norms*; American Council on Education, University of California–Los Angeles.

[9] B. S. Reddy et al., "Nutrition and Its Relationship to Cancer," *Advances in Cancer Research* 32:237, 1980.

51. Suppose you have found the line of best least-squares fit to a collection of points and that you edit the data by adding a point on the line to the data. Will the ex-panded data have the same least-squares line? Explain the rationale for your conclusion, and then experiment to test whether your conclusion is correct.

CHAPTER TEST

1. Graph the linear equation $y = 3x$.

2. Find the y-coordinate of the point having x-coordinate $\frac{1}{2}$ and lying on the line $y = -2x + 6$.

3. Find the equation of the line through the point $(-1, 3)$ and parallel to the line $y = -2x + 6$.

4. Solve the system of linear equations

$$\begin{cases} 2x - 3y = 9 \\ -3x + 7y = -11. \end{cases}$$

5. Show that the lines $3x - y = 1$ and $-\frac{1}{3}x - 4 = y$ are perpendicular.

6. Graph the inequality $x + y \geq 8$ in the plane.

7. Find the equation of the line of slope 2 that passes through the intersection point of the lines $4x + 5y = 11$ and $2x - 3y = 7$.

8. Find the coordinates of the vertices of the feasible set for the following system of inequalities.

$$\begin{cases} x + y \geq 16 \\ -2x + y \leq 10 \\ 3x + y \leq 75 \\ x \geq 0, \ y \geq 0 \end{cases}$$

9. (*Sales Commissions*) Fred sells carpeting. He earns $200 each week plus 5% commission on sales. His sister thinks she has a better job selling carpet because she earns $250 a week plus 3% commission on her sales. For what volume of sales is Fred's sister correct? When is she incorrect?

10. (*Inflation*) The following table gives the average price of a pound of Red Delicious apples in January of the given years. (*Source*: U.S. Bureau of Labor Statistics, Consumer Price Index.)

(a) Use the method of least squares to obtain the straight line that best fits these data.

Year	Price
1995	$0.77
1999	$0.86
2002	$0.88
2005	$0.97

(b) Estimate the average price of a pound of Red Delicious apples in January 2000.

(c) If this trend continues, when will the average price of a pound of Red Delicious apples be $1.05?

CHAPTER 1 PROJECT

Break-Even Analysis

We discussed linear demand curves in Section 1.3. Demand curves normally apply to an entire industry or to a monopolist; that is, a manufacturer that is so large that the quantity it supplies affects the market price of the commodity. We discussed linear cost curves and break-even analysis in Section 1.4. In this chapter project we combine demand curves and cost curves with least-squares lines to extend break-even analysis to a monopolist.

1. Table 1 can be used to obtain the demand curve for a monopolist who manufactures and sells a unique type of camera. The first column gives several production quantities in thousands of cameras and the second column gives the corresponding prices per camera. For instance, in order to sell 200 thousand cameras, the manufacturer must set the price at $79 per camera. Find the least-squares line that best fits these data; that is, find a demand curve for the camera.

TABLE 1	
q (thousands)	p (dollars)
100	90
200	79
300	72
400	59
500	50

2. Use the demand curve from part 1 to estimate the price that must be charged in order to sell 350 thousand cameras. Calculate the revenue for this price and quantity. (*Note*: The revenue is the amount of money received from the sale of the cameras.)

3. Use the demand curve to estimate the quantity that can be sold if the price is $75 per camera. Calculate the revenue for this price and quantity.

4. Determine the expression that gives the revenue from producing and selling q thousand cameras. (*Note*: The number of cameras sold will be $1000q$.)

5. Assuming that the manufacturer has fixed costs of $8,000,000 and that the variable cost of producing each thousand cameras is $25,000, find the equation of the cost curve.

6. Graph the revenue curve from part 4 and the cost curve from part 5 on a graphing calculator and determine the two points of intersection.

7. What is the break-even point; that is, the lowest value of q for which cost equals revenue?

8. For what values of q will the company make a profit?

Matrices

W E begin this chapter by developing a method for solving systems of linear equations in any number of variables. Our discussion of this method will lead naturally into the study of mathematical objects called *matrices*. The arithmetic and applications of matrices are the main topics of the chapter. We discuss in detail the application of matrix arithmetic to input–output analysis, which can be (and is) used to make production decisions for large businesses and entire economies.

2.1 Solving Systems of Linear Equations, I

In Chapter 1 we presented a method for solving systems of linear equations in two variables. The method of Chapter 1 is very efficient for determining the solutions. Unfortunately, it works only for systems of linear equations having *two* variables. In many applications we meet systems having more than two variables, as the following example illustrates.

EXAMPLE 1 **Manufacturing** The Upside Down company specializes in making down jackets, ski vests, and comforters. The requirements for down and labor and the profits earned are given in the following chart.

	Down (pounds)	Time (labor-hours)	Profit ($)
Jacket	3	2	6
Vest	2	1	6
Comforter	4	1	2

Each week the company has available 600 pounds of down and 275 labor-hours. It wants to earn a weekly profit of $1150. How many of each item should the company make each week?

Solution The requirements and earnings can be expressed by a system of equations. Let x be the number of jackets, y the number of vests, and z the number of comforters. If 600 pounds of down are used, then

$$[\text{down in jackets}] + [\text{down in vests}] + [\text{down in comforters}] = 600$$
$$3\,[\text{no. jackets}] + 2\,[\text{no. vests}] + 4\,[\text{no. comforters}] = 600.$$

That is,

$$3x + 2y + 4z = 600.$$

Similarly, the equation for labor is

$$2x + y + z = 275,$$

and the equation for the profit is

$$6x + 6y + 2z = 1150.$$

The numbers x, y, and z must simultaneously satisfy a system of three linear equations in three variables.

$$\begin{cases} 3x + 2y + 4z = 600 \\ 2x + y + z = 275 \\ 6x + 6y + 2z = 1150 \end{cases} \tag{1}$$

Later we present a method for determining the solution to this system. This method yields the solution $x = 50$, $y = 125$, and $z = 50$. It is easy to confirm that these values of x, y, and z satisfy all three equations:

$$3(50) + 2(125) + 4(50) = 600$$
$$2(50) + (125) + (50) = 275$$
$$6(50) + 6(125) + 2(50) = 1150.$$

Thus, the Upside Down company can make a profit of $1150 by producing 50 jackets, 125 vests, and 50 comforters. ∎

In this section we develop a step-by-step procedure for solving systems of linear equations such as (1). The procedure, called the *Gaussian elimination method*, consists of repeatedly simplifying the system, using so-called elementary row operations, until the solution stares us in the face!

In the system of linear equations (1) the equations have been written in such a way that the x-terms, the y-terms, and the z-terms lie in different columns. We shall always be careful to display systems of equations with separate columns for each variable. One of the key ideas of the Gaussian elimination method is to think of the solution as a system of linear equations in its own right. For example, we can write the solution of system (1) as

$$\begin{cases} x & = 50 \\ y & = 125 \\ z = & 50. \end{cases} \tag{2}$$

This is just a system of linear equations in which the coefficients of most terms are zero! Since the only terms with nonzero coefficients are arranged on a diagonal, such a system is said to be in *diagonal form.*

Our method for solving a system of linear equations consists of repeatedly using three operations that alter the system but do not change the solutions. The operations are used to transform the system into a system in diagonal form. Since the operations involve only elementary arithmetic and are applied to entire equations (i.e., rows of the system), they are called *elementary row operations.* Let us begin our study of the Gaussian elimination method by introducing these operations.

Elementary Row Operation 1   Rearrange the equations in any order.

This operation is harmless enough. It certainly does not change the solutions of the system.

Elementary Row Operation 2   Multiply an equation by a nonzero number.

For example, if we are given the system of linear equations

$$\begin{cases} 2x - 3y + 4z = 11 \\ 4x - 19y + z = 31 \\ 5x + 7y - z = 12, \end{cases}$$

then we may replace it by a new system obtained by leaving the last two equations unchanged and multiplying the first equation by 3. To accomplish this, multiply each term of the first equation by 3. The transformed system is

$$\begin{cases} 6x - 9y + 12z = 33 \\ 4x - 19y + z = 31 \\ 5x + 7y - z = 12. \end{cases}$$

The operation of multiplying an equation by a nonzero number does not change the solutions of the system. For if a particular set of values of the variables satisfies the original equation, it satisfies the resulting equation, and vice versa.

Elementary row operation 2 may be used to make the coefficient of a particular variable 1.

EXAMPLE 2   **Demonstrating elementary row operation 2**   Replace the system

$$\begin{cases} -5x + 10y + 20z = 4 \\ x \qquad - 12z = 1 \\ x + y + z = 0 \end{cases}$$

by an equivalent system in which the coefficient of x in the first equation is 1.

Solution The coefficient of x in the first equation is -5, so we use elementary row operation 2 to multiply the first equation by $-\frac{1}{5}$. Multiplying each term of the first equation by $-\frac{1}{5}$ gives

$$\begin{cases} x - 2y - 4z = -\frac{4}{5} \\ x - 12z = 1 \\ x + y + z = 0. \end{cases}$$

Now Try Exercise 1   ■

Another operation that can be performed on a system without changing its solutions is to replace one equation by its sum with some other equation. For example, consider this system of equations:

$$\text{A:} \quad \begin{cases} x + y - 2z = 3 \\ x + 2y - 5z = 4 \\ 5x + 8y - 18z = 14. \end{cases}$$

We can replace the second equation by the sum of the first and the second. Since

$$\begin{array}{r} x + y - 2z = 3 \\ + x + 2y - 5z = 4 \\ \hline 2x + 3y - 7z = 7, \end{array}$$

the resulting system is

$$\text{B:} \quad \begin{cases} x + y - 2z = 3 \\ 2x + 3y - 7z = 7 \\ 5x + 8y - 18z = 14. \end{cases}$$

If a particular choice of x, y, and z satisfies system A, it also satisfies system B. This is because system B results from adding equations. Similarly, system A can be derived from system B by subtracting equations. So any particular solution of system A is a solution of system B, and vice versa.

The operation of adding equations is usually used in conjunction with elementary row operation 2. That is, an equation is changed by adding to it a nonzero multiple of another equation. For example, consider the system

$$\begin{cases} x + y - 2z = 3 \\ x + 2y - 5z = 4 \\ 5x + 8y - 18z = 14. \end{cases}$$

Let us change the second equation by adding to it twice the first. Since

$$\begin{array}{rr} 2(\text{first}) & 2x + 2y - 4z = 6 \\ +(\text{second}) & x + 2y - 5z = 4 \\ \hline & 3x + 4y - 9z = 10, \end{array}$$

the new second equation is

$$3x + 4y - 9z = 10$$

and the transformed system is

$$\begin{cases} x + y - 2z = 3 \\ 3x + 4y - 9z = 10 \\ 5x + 8y - 18z = 14. \end{cases}$$

Since addition of equations and elementary row operation 2 are often used together, let us define a third elementary row operation.

Elementary Row Operation 3   Change an equation by adding to it a multiple of another equation.

For reference, let us summarize the elementary row operations we have just defined.

Elementary Row Operations

1. Rearrange the equations in any order.

2. Multiply an equation by a nonzero number.

3. Change an equation by adding to it a multiple of another equation.

The idea of the Gaussian elimination method is to transform an arbitrary system of linear equations into diagonal form by repeated applications of the three elementary row operations. To see how the method works, consider the following example.

EXAMPLE 3   **Solving a system of equations using Gaussian elimination**   Solve the following system by the Gaussian elimination method:

$$\begin{cases} x - 3y = 7 \\ -3x + 4y = -1. \end{cases}$$

Solution   Let us transform this system into diagonal form by examining one column at a time, starting from the left. Examine the first column:

$$x$$
$$-3x$$

The coefficient of the top x is 1, which is exactly what it should be for the system to be in diagonal form. So we do nothing to this term. Now examine the next term in the column, $-3x$. In diagonal form this term must be absent. In order to accomplish this, we add a multiple of the first equation to the second. Since the coefficient of x in the second is -3, we add three times the first equation to the second equation in order to cancel the x-term. (Abbreviation: $[2] + 3[1]$. The $[2]$ means that we are changing equation 2. The expression $[2] + 3[1]$ means that we are replacing equation 2 by the original equation plus three times equation 1.)

$$\begin{cases} x - 3y = 7 \\ -3x + 4y = -1 \end{cases} \xrightarrow{\;[2]+3[1]\;} \begin{cases} x - 3y = 7 \\ - 5y = 20 \end{cases}$$

The first column now has the proper form, so we proceed to the second column. In diagonal form that column will have one nonzero term, namely the second, and the coefficient of y in that term must be 1. To bring this about, multiply the second equation by $-\frac{1}{5}$ (abbreviation: $-\frac{1}{5}[2]$):

$$\begin{cases} x - 3y = 7 \\ - 5y = 20 \end{cases} \xrightarrow{\;-\frac{1}{5}[2]\;} \begin{cases} x - 3y = 7 \\ y = -4. \end{cases}$$

The second column still does not have the correct form. We must get rid of the $-3y$-term in the first equation. We do this by adding a multiple of the second equation to the first. Since the coefficient of the term to be canceled is -3, we add three times the second equation to the first:

$$\begin{cases} x - 3y = & 7 \\ & y = -4 \end{cases} \xrightarrow{[1] + 3[2]} \begin{cases} x & = -5 \\ y = -4. \end{cases}$$

The system is now in diagonal form and the solution can be read off: $x = -5$, $y = -4$.

Now Try Exercise 25

■

EXAMPLE 4   **Solving a system of equations using Gaussian elimination**   Use the Gaussian elimination method to solve the system

$$\begin{cases} 2x - 6y = -8 \\ -5x + 13y = 1. \end{cases}$$

Solution   We can perform the calculations in a mechanical way, proceeding column by column from the left:

$$\begin{cases} 2x - 6y = -8 \\ -5x + 13y = 1 \end{cases} \xrightarrow{\frac{1}{2}[1]} \begin{cases} x - 3y = -4 \\ -5x + 13y = 1 \end{cases}$$

$$\xrightarrow{[2] + 5[1]} \begin{cases} x - 3y = -4 \\ -2y = -19 \end{cases}$$

$$\xrightarrow{-\frac{1}{2}[2]} \begin{cases} x - 3y = -4 \\ y = \frac{19}{2} \end{cases}$$

$$\xrightarrow{[1] + 3[2]} \begin{cases} x = \frac{49}{2} \\ y = \frac{19}{2}. \end{cases}$$

So the solution of the system is $x = \frac{49}{2}$, $y = \frac{19}{2}$.

Now Try Exercise 33

■

The calculation becomes easier to follow if we omit writing down the variables at each stage and work only with the coefficients. At each stage of the computation, the system is represented by a rectangular array of numbers. For instance, the original system is written[1]

$$\left[\begin{array}{cc|c} 2 & -6 & -8 \\ -5 & 13 & 1 \end{array}\right].$$

The elementary row operations are performed on the rows of this rectangular array just as if the variables were there. So, for example, the first step in the above is to multiply the first equation by $\frac{1}{2}$. This corresponds to multiplying the first row of the array by $\frac{1}{2}$ to get

$$\left[\begin{array}{cc|c} 1 & -3 & -4 \\ -5 & 13 & 1 \end{array}\right].$$

---

[1] The vertical line between the second and third columns is a placemarker that separates the data obtained from the left- and right-hand sides of the equations. It is inserted for visual convenience.

The diagonal form just corresponds to the array

$$\begin{bmatrix} 1 & 0 & \Big| & \frac{49}{2} \\ 0 & 1 & \Big| & \frac{19}{2} \end{bmatrix}.$$

Note that this array has ones down the diagonal and zeros everywhere else on the left. The solution of the system appears on the right.

A rectangular array of numbers is called a *matrix* (plural: *matrices*). In the next example, we use matrices to carry out the Gaussian elimination method.

EXAMPLE 5 **Solving a system of equations using Gaussian elimination** Use the Gaussian elimination method to solve the system

$$\begin{cases} 3x - 6y + 9z = 0 \\ 4x - 6y + 8z = -4 \\ -2x - y + z = 7. \end{cases}$$

Solution The initial array corresponding to the system is

$$\begin{bmatrix} 3 & -6 & 9 & \Big| & 0 \\ 4 & -6 & 8 & \Big| & -4 \\ -2 & -1 & 1 & \Big| & 7 \end{bmatrix}.$$

We must use elementary row operations to transform this array into diagonal form—that is, with ones down the diagonal and zeros everywhere else on the left:

$$\begin{bmatrix} 1 & 0 & 0 & \Big| & * \\ 0 & 1 & 0 & \Big| & * \\ 0 & 0 & 1 & \Big| & * \end{bmatrix}.$$

We proceed one column at a time.

$$\begin{bmatrix} 3 & -6 & 9 & \Big| & 0 \\ 4 & -6 & 8 & \Big| & -4 \\ -2 & -1 & 1 & \Big| & 7 \end{bmatrix} \xrightarrow{\frac{1}{3}[1]} \begin{bmatrix} 1 & -2 & 3 & \Big| & 0 \\ 4 & -6 & 8 & \Big| & -4 \\ -2 & -1 & 1 & \Big| & 7 \end{bmatrix} \xrightarrow{[2]+(-4)[1]}$$

$$\begin{bmatrix} 1 & -2 & 3 & \Big| & 0 \\ 0 & 2 & -4 & \Big| & -4 \\ -2 & -1 & 1 & \Big| & 7 \end{bmatrix} \xrightarrow{[3]+2[1]} \begin{bmatrix} 1 & -2 & 3 & \Big| & 0 \\ 0 & 2 & -4 & \Big| & -4 \\ 0 & -5 & 7 & \Big| & 7 \end{bmatrix} \xrightarrow{\frac{1}{2}[2]}$$

$$\begin{bmatrix} 1 & -2 & 3 & \Big| & 0 \\ 0 & 1 & -2 & \Big| & -2 \\ 0 & -5 & 7 & \Big| & 7 \end{bmatrix} \xrightarrow{[1]+2[2]} \begin{bmatrix} 1 & 0 & -1 & \Big| & -4 \\ 0 & 1 & -2 & \Big| & -2 \\ 0 & -5 & 7 & \Big| & 7 \end{bmatrix} \xrightarrow{[3]+5[2]}$$

$$\begin{bmatrix} 1 & 0 & -1 & \Big| & -4 \\ 0 & 1 & -2 & \Big| & -2 \\ 0 & 0 & -3 & \Big| & -3 \end{bmatrix} \xrightarrow{(-\frac{1}{3})[3]} \begin{bmatrix} 1 & 0 & -1 & \Big| & -4 \\ 0 & 1 & -2 & \Big| & -2 \\ 0 & 0 & 1 & \Big| & 1 \end{bmatrix} \xrightarrow{[1]+1[3]}$$

$$\begin{bmatrix} 1 & 0 & 0 & \Big| & -3 \\ 0 & 1 & -2 & \Big| & -2 \\ 0 & 0 & 1 & \Big| & 1 \end{bmatrix} \xrightarrow{[2]+2[3]} \begin{bmatrix} 1 & 0 & 0 & \Big| & -3 \\ 0 & 1 & 0 & \Big| & 0 \\ 0 & 0 & 1 & \Big| & 1 \end{bmatrix}$$

The last array is in diagonal form, so we just put back the variables and read off the solution:

$$x = -3, \qquad y = 0, \qquad z = 1.$$

Because so much arithmetic has been performed, it is a good idea to check the solution by substituting the values for x, y, and z into each of the equations of the original system. This will uncover any arithmetic errors that may have occurred.

$$\begin{cases} 3x - 6y + 9z = 0 \\ 4x - 6y + 8z = -4 \\ -2x - y + z = 7 \end{cases} \qquad \begin{cases} 3(-3) - 6(0) + 9(1) = 0 \\ 4(-3) - 6(0) + 8(1) = -4 \\ -2(-3) - (0) + (1) = 7 \end{cases}$$

$$\begin{cases} -9 - 0 + 9 = 0 \\ -12 - 0 + 8 = -4 \\ 6 - 0 + 1 = 7 \end{cases}$$

$$\begin{cases} 0 = 0 \\ -4 = -4 \\ 7 = 7 \end{cases}$$

So we have indeed found a solution of the system.

Now Try Exercise 31

▶ *Remark* Note that so far we have not had to use elementary row operation 1, which allows interchange of equations. But in some examples it is definitely needed.

Consider this system:

$$\begin{cases} y + z = 0 \\ 3x - y + z = 6 \\ 6x - z = 3. \end{cases}$$

The first step of the Gaussian elimination method consists of making the x-coefficient 1 in the first equation. But we cannot do this, since the first equation does not involve x. To remedy this difficulty, just interchange the first two equations to guarantee that the first equation involves x. Now proceed as before. Of course, in terms of the matrix of coefficients, interchanging equations corresponds to interchanging rows of the matrix.   ◀

GC Graphing utilities can perform many operations with matrices, including elementary row operations.

TI-83 With a TI-83 Plus or TI-84 Plus, pressing 2ND [MATRX] presents the three menus shown in Fig. 1. (With an ordinary TI-83, press MATRX.) You define a new matrix or alter an existing matrix with the EDIT menu. You place the name of an existing matrix on the home screen with the NAMES menu. You perform operations on existing matrices with the MATH menu.

With the EDIT menu, you first select one of the names **[A]**, **[B]**, **[C]**, ... for the matrix and then specify the size and fill in the entries as shown in Fig. 2. The three elementary row operations are carried out with commands of the following forms from the MATRX/MATH menu.

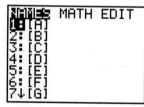

Figure 1

rowSwap(*matrix, rowA, rowB*)	Interchange *rowA* and *rowB* of *matrix*
*row(*value, matrix, row*)	Multiply *row* of *matrix* by *value*
*row+(*value, matrix, rowA, rowB*)	Add *value∗rowA* to *rowB* of *matrix*

```
MATRIX[A] 2 ×3
[ 5    -2    3  ]
[ 20   .4   .6667]



2,3=.6666666666...
```

Figure 2

When one of these commands is carried out, the resulting matrix is displayed but the stored matrix is not changed. Therefore, when a sequence of commands is executed to carry out the Gaussian elimination method, each command should be followed with $\boxed{\text{STO}\ \blacktriangleright}$ *matrix* to change the stored matrix. (*Note*: To display the name of a matrix, press $\boxed{\text{2nd}}\boxed{\text{MATRX}}$ or $\boxed{\text{MATRX}}$, cursor down to the matrix, and press $\boxed{\text{ENTER}}$.) For instance, if in Example 5 the original matrix is named [A], then the first three row operations are carried out with

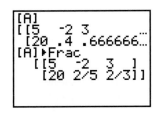

Figure 3

$$\textbf{*row(1/3,[A],1)}\boxed{\text{STO}\ \blacktriangleright}\textbf{[A]}$$

$$\textbf{*row+(-4,[A],1,2)}\boxed{\text{STO}\ \blacktriangleright}\textbf{[A]}$$

$$\textbf{*row+(2,[A],1,3)}\boxed{\text{STO}\ \blacktriangleright}\textbf{[A]}.$$

Matrix entries are normally displayed on the calculator as decimals. From the home screen you can display the entries as fractions with a command such as **[A]** ▶**Frac** or **Ans** ▶**Frac**. (▶**Frac** is displayed by pressing $\boxed{\text{MATH}}$ **1**.) See Fig. 3.

TI-89 Matrices are created and altered with the Data/Matrix editor that is accessed through the $\boxed{\text{APPS}}$ key. See Appendix D for details. The three elementary row operations are carried out with the following commands from the MATH/Matrix/Row ops menu.

rowSwap(*matrix, rowA, rowB*)	Interchange *rowA* and *rowB* of *matrix*
mRow(*value, matrix, row*)	Multiply *row* by *value*
mRowAdd(*value, matrix, rowA, rowB*)	Add *value·rowA* to *rowB*

When one of these commands is carried out, the resulting matrix is displayed but the stored matrix is not changed. Therefore, when a sequence of commands is executed to carry out the Gaussian elimination method, each command should be followed with $\boxed{\text{STO}\ \blacktriangleright}$ *matrix* to change the stored matrix. For instance, if in Example 5 the original matrix is named *a*, then the first three row operations are carried out with

$$\textbf{mRow(1/3,a,1)}\boxed{\text{STO}\ \blacktriangleright}\textbf{a}$$

$$\textbf{mRowAdd(-4,a,1,2)}\boxed{\text{STO}\ \blacktriangleright}\textbf{a}$$

$$\textbf{mRowAdd(2,a,1,3)}\boxed{\text{STO}\ \blacktriangleright}\textbf{a}.$$

ES[2] In an electronic spreadsheet, the entries of an $m \times n$ matrix can be typed directly into an m by n range of cells. (In order to insert a fraction, type in a formula such as **=2/3**.)

A cell or a matrix is most easily used in formulas if it has been given a name, such as **x**, **y**, **A**, **B**. To name a cell or a matrix, select the cell or matrix, click on the Name box (located just above the upper-left corner of the spreadsheet), type the name, and press Enter. Single-letter names can consist of any letter other than C and R. (See Appendix C for further information about naming cells and ranges.) A matrix also can be referred to by an expression of the form

(upper-left cell):(lower-right cell).

For instance, the matrix in Fig. 4 can be referred to as **A** or as **B2:D3**.

The systems of linear equations presented in this section can be solved with a device called Solver that is invoked from the Excel Tools menu. (If Solver does not appear in the Tools menu of your computer, see Appendix C for information on how to install Solver.) The following steps show how to use the Solver to find the solution to Example 5.

   1. Give the cells A1, A2, and A3 the names x, y, and z, respectively. (There is no need to place any values into these cells.)

---

[2]ES is an abbreviation for "Electronic Spreadsheet."

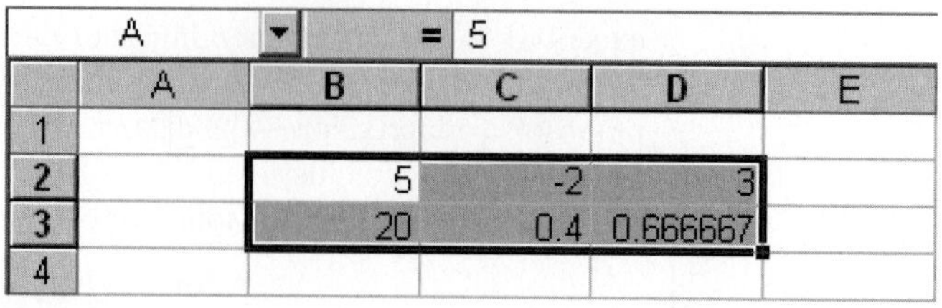

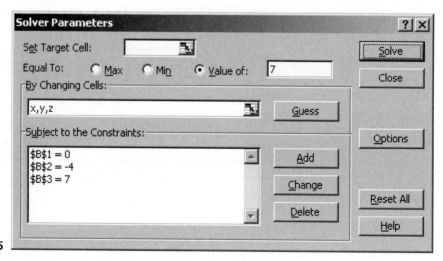

Figure 4

2. In the cells B1, B2, and B3, place the formulas consisting of the left sides of the three equations. For instance, type the formula **=3*x–6*y+9*z** into cell B1.

3. Click on the Tools menu in the toolbar and then click on Solver. A window titled Solver Parameters will appear.

4. Clear the contents, if any, of the Set Target Cell box. (Ignore any entries appearing in the "Equal To" line.)

5. Type **x**, **y**, **z** into the By Changing Cells box.

6. Click the Add button. An Add Constraint dialog box will appear.
   (a) Enter **$B$1,=,** and **0** into the three boxes and then press the Add button.
   (b) Enter **$B$2,=,** and **–4** into the three boxes and then press the Add button.
   (c) Enter **$B$3,=,** and **7** into the three boxes and then press the OK button. The Solver Parameters window will reappear as shown in Fig. 5.

Note: The entries 0, −4, and 7 are the numbers on the right sides of the equations.

Figure 5

7. Click the Solve button. A window titled Solver Results will appear.

8. Click the OK button. The solution to the system of linear equations will appear in the x, y, and z cells. *Note*: The values may differ slightly from the true values due to rounding errors. For instance, the value for y may appear as −6.7E-07 (that is −.00000067) instead of 0.

PRACTICE PROBLEMS 2.1

1. Determine whether the following systems of linear equations are in diagonal form.

(a) $\begin{cases} x \quad\ + z = 3 \\ \quad\ y \quad\ = 2 \\ \quad\quad\ z = 7 \end{cases}$ (b) $\begin{cases} x \quad\quad\ = 3 \\ \quad\ y = 5 \\ \quad\ z \quad = 7 \end{cases}$

(c) $\begin{cases} x \quad\quad\ = -1 \\ \quad\ y \quad = 0 \\ \quad\ 3z = 4 \end{cases}$

2. Perform the indicated elementary row operation.

(a) $\begin{cases} x - 3y = 2 \\ 2x + 3y = 5 \end{cases} \xrightarrow{[2] + (-2)[1]}$

(b) $\begin{cases} x + y = 3 \\ -x + 2y = 5 \end{cases} \xrightarrow{[2] + (1)[1]}$

3. State the next elementary row operation that should be performed when applying the Gaussian elimination method.

(a) $\begin{bmatrix} 0 & 2 & 4 & | & 1 \\ 0 & 3 & -7 & | & 0 \\ 3 & 6 & -3 & | & 3 \end{bmatrix}$

(b) $\begin{bmatrix} 1 & -3 & 4 & | & 5 \\ 0 & 2 & 3 & | & 4 \\ -6 & 5 & -7 & | & 0 \end{bmatrix}$

EXERCISES 2.1

In Exercises 1–8, perform the indicated elementary row operations and give their abbreviations.

1. Operation 2: Multiply the first equation by 2.
$$\begin{cases} \tfrac{1}{2}x - 3y = 2 \\ 5x + 4y = 1 \end{cases}$$

2. Operation 2: Multiply the second equation by -1.
$$\begin{cases} x + 4y = 6 \\ -y = 2 \end{cases}$$

3. Operation 3: Change the second equation by adding to it 5 times the first equation.
$$\begin{cases} x + 2y = 3 \\ -5x + 4y = 1 \end{cases}$$

4. Operation 3: Change the second equation by adding to it $\left(-\tfrac{1}{2}\right)$ times the first equation.
$$\begin{cases} x - 6y = 4 \\ \tfrac{1}{2}x + 2y = 1 \end{cases}$$

5. Operation 3: Change the third equation by adding to it (-4) times the first equation.
$$\begin{cases} x - 2y + z = 0 \\ y - 2z = 4 \\ 4x + y + 3z = 5 \end{cases}$$

6. Operation 3: Change the third equation by adding to it 3 times the second equation.
$$\begin{cases} x + 6y - 4z = 1 \\ y + 3z = 1 \\ -3y + 7z = 2 \end{cases}$$

7. Operation 3: Change the first row by adding to it $\tfrac{1}{2}$ times the second row.
$$\begin{bmatrix} 1 & -\tfrac{1}{2} & | & 3 \\ 0 & 1 & | & 4 \end{bmatrix}$$

8. Operation 3: Change the third row by adding to it (-4) times the second row.
$$\begin{bmatrix} 1 & 0 & 7 & | & 9 \\ 0 & 1 & -2 & | & 3 \\ 0 & 4 & 8 & | & 5 \end{bmatrix}$$

In Exercises 9–12, describe in your own words the meaning of the notation with respect to a matrix.

9. $\left(\tfrac{1}{3}\right)[2]$ **10.** $[2] + (-4)[1]$

11. $[1] + (3)[2]$ **12.** $(-1)[1]$

In Exercises 13–16, carry out the indicated elementary row operation.

13. $\begin{bmatrix} 1 & 2 \\ -3 & 4 \end{bmatrix} \xrightarrow{[2] + 3[1]} \begin{bmatrix} \quad & \quad \\ \quad & \quad \end{bmatrix}$

14. $\begin{bmatrix} -\tfrac{1}{2} & 3 \\ 4 & 5 \end{bmatrix} \xrightarrow{-2[1]} \begin{bmatrix} \quad & \quad \\ \quad & \quad \end{bmatrix}$

15. $\begin{bmatrix} \tfrac{1}{7} & \tfrac{2}{7} \\ 3 & -2 \end{bmatrix} \xrightarrow{7[1]} \begin{bmatrix} \quad & \quad \\ \quad & \quad \end{bmatrix}$

16. $\begin{bmatrix} 1 & 3 \\ 4 & 4 \end{bmatrix} \xrightarrow{[2] + (-4)[1]} \begin{bmatrix} \quad & \quad \\ \quad & \quad \end{bmatrix}$

In Exercises 17–24, state the next elementary row operation that should be performed in order to put the matrix into diagonal form. Do not perform the operation.

17. $\begin{bmatrix} 1 & -5 & | & 1 \\ -2 & 4 & | & 6 \end{bmatrix}$ 18. $\begin{bmatrix} 1 & 3 & | & 4 \\ 0 & 2 & | & 6 \end{bmatrix}$

19. $\begin{bmatrix} 1 & 2 & | & 3 \\ 0 & 1 & | & 4 \end{bmatrix}$ 20. $\begin{bmatrix} 1 & -2 & 5 & | & 7 \\ 0 & -3 & 6 & | & 9 \\ 4 & 5 & -6 & | & 7 \end{bmatrix}$

21. $\begin{bmatrix} 0 & 5 & -3 & | & 6 \\ 2 & -3 & 4 & | & 5 \\ 4 & 1 & -7 & | & 8 \end{bmatrix}$ 22. $\begin{bmatrix} 1 & 4 & -2 & | & 5 \\ 0 & -3 & 6 & | & 9 \\ 0 & 4 & 3 & | & 1 \end{bmatrix}$

23. $\begin{bmatrix} 1 & 0 & 3 & | & 4 \\ 0 & 1 & 2 & | & 5 \\ 0 & 0 & 1 & | & 6 \end{bmatrix}$ 24. $\begin{bmatrix} 1 & 2 & 4 & | & 5 \\ 0 & 0 & 3 & | & 6 \\ 0 & 1 & 1 & | & 7 \end{bmatrix}$

In Exercises 25–38, solve the linear system by using the Gaussian elimination method.

25. $\begin{cases} x + 9y = 8 \\ 2x + 8y = 6 \end{cases}$ 26. $\begin{cases} \frac{1}{3}x + 2y = 1 \\ -2x - 4y = 6 \end{cases}$

27. $\begin{cases} x - 3y + 4z = 1 \\ 4x - 10y + 10z = 4 \\ -3x + 9y - 5z = -6 \end{cases}$

28. $\begin{cases} \frac{1}{2}x + y = 4 \\ -4x - 7y + 3z = -31 \\ 6x + 14y + 7z = 50 \end{cases}$

29. $\begin{cases} 2x - 2y + 4 = 0 \\ 3x + 4y - 1 = 0 \end{cases}$ 30. $\begin{cases} 2x + 3y = 4 \\ -x + 2y = -2 \end{cases}$

31. $\begin{cases} 4x - 4y + 4z = -8 \\ x - 2y - 2z = -1 \\ 2x + y + 3z = 1 \end{cases}$

32. $\begin{cases} x + 2y + 2z - 11 = 0 \\ x - y - z + 4 = 0 \\ 2x + 5y + 9z - 39 = 0 \end{cases}$

33. $\begin{cases} .2x + .3y = 4 \\ .6x + 1.1y = 15 \end{cases}$ 34. $\begin{cases} \frac{3}{2}x + 6y = 9 \\ \frac{1}{2}x - \frac{2}{3}y = 11 \end{cases}$

35. $\begin{cases} x + y + 4z = 3 \\ 4x + y - 2z = -6 \\ -3x + 2z = 1 \end{cases}$

36. $\begin{cases} -2x - 3y + 2z = -2 \\ x + y = 3 \\ -x - 3y + 5z = 8 \end{cases}$

37. $\begin{cases} -x + y = -1 \\ x + z = 4 \\ 6x - 3y + 2z = 10 \end{cases}$ 38. $\begin{cases} x + 2z = 9 \\ y + z = 1 \\ 3x - 2y = 9 \end{cases}$

39. **(PE)** A baked potato smothered with cheddar cheese weighs 180 grams and contains 10.5 grams of protein. If cheddar cheese contains 25% protein and a baked potato contains 2% protein, how many grams of cheddar cheese are there?

(a) 25   (b) 30   (c) 35   (d) 40   (e) 45

40. **(PE)** A high school math department purchased brand A calculators for $80 each and brand B calculators for $95 each. It purchased a total of 20 calculators at a total cost of $1780. How many brand A calculators did the department purchase?

(a) 7   (b) 8   (c) 9   (d) 10   (e) 11

Exercises 41 and 42 are multiple choice exercises with five possible choices. Each exercise consists of a question and two statements that may or may not provide sufficient information to answer the question. Select the response below that best describes the situation.

(**a**) *Statement I alone is sufficient to answer the question, but statement II is not sufficient.*

(**b**) *Statement II alone is sufficient to answer the question, but statement I is not sufficient.*

(**c**) *Both statements together are sufficient to answer the question, but neither alone is sufficient.*

(**d**) *Each statement alone is sufficient to answer the question.*

(**e**) *Both statements together are not sufficient to answer the question.*

41. **(PE)** A box of golf balls and a golf glove cost a total of $20. How much does the box of balls cost?

Statement I: The golf glove costs three times as much as the box of golf balls.

Statement II: The golf glove costs $15.

42. **(PE)** I have four nickels and three pennies in my pocket. What is the total weight of these coins?

Statement I: A nickel weighs twice as much as a penny.

Statement II: The total weight of a nickel and two pennies is 10 grams.

43. **(Sales)** A street vendor has a total of 350 short and long sleeve T-shirts. If she sells the short sleeve shirts for $10 each and the long sleeve shirts for $14 each, how many of each did she sell if she sold all of her stock for $4300?

44. **(Sales)** A grocery store carries two brands of bleach. A 96-ounce bottle of the national brand sells for $1.79, while the same size bottle of the store brand sells for $1.59. How many bottles of each brand were sold if a total of 82 bottles were sold for $142.58?

45. **(Movie Tickets)** A 350-seat movie theater charges $8.50 admission for adults and $5.50 for children. If the theater is full and $2711 is collected, how many adults and how many children are in the audience?

46. **(Batting Average)** A baseball player's batting average is determined by dividing the number of hits by the number of times at bat and multiplying by 1000. (Batting averages are usually, but not necessarily, rounded to the nearest whole number.) For instance, if a player gets 2 hits in 5 times at bat, his batting average is

400: $\left(\frac{2}{5} \times 1000 = 400\right)$. Partway through the season, a player thinks to himself, "If I get a hit in my next time at bat, my average will go up to 250; if I don't get a hit, it will drop to 187.5." How many times has this player batted, how many hits has he had, and what is his current batting average?

47. (*Investment Planning*) A bank wishes to invest a $100,000 trust fund in three sources: bonds paying 8%; certificates of deposit paying 7%; and first mortgages paying 10%. The bank wishes to realize an $8000 annual income from the investment. A condition of the trust is that the total amount invested in bonds and certificates of deposit must be triple the amount invested in mortgages. How much should the bank invest in each possible category? Let x, y, and z, respectively, be the amounts invested in bonds, certificates of deposit, and first mortgages. Solve the system of equations by the Gaussian elimination method.

48. (*Nutrition Planning*) A dietitian wishes to plan a meal around three foods. Each ounce of food I contains 10% of the daily requirements for carbohydrates, 10% for protein, and 15% for vitamin C. Each ounce of food II contains 10% of the daily requirements for carbohydrates, 5% for protein, and 0% for vitamin C. Each ounce of food III contains 10% of the daily requirements for carbohydrates, 25% for protein, and 10% for vitamin C. How many ounces of each food should be served in order to supply exactly the daily requirements for each nutrient? Let x, y, and z, respectively, be the number of ounces of foods I, II, and III.

49. (*Candy Assortments*) A small candy store makes three types of party mixes. The first type contains 40% nonpareils and 60% peanut clusters, while the second type contains 30% peanut clusters and 70% chocolate covered raisins. The third type consists of 40% nonpareils, 30% peanut clusters, and 30% chocolate covered raisins. If the store has 90 pounds of nonpareils, 100 pounds of peanut clusters, and 120 pounds of chocolate covered raisins available, how many pounds of each type of party mix should be made?

50. (*Investment Planning*) New parents Jim and Lucy want to start saving for their son's college education. They have $5000 to invest in three different types of plans. A traditional savings account pays 3% annual interest, a certificate of deposit pays 6% annual interest, and a prepaid college plan pays $7\frac{1}{2}$% annual interest. If they want to invest the same amount in the prepaid college fund as in the other two plans together, how much should they invest in each plan to realize an interest income of $300 for the first year?

Exercises 51–56 require the use of a graphing calculator, spreadsheet, or mathematical software.

51. Enter the matrix corresponding to the equations in Exercise 3, and change the second row by adding to it 5 times the first row.

52. Enter the matrix from Exercise 5, and change the third row by adding to it (-4) times the first row.

53. Enter the matrix in Exercise 7, and interchange the rows.

54. Enter the matrix corresponding to the linear system in Exercise 33, and multiply the first row by 5.

55. Use technology to solve the system of linear equations in Exercise 35.

56. Use technology to solve the system of linear equations in Exercise 36.

In Exercises 57 and 58, solve the system of linear equations by the Gaussian elimination method.

57. $\begin{cases} 2x + 2y + 2z + 4w = -12 \\ \quad\quad y + z + w = -5 \\ \quad\quad\quad\quad z + 2w = -6 \\ x + y + z + 4w = -14 \end{cases}$

58. $\begin{cases} 2x \quad\quad - z + 5w = 19 \\ 2x + 3y - z + 5w = 28 \\ \quad\quad - z + 5w = 21 \\ 6x \quad\quad - 3z = -3 \end{cases}$

SOLUTIONS TO PRACTICE PROBLEMS 2.1

1. (a) Not in diagonal form, since the first equation contains both x and z.

(b) Not in diagonal form, since the variables are not arranged in diagonal fashion.

(c) Not in diagonal form, since the coefficient of z is not 1.

2. (a) Change the system into another system in which the second equation is altered by having (-2)(first equation) added to it. The new sys-

tem is

$$\begin{cases} x - 3y = 2 \\ \quad\quad 9y = 1. \end{cases}$$

The equation $9y = 1$ was obtained as follows:

$$\begin{array}{ll} (-2)(\text{first equation}) & -2x + 6y = -4 \\ + \ (\text{second equation}) & \underline{2x + 3y = \ \ 5} \\ & \quad\quad\ 9y = \ \ 1. \end{array}$$

(b) Change the second equation by adding to it 1 times the first equation. The result is

$$\begin{cases} x + y = 3 \\ 3y = 8. \end{cases}$$

3. (a) The first row should contain a nonzero number as its first entry. This can be accomplished by interchanging the first and third rows.

(b) The first column can be put into proper form by eliminating the -6. To accomplish this, multiply the first row by 6 and add this product to the third row. The notation for this operation is

$$\xrightarrow{[3] + 6[1]}.$$

2.2   Solving Systems of Linear Equations, II

In this section we introduce the operation of pivoting and consider systems of linear equations that do not have unique solutions.

Roughly speaking, the Gaussian elimination method applied to a matrix proceeds as follows: Consider the columns one at a time, from left to right. For each column use the elementary row operations to transform the appropriate entry to a one and the remaining entries in the column to zeros. (The "appropriate" entry is the first entry in the first column, the second entry in the second column, and so forth.) This sequence of elementary row operations performed for each column is called *pivoting*. More precisely,

---

Method   To pivot a matrix about a given nonzero entry,

1. Transform the given entry into a one.

2. Transform all other entries in the same column into zeros.

Pivoting is used in solving problems other than systems of linear equations. As we shall see in Chapter 4, it is the basis for the simplex method of solving linear programming problems.

EXAMPLE 1   **Pivoting**   Pivot the matrix about the circled element.

$$\left[\begin{array}{cc|c} 18 & \boxed{-6} & 15 \\ 5 & -2 & 4 \end{array}\right]$$

Solution   The first step is to transform the -6 to a 1. We do this by multiplying the first row by $-\frac{1}{6}$:

$$\left[\begin{array}{cc|c} 18 & -6 & 15 \\ 5 & -2 & 4 \end{array}\right] \xrightarrow{-\frac{1}{6}[1]} \left[\begin{array}{cc|c} -3 & 1 & -\frac{5}{2} \\ 5 & -2 & 4 \end{array}\right].$$

Next, we transform the -2 (the only remaining entry in column 2) into a 0:

$$\left[\begin{array}{cc|c} -3 & 1 & -\frac{5}{2} \\ 5 & -2 & 4 \end{array}\right] \xrightarrow{[2] + 2[1]} \left[\begin{array}{cc|c} -3 & 1 & -\frac{5}{2} \\ -1 & 0 & -1 \end{array}\right].$$

The last matrix is the result of pivoting the original matrix about the circled entry.

Now Try Exercise 1

∎

In terms of pivoting, we can give the following summary of the Gaussian elimination method.

Gaussian Elimination Method to Transform a System of Linear Equations into Diagonal Form

1. Write down the matrix corresponding to the linear system.

2. Make sure that the first entry in the first column is nonzero. Do this by interchanging the first row with one of the rows below it, if necessary.

3. Pivot the matrix about the first entry in the first column.

4. Make sure that the second entry in the second column is nonzero. Do this by interchanging the second row with one of the rows below it, if necessary.

5. Pivot the matrix about the second entry in the second column.

6. Continue in this manner.

All the systems considered in the preceding section had only a single solution. In this case we say that the solution is *unique*. Let us now use the Gaussian elimination method to study the various possibilities other than a unique solution. We first experiment with an example.

EXAMPLE 2 **Solving a system of equations that has many solutions** Determine all solutions of the system

$$\begin{cases} 2x + 2y + 4z = 8 \\ x - y + 2z = 2 \\ -x + 5y - 2z = 2. \end{cases}$$

Solution We set up the matrix corresponding to the system and perform the appropriate pivoting operations. (The elements pivoted about are circled.)

$$\left[\begin{array}{ccc|c} ② & 2 & 4 & 8 \\ 1 & -1 & 2 & 2 \\ -1 & 5 & -2 & 2 \end{array}\right] \xrightarrow[\substack{[2]+(-1)[1] \\ [3]+(1)[1]}]{\frac{1}{2}[1]} \left[\begin{array}{ccc|c} 1 & 1 & 2 & 4 \\ 0 & ⑳{-2} & 0 & -2 \\ 0 & 6 & 0 & 6 \end{array}\right]$$

$$\xrightarrow[\substack{[1]+(-1)[2] \\ [3]+(-6)[2]}]{\left(-\frac{1}{2}\right)[2]} \left[\begin{array}{ccc|c} 1 & 0 & 2 & 3 \\ 0 & 1 & 0 & 1 \\ 0 & 0 & 0 & 0 \end{array}\right]$$

Note that our method must terminate here, since there is no way to transform the third entry in the third column into a 1 without disturbing the columns already in appropriate form. The equations corresponding to the last matrix read

$$\begin{cases} x \quad + 2z = 3 \\ y \quad\quad = 1 \\ 0 \quad\quad = 0. \end{cases}$$

The last equation does not involve any of the variables and so may be omitted. This leaves the two equations

$$\begin{cases} x & + 2z = 3 \\ & y & = 1. \end{cases}$$

Now, taking the $2z$-term in the first equation to the right side, we can write the equations

$$\begin{cases} x = 3 - 2z \\ y = 1. \end{cases}$$

The value of y is given: $y = 1$. The value of x is given in terms of z. To find a solution to this system, assign any value to z. Then the first equation gives a value for x and thereby a specific solution to the system. For example, if we take $z = 1$, then the corresponding specific solution is

$$z = 1$$
$$x = 3 - 2(1) = 1$$
$$y = 1.$$

If we take $z = -3$, the corresponding specific solution is

$$z = -3$$
$$x = 3 - 2(-3) = 9$$
$$y = 1.$$

Thus, we see that the original system has infinitely many specific solutions, corresponding to the infinitely many possible different choices for z.

We say that the *general solution* of the system is

$$z = \text{any value}$$
$$x = 3 - 2z$$
$$y = 1.$$

Now Try Exercise 15   ■

When a linear system cannot be *completely* diagonalized,

1. Apply the Gaussian elimination method to as many columns as possible. Proceed from left to right, but do not disturb columns that have already been put into proper form.

2. Variables corresponding to columns not in proper form can assume any value.

3. The other variables can be expressed in terms of the variables of step 2.

EXAMPLE 3   **Solving a system of equations that has many solutions**   Find all solutions of the linear system

$$\begin{cases} x + 2y - z + 3w = 5 \\ y + 2z + w = 7. \end{cases}$$

Solution The Gaussian elimination method proceeds as follows:

$$\left[\begin{array}{cccc|c} 1 & 2 & -1 & 3 & 5 \\ 0 & \textcircled{1} & 2 & 1 & 7 \end{array}\right] \quad \text{(The first column is already in proper form.)}$$

$$\xrightarrow{\;[1]+(-2)[2]\;} \left[\begin{array}{cccc|c} 1 & 0 & -5 & 1 & -9 \\ 0 & 1 & 2 & 1 & 7 \end{array}\right].$$

We cannot do anything further with the third and fourth columns (without disturbing the first two columns), so the corresponding variables, z and w, can assume any values. Writing down the equations corresponding to the last matrix yields

$$\begin{cases} x & -5z + w = -9 \\ & y + 2z + w = 7, \end{cases}$$

or

$$z = \text{any value}$$
$$w = \text{any value}$$
$$x = -9 + 5z - w$$
$$y = 7 - 2z - w.$$

To determine a specific solution, let, for example, $z = 1$ and $w = 2$. Then a specific solution of the original system is

$$z = 1$$
$$w = 2$$
$$x = -9 + 5(1) - (2) = -6$$
$$y = 7 - 2(1) - (2) = 3.$$

Now Try Exercise 19   ■

EXAMPLE 4 **Solving a system of equations that has many solutions** Find all solutions of the system of equations

$$\begin{cases} x - 7y + z = 3 \\ 2x - 14y + 3z = 4. \end{cases}$$

Solution The first pivot operation is routine:

$$\left[\begin{array}{ccc|c} \textcircled{1} & -7 & 1 & 3 \\ 2 & -14 & 3 & 4 \end{array}\right] \quad \xrightarrow{\;[2]+(-2)[1]\;} \quad \left[\begin{array}{ccc|c} 1 & -7 & 1 & 3 \\ 0 & 0 & 1 & -2 \end{array}\right].$$

However, it is impossible to pivot about the zero in the second column. So skip the second column and pivot about the second entry in the third column to get

$$\left[\begin{array}{ccc|c} 1 & -7 & 0 & 5 \\ 0 & 0 & 1 & -2 \end{array}\right].$$

This is as far as we can go. The variable corresponding to the second column, namely y, can assume any value, and the general solution of the system is obtained from the equations

$$\begin{cases} x - 7y & = 5 \\ & z = -2. \end{cases}$$

Therefore, the general solution of the system is

$$y = \text{any value}$$
$$x = 5 + 7y$$
$$z = -2.$$

Now Try Exercise 23

We have seen that a linear system may have a unique solution or it may have infinitely many solutions. But another phenomenon can occur: A system may have no solutions at all, as the next example shows.

EXAMPLE 5 **An inconsistent system of equations** Find all solutions of the system

$$\begin{cases} x - y + z = 3 \\ x + y - z = 5 \\ -2x + 4y - 4z = 1. \end{cases}$$

Solution We apply the Gaussian elimination method to the matrix of the system.

$$\begin{bmatrix} ① & -1 & 1 & 3 \\ 1 & 1 & -1 & 5 \\ -2 & 4 & -4 & 1 \end{bmatrix} \xrightarrow[\begin{subarray}{c} [2] + (-1)[1] \\ [3] + 2[1] \end{subarray}]{} \begin{bmatrix} 1 & -1 & 1 & 3 \\ 0 & ② & -2 & 2 \\ 0 & 2 & -2 & 7 \end{bmatrix}$$

$$\xrightarrow[\begin{subarray}{c} \frac{1}{2}[2] \\ [1] + (1)[2] \\ [3] + (-2)[2] \end{subarray}]{} \begin{bmatrix} 1 & 0 & 0 & 4 \\ 0 & 1 & -1 & 1 \\ 0 & 0 & 0 & 5 \end{bmatrix}$$

We cannot pivot about the last zero in the third column, so we have carried the method as far as we can. Let us write out the equations corresponding to the last matrix:

$$\begin{cases} x \quad\quad = 4 \\ \quad y - z = 1 \\ \quad\quad 0 = 5. \end{cases}$$

Note that the last equation is a built-in contradiction. In mathematical terms, the last equation is *inconsistent*. So the last equation can never be satisfied, no matter what the values of x, y, and z. Thus, the original system has no solutions. Systems with no solutions can always be detected by the presence of inconsistent equations in the last matrix resulting from the Gaussian elimination method.

Now Try Exercise 13

At first it might seem strange that some systems have no solutions, some have one, and yet others have infinitely many. The reason for the difference can be explained geometrically. For simplicity, consider the case of systems of two equations in two variables. Each equation in this case has a graph in the xy-plane, and the graph is a straight line. As we have seen, solving the system corresponds to finding the points lying on both lines. There are three possibilities. First, the two lines may intersect. In this case the solution is unique. Second, the two lines may be parallel.

Then the two lines do not intersect, and the system has no solutions. Finally, the two equations may represent the same line, as, for example, the equations $2x + 3y = 1$ and $4x + 6y = 2$ do. In this case every point on the line is a solution of the system; that is, there are infinitely many solutions (Fig. 1).

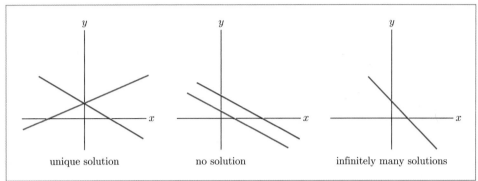

Figure 1

GC   The complete Gaussian elimination method can be carried out in one step as shown in Figs. 2 and 3 for the matrix of Example 4. The command **rref** is found in the MATRIX/MATH menu of the TI-83 and in the MATH/Matrix menu of the TI-89. (*Note*: **rref** stands for "row-reduced echelon form," the name given to the final form of a matrix that has been completely row reduced.)

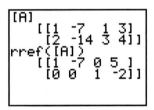

Figure 2.   TI-83   **Figure 3.**   TI-89

ES   When Solver is used with a system of linear equations having infinitely many solutions, only one solution is given. When Solver is used with a system of linear equations having no solutions, the sentence "Solver could not find a feasible solution." is displayed.

PRACTICE PROBLEMS 2.2

1. Find a specific solution to a system of linear equations whose general solution is

$$w = \text{any value}$$
$$y = \text{any value}$$
$$z = 7 + 6w$$
$$x = 26 - 2y + 14w.$$

2. Find all solutions of this system of linear equations.

$$\begin{cases} 2x + 4y - 4z - 4w = 24 \\ -3x - 6y + 10z - 18w = -8 \\ -x - 2y + 4z - 10w = 2 \end{cases}$$

EXERCISES 2.2

In Exercises 1–8, pivot each matrix about the circled element.

1. $\begin{bmatrix} ② & -4 & 6 \\ 3 & 7 & 1 \end{bmatrix}$
 2. $\begin{bmatrix} 1 & 2 & 3 \\ 4 & ⑧ & -12 \end{bmatrix}$

3. $\begin{bmatrix} 7 & 1 & 4 & 5 \\ -1 & 1 & ② & 6 \\ 4 & 0 & 2 & 3 \end{bmatrix}$
 4. $\begin{bmatrix} 5 & 10 & -10 & 12 \\ 4 & 3 & 6 & 12 \\ 4 & ④ & 4 & -16 \end{bmatrix}$

5. $\begin{bmatrix} ② & 3 \\ 6 & 0 \\ 1 & 5 \end{bmatrix}$
 6. $\begin{bmatrix} 2 & 1 \\ ① & 0 \end{bmatrix}$

7. $\begin{bmatrix} 4 & 3 & 0 \\ \frac{2}{3} & 0 & -2 \\ 1 & 3 & ⑥ \end{bmatrix}$
 8. $\begin{bmatrix} 1 & 0 & 2 \\ -1 & 1 & ② \\ 1 & 2 & 6 \end{bmatrix}$

In Exercises 9–22, use the Gaussian elimination method to find all solutions of the systems of linear equations.

9. $\begin{cases} 2x - 4y = 6 \\ -x + 2y = -3 \end{cases}$
 10. $\begin{cases} -\frac{1}{2}x + y = \frac{3}{2} \\ -3x + 6y = 10 \end{cases}$

11. $\begin{cases} x + 2y = 5 \\ 3x - y = 1 \\ -x + 3y = 5 \end{cases}$
 12. $\begin{cases} x - 6y = 12 \\ -\frac{1}{2}x + 3y = -6 \\ \frac{1}{3}x - 2y = 4 \end{cases}$

13. $\begin{cases} x - y + 3z = 3 \\ -2x + 3y - 11z = -4 \\ x - 2y + 8z = 6 \end{cases}$

14. $\begin{cases} x - 3y + z = 5 \\ -2x + 7y - 6z = -9 \\ x - 2y - 3z = 6 \end{cases}$

15. $\begin{cases} x + y + z = -1 \\ 2x + 3y + 2z = 3 \\ 2x + y + 2z = -7 \end{cases}$
 16. $\begin{cases} x - 3y + 2z = 10 \\ -x + 3y - z = -6 \\ -x + 3y + 2z = 6 \end{cases}$

17. $\begin{cases} x + 2y + 3z = 4 \\ 5x + 6y + 7z = 8 \\ x + 2y + 3z = 5 \end{cases}$
 18. $\begin{cases} x + 3y = 7 \\ x + 2y = 5 \\ -x + y = 2 \end{cases}$

19. $\begin{cases} x + y - 2z + 2w = 5 \\ 2x + y - 4z + w = 5 \\ 3x + 4y - 6z + 9w = 20 \\ 4x + 4y - 8z + 8w = 20 \end{cases}$

20. $\begin{cases} 2y + z - w = 1 \\ x - y + z + w = 14 \\ -x - 9y - z + 4w = 11 \\ x + y + z = 9 \end{cases}$

21. $\begin{cases} 6x - 4y = 2 \\ -3x + 3y = 6 \\ 5x + 2y = 39 \end{cases}$
 22. $\begin{cases} 3x + 2y = 5 \\ -x + 3y = 2 \\ 5x + 2y = 6 \\ 6x + y = 39 \end{cases}$

In Exercises 23–25, find three solutions to the systems of equations.

23. $\begin{cases} x + 2y + z = 5 \\ y + 3z = 9 \end{cases}$
 24. $\begin{cases} x + 5y + 3z = 9 \\ 2x + 9y + 7z = 5 \end{cases}$

25. $\begin{cases} x + 7y - 3z = 8 \\ z = 5 \end{cases}$

26. (*Triathlon*) An out-of-shape athlete runs 6 miles per hour, swims 1 mile per hour, and bikes 10 miles per hour. He entered a triathlon, which requires all three events, and finished 5 hours and 40 minutes later. A friend who runs 8 miles per hour, swims 2 miles per hour, and bikes 15 miles per hour finished the same course in 3 hours and 35 minutes. The total course was 32 miles long. How many miles was each segment (running, swimming, and biking)?

27. (*Nutrition Planning*) In a laboratory experiment, a researcher wants to provide a rabbit with exactly 1000 units of vitamin A, exactly 1600 units of vitamin C, and exactly 2400 units of vitamin E. The rabbit is fed a mixture of three foods. Each gram of food 1 contains 2 units of vitamin A, 3 units of vitamin C, and 5 units of vitamin E. Each gram of food 2 contains 4 units of vitamin A, 7 units of vitamin C, and 9 units of vitamin E. Each gram of food 3 contains 6 units of vitamin A, 10 units of vitamin C, and 14 units of vitamin E. How many grams of each food should the rabbit be fed?

28. (*Nutrition Planning*) Rework Exercise 27 with the requirement for vitamin E changed to 2000 units.

29. (*Quilting*) Granny's Custom Quilts receives an order for a patchwork quilt made from square patches of three types: solid green, solid blue, and floral. The quilt is to be 8 squares by 12 squares, and there must be 15 times as many solid squares as floral squares. If Granny's charges $3 per solid square and $5 per floral square, and the customer wishes to spend exactly $300, how many of each type of square may be used in the quilt?

30. (*Purchasing Options*) Amanda is decorating her new home and wants to buy some house plants. She is interested in three types of plants costing $7, $10, and $13. If she has budgeted exactly $150 for the plants and wants to buy exactly 15 of them, what are her options?

31. Find all solutions to the following system of equations.

$$\begin{cases} x^2 + y^2 + z^2 = 14 \\ x^2 - y^2 + 2z^2 = 15 \\ x^2 + 2y^2 + 3z^2 = 36 \end{cases}$$

32. For what value of k will the following system of linear equations have a solution?

$$\begin{cases} 2x + 6y = 4 \\ x + 7y = 10 \\ kx + 8y = 4 \end{cases}$$

33. For what values(s) of k will the following system of linear equations have no solution? Infinitely many solutions?

$$\begin{cases} 2x - 3y = 4 \\ -6x + 9y = k \end{cases}$$

In Exercises 34–42, use a graphing calculator or mathematical software to carry out the tasks.

In Exercises 34–37, graph the three equations together, and determine the number of solutions (exactly one, none, or infinitely many). If there is exactly one solution, estimate the solution.

34. $\begin{cases} 2x + 3y = 5 \\ -3x + 5y = 22 \\ 2x + y = -1 \end{cases}$ **35.** $\begin{cases} x + y = 10 \\ 2x - 3y = 5 \\ -x + 3y = 2 \end{cases}$

36. $\begin{cases} 3x - 2y = 3 \\ -2x + 4y = 14 \\ x + y = 11 \end{cases}$ **37.** $\begin{cases} 2x + y = 12 \\ 3x - y = 2 \\ x + 2y = 16 \end{cases}$

In Exercises 38–41, solve the system of linear equations by the Gaussian elimination method.

38. $\begin{cases} 2x + 3y + 6z = 4 \\ 4x + 7y = 2 \\ 3x + 5y + 3z = 3 \end{cases}$

39. $\begin{cases} 4x - 3y + 2z = 3 \\ -7x + 5y = 2 \\ -10x + 7y + 2z = 4 \end{cases}$

40. $\begin{cases} 8x + 3y - 2z = 5 \\ 12x + 5y + 2z = 3 \\ 5x + 2y = 7 \end{cases}$

41. $\begin{cases} 2x - 5y - 3z = -3 \\ -5x + 12y + 3z = 11 \\ -3x + 7y = 8 \end{cases}$

42. If your calculator or software has **rref** or an analogous command, apply the command to the matrix in Example 5. How does the final matrix differ from the one appearing in the text?

SOLUTIONS TO PRACTICE PROBLEMS 2.2

1. Since w and y can each assume any value, select any numbers, say $w = 1$ and $y = 2$. Then $z = 7 + 6(1) = 13$ and $x = 26 - 2(2) + 14(1) = 36$. So $x = 36$, $y = 2$, $z = 13$, $w = 1$ is a specific solution. There are infinitely many different specific solutions since there are infinitely many different choices for w and y.

2. We apply the Gaussian elimination method to the matrix of the system.

$$\begin{bmatrix} \boxed{2} & 4 & -4 & -4 & | & 24 \\ -3 & -6 & 10 & -18 & | & -8 \\ -1 & -2 & 4 & -10 & | & 2 \end{bmatrix}$$

$$\xrightarrow[\substack{[2]+3[1] \\ [3]+1[1]}]{\frac{1}{2}[1]} \begin{bmatrix} 1 & 2 & -2 & -2 & | & 12 \\ 0 & 0 & \boxed{4} & -24 & | & 28 \\ 0 & 0 & 2 & -12 & | & 14 \end{bmatrix}$$

$$\xrightarrow[\substack{[1]+2[2] \\ [3]+(-2)[2]}]{\frac{1}{4}[2]} \begin{bmatrix} 1 & 2 & 0 & -14 & | & 26 \\ 0 & 0 & 1 & -6 & | & 7 \\ 0 & 0 & 0 & 0 & | & 0 \end{bmatrix}$$

The corresponding system of equations is

$$\begin{cases} x + 2y - 14w = 26 \\ z - 6w = 7. \end{cases}$$

The general solution is

$$w = \text{any value}$$
$$y = \text{any value}$$
$$z = 7 + 6w$$
$$x = 26 - 2y + 14w.$$

2.3 Arithmetic Operations on Matrices

We introduced matrices in Sections 2.1 and 2.2 to display the coefficients of a system of linear equations. For example, the linear system

$$\begin{cases} 5x - 3y = \frac{1}{2} \\ 4x + 2y = -1 \end{cases}$$

is represented by the matrix

$$\left[\begin{array}{cc|c} 5 & -3 & \frac{1}{2} \\ 4 & 2 & -1 \end{array}\right].$$

After we have become accustomed to using such matrices in solving linear systems, we may omit the vertical line that separates the left and right sides of the equations. We need only remember that the right side of the equations is recorded in the right column. So, for example, we would write the preceding matrix in the form

$$\begin{bmatrix} 5 & -3 & \frac{1}{2} \\ 4 & 2 & -1 \end{bmatrix}.$$

A matrix is *any* rectangular array of numbers and may be of any size. Here are some examples of matrices of various sizes:

$$\begin{bmatrix} 3 & 7 \\ 0 & -1 \end{bmatrix}, \quad \begin{bmatrix} 1 \\ 2 \end{bmatrix}, \quad \begin{bmatrix} 2 & 1 \end{bmatrix}, \quad \begin{bmatrix} 6 \end{bmatrix}, \quad \begin{bmatrix} 5 & 7 & -1 \\ 0 & 3 & 5 \\ 6 & 0 & 5 \end{bmatrix}.$$

Examples of matrices abound in everyday life. For example, the newspaper stock market report is a large matrix with several thousand rows, one for each listed stock. The columns of the matrix give the various data about each stock, such as opening and closing price, number of shares traded, and so on. Another example of a matrix is a mileage chart on a road map. The rows and columns are labeled with the names of cities. The number at a given row and column gives the distance between the corresponding cities.

In these everyday examples, matrices are used only to display data. However, the most important applications involve arithmetic operations on matrices—namely, addition, subtraction, and multiplication of matrices. The major goal of this section is to discuss these operations. Before we can do so, however, we need some vocabulary with which to describe matrices.

A matrix is described by the number of rows and columns it contains. For example, the matrix

$$\begin{bmatrix} 7 & 5 \\ \frac{1}{2} & -2 \\ 2 & -11 \end{bmatrix}$$

has three rows and two columns and is referred to as a *3 × 2* (read: "three-by-two") *matrix*. The matrix $\begin{bmatrix} 4 & 5 & 0 \end{bmatrix}$ has one row and three columns and is a *1 × 3 matrix*. A matrix with only one row is often called a *row matrix* (sometimes also called a *row vector*). A matrix, such as

$$\begin{bmatrix} 2 \\ 7 \end{bmatrix},$$

that has only one column is called a *column matrix* or *column vector*. If a matrix has the same number of rows and columns, it is called a *square matrix*. Here are some square matrices of various sizes:

$$\begin{bmatrix} 5 \end{bmatrix}, \quad \begin{bmatrix} 1 & 2 \\ 3 & 4 \end{bmatrix}, \quad \begin{bmatrix} 2 & -1 & 0 \\ 3 & 5 & 4 \\ 0 & 3 & -7 \end{bmatrix}.$$

The rows of a matrix are numbered from the top down, and the columns are numbered from left to right. For example, the first row of the matrix

$$\begin{bmatrix} 1 & -1 & 0 \\ 2 & 1 & 7 \\ -3 & 2 & 4 \end{bmatrix}$$

is $\begin{bmatrix} 1 & -1 & 0 \end{bmatrix}$, and its third column is

$$\begin{bmatrix} 0 \\ 7 \\ 4 \end{bmatrix}.$$

The numbers in a matrix, called *entries*, may be identified in terms of the row and column containing the entry in question. For example, the entry in the first row, third column of the following matrix is 0:

$$\begin{bmatrix} 1 & -1 & \boxed{0} \\ 2 & 1 & 7 \\ -3 & 2 & 4 \end{bmatrix};$$

the entry in the second row, first column is 2:

$$\begin{bmatrix} 1 & -1 & 0 \\ \boxed{2} & 1 & 7 \\ -3 & 2 & 4 \end{bmatrix};$$

and the entry in the third row, third column is 4:

$$\begin{bmatrix} 1 & -1 & 0 \\ 2 & 1 & 7 \\ -3 & 2 & \boxed{4} \end{bmatrix}.$$

We use double-subscripted letters to indicate the locations of the entries of a matrix. We denote the entry in the ith row, jth column by a_{ij}. For instance, in the preceding matrix we have $a_{13} = 0$, $a_{21} = 2$, and $a_{33} = 4$.

We say that two matrices A and B are *equal*, denoted $A = B$, provided that they have the same size and that all their corresponding entries are equal.

Addition and Subtraction of Matrices We define the sum $A + B$ of two matrices A and B only if A and B are two matrices of the same size—that is, if A and B have the same number of rows and the same number of columns. In this case $A + B$ is the matrix formed by adding the corresponding entries of A and B. For example,

$$\begin{bmatrix} 2 & 0 \\ 1 & 1 \\ 5 & 3 \end{bmatrix} + \begin{bmatrix} 5 & 4 \\ 0 & 2 \\ 2 & 6 \end{bmatrix} = \begin{bmatrix} 2+5 & 0+4 \\ 1+0 & 1+2 \\ 5+2 & 3+6 \end{bmatrix} = \begin{bmatrix} 7 & 4 \\ 1 & 3 \\ 7 & 9 \end{bmatrix}.$$

We subtract matrices of the same size by subtracting corresponding entries. Thus, we have

$$\begin{bmatrix} 7 \\ 1 \end{bmatrix} - \begin{bmatrix} 3 \\ 2 \end{bmatrix} = \begin{bmatrix} 7-3 \\ 1-2 \end{bmatrix} = \begin{bmatrix} 4 \\ -1 \end{bmatrix}.$$

Multiplication of Matrices It might seem that to define the product of two matrices, one would start with two matrices of like size and multiply the corresponding entries. But this definition is not useful, since the calculations that arise in applications require a somewhat more complex multiplication. In the interests of simplicity, we start by defining the product of a row matrix times a column matrix.

If A is a row matrix and B is a column matrix, then we can form the product $A \cdot B$ provided that the two matrices have the same length. The product $A \cdot B$ is the 1×1 matrix obtained by multiplying corresponding entries of A and B and then forming the sum.

We may put this definition into algebraic terms as follows. Suppose that A is the row matrix

$$A = \begin{bmatrix} a_1 & a_2 & \cdots & a_n \end{bmatrix},$$

and B is the column matrix

$$B = \begin{bmatrix} b_1 \\ b_2 \\ \vdots \\ b_n \end{bmatrix}.$$

Note that A and B are both of the same length, namely n. Then

$$A \cdot B = \begin{bmatrix} a_1 & a_2 & \cdots & a_n \end{bmatrix} \cdot \begin{bmatrix} b_1 \\ b_2 \\ \vdots \\ b_n \end{bmatrix}$$

is calculated by multiplying corresponding entries of A and B and forming the sum; that is,

$$A \cdot B = \begin{bmatrix} a_1 b_1 + a_2 b_2 + \cdots + a_n b_n \end{bmatrix}.$$

Notice that the product is a 1×1 matrix, namely a single number in brackets.

Here are some examples of the product of a row matrix times a column matrix:

$$\begin{bmatrix} 3 & \frac{1}{2} \end{bmatrix} \cdot \begin{bmatrix} 1 \\ 4 \end{bmatrix} = \begin{bmatrix} 3 \cdot 1 + \frac{1}{2} \cdot 4 \end{bmatrix} = \begin{bmatrix} 5 \end{bmatrix};$$

$$\begin{bmatrix} 2 & 0 & -1 \end{bmatrix} \cdot \begin{bmatrix} 6 \\ 5 \\ 3 \end{bmatrix} = \begin{bmatrix} 2 \cdot 6 + 0 \cdot 5 + (-1) \cdot 3 \end{bmatrix} = \begin{bmatrix} 9 \end{bmatrix}.$$

In multiplying a row matrix times a column matrix, it helps to use both of your hands. Use your left index finger to point to an element of the row matrix and your right to point to the corresponding element of the column. Multiply the elements you are pointing to and keep a running total of the products in your head. After each multiplication move your fingers to the next elements of each matrix. With a little practice you should be able to multiply a row times a column quickly and accurately.

The preceding definition of multiplication may seem strange. But products of this sort occur in many down-to-earth problems. Consider, for instance, the next example.

EXAMPLE 1 **Total revenue as a matrix product** A dairy farm produces three items—milk, eggs, and cheese. The prices of the three items are \$1.50 per gallon, \$0.80 per dozen, and \$2.00 per pound, respectively. In a certain week the dairy farm sells 30,000 gallons of milk, 2000 dozen eggs, and 5000 pounds of cheese. Represent its total revenue as a matrix product.

Solution The total revenue equals

$$(1.50)(30{,}000) + (.80)(2000) + (2)(5000).$$

This suggests that we define two matrices: The first displays the prices of the various items:

$$\begin{bmatrix} 1.50 & .80 & 2 \end{bmatrix}.$$

The second represents the production:

$$\begin{bmatrix} 30{,}000 \\ 2000 \\ 5000 \end{bmatrix}.$$

Then the revenue for the week, when placed in a 1×1 matrix, equals

$$\begin{bmatrix} 1.50 & .80 & 2 \end{bmatrix} \begin{bmatrix} 30{,}000 \\ 2000 \\ 5000 \end{bmatrix} = \begin{bmatrix} 56{,}600 \end{bmatrix}.$$

Now Try Exercise 67 ■

The principle behind Example 1 is this: Any sum of products of the form $a_1 b_1 + a_2 b_2 + \cdots + a_n b_n$, when placed in a 1×1 matrix, can be written as the matrix product

$$\begin{bmatrix} a_1 b_1 + a_2 b_2 + \cdots + a_n b_n \end{bmatrix} = \begin{bmatrix} a_1 & a_2 & \cdots & a_n \end{bmatrix} \cdot \begin{bmatrix} b_1 \\ b_2 \\ \vdots \\ b_n \end{bmatrix}.$$

Let us illustrate the procedure for multiplying more general matrices by working out a typical product:

$$\begin{bmatrix} 2 & 1 \\ 0 & 1 \\ 1 & 0 \end{bmatrix} \cdot \begin{bmatrix} 1 & 1 \\ 4 & 2 \end{bmatrix}.$$

To obtain the entries of the product, we multiply the rows of the left matrix by the columns of the right matrix, taking care to arrange the products in a specific way to yield a matrix, as follows. Start with the first row on the left, $\begin{bmatrix} 2 & 1 \end{bmatrix}$, and the first column on the right, $\begin{bmatrix} 1 \\ 4 \end{bmatrix}$. Their product is $\begin{bmatrix} 6 \end{bmatrix}$, so we enter 6 as the element in the first row, first column of the product:

$$\begin{bmatrix} 2 & 1 \\ 0 & 1 \\ 1 & 0 \end{bmatrix} \cdot \begin{bmatrix} 1 & 1 \\ 4 & 2 \end{bmatrix} = \begin{bmatrix} 6 & \\ & \end{bmatrix}.$$

The product of the first row of the left matrix and the second column of the right matrix is $\begin{bmatrix} 4 \end{bmatrix}$, so we put a 4 in the first row, second column of the product:

$$\begin{bmatrix} 2 & 1 \\ 0 & 1 \\ 1 & 0 \end{bmatrix} \cdot \begin{bmatrix} 1 & 1 \\ 4 & 2 \end{bmatrix} = \begin{bmatrix} 6 & 4 \\ & \\ & \end{bmatrix}.$$

There are no more columns that can be multiplied by the first row, so let us move to the second row and shift back to the first column. Correspondingly, we move down one row in the product:

$$\begin{bmatrix} 2 & 1 \\ 0 & 1 \\ 1 & 0 \end{bmatrix} \cdot \begin{bmatrix} 1 & 1 \\ 4 & 2 \end{bmatrix} = \begin{bmatrix} 6 & 4 \\ 4 & \\ & \end{bmatrix};$$

$$\begin{bmatrix} 2 & 1 \\ 0 & 1 \\ 1 & 0 \end{bmatrix} \cdot \begin{bmatrix} 1 & 1 \\ 4 & 2 \end{bmatrix} = \begin{bmatrix} 6 & 4 \\ 4 & 2 \\ & \end{bmatrix}.$$

We have now exhausted the second row of the left matrix, so we shift to the third row and correspondingly move down one row in the product:

$$\begin{bmatrix} 2 & 1 \\ 0 & 1 \\ 1 & 0 \end{bmatrix} \cdot \begin{bmatrix} 1 & 1 \\ 4 & 2 \end{bmatrix} = \begin{bmatrix} 6 & 4 \\ 4 & 2 \\ 1 & \end{bmatrix};$$

$$\begin{bmatrix} 2 & 1 \\ 0 & 1 \\ 1 & 0 \end{bmatrix} \cdot \begin{bmatrix} 1 & 1 \\ 4 & 2 \end{bmatrix} = \begin{bmatrix} 6 & 4 \\ 4 & 2 \\ 1 & 1 \end{bmatrix}.$$

Note that we have now multiplied every row of the left matrix by every column of the right matrix. This completes the computation of the product:

$$\begin{bmatrix} 2 & 1 \\ 0 & 1 \\ 1 & 0 \end{bmatrix} \cdot \begin{bmatrix} 1 & 1 \\ 4 & 2 \end{bmatrix} = \begin{bmatrix} 6 & 4 \\ 4 & 2 \\ 1 & 1 \end{bmatrix}.$$

EXAMPLE 2 **Matrix multiplication** Calculate the following product:

$$\begin{bmatrix} 1 & 5 \\ 3 & 2 \end{bmatrix} \cdot \begin{bmatrix} 1 & 2 \\ 1 & 0 \end{bmatrix}.$$

Solution

$$\begin{bmatrix} 1 & 5 \\ 3 & 2 \end{bmatrix} \cdot \begin{bmatrix} 1 & 2 \\ 1 & 0 \end{bmatrix} = \begin{bmatrix} 6 & \\ & \end{bmatrix}$$

$$\begin{bmatrix} 1 & 5 \\ 3 & 2 \end{bmatrix} \cdot \begin{bmatrix} 1 & 2 \\ 1 & 0 \end{bmatrix} = \begin{bmatrix} 6 & 2 \\ & \end{bmatrix}$$

$$\begin{bmatrix} 1 & 5 \\ 3 & 2 \end{bmatrix} \cdot \begin{bmatrix} 1 & 2 \\ 1 & 0 \end{bmatrix} = \begin{bmatrix} 6 & 2 \\ 5 & \end{bmatrix}$$

$$\begin{bmatrix} 1 & 5 \\ 3 & 2 \end{bmatrix} \cdot \begin{bmatrix} 1 & 2 \\ 1 & 0 \end{bmatrix} = \begin{bmatrix} 6 & 2 \\ 5 & 6 \end{bmatrix}$$

Thus,

$$\begin{bmatrix} 1 & 5 \\ 3 & 2 \end{bmatrix} \cdot \begin{bmatrix} 1 & 2 \\ 1 & 0 \end{bmatrix} = \begin{bmatrix} 6 & 2 \\ 5 & 6 \end{bmatrix}.$$

Now Try Exercise 25　　■

Notice that we cannot use the preceding method to compute the product $A \cdot B$ of *any* matrices A and B. For the procedure to work, it is crucial that the number of entries of each row of A be the same as the number of entries of each column of B. (Or, to put it another way, the number of columns of the left matrix must equal the number of rows of the right matrix.) Therefore, in order for us to form the product $A \cdot B$, the sizes of A and B must match up in a special way. If A is $m \times n$ and B is $p \times q$, then the product $A \cdot B$ is defined only in case the "inner" dimensions n and p are equal. In that case, the size of the product is determined by the "outer" dimensions m and q. It is an $m \times q$ matrix:

$$A \quad \cdot \quad B \quad = \quad C.$$
$$m \times n \quad p \times q \quad m \times q$$
$$\underset{\text{equal}}{\underbrace{\qquad}}$$

So, for example,

$$\underset{3 \times 4}{\begin{bmatrix} \end{bmatrix}} \underset{4 \times 2}{\begin{bmatrix} \end{bmatrix}} = \underset{3 \times 2}{\begin{bmatrix} \end{bmatrix}}$$

$$\underset{2 \times 2}{\begin{bmatrix} \end{bmatrix}} \underset{2 \times 1}{\begin{bmatrix} \end{bmatrix}} = \underset{2 \times 1}{\begin{bmatrix} \end{bmatrix}}.$$

If the sizes of A and B do not match up in the way just described, the product $A \cdot B$ is not defined.

EXAMPLE 3   **Matrix multiplication**   Calculate the following products, if defined.

(a) $\begin{bmatrix} 3 & -1 \\ 2 & 0 \\ 1 & 5 \end{bmatrix} \begin{bmatrix} 1 & 0 \\ 5 & -4 \\ 2 & -1 \end{bmatrix}$ (b) $\begin{bmatrix} 3 & -1 \\ 2 & 0 \\ 1 & 5 \end{bmatrix} \begin{bmatrix} 5 & 4 \\ -2 & 3 \end{bmatrix}$

Solution (a) The matrices to be multiplied are 3×2 and 3×2. The inner dimensions do not match, so the product is undefined.

(b) We are asked to multiply a 3×2 matrix times a 2×2 matrix. The inner dimensions match, so the product is defined and has size determined by the outer dimensions, that is, 3×2.

$$\begin{bmatrix} 3 & -1 \\ 2 & 0 \\ 1 & 5 \end{bmatrix} \cdot \begin{bmatrix} 5 & 4 \\ -2 & 3 \end{bmatrix} = \begin{bmatrix} 3 \cdot 5 + (-1) \cdot (-2) & 3 \cdot 4 + (-1) \cdot 3 \\ 2 \cdot 5 + 0 \cdot (-2) & 2 \cdot 4 + 0 \cdot 3 \\ 1 \cdot 5 + 5 \cdot (-2) & 1 \cdot 4 + 5 \cdot 3 \end{bmatrix}$$

$$= \begin{bmatrix} 17 & 9 \\ 10 & 8 \\ -5 & 19 \end{bmatrix}$$

Now Try Exercise 27   ∎

Multiplication of matrices has many properties in common with multiplication of ordinary numbers. However, there is at least one important difference. With matrix multiplication, the order of the factors is usually important. For example, the product of a 2×3 matrix times a 3×2 matrix is defined: The product is a 2×2 matrix. If the order is reversed to a 3×2 matrix times a 2×3 matrix, the product is a 3×3 matrix. So reversing the order may change the size of the product. Even when it does not, reversing the order may still change the entries in the product, as the following two products demonstrate:

$$\begin{bmatrix} 1 & 5 \\ 3 & 2 \end{bmatrix} \begin{bmatrix} 1 & 2 \\ 1 & 0 \end{bmatrix} = \begin{bmatrix} 6 & 2 \\ 5 & 6 \end{bmatrix}; \qquad \begin{bmatrix} 1 & 2 \\ 1 & 0 \end{bmatrix} \begin{bmatrix} 1 & 5 \\ 3 & 2 \end{bmatrix} = \begin{bmatrix} 7 & 9 \\ 1 & 5 \end{bmatrix}.$$

EXAMPLE 4   **Investment earnings**   An investment trust has investments in three states. Its deposits in each state are divided among bonds, mortgages, and consumer loans. On January 1 the amount (in millions of dollars) of money invested in each category by state is given by the matrix

	Bonds	Mortgages	Consumer loans
State A	10	5	20
State B	30	12	10
State C	15	6	25

The current average yields are 7% for bonds, 9% for mortgages, and 15% for consumer loans. Determine the earnings of the trust from its investments in each state.

Solution Define the matrix of investment yields by

$$\begin{bmatrix} .07 \\ .09 \\ .15 \end{bmatrix} \begin{matrix} \text{Bonds} \\ \text{Mortgages} \\ \text{Consumer loans.} \end{matrix}$$

The amount earned in state A, for instance, is

[amount of bonds][yield of bonds]

+ [amount of mortgages][yield of mortgages]

+ [amount of consumer loans][yield of consumer loans]

$$= (10)(.07) + (5)(.09) + (20)(.15).$$

And this is just the first entry of the product:

$$\begin{bmatrix} 10 & 5 & 20 \\ 30 & 12 & 10 \\ 15 & 6 & 25 \end{bmatrix} \begin{bmatrix} .07 \\ .09 \\ .15 \end{bmatrix}.$$

Similarly, the earnings for the other states are the second and third entries of the product. Carrying out the arithmetic, we find that

$$\begin{bmatrix} 10 & 5 & 20 \\ 30 & 12 & 10 \\ 15 & 6 & 25 \end{bmatrix} \begin{bmatrix} .07 \\ .09 \\ .15 \end{bmatrix} = \begin{bmatrix} 4.15 \\ 4.68 \\ 5.34 \end{bmatrix}.$$

Therefore, the trust earns $4.15 million in state A, $4.68 million in state B, and $5.34 million in state C.

Now Try Exercise 51(a)

■

EXAMPLE 5 **Manufacturing revenue** A clothing manufacturer has factories in Los Angeles, San Antonio, and Newark. Sales (in thousands) during the first quarter of last year are summarized in the production matrix

	Los Angeles	San Antonio	Newark
Coats	12	13	38
Shirts	25	5	26
Sweaters	11	8	8
Ties	5	0	12

During this period the selling price of a coat was $100, of a shirt $10, of a sweater $25, and of a tie $5.

(a) Use a matrix calculation to determine the total revenue produced by each of the factories.

(b) Suppose that the prices had been $110, $8, $20, and $10, respectively. How would this have affected the revenue of each factory?

Solution (a) For each factory, we wish to multiply the price of each item by the number produced to arrive at revenue. Since the production figures for the various items of clothing are arranged down the columns, we arrange the prices in a row matrix, ready for multiplication. The price matrix is

$$\begin{bmatrix} 100 & 10 & 25 & 5 \end{bmatrix}.$$

The revenues of the various factories are then the entries of the product

$$\begin{bmatrix} 100 & 10 & 25 & 5 \end{bmatrix} \begin{bmatrix} 12 & 13 & 38 \\ 25 & 5 & 26 \\ 11 & 8 & 8 \\ 5 & 0 & 12 \end{bmatrix} = \begin{matrix} \text{Los Angeles} & \text{San Antonio} & \text{Newark} \\ \begin{bmatrix} 1750 & 1550 & 4320 \end{bmatrix} \end{matrix}.$$

Since the production figures are in thousands, the revenue figures are in thousands of dollars. That is, the Los Angeles factory has revenues of $1,750,000, the San Antonio factory $1,550,000, and the Newark factory $4,320,000.

(b) In a similar way, we determine the revenue of each factory if the price matrix had been $\begin{bmatrix} 110 & 8 & 20 & 10 \end{bmatrix}$.

$$\begin{bmatrix} 110 & 8 & 20 & 10 \end{bmatrix} \begin{bmatrix} 12 & 13 & 38 \\ 25 & 5 & 26 \\ 11 & 8 & 8 \\ 5 & 0 & 12 \end{bmatrix} = \begin{matrix} \text{Los Angeles} & \text{San Antonio} & \text{Newark} \\ \begin{bmatrix} 1790 & 1630 & 4668 \end{bmatrix} \end{matrix}.$$

The change in revenue at each factory can be read from the difference of the revenue matrices:

$$\begin{bmatrix} 1790 & 1630 & 4668 \end{bmatrix} - \begin{bmatrix} 1750 & 1550 & 4320 \end{bmatrix} = \begin{bmatrix} 40 & 80 & 348 \end{bmatrix}.$$

If prices had been as given in (b), revenues of the Los Angeles factory would have increased by $40,000, revenues at San Antonio would have increased by $80,000, and revenues at Newark would have increased by $348,000.

Now Try Exercise 55   ■

There are special matrices analogous to the number 1. Such matrices are called *identity matrices*. The identity matrix I_n of size n is the $n \times n$ square matrix with all zeros except for ones down the upper-left-to-lower-right diagonal. Here are the identity matrices of sizes 2, 3, and 4:

$$I_2 = \begin{bmatrix} 1 & 0 \\ 0 & 1 \end{bmatrix}; \quad I_3 = \begin{bmatrix} 1 & 0 & 0 \\ 0 & 1 & 0 \\ 0 & 0 & 1 \end{bmatrix}; \quad I_4 = \begin{bmatrix} 1 & 0 & 0 & 0 \\ 0 & 1 & 0 & 0 \\ 0 & 0 & 1 & 0 \\ 0 & 0 & 0 & 1 \end{bmatrix}.$$

The characteristic property of an identity matrix is that it plays the role of the number 1; that is,

$$I_n \cdot A = A \cdot I_n = A$$

for all $n \times n$ matrices A.

One of the principal uses of matrices is in dealing with systems of linear equations. Matrices provide a compact way of writing systems, as the next example shows.

EXAMPLE 6 **Representing a system of equations as a matrix equation** Write the system of linear equations

$$\begin{cases} -2x + 4y = 2 \\ -3x + 7y = 7 \end{cases}$$

as a matrix equation.

Solution The system of equations can be written in the form

$$\begin{bmatrix} -2x + 4y \\ -3x + 7y \end{bmatrix} = \begin{bmatrix} 2 \\ 7 \end{bmatrix}.$$

So consider the matrices

$$A = \begin{bmatrix} -2 & 4 \\ -3 & 7 \end{bmatrix}, \qquad X = \begin{bmatrix} x \\ y \end{bmatrix}, \qquad B = \begin{bmatrix} 2 \\ 7 \end{bmatrix}.$$

Notice that

$$AX = \begin{bmatrix} -2 & 4 \\ -3 & 7 \end{bmatrix} \begin{bmatrix} x \\ y \end{bmatrix} = \begin{bmatrix} -2x + 4y \\ -3x + 7y \end{bmatrix}.$$

Thus, AX is a 2×1 column matrix whose entries correspond to the left side of the given system of linear equations. Since the entries of B correspond to the right side of the system of equations, we can rewrite the given system in the form

$$AX = B$$

—that is,

$$\begin{bmatrix} -2 & 4 \\ -3 & 7 \end{bmatrix} \begin{bmatrix} x \\ y \end{bmatrix} = \begin{bmatrix} 2 \\ 7 \end{bmatrix}.$$

Now Try Exercise 39 ■

The matrix A of the preceding example displays the coefficients of the variables x and y, and so it is called the *coefficient matrix* of the system.

GC The arithmetic operations $+$, $-$, and $*$ can be applied to matrices in much the same way as to numbers, as Figs. 1, 2, 4, and 5 show. (*Note*: With the TI-83 **[A]**∗**[B]** also can be written as **[A][B]**.) Identity matrices are placed on the home screen with the command **identity** (in the MATRIX/MATH menu of the TI-83 and in the MATH/Matrix menu of the TI-89), as shown in Figures 3 and 6. The value in the ith row, jth column of the matrix **[A]** can be displayed with **[A](i,j)** on the TI-83 and with **a[i,j]** on the TI-89.

```
[A]
            [[1 5]
             [3 2]]
[B]
            [[1 2]
             [1 0]]
```

Figure 1

```
[A]+[B]
            [[2 7]
             [4 2]]
[A]*[B]
            [[6 2]
             [5 6]]
```

Figure 2

```
identity(2)
            [[1 0]
             [0 1]]
identity(3)
            [[1 0 0]
             [0 1 0]
             [0 0 1]]
```

Figure 3

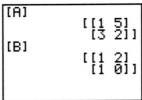

Figure 4

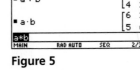

Figure 5

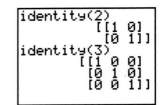

Figure 6

ES Matrices are added, subtracted, and multiplied with the + operator, the − operator, and the MMULT function, respectively. The following steps carry out these operations on the matrices A and B:

1. Highlight a range of cells having the size of the matrix to be computed.
2. Type **=A+B**, **=A−B**, or **=MMULT(A,B)**.
3. Press Ctrl+Shift+Enter. (*Note:* The plus signs indicate that the three keys should be held down together.) The computed matrix will be displayed in the highlighted block.

PRACTICE PROBLEMS 2.3

1. Compute

$$\begin{bmatrix} 3 & 1 & 2 \\ -1 & 0 & \frac{1}{2} \\ 0 & 4 & 1 \end{bmatrix} \begin{bmatrix} 7 & -1 & 0 \\ 5 & 4 & 2 \\ -6 & 0 & 4 \end{bmatrix}.$$

2. Give the system of linear equations that is equivalent to the matrix equation

$$\begin{bmatrix} 3 & -6 \\ 2 & 1 \end{bmatrix} \begin{bmatrix} x \\ y \end{bmatrix} = \begin{bmatrix} 5 \\ 0 \end{bmatrix}.$$

3. Give a matrix equation equivalent to this system of equations:

$$\begin{cases} 8x + 3y = 7 \\ 9x - 2y = -5. \end{cases}$$

EXERCISES 2.3

In Exercises 1–6, give the size and special characteristics of the given matrix (such as square, column, row, identity).

1. $\begin{bmatrix} 3 & 2 & 4 \\ \frac{1}{2} & 0 & 6 \end{bmatrix}$

2. $\begin{bmatrix} 3 \\ -1 \end{bmatrix}$

3. $\begin{bmatrix} 2 & \frac{1}{3} & 0 \end{bmatrix}$

4. $\begin{bmatrix} 1 & 0 \\ 0 & 1 \end{bmatrix}$

5. $\begin{bmatrix} 1 & 0 \\ 0 & 0 \end{bmatrix}$

6. $\begin{bmatrix} 0 & 0 & 0 & 0 \\ 0 & 0 & 0 & 0 \end{bmatrix}$

Exercises 7–10 refer to the 2×3 matrix $A = \begin{bmatrix} 2 & -4 & 6 \\ 0 & 3 & -1 \end{bmatrix}$.

7. Find a_{12} and a_{21}.

8. Find a_{23} and a_{11}.

9. For what values of i and j does $a_{ij} = 6$?

10. For what values of i and j does $a_{ij} = 3$?

In Exercises 11–18, perform the indicated matrix calculations.

11. $\begin{bmatrix} 4 & -2 \\ 3 & 0 \end{bmatrix} + \begin{bmatrix} 5 & 5 \\ 4 & -1 \end{bmatrix}$

12. $\begin{bmatrix} 8 \\ -3 \end{bmatrix} + \begin{bmatrix} 5 \\ 6 \end{bmatrix}$

13. $\begin{bmatrix} 2 & 8 \\ \frac{4}{3} & 4 \\ 1 & -2 \end{bmatrix} - \begin{bmatrix} 1 & 5 \\ \frac{1}{3} & 2 \\ -3 & 0 \end{bmatrix}$

14. $\begin{bmatrix} 1 & 0 \\ 0 & 1 \end{bmatrix} - \begin{bmatrix} .8 & .5 \\ .2 & .5 \end{bmatrix}$

15. $\begin{bmatrix} 5 & 3 \end{bmatrix} \begin{bmatrix} 1 \\ 2 \end{bmatrix}$

16. $\begin{bmatrix} 1 & 0 & 0 \end{bmatrix} \begin{bmatrix} \frac{1}{2} \\ 6 \\ 2 \end{bmatrix}$

17. $\begin{bmatrix} 6 & 1 & 5 \end{bmatrix} \begin{bmatrix} \frac{1}{2} \\ -3 \\ 2 \end{bmatrix}$

18. $\begin{bmatrix} 0 & 0 \end{bmatrix} \begin{bmatrix} 5 \\ -3 \end{bmatrix}$

In Exercises 19–24, the sizes of two matrices are given. Tell whether or not the product AB is defined. If so, give its size.

19. A, 3×4; B, 4×5

20. A, 3×3; B, 3×4

21. A, 3×2; B, 3×2

22. A, 1×1; B, 1×1

23. A, 3×3; B, 3×1

24. A, 4×2; B, 3×4

In Exercises 25–34, perform the multiplication.

25. $\begin{bmatrix} 3 & 1 \\ 0 & 2 \end{bmatrix} \begin{bmatrix} 1 & 4 \\ 3 & 5 \end{bmatrix}$

26. $\begin{bmatrix} 4 & -1 \\ 2 & \frac{1}{2} \end{bmatrix} \begin{bmatrix} 3 \\ 2 \end{bmatrix}$

27. $\begin{bmatrix} 4 & 1 & 0 \\ -2 & 0 & 3 \\ 1 & 5 & -1 \end{bmatrix} \begin{bmatrix} 5 \\ 1 \\ 2 \end{bmatrix}$

28. $\begin{bmatrix} 0 & 0 \\ 0 & 0 \\ 0 & 0 \end{bmatrix} \begin{bmatrix} 1 & 2 \\ 3 & 4 \end{bmatrix}$

29. $\begin{bmatrix} 1 & 0 \\ 0 & 1 \end{bmatrix} \begin{bmatrix} 5 & 6 \\ 7 & 8 \end{bmatrix}$ **30.** $\begin{bmatrix} 1 & 2 \\ 1 & 3 \end{bmatrix} \begin{bmatrix} 3 & -2 \\ -1 & 1 \end{bmatrix}$

31. $\begin{bmatrix} .6 & .3 \\ .4 & .7 \end{bmatrix} \begin{bmatrix} .6 & .3 \\ .4 & .7 \end{bmatrix}$

32. $\begin{bmatrix} 0 & 1 & 2 \\ -1 & 4 & \frac{1}{2} \\ 1 & 3 & 0 \end{bmatrix} \begin{bmatrix} 3 & -1 & 5 \\ 0 & 2 & 2 \\ 4 & -6 & 0 \end{bmatrix}$

33. $\begin{bmatrix} 2 & -1 & 4 \\ 0 & 1 & 0 \\ \frac{1}{2} & 3 & -2 \end{bmatrix} \begin{bmatrix} 4 & 8 & 0 \\ 3 & -1 & 2 \\ 5 & 0 & 1 \end{bmatrix}$

34. $\begin{bmatrix} 1 & 0 & 0 \\ 0 & 1 & 0 \\ 0 & 0 & 1 \end{bmatrix} \begin{bmatrix} 1 \\ 2 \\ 3 \end{bmatrix}$

In Exercises 35–38, give the system of linear equations that is equivalent to the matrix equation. Do not solve.

35. $\begin{bmatrix} 2 & 3 \\ 4 & 5 \end{bmatrix} \begin{bmatrix} x \\ y \end{bmatrix} = \begin{bmatrix} 6 \\ 7 \end{bmatrix}$ **36.** $\begin{bmatrix} -3 & 4 \\ 0 & 1 \end{bmatrix} \begin{bmatrix} x \\ y \end{bmatrix} = \begin{bmatrix} 1 \\ 1 \end{bmatrix}$

37. $\begin{bmatrix} 1 & 2 & 3 \\ 4 & 5 & 6 \\ 7 & 8 & 9 \end{bmatrix} \begin{bmatrix} x \\ y \\ z \end{bmatrix} = \begin{bmatrix} 10 \\ 11 \\ 12 \end{bmatrix}$

38. $\begin{bmatrix} 1 & 0 & 0 \\ 0 & 1 & 0 \\ 0 & 0 & 1 \end{bmatrix} \begin{bmatrix} x \\ y \\ z \end{bmatrix} = \begin{bmatrix} 1 \\ 2 \\ 3 \end{bmatrix}$

In Exercises 39–42, write the given system of linear equations in matrix form.

39. $\begin{cases} 3x + 2y = -1 \\ 7x - y = 2 \end{cases}$ **40.** $\begin{cases} 5x - 2y = 6 \\ -2x + 4y = 0 \end{cases}$

41. $\begin{cases} x - 2y + 3z = 5 \\ y + z = 6 \\ z = 2 \end{cases}$ **42.** $\begin{cases} -2x + 4y - z = 5 \\ x + 6y + 3z = -1 \\ 7x + 4z = 8 \end{cases}$

The distributive law says that $(A+B)C = AC + BC$. That is, adding A and B and then multiplying on the right by C gives the same result as first multiplying each of A and B on the right by C and then adding. In Exercises 43 and 44, verify the distributive law for the given matrices.

43. $A = \begin{bmatrix} 1 & 2 \\ 0 & 3 \end{bmatrix}$, $B = \begin{bmatrix} 3 & -2 \\ 4 & 5 \end{bmatrix}$, $C = \begin{bmatrix} 1 & 6 \\ 2 & 0 \end{bmatrix}$

44. $A = \begin{bmatrix} 1 & 0 & 0 \\ 0 & 1 & 0 \\ 0 & 0 & 1 \end{bmatrix}$, $B = \begin{bmatrix} 2 & 1 & 3 \\ 0 & 5 & -1 \\ 3 & 6 & 0 \end{bmatrix}$, $C = \begin{bmatrix} 0 \\ 3 \\ -4 \end{bmatrix}$

Two $n \times n$ matrices A and B are called inverses *(of one another) if both products AB and BA equal I_n. Check that the pairs of matrices in Exercises 45 and 46 are inverses.*

45. $\begin{bmatrix} 3 & -1 \\ -1 & \frac{1}{2} \end{bmatrix}$, $\begin{bmatrix} 1 & 2 \\ 2 & 6 \end{bmatrix}$

46. $\begin{bmatrix} 2 & 8 & -11 \\ -1 & -5 & 7 \\ 1 & 2 & -3 \end{bmatrix}$, $\begin{bmatrix} 1 & 2 & 1 \\ 4 & 5 & -3 \\ 3 & 4 & -2 \end{bmatrix}$

47. (*Wardrobe Costs*) The quantities of pants, shirts, and jackets owned by Mike and Don are given by the matrix A, and the costs of these items are given by matrix B.

	Pants	Shirts	Jackets
Mike	6	8	2
Don	2	5	3

$= A$

Pants	20
Shirts	15
Jackets	50

$= B$

(a) Calculate the matrix AB.

(b) Interpret the entries of the matrix AB.

48. (*Retail Sales*) Two stores sell the exact same brand and style of a dresser, a nightstand, and a bookcase. Matrix A gives the retail prices (in dollars) for the items. Matrix B gives the number of each item sold at each store in one month.

	Dresser	Nightstand	Bookcase
$A =$	250	80	60

$B =$	Store 1	Store 2	
	40	35	Dresser
	30	35	Nightstand
	50	75	Bookcase

(a) Calculate AB.

(b) Interpret the entries of AB.

49. (*Retail Sales*) A candy shop sells various items for the price per pound (in dollars) indicated in matrix A. Matrix B gives the number of pounds of peanuts, raisins, and coffee beans used in a week. Matrix C gives the total number of pounds of plain, chocolate-covered, and yogurt-covered items sold each week.

	Plain	Chocolate	Yogurt	
	3	4	3.50	Peanuts
$A =$	2	3	2.75	Raisins
	6	12	9.25	Coffee beans

	Peanuts	Raisins	Coffee beans
$B =$	100	175	80

$C =$		
	250	Plain
	400	Chocolate
	175	Yogurt

Determine and interpret the following matrices.

(a) BA (b) AC

50. (*Wholesale and Retail Sales*) A company has three appliance stores that sell washers, dryers, and stoves. Matrices A and B give the wholesale and retail prices of these items, respectively. Matrices C and D give the quantities of these items sold by the three stores in September and October, respectively.

$$A = \begin{bmatrix} \overset{\text{Washers}}{300} & \overset{\text{Dryers}}{250} & \overset{\text{Stoves}}{450} \end{bmatrix}$$

$$B = \begin{bmatrix} \overset{\text{Washers}}{500} & \overset{\text{Dryers}}{450} & \overset{\text{Stoves}}{750} \end{bmatrix}$$

	Store 1	Store 2	Store 3	
$C =$	30	40	20	Washers
	20	30	10	Dryers
	10	5	35	Stoves

	Store 1	Store 2	Store 3	
$D =$	20	50	30	Washers
	30	10	20	Dryers
	10	20	30	Stoves

Determine and interpret the following matrices.

(a) AC (b) AD

(c) BC (d) BD

(e) $B - A$ (f) $(B - A)C$

(g) $(B - A)D$ (h) $C + D$

(i) $(B - A)(C + D)$

51. (*Course Grades*) Three professors teaching the same course have entirely different grading policies. The percentage of students given each grade by the professors is summarized in the following matrix:

	Grade				
	A	B	C	D	F
Prof. I	25	35	30	10	0
Prof. II	10	20	40	20	10
Prof. III	5	10	20	40	25

(a) The point values of the grades are A = 4, B = 3, C = 2, D = 1, and F = 0. Use matrix multiplication to determine the average grade given by each professor.

(b) Professor I has 240 students, professor II has 120 students, and professor III has 40 students. Use matrix multiplication to determine the numbers of A's, B's, C's, D's, and F's given.

52. (*Semester Grades*) A professor bases semester grades on four 100-point items: homework, quizzes, a midterm exam, and a final exam. Students may choose one of three schemes summarized in the following matrix for weighting the points from the four items. Use matrix multiplication to determine the most advantageous weighting scheme for a student who earned 97 points on homework, 72 points on the quizzes, 83 points on the midterm exam, and 75 points on the final exam.

	Items			
	Hw	Qu	ME	FE
Scheme I	.10	.10	.30	.50
Scheme II	.10	.20	.30	.40
Scheme III	.15	.15	.35	.35

53. (*Voter Analysis*) In a certain town the proportions of voters voting Democratic and Republican by various age groups is summarized by this matrix:

	Dem.	Rep.	
Under 30	.65	.35	
30–50	.55	.45	$= A.$
Over 50	.45	.55	

The population of voters in the town by age group is given by the matrix

$$B = \begin{bmatrix} \underset{\substack{\text{Under} \\ 30}}{6000} & \underset{\substack{\text{30–50}}}{8000} & \underset{\substack{\text{Over} \\ 50}}{4000} \end{bmatrix}.$$

Interpret the entries of the matrix product BA.

54. (*Voter Analysis*) Refer to Exercise 53.

(a) Using the given data, which party would win and what would be the percentage of the winning vote?

(b) Suppose that the population of the town shifted toward older residents as reflected in the population matrix $B = \begin{bmatrix} 2000 & 4000 & 12{,}000 \end{bmatrix}$. What would be the result of the election now?

55. (*Labor Costs*) Suppose that a contractor employs carpenters, bricklayers, and plumbers, working three shifts per day. The number of labor-hours employed in each of the shifts is summarized in the following matrix:

	Shift		
	1	2	3
Carpenters	50	20	10
Bricklayers	30	30	15
Plumbers	20	20	5

Labor in shift 1 costs $10 per hour, in shift 2 $15 per hour, and in shift 3 $20 per hour. Use matrix multiplication to compute the amount spent on each type of labor.

56. (*Epidemiology*) A flu epidemic hits a large city. Each resident of the city is either sick, well, or a carrier. The proportion of people in each of the categories is expressed by the following matrix:

$$\begin{array}{l} \\ \\ \text{Well} \\ \text{Sick} \\ \text{Carrier} \end{array} \begin{array}{c} \text{Age} \\ \hline \begin{array}{ccc} \text{0–10} & \text{10–30} & \text{Over 30} \end{array} \\ \begin{bmatrix} .70 & .70 & .60 \\ .10 & .20 & .30 \\ .20 & .10 & .10 \end{bmatrix} \end{array} = A.$$

The population of the city is distributed by age and sex as follows:

$$\text{Age} \left| \begin{array}{l} \text{0–10} \\ \text{10–30} \\ \text{Over 30} \end{array} \right. \begin{array}{c} \begin{array}{cc} \text{Male} & \text{Female} \end{array} \\ \begin{bmatrix} 60{,}000 & 65{,}000 \\ 100{,}000 & 110{,}000 \\ 200{,}000 & 230{,}000 \end{bmatrix} \end{array} = B.$$

(a) Compute AB.

(b) How many sick males are there?

(c) How many female carriers are there?

57. (*Nutrition Analysis*) Mikey's diet consists of food X and food Y. The matrix N represents the number of units of nutrients 1, 2, and 3 per ounce for each of the foods.

$$N = \begin{array}{c} \begin{array}{ccc} 1 & 2 & 3 \end{array} \\ \begin{bmatrix} 60 & 50 & 38 \\ 42 & 50 & 67 \end{bmatrix} \end{array} \begin{array}{c} \\ X \\ Y \end{array}$$

The matrices B, L, and D represent the number of ounces of each food that Mikey eats each day for breakfast, lunch, and dinner, respectively.

$$B = \begin{array}{c} \begin{array}{cc} X & Y \end{array} \\ \begin{bmatrix} 2 & 1 \end{bmatrix} \end{array} \quad L = \begin{array}{c} \begin{array}{cc} X & Y \end{array} \\ \begin{bmatrix} 1 & 3 \end{bmatrix} \end{array} \quad D = \begin{array}{c} \begin{array}{cc} X & Y \end{array} \\ \begin{bmatrix} 2 & 4 \end{bmatrix} \end{array}$$

Calculate and interpret the following.

(a) BN (b) LN (c) DN

(d) $B+L+D$ (e) $(B+L+D)N$

58. (*Bakery Sales*) A bakery makes three types of cookies, I, II, and III. Each type of cookie is made from the four ingredients A, B, C, and D. The number of units of each ingredient used in each type of cookie is given by the matrix M. The cost per unit of each of the four ingredients (in cents) is given by the matrix N. The selling price for each of the cookies (in cents) is given by the matrix S. The baker receives an order for 10 type I cookies, 20 type II cookies, and 15 type III cookies, as represented by the matrix R.

$$M = \begin{array}{c} \begin{array}{cccc} A & B & C & D \end{array} \\ \begin{bmatrix} 1 & 0 & 2 & 4 \\ 3 & 2 & 1 & 1 \\ 2 & 5 & 3 & 1 \end{bmatrix} \end{array} \begin{array}{c} \\ \text{I} \\ \text{II} \\ \text{III} \end{array} \qquad N = \begin{bmatrix} 10 \\ 20 \\ 15 \\ 17 \end{bmatrix} \begin{array}{c} A \\ B \\ C \\ D \end{array}$$

$$S = \begin{bmatrix} 175 \\ 150 \\ 225 \end{bmatrix} \begin{array}{c} \text{I} \\ \text{II} \\ \text{III} \end{array} \qquad R = \begin{array}{c} \begin{array}{ccc} \text{I} & \text{II} & \text{III} \end{array} \\ \begin{bmatrix} 10 & 20 & 15 \end{bmatrix} \end{array}$$

Calculate and interpret the following.

(a) RM (b) MN (c) RMN

(d) $S-MN$ (e) $R(S-MN)$ (f) RS

59. (*Revenue*) A community fitness center has a pool and a weight room. The admission prices (in dollars) for residents and nonresidents are given by the matrix

$$P = \begin{bmatrix} 4.50 \\ 5.00 \end{bmatrix} \begin{array}{l} \text{Residents} \\ \text{Nonresidents.} \end{array}$$

The average daily numbers of customers for the fitness center are given by the matrix

$$A = \begin{array}{c} \begin{array}{cc} \text{Residents} & \text{Nonresidents} \end{array} \\ \begin{bmatrix} 90 & 63 \\ 78 & 59 \end{bmatrix} \end{array} \begin{array}{l} \\ \text{Pool} \\ \text{Weight room.} \end{array}$$

(a) Compute AP.

(b) What is the average amount of money taken in by the pool each day?

60. (*Production Planning*) A company makes radios and TV sets. Each radio requires 3 hours of assembly and $\frac{1}{2}$ hour of packaging, while each TV set requires 5 hours of assembly and 1 hour of packaging.

(a) Write a matrix T representing the required time for assembly and packaging of radios and TV sets.

(b) The company receives an order from a retail outlet for 30 radios and 20 TV sets. Find a matrix S so that either ST or TS gives the total assembly time and the total packaging time required to fill the order. What is the total assembly time? What is the total packaging time?

61. Make up an application whose answer is that the total cost is given by

$$\begin{bmatrix} 20 & 30 \end{bmatrix} \begin{bmatrix} 600 \\ 700 \end{bmatrix}.$$

62. Find the values of a and b for which $A \cdot B = I_3$, where

$$A = \begin{bmatrix} 3 & 2 & 0 \\ 1 & 1 & 0 \\ 0 & 0 & 1 \end{bmatrix} \quad \text{and} \quad B = \begin{bmatrix} a & b & 0 \\ -1 & 3 & 0 \\ 0 & 0 & 1 \end{bmatrix}.$$

In Exercises 63 and 64, determine the matrix B based on the screen shown.

63.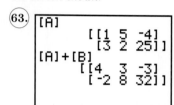

```
[A]
        [[1  5  -4]
         [3  2  25]]
[A]+[B]
        [[4   3  -3]
         [-2  8  32]]
```

64.

```
[A]
        [[1  5  -4]
         [3  2  25]]
[A]-[B]
        [[-4  1  -1]
         [3   3  23]]
```

65. If A is a 3×4 matrix and $A(BB)$ is defined, what is the size of matrix B?

66. If B is a 3×5 matrix and $(AA)B$ is defined, what is the size of matrix A?

67. Table 1 gives the number of public school teachers (elementary and secondary) and the average number of pupils per teacher for three mid-Atlantic states. Set up a product of two matrices that gives the total number of pupils in the three states. (*Source: The World Almanac and Book of Facts*, 2005)

TABLE 1 Teachers and Pupils

	Delaware	Maryland	Virginia
Teachers	7698	55,382	99,919
Pupils per teacher	15.1	15.7	11.8

68. Table 2 gives the area and population density for three West-Coast states. Set up a product of two matrices that gives total population of the three states. (*Source: The World Almanac and Book of Facts*, 2005)

TABLE 2 State Areas and Densities

	California	Oregon	Washington
Land Area (sq. mi.)	155,959	95,997	66,544
Pop. Density (per sq. mile)	227.5	37.1	92.1

In Exercises 69–76, use a graphing calculator, spreadsheet, or mathematical software to carry out the matrix operations.

In Exercises 69–76, calculate the given expression, where

$$A = \begin{bmatrix} .4 & 7 & -3 \\ 19 & .5 & 1.6 \\ -9 & 11 & 2 \end{bmatrix}, \qquad B = \begin{bmatrix} 6 & -9 & .3 \\ 1.5 & 22 & -4 \\ -5 & 6.6 & 14 \end{bmatrix},$$

and

$$C = \begin{bmatrix} 2.4 & 8 & -.2 \\ -11 & .3 & 6 \\ 7 & -4 & 5.1 \end{bmatrix}.$$

69. AB

70. BA

71. $A(B+C)$

72. $AB + AC$

73. A^2 (that is, AA) **74.** A^3 (that is, AAA)

75. Calculate `[A]^2` and `[A]^3`. Compare your results with your answers in Exercises 73 and 74.

76. Try multiplying matrices A and B, where the number of columns of A differs from the number of rows of B. How does your calculator or computer respond?

SOLUTIONS TO PRACTICE PROBLEMS 2.3

1. Answer:

$$\begin{bmatrix} 3 & 1 & 2 \\ -1 & 0 & \frac{1}{2} \\ 0 & 4 & 1 \end{bmatrix} \begin{bmatrix} 7 & -1 & 0 \\ 5 & 4 & 2 \\ -6 & 0 & 4 \end{bmatrix}$$

$$= \begin{bmatrix} 14 & 1 & 10 \\ -10 & 1 & 2 \\ 14 & 16 & 12 \end{bmatrix}.$$

The systematic steps to be taken are as follows:

(a) Determine the size of the product matrix. Since we have a

③× 3   times a   3 ×③,
└─ outer dimensions ─┘

the size of the product is given by the outer dimensions or 3×3. Begin by drawing a 3×3 rectangular array.

SOLUTIONS TO PRACTICE PROBLEMS 2.3 (CONTINUED)

(b) Find the entries one at a time. To find the entry in the first row, first column of the product, look at the first row of the left matrix and the first column of the right matrix and form their product.

$$\begin{bmatrix} 3 & 1 & 2 \\ -1 & 0 & \frac{1}{2} \\ 0 & 4 & 1 \end{bmatrix} \begin{bmatrix} 7 & -1 & 0 \\ 5 & 4 & 2 \\ -6 & 0 & 4 \end{bmatrix}$$

$$= \begin{bmatrix} 14 & & \\ & & \\ & & \end{bmatrix}$$

since $3 \cdot 7 + 1 \cdot 5 + 2(-6) = 14$. In general, to find the entry in the ith row, jth column of the product, put one finger on the ith row of the left matrix and another finger on the jth column of the right matrix. Then multiply the row matrix times the column matrix to get the desired entry.

2. Denote the three matrices by A, X, and B, respectively. Since b_{11} (the entry of the first row, first col-

umn of B) is 5, this means that

$$\begin{bmatrix} \text{first row of } A \end{bmatrix} \begin{bmatrix} \text{first} \\ \text{column} \\ \text{of } X \end{bmatrix} = \begin{bmatrix} b_{11} \end{bmatrix}.$$

That is,

$$\begin{bmatrix} 3 & -6 \end{bmatrix} \begin{bmatrix} x \\ y \end{bmatrix} = \begin{bmatrix} 5 \end{bmatrix} \quad \text{or} \quad 3x - 6y = 5.$$

Similarly, $b_{21} = 0$ says that $2x + y = 0$. Therefore, the corresponding system of linear equations is

$$\begin{cases} 3x - 6y = 5 \\ 2x + \ y = 0. \end{cases}$$

3. The coefficient matrix is

$$\begin{bmatrix} 8 & 3 \\ 9 & -2 \end{bmatrix}.$$

So the system is equivalent to the matrix equation

$$\begin{bmatrix} 8 & 3 \\ 9 & -2 \end{bmatrix} \begin{bmatrix} x \\ y \end{bmatrix} = \begin{bmatrix} 7 \\ -5 \end{bmatrix}.$$

2.4 The Inverse of a Matrix

In Section 2.3 we introduced the operations of addition, subtraction, and multiplication of matrices. In this section let us pursue the algebra of matrices a bit further and consider equations involving matrices. Specifically, we consider equations of the form

$$AX = B, \tag{1}$$

where A and B are given matrices and X is an unknown matrix whose entries are to be determined. Such equations among matrices are intimately bound up with the theory of systems of linear equations. Indeed, we described the connection in a special case in Example 6 of Section 2.3. In that example we wrote the system of linear equations

$$\begin{cases} -2x + 4y = 2 \\ -3x + 7y = 7 \end{cases}$$

as a matrix equation of the form (1), where

$$A = \begin{bmatrix} -2 & 4 \\ -3 & 7 \end{bmatrix}, \qquad B = \begin{bmatrix} 2 \\ 7 \end{bmatrix}, \qquad X = \begin{bmatrix} x \\ y \end{bmatrix}.$$

Note that by determining the entries (x and y) of the unknown matrix X, we solve the system of linear equations. We will return to this example after we have made a complete study of the matrix equation (1).

As motivation for our solution of equation (1), let us consider the analogous equation among numbers:

$$ax = b,$$

where a and b are given numbers[1] and x is to be determined. Let us examine its solution in great detail. Multiply both sides by $1/a$. (Note that $1/a$ makes sense, since $a \neq 0$.)

$$\left(\frac{1}{a}\right) \cdot (ax) = \frac{1}{a} \cdot b$$

$$\left(\frac{1}{a} \cdot a\right) \cdot x = \frac{1}{a} \cdot b$$

$$1 \cdot x = \frac{1}{a} \cdot b$$

$$x = \frac{1}{a} \cdot b$$

Let us model our solution of equation (1) on the preceding calculation. To do so, we need to multiply both sides of the equation by a matrix that plays the same role in matrix arithmetic as $1/a$ plays in ordinary arithmetic. Our first task then will be to introduce this matrix and study its properties.

The number $1/a$ has the following relationship to the number a:

$$\frac{1}{a} \cdot a = a \cdot \frac{1}{a} = 1. \tag{2}$$

The matrix analog of the number 1 is an identity matrix I. This prompts us to generalize equation (2) to matrices as follows. Suppose that we are given a square matrix A. Then the *inverse* of A, denoted A^{-1}, is a square matrix with the property

$$A^{-1}A = I \quad \text{and} \quad AA^{-1} = I, \tag{3}$$

where I is an identity matrix of the same size as A. The matrix A^{-1} is the matrix analog of the number $1/a$. It can be shown that a matrix A has at most one inverse. (However, A may not have an inverse at all; see Example 3.)

If we are given a matrix A, then it is easy to determine whether or not a given matrix is its inverse. Merely check equation (3) with the given matrix substituted for A^{-1}. For example, if

$$A = \begin{bmatrix} -2 & 4 \\ -3 & 7 \end{bmatrix}, \quad \text{then} \quad A^{-1} = \begin{bmatrix} -\frac{7}{2} & 2 \\ -\frac{3}{2} & 1 \end{bmatrix}.$$

Indeed, we have

$$\underset{A^{-1}}{\begin{bmatrix} -\frac{7}{2} & 2 \\ -\frac{3}{2} & 1 \end{bmatrix}} \underset{A}{\begin{bmatrix} -2 & 4 \\ -3 & 7 \end{bmatrix}} = \begin{bmatrix} 7-6 & -14+14 \\ 3-3 & -6+7 \end{bmatrix} = \underset{I_2}{\begin{bmatrix} 1 & 0 \\ 0 & 1 \end{bmatrix}}$$

and

$$\underset{A}{\begin{bmatrix} -2 & 4 \\ -3 & 7 \end{bmatrix}} \underset{A^{-1}}{\begin{bmatrix} -\frac{7}{2} & 2 \\ -\frac{3}{2} & 1 \end{bmatrix}} = \begin{bmatrix} 7-6 & -4+4 \\ \frac{21}{2}-\frac{21}{2} & -6+7 \end{bmatrix} = \underset{I_2}{\begin{bmatrix} 1 & 0 \\ 0 & 1 \end{bmatrix}}.$$

The inverse of a matrix can be calculated using Gaussian elimination, as the next example illustrates.

---

[1] We may as well assume that $a \neq 0$. Otherwise, x does not occur.

EXAMPLE 1   **Finding the inverse of a matrix** Let

$$A = \begin{bmatrix} 3 & 1 \\ 5 & 2 \end{bmatrix}.$$

Determine A^{-1}.

Solution Since A is a 2×2 matrix, A^{-1} is also a 2×2 matrix and satisfies

$$AA^{-1} = I_2 \quad \text{and} \quad A^{-1}A = I_2, \tag{4}$$

where $I_2 = \begin{bmatrix} 1 & 0 \\ 0 & 1 \end{bmatrix}$ is the 2×2 identity matrix. Suppose that

$$A^{-1} = \begin{bmatrix} x & y \\ z & w \end{bmatrix}.$$

Then the first equation of (4) reads

$$\begin{bmatrix} 3 & 1 \\ 5 & 2 \end{bmatrix} \begin{bmatrix} x & y \\ z & w \end{bmatrix} = \begin{bmatrix} 1 & 0 \\ 0 & 1 \end{bmatrix}.$$

Multiplying out the matrices on the left gives

$$\begin{bmatrix} 3x + z & 3y + w \\ 5x + 2z & 5y + 2w \end{bmatrix} = \begin{bmatrix} 1 & 0 \\ 0 & 1 \end{bmatrix}.$$

Now equate corresponding elements in the two matrices to obtain the equations

$$\begin{cases} 3x + z = 1 \\ 5x + 2z = 0, \end{cases} \qquad \begin{cases} 3y + w = 0 \\ 5y + 2w = 1. \end{cases}$$

Notice that the equations break up into two pairs of linear equations, each pair involving only two variables. Solving these two systems of linear equations yields $x = 2$, $z = -5$, $y = -1$, and $w = 3$. Therefore,

$$A^{-1} = \begin{bmatrix} 2 & -1 \\ -5 & 3 \end{bmatrix}.$$

Indeed, we may readily verify that

$$\begin{bmatrix} 3 & 1 \\ 5 & 2 \end{bmatrix} \begin{bmatrix} 2 & -1 \\ -5 & 3 \end{bmatrix} = \begin{bmatrix} 1 & 0 \\ 0 & 1 \end{bmatrix}$$

$$\begin{bmatrix} 2 & -1 \\ -5 & 3 \end{bmatrix} \begin{bmatrix} 3 & 1 \\ 5 & 2 \end{bmatrix} = \begin{bmatrix} 1 & 0 \\ 0 & 1 \end{bmatrix}.$$

Now Try Exercise 3   ■

The preceding method can be used to calculate the inverse of matrices of any size, although it involves considerable calculation. We provide a rather efficient computational method for calculating A^{-1} in the next section. For now, however, let us be content with the above method. Using it, we can derive a general formula for A^{-1} in the case where A is a 2×2 matrix.

To determine the inverse of a 2×2 matrix, let

$$A = \begin{bmatrix} a & b \\ c & d \end{bmatrix}.$$

Let $\Delta = ad - bc$, and assume that $\Delta \neq 0$. Then A^{-1} is given by the formula

$$A^{-1} = \begin{bmatrix} \dfrac{d}{\Delta} & -\dfrac{b}{\Delta} \\ -\dfrac{c}{\Delta} & \dfrac{a}{\Delta} \end{bmatrix}. \tag{5}$$

We will omit the derivation of this formula. It proceeds along lines similar to those of Example 1. Notice that formula (5) involves division by Δ. Since division by 0 is not permissible, it is necessary that $\Delta \neq 0$ for formula (5) to be applied. We discuss the case $\Delta = 0$ in Example 3.

Obtaining equation (5) can be reduced to a simple step-by-step procedure.

To determine the inverse of $\begin{bmatrix} a & b \\ c & d \end{bmatrix}$ if $\Delta = ad - bc \neq 0$,

1. Interchange a and d to get $\begin{bmatrix} d & b \\ c & a \end{bmatrix}$.

2. Change the signs of b and c to get $\begin{bmatrix} d & -b \\ -c & a \end{bmatrix}$.

3. Divide all entries by Δ to get $\begin{bmatrix} \dfrac{d}{\Delta} & -\dfrac{b}{\Delta} \\ -\dfrac{c}{\Delta} & \dfrac{a}{\Delta} \end{bmatrix}.$

EXAMPLE 2   **Using the formula for the inverse of a 2 x 2 matrix**   Calculate the inverse of

$$\begin{bmatrix} -2 & 4 \\ -3 & 7 \end{bmatrix}.$$

Solution $\Delta = (-2) \cdot 7 - 4 \cdot (-3) = -2$, so $\Delta \neq 0$, and we may use the preceding computation.

1. Interchange a and d:

$$\begin{bmatrix} 7 & 4 \\ -3 & -2 \end{bmatrix}.$$

2. Change the signs of b and c:

$$\begin{bmatrix} 7 & -4 \\ 3 & -2 \end{bmatrix}.$$

3. Divide all entries by $\Delta = -2$:

$$\begin{bmatrix} -\dfrac{7}{2} & 2 \\ -\dfrac{3}{2} & 1 \end{bmatrix}.$$

Thus

$$\begin{bmatrix} -2 & 4 \\ -3 & 7 \end{bmatrix}^{-1} = \begin{bmatrix} -\frac{7}{2} & 2 \\ -\frac{3}{2} & 1 \end{bmatrix}.$$

Now Try Exercise 5   ■

Not every square matrix has an inverse. Indeed, it may be impossible to satisfy equation (3) for any choice of A^{-1}. This phenomenon can even occur in the case of 2×2 matrices. Here one can show that *if* $\Delta = 0$, *then the matrix does not have an inverse.* The next example illustrates this phenomenon in a special case.

EXAMPLE 3   **Showing that a 2 x 2 matrix does not have an inverse**   Show that

$$\begin{bmatrix} 1 & 1 \\ 1 & 1 \end{bmatrix}$$

does not have an inverse.

Solution   Note first that $\Delta = 1 \cdot 1 - 1 \cdot 1 = 0$, so the inverse cannot be computed via equation (5). Suppose that the given matrix did have an inverse, say

$$\begin{bmatrix} s & t \\ u & v \end{bmatrix}.$$

Then the following equation would hold:

$$\begin{bmatrix} s & t \\ u & v \end{bmatrix} \begin{bmatrix} 1 & 1 \\ 1 & 1 \end{bmatrix} = \begin{bmatrix} 1 & 0 \\ 0 & 1 \end{bmatrix}.$$

On multiplying out the two matrices on the left, we get the equation

$$\begin{bmatrix} s+t & s+t \\ u+v & u+v \end{bmatrix} = \begin{bmatrix} 1 & 0 \\ 0 & 1 \end{bmatrix},$$

so that, on equating entries in the first row:

$$s+t = 1, \qquad s+t = 0.$$

But $s + t$ cannot equal both 1 and 0. So we reach a contradiction, and therefore the original matrix cannot have an inverse.

Now Try Exercise 23   ■

We were led to introduce the inverse of a matrix from a discussion of the matrix equation $AX = B$. Let us now return to that discussion. Suppose that A and B are given matrices and that we wish to solve the matrix equation

$$AX = B$$

for the unknown matrix X. Suppose further that A has an inverse A^{-1}. Multiply both sides of the equation on the left by A^{-1} to obtain

$$A^{-1} \cdot AX = A^{-1}B.$$

Because $A^{-1} \cdot A = I$, we have

$$IX = A^{-1}B$$
$$X = A^{-1}B.$$

Thus the matrix X is found by simply multiplying B on the left by A^{-1}, and we can summarize our findings as follows:

> **Solving a Matrix Equation**   If the matrix A has an inverse, then the solution of the matrix equation
>
> $$AX = B \quad \text{is given by} \quad X = A^{-1}B.$$

Matrix equations can be used to solve systems of linear equations, as illustrated in the next example.

EXAMPLE 4   **Using a matrix inverse to solve a system of equations**   Use a matrix equation to solve the system of linear equations

$$\begin{cases} -2x + 4y = 2 \\ -3x + 7y = 7. \end{cases}$$

Solution   In Example 6 of Section 2.3 we saw that the system could be written as a matrix equation:

$$\underset{A}{\begin{bmatrix} -2 & 4 \\ -3 & 7 \end{bmatrix}} \underset{X}{\begin{bmatrix} x \\ y \end{bmatrix}} = \underset{B}{\begin{bmatrix} 2 \\ 7 \end{bmatrix}}.$$

We happen to know A^{-1} from Example 2, namely

$$A^{-1} = \begin{bmatrix} -\frac{7}{2} & 2 \\ -\frac{3}{2} & 1 \end{bmatrix}.$$

So we may compute the matrix $X = A^{-1}B$:

$$X = \begin{bmatrix} x \\ y \end{bmatrix} = \begin{bmatrix} -\frac{7}{2} & 2 \\ -\frac{3}{2} & 1 \end{bmatrix} \begin{bmatrix} 2 \\ 7 \end{bmatrix} = \begin{bmatrix} 7 \\ 4 \end{bmatrix}.$$

Thus, the solution of the system is $x = 7$, $y = 4$.

Now Try Exercise 11   ■

EXAMPLE 5   **Analyzing marriage trends**   Let x and y denote the number of married and single adults in a certain town as of January 1. Let m and s denote the corresponding numbers for the following year. A statistical survey shows that x, y, m, and s are related by the equations

$$.9x + .2y = m$$
$$.1x + .8y = \; s.$$

In a given year there were found to be 490,000 married adults and 147,000 single adults.

(a) How many married adults were there in the preceding year?

(b) How many married adults were there two years ago?

Solution   (a) The given equations can be written in the matrix form

$$AX = B,$$

where

$$A = \begin{bmatrix} .9 & .2 \\ .1 & .8 \end{bmatrix}, \qquad X = \begin{bmatrix} x \\ y \end{bmatrix}, \qquad B = \begin{bmatrix} m \\ s \end{bmatrix}.$$

We are given that $B = \begin{bmatrix} 490{,}000 \\ 147{,}000 \end{bmatrix}$. So, since

$$X = A^{-1}B \quad \text{and} \quad A^{-1} = \begin{bmatrix} \frac{8}{7} & -\frac{2}{7} \\ -\frac{1}{7} & \frac{9}{7} \end{bmatrix},$$

we have

$$X = \begin{bmatrix} \frac{8}{7} & -\frac{2}{7} \\ -\frac{1}{7} & \frac{9}{7} \end{bmatrix} \begin{bmatrix} 490{,}000 \\ 147{,}000 \end{bmatrix} = \begin{bmatrix} 518{,}000 \\ 119{,}000 \end{bmatrix}.$$

Thus last year there were 518,000 married adults and 119,000 single adults.

(b) We deduce x and y for two years ago from the values of m and s for last year, namely $m = 518{,}000$, $s = 119{,}000$.

$$X = A^{-1}B = \begin{bmatrix} \frac{8}{7} & -\frac{2}{7} \\ -\frac{1}{7} & \frac{9}{7} \end{bmatrix} \begin{bmatrix} 518{,}000 \\ 119{,}000 \end{bmatrix} = \begin{bmatrix} 558{,}000 \\ 79{,}000 \end{bmatrix}.$$

That is, two years ago there were 558,000 married adults and 79,000 single adults.

Now Try Exercise 17   ■

EXAMPLE 6   **Using a matrix inverse to solve systems of equations**   In Section 2.5 we will show that if

$$A = \begin{bmatrix} 4 & -2 & 3 \\ 8 & -3 & 5 \\ 7 & -2 & 4 \end{bmatrix}, \quad \text{then} \quad A^{-1} = \begin{bmatrix} -2 & 2 & -1 \\ 3 & -5 & 4 \\ 5 & -6 & 4 \end{bmatrix}.$$

(a) Use this fact to solve the system of linear equations

$$\begin{cases} 4x - 2y + 3z = 1 \\ 8x - 3y + 5z = 4 \\ 7x - 2y + 4z = 5. \end{cases}$$

(b) Solve the system of equations

$$\begin{cases} 4x - 2y + 3z = 4 \\ 8x - 3y + 5z = 7 \\ 7x - 2y + 4z = 6. \end{cases}$$

Solution   (a) The system can be written in the matrix form

$$\underbrace{\begin{bmatrix} 4 & -2 & 3 \\ 8 & -3 & 5 \\ 7 & -2 & 4 \end{bmatrix}}_{A} \underbrace{\begin{bmatrix} x \\ y \\ z \end{bmatrix}}_{X} = \underbrace{\begin{bmatrix} 1 \\ 4 \\ 5 \end{bmatrix}}_{B}.$$

The solution of this matrix equation is $X = A^{-1}B$, or

$$\begin{bmatrix} x \\ y \\ z \end{bmatrix} = \begin{bmatrix} -2 & 2 & -1 \\ 3 & -5 & 4 \\ 5 & -6 & 4 \end{bmatrix} \begin{bmatrix} 1 \\ 4 \\ 5 \end{bmatrix} = \begin{bmatrix} 1 \\ 3 \\ 1 \end{bmatrix}.$$

Thus the solution of the system is $x = 1$, $y = 3$, $z = 1$.

(b) This system has the same left-hand side as the preceding system, so its solution is

$$\begin{bmatrix} x \\ y \\ z \end{bmatrix} = \begin{bmatrix} -2 & 2 & -1 \\ 3 & -5 & 4 \\ 5 & -6 & 4 \end{bmatrix} \begin{bmatrix} 4 \\ 7 \\ 6 \end{bmatrix} = \begin{bmatrix} 0 \\ 1 \\ 2 \end{bmatrix}.$$

That is, the solution of the system is $x = 0$, $y = 1$, $z = 2$.

Now Try Exercise 19   ■

Using the method of matrix equations to solve a system of linear equations is especially efficient if one wishes to solve a number of systems all having the same left-hand sides but different right-hand sides. For then A^{-1} must be computed only once for all the systems under consideration. (This point is useful in Exercises 19–22.)

GC TI-83   The inverse of a square matrix can be obtained directly with the inverse key $\boxed{x^{-1}}$. See Fig. 1. (*Note*: Do not use **[A]^-1**.)

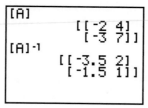

Figure 1

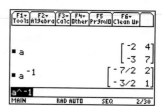

Figure 2

TI-89   The inverse of a square matrix can be obtained directly by raising the matrix to the -1 power. See Fig. 2.

ES   To obtain the inverse of the $n \times n$ matrix named A, select an n by n square of cells where you would like the inverse matrix to be displayed, type **=MINVERSE(A)**, and press Crtl+Shift+Enter.

PRACTICE PROBLEMS 2.4

1. Show that the inverse of

$$\begin{bmatrix} -4 & 1 & 2 \\ 7 & -1 & -4 \\ -\frac{1}{2} & 0 & \frac{1}{2} \end{bmatrix} \text{ is } \begin{bmatrix} 1 & 1 & 4 \\ 3 & 2 & 4 \\ 1 & 1 & 6 \end{bmatrix}.$$

2. Use the method of this section to solve the system of linear equations

$$\begin{cases} .8x + .6y = 5 \\ .2x + .4y = 2. \end{cases}$$

EXERCISES 2.4

In Exercises 1 and 2, use the fact that

$$\begin{bmatrix} 2 & 2 \\ \frac{1}{2} & 1 \end{bmatrix}^{-1} = \begin{bmatrix} 1 & -2 \\ -\frac{1}{2} & 2 \end{bmatrix}.$$

1. Solve $\begin{cases} 2x + 2y = 4 \\ \frac{1}{2}x + y = 1. \end{cases}$

2. Solve $\begin{cases} 2x + 2y = 14 \\ \frac{1}{2}x + y = 4. \end{cases}$

In Exercises 3–10, find the inverse of the given matrix.

3. $\begin{bmatrix} 7 & 2 \\ 3 & 1 \end{bmatrix}$ **4.** $\begin{bmatrix} 2 & 3 \\ 5 & 7 \end{bmatrix}$

5. $\begin{bmatrix} 6 & 2 \\ 5 & 2 \end{bmatrix}$ **6.** $\begin{bmatrix} 1 & .5 \\ 0 & .5 \end{bmatrix}$

7. $\begin{bmatrix} .7 & .2 \\ .3 & .8 \end{bmatrix}$ **8.** $\begin{bmatrix} 0 & 1 \\ 1 & 0 \end{bmatrix}$

9. $\begin{bmatrix} 3 \end{bmatrix}$ **10.** $\begin{bmatrix} .2 \end{bmatrix}$

In Exercises 11–14, use the method of this section to solve the system of linear equations.

11. $\begin{cases} x + 2y = 3 \\ 2x + 6y = 5 \end{cases}$ **12.** $\begin{cases} 5x + 3y = 1 \\ 7x + 4y = 2 \end{cases}$

13. $\begin{cases} \frac{1}{2}x + 2y = 4 \\ 3x + 16y = 0 \end{cases}$ **14.** $\begin{cases} .8x + .6y = 2 \\ .2x + .4y = 1 \end{cases}$

15. (*Marriage Trends*) It is found that the number of married and single adults in a certain town are subject to the following statistics. Suppose that x and y denote the number of married and single adults, respectively, in a given year (say as of January 1) and let m, s denote the corresponding numbers for the following year. Then

$$.8x + .3y = m$$
$$.2x + .7y = s.$$

(a) Write this system of equations in matrix form.

(b) Solve the resulting matrix equation for $X = \begin{bmatrix} x \\ y \end{bmatrix}$.

(c) Suppose that in a given year there were found to be 100,000 married adults and 50,000 single adults. How many married (respectively, single) adults were there the preceding year?

(d) How many married (respectively, single) adults were there two years ago?

16. (*Epidemiology*) A flu epidemic is spreading through a town of 48,000 people. It is found that if x and y denote the numbers of people sick and well in a given week,

respectively, and if s and w denote the corresponding numbers for the following week, then

$$\tfrac{1}{3}x + \tfrac{1}{4}y = s$$
$$\tfrac{2}{3}x + \tfrac{3}{4}y = w.$$

(a) Write this system of equations in matrix form.

(b) Solve the resulting matrix equation for $X = \begin{bmatrix} x \\ y \end{bmatrix}$.

(c) Suppose that 13,000 people are sick in a given week. How many were sick the preceding week?

(d) Same question as part (c), except assume that 14,000 are sick.

17. (*Housing Trends*) Statistics show that at a certain university, 70% of the students who live on campus during a given semester will remain on campus the following semester, and 90% of students living off campus during a given semester will remain off campus the following semester. Let x and y denote the number of students who live on and off campus this semester, and let u and v be the corresponding numbers for the next semester. Then

$$.7x + .1y = u$$
$$.3x + .9y = v.$$

(a) Write this system of equations in matrix form.

(b) Solve the resulting matrix equation for $\begin{bmatrix} x \\ y \end{bmatrix}$.

(c) Suppose that out of a group of 9000 students, 6000 currently live on campus and 3000 live off campus. How many lived on campus last semester? How many will live off campus next semester?

18. (*Performance on Tests*) A teacher estimates that of the students who pass a test, 80% will pass the next test, while of the students who fail a test, 50% will pass the next test. Let x and y denote the number of students who pass and fail a given test, and let u and v be the corresponding numbers for the following test.

(a) Write a matrix equation relating $\begin{bmatrix} x \\ y \end{bmatrix}$ to $\begin{bmatrix} u \\ v \end{bmatrix}$.

(b) Suppose that 25 of the teacher's students pass the third test and 8 fail the third test. How many students will pass the fourth test? Approximately how many passed the second test?

In Exercises 19 and 20, use the fact that

$$\begin{bmatrix} 1 & 2 & 2 \\ 1 & 3 & 2 \\ 1 & 2 & 3 \end{bmatrix}^{-1} = \begin{bmatrix} 5 & -2 & -2 \\ -1 & 1 & 0 \\ -1 & 0 & 1 \end{bmatrix}.$$

19. Solve $\begin{cases} x + 2y + 2z = 1 \\ x + 3y + 2z = -1 \\ x + 2y + 3z = -1. \end{cases}$

20. Solve $\begin{cases} x + 2y + 2z = 1 \\ x + 3y + 2z = 0 \\ x + 2y + 3z = 0. \end{cases}$

In Exercises 21 and 22, use the fact that

$$\begin{bmatrix} 9 & 0 & 2 & 0 \\ -20 & -9 & -5 & 5 \\ 4 & 0 & 1 & 0 \\ -4 & -2 & -1 & 1 \end{bmatrix}^{-1} = \begin{bmatrix} 1 & 0 & -2 & 0 \\ 0 & 1 & 0 & -5 \\ -4 & 0 & 9 & 0 \\ 0 & 2 & 1 & -9 \end{bmatrix}.$$

21. Solve $\begin{cases} 9x + 2z = 1 \\ -20x - 9y - 5z + 5w = 0 \\ 4x + z = 0 \\ -4x - 2y - z + w = -1. \end{cases}$

22. Solve $\begin{cases} 9x + 2z = 2 \\ -20x - 9y - 5z + 5w = 1 \\ 4x + z = 3 \\ -4x - 2y - z + w = 0. \end{cases}$

23. Without computing Δ, show that the matrix $\begin{bmatrix} 6 & 3 \\ 2 & 1 \end{bmatrix}$ does not have an inverse.

24. If $A^{-1} = \begin{bmatrix} 2 & 7 \\ 1 & -3 \end{bmatrix}$, what is the matrix A?

25. (*Age Distribution*) There are two age groups for a particular species of organism. Group I consists of all organisms aged under 1 year, while group II consists of all organisms aged from 1 to 2 years. No organism survives more than 2 years. The average number of offspring per year born to each member of group I is 1, while the average number of organisms per year born to each member of group II is 2. Nine-tenths of group I survive to enter group II each year.

    (a) Let x and y represent the initial number of organisms in groups I and II, respectively. Let a and b represent the number of organisms in groups I and II, respectively, after one year. Write a matrix equation relating $\begin{bmatrix} x \\ y \end{bmatrix}$ to $\begin{bmatrix} a \\ b \end{bmatrix}$.

    (b) If there are initially 450,000 organisms in group I and 360,000 organisms in group II, calculate the number of organisms in each of the groups after 1 year and after 2 years.

    (c) Suppose that at a certain time there were 810,000 organisms in group I and 630,000 organisms in

group II. Determine the population of each group 1 year earlier.

26. If $A^2 = \begin{bmatrix} -2 & -1 \\ 2 & -1 \end{bmatrix}$ and $A^3 = \begin{bmatrix} -2 & 1 \\ -2 & -3 \end{bmatrix}$, what is A?

27. Show that if AB is a matrix of all zeros and A has an inverse, then B is a matrix of all zeros.

28. Consider the matrices $A = \begin{bmatrix} 3 & 1 \\ 5 & 2 \end{bmatrix}$ and $B = \begin{bmatrix} 6 & 2 \\ 5 & 2 \end{bmatrix}$. Show that $(AB)^{-1} = B^{-1}A^{-1}$.

29. Find a 2×2 matrix A and a 2×1 column matrix B for which $AX = B$ has no solution.

30. Find a 2×2 matrix A and a 2×1 column matrix B for which $AX = B$ has infinitely many solutions.

Exercises 31–39 require the use of a graphing calculator or a computer.

In Exercises 31–34, use the inverse operation to find the inverse of the given matrix. Display the entries as fractions.

31. $\begin{bmatrix} .2 & 3 \\ 4 & 1.6 \end{bmatrix}$ **32.** $\begin{bmatrix} -12 & 3.3 \\ 6 & .4 \end{bmatrix}$

33. $\begin{bmatrix} .6 & 3 & -7 \\ 2.5 & -1 & 4 \\ -2 & .3 & 9 \end{bmatrix}$ **34.** $\begin{bmatrix} 5 & 2.3 & 6 \\ 1.2 & 5 & -7 \\ -3 & -4 & 6.5 \end{bmatrix}$

In Exercises 35–38, calculate the answer using $[A]^{-1} * [B]$ and give the answer using fractions.

35. $\begin{cases} 2x - 4y + 7z = 11 \\ x + 3y - 5z = -9 \\ 3x - y + 3z = 7 \end{cases}$

36. $\begin{cases} 5x + 2y - 3z = 1 \\ 4x - y + z = 22 \\ -x + 5y - 6z = 4 \end{cases}$

37. $\begin{cases} 2x + 7z + 5w = 10 \\ 5x - y + 3z = -2 \\ x + 2y - 2w = 0 \\ 3x - 4y + 2z - 5w = -18 \end{cases}$

38. $\begin{cases} x + 4y - z + 2w = 9 \\ 3x - 8y + 2z + 4w = -2 \\ -x - 3y + 7z - 6w = 10 \\ 4x + 2y - 3z + w = -6 \end{cases}$

39. Try finding the inverse of a matrix that does not have an inverse. How does your calculator or computer respond?

SOLUTIONS TO PRACTICE PROBLEMS 2.4

1. To see if this matrix is indeed the inverse, multiply it by the original matrix and find out if the products are identity matrices.

$$\begin{bmatrix} 1 & 1 & 4 \\ 3 & 2 & 4 \\ 1 & 1 & 6 \end{bmatrix} \begin{bmatrix} -4 & 1 & 2 \\ 7 & -1 & -4 \\ -\frac{1}{2} & 0 & \frac{1}{2} \end{bmatrix} = \begin{bmatrix} 1 & 0 & 0 \\ 0 & 1 & 0 \\ 0 & 0 & 1 \end{bmatrix},$$

(an identity matrix)

$$\begin{bmatrix} -4 & 1 & 2 \\ 7 & -1 & -4 \\ -\frac{1}{2} & 0 & \frac{1}{2} \end{bmatrix} \begin{bmatrix} 1 & 1 & 4 \\ 3 & 2 & 4 \\ 1 & 1 & 6 \end{bmatrix} = \begin{bmatrix} 1 & 0 & 0 \\ 0 & 1 & 0 \\ 0 & 0 & 1 \end{bmatrix}.$$

2. The matrix form of this system is

$$\begin{bmatrix} .8 & .6 \\ .2 & .4 \end{bmatrix} \begin{bmatrix} x \\ y \end{bmatrix} = \begin{bmatrix} 5 \\ 2 \end{bmatrix}.$$

Therefore, the solution is

$$\begin{bmatrix} x \\ y \end{bmatrix} = \begin{bmatrix} .8 & .6 \\ .2 & .4 \end{bmatrix}^{-1} \begin{bmatrix} 5 \\ 2 \end{bmatrix}.$$

To compute the inverse of the 2×2 matrix, first compute Δ.

$$\Delta = ad - bc = (.8)(.4) - (.6)(.2) = .32 - .12 = .2$$

Thus,

$$\begin{bmatrix} .8 & .6 \\ .2 & .4 \end{bmatrix}^{-1} = \begin{bmatrix} .4/.2 & -.6/.2 \\ -.2/.2 & .8/.2 \end{bmatrix} = \begin{bmatrix} 2 & -3 \\ -1 & 4 \end{bmatrix}.$$

Therefore,

$$\begin{bmatrix} x \\ y \end{bmatrix} = \begin{bmatrix} 2 & -3 \\ -1 & 4 \end{bmatrix} \begin{bmatrix} 5 \\ 2 \end{bmatrix} = \begin{bmatrix} 4 \\ 3 \end{bmatrix},$$

so the solution is $x = 4$, $y = 3$.

2.5 The Gauss–Jordan Method for Calculating Inverses

Of the several popular methods for finding the inverse of a matrix, the Gauss–Jordan method is probably the easiest to describe. It can be used on square matrices of any size. Also, the mechanical nature of the computations allows this method to be programmed for a computer with relative ease. We shall illustrate the procedure with a 2×2 matrix, whose inverse can also be calculated using the method of the previous section. Let

$$A = \begin{bmatrix} \frac{1}{2} & 1 \\ 1 & 3 \end{bmatrix}.$$

It is simple to check that

$$A^{-1} = \begin{bmatrix} 6 & -2 \\ -2 & 1 \end{bmatrix}.$$

Let us now derive this result using the Gauss–Jordan method.

> **Step 1** Write down the matrix A, and on its right write an identity matrix of the same size.

This is most conveniently done by placing I_2 beside A in a single matrix.

$$\left[\begin{array}{cc|cc} \frac{1}{2} & 1 & 1 & 0 \\ 1 & 3 & 0 & 1 \end{array} \right]$$

$$\underbrace{}_{A} \quad \underbrace{}_{I_2}$$

> **Step 2** Perform elementary row operations on the left-hand matrix so as to transform it into an identity matrix. Each operation performed on the left-hand matrix is also performed on the right-hand matrix.

This step proceeds exactly like the Gaussian elimination method and may be most conveniently expressed in terms of pivoting.

$$\left[\begin{array}{cc|cc} \tiny{\texttextcircled{$\frac{1}{2}$}} & 1 & 1 & 0 \\ 1 & 3 & 0 & 1 \end{array}\right], \quad \left[\begin{array}{cc|cc} 1 & 2 & 2 & 0 \\ 0 & \textcircled{1} & -2 & 1 \end{array}\right], \quad \left[\begin{array}{cc|cc} 1 & 0 & 6 & -2 \\ 0 & 1 & -2 & 1 \end{array}\right]$$

> **Step 3** When the matrix on the left becomes an identity matrix, the matrix on the right is the desired inverse.

So, from the last matrix of our calculation above, we have

$$A^{-1} = \begin{bmatrix} 6 & -2 \\ -2 & 1 \end{bmatrix}.$$

This is the same result obtained earlier.

We will demonstrate why the preceding method works after some further examples.

EXAMPLE 1 **Finding the inverse of a matrix using the Gauss–Jordan method** Find the inverse of the matrix

$$A = \begin{bmatrix} 4 & -2 & 3 \\ 8 & -3 & 5 \\ 7 & -2 & 4 \end{bmatrix}.$$

Solution

$$\left[\begin{array}{ccc|ccc} \textcircled{4} & -2 & 3 & 1 & 0 & 0 \\ 8 & -3 & 5 & 0 & 1 & 0 \\ 7 & -2 & 4 & 0 & 0 & 1 \end{array}\right]$$

$$\left[\begin{array}{ccc|ccc} 1 & -\frac{1}{2} & \frac{3}{4} & \frac{1}{4} & 0 & 0 \\ 0 & \textcircled{1} & -1 & -2 & 1 & 0 \\ 0 & \frac{3}{2} & -\frac{5}{4} & -\frac{7}{4} & 0 & 1 \end{array}\right]$$

$$\left[\begin{array}{ccc|ccc} 1 & 0 & \frac{1}{4} & -\frac{3}{4} & \frac{1}{2} & 0 \\ 0 & 1 & -1 & -2 & 1 & 0 \\ 0 & 0 & \textcircled{$\frac{1}{4}$} & \frac{5}{4} & -\frac{3}{2} & 1 \end{array}\right]$$

$$\left[\begin{array}{ccc|ccc} 1 & 0 & 0 & -2 & 2 & -1 \\ 0 & 1 & 0 & 3 & -5 & 4 \\ 0 & 0 & 1 & 5 & -6 & 4 \end{array}\right]$$

Therefore,

$$A^{-1} = \begin{bmatrix} -2 & 2 & -1 \\ 3 & -5 & 4 \\ 5 & -6 & 4 \end{bmatrix}.$$

Now Try Exercise 7   ■

Not all square matrices have inverses. If a matrix does not have an inverse, this will become apparent when applying the Gauss–Jordan method. At some point there will be no way to continue transforming the left-hand matrix into an identity matrix. This is illustrated in the next example.

EXAMPLE 2 **Demonstrating that a matrix does not have an inverse** Find the inverse of the matrix

$$A = \begin{bmatrix} 1 & 3 & 2 \\ 0 & 1 & 4 \\ 1 & 5 & 10 \end{bmatrix}.$$

Solution

$$\left[\begin{array}{ccc|ccc} ① & 3 & 2 & 1 & 0 & 0 \\ 0 & 1 & 4 & 0 & 1 & 0 \\ 1 & 5 & 10 & 0 & 0 & 1 \end{array}\right]$$

$$\left[\begin{array}{ccc|ccc} 1 & 3 & 2 & 1 & 0 & 0 \\ 0 & ① & 4 & 0 & 1 & 0 \\ 0 & 2 & 8 & -1 & 0 & 1 \end{array}\right]$$

$$\left[\begin{array}{ccc|ccc} 1 & 0 & -10 & 1 & -3 & 0 \\ 0 & 1 & 4 & 0 & 1 & 0 \\ 0 & 0 & 0 & -1 & -2 & 1 \end{array}\right]$$

Since the third row of the left-hand matrix has only zero entries, it is impossible to complete the Gauss–Jordan method. Therefore, the matrix A has no inverse matrix.

Now Try Exercise 9   ■

Verification of the Gauss–Jordan Method for Calculating Inverses
In Section 2.4 we showed how to calculate the inverse by solving several systems of linear equations. Actually, the Gauss–Jordan method is just an organized way of going about the calculation. To see why, let us consider a concrete example:

$$A = \begin{bmatrix} 4 & -2 & 3 \\ 8 & -3 & 5 \\ 7 & -2 & 4 \end{bmatrix}.$$

We wish to determine A^{-1}, so regard it as a matrix of unknowns:

$$A^{-1} = \begin{bmatrix} x_1 & x_2 & x_3 \\ y_1 & y_2 & y_3 \\ z_1 & z_2 & z_3 \end{bmatrix}.$$

The statement $AA^{-1} = I_3$ is

$$\begin{bmatrix} 4 & -2 & 3 \\ 8 & -3 & 5 \\ 7 & -2 & 4 \end{bmatrix} \begin{bmatrix} x_1 & x_2 & x_3 \\ y_1 & y_2 & y_3 \\ z_1 & z_2 & z_3 \end{bmatrix} = \begin{bmatrix} 1 & 0 & 0 \\ 0 & 1 & 0 \\ 0 & 0 & 1 \end{bmatrix}.$$

Multiplying out the matrices on the left and comparing the result with the matrix on the right give us nine equations, namely

$$\begin{cases} 4x_1 - 2y_1 + 3z_1 = 1 \\ 8x_1 - 3y_1 + 5z_1 = 0 \\ 7x_1 - 2y_1 + 4z_1 = 0 \end{cases}$$

$$\begin{cases} 4x_2 - 2y_2 + 3z_2 = 0 \\ 8x_2 - 3y_2 + 5z_2 = 1 \\ 7x_2 - 2y_2 + 4z_2 = 0 \end{cases}$$

$$\begin{cases} 4x_3 - 2y_3 + 3z_3 = 0 \\ 8x_3 - 3y_3 + 5z_3 = 0 \\ 7x_3 - 2y_3 + 4z_3 = 1. \end{cases}$$

Notice that each system of equations corresponds to one column of unknowns in A^{-1}. More precisely, if we set

$$X_1 = \begin{bmatrix} x_1 \\ y_1 \\ z_1 \end{bmatrix}, \qquad X_2 = \begin{bmatrix} x_2 \\ y_2 \\ z_2 \end{bmatrix}, \qquad X_3 = \begin{bmatrix} x_3 \\ y_3 \\ z_3 \end{bmatrix},$$

then the preceding three systems have the respective matrix forms

$$AX_1 = \begin{bmatrix} 1 \\ 0 \\ 0 \end{bmatrix}, \qquad AX_2 = \begin{bmatrix} 0 \\ 1 \\ 0 \end{bmatrix}, \qquad AX_3 = \begin{bmatrix} 0 \\ 0 \\ 1 \end{bmatrix}.$$

Now imagine the process of applying Gaussian elimination to solve these three systems. We apply elementary row operations to the matrices

$$\left[A \,\middle|\, \begin{matrix} 1 \\ 0 \\ 0 \end{matrix} \right], \qquad \left[A \,\middle|\, \begin{matrix} 0 \\ 1 \\ 0 \end{matrix} \right], \qquad \left[A \,\middle|\, \begin{matrix} 0 \\ 0 \\ 1 \end{matrix} \right].$$

The process ends when we convert A into the identity matrix, at which point the solutions may be read off the right column. So the procedure ends with the matrices

$$\begin{bmatrix} I_3 & | & X_1 \end{bmatrix}, \qquad \begin{bmatrix} I_3 & | & X_2 \end{bmatrix}, \qquad \begin{bmatrix} I_3 & | & X_3 \end{bmatrix}.$$

Realize, however, that at each step of the three Gaussian eliminations we are performing the same operations, since all three start with the matrix A on the left. So, in order to save calculations, perform the three Gaussian eliminations simultaneously by performing the row operations on the composite matrix

$$\left[A \,\middle|\, \begin{matrix} 1 & 0 & 0 \\ 0 & 1 & 0 \\ 0 & 0 & 1 \end{matrix} \right] = \begin{bmatrix} A & | & I_3 \end{bmatrix}.$$

The procedure ends when this matrix is converted into

$$\begin{bmatrix} I_3 & | & X_1 & X_2 & X_3 \end{bmatrix}.$$

That is, since $A^{-1} = \begin{bmatrix} X_1 & X_2 & X_3 \end{bmatrix}$, the procedure ends with A^{-1} on the right. This is the reasoning behind the Gauss–Jordan method of calculating inverses. ■

PRACTICE PROBLEMS 2.5

1. Use the Gauss–Jordan method to calculate the inverse of the matrix

$$\begin{bmatrix} 1 & 0 & 2 \\ 0 & 1 & -4 \\ 0 & 0 & 2 \end{bmatrix}.$$

2. Solve the system of linear equations

$$\begin{cases} x & + 2z = 4 \\ y - 4z = 6 \\ 2z = 9. \end{cases}$$

EXERCISES 2.5

In Exercises 1–12, use the Gauss–Jordan method to compute the inverse of the matrix.

1. $\begin{bmatrix} 7 & 3 \\ 5 & 2 \end{bmatrix}$

2. $\begin{bmatrix} 5 & -2 \\ 6 & 2 \end{bmatrix}$

3. $\begin{bmatrix} 10 & 12 \\ 3 & -4 \end{bmatrix}$

4. $\begin{bmatrix} 1 & -3 \\ 0 & 1 \end{bmatrix}$

5. $\begin{bmatrix} 2 & -4 \\ -1 & 2 \end{bmatrix}$

6. $\begin{bmatrix} 1 & 3 & 1 \\ -1 & 2 & 0 \\ 2 & 11 & 3 \end{bmatrix}$

7. $\begin{bmatrix} 1 & 2 & -2 \\ 1 & 1 & 1 \\ 0 & 0 & 1 \end{bmatrix}$

8. $\begin{bmatrix} 2 & 2 & 0 \\ 0 & -2 & 0 \\ 3 & 0 & 1 \end{bmatrix}$

9. $\begin{bmatrix} -2 & 5 & 2 \\ 1 & -3 & -1 \\ -1 & 2 & 1 \end{bmatrix}$

10. $\begin{bmatrix} 1 & 0 & 0 \\ 2 & 1 & -2 \\ -1 & 2 & 1 \end{bmatrix}$

11. $\begin{bmatrix} 1 & 6 & 0 & 0 \\ 1 & 5 & 0 & 0 \\ 0 & 0 & 4 & 2 \\ 0 & 0 & 50 & 2 \end{bmatrix}$

12. $\begin{bmatrix} 6 & 0 & 2 & 0 \\ -6 & 1 & 0 & 1 \\ 1 & 0 & 1 & 0 \\ -9 & 0 & -1 & 1 \end{bmatrix}$

In Exercises 13–16, use matrix inversion to solve the systems of linear equations.

13. $\begin{cases} x + y + 2z = 3 \\ 3x + 2y + 2z = 4 \\ x + y + 3z = 5 \end{cases}$

14. $\begin{cases} x + 2y + 3z = 4 \\ 3x + 5y + 5z = 3 \\ 2x + 4y + 2z = 4 \end{cases}$

15. $\begin{cases} x & - 2z - 2w = 0 \\ y & - 5w = 1 \\ -4x & + 9z + 9w = 2 \\ 2y + z - 8w = 3 \end{cases}$

16. $\begin{cases} y + 2z = 1 \\ 2x + y + 3z = 2 \\ x + y + 2z = 3 \end{cases}$

17. If $A^{-1} = \begin{bmatrix} 2 & 7 \\ 1 & -3 \end{bmatrix}$, what is the matrix A?

18. Find the 2×2 matrix C for which $A \cdot C = B$, where

$$A = \begin{bmatrix} 3 & 2 \\ 4 & 3 \end{bmatrix} \quad \text{and} \quad B = \begin{bmatrix} 5 & 6 \\ 1 & 7 \end{bmatrix}.$$

19. Find a 2×2 matrix A for which

$$A \cdot \begin{bmatrix} 2 \\ 1 \end{bmatrix} = \begin{bmatrix} -1 \\ 4 \end{bmatrix} \quad \text{and} \quad A \cdot \begin{bmatrix} 5 \\ 3 \end{bmatrix} = \begin{bmatrix} 0 \\ 2 \end{bmatrix}.$$

20. Let

$$A = \begin{bmatrix} 7 & 4 \\ 3 & 2 \end{bmatrix} \quad \text{and} \quad B = \begin{bmatrix} 1 & 5 \\ -3 & 4 \end{bmatrix}.$$

Each of the equations $AX = B$ and $XA = B$ has a 2×2 matrix X as a solution. Find X in each case and explain why the two answers are different.

In Exercises 21–24, use a graphing calculator or mathematical software.

In Exercises 21–24, use the Gauss–Jordan method to compute the inverse of the matrix.

21. $\begin{bmatrix} 4 & 1 \\ 7 & 2 \end{bmatrix}$

22. $\begin{bmatrix} 8 & 3 \\ 13 & 5 \end{bmatrix}$

23. $\begin{bmatrix} 2 & 5 & 3 \\ 1 & 3 & 0 \\ 2 & 3 & 4 \end{bmatrix}$

24. $\begin{bmatrix} 4 & -2 & -8 \\ -3 & 1 & 3 \\ 5 & 0 & 6 \end{bmatrix}$

SOLUTIONS TO PRACTICE PROBLEMS 2.5

1. First write the given matrix beside an identity matrix of the same size

$$\begin{bmatrix} 1 & 0 & 2 & | & 1 & 0 & 0 \\ 0 & 1 & -4 & | & 0 & 1 & 0 \\ 0 & 0 & 2 & | & 0 & 0 & 1 \end{bmatrix}.$$

The object is to use elementary row operations to transform the 3×3 matrix on the left into the identity matrix. The first two columns are already in the correct form.

SOLUTIONS TO PRACTICE PROBLEMS 2.5 (CONTINUED)

$$\begin{bmatrix} 1 & 0 & 2 & | & 1 & 0 & 0 \\ 0 & 1 & -4 & | & 0 & 1 & 0 \\ 0 & 0 & 2 & | & 0 & 0 & 1 \end{bmatrix}$$

$$\xrightarrow{\frac{1}{2}[3]} \begin{bmatrix} 1 & 0 & 2 & | & 1 & 0 & 0 \\ 0 & 1 & -4 & | & 0 & 1 & 0 \\ 0 & 0 & 1 & | & 0 & 0 & \frac{1}{2} \end{bmatrix}$$

$$\xrightarrow{[1]+(-2)[3]} \begin{bmatrix} 1 & 0 & 0 & | & 1 & 0 & -1 \\ 0 & 1 & -4 & | & 0 & 1 & 0 \\ 0 & 0 & 1 & | & 0 & 0 & \frac{1}{2} \end{bmatrix}$$

$$\xrightarrow{[2]+(4)[3]} \begin{bmatrix} 1 & 0 & 0 & | & 1 & 0 & -1 \\ 0 & 1 & 0 & | & 0 & 1 & 2 \\ 0 & 0 & 1 & | & 0 & 0 & \frac{1}{2} \end{bmatrix}$$

Thus the inverse of the given matrix is

$$\begin{bmatrix} 1 & 0 & -1 \\ 0 & 1 & 2 \\ 0 & 0 & \frac{1}{2} \end{bmatrix}.$$

2. The matrix form of this system of equations is $AX = B$, where A is the matrix whose inverse was found in Problem 1, and

$$B = \begin{bmatrix} 4 \\ 6 \\ 9 \end{bmatrix}.$$

Therefore, $X = A^{-1}B$, so that

$$\begin{bmatrix} x \\ y \\ z \end{bmatrix} = \begin{bmatrix} 1 & 0 & -1 \\ 0 & 1 & 2 \\ 0 & 0 & \frac{1}{2} \end{bmatrix} \begin{bmatrix} 4 \\ 6 \\ 9 \end{bmatrix} = \begin{bmatrix} -5 \\ 24 \\ \frac{9}{2} \end{bmatrix}.$$

So the solution of the system is $x = -5$, $y = 24$, $z = \frac{9}{2}$.

2.6 Input–Output Analysis

In recent years matrix arithmetic has played an ever-increasing role in economics, especially in that branch of economics called *input–output analysis*. Pioneered by the Harvard economist Vassily Leontieff, input–output analysis is used to analyze an economy in order to meet given consumption and export demands. As we shall see, such analysis leads to matrix calculations and in particular to inverses. Input–output analysis has been of such great significance that Leontieff was awarded the 1973 Nobel Prize in economics for his fundamental work in the subject.

Suppose that we divide an economy into a number of industries—transportation, agriculture, steel, and so on. Each industry produces a certain output using certain raw materials (or input). The input of each industry is made up in part by the outputs of other industries. For example, in order to produce food, agriculture uses as input the output of many industries, such as transportation (tractors and trucks) and oil (gasoline and fertilizers). This interdependence among the industries of the economy is summarized in a matrix—an *input–output matrix*. There is one column for each industry's input requirements. The entries in the column reflect the amount of input required from each of the industries. A typical input–output matrix looks like this:

		Input requirements of:		
		Industry 1	Industry 2	Industry 3 ...
From	Industry 1			
	Industry 2			
	Industry 3			
	⋮			

It is most convenient to express the entries of this matrix in monetary terms. That is, each column gives the dollar values of the various inputs needed by an industry in order to produce $1 worth of output.

There are consumers (other than the industries themselves) who want to purchase some of the output of these industries. The quantity of goods that these consumers want (or demand) is called the *final demand* on the economy. The final demand can be represented by a column matrix, with one entry for each industry, indicating the amount of consumable output demanded from the industry:

$$[\text{final demand}] = \begin{bmatrix} \text{amount from industry 1} \\ \text{amount from industry 2} \\ \vdots \end{bmatrix}.$$

We shall consider the situation in which the final-demand matrix is given and it is necessary to determine how much output should be produced by each industry in order to provide the needed inputs of the various industries and also to satisfy the final demand. The proper level of output can be computed using matrix calculations, as illustrated in the next example.

EXAMPLE 1 **Determining industrial production** Suppose that an economy is composed of only three industries—coal, steel, and electricity. Each of these industries depends on the others for some of its raw materials. Suppose that to make $1 of coal, it takes no coal, but $.02 of steel and $.01 of electricity; to make $1 of steel, it takes $.15 of coal, $.03 of steel, and $.08 of electricity; and to make $1 of electricity, it takes $.43 of coal, $.20 of steel, and $.05 of electricity. How much should each industry produce to allow for consumption (not used for production) at these levels: $2 billion coal, $1 billion steel, $3 billion electricity?

Solution Put all the data indicating the interdependence of the industries in a matrix. In each industry's column, put the amount of input from each of the industries needed to produce $1 of output in that particular industry:

$$\begin{array}{c} \\ \text{Coal} \\ \text{Steel} \\ \text{Electricity} \end{array} \begin{array}{ccc} \text{Coal} & \text{Steel} & \text{Electricity} \\ \begin{bmatrix} 0 & .15 & .43 \\ .02 & .03 & .20 \\ .01 & .08 & .05 \end{bmatrix} \end{array} = A.$$

This matrix is the *input–output matrix* corresponding to the economy. Let D denote the final-demand matrix. Then, letting the numbers in D stand for billions of dollars, we have

$$D = \begin{bmatrix} 2 \\ 1 \\ 3 \end{bmatrix}.$$

Suppose that the coal industry produces x billion dollars of output, the steel industry y billion dollars, and the electrical industry z billion dollars. Our problem is to determine the x, y, and z that yield the desired amounts left over from the production process. As an example, consider coal. The amount of coal that can be consumed or exported is just

$$x - [\text{amount of coal used in production}].$$

To determine the amount of coal used in production, refer to the input–output matrix. Production of x billion dollars of coal takes $0 \cdot x$ billion dollars of coal; production of y billion dollars of steel takes $.15y$ billion dollars of coal; and production of z billion dollars of electricity takes $.43z$ billion dollars of coal. Thus,

$$\left[\text{amount of coal used in production}\right] = 0 \cdot x + .15y + .43z.$$

This quantity should be recognized as the first entry of a matrix product. Namely, if we let

$$X = \begin{bmatrix} x \\ y \\ z \end{bmatrix},$$

then

$$\begin{bmatrix} \text{coal} \\ \text{steel} \\ \text{electricity} \end{bmatrix}_{\text{used in production}} = \begin{bmatrix} 0 & .15 & .43 \\ .02 & .03 & .20 \\ .01 & .08 & .05 \end{bmatrix} \begin{bmatrix} x \\ y \\ z \end{bmatrix} = AX.$$

But then the amount of each output available for purposes other than production is $X - AX$. That is, we have the matrix equation

$$X - AX = D.$$

To solve this equation for X, proceed as follows. Since $IX = X$, write the equation in the form

$$IX - AX = D$$

$$(I - A)X = D$$

$$X = (I - A)^{-1}D. \tag{1}$$

So, in other words, X may be found by multiplying D on the left by $(I - A)^{-1}$. Let us now do the arithmetic.

$$I - A = \begin{bmatrix} 1 & 0 & 0 \\ 0 & 1 & 0 \\ 0 & 0 & 1 \end{bmatrix} - \begin{bmatrix} 0 & .15 & .43 \\ .02 & .03 & .20 \\ .01 & .08 & .05 \end{bmatrix} = \begin{bmatrix} 1 & -.15 & -.43 \\ -.02 & .97 & -.20 \\ -.01 & -.08 & .95 \end{bmatrix}$$

Applying the Gauss–Jordan method, we find that

$$(I - A)^{-1} = \begin{bmatrix} 1.01 & .20 & .50 \\ .02 & 1.05 & .23 \\ .01 & .09 & 1.08 \end{bmatrix},$$

where all figures are carried to two decimal places (Exercise 9). Therefore,

$$X = (I - A)^{-1}D = \begin{bmatrix} 1.01 & .20 & .50 \\ .02 & 1.05 & .23 \\ .01 & .09 & 1.08 \end{bmatrix} \begin{bmatrix} 2 \\ 1 \\ 3 \end{bmatrix} = \begin{bmatrix} 3.72 \\ 1.78 \\ 3.35 \end{bmatrix}.$$

In other words, coal should produce \$3.72 billion worth of output, steel \$1.78 billion, and electricity \$3.35 billion. This output will meet the required final demands from each industry.

Now Try Exercise 5   ■

The preceding analysis is useful in studying not only entire economies but also segments of economies and even individual companies.

EXAMPLE 2 **Determining production for a conglomerate** A conglomerate has three divisions, which produce computers, semiconductors, and business forms. For each $1 of output, the computer division needs $.02 worth of computers, $.20 worth of semiconductors, and $.10 worth of business forms. For each $1 of output, the semiconductor division needs $.02 worth of computers, $.01 worth of semiconductors, and $.02 worth of business forms. For each $1 of output, the business forms division requires $.10 worth of computers and $.01 worth of business forms. The conglomerate estimates the sales demand to be $300,000,000 for the computer division, $100,000,000 for the semiconductor division, and $200,000,000 for the business forms division. At what level should each division produce in order to satisfy this demand?

Solution The conglomerate can be viewed as a miniature economy and its sales as the final demand. The input–output matrix for this "economy" is

$$
\begin{array}{c}
\\
\text{Computers} \\
\text{Semiconductors} \\
\text{Business forms}
\end{array}
\begin{array}{ccc}
\text{Computers} & \text{Semiconductors} & \begin{array}{c}\text{Business}\\\text{forms}\end{array}
\end{array}
\begin{bmatrix}
.02 & .02 & .10 \\
.20 & .01 & 0 \\
.10 & .02 & .01
\end{bmatrix} = A.
$$

The final-demand matrix is

$$
D = \begin{bmatrix} 3 \\ 1 \\ 2 \end{bmatrix},
$$

where the demand is expressed in hundreds of millions of dollars. By equation (1) the matrix X, giving the desired levels of production for the various divisions, is given by

$$
X = (I - A)^{-1}D.
$$

But

$$
I - A = \begin{bmatrix}
.98 & -.02 & -.10 \\
-.20 & .99 & 0 \\
-.10 & -.02 & .99
\end{bmatrix},
$$

so that (Exercise 10)

$$
(I - A)^{-1} = \begin{bmatrix}
1.04 & .02 & .10 \\
.21 & 1.01 & .02 \\
.11 & .02 & 1.02
\end{bmatrix} \quad \text{and} \quad (I - A)^{-1}D = \begin{bmatrix} 3.34 \\ 1.68 \\ 2.39 \end{bmatrix}.
$$

Therefore,

$$
X = \begin{bmatrix} 3.34 \\ 1.68 \\ 2.39 \end{bmatrix}.
$$

That is, the computer division should produce $334,000,000, the semiconductor division $168,000,000, and the business forms division $239,000,000.

Now Try Exercise 7 ■

Input–output analysis is usually applied to the entire economy of a country having hundreds of industries. The resulting matrix equation $(I - A)X = D$ could be solved by the Gaussian elimination method. However, it is best to find the inverse of $I - A$ and solve for X as we have done in the examples of this section.

Over a short period, D might change but A is unlikely to change. Therefore, the proper outputs to satisfy the new demand can easily be determined by using the already computed inverse of $I - A$.

The Closed Leontieff Model   The foregoing description of an economy is usually called the *Leontieff open model* since it views exports as an activity that takes place external to the economy. However, it is possible to consider exports as yet another industry in the economy. Instead of describing exports by a demand column D, we describe it by a column in the input–output matrix. That is, the export column describes how each dollar of exports is divided among the various industries. Since exports are now regarded as another industry, each of the original columns has an additional entry, namely the amount of output from the export industry (that is, imports) used to produce \$1 of goods (of the industry corresponding to the column). If A denotes the expanded input–output matrix and X the production matrix (as before), then AX is the matrix describing the total demand experienced by each of the industries. In order for the economy to function efficiently, the total amount demanded by the various industries should equal the amount produced. That is, the production matrix must satisfy the equation

$$AX = X.$$

By studying the solutions to this equation, it is possible to determine the equilibrium states of the economy—that is, the production matrices X for which the amounts produced exactly equal the amounts needed by the various industries. The model just described is called *Leontieff's closed model.*

We may expand the Leontieff closed model to include the effects of labor and monetary phenomena by considering labor and banking as yet further industries to be incorporated in the input–output matrix.

GC   With a graphing calculator, the matrix $(I - A)^{-1}D$ can be calculated with a single press of the ENTER key. Figures 1 and 2 show the calculation for the matrices of Example 2.

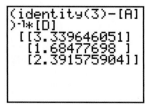

Figure 1

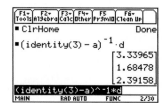

Figure 2

ES   To display the matrix $(I - A)^{-1}D$ on an Excel spreadsheet, select a column of n cells, type the formula

$$\texttt{=MMULT(MINVERSE(I-A),D)}$$

and press Ctrl+Shift+Enter. (Here D is an $n \times 1$ matrix, I is the $n \times n$ identity matrix, and A is an $n \times n$ square matrix.)

PRACTICE PROBLEMS 2.6

1. Let

$$I = \begin{bmatrix} 1 & 0 & 0 \\ 0 & 1 & 0 \\ 0 & 0 & 1 \end{bmatrix}, \quad A = \begin{bmatrix} .1 & 0 & .1 \\ .2 & .1 & .1 \\ .1 & .2 & 0 \end{bmatrix},$$

$$X = \begin{bmatrix} x \\ y \\ z \end{bmatrix}, \quad D = \begin{bmatrix} 100 \\ 200 \\ 50 \end{bmatrix}.$$

Solve the matrix equation

$$(I - A)X = D.$$

2. Let I, A, and X be as in Problem 1, but let

$$D = \begin{bmatrix} 300 \\ 100 \\ 100 \end{bmatrix}.$$

Solve the matrix equation $(I - A)X = D$.

EXERCISES 2.6

(Three-Sector Economy) In Exercises 1–4, suppose a simplified economy consisting of the three sectors Manufacturing, Energy, and Services has the input-output matrix

$$\begin{array}{c} \\ M \\ E \\ S \end{array} \begin{array}{ccc} M & E & S \\ \begin{bmatrix} .3 & .1 & .2 \\ .2 & .25 & .15 \\ .1 & .2 & .15 \end{bmatrix} \end{array}.$$

1. How many cents of energy are required to produce $1 worth of manufactured goods?

2. How many cents of energy are required to produce $1 worth of services?

3. Which sector of the economy requires the greatest amount of services in order to produce $1 worth of output?

4. What is the dollar amount of the energy costs needed to produce 10 million dollars worth of goods from each sector?

5. *(Industrial Production)* Suppose that in the economy of Example 1 the demand for electricity triples and the demand for coal doubles, whereas the demand for steel increases only by 50%. At what levels should the various industries produce in order to satisfy the new demand?

6. *(Conglomerate)* Suppose that the conglomerate of Example 2 is faced with an increase of 50% in demand for computers, a doubling in demand for semiconductors, and a decrease of 50% in demand for business forms. At what levels should the various divisions produce in order to satisfy the new demand?

7. *(Conglomerate)* Suppose that the conglomerate of Example 2 experiences a doubling in the demand for business forms. At what levels should the computer and semiconductor divisions produce?

8. *(Multinational Corporation)* A multinational corporation does business in the United States, Canada, and England. Its branches in one country purchase goods

from the branches in other countries according to the matrix

Purchase from:	Branch in:		
	United States	Canada	England
United States	.02	0	.02
Canada	.01	.03	.01
England	.03	0	.01

where the entries in the matrix represent proportions of total sales by the respective branch. The external sales by each of the offices are $800,000,000 for the U.S. branch, $300,000,000 for the Canadian branch, and $1,400,000,000 for the English branch. At what level should each of the branches produce in order to satisfy the total demand?

9. Show that to two decimal places

$$\begin{bmatrix} 1 & -.15 & -.43 \\ -.02 & .97 & -.20 \\ -.01 & -.08 & .95 \end{bmatrix}^{-1} = \begin{bmatrix} 1.01 & .20 & .50 \\ .02 & 1.05 & .23 \\ .01 & .09 & 1.08 \end{bmatrix}.$$

10. Show that to two decimal places

$$\begin{bmatrix} .98 & -.02 & -.10 \\ -.20 & .99 & 0 \\ -.10 & -.02 & .99 \end{bmatrix}^{-1} = \begin{bmatrix} 1.04 & .02 & .10 \\ .21 & 1.01 & .02 \\ .11 & .02 & 1.02 \end{bmatrix}.$$

11. *(Two-Sector Economy)* A simplified economy consists of the two sectors Transportation and Energy. For each $1 worth of output, the transportation sector requires $.25 worth of input from the transportation sector and $.20 of input from the energy sector. For each $1 worth of output, the energy sector requires $.30 from the transportation sector and $.15 from the energy sector.

   (a) Give the input-output matrix A for this economy.

   (b) Determine the matrix $(I - A)^{-1}$. (Round entries to two decimal places.)

   (c) At what level of output should each sector produce to meet a demand for $5 billion worth of transportation and $3 billion worth of energy?

12. (*Three-Sector Economy*) An economy consists of the three sectors agriculture, energy, and manufacturing. For each \$1 worth of output, the agriculture sector requires \$.08 worth of input from the agriculture sector, \$.10 worth of input from the energy sector, and \$.20 worth of input from the manufacturing sector. For each \$1 worth of output, the energy sector requires \$.15 worth of input from the agriculture sector, \$.14 worth of input from the energy sector, and \$.10 worth of input from the manufacturing sector. For each \$1 worth of output, the manufacturing sector requires \$.25 worth of input from the agriculture sector, \$.12 worth of input from the energy sector, and \$.05 worth of input from the manufacturing sector.

(a) Give the input-output matrix A for this economy.

(b) Determine the matrix $(I - A)^{-1}$. (Round entries to two decimal places.)

(c) At what level of output should each sector produce to meet a demand for \$4 billion worth of agriculture, \$3 billion worth of energy, and \$2 billion worth of manufacturing?

13. (*Two-Product Corporation*) A corporation has a plastics division and an industrial equipment division. For each \$1 worth of output, the plastics division needs \$.02 worth of plastics and \$.10 worth of equipment. For each \$1 worth of output, the industrial equipment division needs \$.01 worth of plastics and \$.05 worth of equipment. At what level should the divisions produce to meet a demand for \$930,000 worth of plastics and \$465,000 worth of industrial equipment?

14. (*Two-Product Corporation*) Rework Exercise 13 under the condition that the demand for plastics is \$1,860,000 and the demand for industrial equipment is \$2,790,000.

15. (*Three-Sector Industry*) An industrial system involves manufacturing, transportation, and agriculture. The interdependence of the three industries is given by the input–output matrix

$$\begin{array}{ccc} M & T & A \end{array}$$
$$\begin{bmatrix} .4 & .3 & .1 \\ .2 & .2 & .2 \\ .1 & .1 & .4 \end{bmatrix} \begin{array}{c} M \\ T \\ A \end{array}$$

At what levels must the industries produce to satisfy a demand for \$100 million worth of manufactured goods,

\$80 million of transportation, and \$200 million worth of agricultural products?

16. (*Localized Economy*) A town has a merchant, a baker, and a farmer. To produce \$1 worth of output, the merchant requires \$.30 worth of baked goods and \$.40 worth of the farmer's products. To produce \$1 worth of output, the baker requires \$.50 worth of the merchant's goods, \$.10 worth of his own goods, and \$.30 worth of the farmer's goods. To produce \$1 worth of output, the farmer requires \$.30 worth of the merchant's goods, \$.20 worth of baked goods, and \$.30 worth of his own products. How much should the merchant, baker, and farmer produce to meet a demand for \$20,000 worth of output from the merchant, \$15,000 worth of output from the baker, and \$18,000 worth of output from the farmer?

17. The matrix $(I - A)^{-1}$ from Example 1 is given in Exercise 9. Show that if the final demand for coal is increased by one billion dollars, then the additional amounts (in billions of dollars) that must be produced by each of the three industries is given by the first column of $(I - A)^{-1}$.

Hint: $(I - A)^{-1} \begin{bmatrix} 3 \\ 1 \\ 3 \end{bmatrix} = (I - A)^{-1} \left(\begin{bmatrix} 2 \\ 1 \\ 3 \end{bmatrix} + \begin{bmatrix} 1 \\ 0 \\ 0 \end{bmatrix} \right).$

18. Refer to Exercise 17. Interpret the significance of the second and third columns of $(I - A)^{-1}$.

In the following exercises, use a graphing calculator, a spreadsheet, or mathematical software to carry out the matrix operations.

In Exercises 19 and 20, use the input–output matrix A and the final-demand matrix D to find the production matrix X for the Leontieff open model. Round your answers to two decimal places.

19. $A = \begin{bmatrix} .1 & .2 & .4 \\ .05 & .3 & .25 \\ .15 & .1 & .2 \end{bmatrix}, D = \begin{bmatrix} 3 \\ 7 \\ 4 \end{bmatrix}.$

20. $A = \begin{bmatrix} .2 & .35 & .15 & .05 \\ .1 & .1 & .3 & .2 \\ .075 & .2 & .05 & .1 \\ .3 & .04 & .1 & .15 \end{bmatrix}, D = \begin{bmatrix} 5 \\ 2 \\ 1 \\ 7 \end{bmatrix}.$

SOLUTIONS TO PRACTICE PROBLEMS 2.6

1. The equation $(I - A)X = D$ has the form $CX = D$, where C is the matrix $I - A$. From Section 2.4 we know that $X = C^{-1}D$. That is, $X = (I - A)^{-1}D$. Now

$$I - A = \begin{bmatrix} 1 & 0 & 0 \\ 0 & 1 & 0 \\ 0 & 0 & 1 \end{bmatrix} - \begin{bmatrix} .1 & 0 & .1 \\ .2 & .1 & .1 \\ .1 & .2 & 0 \end{bmatrix}$$

$$= \begin{bmatrix} .9 & 0 & -.1 \\ -.2 & .9 & -.1 \\ -.1 & -.2 & 1 \end{bmatrix}.$$

Using the Gauss–Jordan method to find the inverse of this matrix, we have (to two decimal places)

$$(I - A)^{-1} = \begin{bmatrix} 1.13 & .03 & .12 \\ .27 & 1.14 & .14 \\ .17 & .23 & 1.04 \end{bmatrix}.$$

Therefore, rounding to the nearest integer, we have

$$X = (I - A)^{-1}D = \begin{bmatrix} 1.13 & .03 & .12 \\ .27 & 1.14 & .14 \\ .17 & .23 & 1.04 \end{bmatrix} \begin{bmatrix} 100 \\ 200 \\ 50 \end{bmatrix}$$

$$= \begin{bmatrix} 125 \\ 262 \\ 115 \end{bmatrix}.$$

2. We have $X = (I - A)^{-1}D$, where $(I - A)^{-1}$ is as computed in Problem 1. So

$$X = (I - A)^{-1}D = \begin{bmatrix} 1.13 & .03 & .12 \\ .27 & 1.14 & .14 \\ .17 & .23 & 1.04 \end{bmatrix} \begin{bmatrix} 300 \\ 100 \\ 100 \end{bmatrix}$$

$$= \begin{bmatrix} 354 \\ 209 \\ 178 \end{bmatrix}.$$

CHAPTER 2 SUMMARY

1. The three elementary row operations for a system of linear equations (or a matrix) are as follows:

 (a) Rearrange the equations (rows) in any order.

 (b) Multiply an equation (row) by a nonzero number.

 (c) Change an equation (row) by adding to it a multiple of another equation (row).

2. When an elementary row operation is applied to a system of linear equations (or an augmented matrix) the solutions remain the same. The Gaussian elimination method is a systematic process that applies a sequence of elementary row operations until the solutions can be easily obtained.

3. The process of pivoting on a specific element of a matrix is to apply a sequence of elementary row operations so that the specific element becomes 1 and the other elements in its column become 0. To apply the Gaussian elimination method, proceed from left to right and perform pivots on as many columns to the left of the vertical line as possible, with the specific elements for the pivots coming from different rows.

4. After an augmented matrix has been completely reduced with the Gaussian elimination method, all the solutions to the corresponding system of linear equations can be obtained. If the matrix on the left of the vertical line is an identity matrix, then there is a unique solution. If one row of the augmented matrix is of the form $0 \ 0 \ 0 \ \cdots \ 0 \mid a$, where a is a nonzero number, then there are no solutions. Otherwise, there are infinitely many solutions. In this case, variables corresponding to columns that have not been pivoted can assume any values, and the values of the other variables can be expressed in terms of those variables.

5. A matrix of size $m \times n$ has m rows and n columns.

6. Matrices of the same size can be added (or subtracted) by adding (or subtracting) corresponding elements.

7. The product of an $m \times n$ matrix and an $n \times r$ matrix is the $m \times r$ matrix whose ijth element is obtained by multiplying the ith row of the first matrix by the jth column of the second matrix. (The product of each row and column is calculated as the sum of the products of successive entries.)

8. The inverse of a square matrix A is a square matrix A^{-1} with the property that $A^{-1}A = I$ and $AA^{-1} = I$, where I is an identity matrix.

9. A 2×2 matrix $\begin{bmatrix} a & b \\ c & d \end{bmatrix}$ has an inverse if $\Delta = ad - bc \neq 0$.

If so, the inverse matrix is

$$\begin{bmatrix} \dfrac{d}{\Delta} & -\dfrac{b}{\Delta} \\[2ex] -\dfrac{c}{\Delta} & \dfrac{a}{\Delta} \end{bmatrix}.$$

10. A system of linear equations can be written in the form $AX = B$, where A is a rectangular matrix of coefficients of the variables, X is a column of variables, and B is a column matrix of the constants from the right side of the system. If the matrix A has an inverse, then the solution of the equation is given by $X = A^{-1}B$.

11. To calculate the inverse of a matrix by the Gauss–Jordan method, append an identity matrix to the right of the original matrix and perform pivots to reduce the original matrix to an identity matrix. The matrix on the right will then be the inverse of the original matrix. (If the original matrix cannot be reduced to an identity matrix, then the original matrix does not have an inverse.)

12. An input–output matrix has rows and columns labeled with the different industries in an economy. The ijth entry of the matrix gives the cost of the input from the industry in row i used in the production of \$1 worth of the output of industry in column j.

13. If A is an input–output matrix and D is a demand matrix giving the dollar values of the outputs from the various industries to be supplied to outside customers, then the matrix $X = (I - A)^{-1}D$ gives the amounts that must be produced by the various industries in order to meet the demand.

REVIEW OF FUNDAMENTAL CONCEPTS

1. What is meant by a solution to a system of linear equations?

2. What is a matrix?

3. State the three elementary row operations on equations or matrices.

4. What does it mean for a system of equations or a matrix to be in diagonal form?

5. What is meant by pivoting a matrix about a nonzero entry?

6. State the Gaussian elimination method for transforming a system of linear equations into diagonal form.

7. What is a row matrix? Column matrix? Square matrix? Identity matrix, I_n?

8. What is meant by a_{ij}, the ijth entry of a matrix?

9. Define the sum and difference of two matrices.

10. Define the product of two matrices.

11. Define the inverse of a matrix, A^{-1}.

12. Give the formula for the inverse of a 2×2 matrix.

13. Explain how to use the inverse of a matrix to solve a system of linear equations.

14. Describe the steps of the Gauss–Jordan method for calculating the inverse of a matrix.

15. What is an input–output matrix and a final-demand matrix?

16. Explain how to solve an input–output analysis problem.

KEY FORMULAS

Inverse of $\begin{bmatrix} a & b \\ c & d \end{bmatrix} = \begin{bmatrix} \dfrac{d}{\Delta} & -\dfrac{b}{\Delta} \\[2ex] -\dfrac{c}{\Delta} & \dfrac{a}{\Delta} \end{bmatrix}$, where $\Delta = ad - bc \neq 0$.

Solution of $AX = B$, where A has an inverse is $X = A^{-1}B$. If A is the input–output matrix for an economy and D is the demand matrix, then the proper production amounts for each sector are given by $X = (I - A)^{-1}D$.

SUPPLEMENTARY EXERCISES

In Exercises 1 and 2, pivot each matrix around the circled element.

1. $\begin{bmatrix} ③ & -6 & 1 \\ 2 & 4 & 6 \end{bmatrix}$

2. $\begin{bmatrix} -5 & -3 & 1 \\ 4 & ② & 0 \\ 0 & 6 & 7 \end{bmatrix}$

In Exercises 3–8, use the Gaussian elimination method to find all solutions of the systems of linear equations.

3. $\begin{cases} \frac{1}{2}x - y = -3 \\ 4x - 5y = -9 \end{cases}$

4. $\begin{cases} 3x \qquad\; + 9z = 42 \\ 2x + y + 6z = 30 \\ -x + 3y - 2z = -20 \end{cases}$

5. $\begin{cases} 3x - 6y + 6z = -5 \\ -2x + 3y - 5z = \frac{7}{3} \\ x + y + 10z = 3 \end{cases}$

6. $\begin{cases} 3x + 6y - 9z = 1 \\ 2x + 4y - 6z = 1 \\ 3x + 4y + 5z = 0 \end{cases}$

7. $\begin{cases} x + 2y - 5z + 3w = 16 \\ -5x - 7y + 13z - 9w = -50 \\ -x + y - 7z + 2w = 9 \\ 3x + 4y - 7z + 6w = 33 \end{cases}$

8. $\begin{cases} 5x - 10y = 5 \\ 3x - 8y = -3 \\ -3x + 7y = 0 \end{cases}$

In Exercises 9 and 10, perform the indicated matrix operations.

9. $\begin{bmatrix} 2 \\ -1 \\ 0 \end{bmatrix} + \begin{bmatrix} 3 \\ 4 \\ 7 \end{bmatrix}$

10. $\begin{bmatrix} 1 & 3 & -2 \\ 4 & 0 & -1 \end{bmatrix} \begin{bmatrix} 3 & 5 \\ 1 & 0 \\ 0 & -6 \end{bmatrix}$

11. Find the inverse of the appropriate matrix, and use it to solve the system of equations

$$\begin{cases} 3x + 2y = 0 \\ 5x + 4y = 2. \end{cases}$$

12. The matrices

$$\begin{bmatrix} 4 & -2 & 3 \\ 8 & -3 & 5 \\ 7 & -2 & 4 \end{bmatrix} \text{ and } \begin{bmatrix} -2 & 2 & -1 \\ 3 & -5 & 4 \\ 5 & -6 & 4 \end{bmatrix}$$

are inverses of each other. Use these matrices to solve the following systems of linear equations.

(a) $\begin{cases} -2x + 2y - z = 1 \\ 3x - 5y + 4z = 0 \\ 5x - 6y + 4z = 3 \end{cases}$

(b) $\begin{cases} 4x - 2y + 3z = 0 \\ 8x - 3y + 5z = -1 \\ 7x - 2y + 4z = 2 \end{cases}$

In Exercises 13 and 14, use the Gauss–Jordan method to calculate the inverses of the following matrices.

13. $\begin{bmatrix} 2 & 6 \\ 1 & 2 \end{bmatrix}$

14. $\begin{bmatrix} 1 & 1 & 1 \\ 3 & 4 & 3 \\ 1 & 1 & 2 \end{bmatrix}$

15. (*Crop Allocation*) Farmer Brown has 1000 acres of land on which he plans to grow corn, wheat, and soybeans. The cost of cultivating these crops is $160 per acre for corn, $60 per acre for wheat, and $82 per acre for soybeans. If Farmer Brown wishes to use all his available land and his entire budget of $121,000, and if he wishes to plant the same number of acres of corn as wheat and soybeans combined, how many acres of each crop can he grow?

16. (*Equipment Sales*) A company makes backyard playground equipment like swing sets, slides, and jungle gyms. The cost (in dollars) to make a specific style of each piece is given in matrix C. The sales price (in dollars) for each piece is given in matrix S. Two stores sell these specific pieces and matrix A gives the quantities sold during one month.

$$C = \begin{bmatrix} 60 \\ 30 \\ 45 \end{bmatrix} \begin{matrix} \text{Swing set} \\ \text{Slide} \\ \text{Gym} \end{matrix}$$

$$S = \begin{bmatrix} 135 \\ 75 \\ 110 \end{bmatrix} \begin{matrix} \text{Swing set} \\ \text{Slide} \\ \text{Gym} \end{matrix}$$

$$A = \begin{array}{ccc} \text{Swing set} & \text{Slide} & \text{Gym} \\ \begin{bmatrix} 15 & 20 & 8 \\ 10 & 17 & 12 \end{bmatrix} \end{array}$$

Determine and interpret the following matrices.

(a) AC **(b)** AS **(c)** $S - C$ **(d)** $A(S - C)$

17. (*Fruit Baskets*) The produce department at a grocery store makes fruit baskets of apples, bananas, and oranges. The cost for each apple is $.50, for each banana is $.15, and for each orange is $.33. Suppose each basket needs to have 18 pieces of fruit and costs $5 to make. How many apples, bananas, and oranges are in each basket if the number of bananas in each basket is the same as the number of apples and oranges together?

18. (*Nuclear Disarmament*) In an arms race between two superpowers, each nation takes stock of its own and its enemy's nuclear arsenal each year. Each nation has the policy of dismantling a certain percentage of its stockpile each year and adding that same percentage of its competitor's stockpile. Nation A uses 20% and nation B uses 10%. Suppose that the current stockpiles of nations A and B are 10,000 and 7000 weapons, respectively.

(a) What will the stockpiles be in each of the next two years?

(b) What were the stockpiles in each of the preceding two years?

(c) Show that the "missile gap" between the superpowers decreases by 30% each year under these policies. Show that the total number of weapons decreases each year if nation A begins with the most weapons, and the total number of weapons increases if nation B begins with the most weapons.

19. (*Two-Sector Economy*) The economy of a small country can be regarded as consisting of two industries, I and II, whose input–output matrix is

$$A = \begin{bmatrix} .4 & .2 \\ .1 & .3 \end{bmatrix}.$$

How many units should be produced by each industry in order to meet a demand for 8 units from industry I and 12 units from industry II?

20. (*Coins*) (**PE**) Joe has $3.30 in his pocket made up of nickels, dimes, and quarters. There are 30 coins and there are five times as many dimes as quarters. How many quarters does Joe have?

(**a**) 4   (**b**) 5   (**c**) 6   (**d**) 7   (**e**) 8

Conceptual Exercises

21. Identify each statement as true or false.

(**a**) If a system of linear equations has two different solutions, it must have infinitely many solutions.

(**b**) If a system of linear equations has more equations than variables, it cannot have a unique solution.

(**c**) If a system of linear equations has more variables than equations, then it must have infinitely many solutions.

22. Make up a system of two linear equations that has infinitely many solutions. No solution.

23. If the product of two numbers is zero, then one of the numbers must be zero. If the two matrices A and B are such that AB is a matrix of all zeros, must A or B be a matrix of all zeros?

24. Suppose that we try to solve the matrix equation $AX = B$ by using matrix inversion but find that even though the matrix A is a square matrix, it has no inverse. What can be said about the outcome from solving the associated system of linear equations by the Gaussian elimination method?

25. Why should the numbers in a single column of an input–output matrix have a sum that is less than 1?

CHAPTER TEST

1. Solve the following system of linear equations by using the Gaussian elimination method.

$$\begin{cases} x + 2y + z = 5 \\ 2x - y + z = -5 \\ -3x + y - 2z = 8 \end{cases}$$

2. Each of the following is the final matrix of a Gaussian elimination process. Give the solutions to the corresponding systems of linear equations.

(**a**) $\begin{bmatrix} 1 & 0 & 0 & | & 4 \\ 0 & 1 & 0 & | & -3 \\ 0 & 0 & 1 & | & 6 \end{bmatrix}$ (**b**) $\begin{bmatrix} 1 & 0 & 0 & | & 2 \\ 0 & 1 & 0 & | & 3 \\ 0 & 0 & 1 & | & 5 \end{bmatrix}$

(**c**) $\begin{bmatrix} 1 & 0 & 0 & | & 1 \\ 0 & 1 & 0 & | & -3 \\ 0 & 0 & 1 & | & 4 \\ 0 & 0 & 0 & | & 2 \end{bmatrix}$ (**d**) $\begin{bmatrix} 1 & 0 & -1 & | & 2 \\ 0 & 1 & 2 & | & 2 \end{bmatrix}$

(**e**) $\begin{bmatrix} 1 & 2 & 0 & | & 0 \\ 0 & 0 & 1 & | & 0 \\ 0 & 0 & 0 & | & 0 \end{bmatrix}$

3. The following is the general solution to a system of linear equations. Find a *specific* solution.

$$\begin{cases} w = \text{any value} \\ z = \text{any value} \\ y = 2x + w - z \\ x = w + z \end{cases}$$

4. Find three solutions to the system

$$\begin{cases} x + y - 2z = 15 \\ y + 4z = 6. \end{cases}$$

5. Let matrices A, B, and C be defined as shown. Calculate the following matrices whenever they are defined: $A + B$, $A + C$, AB, AC, BC.

$$A = \begin{bmatrix} 2 & 1 & 0 \\ 3 & 2 & 1 \end{bmatrix} \quad B = \begin{bmatrix} -1 & 1 \\ 0 & 1 \\ 1 & -1 \end{bmatrix} \quad C = \begin{bmatrix} 0 & 1 & 2 \\ -1 & 1 & 1 \end{bmatrix}$$

6. (*Resource Allocation*) Making a large decorative plate for retail sale takes $\frac{1}{2}$ hour molding time, 3 hours oven time, and 4 hours painting time. Vases require 1 hour molding time, 2.5 hours oven time, and 3 hours painting time. Bowls require 1 hour molding time, 2 hours oven time, and 2 hours painting time. We intend to produce x plates, y vases, and z bowls.

(**a**) Write three equations expressing molding time (m), oven time (v), and painting time (p) needed in terms of x, y, and z.

(**b**) Write the system as a matrix equation.

(**c**) Determine the number of hours each of molding time, oven time, and painting time to produce 100 plates, 300 vases, and 200 bowls.

7. Find A^{-1} if it exists, where

$$A = \begin{bmatrix} 1 & 2 & 1 \\ 0 & 1 & 1 \\ 1 & -1 & 0 \end{bmatrix}.$$

8. (*College Tuition*) The freshman class at State University has 1500 students. Out-of-state students pay $10,000 tuition, while in-state students are charged $4500 for tuition. The school collected $12,800,000 in tuition. How many students are in-state?

9. (*Three-Sector Economy*) In an economic system, each of three industries depends on the others for raw materials. To make $1 of processed wood requires 30¢ wood, 20¢ steel, and 10¢ coal. To make $1 of steel requires no wood, 30¢ steel, and 20¢ coal. To make $1 of coal requires 10¢ wood, 20¢ steel, and 5¢ coal. To allow for $1 consumption in wood, $4 consumption in steel, and $2 consumption in coal, what levels of production for wood, steel, and coal are required?

Population Dynamics

In 1991 the U.S. Fish and Wildlife Service proposed logging restrictions on nearly 12 million acres of Pacific Northwest forest to help save the endangered northern spotted owl. This decision caused considerable controversy between the logging industry and environmentalists.

Mathematical ecologists created a mathematical model to analyze the population dynamics of the spotted owl.[1] They divided the female owl population into three categories—juvenile (up to one year old), subadult (1 to 2 years old), and adult (over 2 years old). Suppose that in a certain region there are currently 2950 female spotted owls made up of 650 juveniles, 200 subadults, and 2100 adults. The ecologists used matrices to project the changes in the population from year to year. The original numbers can be displayed in the column matrix

$$X_0 = \begin{bmatrix} 650 \\ 200 \\ 2100 \end{bmatrix}_0.$$

The populations after one year are given by the column matrix

$$X_1 = \begin{bmatrix} 693 \\ 117 \\ 2116 \end{bmatrix}_1.$$

The subscript 1 tells us that the matrix gives the population after one year. The column matrices for subsequent years will have subscripts 2, 3, 4, etc.

1. How many subadult females are there after one year?

2. Did the total population of females increase or decrease during the year?

Let A denote the matrix

$$\begin{bmatrix} 0 & 0 & .33 \\ .18 & 0 & 0 \\ 0 & .71 & .94 \end{bmatrix}.$$

According to the mathematical model, subsequent population distributions are generated by multiplication on the left by A. That is,

$$A \cdot X_0 = X_1, \quad A \cdot X_1 = X_2, \quad A \cdot X_2 = X_3, \dots.$$

If the population distribution at any time is given by $\begin{bmatrix} j \\ s \\ a \end{bmatrix}$, then the distribution one year later is $A \cdot \begin{bmatrix} j \\ s \\ a \end{bmatrix}$.

3. Fill in the blanks in the following statements.

(a) Each year _____ juvenile females are born for each 100 adult females.

(b) Each year _____% of the juvenile females survive to become subadults.

(c) Each year _____% of the subadults survive to become adults and _____% of the adults survive.

4. Calculate the column matrices X_2, X_3, X_4, and X_5. (*Note:* With a graphing calculator you can specify matrix **[A]** to be the 3×3 matrix A and specify **[B]** to be the initial 3×1 population matrix X_0. Display **[B]** in the home screen and then enter the command **[A]∗Ans**. Each press of the ENTER key will generate the next population

---

[1]Lamberson, R. H., R. McKelvey, B. R. Noon, and C. Voss, "A Dynamic Analysis of Northern Spotted Owl Viability in a Fragmented Forest Landscape," *Conservation Biology*, Vol. 6, No. 4, December 1992; 505–512.

distribution matrix. *Tip*: Prior to generating the matrices invoke the MODE list and set Float to 0 so that all numbers in the population matrices will be rounded to whole numbers.)

5. Refer to item 4. Is the total female population increasing, decreasing, or neither during the first five years?

6. Explain why calculating `[A]`$^{\wedge}$`50`*`[B]` gives the column matrix for the population distribution after 50 years.

7. Find the projected population matrix after 50 years; after 100 years; after 150 years. Based on this mathematical model, what do you conclude about the prospects for the northern spotted owl?

In this model, the main impediment to the survival of the owl is the number .18 in the second row of matrix A. This number is low for two reasons. The first year of life is precarious for most animals living in the wild. In addition, juvenile owls must eventually leave the nest and establish their own territory. If much of the forest near their original home has been cleared, then they are vulnerable to predators while searching for a new home.

8. Suppose that due to better forest management, the number .18 can be increased to .26. Find the total female population for the first five years under this new assumption. Repeat item 7 and determine if extinction will be avoided under the new assumption.

CHAPTER 3

Linear Programming, A Geometric Approach

LINEAR programming is a method for solving problems in which a linear function (representing cost, profit, distance, weight, or the like) is to be maximized or minimized. Such problems are called *optimization problems*. As we shall see, these problems, when translated into mathematical language, involve systems of linear inequalities, systems of linear equations, and eventually (in Chapter 4) matrices.

3.1 A Linear Programming Problem

Let us begin with a detailed discussion of a typical problem that can be solved by linear programming.

Furniture Manufacturing Problem A furniture manufacturer makes two types of furniture—chairs and sofas. For simplicity, divide the production process into three distinct operations—carpentry, finishing, and upholstery. The amount of labor required for each operation varies. Manufacture of a chair requires 6 hours of carpentry, 1 hour of finishing, and 2 hours of upholstery. Manufacture of a sofa requires 3 hours of carpentry, 1 hour of finishing, and 6 hours of upholstery. Due to limited availability of skilled labor as well as of tools and equipment, the factory has available each day 96 labor-hours for carpentry, 18 labor-hours for finishing, and 72 labor-hours for upholstery. The profit per chair is $80 and the profit per sofa $70. How many chairs and how many sofas should be produced each day to maximize the profit?

It is often helpful to tabulate data given in verbal problems. Our first step, then, is to construct a chart.

	Chair	Sofa	Available time
Carpentry	6 hours	3 hours	96 labor-hours
Finishing	1 hour	1 hour	18 labor-hours
Upholstery	2 hours	6 hours	72 labor-hours
Profit	$80	$70	

The next step is to translate the problem into mathematical language. As you know, this is done by identifying what is unknown and denoting the unknown quantities by letters. Since the problem asks for the optimal number of chairs and sofas to be produced each day, there are two unknowns—the number of chairs produced each day and the number of sofas produced each day. Let x denote the number of chairs and y the number of sofas.

To achieve a large profit, one need only manufacture a large number of chairs and sofas. But due to restricted availability of tools and labor, the factory cannot manufacture an unlimited quantity of furniture. Let us translate the restrictions into mathematical language. Each of the first three rows of the chart gives one restriction. The first row says that the amount of carpentry required is 6 hours for each chair and 3 hours for each sofa. Also, there are available only 96 labor-hours of carpentry per day. We can compute the total number of labor-hours of carpentry required per day to produce x chairs and y sofas as follows:

$\left[\text{number of labor-hours per day of carpentry}\right]$

$\qquad = (\text{number of hours carpentry per chair}) \cdot (\text{number of chairs per day})$

$\qquad\quad + (\text{number of hours carpentry per sofa}) \cdot (\text{number of sofas per day})$

$\qquad = 6 \cdot x + 3 \cdot y.$

The requirement that at most 96 labor-hours of carpentry be used per day means that x and y must satisfy the inequality

$$6x + 3y \leq 96. \tag{1}$$

The second row of the chart gives a restriction imposed by finishing. Since 1 hour of finishing is required for each chair and sofa, and since at most 18 labor-hours of finishing are available per day, the same reasoning as used to derive inequality (1) yields

$$x + y \leq 18. \tag{2}$$

Similarly, the third row of the chart gives the restriction due to upholstery:

$$2x + 6y \leq 72. \tag{3}$$

Further restrictions are given by the fact that the numbers of chairs and sofas must be nonnegative:

$$x \geq 0, \qquad y \geq 0. \tag{4}$$

Now that we have written down the restrictions constraining x and y, let us express the profit (which is to be maximized) in terms of x and y. The profit comes from two sources—chairs and sofas. Therefore,

$$\begin{aligned}
[\text{profit}] &= [\text{profit from chairs}] + [\text{profit from sofas}] \\
&= [\text{profit per chair}] \cdot [\text{number of chairs}] \\
&\quad + [\text{profit per sofa}] \cdot [\text{number of sofas}] \\
&= 80x + 70y.
\end{aligned} \tag{5}$$

Since the objective of the problem is to optimize profit, the expression $80x + 70y$ is called the *objective function*. Combining (1) to (5), we arrive at the following:

Furniture Manufacturing Problem—Mathematical Formulation

Find numbers x and y for which the objective function $80x + 70y$ is as large as possible, and for which all the following inequalities hold simultaneously:

$$\begin{cases}
6x + 3y \leq 96 \\
x + y \leq 18 \\
2x + 6y \leq 72 \\
x \geq 0, \quad y \geq 0.
\end{cases} \tag{6}$$

We may describe this mathematical problem in the following general way. We are required to maximize an expression in a certain number of variables, where the variables are subject to restrictions in the form of one or more inequalities. Problems of this sort are called *mathematical programming problems*. Actually, general mathematical programming problems can be quite involved, and their solutions may require very sophisticated mathematical ideas. However, this is not the case with the furniture manufacturing problem. What makes it a rather simple mathematical programming problem is that both the expression to be maximized and the inequalities are linear. For this reason the furniture manufacturing problem is called a *linear programming problem*. The theory of linear programming is a fairly recent advance in mathematics. It was developed over the last 65 years to deal with the increasingly more complicated problems of our technological society. The 1975 Nobel Prize in economics was awarded to Kantorovich and Koopmans for their pioneering work in the field of linear programming.

We will solve the furniture manufacturing problem in Section 3.2, where we develop a general technique for handling similar linear programming problems. At this point it is worthwhile to attempt to gain some insights into the problem and possible methods for attacking it.

It seems clear that a factory will operate most efficiently when its labor is fully utilized. Let us therefore take the operations one at a time and determine the conditions on x and y that fully utilize the three kinds of labor. The restriction on carpentry asserts that

$$6x + 3y \leq 96.$$

If x and y were chosen so that $6x + 3y$ is actually *less* than 96, we would leave the carpenters idle some of the time, a waste of labor. Thus it would seem reasonable to choose x and y to satisfy

$$6x + 3y = 96.$$

Similarly, to utilize all the finishers' time, x and y must satisfy

$$x + y = 18,$$

and to utilize all the upholsterers' time, we must have

$$2x + 6y = 72.$$

Thus, if no labor is to be wasted, then x and y must satisfy the system of equations

$$\begin{cases} 6x + 3y = 96 \\ x + y = 18 \\ 2x + 6y = 72. \end{cases} \qquad (7)$$

Let us now graph the three equations of (7), which represent the conditions for full utilization of all forms of labor. (See the following chart and Fig. 1.)

Equation	Standard form	x-Intercept	y-Intercept
$6x + 3y = 96$	$y = -2x + 32$	$(16, 0)$	$(0, 32)$
$x + y = 18$	$y = -x + 18$	$(18, 0)$	$(0, 18)$
$2x + 6y = 72$	$y = -\frac{1}{3}x + 12$	$(36, 0)$	$(0, 12)$

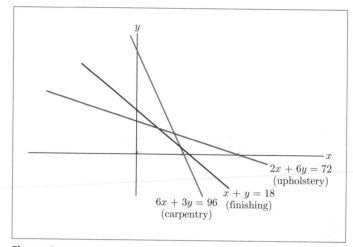

Figure 1

What does Fig. 1 say about the furniture manufacturing problem? Each particular pair of numbers (x, y) is called a *production schedule*. Each of the lines in Fig. 1 gives the production schedules that fully utilize one of the types of labor. Notice that the three lines do not have a common intersection point. This means that there is *no* production schedule that *simultaneously* makes full use of all three types of labor. In any production schedule at least some of the labor-hours must be wasted. This is not a solution to the furniture manufacturing problem, but it is a valuable insight. It says that in the inequalities of (6) not all of the corresponding equations can hold. This suggests that we take a closer look at the system of inequalities.

The standard forms of the inequalities (6) are

$$\begin{cases} y \le -2x + 32 \\ y \le -x + 18 \\ y \le -\frac{1}{3}x + 12 \\ x \ge 0, \quad y \ge 0. \end{cases}$$

By using the techniques of Section 1.2, we arrive at a feasible set for the preceding system of inequalities, as shown in Fig. 2.

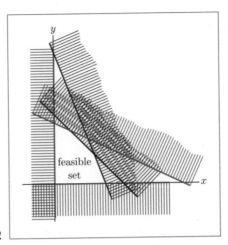

Figure 2

The feasible set for the furniture manufacturing problem is a bounded, five-sided region. The points on and inside the boundary of this feasible set give the production schedules that satisfy all the restrictions. In Section 3.2 we show how to pick out the particular point of the feasible set that corresponds to a maximum profit.

PRACTICE PROBLEMS 3.1

1. Determine whether the following points are in the feasible set of the furniture manufacturing problem: (a) $(10, 9)$; (b) $(14, 4)$.

2. A physical fitness enthusiast decides to devote her exercise time to a combination of jogging and cycling. She wants to earn aerobic points (a measure of the benefit of the exercise to strengthening the heart and lungs) and also to achieve relaxation and enjoyment. She jogs at 6 miles per hour and cycles at 18 miles per hour. An hour of jogging earns 12 aerobic points, and an hour of cycling earns 9 aerobic points. Each week she would like to earn at least 36 aerobic points, cover at least 54 miles, and cycle at least as much as she jogs.

(a) Fill in the following chart.

(b) Let x be the number of hours of jogging and y the number of hours of cycling each week. Referring to the chart, give the inequalities that x and y must satisfy due to miles covered and aerobic points.

(c) Give the inequalities that x and y must satisfy due to her preference for cycling and also due to the fact that x and y cannot be negative.

(d) Express the time required as a linear function of x and y.

(e) Graph the feasible set for the system of linear inequalities.

	One hour of jogging	One hour of cycling	Requirement
Miles covered			
Aerobic points			
Time required			

EXERCISES 3.1

In Exercises 1–4, determine whether the given point is in the feasible set of the furniture manufacturing problem. (The inequalities are as follows.)

$$\begin{cases} 6x + 3y \leq 96; & 2x + 6y \leq 72 \\ x + y \leq 18; & x \geq 0,\ y \geq 0 \end{cases}$$

1. $(8, 7)$

2. $(14, 3)$

3. $(9, 10)$

4. $(16, 0)$

5. (*Shipping*) A truck traveling from New York to Baltimore is to be loaded with two types of cargo. Each crate of cargo A is 4 cubic feet in volume, weighs 100 pounds, and earns \$13 for the driver. Each crate of cargo B is 3 cubic feet in volume, weighs 200 pounds, and earns \$9 for the driver. The truck can carry no more than 300 cubic feet of crates and no more than 10,000 pounds. Also, the number of crates of cargo B must be less than or equal to twice the number of crates of cargo A.

(a) Fill in the following chart.

	A	B	Truck capacity
Volume			
Weight			
Earnings			

(b) Let x be the number of crates of cargo A and y the number of crates of cargo B. Referring to the chart, give the two inequalities that x and y must satisfy because of the truck's capacity for volume and weight.

(c) Give the inequalities that x and y must satisfy because of the last sentence of the problem and also because x and y cannot be negative.

(d) Express the earnings from carrying x crates of cargo A and y crates of cargo B.

(e) Graph the feasible set for the shipping problem.

6. (*Mining*) A coal company owns mines in two different locations. Each day mine 1 produces 4 tons of anthracite (hard coal), 4 tons of ordinary coal, and 7 tons of bituminous (soft) coal. Each day mine 2 produces 10 tons of anthracite, 5 tons of ordinary coal, and 5 tons of bituminous coal. It costs the company \$150 per day to operate mine 1 and \$200 per day to operate mine 2. An order is received for 80 tons of anthracite, 60 tons of ordinary coal, and 75 tons of bituminous coal.

(a) Fill in the following chart.

	Mine 1	Mine 2	Ordered
Anthracite			
Ordinary			
Bituminous			
Daily cost			

(b) Let x be the number of days mine 1 should be operated and y the number of days mine 2 should be operated. Refer to the chart and give three inequalities that x and y must satisfy to fill the order.

(c) Give other requirements that x and y must satisfy.

(d) Find the cost of operating mine 1 for x days and mine 2 for y days.

(e) Graph the feasible set for the mining problem.

7. (*Exam Strategy*) A student is taking an exam consisting of 10 essay questions and 50 short-answer questions. He has 90 minutes to take the exam and knows he cannot possibly answer every question. The essay questions are worth 20 points each and the short-answer questions are worth 5 points each. An essay question takes 10 minutes to answer and a short-answer question takes 2 minutes. The student must do at least 3 essay questions and at least 10 short-answer questions.

(a) Fill in the following chart.

	Essay questions	Short-answer questions	Available
Time to answer			
Quantity			
Required			
Worth			

(b) Let x be the number of essay questions to be answered and y the number of short-answer questions to be answered. Refer to the chart and give the inequality that x and y must satisfy due to the amount of time available.

(c) Give the inequalities that x and y must satisfy because of the numbers of each type of question and also because of the minimum number of each type of question that must be answered.

(d) Give an expression for the score obtained from answering x essay questions and y short-answer questions.

(e) Graph the feasible set for the exam strategy problem.

8. (*Political Campaign—Resource Allocation*) A local politician has budgeted at most $80,000 for her media campaign. She plans to distribute these funds between TV ads and radio ads. Each 1-minute TV ad is expected to be seen by 20,000 viewers, and each 1-minute radio ad is expected to be heard by 4000 listeners. Each minute of TV time costs $8000 and each minute of radio time costs $2000. She has been advised to use at most 90% of her media campaign budget on television ads.

(a) Fill in the following chart. (*Note*: Fill in only the first entry of the last column.)

	One-minute TV ads	One-minute radio ads	Money available
Cost			
Audience reached			

(b) Let x be the number of minutes of TV ads and let y be the number of minutes of radio ads. Refer to the chart and give an inequality that x and y must satisfy due to the amount of money available.

(c) Give the inequality that x must satisfy due to the limitation on the amount of money to be spent on TV ads. Also, give the inequalities that x and y must satisfy because x and y cannot be negative.

(d) Give an expression for the audience reached by x minutes of TV ads and y minutes of radio ads.

(e) Graph the feasible set for the political campaign problem.

9. (*Nutrition—Dairy Cows*) A dairy farmer concludes that his small herd of cows will need at least 4550 pounds of protein in their winter feed, at least 26,880 pounds of total digestible nutrients (TDN), and at least 43,200 international units (IUs) of vitamin A. Each pound of alfalfa hay provides .13 pound of protein, .48 pound of TDN, and 2.16 IUs of vitamin A. Each pound of ground ears of corn supplies .065 pound of protein, .96 pound of TDN, and no vitamin A. Alfalfa hay costs $1 per 100-pound sack. Ground ears corn costs $1.60 per 100-pound sack.

(a) Fill in the following chart.

	Alfalfa	Corn	Requirements
Protein			
TDN			
Vitamin A			
Cost/lb			

(b) Let x be the number of pounds of alfalfa hay and y be the number of pounds of ground ears of corn to be bought. Give the inequalities that x and y must satisfy.

(c) Graph the feasible set for the system of linear inequalities.

(d) Express the cost of buying x pounds of alfalfa hay and y pounds of ground ears of corn.

10. (*Manufacturing—Resource Allocation*) A clothing manufacturer makes denim and hooded fleece jackets. Each denim jacket requires 2 labor-hours for cutting the pieces, 2 labor-hours for sewing, and 1 labor-hour for finishing. Each hooded fleece jacket requires 1 labor-hour for cutting, 4 labor-hours for sewing, and 1 labor-hour for finishing. There are 42 labor-hours for cutting, 90 labor-hours for sewing, and 27 labor-hours for finishing available each day. The profit is $9 per denim jacket and $5 per hooded fleece jacket.

(a) Fill in the following chart.

	Denim	Hooded fleece	Available hours
Cutting			
Sewing			
Finishing			
Profit			

(b) Let x be the number of denim jackets made each day. Let y be the number of hooded fleece jackets made each day. Refer to the chart and give three inequalities that x and y must satisfy due to the available labor-hours.

(c) Give other requirements that x and y must satisfy.

(d) Find the profit in making x denim jackets and y hooded fleece jackets.

(e) Graph the feasible set for the clothing problem.

SOLUTIONS TO PRACTICE PROBLEMS 3.1

1. A point is in the feasible set of a system of inequalities if it satisfies every inequality. Either the original form or the standard form of the inequalities may be used. The original forms of the inequalities of the furniture manufacturing problem are

$$\begin{cases} 6x + 3y \leq 96 \\ x + y \leq 18 \\ 2x + 6y \leq 72 \\ x \geq 0, \quad y \geq 0. \end{cases}$$

(a) $(10, 9)$
$$\begin{cases} 6(10) + 3(9) \leq 96 \\ 10 + 9 \leq 18 \\ 2(10) + 6(9) \leq 72 \\ 10 \geq 0, \quad 9 \geq 0; \end{cases} \quad \begin{cases} 87 \leq 96 & \text{true} \\ 19 \leq 18 & \text{false} \\ 74 \leq 72 & \text{false} \\ 10 \geq 0, \quad 9 \geq 0 & \text{true} \end{cases}$$

(b) $(14, 4)$
$$\begin{cases} 6(14) + 3(4) \leq 96 \\ 14 + 4 \leq 18 \\ 2(14) + 6(4) \leq 72 \\ 14 \geq 0, \quad 4 \geq 0; \end{cases} \quad \begin{cases} 96 \leq 96 & \text{true} \\ 18 \leq 18 & \text{true} \\ 52 \leq 72 & \text{true} \\ 14 \geq 0, \quad 4 \geq 0 & \text{true} \end{cases}$$

Therefore, $(14, 4)$ is in the feasible set and $(10, 9)$ is not.

2. **(a)**

	One hour of jogging	One hour of cycling	Requirement
Miles covered	6	18	54
Aerobic points	12	9	36
Time required	1	1	

(b) Miles covered: $6x + 18y \geq 54$
Aerobic points: $12x + 9y \geq 36$

(c) $y \geq x$, $x \geq 0$. It is not necessary to list $y \geq 0$ since this is automatically assured if the other two inequalities hold.

(d) $x + y$. (An objective of the exercise program might be to minimize $x + y$.)

(e)

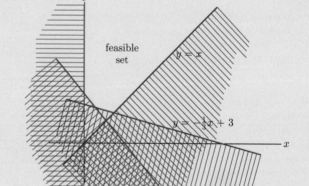

3.2 Linear Programming I

We have shown that the feasible set for the furniture manufacturing problem—that is, the set of points corresponding to production schedules satisfying all five restriction inequalities—consists of the points in the interior and on the boundary of the five-sided region drawn in Fig. 1. For reference, we have labeled each line

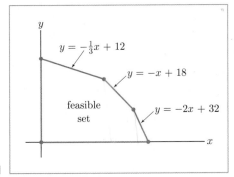

Figure 1

segment with the equation of the line to which it belongs. The line segments intersect in five points, each of which is a corner of the feasible set. Such a corner is called a *vertex*. Somehow, we must pick out of the feasible set an *optimal point*— that is, a point corresponding to a production schedule that yields a maximum profit. To assist us in this task, we have the following result[1]:

Fundamental Theorem of Linear Programming   The maximum (or minimum) value of the objective function is achieved at one of the vertices of the feasible set.

This result does not completely solve the furniture manufacturing problem for us, but it comes close. It tells us that an optimal production schedule (a, b) corresponds to one of the five points labeled A–E in Fig. 2. So to complete the solution of the furniture manufacturing problem, it suffices to find the coordinates of the five points, evaluate the profit at each, and then choose the point corresponding to the maximum profit.

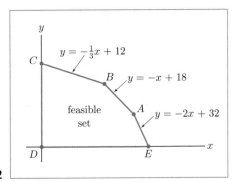

Figure 2

Solution of the Furniture Manufacturing Problem   Let us begin by determining the coordinates of the points A–E in Fig. 2. Remembering that the x-axis has the equation $y = 0$ and the y-axis the equation $x = 0$, we see from Fig. 2 that the coordinates of A–E can be found as intersections of the following lines:

$$A: \quad \begin{cases} y = -x + 18 \\ y = -2x + 32 \end{cases}$$

---

[1] For a verification, see Section 3.3.

$$B: \begin{cases} y = -x + 18 \\ y = -\frac{1}{3}x + 12 \end{cases}$$

$$C: \begin{cases} y = -\frac{1}{3}x + 12 \\ x = 0 \end{cases}$$

$$D: \begin{cases} y = 0 \\ x = 0 \end{cases}$$

$$E: \begin{cases} y = 0 \\ y = -2x + 32. \end{cases}$$

The point D is clearly $(0,0)$, and C is clearly the point $(0,12)$. We obtain A from

$$-x + 18 = -2x + 32$$
$$x = 14$$
$$y = -14 + 18 = 4.$$

Hence $A = (14, 4)$. Similarly, we obtain B from

$$-x + 18 = -\frac{1}{3}x + 12$$
$$-\frac{2}{3}x = -6$$
$$x = 9$$
$$y = -9 + 18 = 9,$$

so $B = (9, 9)$. Finally, E is obtained from

$$0 = -2x + 32$$
$$2x = 32$$
$$x = 16$$
$$y = 0,$$

and thus $E = (16, 0)$. We have displayed the vertices in Fig. 3 and listed them in Table 1. In the second column we have evaluated the profit, which is given by $80x + 70y$, at each of the vertices. Note that the largest profit occurs at the vertex $(14, 4)$, so the solution of the linear programming problem is $x = 14$, $y = 4$. In other words, the factory should produce 14 chairs and 4 sofas each day in order to achieve maximum profit, and the maximum profit is $1400 per day.

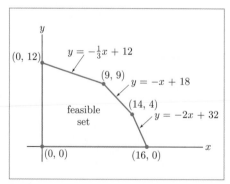

Figure 3

TABLE 1	
Vertex	**Profit $= 80x + 70y$**
$(14, 4)$	$80(14) + 70(4) = 1400$
$(9, 9)$	$80(9) + 70(9) = 1350$
$(0, 12)$	$80(0) + 70(12) = 840$
$(0, 0)$	$80(0) + 70(0) = 0$
$(16, 0)$	$80(16) + 70(0) = 1280$

The furniture manufacturing problem is one particular example of a linear programming problem. Generally, such problems involve finding the values of x and y that maximize (or minimize) a particular linear expression in x and y and where x and y are chosen so as to satisfy one or more restrictions in the form of linear inequalities. The expression that is to be maximized (or minimized) is called the *objective function*. On the basis of our experience with the furniture manufacturing problem, we can summarize the steps to be followed in approaching *any* linear programming problem.

Step 1   Translate the problem into mathematical language.

   A. Organize the data.

   B. Identify the unknown quantities and define corresponding variables.

   C. Translate the restrictions into linear inequalities.

   D. Form the objective function.

Step 2   Graph the feasible set.

   A. Put the inequalities in standard form.

   B. Graph the straight line corresponding to each inequality.

   C. Determine the side of the line belonging to the graph of each inequality. Cross out the other side. The remaining region is the feasible set.

Step 3   Determine the vertices of the feasible set.

Step 4   Evaluate the objective function at each vertex. Determine the optimal point.

Linear programming can be applied to many problems. The Army Corps of Engineers has used linear programming to plan the location of a series of dams so as to maximize the resulting hydroelectric power production. The restrictions were to provide adequate flood control and irrigation. Public transit companies have used linear programming to plan routes and schedule buses in order to maximize services. The restrictions in this case arose from the limitations on labor, equipment, and funding. The petroleum industry uses linear programming in the refining and blending of gasoline. Profit is maximized subject to restrictions on availability of raw materials, refining capacity, and product specifications. Some large advertising firms have used linear programming in media selection. The problem consists of determining how much to spend in each medium in order to maximize the number of consumers reached. The restrictions come from limitations on the budget and the relative costs of different media. Linear programming has also been used by psychologists to design an optimal battery of tests. The problem is to maximize the correlation between test scores and the characteristic that is to be predicted. The restrictions are imposed by the length and cost of the testing.

Linear programming is also used by dietitians in planning meals for large numbers of people. The object is to minimize the cost of the diet, and the restrictions reflect the minimum daily requirements of the various nutrients considered in the diet. The next example is representative of this type of problem. Whereas in actual practice many nutritional factors are considered, we shall simplify the problem by considering only three: protein, calories, and riboflavin.

EXAMPLE 1 **Nutrition—People** Suppose a person decides to make rice and soybeans part of his staple diet. The object is to design a lowest-cost diet that provides certain minimum levels of protein, calories, and vitamin B_2 (riboflavin). Suppose that one cup of uncooked rice costs 21 cents and contains 15 grams of protein, 810 calories, and $\frac{1}{9}$ milligram of riboflavin. On the other hand, one cup of uncooked soybeans costs 14 cents and contains 22.5 grams of protein, 270 calories, and $\frac{1}{3}$ milligram of riboflavin. Suppose that the minimum daily requirements are 90 grams of protein, 1620 calories, and 1 milligram of riboflavin. Design the lowest-cost diet meeting these specifications.

Solution We solve the problem by following steps 1–4. The first step is to translate the problem into mathematical language, and the first part of this step is to organize the data, preferably into a chart (Table 2).

TABLE 2			
	Rice	**Soybeans**	**Required level per day**
Protein (grams/cup)	15	22.5	90
Calories (per cup)	810	270	1620
Riboflavin (milligrams/cup)	$\frac{1}{9}$	$\frac{1}{3}$	1
Cost (cents/cup)	21	14	

Now that we have organized the data, we ask for the unknowns. We wish to know how many cups each of rice and soybeans should comprise the diet, so we identify appropriate variables:

$$x = \text{number of cups of rice per day}$$
$$y = \text{number of cups of soybeans per day.}$$

Next, we obtain the restrictions on the variables. There is one restriction corresponding to each nutrient. That is, there is one restriction for each of the first three rows of the chart. If x cups of rice and y cups of soybeans are consumed, the amount of protein is $15x + 22.5y$ grams. Thus, from the first row of the chart, $15x + 22.5y \geq 90$, a restriction expressing the fact that there must be at least 90 grams of protein per day. Similarly, the restrictions for calories and riboflavin lead to the inequalities $810x + 270y \geq 1620$ and $\frac{1}{9}x + \frac{1}{3}y \geq 1$, respectively. As in the furniture manufacturing problem, x and y cannot be negative, so there are two further restrictions: $x \geq 0$, $y \geq 0$. In all, there are five restrictions:

$$\begin{cases} 15x + 22.5y \geq & 90 \\ 810x + 270y \geq 1620 \\ \frac{1}{9}x + \frac{1}{3}y \geq & 1 \\ x \geq 0, \quad y \geq 0. \end{cases} \qquad (1)$$

Now that we have the restrictions, we form the objective function, which tells us what we want to maximize or minimize. Since we wish to minimize cost, we express cost in terms of x and y. Now x cups of rice cost $21x$ cents and y cups of soybeans cost $14y$ cents, so the objective function is given by

$$[\text{cost}] = 21x + 14y. \qquad (2)$$

The problem can finally be stated in mathematical form: Minimize the objective function (2) subject to the restrictions (1). This completes the first step of the solution process.

The second step requires that we graph each of the inequalities (1). In Table 3 we have summarized all the steps necessary to obtain the information from which to draw the graphs.

TABLE 3

Inequality	Standard form	Line	Intercepts x	Intercepts y	Graph
$15x + 22.5y \geq 90$	$y \geq -\frac{2}{3}x + 4$	$y = -\frac{2}{3}x + 4$	$(6, 0)$	$(0, 4)$	above
$810x + 270y \geq 1620$	$y \geq -3x + 6$	$y = -3x + 6$	$(2, 0)$	$(0, 6)$	above
$\frac{1}{9}x + \frac{1}{3}y \geq 1$	$y \geq -\frac{1}{3}x + 3$	$y = -\frac{1}{3}x + 3$	$(9, 0)$	$(0, 3)$	above
$x \geq 0$	$x \geq 0$	$x = 0$	$(0, 0)$	—	right
$y \geq 0$	$y \geq 0$	$y = 0$	—	$(0, 0)$	above

We have sketched the graphs in Fig. 4. From Fig. 4(b) we see that the feasible set is an unbounded, five-sided region. There are four vertices, two of which are known from Table 3, since they are intercepts of boundary lines. Label the remaining two vertices A and B (Fig. 5).

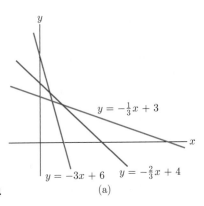

$y = -\frac{1}{3}x + 3$

$y = -3x + 6$ $y = -\frac{2}{3}x + 4$

(a)

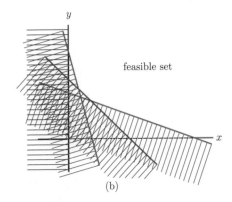

feasible set

(b)

Figure 4

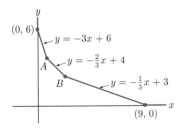

$(0, 6)$

$y = -3x + 6$

$y = -\frac{2}{3}x + 4$

A

B

$y = -\frac{1}{3}x + 3$

$(9, 0)$

Figure 5

The third step of the solution process consists of determining the coordinates of A and B. From Fig. 5, these coordinates can be found by solving the following systems of equations:

$$A: \begin{cases} y = -3x + 6 \\ y = -\frac{2}{3}x + 4 \end{cases} \qquad B: \begin{cases} y = -\frac{2}{3}x + 4 \\ y = -\frac{1}{3}x + 3. \end{cases}$$

To solve the first system, equate the two expressions for y:

$$-\frac{2}{3}x + 4 = -3x + 6$$
$$3x - \frac{2}{3}x = 6 - 4$$
$$\frac{7}{3}x = 2$$
$$x = \frac{6}{7}$$
$$y = -3x + 6 = -3\left(\frac{6}{7}\right) + 6 = \frac{24}{7}$$
$$A = \left(\frac{6}{7}, \frac{24}{7}\right).$$

Similarly, we find B:

$$-\tfrac{2}{3}x + 4 = -\tfrac{1}{3}x + 3$$
$$-\tfrac{1}{3}x = -1$$
$$x = 3$$
$$y = -\tfrac{2}{3}(3) + 4 = 2$$
$$B = (3, 2).$$

TABLE 4	
Vertex	**Cost $= 21x + 14y$**
$(0, 6)$	$21 \cdot 0 + 14 \cdot 6 = 84$
$\left(\tfrac{6}{7}, \tfrac{24}{7}\right)$	$21 \cdot \tfrac{6}{7} + 14 \cdot \tfrac{24}{7} = 66$
$(3, 2)$	$21 \cdot 3 + 14 \cdot 2 = 91$
$(9, 0)$	$21 \cdot 9 + 14 \cdot 0 = 189$

The fourth step consists of evaluating the objective function, in this case $21x + 14y$, at each vertex. From Table 4 we see that the minimum cost is achieved at the vertex $\left(\tfrac{6}{7}, \tfrac{24}{7}\right)$. So the optimal diet—that is, the one that gives nutrients at the desired levels but at minimum cost—is the one that has $\tfrac{6}{7}$ cup of rice per day and $\tfrac{24}{7}$ cups of soybeans per day.

Now Try Exercise 31   ■

▶ *Note*   We have assumed that all linear programming problems have solutions. Although every linear programming problem presented in this text has a solution, there are problems that have no optimal feasible solution. This can happen in two ways. First, there might be no points in the feasible set. Second, feasible solutions to the system of inequalities might exist, but the objective function might not have a maximum (or minimum) value within the feasible set. See Exercises 38 and 39. ◀

GC   Graphing calculators can draw a feasible set, determine the vertices of the feasible set, and evaluate the objective function at the vertices. Figures 6 to 8 show how some of these tasks can be carried out on a TI-83 for the furniture manufacturing problem.

	A	B
1		=6*x + 3*y
2		=x + y
3		=2*x + 6*y
4		=x
5		=y
6		=80*x + 70*y

Figure 9

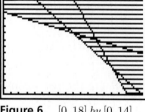

Figure 6. $[0, 18]$ *by* $[0, 14]$

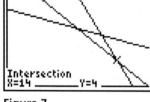

Figure 7

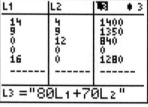

Figure 8

ES   Solver can be used to solve linear programming problems that have a unique solution. Each linear inequality should be written with the variables on the left and a constant on the right. For instance, the linear inequalitites for the furniture manufacturing problem should be written as near the top of page 125. In Fig. 9, the cells A1 and A2 have been named x and y, respectively. The figure shows the formulas that are typed into column B. The formulas entered into the first five cells of column B are the left sides of the inequalities and the formula entered into cell B6 is the objective function.

To obtain the solution of the linear programming problem, click on Solver in the Tools menu and fill in the Solver Parameters window as shown in Fig. 10. The cell containing the objective function is the target cell, and the Max button is selected since the objective function is to be maximized. The rest of the window is completed in much the same way as when solving a system of linear equations. Fig. 11 shows the answer generated after the Solve and then the OK buttons have been clicked.

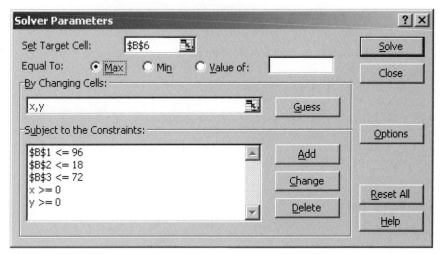

Figure 10

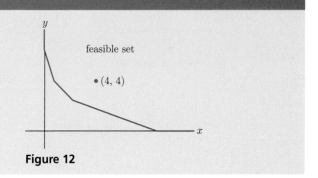

Figure 11

PRACTICE PROBLEMS 3.2

1. The feasible set for the nutrition problem is shown in Fig. 12. The cost is $21x + 14y$. *Without* using the fundamental theorem of linear programming, explain why the cost could not possibly be minimized at the point $(4, 4)$.

2. Rework the nutrition problem assuming that the cost of rice is changed to 7 cents per cup.

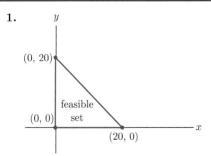

Figure 12

EXERCISES 3.2

For each of the feasible sets in Exercises 1–4, determine x and y so that the objective function $4x + 3y$ is maximized.

1.

2.

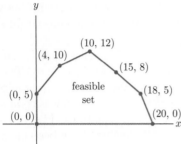

3.

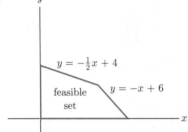

4.

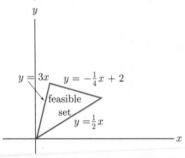

In Exercises 5–8, find the values of x and y that maximize the given objective function for the feasible set in Fig. 13.

5. $x + 2y$ **6.** $x + y$

7. $2x + y$ **8.** $3 - x - y$

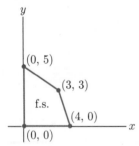

Figure 13   **Figure 14**

In Exercises 9–12, find the values of x and y that minimize the given objective function for the feasible set in Fig. 14.

9. $8x + y$ **10.** $3x + 2y$

11. $2x + 3y$ **12.** $x + 8y$

13. (*Shipping*) Refer to Exercises 3.1, Problem 5. How many crates of each cargo should be shipped in order to satisfy the shipping requirements and yield the

greatest earnings? (See the graph of the feasible set in Fig. 15.)

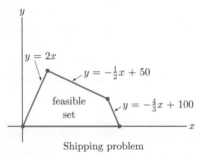

Shipping problem

Figure 15

14. (*Mining*) Refer to Exercises 3.1, Problem 6. Find the number of days that each mine should be operated in order to fill the order at the least cost. (See the graph of the feasible set in Fig. 16.)

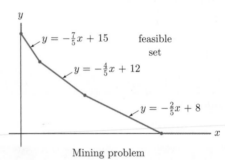

Mining problem

Figure 16

15. (*Exam Strategy*) Refer to Exercises 3.1, Problem 7. How many of each type of question should the student do to maximize the total score? (See the graph of the feasible set in Fig. 17.)

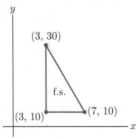

Exam strategy problem

Figure 17

16. (*Political Campaign—Resource Allocation*) Refer to Exercises 3.1, Problem 8. How should the media funds be allocated so as to maximize the total audience? (See the graph of the feasible set in Fig. 18.)

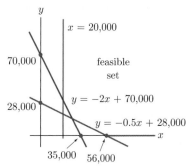

Political campaign problem

Figure 18

17. (*Nutrition—Dairy Cows*) Refer to Exercises 3.1, Problem 9. How many pounds of each food should be purchased in order to meet the nutritional requirements at the least cost? (See the graph of the feasible set in Fig. 19.)

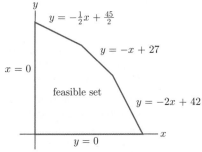

Figure 19

18. (*Manufacturing-Resource Allocation*) Refer to Exercises 3.1, Problem 10. How many of each type of jacket should be made to maximize the profit? (See the graph of the feasible set in Fig. 20.)

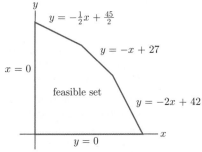

Figure 20

(Furniture Manufacturing) In Exercises 19 and 20, rework the furniture manufacturing problem, where everything is the same except that the profit per chair is changed to the given value. (See Table 1 for vertices.)

19. $150   **20.** $60

21. Minimize the objective function $3x + 4y$ subject to the

constraints

$$\begin{cases} 2x + y \geq 10 \\ x + 2y \geq 14 \\ x \geq 0, \quad y \geq 0. \end{cases}$$

22. Maximize the objective function $7x + 4y$ subject to the constraints

$$\begin{cases} 3x + 2y \leq 36 \\ x + 4y \leq 32 \\ x \geq 0, \quad y \geq 0. \end{cases}$$

23. Maximize the objective function $2x + 5y$ subject to the constraints

$$\begin{cases} x + 2y \leq 20 \\ 3x + 2y \geq 24 \\ x \leq 6 \\ x \geq 0, \quad y \geq 0. \end{cases}$$

24. Minimize the objective function $2x + 3y$ subject to the constraints

$$\begin{cases} x + y \geq 10 \\ y \leq x \\ x \leq 8. \end{cases}$$

25. Maximize the objective function $100x + 150y$ subject to the constraints

$$\begin{cases} x + 3y \leq 120 \\ 35x + 10y \leq 780 \\ x \leq 20 \\ x \geq 0, \quad y \geq 0. \end{cases}$$

26. Minimize the objective function $\frac{1}{2}x + \frac{3}{4}y$ subject to the constraints

$$\begin{cases} 2x + 2y \geq 8 \\ 3x + 5y \geq 16 \\ x \geq 0, \quad y \geq 0. \end{cases}$$

27. Minimize the objective function $7x + 4y$ subject to the constraints

$$\begin{cases} y \geq -2x + 11 \\ y \leq -x + 10 \\ y \leq -\frac{1}{3}x + 6 \\ y \geq -\frac{1}{4}x + 4. \end{cases}$$

28. Maximize the objective function $x + 2y$ subject to the constraints

$$\begin{cases} y \leq -x + 100 \\ y \geq \frac{1}{3}x + 20 \\ y \leq x. \end{cases}$$

29. (*Manufacturing—Resource Allocation*) Infotron Inc. makes electronic hockey and soccer games. Each hockey game requires 2 labor-hours of assembly and 2 labor-hours of testing. Each soccer game requires 3 labor-hours of assembly and 1 labor-hour of testing. Each day there are 42 labor-hours available for assembly and 26 labor-hours available for testing. How many of each game should Infotron produce each day to maximize its total daily output?

30. (*Manufacturing—Production Planning*) An electronics company has factories in Cleveland and Toledo that manufacture three-head and four-head VCRs. Each day the Cleveland factory produces 500 three-head VCRs and 300 four-head VCRs at a cost of $18,000. Each day the Toledo factory produces 300 of each type of VCR at a cost of $15,000. An order is received for 25,000 three-head VCRs and 21,000 four-head VCRs. For how many days should each factory operate to fill the order at the least cost?

31. (*Nutrition—People*) A nutritionist, working for NASA, must meet certain nutritional requirements and yet keep the weight of the food at a minimum. He is considering a combination of two foods, which are packaged in tubes. Each tube of food A contains 4 units of protein, 2 units of carbohydrates, and 2 units of fat and weighs 3 pounds. Each tube of food B contains 3 units of protein, 6 units of carbohydrates, and 1 unit of fat and weighs 2 pounds. The requirement calls for 42 units of protein, 30 units of carbohydrates, and 18 units of fat. How many tubes of each food should be supplied to the astronauts?

32. (*Construction—Resource Allocation*) A contractor builds two types of homes. The first type requires one lot, $12,000 capital, and 150 labor-days to build and is sold for a profit of $2400. The second type of home requires one lot, $32,000 capital, and 200 labor-days to build and is sold for a profit of $3400. The contractor owns 150 lots and has available for the job $2,880,000 capital and 24,000 labor-days. How many homes of each type should she build to realize the greatest profit?

33. (*Packaging—Product Mix*) The Beautiful Day Fruit Juice Company makes two varieties of fruit drink. Each can of Fruit Delight contains 10 ounces of pineapple juice, 3 ounces of orange juice, and 1 ounce of apricot juice and makes a profit of 20 cents. Each can of Heavenly Punch contains 10 ounces of pineapple juice, 2 ounces of orange juice, and 2 ounces of apricot juice and makes a profit of 30 cents. Each week the company has available 9000 ounces of pineapple juice, 2400 ounces of orange juice, and 1400 ounces of apricot juice. How many cans of Fruit Delight and of Heavenly Punch should be produced each week to maximize profits?

34. (*Manufacturing—Resource Allocation*) The Bluejay Lacrosse Stick Company makes two kinds of lacrosse sticks. Type A sticks require 2 labor-hours for cutting, 1 labor-hour for stringing, and 2 labor-hours for finishing and are sold for a profit of $8. Type B sticks require 1 labor-hour for cutting, 3 labor-hours for stringing, and 2 labor-hours for finishing and are sold for a profit of $10. Each day the company has available 120 labor-hours for cutting, 150 labor-hours for stringing, and 140 labor-hours for finishing. How many lacrosse sticks of each kind should be manufactured each day to maximize profits?

35. (*Agriculture—Crop Planning*) A farmer has 100 acres on which to plant oats or corn. Each acre of oats requires $18 capital and 2 hours of labor. Each acre of corn requires $36 capital and 6 hours of labor. Labor costs are $8 per hour. The farmer has $2100 available for capital and $2400 available for labor. If the revenue is $55 from each acre of oats and $125 from each acre of corn, what planting combination will produce the greatest total profit? (Profit here is revenue plus leftover capital and labor cash reserve.) What is the maximum profit?

36. (*Manufacturing—Resource Allocation*) A company makes two items, I_1 and I_2, from three raw materials, M_1, M_2, and M_3. Item I_1 uses 3 ounces of M_1, 2 ounces of M_2, and 2 ounces of M_3. Item I_2 uses 4 ounces of M_1, 1 ounce of M_2, and 3 ounces of M_3. The profit on item I_1 is $8 and on item I_2 is $6. The company has a daily supply of 40 ounces of M_1, 20 ounces of M_2, and 60 ounces of M_3.

   (a) How many of items I_1 and I_2 should be made each day to maximize profit?

   (b) What is the maximum profit?

   (c) How many ounces of each raw material are used?

   (d) If the profit on item I_1 increases to $13, how many of items I_1 and I_2 should be made each day to maximize profit?

37. (*Agriculture—Crop Planning*) Suppose that the farmer of Exercise 35 can allocate the $4500 available for capital and labor however he or she wants.

   (a) Without solving the linear programming problem, explain why the optimal profit cannot be less than what was found in Exercise 35.

   (b) Find the optimal solution in the new situation. Does it provide more profit than in Exercise 35?

38. Consider the following linear programming problem: Maximize $M = 10x + 6y$ subject to the constraints

$$\begin{cases} x + y \geq 6 \\ x \geq 0, \quad y \geq 0. \end{cases}$$

   (a) Sketch the feasible set.

   (b) Determine three points in the feasible set and calculate M at each of them.

   (c) Show that the objective function attains no maximum value for points in the feasible set.

39. Consider the following linear programming problem: Minimize $M = 10x + 6y$ subject to the constraints

$$\begin{cases} x + y \geq 6 \\ 4x + 3y \leq 4 \\ x \geq 0, \quad y \geq 0. \end{cases}$$

(a) Sketch the feasible set for the linear programming problem.

(b) Determine a point of the feasible set.

Exercises 40 and 41 require a spreadsheet.

40. Use an electronic spreadsheet such as Excel to solve Exercise 21.

41. Use an electronic spreadsheet such as Excel to solve Exercise 22.

SOLUTIONS TO PRACTICE PROBLEMS 3.2

1. The point P in Fig. 21 has a smaller value of x and a smaller value of y than $(4, 4)$ and is still in the feasible set. It therefore corresponds to a lower cost than $(4, 4)$ and still meets the requirements. We conclude that no interior point of the feasible set could possibly be an optimal point. This geometric argument indicates that an optimal point might be one that juts out far—that is, a vertex.

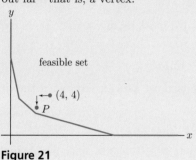

Figure 21

2. The system of linear inequalities, feasible set, and vertices will all be the same as before. Only the objective function changes. The new objective function is $7x + 14y$. The minimum cost occurs when using 3 cups of rice and 2 cups of soybeans.

Vertex	Cost $= 7x + 14y$
$(0, 6)$	84
$\left(\frac{6}{7}, \frac{24}{7}\right)$	54
$(3, 2)$	49
$(9, 0)$	63

3.3 Linear Programming II

In this section we apply the technique of linear programming to the design of a portfolio for a retirement fund and to the transportation of goods from warehouses to retail outlets. The significant new feature of each of these problems is that, on the surface, they appear to involve more than two variables. However, they can be translated into mathematical language so that only two variables are required.

EXAMPLE 1 **Investment planning** A pension fund has $30 million to invest. The money is to be divided among Treasury notes, bonds, and stocks. The rules for administration of the fund require that at least $3 million be invested in each type of investment, that at least half the money be invested in Treasury notes and bonds, and that the amount invested in bonds not exceed twice the amount invested in Treasury notes. The annual yields for the various investments are 7% for Treasury notes, 8% for bonds, and 9% for stocks. How should the money be allocated among the various investments to produce the largest return?

Solution First, let us agree that all numbers stand for millions. That is, we write 30 to stand for 30 million dollars. This will save us from writing many zeros. In examining the problem, we find that very little organization needs to be done. The rules for administration of the fund are written in a form from which inequalities can be read right off. Let us just summarize the remaining data in the first row of a chart (Table 1).

TABLE 1			
	Treasury notes	**Bonds**	**Stocks**
Yield	.07	.08	.09
Variables	x	y	$30 - (x+y)$

There appear to be three variables—the amounts to be invested in each of the three categories. However, since the three investments must total 30, we need only two variables. Let $x =$ the amount to be invested in Treasury notes and $y =$ the amount to be invested in bonds. Then the amount to be invested in stocks is $30 - (x + y)$. We have displayed the variables in Table 1.

Now for the restrictions. Since at least 3 (million dollars) must be invested in each category, we have the three inequalities

$$x \geq 3$$
$$y \geq 3$$
$$30 - (x + y) \geq 3.$$

Moreover, since at least half the money, or 15, must be invested in Treasury notes and bonds, we must have

$$x + y \geq 15.$$

Finally, since the amount invested in bonds must not exceed twice the amount invested in Treasury notes, we must have

$$y \leq 2x.$$

(In this example we do not need to state that $x \geq 0$, $y \geq 0$, since we have already required that they be greater than or equal to 3.) Thus there are five restriction inequalities:

$$\begin{cases} x \geq 3, \quad y \geq 3 \\ 30 - (x + y) \geq \quad 3 \\ \qquad\qquad x + y \geq 15 \\ \qquad\qquad\quad y \leq 2x. \end{cases} \tag{1}$$

Next, we form the objective function, which in this case equals the total return on the investment. Since x dollars is invested at 7%, y dollars at 8%, and $30-(x+y)$ dollars at 9%, the total return is

$$[\text{return}] = .07x + .08y + .09[30 - (x + y)]$$
$$= .07x + .08y + 2.7 - .09x - .09y$$
$$= 2.7 - .02x - .01y. \tag{2}$$

So the mathematical statement of the problem is: Maximize the objective function (2) subject to the restrictions (1).

The next step of the solution is to graph the inequalities (1). The necessary information is tabulated in Table 2.

One point about the chart is worth noting: It contains enough data to graph each of the lines, with the exception of $y = 2x$. The reason is that the x- and y-intercepts of this line are the same, $(0,0)$. So to graph $y = 2x$, we must find an additional point on the line. For example, if we set $x = 2$, then $y = 4$, so $(2, 4)$ is on the line. In Fig. 1(a) we have drawn the various lines, and in Fig. 1(b) we

TABLE 2

Inequality	Standard form	Line	Intercepts x	y	Graph
$x \geq 3$	$x \geq 3$	$x = 3$	$(3,0)$	—	Right of line
$y \geq 3$	$y \geq 3$	$y = 3$	—	$(0,3)$	Above line
$30 - (x+y) \geq 3$	$y \leq -x + 27$	$y = -x + 27$	$(27,0)$	$(0,27)$	Below line
$x + y \geq 15$	$y \geq -x + 15$	$y = -x + 15$	$(15,0)$	$(0,15)$	Above line
$y \leq 2x$	$y \leq 2x$	$y = 2x$	$(0,0)$	$(0,0)$	Below line

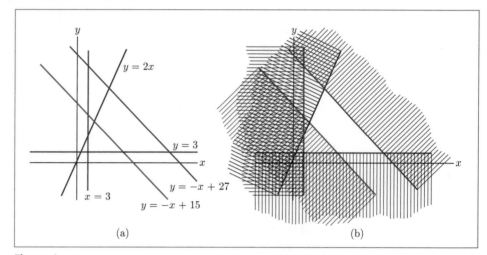

(a)   (b)

Figure 1

have crossed out the appropriate regions to produce the graph of the system. The feasible set, as well as the equations of the various lines that make up its boundary, are shown in Fig. 2. From Fig. 2 we find the pairs of equations that determine each of the vertices A–D. This is the third step of the solution procedure.

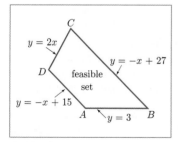

Figure 2

$$A: \begin{cases} y = 3 \\ y = -x + 15 \end{cases} \qquad B: \begin{cases} y = 3 \\ y = -x + 27 \end{cases}$$

$$C: \begin{cases} y = -x + 27 \\ y = 2x \end{cases} \qquad D: \begin{cases} y = 2x \\ y = -x + 15 \end{cases}$$

A and B are the easiest to determine. To find A, we must solve

$$3 = -x + 15$$
$$x = 12$$
$$y = 3$$
$$A = (12, 3).$$

Similarly, $B = (24, 3)$. To find C, we must solve

$$2x = -x + 27$$
$$3x = 27$$
$$x = 9$$
$$y = 2(9) = 18$$
$$C = (9, 18).$$

Similarly, $D = (5, 10)$.

Finally, we list the four vertices and evaluate the objective function (2) at each one. The results are summarized in Table 3.

TABLE 3	
Vertex	**Return = 2.7 − .02x − .01y**
$(5, 10)$	$2.7 - .02(5) - .01(10) = \$2.5$ million
$(9, 18)$	$2.7 - .02(9) - .01(18) = \$2.34$ million
$(24, 3)$	$2.7 - .02(24) - .01(3) = \$2.19$ million
$(12, 3)$	$2.7 - .02(12) - .01(3) = \$2.43$ million

It is clear that the largest return occurs when $x = 5$, $y = 10$. In other words, $5 million should be invested in Treasury notes, $10 million in bonds, and $30 - (x + y) = 30 - (5 + 10) = \$15$ million in stocks.

Now Try Exercise 23   ■

Linear programming is of use not only in analyzing investments but in the fields of transportation and shipping. It is often used to plan routes, determine locations of warehouses, and develop efficient procedures for getting goods to people. Many linear programming problems of this variety can be formulated as *transportation problems*. A typical transportation problem involves determining the least-cost scheme for delivering a commodity stocked in a number of different warehouses to a number of different locations, say retail stores. Of course, in practical applications, it is necessary to consider problems involving perhaps dozens or even hundreds of warehouses, and possibly just as many delivery locations. For problems on such a grand scale, the methods developed so far are inadequate. For one thing, the number of variables required is usually more than two. We must wait until Chapter 4 for methods that apply to such problems. However, the next example gives an instance of a transportation problem that does not involve too many warehouses or too many delivery points. It gives the flavor of general transportation problems.

EXAMPLE 2 **Transportation—Shipping** Suppose that a Maryland TV dealer has stores in Annapolis and Rockville and warehouses in College Park and Baltimore. The cost of shipping a TV set from College Park to Annapolis is $6; from College Park to Rockville, $3; from Baltimore to Annapolis, $9; and from Baltimore to Rockville, $5. Suppose that the Annapolis store orders 25 TV sets and the Rockville store 30. Suppose further that the College Park warehouse has a stock of 45 sets and the Baltimore warehouse 40. What is the most economical way to supply the requested TV sets to the two stores?

Solution The first step in solving a linear programming problem is to translate it into mathematical language. And the first part of this step is to organize the information given, preferably in the form of a chart. In this case, since the problem is geographic, we draw a schematic diagram, as in Fig. 3, which shows the flow of goods between warehouses and retail stores. By each route, we have written the cost. Below each warehouse we have written down its stock and below each retail store the number of TV sets it ordered.

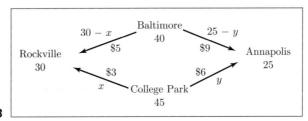

Figure 3

Next, let us determine the variables. It appears initially that four variables are required, namely the number of TV sets to be shipped over each route. However, a closer look shows that only two variables are required. For if x denotes the number of TV sets to be shipped from College Park to Rockville, then since Rockville ordered 30 sets, the number shipped from Baltimore to Rockville is $30 - x$. Similarly, if y denotes the number of sets shipped from College Park to Annapolis, then the number shipped from Baltimore to Annapolis is $25 - y$. We have written the appropriate shipment sizes beside the various routes in Fig. 3.

As the third part of the translation process, let us write down the restrictions on the variables. Basically, there are two kinds of restrictions: None of x, y, $30 - x$, $25 - y$ can be negative, and a warehouse cannot ship more TV sets than it has in stock. Referring to Fig. 3, we see that College Park ships $x + y$ sets, so that $x + y \leq 45$. Similarly, Baltimore ships $(30 - x) + (25 - y)$ sets, so that $(30 - x) + (25 - y) \leq 40$. Simplifying this inequality, we get

$$
\begin{aligned}
55 - x - y &\leq 40 \\
-x - y &\leq -15 \\
x + y &\geq 15.
\end{aligned}
$$

The inequality $30 - x \geq 0$ can be simplified to $x \leq 30$, and the inequality $25 - y \geq 0$ can be written $y \leq 25$. So our restriction inequalities are these:

$$
\begin{cases}
x \geq 0, & y \geq 0 \\
x \leq 30, & y \leq 25 \\
x + y \geq 15 \\
x + y \leq 45.
\end{cases}
\tag{3}
$$

The final step in the translation process is to form the objective function. In this problem we are attempting to minimize cost, so the objective function must express the cost in terms of x and y. Refer again to Fig. 3. There are x sets going from College Park to Rockville, and each costs \$3 to transport, so the cost of delivering these x sets is $3x$. Similarly, the costs of making the other deliveries are $6y$, $5(30 - x)$, and $9(25 - y)$. Thus the objective function is

$$
\begin{aligned}
[\text{cost}] &= 3x + 6y + 5(30 - x) + 9(25 - y) \\
&= 3x + 6y + 150 - 5x + 225 - 9y \\
&= 375 - 2x - 3y.
\end{aligned}
\tag{4}
$$

So the mathematical problem we must solve is as follows: Find x and y that minimize the objective function (4) and satisfy the restrictions (3).

To solve the mathematical problem, we must graph the system of inequalities in (3). Four of the inequalities have graphs determined by horizontal or vertical lines. The only inequalities involving any work are $x + y \geq 15$ and $x + y \leq 45$. And even these are very easy to graph. The result is the graph in Fig. 4.

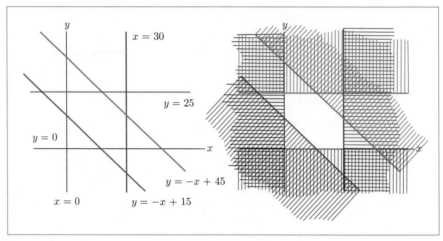

Figure 4

In Fig. 5 we have drawn the feasible set and have labeled each boundary line with its equation. The vertices A–F are now simple to determine. First, A and F are the intercepts of the line $y = -x + 15$. Therefore, $A = (15, 0)$ and $F = (0, 15)$. Since B is the x-intercept of the line $x = 30$, we have $B = (30, 0)$. Similarly, $E = (0, 25)$. Since C is on the line $x = 30$, its x-coordinate is 30. Its y-coordinate is $y = -30 + 45 = 15$, so $C = (30, 15)$. Similarly, since D has y-coordinate 25, its x-coordinate is given by $25 = -x + 45$ or $x = 20$. Thus $D = (20, 25)$.

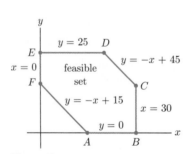

Figure 5

TABLE 4	
Vertex	**Cost = 375 − 2x − 3y**
$(0, 25)$	300
$(0, 15)$	330
$(15, 0)$	345
$(30, 0)$	315
$(30, 15)$	270
$(20, 25)$	260

We have listed in Table 4 the vertices A–F as well as the cost corresponding to each one. The minimum cost of \$260 occurs at the vertex $(20, 25)$. So $x = 20$, $y = 25$ yields the minimum of the objective function. In other words, 20 TV sets should be shipped from College Park to Rockville and 25 from College Park to Annapolis, $30 - x = 10$ from Baltimore to Rockville, and $25 - y = 0$ from Baltimore to Annapolis. This solves our problem.

Now Try Exercise 25   ∎

▶ **Remarks Concerning** Note that the highest-cost route is the one from Baltimore to Annapolis. The
the Transportation solution we have obtained eliminates any shipments over this route. One might
Problem infer from this that one should always avoid the most expensive route. But this is

not correct reasoning. To see why, reconsider Example 2, except change the cost of transporting a TV set from Baltimore to Annapolis from $9 to $7. The Baltimore–Annapolis route is still the most expensive. However, in this case the minimum cost is not obtained by eliminating the Baltimore–Annapolis route. For the revised problem, the linear inequalities stay the same. So the feasible set and the vertices remain the same. The only change is in the objective function, which now is given by

$$[\text{cost}] = 3x + 6y + 5(30 - x) + 7(25 - y) = 325 - 2x - y.$$

The costs at the various vertices are given in Table 5. So the minimum cost of $250 is achieved when $x = 30$, $y = 15$, $30 - x = 0$, and $25 - y = 10$. Note that 10 sets are being shipped from Baltimore to Annapolis, even though this is the most expensive route.

TABLE 5

Vertex	Cost = $325 - 2x - y$
$(0, 25)$	300
$(0, 15)$	310
$(15, 0)$	295
$(30, 0)$	265
$(30, 15)$	250
$(20, 25)$	260

It is even possible for the cost function to be optimized simultaneously at two different vertices. For example, if the cost from Baltimore to Annapolis is $8 and all other data are the same as in Example 2, then the optimal cost is $260 and is achieved at both vertices $(30, 15)$ and $(20, 25)$.   ◀

Verification of the Fundamental Theorem   The fundamental theorem of linear programming asserts that the objective function assumes its optimal value at a vertex of the feasible set. Let us verify this fact. For simplicity, we give the argument only in a special case, namely for the furniture manufacturing problem. However, this is for convenience of exposition only. The same argument given in the following example may be used to prove the fundamental theorem in general. Our argument relies on the parallel property for straight lines, which asserts that parallel lines have the same slope.

EXAMPLE 3   **Proof of the Fundamental Theorem of Linear Programming**   Prove the fundamental theorem of linear programming in the special case of the furniture manufacturing problem.

Solution   The profit derived from producing x chairs and y sofas is $80x + 70y$ dollars. Let us examine all those production schedules having a given profit. As an example, consider a profit of $2800. Then x and y must satisfy $80x + 70y = 2800$. That is, (x, y) must lie on the line whose equation is $80x + 70y = 2800$, or in standard form, $y = -\frac{8}{7}x + 40$. The slope of this line is $-\frac{8}{7}$ and its y-intercept is $(0, 40)$. We have drawn this line in Fig. 6(a), in which we have also drawn the feasible set for the furniture manufacturing problem. Note two fundamental facts: (1) Every production schedule on the line corresponds to a profit of $2800. (2) The line lies above the feasible set. In particular, no production schedule on the line satisfies all

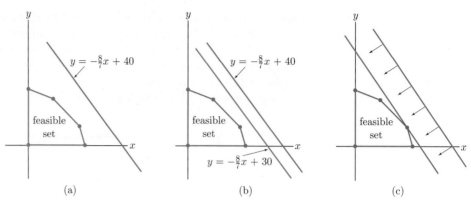

Figure 6   (a)   (b)   (c)

the restrictions of the problem. The difficulty is that $2800 is too high a profit for which to ask.

So now lower the profit, say, to $2100. In this case the production schedule (x, y) lies on the line $80x + 70y = 2100$, or in standard form, $y = -\frac{8}{7}x + 30$. This line is drawn in Fig. 6(b). Note that since both lines have slope $-\frac{8}{7}$, they are parallel by the parallel property. Actually, if we look at the production schedules yielding any fixed profit p, then they will lie along a line of slope $-\frac{8}{7}$, which is parallel to the two lines already drawn. For if the production schedule (x, y) yields a profit p, then

$$80x + 70y = p \quad \text{or} \quad y = -\frac{8}{7}x + \frac{p}{70}.$$

In other words, (x, y) lies on a line of slope $-\frac{8}{7}$ and y-intercept $(0, p/70)$. In particular, all the "lines of constant profit" are parallel to one another. So let us go back to the line of $2800 profit. It does not touch the feasible set. So now lower the profit and therefore translate the line downward parallel to itself. Next lower the profit until we first touch the feasible set. This line now touches the feasible set at a vertex [Fig. 6(c)]. And this vertex corresponds to the optimal production schedule, since any other point of the feasible set lies on a "line of constant profit" corresponding to an even-lower profit. This shows why the fundamental theorem of linear programming is true.

Now Try Exercise 1   ∎

Range of Optimality   The objective function for the furniture manufacturing problem is $80x + 70y$, and the optimal production schedule is $(14, 4)$. The numbers 80 and 70 in the objective function were the profits per chair and sofa, respectively. If the profit per chair is increased to $90 and all other numbers in the problem remain the same, then the optimal production schedule will still be $(14, 4)$. Actually, if

$$70 \leq [\text{profit per chair}] \leq 140,$$

then the production schedule $(14, 4)$ will still provide an optimal solution. The interval $[70, 140]$ is called the *range of optimality* for the profit per chair. (Exercise 34 shows how to use the concept developed in Example 3 to calculate ranges of optimality.) Since $80 - 70 = 10$ and $140 - 80 = 60$, the profit per chair can be decreased by as much as $10 or increased by as much as $60 while still keeping $(14, 4)$ as an optimal production schedule. We say that the *allowable decrease* for the profit per chair is 10 and the *allowable increase* is 60. The *range of optimality* for the profit per sofa is $[40, 80]$. Therefore, the profit per sofa has an allowable decrease of 30 and an allowable increase of 10.

If the objective function is parallel to one of the boundary lines of the feasible set, there will be infinitely many solutions—all points on that boundary line segment provide optimal values for the objective function. At least one such point is a vertex of the feasible set.

EXAMPLE 4   **Transportation—Shipping**   Reconsider the TV shipping problem discussed in Example 2 with the cost of shipping from Baltimore to Annapolis now \$8 and all other data the same as in Example 2. Minimize the shipping costs.

Solution   The cost function to be minimized is now

$$[\text{cost}] = 350 - 2x - 2y.$$

Note that for any fixed value of the cost, the slope of the objective function is -1. The feasible set is identical to that of Example 2 (Fig. 5). The slope of the boundary line $y = -x + 45$ is also -1. A check of the vertices of the feasible set (Table 6) shows that the minimal cost of \$260 is achieved at the two vertices $C = (30, 15)$ and $D = (20, 25)$. In fact, the cost of \$260 is achieved at every point on the boundary line joining these two vertices. Let's verify this in two cases. The points $(25, 20)$ and $(28, 17)$ both lie on the line $y = -x + 45$ and produce a cost of \$260. Figure 7 illustrates the result. ■

TABLE 6	
Vertex	**Cost = $350 - 2x - 2y$**
(0, 25)	300
(0, 15)	320
(15, 0)	320
(30, 0)	290
(30,15)	260
(20,25)	260

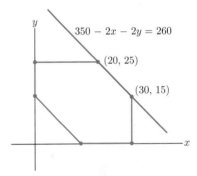

Figure 7

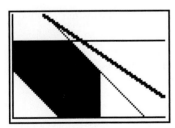

Figure 8

GC   Graphing calculators can draw accurate feasible sets and accurate lines of constant profit (or cost). These precise drawings often permit us to select the optimal vertex by inspection. Figure 8 shows the TI-83 screen for the shipping problem of Example 2. The thick line is the line of constant cost, \$253. As the thick line is lowered, the point $(20, 25)$ is selected. A similar screen can be produced with the TI-89.

ES   Solver can be asked to provide allowable increases and decreases for the coefficients of the objective function along with the solution. After the Solver Parameters window has been filled in, click the Options button to invoke the Solver Options window. In the Solver Options window, click on the box labeled "Assume Linear Model" and then click the OK button to return to the Solver Parameters window. As before, click on the Solve button to invoke the Solver Results window. In this window, click on Sensitivity in the Reports list, and then click on the OK button. The requested information will be contained in a table in a sheet named "Sensitivity Report 1." See Table 7. The sheet contains a second table that is discussed at the end of Section 4.4.

TABLE 7	The first table of the sensitivity sheet					
Adjustable Cells						
Cell	Name	Final Value	Reduced Cost	Objective Coefficient	Allowable Increase	Allowable Decrease
$A$1	x	14	0	80	60	10
$A$2	y	4	0	70	10	30

PRACTICE PROBLEMS 3.3

Problems 1–3 refer to Example 1. Translate the statement into an inequality.

1. The amount to be invested in bonds is at most $5 million more than the amount to be invested in Treasury notes.

2. No more than $25 million should be invested in stocks and bonds.

3. Rework Example 1, assuming that the yield for Treasury notes goes up to 8%.

4. A linear programming problem has objective function $\left[\text{cost}\right] = 5x + 10y$, which is to be minimized. Figure 9 shows the feasible set and the straight line of all combinations of x and y for which $\left[\text{cost}\right] = \$20$.

   (a) Give the linear equation (in standard form) of the line of constant cost c.

   (b) As c increases, does the line of constant cost c move up or down?

   (c) By inspection, find the vertex of the feasible set that gives the optimal solution.

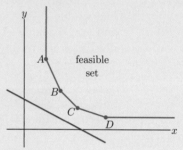

Figure 9

EXERCISES 3.3

1. Figure 10(a) shows the feasible set of the nutrition problem of Section 3.2 and the straight line of all combinations of rice and soybeans for which the cost is 42 cents.

   (a) The objective function is $21x + 14y$. Give the linear equation (in standard form) of the line of constant cost c.

   (b) As c increases, does the line of constant cost move up or down?

   (c) By inspection, find the vertex of the feasible set that gives the optimal solution.

2. Figure 10(b) shows the feasible set of the transportation problem of Example 2 and the straight line of all combinations of shipments for which the transportation cost is $240.

   (a) The objective function is $\left[\text{cost}\right] = 375 - 2x - 3y$. Give the linear equation (in standard form) of the line of constant cost c.

   (b) As c increases, does the line of constant cost move up or down?

   (c) By inspection, find the vertex of the feasible set that gives the optimal solution.

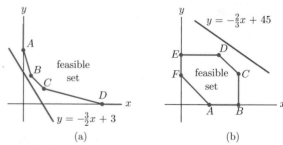

(a)   (b)

Figure 10

Consider the feasible set in Fig. 11(a). In Exercises 3–6, find an objective function of the form $ax + by$ that has its greatest value at the given point.

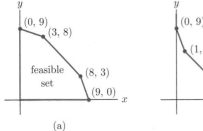

Figure 11

(a)

(b)

3. $(9, 0)$ **4.** $(3, 8)$ **5.** $(8, 3)$ **6.** $(0, 9)$

Consider the feasible set in Fig. 11(b). In Exercises 7–10, find an objective function of the form $ax + by$ that has its least value at the given point.

7. $(9, 0)$ **8.** $(1, 6)$ **9.** $(6, 1)$ **10.** $(0, 9)$

Consider the feasible set in Fig. 12(a), where three of the boundary lines are labeled with their slopes. In Exercises 11–14, find the point at which the given objective function has its greatest value.

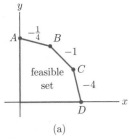

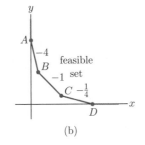

Figure 12

(a)

(b)

11. $3x + 2y$ **12.** $2x + 10y$

13. $10x + 2y$ **14.** $2x + 3y$

Consider the feasible set in Fig. 12(b) on the previous page, where three of the boundary lines are labeled with their slopes. In Exercises 15–18, find the point at which the given objective function has its least value.

15. $2x + 10y$ **16.** $10x + 2y$

17. $2x + 3y$ **18.** $3x + 2y$

19. Consider the feasible set in Fig. 13. For what values of k will the objective function $x + ky$ be maximized at the vertex $(3, 4)$?

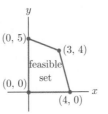

Figure 13

20. Consider the feasible set in Fig. 14. Explain why the objective function $ax + by$, with a and b positive, must have its maximum value at point E.

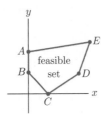

Figure 14

21. (*Nutrition—Animal*) Mr. Smith decides to feed his pet Doberman pinscher a combination of two dog foods. Each can of brand A contains 3 units of protein, 1 unit of carbohydrates, and 2 units of fat and costs 80 cents. Each can of brand B contains 1 unit of protein, 1 unit of carbohydrates, and 6 units of fat and costs 50 cents. Mr. Smith feels that each day his dog should have at least 6 units of protein, 4 units of carbohydrates, and 12 units of fat. How many cans of each dog food should he give to his dog each day to provide the minimum requirements at the least cost?

22. (*Oil Production*) An oil company owns two refineries. Refinery I produces each day 100 barrels of high-grade oil, 200 barrels of medium-grade oil, and 300 barrels of low-grade oil and costs \$10,000 to operate. Refinery II produces each day 200 barrels of high-grade, 100 barrels of medium-grade, and 200 barrels of low-grade oil and costs \$9000 to operate. An order is received for 1000 barrels of high-grade oil, 1000 barrels of medium-grade oil, and 1800 barrels of low-grade oil. How many days should each refinery be operated to fill the order at the least cost?

23. (*Investment Planning*) Mr. Jones has \$9000 to invest in three types of stocks: low-risk, medium-risk, and high-risk. He invests according to three principles. The amount invested in low-risk stocks will be at most \$1000 more than the amount invested in medium-risk stocks. At least \$5000 will be invested in low- and medium-risk stocks. No more than \$7000 will be invested in medium- and high-risk stocks. The expected yields are 6% for low-risk stocks, 7% for medium-risk stocks, and 8% for high-risk stocks. How much money should Mr. Jones invest in each type of stock to maximize his total expected yield?

24. (*Shipping—Product Mix*) A produce dealer in Florida ships oranges, grapefruits, and avocados to New York by truck. Each truckload consists of 100 crates, of which at least 20 crates must be oranges, at least 10 crates must be grapefruits, at least 30 crates must be avocados, and there must be at least as many crates of oranges as grapefruits. The profit per crate is $5 for oranges, $6 for grapefruits, and $4 for avocados. How many crates of each type should be shipped to maximize the profit? [*Hint*: Let x = number of crates of oranges, y = number of crates of grapefruit. Then $100 - x - y$ = number of crates of avocados.]

25. (*Transportation—Shipping*) A foreign-car dealer with warehouses in New York and Baltimore receives orders from dealers in Philadelphia and Trenton. The dealer in Philadelphia needs 4 cars and the dealer in Trenton needs 7. The New York warehouse has 6 cars and the Baltimore warehouse has 8. The cost of shipping cars from Baltimore to Philadelphia is $120 per car, from Baltimore to Trenton $90 per car, from New York to Philadelphia $100 per car, and from New York to Trenton $70 per car. Find the number of cars to be shipped from each warehouse to each dealer to minimize the shipping cost.

26. (*Manufacturing—Production Planning*) An automobile manufacturer has assembly plants in Detroit and Cleveland, each of which can assemble cars and trucks. The Detroit plant can assemble at most 800 vehicles in one day at a cost of $1200 per car and $2100 per truck. The Cleveland plant can assemble at most 500 vehicles in one day at a cost of $1000 per car and $2000 per truck. A rush order is received for 600 cars and 300 trucks. How many vehicles of each type should each plant produce to fill the order at the least cost? (*Hint*: Let x = number of cars to be produced in Detroit, y = number of trucks to be produced in Detroit, $600 - x$ = number of cars to be produced in Cleveland, and $300 - y$ = number of trucks to be produced in Cleveland.)

27. (*Manufacturing—Production Planning*) An oil refinery produces gasoline, jet fuel, and diesel fuel. The profits per gallon from the sale of these fuels are $.15, $.12, and $.10, respectively. The refinery has a contract with an airline to deliver a minimum of 20,000 gallons per day of jet fuel and/or gasoline (or some of each). It has a contract with a trucking firm to deliver a minimum of 50,000 gallons per day of diesel fuel and/or gasoline (or some of each). The refinery can produce 100,000 gallons of fuel per day, distributed among the fuels in any fashion. It wishes to produce at least 5000 gallons per day of each fuel. How many gallons of each should be produced to maximize the profit?

28. (*Manufacturing—Production Planning*) Suppose that a price war reduces the profits of gasoline in Exercise 27 to $.05 per gallon and that the profits on jet fuel and diesel fuel are unchanged. How many gallons of each fuel should now be produced to maximize the profit?

29. (*Shipping—Resource Allocation*) A shipping company is buying new trucks. The high-capacity trucks cost $50,000 and hold 320 cases of merchandise. The low-capacity trucks cost $30,000 and hold 200 cases of merchandise. The company has budgeted $1,080,000 for the new trucks and has a maximum of 30 people qualified to drive the trucks. Due to availability limitations, the company can purchase at most 15 high-capacity trucks. How many of each type of truck should the company purchase to maximize the number of cases of merchandise that can be shipped simultaneously?

30. (*Shipping—Resource Allocation*) Suppose that the shipping company of Exercise 29 needs to buy enough new trucks to be able to ship 11,200 cases of merchandise. Of course, the company is willing to increase its budget.

    (a) How many of each type of truck should the company purchase to minimize cost?

    (b) What if the company hires 23 additional qualified drivers?

31. (*Transportation—Shipping*) A major coffee supplier has warehouses in Seattle and San José. The coffee supplier receives orders from coffee retailers in Salt Lake City and Reno. The retailer in Salt Lake City needs 400 pounds of coffee, and the retailer in Reno needs 350 pounds of coffee. The Seattle warehouse has 700 pounds available, and the warehouse in San José has 500 pounds available. The cost of shipping from Seattle to Salt Lake City is $2.50 per pound, from Seattle to Reno $3 per pound, from San José to Salt Lake City $4 per pound, and from San José to Reno $2 per pound. Find the number of pounds to be shipped from each warehouse to each retailer to minimize the cost.

32. (*Transportation—Shipping*) Suppose that the coffee supplier's costs in Exercise 31 change so that the cost of shipping from Seattle to Salt Lake City is $3 per pound, from Seattle to Reno $3 per pound, from San José to Salt Lake City $3.50 per pound, and from San José to Reno $2.50 per pound. How many pounds should be shipped from each warehouse to each retailer to minimize the cost?

33. (*Packaging—Product Mix*) A pet store sells three different starter kits for 10-gallon aquariums. The Basic I package contains a filter, 2 pounds of gravel, and 1 package of fish food and gives a profit of $7. The Basic II package contains 2 filters, 2 pounds of gravel, and no packages of fish food and gives a profit of $10. The Deluxe package contains a filter, 3 pounds of gravel, and 2 packages of fish food and gives a profit of $13. The store has 54 filters, 100 pounds of gravel, and 53 packages of fish food. If the store makes as many Basic I packages as Basic II packages and Deluxe packages

together, how many of each package should be created to maximize profit?

34. Refer to Fig. 6 in Example 3. As the lines of constant profit were lowered, the final line had slope $-\frac{8}{7}$ and passed through the optimal vertex of the feasible set. Figure 15 shows that as long as the slope of the final line is between -2 and -1, the optimal solution would still be $(14, 4)$.

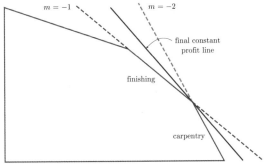

Figure 15

(a) Suppose that the profit per chair is changed from \$80 to \$$A$, but that the profit per sofa remains at \$70. Show that the lines of constant profit will each have slope $-\frac{A}{70}$.

(b) Show that when

$$-2 \leq -\frac{A}{70} \leq -1,$$

then $70 \leq A \leq 140$. Conclude that the interval $[70, 140]$ is the range of optimality for the profit per chair.

(c) Use similar reasoning to show that the range of optimality for the profit per sofa is $[40, 80]$.

Exercise 35 requires a spreadsheet.

35. Use an electronic spreadsheet such as Excel to create a sensitivity report for the nutrition problem of Example 1 of Section 3.2. Use the report to determine the range of optimality for the cost of rice and for the cost of soybeans.

SOLUTIONS TO PRACTICE PROBLEMS 3.3

1. Amount invested in bonds $= y$. Five million dollars more than the amount invested in Treasury notes is $x + 5$. Therefore, $y \leq x + 5$.

2. Amount invested in stocks $= 30 - (x + y)$.
Amount invested in bonds $= y$.
Therefore,

$$30 - (x + y) + y \leq 25$$
$$30 - x \leq 25$$
$$x \geq 5.$$

3. The feasible set stays the same but the return becomes

$$\left[\text{return}\right] = .08x + .08y + .09[30 - (x + y)]$$
$$= .08x + .08y + 2.7 - .09x - .09y$$
$$= 2.7 - .01x - .01y.$$

When the return is evaluated at each of the vertices of the feasible set, the greatest return is achieved at two vertices. Either of these vertices yields an optimal solution.

(x, y)	$2.7 - .01x - .01y$
$(5, 10)$	2.55
$(12, 3)$	2.55
$(24, 3)$	2.43
$(9, 18)$	2.43

4. (a) The values of x and y for which the cost is c dollars satisfy $5x + 10y = c$. The standard form of this linear equation is $y = -\frac{1}{2}x + \frac{c}{10}$.

(b) The line $y = -\frac{1}{2}x + \frac{c}{10}$ has slope $-\frac{1}{2}$ and y-intercept $(0, c/10)$. As c increases, the slope stays the same, but the y-intercept moves up. Therefore, the line moves up.

(c) The line of constant cost \$20 does not contain any points of the feasible set, so such a low cost cannot be achieved. Increase the cost until the line of constant cost just touches the feasible set. As c increases, the line moves up (keeping the same slope) and first touches the feasible set at vertex C. Therefore, taking x and y to be the coordinates of C yields the minimum cost.

CHAPTER 3 SUMMARY

1. A linear programming problem asks us to find the point (or points) in the feasible set of a system of linear inequalities at which the value of a linear expression involving the variables, called the objective function, is either maximized or minimized.

2. The fundamental theorem of linear programming states that the optimal value of the objective function for a linear programming problem occurs at a vertex of the feasible set.

3. To solve a linear programming word problem, assign variables to the unknown quantities, translate the restrictions into a system of linear inequalities involving no more than two variables, form an objective function for the quantity to be optimized, graph the feasible set, determine the vertices of the feasible set, evaluate the objective function at each vertex, and identify the vertex that gives the optimal value.

REVIEW OF FUNDAMENTAL CONCEPTS

1. What is the nature of a linear programming problem?

2. What is the role of the objective function in a linear programming problem?

3. State the fundamental theorem of linear programming.

4. Give the four-step procedure for solving a linear programming problem.

SUPPLEMENTARY EXERCISES

1. (*Travel—Resource Allocation*) Terrapin Airlines wants to fly 1400 members of a ski club to Colorado. The airline owns two types of planes. Type A can carry 50 passengers, requires 3 flight attendants, and costs $14,000 for the trip. Type B can carry 300 passengers, requires 4 flight attendants, and costs $90,000 for the trip. If the airline must use at least as many type A planes as type B and has available only 42 flight attendants, how many planes of each type should be used to minimize the cost for the trip?

2. (*Nutrition—People*) A nutritionist is designing a new breakfast cereal using wheat germ and enriched oat flour as the basic ingredients. Each ounce of wheat germ contains 2 milligrams of niacin, 3 milligrams of iron, and .5 milligram of thiamin and costs 3 cents. Each ounce of enriched oat flour contains 3 milligrams of niacin, 3 milligrams of iron, and .25 milligram of thiamin and costs 4 cents. The nutritionist wants each serving of the cereal to have at least 7 milligrams of niacin, 9 milligrams of iron, and 1 milligram of thiamin. How many ounces of wheat germ and how many ounces of enriched oat flour should be used in each serving to meet the nutritional requirements at the least cost?

3. (*Manufacturing—Resource Allocation*) An automobile manufacturer makes hardtops and sports cars. Each hardtop requires 8 labor-hours to assemble, 2 labor-hours to paint, and 2 labor-hours to upholster and is sold for a profit of $90. Each sports car requires 18 labor-hours to assemble, 2 labor-hours to paint, and 1 labor-hour to upholster and is sold for a profit of $100. During each day 360 labor-hours are available to as-semble, 50 labor-hours to paint, and 40 labor-hours to upholster automobiles. How many hardtops and sports cars should be produced each day to maximize the profit?

4. (*Packaging—Product Mix*) A confectioner makes two raisin–nut mixtures. A box of mixture A contains 6 ounces of peanuts, 1 ounce of raisins, and 4 ounces of cashews and sells for $2.74. A box of mixture B contains 12 ounces of peanuts, 3 ounces of raisins, and 2 ounces of cashews and sells for $4.24. He has available 5400 ounces of peanuts, 1200 ounces of raisins, and 2400 ounces of cashews. How many boxes of each mixture should he make to maximize revenue?

5. (*Publishing—Product Mix*) A textbook publisher puts out 72 new books each year, which are classified as elementary, intermediate, and advanced. The company's policy for new books is to publish at least four advanced books, at least three times as many elementary books as intermediate books, and at least twice as many intermediate books as advanced books. On the average, the annual profits are $8000 for each elementary book, $7000 for each intermediate book, and $1000 for each advanced book. How many new books of each type should be published to maximize the annual profit while conforming to company policy?

6. (*Shipping—Resource Allocation*) A computer company has two manufacturing plants, one in Rochester and one in Queens. Transporting a computer from Rochester to the retail outlet takes 15 hours and costs $15, while transporting a computer from Queens to

the retail outlet takes 20 hours and costs $30. The profit on each computer manufactured in Rochester is $40, and the profit on each computer manufactured in Queens is $30. The Rochester plant has 80 computers available, and the Queens plant has 120 computers available. If there are 2100 hours and $3000 allotted for transporting the computers, how many computers should be sent to the retail outlet from each of the two plants to maximize the company's profits?

7. (*Transportation—Shipping*) An appliance company has two warehouses and two retail outlets. Warehouse A has 400 refrigerators, and warehouse B has 300 refrigerators. Outlet I needs 200 refrigerators, and outlet II needs 300 refrigerators. It costs $36 to ship a refrigerator from warehouse A to outlet I and $30 to ship a refrigerator from warehouse A to outlet II. It costs $30 to ship a refrigerator from warehouse B to outlet I and $25 to ship a refrigerator from warehouse B to outlet II. How should the company ship the refrigerators to minimize the cost?

8. (*Investment Planning*) Portia has $10,000 to invest. She is considering a certificate of deposit (CD) that is expected to yield 5%, a mutual fund expected to yield 7%, and stocks expected to yield 9%. The amount invested in the mutual fund can be no more than the amount invested in the CD and stocks together. The

amount in the mutual fund and stocks must be no more than $8000. How much should Portia invest in each investment vehicle to maximize her total expected yield?

Conceptual Exercises

9. Suppose a constraint is added to a cost minimization problem. Is it possible for the new optimal cost to be greater than the original optimal cost? Is it possible for the new optimal cost to be less than the original optimal cost?

10. Suppose a constraint is removed from a profit maximization problem. Is it possible for the new optimal profit to be greater than the original optimal profit? Is it possible for the new optimal profit to be less than the original optimal profit?

11. Make up a linear programming problem for which the objective function assumes neither a minimum nor a maximum value.

12. Explain why a linear programming problem will always have a solution if the feasible set is bounded.

13. Suppose the maximum value of an objective function occurs at two vertices. Explain why every point on the line segment between the two vertices yields a maximum value of the objective function.

CHAPTER TEST

1. Describe the four main steps for solving a linear programming problem graphically.

2. Determine the objective function and the constraints for the following linear programming problems. *Do not solve the problems.*

   (a) (*Manufacturing—Resource Allocation*) A video game company makes two types of hockey games at two different manufacturing plants. Plant A can produce 20 type I games and 10 type II games per hour. Plant B can produce 30 type I games and 20 type II games per hour. It costs $70 per hour to operate plant A and $90 per hour to operate Plant B. The company needs 300 type I hockey games and 200 type II hockey games per day. How many hours per day should each plant spend on producing these games in order to meet the company's need and minimize production costs?

   (b) (*Investment Planning*) Mabel plans to invest a total of $500,000 in mutual funds, bonds, and certificates of deposit (CDs). The annual yields for mutual funds, bonds, and CDs are 7%, 6%, and 4.5%, respectively. She wishes to invest at least $150,000 in bonds and no more than $200,000 in CDs. She will invest at least half as much money

in mutual funds as she does in bonds. How should she allocate the $500,000 so as to maximize her total annual yield?

3. (a) Graph the feasible set for the following linear programming problem. Maximize $3x + 2y$ subject to the constraints

$$\begin{cases} x + 2y \geq 10 \\ 2x + y \geq 8 \\ x \geq 0, \quad y \geq 0. \end{cases}$$

   (b) Does the linear programming problem in part (a) have a solution? Explain.

4. Find the coordinates of the vertices of the following feasible set. (See next page.)

$$\begin{cases} x + y \leq 12 \\ 5x + 7y \leq 70 \\ x \leq 9 \\ x \geq 0, \quad y \geq 0 \end{cases}$$

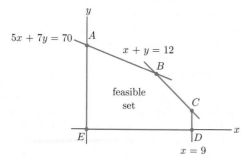

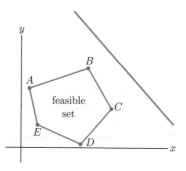

5. For the following feasible set, determine x and y so that the objective function $20x - 10y$ is

  (a) maximized

  (b) minimized.

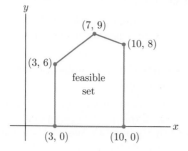

6. Suppose the objective function for a linear programming problem is $[\text{cost}] = 8x + 5y$. The feasible set for the linear programming problem is shown, along with the straight line of all combinations of x and y for which the cost is \$3200.

  (a) Give the linear equation (in standard form) of the line of constant cost shown in the figure.

  (b) Give the linear equation (in standard form) of the line of constant cost c.

  (c) As c decreases, does the line of constant cost move up or down?

  (d) By inspection, determine the vertex of the feasible set that results in the minimum cost.

7. Consider the feasible set shown, where three of the boundary lines are labeled with their slopes. Find the point at which the objective function $5x + 8y$ has its least value.

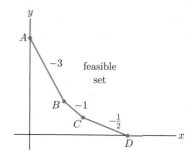

8. (*Resource Allocation*) A small craft shop makes toy stuffed bears and silk flower wreaths. Each bear requires 3 hours of preparation time, 6 hours of assembly time, and 2 hours of finishing time. Each wreath requires 2 hours of preparation time, 5 hours of assembly time, and 3 hours of finishing time. There are 36 hours of preparation time, 78 hours of assembly time, and 42 hours of finishing time available each week. If each bear earns a profit of \$10 and each wreath earns a profit of \$8, how many of each should be crafted to maximize the profit?

CHAPTER 3 PROJECT

Shadow Prices

When a mathematician is presented with a linear programming problem, he or she will not only determine the optimal solution but will also supply what are called *shadow prices* for each constraint. This chapter project develops the concept of a shadow price.

Consider the furniture manufacturing problem. The constraint for finishing is $x + y \leq 18$. The number 18 came from the fact that 18 hours are available for finishing each day. Suppose you could increase the number of hours available for finishing by one hour. The shadow price for the finishing constraint is the maximum price you would be willing to pay for that additional hour.

1. What is the new inequality for the finishing constraint? What is its corresponding linear equation?

2. The figure shows the graph of the original feasible set for the furniture manufacturing problem drawn with a red boundary. The blue line segments show the change in the feasible set when the new finishing constraint is used. Find the coordinates of the points A and B.

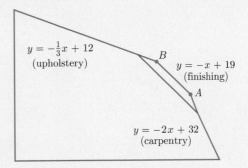

3. Determine the optimal solution for the revised linear programming problem. What is the new maximum profit? By how much was the profit increased due to the additional hour for finishing?

4. What is the shadow price for the finishing constraint?

5. Return to the original furniture manufacturing problem and assume that one additional hour is available for carpentry. Solve the altered problem and determine the shadow price for the carpentry constraint.

6. Use your knowledge of the solution of the original furniture manufacturing problem and the definition of shadow price to explain why the shadow price for the upholstery constraint is 0. (*Hint*: No computation is necessary.)

7. Fill in the blanks in the following sentence. The shadow price associated with a constraint can be interpreted as the change in value of the _____ _____ per unit change of the constraint's right-hand-side resource.

The Simplex Method

IN Chapter 3 we introduced a graphical method for solving linear programming problems. This method, although very simple, is of limited usefulness since it applies only to problems that involve (or can be reduced to) two variables. On the other hand, linear programming applications in business and economics can involve dozens or even hundreds of variables. In this chapter we describe a method for handling such applications. This method, called the *simplex method* (or *simplex algorithm*), was developed by the mathematician George B. Dantzig in the late 1940s and today is the principal method used in solving complex linear programming problems. The simplex method can be used for problems in any number of variables and is easily adapted to computer calculations.

4.1 Slack Variables and the Simplex Tableau

In this section and the next we explain how the simplex method can be used to solve linear programming problems. Let us reconsider the furniture manufacturing problem of Chapter 3. You may recall that the problem is to determine the number of chairs and the number of sofas that should be produced each day in order to maximize the profit. The requirements and availability of resources for carpentry, finishing, and upholstery determine the constraints on the production schedule. Thus, we try to find numbers x and y for which

$$80x + 70y$$

is as large as possible subject to the constraints

$$\begin{cases} 6x + 3y \leq 96 \\ x + y \leq 18 \\ 2x + 6y \leq 72 \\ x \geq 0, \quad y \geq 0. \end{cases}$$

Here x is the number of chairs to be produced each day and y is the number of sofas to be produced each day.

This problem exhibits certain features that make it particularly convenient to work with.

1. The objective function is to be maximized.
2. Each variable is constrained to be greater than or equal to 0.
3. All other constraints are of the form

$$[\text{linear polynomial}]^* \leq [\text{nonnegative constant}].$$

A linear programming problem satisfying these conditions is said to be in *standard form*. Our initial discussion of the simplex method will involve only such problems. Then, in Section 4.3, we consider problems in nonstandard form.

The essential feature of the simplex method is that it provides a systematic method of testing selected vertices of the feasible set until an optimal vertex is reached. The method usually begins at the origin, if it is in the feasible set, and then considers the adjacent vertex that most improves the value of the objective function. This process continues until the optimal vertex is found.

The first step of the simplex method is to convert the given linear programming problem into a system of linear *equations*. To see how this is done, consider the furniture manufacturing problem. It specifies that the variables x and y are subject to the constraint

$$6x + 3y \leq 96.$$

Let us introduce another variable, u, which turns the inequality into an equation:

$$6x + 3y + u = 96.$$

The variable u "takes up the slack" between $6x + 3y$ and 96 and is therefore called a *slack variable*. Moreover, since $6x + 3y$ is at most 96, the variable u must be greater than or equal to 0. In a similar way, the constraint

$$x + y \leq 18$$

can be turned into the equation

$$x + y + v = 18,$$

where v is a slack variable and $v \geq 0$. The third constraint,

$$2x + 6y \leq 72,$$

---

*A linear polynomial is an expression of the form $ax + by + cz + \cdots + dw$, where $a, b, c, \ldots, d$ are specific numbers and $x, y, z, \ldots, w$ are variables. Some examples are $2x - 3y + z$, $x + 2y + 3z - 4w$, and $-x + 3z - 2w$.

becomes the equation

$$2x + 6y + w = 72,$$

where w is also a slack variable and $w \geq 0$. Let us even turn our objective function $80x + 70y$ into an equation by introducing the new variable M defined by $M = 80x + 70y$. Then M is the variable we want to maximize. Moreover, it satisfies the equation

$$-80x - 70y + M = 0.$$

Thus, the furniture manufacturing problem can be restated in terms of a system of linear equations.

Furniture Manufacturing Problem   Among all the solutions of the system of linear equations

$$\left\{ \begin{array}{rcrcrcrcrcr} 6x &+& 3y &+& u & & & & & =& 96 \\ x &+& y & & &+& v & & & =& 18 \\ 2x &+& 6y & & & & &+& w &=& 72 \\ -80x &-& 70y & & & & & &+ M &=& 0, \end{array} \right.$$

find one for which $x \geq 0$, $y \geq 0$, $u \geq 0$, $v \geq 0$, $w \geq 0$, and for which M is as large as possible.

In a similar way, any linear programming problem in standard form can be reduced to that of determining a certain type of solution of a system of linear equations.

EXAMPLE 1   **Using linear equations in a linear programming problem**   Formulate the following linear programming problem in terms of a system of linear equations.

Maximize the objective function $3x + 4y$ subject to the constraints

$$\left\{ \begin{array}{l} x + y \leq 20 \\ x + 2y \leq 25 \\ x \geq 0 \\ y \geq 0. \end{array} \right.$$

Solution   The two constraints $x + y \leq 20$ and $x + 2y \leq 25$ yield the equations

$$\begin{array}{rcrcrcr} x &+& y &+& u & =& 20 \\ x &+& 2y & & &+ v =& 25, \end{array}$$

where u and v are slack variables and $u \geq 0$ and $v \geq 0$. The objective function gives the equation $M = 3x + 4y$, or

$$-3x - 4y + M = 0.$$

So the problem can be reformulated: Among all the solutions of the system of linear equations

$$\left\{ \begin{array}{rcrcrcrcr} x &+& y &+& u & & & =& 20 \\ x &+& 2y & & &+& v & =& 25 \\ -3x &-& 4y & & & & &+ M =& 0, \end{array} \right.$$

find one for which $x \geq 0$, $y \geq 0$, $u \geq 0$, $v \geq 0$, and M is as large as possible.   ∎

Now Try Exercise 1

EXAMPLE 2 **Using linear equations in a linear programming problem** Formulate the following linear programming problem in terms of a system of linear equations.

Maximize the objective function $x + 2y + z$ subject to the constraints

$$\begin{cases} x - y + 2z \leq 10 \\ 2x + y + 3z \leq 12 \\ x \geq 0, \quad y \geq 0, \quad z \geq 0. \end{cases}$$

Solution The two constraints $x - y + 2z \leq 10$ and $2x + y + 3z \leq 12$ yield the equations

$$\begin{aligned} x - y + 2z + u &= 10 \\ 2x + y + 3z + v &= 12. \end{aligned}$$

The objective function yields the equation $M = x + 2y + z$—that is,

$$-x - 2y - z + M = 0.$$

So the problem can be reformulated: Among all solutions of the system of linear equations

$$\begin{cases} x - y + 2z + u = 10 \\ 2x + y + 3z + v = 12 \\ -x - 2y - z + M = 0, \end{cases}$$

find one for which $x \geq 0$, $y \geq 0$, $z \geq 0$, $u \geq 0$, $v \geq 0$, and M is as large as possible.

Now Try Exercise 3

■

We shall now discuss a scheme for solving systems of equations like those just encountered. For the moment, we will not worry about maximizing M or keeping the variables ≥ 0. Rather, let us concentrate on a particular method for determining solutions. In order to be concrete, consider the system of linear equations from the furniture manufacturing problem:

$$\begin{cases} 6x + 3y + u = 96 \\ x + y + v = 18 \\ 2x + 6y + w = 72 \\ -80x - 70y + M = 0. \end{cases} \qquad (1)$$

This system of equations has an infinite number of solutions. We can rewrite the equations as

$$\begin{aligned} u &= 96 - 6x - 3y \\ v &= 18 - x - y \\ w &= 72 - 2x - 6y \\ M &= 80x + 70y. \end{aligned}$$

Given any values of x and y, we can determine corresponding values for u, v, w, and M. For example, if $x = 0$ and $y = 0$, then $u = 96$, $v = 18$, $w = 72$, and $M = 0$. These values for u, v, w, and M are precisely the numbers that appear to the right of the equality signs in our original system of linear equations. Therefore, this particular solution could have been read directly from system (1) without any computation. This method of generating solutions is used in the simplex method, so let us further explore the special properties of the system that allowed us to read off a specific solution so easily.

Note that the system of linear equations has six variables: x, y, u, v, w, and M. These variables can be divided into two groups. Group I consists of those that were set equal to 0, namely, x and y. Group II consists of those whose particular values were read from the right-hand sides of the equations, namely, u, v, w, and M. Note also that the system has a special form that allows the particular values of the group II variables to be read off: Each of the equations involves exactly one of the group II variables, and these variables always appear with coefficient 1. Thus, for example, the first equation involves the group II variable u:

$$6x + 3y + u = 96.$$

Therefore, when all group I variables (x and y) are set equal to 0, only the u-term remains on the left, and the particular value of u can be read off the right-hand side.

The special form of the system can best be described in matrix form. Write the system in the usual way as a matrix, but add column headings corresponding to the variables:

$$
\begin{array}{cccccc}
x & y & u & v & w & M \\
\end{array}
$$
$$
\left[
\begin{array}{cccccc|c}
6 & 3 & 1 & 0 & 0 & 0 & 96 \\
1 & 1 & 0 & 1 & 0 & 0 & 18 \\
2 & 6 & 0 & 0 & 1 & 0 & 72 \\
-80 & -70 & 0 & 0 & 0 & 1 & 0 \\
\end{array}
\right].
$$

Note closely the columns corresponding to the group II variables u, v, w, and M:

$$
\begin{array}{cccccc}
x & y & u & v & w & M \\
\end{array}
$$
$$
\left[
\begin{array}{cccccc|c}
6 & 3 & 1 & 0 & 0 & 0 & 96 \\
1 & 1 & 0 & 1 & 0 & 0 & 18 \\
2 & 6 & 0 & 0 & 1 & 0 & 72 \\
-80 & -70 & 0 & 0 & 0 & 1 & 0 \\
\end{array}
\right].
$$

The presence of these columns gives the system the special form discussed previously. Indeed, the u column asserts that u appears only in the first equation and its coefficient there is 1, and similarly for the v, w, and M columns.

The property of allowing us to read off a particular solution from the right-hand column is shared by all linear systems whose matrices contain the columns

$$
\begin{array}{cccccc}
1 & 0 & 0 & \cdots & 0 \\
0 & 1 & 0 & \cdots & 0 \\
0 & 0 & 1 & \cdots & 0 \\
\vdots & \vdots & \vdots & & \vdots \\
0 & 0 & 0 & \cdots & 1.
\end{array}
$$

(These columns need not appear in exactly the order shown.) The variables corresponding to these columns are called the group II variables. The group I variables consist of all the others. To get one particular solution to the system, set all the group I variables equal to zero and read off the values of the group II variables from the right-hand side of the system. This procedure is illustrated in the following example.

EXAMPLE 3 **Finding group II variables from the matrix form** Determine by inspection one set of solutions to each of these systems of linear equations:

(a) $\begin{cases} x - 5y + u & = 3 \\ -2x + 8y & + v & = 11 \\ -\frac{1}{2}x & + M = 0 \end{cases}$ (b) $\begin{cases} - y + 2u + v & = 12 \\ x + \frac{1}{2}y - 6u & = -1 \\ 3y + 8u & + M = 4. \end{cases}$

Solution (a) The matrix of the system is

$$
\begin{array}{ccccc}
x & y & u & v & M \\
\end{array}
$$
$$
\left[\begin{array}{ccccc|c}
1 & -5 & 1 & 0 & 0 & 3 \\
-2 & 8 & 0 & 1 & 0 & 11 \\
-\frac{1}{2} & 0 & 0 & 0 & 1 & 0
\end{array} \right].
$$

We look for each variable whose column contains one entry of 1 and all the other entries 0.

$$
\begin{array}{ccccc}
x & y & u & v & M \\
\end{array}
$$
$$
\left[\begin{array}{ccccc|c}
1 & -5 & 1 & 0 & 0 & 3 \\
-2 & 8 & 0 & 1 & 0 & 11 \\
-\frac{1}{2} & 0 & 0 & 0 & 1 & 0
\end{array} \right]
$$

The group II variables should be u, v, and M, with x, y as the group I variables. Set all group I variables equal to 0. The corresponding values of the group II variables may then be read off the last column: $u = 3$, $v = 11$, $M = 0$. So one solution of the system is

$$x = 0, \qquad y = 0, \qquad u = 3, \qquad v = 11, \qquad M = 0.$$

(b) The matrix of the system is

$$
\begin{array}{ccccc}
x & y & u & v & M \\
\end{array}
$$
$$
\left[\begin{array}{ccccc|c}
0 & -1 & 2 & 1 & 0 & 12 \\
1 & \frac{1}{2} & -6 & 0 & 0 & -1 \\
0 & 3 & 8 & 0 & 1 & 4
\end{array} \right].
$$

The shaded columns show that the group II variables should be v, x, and M, with y, u as the group I variables. So the corresponding solution is

$$x = -1, \qquad y = 0, \qquad u = 0, \qquad v = 12, \qquad M = 4.$$

Now Try Exercise 9 ■

A *simplex tableau* is a matrix (corresponding to a linear system) in which each of the columns

$$
\begin{array}{cccc}
1 & 0 & \cdots & 0 \\
0 & 1 & \cdots & 0 \\
\vdots & \vdots & & \vdots \\
0 & 0 & \cdots & 1
\end{array}
$$

is present (in some order) to the left of the vertical line. We have seen how to construct a simplex tableau corresponding to a linear programming problem in standard form. From this initial tableau we can read off one particular solution of the linear system by using the method described previously. This particular solution may or may not correspond to the solution of the original optimization problem. If it does not, we replace the initial tableau with another one whose corresponding solution is "closer" to the optimum. How do we replace the initial simplex tableau with another? Just pivot it about a nonzero entry! Indeed, one of the key reasons the simplex method works is that pivoting transforms one simplex tableau into another. Note also that since pivoting consists of elementary row operations, a solution corresponding to a transformed tableau is a solution of the original linear system. The next example illustrates how pivoting transforms a tableau into another one.

EXAMPLE 4 **Finding a feasible solution by pivoting** Consider the simplex tableau obtained from the furniture manufacturing problem:

$$
\begin{array}{cccccc}
x & y & u & v & w & M \\
\end{array}
$$
$$
\left[
\begin{array}{cccccc|c}
⑥ & 3 & 1 & 0 & 0 & 0 & 96 \\
1 & 1 & 0 & 1 & 0 & 0 & 18 \\
2 & 6 & 0 & 0 & 1 & 0 & 72 \\
-80 & -70 & 0 & 0 & 0 & 1 & 0 \\
\end{array}
\right].
$$

(a) Pivot this tableau around the circled entry, 6.

(b) Calculate the particular solution corresponding to the transformed tableau that results from setting the new group I variables equal to 0.

Solution (a) The first step in pivoting is to replace the pivot element 6 by a 1. To do this, multiply the first row of the tableau by $\frac{1}{6}$ to get

$$
\begin{array}{cccccc}
x & y & u & v & w & M \\
\end{array}
$$
$$
\left[
\begin{array}{cccccc|c}
1 & \frac{1}{2} & \frac{1}{6} & 0 & 0 & 0 & 16 \\
1 & 1 & 0 & 1 & 0 & 0 & 18 \\
2 & 6 & 0 & 0 & 1 & 0 & 72 \\
-80 & -70 & 0 & 0 & 0 & 1 & 0 \\
\end{array}
\right].
$$

Next we must replace all nonpivot elements in the first column by zeros. Do this by adding to the second row (-1) times the first row:

$$
\begin{array}{cccccc}
x & y & u & v & w & M \\
\end{array}
$$
$$
\left[
\begin{array}{cccccc|c}
1 & \frac{1}{2} & \frac{1}{6} & 0 & 0 & 0 & 16 \\
0 & \frac{1}{2} & -\frac{1}{6} & 1 & 0 & 0 & 2 \\
2 & 6 & 0 & 0 & 1 & 0 & 72 \\
-80 & -70 & 0 & 0 & 0 & 1 & 0 \\
\end{array}
\right],
$$

and by adding to the third row (-2) times the first row:

$$
\left[
\begin{array}{cccccc|c}
x & y & u & v & w & M & \\
1 & \frac{1}{2} & \frac{1}{6} & 0 & 0 & 0 & 16 \\
0 & \frac{1}{2} & -\frac{1}{6} & 1 & 0 & 0 & 2 \\
0 & 5 & -\frac{1}{3} & 0 & 1 & 0 & 40 \\
-80 & -70 & 0 & 0 & 0 & 1 & 0
\end{array}
\right],
$$

and, finally, by adding to the fourth row 80 times the first row:

$$
\left[
\begin{array}{cccccc|c}
x & y & u & v & w & M & \\
1 & \frac{1}{2} & \frac{1}{6} & 0 & 0 & 0 & 16 \\
0 & \frac{1}{2} & -\frac{1}{6} & 1 & 0 & 0 & 2 \\
0 & 5 & -\frac{1}{3} & 0 & 1 & 0 & 40 \\
0 & -30 & \frac{40}{3} & 0 & 0 & 1 & 1280
\end{array}
\right].
$$

Note that we indeed get a new simplex tableau. The new group II variables are x, v, w, and M. The group I variables are y and u.

$$
\left[
\begin{array}{cccccc|c}
x & y & u & v & w & M & \\
1 & \frac{1}{2} & \frac{1}{6} & 0 & 0 & 0 & 16 \\
0 & \frac{1}{2} & -\frac{1}{6} & 1 & 0 & 0 & 2 \\
0 & 5 & -\frac{1}{3} & 0 & 1 & 0 & 40 \\
0 & -30 & \frac{40}{3} & 0 & 0 & 1 & 1280
\end{array}
\right].
$$

(b) Set the group I variables equal to 0:

$$y = 0, \qquad u = 0.$$

Read off the particular values of the group II variables from the right-hand column:

$$x = 16, \qquad v = 2, \qquad w = 40, \qquad M = 1280.$$

So the particular solution corresponding to the transformed tableau is

$$x = 16, \qquad y = 0, \qquad u = 0, \qquad v = 2, \qquad w = 40, \qquad M = 1280.$$

We see that the simplex tableau leads to the vertex $(16, 0)$ that was listed and tested in the graphical solution of the problem presented in Chapter 3.

Now Try Exercise 13 ■

PRACTICE PROBLEMS 4.1

1. Determine by inspection a particular solution of the following system of linear equations:

$$
\begin{cases}
x + 2y + 3u & = 6 \\
y \quad\;\; + v & = 4 \\
5y + 2u \;\;\; + M = 0.
\end{cases}
$$

2. Pivot the simplex tableau about the circled element.

$$
\left[
\begin{array}{ccccc|c}
2 & 4 & 1 & 0 & 0 & 6 \\
3 & ① & 0 & 1 & 0 & 0 \\
1 & 1 & 0 & 0 & 1 & 1
\end{array}
\right]
$$

EXERCISES 4.1

For each of the following linear programming problems, determine the corresponding linear system and restate the linear programming problem in terms of the linear system.

1. Maximize $8x + 13y$ subject to the constraints

$$\begin{cases} 20x + 30y \le 3500 \\ 50x + 10y \le 5000 \\ x \ge 0 \\ y \ge 0. \end{cases}$$

2. Maximize $x + 15y$ subject to the constraints

$$\begin{cases} 3x + 2y \le 10 \\ x \quad\;\; \le 15 \\ \quad y \le 3 \\ x + y \le 5 \\ x \ge 0 \\ y \ge 0. \end{cases}$$

3. Maximize $x + 2y - 3z$ subject to the constraints

$$\begin{cases} x + y + z \le 100 \\ 3x \quad\;\; + z \le 200 \\ 5x + 10y \quad\;\; \le 100 \\ x \ge 0 \\ y \ge 0 \\ z \ge 0. \end{cases}$$

4. Maximize $2x + y + 50$ subject to the constraints

$$\begin{cases} x + 3y \le 24 \\ \quad y \le 5 \\ x + 7y \le 10 \\ x \ge 0 \\ y \ge 0. \end{cases}$$

5. Maximize $3x + 5y + 12z$ subject to the constraints

$$\begin{cases} 4x + 6y - 7z \le 16 \\ 3x + 2y \quad\;\; \le 11 \\ \quad 9y + 3z \le 21 \\ x \ge 0 \\ y \ge 0 \\ z \ge 0. \end{cases}$$

6–10. For each of the linear programming problems in Exercises 1–5,

(a) Set up the simplex tableau.

(b) Determine the particular solution corresponding to the initial tableau.

In Exercises 11–14, find the particular solution corresponding to the tableau.

11.
$$\begin{array}{ccccc} x & y & u & v & M \\ \left[\begin{array}{ccccc|c} 0 & 2 & 1 & 0 & 0 & 10 \\ 1 & 3 & 0 & 12 & 0 & 15 \\ 0 & -1 & 0 & 17 & 1 & 20 \end{array}\right] \end{array}$$

12.
$$\begin{array}{ccccc} x & y & u & v & M \\ \left[\begin{array}{ccccc|c} 1 & 0 & 3 & 11 & 0 & 6 \\ 0 & 1 & 10 & 17 & 0 & 16 \\ 0 & 0 & 5 & -1 & 1 & 3 \end{array}\right] \end{array}$$

13.
$$\begin{array}{ccccccc} x & y & z & u & v & w & M \\ \left[\begin{array}{ccccccc|c} 0 & 3 & 1 & 0 & 1 & 15 & 0 & 15 \\ 1 & -1 & 0 & 0 & 2 & -5 & 0 & 10 \\ 0 & 2 & 0 & 1 & -5 & 4 & 0 & 23 \\ 0 & 11 & 0 & 0 & 11 & 6 & 1 & -11 \end{array}\right] \end{array}$$

14.
$$\begin{array}{ccccccc} x & y & z & u & v & w & M \\ \left[\begin{array}{ccccccc|c} 6 & 0 & 1 & 0 & 5 & -1 & 0 & \frac{1}{4} \\ 5 & 1 & 0 & 0 & 3 & \frac{1}{3} & 0 & 100 \\ 4 & 0 & 0 & 1 & 8 & \frac{1}{2} & 0 & 11 \\ 2 & 0 & 0 & 0 & 6 & \frac{1}{7} & 1 & -\frac{1}{2} \end{array}\right] \end{array}$$

15. Pivot the simplex tableau

$$\begin{array}{ccccc} x & y & u & v & M \\ \left[\begin{array}{ccccc|c} 2 & 3 & 1 & 0 & 0 & 12 \\ 1 & 1 & 0 & 1 & 0 & 10 \\ -10 & -20 & 0 & 0 & 1 & 0 \end{array}\right] \end{array}$$

about the indicated element and compute the particular solution corresponding to the new tableau.

(a) 2   (b) 3

(c) 1 (second row, first column)

(d) 1 (second row, second column)

16. Pivot the simplex tableau

$$\begin{array}{ccccc} x & y & u & v & M \\ \left[\begin{array}{ccccc|c} 5 & 4 & 1 & 0 & 0 & 100 \\ 10 & 6 & 0 & 1 & 0 & 1200 \\ -1 & 2 & 0 & 0 & 1 & 0 \end{array}\right] \end{array}$$

about the indicated element and compute the solution corresponding to the new tableau.

(a) 5   (b) 4   (c) 10   (d) 6

17. Determine which of the pivot operations in Exercise 15 increases M the most.

18. Determine which of the pivot operations in Exercise 16 increases M the most.

19. **(a)** Name the group I and group II variables in the tableau as given.

   (b) Pivot the simplex tableau about the indicated element and compute the solution corresponding to the new tableau. Which solutions are feasible (that is, have all values ≥ 0)? Which variables are now in group I, which are in group II?

$$\begin{array}{ccccc} x & y & u & v & M \\ \end{array}$$
$$\left[\begin{array}{ccccc|c} 2 & 5 & 1 & 0 & 0 & 100 \\ 3 & 1 & 0 & 1 & 0 & 300 \\ -10 & -7 & 0 & 0 & 1 & 0 \end{array}\right]$$

(i) 2   (ii) 5   (iii) 3   (iv) 1 (row 2, column 2)

   (c) Which of the solutions is feasible and also increases the value of M the most?

SOLUTIONS TO PRACTICE PROBLEMS 4.1

1. The matrix of the system is

$$\begin{array}{ccccc} x & y & u & v & M \\ \end{array}$$
$$\left[\begin{array}{ccccc|c} 1 & 2 & 3 & 0 & 0 & 6 \\ 0 & 1 & 0 & 1 & 0 & 4 \\ 0 & 5 & 2 & 0 & 1 & 0 \end{array}\right],$$

from which we see that the group II variables are x, v, and M, and the group I variables y and u. To obtain a solution, we set the group I variables equal to 0. We obtain from the first equation that $x = 6$, from the second that $v = 4$, and from the third that $M = 0$. Thus a solution of the system is $x = 6$, $y = 0$, $u = 0$, $v = 4$, $M = 0$.

2. We must use elementary row operations to transform the second column into $\begin{bmatrix} 0 \\ 1 \\ 0 \end{bmatrix}$.

$$\begin{bmatrix} 2 & 4 & 1 & 0 & 0 & 6 \\ 3 & \textcircled{1} & 0 & 1 & 0 & 0 \\ 1 & 1 & 0 & 0 & 1 & 1 \end{bmatrix}$$

$$\xrightarrow{[1]+(-4)[2]} \begin{bmatrix} -10 & 0 & 1 & -4 & 0 & 6 \\ 3 & 1 & 0 & 1 & 0 & 0 \\ 1 & 1 & 0 & 0 & 1 & 1 \end{bmatrix}$$

$$\xrightarrow{[3]+(-1)[2]} \begin{bmatrix} -10 & 0 & 1 & -4 & 0 & 6 \\ 3 & 1 & 0 & 1 & 0 & 0 \\ -2 & 0 & 0 & -1 & 1 & 1 \end{bmatrix}$$

4.2 The Simplex Method I: Maximum Problems

We can now describe the simplex method for solving linear programming problems. The procedure will be illustrated as we solve the furniture manufacturing problem of Section 4.1. Recall that we must maximize the objective function $80x + 70y$ subject to the constraints

$$\begin{cases} 6x + 3y \leq 96 \\ x + y \leq 18 \\ 2x + 6y \leq 72 \\ x \geq 0, \quad y \geq 0. \end{cases}$$

Step 1 Introduce slack variables and state the problem in terms of a system of linear equations.

We carried out this step in Section 4.1. The result was the following restatement of the problem.

Furniture Manufacturing Problem Among all the solutions of the system of linear equations

$$\begin{cases} 6x + 3y + u = 96 \\ x + y + v = 18 \\ 2x + 6y + w = 72 \\ -80x - 70y + M = 0, \end{cases}$$

find one for which $x \geq 0$, $y \geq 0$, $u \geq 0$, $v \geq 0$, $w \geq 0$, and for which M is as large as possible.

  Step 2 Construct the simplex tableau corresponding to the linear system.

This step was also carried out in Section 4.1. The tableau is

	x	y	u	v	w	M	
u	6	3	1	0	0	0	96
v	1	1	0	1	0	0	18
w	2	6	0	0	1	0	72
M	-80	-70	0	0	0	1	0

Note that we have made two additions to the previously found tableau. First, we have separated the last row from the others by means of a horizontal line. This is because the last row, which corresponds to the objective function in the original problem, will play a special role in what follows. The second addition is that we have labeled each row with one of the group II variables—namely, the variable whose value is determined by the row. Thus, for example, the first row gives the particular value of u, which is 96, so the row is labeled with a u. We will find these labels convenient.

  Corresponding to this tableau, there is a particular solution to the linear system, namely the one obtained by setting all group I variables equal to 0. Reading the values of the group II variables from the last column, we obtain

$$x = 0, \qquad y = 0, \qquad u = 96, \qquad v = 18, \qquad w = 72, \qquad M = 0.$$

Our objective is to make M as large as possible. How can the value of M be increased? Look at the equation corresponding to the last row of the tableau. It reads

$$-80x - 70y + M = 0.$$

Note that two of the coefficients, -80 and -70, are negative. Or, what amounts to the same thing, if we solve for M and get

$$M = 80x + 70y,$$

then the coefficients on the right-hand side are *positive*. This fact is significant. It says that M can be increased by increasing either the value of x or the value of y. A unit change in x will increase M by 80 units, whereas a unit change in y will increase M by 70 units. And since we wish to increase M by as much as possible, it is reasonable to attempt to increase the value of x. Let us indicate this by drawing an arrow pointing to the x column of the tableau:

$$
\begin{array}{c}
\begin{array}{ccccccc}
 & x & y & u & v & w & M \\
\end{array} \\
\begin{array}{c}
u \\
v \\
w \\
M
\end{array}
\left[
\begin{array}{cccccc|c}
6 & 3 & 1 & 0 & 0 & 0 & 96 \\
1 & 1 & 0 & 1 & 0 & 0 & 18 \\
2 & 6 & 0 & 0 & 1 & 0 & 72 \\
\hline
-80 & -70 & 0 & 0 & 0 & 1 & 0
\end{array}
\right]
\end{array}
\qquad (1)
$$

$$\uparrow$$

To increase x (from its present value, zero), we will pivot about one of the entries (above the horizontal line) in the x column. In this way, x will become a group II variable and hence will not necessarily be zero in our next particular solution. But around which entry should we pivot? To find out, let us experiment. The results from pivoting about the 6, the 1, and the 2 in the x column are, respectively,

$$
\begin{array}{c}
\begin{array}{ccccccc}
 & x & y & u & v & w & M \\
\end{array} \\
\begin{array}{c}
x \\
v \\
w \\
M
\end{array}
\left[
\begin{array}{cccccc|c}
1 & \frac{1}{2} & \frac{1}{6} & 0 & 0 & 0 & 16 \\
0 & \frac{1}{2} & -\frac{1}{6} & 1 & 0 & 0 & 2 \\
0 & 5 & -\frac{1}{3} & 0 & 1 & 0 & 40 \\
\hline
0 & -30 & \frac{40}{3} & 0 & 0 & 1 & 1280
\end{array}
\right]
\end{array}
$$
Pivot about 6

$$
\begin{array}{c}
\begin{array}{ccccccc}
 & x & y & u & v & w & M \\
\end{array} \\
\begin{array}{c}
u \\
x \\
w \\
M
\end{array}
\left[
\begin{array}{cccccc|c}
0 & -3 & 1 & -6 & 0 & 0 & -12 \\
1 & 1 & 0 & 1 & 0 & 0 & 18 \\
0 & 4 & 0 & -2 & 1 & 0 & 36 \\
\hline
0 & 10 & 0 & 80 & 0 & 1 & 1440
\end{array}
\right]
\end{array}
$$
Pivot about 1

$$
\begin{array}{c}
\begin{array}{ccccccc}
 & x & y & u & v & w & M \\
\end{array} \\
\begin{array}{c}
u \\
v \\
x \\
M
\end{array}
\left[
\begin{array}{cccccc|c}
0 & -15 & 1 & 0 & -3 & 0 & -120 \\
0 & -2 & 0 & 1 & -\frac{1}{2} & 0 & -18 \\
1 & 3 & 0 & 0 & \frac{1}{2} & 0 & 36 \\
\hline
0 & 170 & 0 & 0 & 40 & 1 & 2880
\end{array}
\right]
\end{array}
$$
Pivot about 2

Note that the labels on the rows have *changed* because the group II variables are now *different*. The solutions corresponding to these tableaux are, respectively,

$$x = 16, \quad y = 0, \quad u = 0, \quad v = 2, \quad w = 40, \quad M = 1280,$$

$$x = 18, \quad y = 0, \quad u = -12, \quad v = 0, \quad w = 36, \quad M = 1440,$$

$$x = 36, \quad y = 0, \quad u = -120, \quad v = -18, \quad w = 0, \quad M = 2880.$$

The second and third solutions violate the requirement that all variables be ≥ 0. Thus, we use the first solution, in which we pivoted about 6. Using this solution, we have increased the value of M to 1280 and have replaced our original tableau by

$$
\begin{array}{c}
\begin{array}{cccccc}
x & y & u & v & w & M
\end{array} \\
\begin{array}{c}
x \\
v \\
w \\
M
\end{array}
\left[
\begin{array}{cccccc|c}
1 & \frac{1}{2} & \frac{1}{6} & 0 & 0 & 0 & 16 \\
0 & \frac{1}{2} & -\frac{1}{6} & 1 & 0 & 0 & 2 \\
0 & 5 & -\frac{1}{3} & 0 & 1 & 0 & 40 \\
\hline
0 & -30 & \frac{40}{3} & 0 & 0 & 1 & 1280
\end{array}
\right].
\end{array}
$$

Can M be increased further? To answer this question, look at the last row of the tableau, which corresponds to the equation

$$-30y + \tfrac{40}{3}u + M = 1280.$$

There is a negative coefficient for the variable y in this equation. Correspondingly, when the equation is solved for M, there is a positive coefficient for y:

$$M = 1280 + 30y - \tfrac{40}{3}u.$$

Now it is clear that we should try to increase y. So we pivot about one of the entries in the y column. A calculation for each of the possible pivots shows that pivoting about the first or the third entries leads to solutions having some negative values. Therefore, we pivot about the second entry in the y column. The result is

$$
\begin{array}{c}
\begin{array}{cccccc}
x & y & u & v & w & M
\end{array} \\
\begin{array}{c}
x \\
y \\
w \\
M
\end{array}
\left[
\begin{array}{cccccc|c}
1 & 0 & \frac{1}{3} & -1 & 0 & 0 & 14 \\
0 & 1 & -\frac{1}{3} & 2 & 0 & 0 & 4 \\
0 & 0 & \frac{4}{3} & -10 & 1 & 0 & 20 \\
\hline
0 & 0 & \frac{10}{3} & 60 & 0 & 1 & 1400
\end{array}
\right].
\end{array}
$$

The corresponding solution is

$$x = 14, \qquad y = 4, \qquad u = 0, \qquad v = 0, \qquad w = 20, \qquad M = 1400.$$

Note that with this pivot operation we have increased M from 1280 to 1400.

Can we increase M any further? Let us reason as before. Use the last row of the current tableau to write M in terms of the other variables:

$$\tfrac{10}{3}u + 60v + M = 1400, \qquad M = 1400 - \tfrac{10}{3}u - 60v.$$

Note, however, that in contrast to the previous expressions for M, this one has *no positive coefficients*. And since u and v are ≥ 0, this means that M can be *at most* 1400. But M is already 1400. So M cannot be increased further. Thus, we have shown that the maximum value of M is 1400, and this occurs when $x = 14$ and $y = 4$. Thus, to maximize profits, the furniture manufacturer should be making 14 chairs and 4 sofas each day. The maximum profit is $1400. From the tableau we can read off the values of the slack variables: $u = 0$, $v = 0$, and $w = 20$. This shows that we have no slack resulting from the first inequality, so we have used all the labor-hours available for carpentry. Similarly, since $v = 0$, we have used all of the labor-hours available for finishing. But since $w = 20$, we have 20 labor-hours of upholstery remaining when we manufacture the optimal number of chairs and sofas.

Let us compare the simplex method solution of the furniture manufacturing problem with the geometric solution carried out in Chapter 3. Both solutions yield the same optimal production schedule. In the geometric solution, we found *all* of the vertices of the feasible set and then evaluated the objective function at every one of these vertices. The following table was obtained:

Vertex	Profit $= 80x + 70y$
$(14, 4)$	$80(14) + 70(4) = 1400$
$(9, 9)$	$80(9) + 70(9) = 1350$
$(0, 12)$	$80(0) + 70(12) = 840$
$(0, 0)$	$80(0) + 70(0) = 0$
$(16, 0)$	$80(16) + 70(0) = 1280$

We selected the optimal solution ($x = 14$, $y = 4$) because it produced the greatest profit.

With the simplex method, we had to consider only *some* of the vertices. In the initial tableau we first considered the vertex $(0, 0)$—that is, both x and y were 0. M was also 0. In the second tableau we looked at the vertex $(16, 0)$—that is, $x = 16$ and $y = 0$, and the tableau showed that $M = 1280$. Finally, as a result of the last pivot operation, we came to the vertex $(14, 4)$. This meant that $x = 14$ and $y = 4$. The value of the objective function was read from the tableau: $M = 1400$. Since we could not increase M any more, we did not have to consider any other vertices. In larger linear programming problems, the time saved from looking at just *some* of the vertices, rather than *all* of the vertices, can be substantial.

Based on the preceding discussion, we can state several general principles. First of all, the following criterion determines when a simplex tableau yields a maximum.

Condition for a Maximum   The particular solution derived from a simplex tableau is a maximum if and only if the bottom row contains no negative entries except perhaps the entry in the last column.[1]

We saw this condition illustrated in the previous example. Each of the first two tableaux had negative entries in the last row, and as we showed, their corresponding solutions were not maxima. However, the third tableau, with no negative entries in the last row, did yield a maximum.

The crucial point of the simplex method is the correct choice of a pivot element. In the preceding example we decided to choose a pivot element from the column corresponding to the most-negative entry in the last row. It can be proved that this is the proper choice in general; that is, we have the following rule:

Choosing the Pivot Column   The pivot element should be chosen from that column to the left of the vertical line that has the most-negative entry in the last row.[2]

---

[1] In Section 4.3 we shall encounter maximum problems whose final tableaux have a negative number in the lower right-hand corner.

[2] In case two or more columns are tied for the honor of being the pivot column, an arbitrary choice among them may be made.

Choosing the correct pivot element from the designated column is somewhat more complicated. Our approach before was to calculate the tableau associated with each element and observe that only one corresponded to a solution with nonnegative elements. However, there is a simpler way to make the choice. As an illustration, let us reconsider tableau (1). We have already decided to pivot around some entry in the first column. For each *positive* entry in the pivot column we compute a ratio: the corresponding entry in the right-hand column divided by the entry in the pivot column. So, for example, for the first entry the ratio is $\frac{96}{6}$; for the second entry the ratio is $\frac{18}{1}$; and for the third entry the ratio is $\frac{72}{2}$. We write these ratios to the right of the matrix as follows:

$$
\begin{array}{c}
\\
u \\
v \\
w \\
M
\end{array}
\begin{array}{c}
\begin{array}{cccccc}
x & y & u & v & w & M
\end{array} \\
\left[\begin{array}{cccccc|c}
6 & 3 & 1 & 0 & 0 & 0 & 96 \\
1 & 1 & 0 & 1 & 0 & 0 & 18 \\
2 & 6 & 0 & 0 & 1 & 0 & 72 \\
\hline
-80 & -70 & 0 & 0 & 0 & 1 & 0
\end{array}\right]
\end{array}
\begin{array}{l}
\frac{96}{6} = 16 \\
\frac{18}{1} = 18 \\
\frac{72}{2} = 36
\end{array}
$$

It is possible to prove the following rule, which allows us to determine the pivot element from the preceding display:

Choosing the Pivot Element   For each positive entry of the pivot column, compute the appropriate ratio. Choose as pivot element the one corresponding to the smallest nonnegative ratio.

For instance, consider the choice of pivot element in the preceding example. The least of the ratios is 16. So we choose 6 as the pivot element.

At first, this method for choosing the pivot element might seem very odd. However, it is just a way of guaranteeing that the last column of the new tableau will have entries ≥ 0. And that is just the basis on which we chose the pivot element earlier. To obtain further insight, let us analyze the preceding example yet further.

Suppose that we pivot our tableau about the 6 in column 1. The first step in pivoting is to divide the pivot row by the pivot element (in this case, 6). This gives the array

$$
\begin{array}{c}
\begin{array}{cccccc}
x & y & u & v & w & M
\end{array} \\
\left[\begin{array}{cccccc|c}
1 & \frac{1}{2} & \frac{1}{6} & 0 & 0 & 0 & \frac{96}{6} \\
1 & 1 & 0 & 1 & 0 & 0 & 18 \\
2 & 6 & 0 & 0 & 1 & 0 & 72 \\
\hline
-80 & -70 & 0 & 0 & 0 & 1 & 0
\end{array}\right]
\end{array},
$$

where we have written $\frac{96}{6}$ rather than 16 to emphasize that we have divided by the pivot element. The next step in the pivot procedure is to replace the second row by $[2] + (-1)[1]$. The result is

$$
\left[
\begin{array}{cccccc|c}
x & y & u & v & w & M & \\
1 & \frac{1}{2} & \frac{1}{6} & 0 & 0 & 0 & \frac{96}{6} \\
0 & \frac{1}{2} & -\frac{1}{6} & 1 & 0 & 0 & 18 - \frac{96}{6} \\
2 & 6 & 0 & 0 & 1 & 0 & 72 \\
\hline
-80 & -70 & 0 & 0 & 0 & 1 & 0
\end{array}
\right].
$$

The next step in the pivot process is to replace the third row by $[3] + (-2)[1]$ to obtain

$$
\left[
\begin{array}{cccccc|c}
x & y & u & v & w & M & \\
1 & \frac{1}{2} & \frac{1}{6} & 0 & 0 & 0 & \frac{96}{6} \\
0 & \frac{1}{2} & -\frac{1}{6} & 1 & 0 & 0 & 18 - \frac{96}{6} \\
0 & 5 & -\frac{1}{3} & 0 & 1 & 0 & 72 - 2\left(\frac{96}{6}\right) \\
\hline
-80 & -70 & 0 & 0 & 0 & 1 & 0
\end{array}
\right].
$$

The final step of the pivot process is to replace the fourth row by $[4] + 80[1]$ to obtain

$$
\left[
\begin{array}{cccccc|c}
x & y & u & v & w & M & \\
1 & \frac{1}{2} & \frac{1}{6} & 0 & 0 & 0 & \frac{96}{6} \\
0 & \frac{1}{2} & -\frac{1}{6} & 1 & 0 & 0 & 18 - \frac{96}{6} \\
0 & 5 & -\frac{1}{3} & 0 & 1 & 0 & 72 - 2\left(\frac{96}{6}\right) \\
\hline
0 & -30 & \frac{40}{3} & 0 & 0 & 1 & 1280
\end{array}
\right].
$$

The entries in the upper part of the right-hand column may be written

$$
\frac{96}{6}, \qquad \frac{18}{1} - \frac{96}{6}, \qquad 2\left(\frac{72}{2} - \frac{96}{6}\right).
$$

If we had pivoted about the 1 or 2 in the first column of the original tableau, the upper entries in the last column of the tableau would have been

$$
6\left(\frac{96}{6} - \frac{18}{1}\right), \qquad \frac{18}{1}, \qquad 2\left(\frac{72}{2} - \frac{18}{1}\right),
$$

or

$$
6\left(\frac{96}{6} - \frac{72}{2}\right), \qquad \frac{18}{1} - \frac{72}{2}, \qquad \frac{72}{2},
$$

respectively. Notice that all of the combinations of the differences of the pairs of ratios appear in these triples. In the first case, the ratio $\frac{96}{6}$ is subtracted from each of the other two ratios, whereas in the next two cases, the ratios $\frac{18}{1}$ and $\frac{72}{2}$ are subtracted. In order for the upper entries in the last column to be nonnegative, we must subtract off the smallest of the ratios. That is, we should pivot about the entry corresponding to the smallest ratio. This is the rationale governing our choice of pivot element!

Now that we have assembled all the components of the simplex method, we can summarize it as follows:

The Simplex Method for Problems in Standard Form

1. Introduce slack variables and state the problem in terms of a system of linear equations.

2. Construct the simplex tableau corresponding to the system.

3. Determine if the left part of the bottom row contains negative entries. If none are present, the solution corresponding to the tableau yields a maximum and the problem is solved.

4. If the left part of the bottom row contains negative entries, construct a new simplex tableau.

   (a) Choose the pivot column by inspecting the entries of the last row of the current tableau, excluding the right-hand entry. The pivot column is the one containing the most negative of these entries.

   (b) Choose the pivot element by computing ratios associated with the positive entries of the pivot column. The pivot element is the one corresponding to the smallest nonnegative ratio.

   (c) Construct the new simplex tableau by pivoting around the selected element.

5. Return to step 3. Steps 3 and 4 are repeated as many times as necessary to find a maximum.

Let us now work some problems to see how this method is applied.

EXAMPLE 1 **Using the simplex tableau** Maximize the objective function $10x + y$ subject to the constraints

$$\begin{cases} x + 2y \leq 10 \\ 3x + 4y \leq 6 \\ x \geq 0, \quad y \geq 0. \end{cases}$$

Solution The corresponding system of linear equations with slack variables is

$$\begin{cases} x + 2y + u & = 10 \\ 3x + 4y & + v & = 6 \\ -10x - y & + M = 0, \end{cases}$$

and we must find that solution of the system for which $x \geq 0$, $y \geq 0$, $u \geq 0$, $v \geq 0$, and M is as large as possible. Here is the initial simplex tableau:

$$\begin{array}{c} & \begin{array}{ccccc} x & y & u & v & M \end{array} \\ \begin{array}{c} u \\ v \\ M \end{array} & \left[\begin{array}{ccccc|c} 1 & 2 & 1 & 0 & 0 & 10 \\ 3 & 4 & 0 & 1 & 0 & 6 \\ -10 & -1 & 0 & 0 & 1 & 0 \end{array}\right] \\ & \hspace{0.3em}\uparrow \end{array}.$$

Note that this tableau does not correspond to a maximum, since the left part of the bottom row has negative entries. So we pivot to create a new tableau. Since -10

is the most-negative entry in the last row, we choose the first column as the pivot column. To determine the pivot element, we compute ratios

$$
\begin{array}{c}
\quad\quad x \quad y \quad u \quad v \quad M \quad\quad\quad \text{Ratios} \\
\begin{array}{c} u \\ v \\ M \end{array}
\left[
\begin{array}{ccccc|c}
1 & 2 & 1 & 0 & 0 & 10 \\
\boxed{3} & 4 & 0 & 1 & 0 & 6 \\
\hline
-10 & -1 & 0 & 0 & 1 & 0
\end{array}
\right]
\begin{array}{c} 10/1 = 10 \\ 6/3 = 2. \\ \\ \end{array}
\\
\quad\quad \uparrow
\end{array}
$$

The smallest ratio is 2, so we pivot about 3, which we have circled. The new tableau is therefore

$$
\begin{array}{c}
\quad\quad x \quad\quad y \quad\quad u \quad\quad v \quad\quad M \\
\begin{array}{c} u \\ x \\ M \end{array}
\left[
\begin{array}{ccccc|c}
0 & \frac{2}{3} & 1 & -\frac{1}{3} & 0 & 8 \\
1 & \frac{4}{3} & 0 & \frac{1}{3} & 0 & 2 \\
\hline
0 & \frac{37}{3} & 0 & \frac{10}{3} & 1 & 20
\end{array}
\right] .
\end{array}
$$

Note that this tableau corresponds to a maximum, since there are no negative entries in the left part of the last row. The solution corresponding to the tableau is

$$x = 2, \quad y = 0, \quad u = 8, \quad v = 0, \quad M = 20.$$

Therefore, the objective function assumes its maximum value of 20 when $x = 2$ and $y = 0$.

Now Try Exercise 1   ■

The simplex method can be used to solve problems in any number of variables. Let us illustrate the method for three variables.

EXAMPLE 2 **Using the simplex tableau** Maximize the objective function $x + 2y + z$ subject to the constraints

$$
\begin{cases}
x - y + 2z \le 10 \\
2x + y + 3z \le 12 \\
x \ge 0, \quad y \ge 0, \quad z \ge 0.
\end{cases}
$$

Solution We determined the corresponding linear system in Example 2 of Section 4.1:

$$
\begin{cases}
x - y + 2z + u \quad\quad\quad = 10 \\
2x + y + 3z \quad\quad + v \quad\quad = 12 \\
-x - 2y - z \quad\quad\quad\quad + M = 0.
\end{cases}
$$

So the simplex method works as follows:

$$
\begin{array}{c}
\quad\quad x \quad y \quad z \quad u \quad v \quad M \\
\begin{array}{c} u \\ v \\ M \end{array}
\left[
\begin{array}{cccccc|c}
1 & -1 & 2 & 1 & 0 & 0 & 10 \\
2 & \boxed{1} & 3 & 0 & 1 & 0 & 12 \\
\hline
-1 & -2 & -1 & 0 & 0 & 1 & 0
\end{array}
\right]
\begin{array}{l} \\ 12/1 = 12 \\ \text{(smallest} \\ \text{nonnegative ratio)} \end{array}
\\
\quad\quad\quad\quad \uparrow
\end{array}
$$

	x	y	z	u	v	M	
u	3	0	5	1	1	0	22
y	2	1	3	0	1	0	12
M	3	0	5	0	2	1	24

Thus, the solution of the original problem ($x = 0$, $y = 12$, $z = 0$) yields the maximum value of the objective function $x + 2y + z$. The maximum value is 24.

Now Try Exercise 9

PRACTICE PROBLEMS 4.2

1. Which of these simplex tableaux has a solution that corresponds to a maximum for the associated linear programming problem?

(a)

	x	y	u	v	M	
	3	1	0	1	0	5
	2	0	0	0	1	0
	−1	−2	1	0	0	3

(b)

	x	y	u	v	M	
	2	1	0	11	0	10
	1	0	1	7	0	1
	1	0	0	4	1	−2

2. Suppose that in the solution of a linear programming problem by the simplex method, we encounter the following simplex tableau. What is the next step in the solution?

	x	y	u	v	M	
	0	4	1	2	0	4
	1	5	0	1	0	9
	0	2	0	−3	1	6

EXERCISES 4.2

For each of the simplex tableaux in Exercises 1–4,

(a) Compute the next pivot element.

(b) Determine the next tableau.

(c) Determine the particular solution corresponding to the tableau of part (b).

1.

	x	y	u	v	M	
	6	2	1	0	0	10
	1	3	0	1	0	6
	−4	−12	0	0	1	0

2.

	x	y	u	v	M	
	1	0	3	1	0	5
	0	1	2	0	0	12
	−6	0	5	0	1	10

3.

	x	y	u	v	M	
	5	12	1	0	0	12
	15	10	0	1	0	5
	4	−2	0	0	1	0

4.

	x	y	u	v	M	
	0	6	3	1	0	5
	1	−5	2	0	0	8
	0	20	−10	0	1	22

In Exercises 5–14, solve the linear programming problems using the simplex method.

✓ **5.** Maximize $x + 3y$ subject to the constraints

$$\begin{cases} x + y \le 7 \\ x + 2y \le 10 \\ x \ge 0, \quad y \ge 0. \end{cases}$$

6. Maximize $x + 2y$ subject to the constraints

$$\begin{cases} -x + y \le 100 \\ 6x + 6y \le 1200 \\ x \ge 0, \quad y \ge 0. \end{cases}$$

7. Maximize $4x + 2y$ subject to the constraints

$$\begin{cases} 5x + y \le 80 \\ 3x + 2y \le 76 \\ x \ge 0, \quad y \ge 0. \end{cases}$$

8. Maximize $2x + 6y$ subject to the constraints

$$\begin{cases} -x + 8y \le 160 \\ 3x - y \le 3 \\ x \ge 0, \quad y \ge 0. \end{cases}$$

9. Maximize $x + 3y + 5z$ subject to the constraints

$$\begin{cases} x + 2z \le 10 \\ 3y + z \le 24 \\ x \ge 0, \quad y \ge 0, \quad z \ge 0. \end{cases}$$

10. Maximize $-x + 8y + z$ subject to the constraints

$$\begin{cases} x - 2y + 9z \le 10 \\ y + 4z \le 12 \\ x \ge 0, \quad y \ge 0, \quad z \ge 0. \end{cases}$$

11. Maximize $2x + 3y$ subject to the constraints

$$\begin{cases} 5x + y \le 30 \\ 3x + 2y \le 60 \\ x + y \le 50 \\ x \ge 0, \quad y \ge 0. \end{cases}$$

12. Maximize $10x + 12y + 10z$ subject to the constraints

$$\begin{cases} x - 2y \le 6 \\ 3x + z \le 9 \\ y + 3z \le 12 \\ x \ge 0, \quad y \ge 0, \quad z \ge 0. \end{cases}$$

13. Maximize $6x + 7y + 300$ subject to the constraints

$$\begin{cases} 2x + 3y \le 400 \\ x + y \le 150 \\ x \ge 0, \quad y \ge 0. \end{cases}$$

14. Maximize $10x + 20y + 50$ subject to the constraints

$$\begin{cases} x + y \le 10 \\ 5x + 2y \le 20 \\ x \ge 0, \quad y \ge 0. \end{cases}$$

15. (*Toy Factory*) A toy manufacturer makes lightweight balls for indoor play. The large basketball uses 4 ounces of foam and 20 minutes of labor and brings a profit of $2.50. The football uses 3 ounces of foam and 30 minutes of labor and brings a profit of $2. The manufacturer has 48 pounds of foam and 120 labor-hours a week. Use the simplex method to determine the optimal production schedule so as to maximize profits. Show that the geometric method gives the same solution.

16. (*Agriculture*) A large agricultural firm has 250 acres and $8000 available for cultivating three crops: barley, oats, and wheat. Barley requires $10 per acre for cultivation, oats require $15 per acre for cultivation and wheat requires $12 per acre of cultivation. Barley requires 7 hours of labor per acre, oats require 9 hours of labor per acre, and wheat requires 8 hours of labor per acre. The firm has 2100 hours of labor available. The profits per acre of each crop are: barley $60, oats $75, wheat $70. How many acres of each crop should be planted to maximize profit?

17. (*Furniture Factory*) Suppose that a furniture manufacturer makes chairs, sofas, and tables. The amounts of labor of various types as well as the relative availability of each type are summarized by the following chart:

	Chair	Sofa	Table	Daily labor available (labor-hours)
Carpentry	6	3	8	768
Finishing	1	1	2	144
Upholstery	2	5	0	216

The profit per chair is $80, per sofa $70, and per table $120. How many pieces of each type of furniture should be manufactured each day to maximize the profit?

18. (*Stereo Store*) A stereo store sells three brands of stereo systems, brands A, B, and C. It can sell a total of 100 stereo systems per month. Brands A, B, and C take up, respectively, 5, 4, and 4 cubic feet of warehouse space, and a maximum of 480 cubic feet of warehouse space is available. Brands A, B, and C generate sales commissions of $40, $20, and $30, respectively, and $3200 is available to pay the sales commissions. The profit generated from the sale of each brand is $70, $210, and $140, respectively. How many of each brand of stereo system should be sold to maximize the profit?

19. (*Weight Loss and Exercise*) As part of a weight-reduction program, a man designs a monthly exercise program consisting of bicycling, jogging, and swimming. He would like to exercise at most 30 hours, devote at most 4 hours to swimming, and jog for no more

than the total number of hours bicycling and swimming. The calories burned per hour by bicycling, jogging, and swimming are 200, 475, and 275, respectively. How many hours should be allotted to each activity to maximize the number of calories burned? If he loses 1 pound of weight for each 3500 calories burned, how many pounds will he lose each month exercising?

20. (*Furniture Factory*) A furniture manufacturer produces small sofas, large sofas, and chairs. The profits per item are, respectively, $60, $60, and $50. The pieces of furniture require the following numbers of labor-hours for their manufacture:

	Carpentry	Upholstery	Finishing
Small sofas	10	30	20
Large sofas	10	30	0
Chairs	10	10	10

The following amounts of labor are available per month: carpentry, at most 1200 hours; upholstery, at most 3000 hours; and finishing, at most 1800 hours. How many each of small sofas, large sofas, and chairs should be manufactured to maximize the profit?

21. (*Fast-Food Restaurants*) The XYZ Corporation plans to open three different types of fast-food restaurants. Type A restaurants require an initial cash outlay of $600,000, need 15 employees, and are expected to make an annual profit of $40,000. Type B restaurants require an initial cash outlay of $400,000, need 9 employees, and are expected to make an annual profit of $30,000. Type C restaurants require an initial cash outlay of $300,000, need 5 employees, and are expected to make an annual profit of $25,000. The XYZ Corporation has $48,000,000 available for initial outlays, does not want to hire more than 1000 new employees, and would like to open at most 70 restaurants. How many restaurants of each type should be opened to maximize the expected annual profit?

22. (*Baby Products*) A baby products company makes car seats, strollers, and travel yards. A particular model of car seat requires 5 labor-hours for parts creation, 4 labor-hours of assembly, and 2 labor-hours of finishing and packaging. A particular model of stroller requires 3 labor-hours for parts creation, 3 labor-hours of assembly, and 1 labor-hour of finishing and packaging. A particular model of travel yard requires 2 labor-hours for parts creation, 3 labor-hours of assembly, and 1 labor-hour of finishing and packaging. There are 450 labor-hours available for parts creation, 380 for assembly, and 140 for finishing and packaging. If the profit per car seat is $30, per stroller is $10, and per travel yard is $18, how many of each should be produced to maximize profit? What advice would you give this company?

23. (*Potting Soil Mixes*) A lawn and garden store creates three different potting mixes sold in 20-pound bags.

Mix A contains 12 pounds of peat, 5 pounds of perlite, and 3 pounds of organic material and earns a profit of $3 per 20-pound bag. Mix B contains 10 pounds of peat, 6 pounds of perlite, and 4 pounds of organic material and earns a profit of $5 per 20-pound bag. Mix C contains 8 pounds of peat, 8 pounds of perlite, and 4 pounds of organic material and earns a profit of $6 per 20-pound bag. There are 1200 pounds of peat, 800 pounds of perlite, and 600 pounds of organic material available every week to make these potting mixes. How many 20-pound bags of each mix should be made in order to maximize weekly profit?

24. Maximize $60x + 90y + 300z$ subject to the constraints

$$\begin{cases} x + y + z \le 600 \\ x + 3y \quad\;\; \le 600 \\ 2x \quad\;\; + z \le 900 \\ x \ge 0, \quad y \ge 0, \quad z \ge 0. \end{cases}$$

25. Maximize $200x + 500y$ subject to the constraints

$$\begin{cases} x + 4y \le 300 \\ x + 2y \le 200 \\ x \ge 0, \quad y \ge 0. \end{cases}$$

In Exercises 26–29, use a graphing calculator, a spreadsheet, or mathematical software to solve the linear programming problem.

26. Maximize $2x + 4y$ subject to the constraints

$$\begin{cases} 5x + y \le 8 \\ x + 2y \le 10 \\ x \ge 0, \quad y \ge 0. \end{cases}$$

27. Maximize $4x + 6y$ subject to the constraints

$$\begin{cases} x + 4y \le 4 \\ 3x + 2y \le 6 \\ x \ge 0, \quad y \ge 0. \end{cases}$$

28. Maximize $3x + 2y + 2z$ subject to the constraints

$$\begin{cases} x + y + 2z \le 6 \\ x + 5y + 2z \le 20 \\ 2x + y + z \le 4 \\ x \ge 0, \quad y \ge 0, \quad z \ge 0. \end{cases}$$

29. Maximize $16x + 4y - 20z$ subject to the constraints

$$\begin{cases} 4x + y + 5z \le 20 \\ x + 2y + 4z \le 50 \\ 4x + 10y + z \le 32 \\ x \ge 0, \quad y \ge 0, \quad z \ge 0. \end{cases}$$

SOLUTIONS TO PRACTICE PROBLEMS 4.2

1. (a) The tableau does not correspond to a maximum, since among the entries $-1, -2, 1, 0, 0$ in the last row, at least one is negative.

   (b) The tableau corresponds to a maximum since none of the entries $1, 0, 0, 4, 1$ of the last row is negative. Note that it does not matter that the entry -2 in the lower right-hand corner of the matrix is negative. This number gives the value of M. In this example -2 is as large as M can become.

2. First choose the column corresponding to the most-negative entry of the final row, that is, the fourth column. For each positive entry in the fourth col-

umn that is above the horizontal line, compute the ratio with the sixth column. The smallest ratio is 2 and appears in the first row, so the next operation is to pivot around the 2 in the first row of the fourth column.

$$\left[\begin{array}{ccccc|c} 0 & 4 & 1 & ② & 0 & 4 \\ 1 & 5 & 0 & 1 & 0 & 9 \\ \hline 0 & 2 & 0 & -3 & 1 & 6 \end{array}\right] \begin{array}{l} \frac{4}{2} = 2 \\ \frac{9}{1} = 9 \\ \\ \end{array}$$

$$\uparrow$$

4.3   The Simplex Method II: Minimum Problems

In the preceding section we developed the simplex method and applied it to a number of problems. However, throughout we restricted ourselves to linear programming problems in standard form. Recall that such problems satisfied three properties: (1) the objective function is to be maximized; (2) each variable must be ≥ 0; and (3) all constraints other than those implied by (2) must be of the form

$$[\text{linear polynomial}] \leq [\text{nonnegative constant}].$$

In this section we do what we can to relax these restrictions.

Let us begin with restriction (3). This could be violated in two ways. First, the constant on the right-hand side of one or more constraints could be negative. Thus, for example, one constraint might be

$$x - y \leq -2.$$

A second way in which restriction (3) can be violated is for some constraints to involve $\geq$ rather than $\leq$. An example of such a constraint is

$$2x + 3y \geq 5.$$

However, we can convert such a constraint into one involving $\leq$ by multiplying both sides of the inequality by -1:

$$-2x - 3y \leq -5.$$

Of course, the right-hand constant is no longer nonnegative. Thus, if we allow negative constants on the right, we can write all constraints in the form

$$[\text{linear polynomial}] \leq [\text{constant}].$$

Henceforth, the first step in solving a linear programming problem will be to write the constraints in this form. Let us now see how to deal with the phenomenon of negative constants.

EXAMPLE 1   **Working with non-standard inequalities**   Maximize the objective function $5x + 10y$ subject to the constraints

$$\begin{cases} x + y \leq 20 \\ 2x - y \geq 10 \\ x \geq 0, \quad y \geq 0. \end{cases}$$

Solution The first step is to put the second constraint into $\leq$ form. Multiply the second inequality by -1 to obtain

$$\begin{cases} x + y \leq 20 \\ -2x + y \leq -10 \\ x \geq 0, \quad y \geq 0. \end{cases}$$

Just as before, write the linear programming problem as a linear system:

$$\begin{cases} x + \ \ y + u \ \ \ \ \ \ \ \ \ \ \ \ = \ \ \ 20 \\ -2x + \ \ y \ \ \ \ \ \ + v \ \ \ \ \ \ = -10 \\ -5x - 10y \ \ \ \ \ \ \ \ \ \ + M = \ \ \ \ 0. \end{cases}$$

From the linear system, construct the simplex tableau:

$$\begin{array}{c} \\ u \\ v \\ M \end{array} \begin{array}{c} \begin{matrix} x & y & u & v & M \end{matrix} \\ \left[\begin{array}{ccccc|c} 1 & 1 & 1 & 0 & 0 & 20 \\ -2 & 1 & 0 & 1 & 0 & -10 \\ \hline -5 & -10 & 0 & 0 & 1 & 0 \end{array} \right]. \end{array}$$

Everything would proceed exactly as before, except that the right-hand column has a -10 in it. This means that the initial value for v is -10, which violates the condition that all variables be ≥ 0. Before we can apply the simplex method of Section 4.2, we must first put the tableau into standard form. This can be done by pivoting to remove the negative entry in the right column.

We choose the pivot element as follows. Look along the left side of the -10 row of the tableau and locate any negative entry. There is only one: -2. Use the column containing the -2 as the pivot column (that is, use column 1). Now compute ratios as before:[1]

$$\begin{array}{c} \\ u \\ v \\ M \end{array} \begin{array}{c} \begin{matrix} x & y & u & v & M \end{matrix} \\ \left[\begin{array}{ccccc|c} 1 & 1 & 1 & 0 & 0 & 20 \\ \boxed{-2} & 1 & 0 & 1 & 0 & -10 \\ \hline -5 & -10 & 0 & 0 & 1 & 0 \end{array} \right] \end{array} \begin{array}{l} \frac{20}{1} = 20 \\ \frac{-10}{-2} = \ 5. \end{array}$$

$\uparrow$

The smallest positive ratio is 5, so we choose -2 as the pivot element. The new tableau is

$$\begin{array}{c} \\ u \\ x \\ M \end{array} \begin{array}{c} \begin{matrix} x & y & u & v & M \end{matrix} \\ \left[\begin{array}{ccccc|c} 0 & \frac{3}{2} & 1 & \frac{1}{2} & 0 & 15 \\ 1 & -\frac{1}{2} & 0 & -\frac{1}{2} & 0 & 5 \\ \hline 0 & -\frac{25}{2} & 0 & -\frac{5}{2} & 1 & 25 \end{array} \right]. \end{array}$$

Note that all entries in the right-hand column are now nonnegative;[2] that is, the corresponding solution has all variables ≥ 0. From here on we follow the simplex

---

[1]Note, however, that in this circumstance we compute ratios corresponding to both positive *and* negative entries (except the last) in the pivot column, considering further only positive ratios.

[2]In general, it may be necessary to pivot several times before all entries in the last column are ≥ 0.

method for tableaux in standard form:

$$
\begin{array}{c c}
 & \begin{array}{c c c c c} x & y & u & v & M \end{array} \\
\begin{array}{c} u \\ x \\ M \end{array} &
\left[\begin{array}{c c c c c|c}
0 & \boxed{\tfrac{3}{2}} & 1 & \tfrac{1}{2} & 0 & 15 \\
1 & -\tfrac{1}{2} & 0 & -\tfrac{1}{2} & 0 & 5 \\
\hline
0 & -\tfrac{25}{2} & 0 & -\tfrac{5}{2} & 1 & 25
\end{array}\right] & 15/\tfrac{3}{2}=10
\end{array}
$$

$$
\begin{array}{c c}
 & \begin{array}{c c c c c} x & y & u & v & M \end{array} \\
\begin{array}{c} y \\ x \\ M \end{array} &
\left[\begin{array}{c c c c c|c}
0 & 1 & \tfrac{2}{3} & \tfrac{1}{3} & 0 & 10 \\
1 & 0 & \tfrac{1}{3} & -\tfrac{1}{3} & 0 & 10 \\
\hline
0 & 0 & \tfrac{25}{3} & \tfrac{5}{3} & 1 & 150
\end{array}\right]
\end{array}.
$$

So the maximum value of M is 150, which is attained for $x = 10$, $y = 10$.

Now Try Exercise 1   ■

In summary,

The Simplex Method for Problems in Nonstandard Form

1. If necessary, convert all inequalities (except $x \geq 0$, $y \geq 0$) into the form
$$[\text{linear polynomial}] \leq [\text{constant}].$$

2. If a negative number appears in the upper part of the last column of the simplex tableau, remove it by pivoting.
 (a) Select one of the negative entries in its row. The column containing the entry will be the pivot column.
 (b) Select the pivot element by determining the least of the positive ratios associated with entries in the pivot column (except the bottom entry).
 (c) Pivot.

3. Repeat step 2 until there are no negative entries in the upper part of the right-hand column of the simplex tableau.

4. Proceed to apply the simplex method for tableaux in standard form.

The method we have just developed can be used to solve *minimum* problems as well as maximum problems. Minimizing the objective function f is the same as maximizing $(-1) \cdot f$. This is so since multiplying an inequality by -1 reverses the direction of the inequality sign. Thus, to apply our method to a minimum problem, we merely multiply the objective function by -1 and turn the problem into a maximum problem.

■ **EXAMPLE 2**   **Minimizing the objective function**   Minimize the objective function $3x + 2y$ subject to the constraints

$$\begin{cases} x + y \geq 10 \\ x - y \leq 15 \\ x \geq 0, \quad y \geq 0. \end{cases}$$

Solution   First transform the problem so that the first two constraints are in $\leq$ form:

$$\begin{cases} -x - y \leq -10 \\ x - y \leq 15 \\ x \geq 0, \quad y \geq 0. \end{cases}$$

Instead of minimizing $3x + 2y$, let us maximize $-3x - 2y$. Let $M = -3x - 2y$. Then our initial simplex tableau reads

$$
\begin{array}{c}
 \\ u \\ v \\ M
\end{array}
\begin{array}{ccccc}
x & y & u & v & M \\
\end{array}
\left[\begin{array}{ccccc|c}
-1 & \boxed{-1} & 1 & 0 & 0 & -10 \\
1 & -1 & 0 & 1 & 0 & 15 \\
\hline
3 & 2 & 0 & 0 & 1 & 0
\end{array} \right]
\quad \frac{-10}{-1} = 10.
$$

We first eliminate the -10 in the right-hand column. We have a choice of two negative entries in the -10 row. Let us choose the one in the y column. The ratios are then tabulated as before, and we pivot around the circled element. The new tableau is[3]

$$
\begin{array}{c}
 \\ y \\ v \\ M
\end{array}
\begin{array}{ccccc}
x & y & u & v & M \\
\end{array}
\left[\begin{array}{ccccc|c}
1 & 1 & -1 & 0 & 0 & 10 \\
2 & 0 & -1 & 1 & 0 & 25 \\
\hline
1 & 0 & 2 & 0 & 1 & -20
\end{array} \right].
$$

Since all entries in the bottom row, except the last, are positive, this tableau corresponds to a maximum. Thus, the maximum value of $-3x - 2y$ (subject to the constraints) is -20, and this value occurs for $x = 0$, $y = 10$. Thus, the *minimum* value of $3x + 2y$ subject to the constraints is 20.

Now Try Exercise 3   ■

Let us now rework an applied problem previously treated (see Example 2 in Section 3.3), this time using the simplex method. For easy reference we restate the problem.

■ **EXAMPLE 3**   **Transportation problem**   Suppose that a TV dealer has stores in Annapolis and Rockville and warehouses in College Park and Baltimore. The cost of shipping sets from College Park to Annapolis is $6 per set; from College Park to Rockville, $3; from Baltimore to Annapolis, $9; and from Baltimore to Rockville, $5. Suppose

---

[3]Note that we do not need the last entry in the last column to be positive. We require *only* that x, y, u, and v be ≥ 0.

that the Annapolis store orders 25 TV sets and the Rockville store 30. Further suppose that the College Park warehouse has a stock of 45 sets, and the Baltimore warehouse 40. What is the most economical way to supply the requested TV sets to the two stores?

Solution via the Simplex Method As in the previous solution, let x be the number of sets shipped from College Park to Rockville, and y the number shipped from College Park to Annapolis. The flow of sets is depicted in Fig. 1.

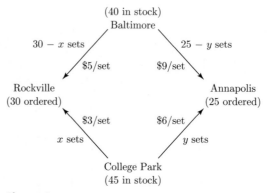

Figure 1

Exactly as in our previous solution, we reduce the problem to the following algebraic form: Minimize $375 - 2x - 3y$ subject to the constraints

$$\begin{cases} x \le 30, \quad y \le 25 \\ x + y \ge 15 \\ x + y \le 45 \\ x \ge 0, \quad y \ge 0. \end{cases}$$

Two changes are needed. First, instead of minimizing $375 - 2x - 3y$, we maximize $-(375 - 2x - 3y) = 2x + 3y - 375$. Second, we write the constraint $x + y \ge 15$ in the form

$$-x - y \le -15.$$

With these changes made, we can write down the linear system:

$$\begin{cases} x & +t & & & & = & 30 \\ & y & +u & & & = & 25 \\ -x - & y & & +v & & = & -15 \\ x + & y & & & +w & = & 45 \\ -2x - & 3y & & & & +M & = & -375. \end{cases}$$

From here on we follow our routine procedure in a mechanical way:

$$
\begin{array}{c}
\begin{array}{ccccccc}
x & y & t & u & v & w & M
\end{array}\\
\begin{array}{c}
t\\u\\v\\w\\M
\end{array}
\left[
\begin{array}{ccccccc|c}
1 & 0 & 1 & 0 & 0 & 0 & 0 & 30\\
0 & 1 & 0 & 1 & 0 & 0 & 0 & 25\\
\boxed{-1} & -1 & 0 & 0 & 1 & 0 & 0 & -15\\
1 & 1 & 0 & 0 & 0 & 1 & 0 & 45\\
\hline
-2 & -3 & 0 & 0 & 0 & 0 & 1 & -375
\end{array}
\right]
\end{array}
\qquad
\begin{array}{l}
\frac{30}{1}=30\\[1em]
\\
\frac{-15}{-1}=15\\[0.5em]
\frac{45}{1}=45
\end{array}
$$

$\uparrow$

$$
\begin{array}{c}
\begin{array}{ccccccc}
x & y & t & u & v & w & M
\end{array}\\
\begin{array}{c}
t\\u\\x\\w\\M
\end{array}
\left[
\begin{array}{ccccccc|c}
0 & -1 & 1 & 0 & \boxed{1} & 0 & 0 & 15\\
0 & 1 & 0 & 1 & 0 & 0 & 0 & 25\\
1 & 1 & 0 & 0 & -1 & 0 & 0 & 15\\
0 & 0 & 0 & 0 & 1 & 1 & 0 & 30\\
\hline
0 & -1 & 0 & 0 & -2 & 0 & 1 & -345
\end{array}
\right]
\end{array}
\qquad
\begin{array}{l}
\frac{15}{1}=15\\[1em]
\\
\frac{30}{1}=30
\end{array}
$$

$\uparrow$

$$
\begin{array}{c}
\begin{array}{ccccccc}
x & y & t & u & v & w & M
\end{array}\\
\begin{array}{c}
v\\u\\x\\w\\M
\end{array}
\left[
\begin{array}{ccccccc|c}
0 & -1 & 1 & 0 & 1 & 0 & 0 & 15\\
0 & 1 & 0 & 1 & 0 & 0 & 0 & 25\\
1 & 0 & 1 & 0 & 0 & 0 & 0 & 30\\
0 & \boxed{1} & -1 & 0 & 0 & 1 & 0 & 15\\
\hline
0 & -3 & 2 & 0 & 0 & 0 & 1 & -315
\end{array}
\right]
\end{array}
\qquad
\begin{array}{l}
\\
\frac{25}{1}=25\\[1em]
\\
\frac{15}{1}=15
\end{array}
$$

$\uparrow$

$$
\begin{array}{c}
\begin{array}{ccccccc}
x & y & t & u & v & w & M
\end{array}\\
\begin{array}{c}
v\\u\\x\\y\\M
\end{array}
\left[
\begin{array}{ccccccc|c}
0 & 0 & 0 & 0 & 1 & 1 & 0 & 30\\
0 & 0 & \boxed{1} & 1 & 0 & -1 & 0 & 10\\
1 & 0 & 1 & 0 & 0 & 0 & 0 & 30\\
0 & 1 & -1 & 0 & 0 & 1 & 0 & 15\\
\hline
0 & 0 & -1 & 0 & 0 & 3 & 1 & -270
\end{array}
\right]
\end{array}
\qquad
\begin{array}{l}
\\
\frac{10}{1}=10\\[0.5em]
\frac{30}{1}=30
\end{array}
$$

$\uparrow$

$$
\begin{array}{c}
\begin{array}{ccccccc}
x & y & t & u & v & w & M
\end{array}\\
\begin{array}{c}
v\\t\\x\\y\\M
\end{array}
\left[
\begin{array}{ccccccc|c}
0 & 0 & 0 & 0 & 1 & 1 & 0 & 30\\
0 & 0 & 1 & 1 & 0 & -1 & 0 & 10\\
1 & 0 & 0 & -1 & 0 & 1 & 0 & 20\\
0 & 1 & 0 & 1 & 0 & 0 & 0 & 25\\
\hline
0 & 0 & 0 & 1 & 0 & 2 & 1 & -260
\end{array}
\right]
\end{array}\ .
$$

The last tableau corresponds to a maximum. So $2x+3y-375$ has a maximum value -260, and therefore $375 - 2x - 3y$ has a minimum value 260. This value occurs when $x = 20$ and $y = 25$. This is in agreement with our previous graphical solution of the problem.

Now Try Exercise 7   ■

The calculations used in Example 3 are not that much simpler than those in the original solution. Why, then, should we concern ourselves with the simplex method? For one thing, the simplex method is so mechanical in its execution that it is much easier to program for a computer. For another, our previous method was restricted to problems in two variables. However, suppose that the two warehouses were to deliver their TV sets to three or four or perhaps even 100 stores. Our previous method could not be applied. However, the simplex method, although yielding very large matrices and very tedious calculations, is applicable. Indeed, this is the method many industries use to optimize distribution of their products.

Some Further Comments on the Simplex Method   Our discussion has omitted some of the technical complications arising in the simplex method. A complete discussion of these is beyond the scope of this book. However, let us mention three. First, it is possible that a given linear programming problem has more than one solution. This can occur, for example, if there are ties for the choice of pivot column. For instance, if the bottom of the simplex tableau is

$$\begin{bmatrix} -3 & -7 & 4 & -7 & 1 & 3 \end{bmatrix},$$
$$\qquad\quad \uparrow \qquad\qquad \uparrow$$

then -7 is the most-negative entry and we may choose as pivot column either the second or fourth. In such a circumstance the pivot column may be chosen arbitrarily. Different choices, however, may lead to different solutions of the problem.

A second difficulty is that a given linear programming problem may have no solution at all. In this case the method will break down at some point. For example, among the ratios at a given stage there may be no nonnegative ones to consider. Then we cannot choose a pivot element. Such a breakdown of the method indicates that the associated linear programming problem has no solution.

Finally, whenever there is a tie for the choice of a pivot element, we choose one of the candidates arbitrarily. Occasionally, this may lead to a loop in which the simplex algorithm leads back to a previously encountered tableau. To prevent the loop from recurring, we should then make a different selection from the tied pivot possibilities.

PRACTICE PROBLEMS 4.3

1. Convert the following minimum problem into a maximum problem in standard form: Minimize $3x + 4y$ subject to the constraints

$$\begin{cases} x - y \geq 0 \\ 3x - 4y \geq 0 \\ x \geq 0, \quad y \geq 0. \end{cases}$$

2. Suppose that the solution of a minimum problem yields the final simplex tableau

x	y	u	v	M	
1	6	−1	0	0	11
0	5	3	1	0	16
0	2	4	0	1	−40

What is the minimum value sought in the original problem?

EXERCISES 4.3

In Exercises 1–8, solve the linear programming problems by the simplex method.

1. Maximize $40x + 30y$ subject to the constraints

$$\begin{cases} x + y \leq 5 \\ -2x + 3y \geq 12 \\ x \geq 0, \quad y \geq 0. \end{cases}$$

2. Maximize $3x - y$ subject to the constraints

$$\begin{cases} 2x + 5y \leq 100 \\ x \quad\quad \geq 10 \\ y \geq 0. \end{cases}$$

3. Minimize $3x + y$ subject to the constraints

$$\begin{cases} x + y \geq 3 \\ 2x \quad \geq 5 \\ x \geq 0, \quad y \geq 0. \end{cases}$$

4. Minimize $3x + 5y + z$ subject to the constraints

$$\begin{cases} x + y + z \geq 20 \\ y + 2z \geq 10 \\ x \geq 0, \quad y \geq 0, \quad z \geq 0. \end{cases}$$

5. Minimize $13x + 4y$ subject to the constraints

$$\begin{cases} y \geq -2x + 11 \\ y \leq -x + 10 \\ y \leq -\frac{1}{3}x + 6 \\ y \geq -\frac{1}{4}x + 4 \\ x \geq 0, \quad y \geq 0. \end{cases}$$

6. Minimize $500 - 10x - 3y$ subject to the constraints

$$\begin{cases} x + y \leq 20 \\ 3x + 2y \geq 50 \\ x \geq 0, \quad y \geq 0. \end{cases}$$

7. Minimize $2x + 7y$ subject to the constraints

$$\begin{cases} 2x + 5y \geq 30 \\ -3x + 5y \geq 5 \\ 8x + 3y \leq 101 \\ -9x + 7y \leq 42 \\ x \geq 0, \quad y \geq 0. \end{cases}$$

8. Minimize $10x + y$ subject to the constraints

$$\begin{cases} 3x + y \geq 16 \\ x + 2y \geq 12 \\ x \geq 2 \\ x \geq 0, \quad y \geq 0. \end{cases}$$

9. (*Nutrition*) A dietitian is designing a daily diet that is to contain at least 60 units of protein, 40 units of carbohydrates, and 120 units of fat. The diet is to consist of two types of foods. One serving of food A contains 30 units of protein, 10 units of carbohydrates, and 20 units of fat and costs \$3. One serving of food B contains 10 units of protein, 10 units of carbohydrates, and 60 units of fat and costs \$1.50. Design the diet that provides the daily requirements at the least cost.

10. (*Electronics Manufacture*) A manufacturing company has two plants, each capable of producing radios, TV sets, and stereo systems. The daily production capacities of each plant are as follows:

	Plant I	Plant II
Radios	10	20
TV sets	30	20
Stereo systems	20	10

Plant I costs \$1500 per day to operate, whereas plant II costs \$1200. How many days should each plant be operated to fill an order for 1000 radios, 1800 TV sets, and 1000 stereo systems at the minimum cost?

11. (*Supply and Demand*) An appliance store sells three brands of TV sets, brands A, B, and C. The profit per set is \$30 for brand A, \$50 for brand B, and \$60 for brand C. The total warehouse space allotted to all brands is sufficient for 600 sets, and the inventory is delivered only once per month. At least 100 customers per month will demand brand A, at least 50 will demand brand B, and at least 200 will demand either brand B or brand C. How can the appliance store satisfy all these constraints and earn maximum profit?

12. (*Political Campaign*) A citizen decides to campaign for the election of a candidate for city council. Her goal is to generate at least 210 votes by a combination of door-to-door canvassing, letter writing, and phone calls. She figures that each hour of door-to-door canvassing will generate four votes, each hour of letter writing will generate two votes, and each hour on the phone will generate three votes. She would like to devote at least seven hours to phone calls and spend at most half her time at door-to-door canvassing. How much time should she allocate to each task in order to achieve her goal in the least amount of time?

13. (*Inventory*) A manufacturer of computers must fill orders from two dealers. The computers are stored in two warehouses located at two airports, one in Boston (BOS) and one in Chicago (MDW). The dealers are located in Detroit, Michigan and Fletcher, North Carolina. There are 50 computers in stock in Boston and 80 in stock in Chicago. The dealer in Detroit orders 40 computers and the dealer in Fletcher orders 30 computers. The following table shows the costs of shipping one computer from each warehouse to each dealer. Find the shipping schedule with the minimum cost. What is the minimum cost?

	Detroit	Fletcher
Boston	$125	$180
Chicago	$100	$160

14. (*Inventory*) The manufacturer of computers in Exercise 13 gets a revised order from the dealer in Fletcher, now requiring 50 computers. How should the manufacturer adjust the schedule? What is the shipping schedule with the minimum cost? What is the cost?

In Exercises 15 and 16, use a graphing calculator or mathematical software to solve the linear programming problems by the simplex method.

15. Maximize $x - 2y$ subject to the constraints

$$\begin{cases} 4x + y \le 5 \\ x + 3y \ge 4 \\ x \ge 0, \quad y \ge 0. \end{cases}$$

16. Minimize $30x + 20y$ subject to the constraints

$$\begin{cases} 5x + 10y \ge 3 \\ 3x + 2y \ge 2 \\ x \ge 0, \quad y \ge 0. \end{cases}$$

SOLUTIONS TO PRACTICE PROBLEMS 4.3

1. To minimize $3x + 4y$, we maximize $-(3x + 4y) = -3x - 4y$. So the associated maximum problem is as follows: Maximize $-3x - 4y$ subject to the constraints

$$\begin{cases} -x + y \le 0 \\ -3x + 4y \le 0 \\ x \ge 0, \quad y \ge 0. \end{cases}$$

2. The value -40 in the lower right corner gives the solution of the associated *maximum* problem. The minimum value originally sought is the negative of the maximum value—that is, $-(-40) = 40$.

4.4 Sensitivity Analysis and Matrix Formulations of Linear Programming Problems

Sensitivity Analysis[1] **of Linear Programming Problems**   The simplex method not only provides the optimal solutions to linear programming problems but also gives other useful information. As a matter of fact, each number appearing in the final simplex tableau has an interpretation that not only sheds light on the current situation but also can be used to analyze the benefits of small changes in the available resources.

Consider the final simplex tableau of the furniture manufacturing problem. Since the variable u was introduced to take up the slack in the carpentry inequality, the u column has been labeled *Carpentry*. Similarly, the v column and the w column have been labeled *Finishing* and *Upholstery*, respectively:

	x	y	(*Carpentry*) u	(*Finishing*) v	(*Upholstery*) w	M	
x	1	0	$\frac{1}{3}$	-1	0	0	14
y	0	1	$-\frac{1}{3}$	2	0	0	4
w	0	0	$\frac{4}{3}$	-10	1	0	20
M	0	0	$\frac{10}{3}$	60	0	1	1400

The optimum profit is $1400, which occurs when $x = 14$ chairs and $y = 4$ sofas.

---

[1]Sensitivity analysis is also known as *postoptimality analysis*, *optimality analysis*, and *marginal analysis*.

We now consider the following question. If additional labor becomes available, how will this change the production level and the profit? To be specific, suppose that we had 3 more labor-hours available for carpentry. The initial tableau of the furniture manufacturing problem would become

		(Carpentry)	(Finishing)	(Upholstery)		
	x	y	u	v	w	M	
u	6	3	1	0	0	0	$96 + 3$
v	1	1	0	1	0	0	$18 + 0$
w	2	6	0	0	1	0	$72 + 0$
M	-80	-70	0	0	0	1	$0 + 0$

Note that only the first entry of the right-hand column has been changed. The increment to the right-hand column can be written as the column

$$\begin{matrix} 3 \\ 0 \\ 0 \\ \hline 0, \end{matrix}$$

which is three times the u column and is referred to as the *increment column*. Now, when the simplex method is performed on the new initial tableau, all of the row operations will affect the increment column exactly as they affected the u column in the original initial tableau. Therefore, the final increment column will be three times the final u column. Hence, the new final tableau will be

		(Carpentry)	(Finishing)	(Upholstery)		
	x	y	u	v	w	M	
x	1	0	$\frac{1}{3}$	-1	0	0	$14 + 3\left(\frac{1}{3}\right)$
y	0	1	$-\frac{1}{3}$	2	0	0	$4 + 3\left(-\frac{1}{3}\right)$
w	0	0	$\frac{4}{3}$	-10	1	0	$20 + 3\left(\frac{4}{3}\right)$
M	0	0	$\frac{10}{3}$	60	0	1	$1400 + 3\left(\frac{10}{3}\right)$

Thus, when 3 additional labor-hours of carpentry are available,

$$x = 14 + 3\left(\tfrac{1}{3}\right) = 15 \text{ chairs} \quad \text{and} \quad y = 4 + 3\left(-\tfrac{1}{3}\right) = 3 \text{ sofas}$$

should be produced. The maximum profit will increase to the new value of

$$M = 1400 + 3\left(\tfrac{10}{3}\right) = 1410 \text{ dollars.}$$

The number 3 was arbitrary. If h is a suitable number, then adding h labor-hours of labor for carpentry to the original problem results in the final tableau:

		(Carpentry)	(Finishing)	(Upholstery)		
	x	y	u	v	w	M	
x	1	0	$\frac{1}{3}$	-1	0	0	$14 + h\left(\frac{1}{3}\right)$
y	0	1	$-\frac{1}{3}$	2	0	0	$4 + h\left(-\frac{1}{3}\right)$
w	0	0	$\frac{4}{3}$	-10	1	0	$20 + h\left(\frac{4}{3}\right)$
M	0	0	$\frac{10}{3}$	60	0	1	$1400 + h\left(\frac{10}{3}\right)$

The number h can be positive or negative. For instance, if three fewer labor-hours are available for carpentry, then setting $h = -3$ yields the optimal production schedule $x = 13$, $y = 5$ and a profit of \$1390. The only restriction on h is that the numbers in the upper part of the right-hand column of the final tableau must all be nonnegative. These three numbers will be nonnegative provided that $-15 \leq h \leq 12$; that is, provided that the right-hand side of the carpentry inequality is between 81 and 108. The interval $[81, 108]$ is called the *range of feasibility* for the carpentry constraint.

Similarly, if in the original problem, the amount of labor available for finishing is increased by h labor-hours, the initial tableau becomes

	x	y	u	v	w	M	
u	6	3	1	0	0	0	$96+0$
v	1	1	0	1	0	0	$18+h$
w	2	6	0	0	1	0	$72+0$
M	-80	-70	0	0	0	1	$0+0$

with headers (Carpentry), (Finishing), (Upholstery).

The right-hand column of the original tableau was changed by adding an increment column that is h times the v column, and therefore the right-hand column of the new final tableau will be the original final right-hand column plus h times the final v column:

	x	y	u	v	w	M	
x	1	0	$\frac{1}{3}$	-1	0	0	$14+h(-1)$
y	0	1	$-\frac{1}{3}$	2	0	0	$4+h(2)$
w	0	0	$\frac{4}{3}$	-10	1	0	$20+h(-10)$
M	0	0	$\frac{10}{3}$	60	0	1	$1400+h(60)$

Hence, if one additional labor-hour were available for finishing, the optimal production schedule would be $x = 13$ chairs, $y = 6$ sofas, and the profit would be \$1460.

Finally, if in the original problem, the amount of labor available for upholstery is increased by h labor-hours, the initial and final tableaux become

	x	y	u	v	w	M	
u	6	3	1	0	0	0	$96+0$
v	1	1	0	1	0	0	$18+0$
w	2	6	0	0	1	0	$72+h$
M	-80	-70	0	0	0	1	$0+0$

	(Carpentry)	(Finishing)	(Upholstery)			
	x	y	u	v	w	M
x	1	0	$\frac{1}{3}$	-1	0	0	$14 + h(0)$
y	0	1	$-\frac{1}{3}$	2	0	0	$4 + h(0)$
w	0	0	$\frac{4}{3}$	-10	1	0	$20 + h(1)$
M	0	0	$\frac{10}{3}$	60	0	1	$1400 + h(0)$

Therefore, a change in the amount of labor available for upholstery has no effect on the production schedule or the profit. This makes sense, since we had excess labor available for upholstery in the solution to the original problem. The slack in carpentry and finishing was used up (u and v were 0), but there was slack in the labor available for upholstery (w was 20).

In summary, each of the slack variable columns in the final tableau of the original furniture manufacturing problem gives the sensitivity to change in the production schedule and in the profit due to a suitable change in one of the factors of production. The final values in each of these columns ($u = \frac{10}{3}$, $v = 60$, and $w = 0$) are called the *shadow prices* or *marginal values* of the three factors of production—carpentry, finishing, and upholstery.

The following example shows a complete analysis of a new linear programming problem.

EXAMPLE 1 **A production problem—adjusting to changing resources** The Cutting Edge Knife Company manufactures paring knives and pocket knives. Each paring knife requires 3 labor-hours, 7 units of steel, and 4 units of wood. Each pocket knife requires 6 labor-hours, 5 units of steel, and 3 units of wood. The profit on each paring knife is $3, and the profit on each pocket knife is $5. Each day the company has available 90 labor-hours, 138 units of steel, and 120 units of wood.

(a) How many of each type of knife should the Cutting Edge Knife Company manufacture daily to maximize its profits?

(b) Suppose that an additional 18 units of steel were available each day. What effect would this have on the optimal solution?

(c) Generalize the result in part (b) to the case where the increase in the number of units of steel available each day is h. (The value of h can be positive or negative.) For what range of values will the result be valid?

Solution We need to find the number of paring knives, x, and pocket knives, y, that will maximize the profit, $M = 3x + 5y$, subject to the constraints

$$\begin{cases} 3x + 6y \leq 90 \\ 7x + 5y \leq 138 \\ 4x + 3y \leq 120 \\ x \geq 0, \quad y \geq 0. \end{cases}$$

The initial tableau with slack variables u, v, and w added for labor, steel, and wood, respectively, is

		(Labor)	(Steel)	(Wood)			
	x	y	u	v	w	M	

$$
\begin{array}{c}
u \\
v \\
w \\
M
\end{array}
\left[
\begin{array}{cccccc|c}
3 & ⑥ & 1 & 0 & 0 & 0 & 90 \\
7 & 5 & 0 & 1 & 0 & 0 & 138 \\
4 & 3 & 0 & 0 & 1 & 0 & 120 \\
\hline
-3 & -5 & 0 & 0 & 0 & 1 & 0
\end{array}
\right]
\begin{array}{l}
\frac{90}{6} = 15 \\
\frac{138}{5} = 27.6 \\
\frac{120}{3} = 40.
\end{array}
$$

The proper pivot element is the entry 6 in the y column. The next tableau is

		(Labor)	(Steel)	(Wood)			
	x	y	u	v	w	M	

$$
\begin{array}{c}
y \\
v \\
w \\
M
\end{array}
\left[
\begin{array}{cccccc|c}
\frac{1}{2} & 1 & \frac{1}{6} & 0 & 0 & 0 & 15 \\
⑨⁄₂ & 0 & -\frac{5}{6} & 1 & 0 & 0 & 63 \\
\frac{5}{2} & 0 & -\frac{1}{2} & 0 & 1 & 0 & 75 \\
\hline
-\frac{1}{2} & 0 & \frac{5}{6} & 0 & 0 & 1 & 75
\end{array}
\right]
\begin{array}{l}
15/\frac{1}{2} = 30 \\
63/\frac{9}{2} = 14 \\
75/\frac{5}{2} = 30.
\end{array}
$$

The proper pivot element is the entry $\frac{9}{2}$ in the x column. The next tableau is

		(Labor)	(Steel)	(Wood)			
	x	y	u	v	w	M	

$$
\begin{array}{c}
y \\
x \\
w \\
M
\end{array}
\left[
\begin{array}{cccccc|c}
0 & 1 & \frac{7}{27} & -\frac{1}{9} & 0 & 0 & 8 \\
1 & 0 & -\frac{5}{27} & \frac{2}{9} & 0 & 0 & 14 \\
0 & 0 & -\frac{1}{27} & -\frac{5}{9} & 1 & 0 & 40 \\
\hline
0 & 0 & \frac{20}{27} & \frac{1}{9} & 0 & 1 & 82
\end{array}
\right].
$$

Since there are no negative entries in the last row of this tableau, the simplex method is complete.

(a) The Cutting Edge Knife Company should produce 14 paring knives and 8 pocket knives each day for a profit of \$82. (Since the slack variable w has the value 40, there will be 40 excess units of wood each day.)

(b) Since 18 additional units of steel are available, the final tableau of the revised problem can be obtained from the final tableau of the original problem by adding 18 times the v column to the right-hand column:

		(Labor)	(Steel)	(Wood)			
	x	y	u	v	w	M	

$$
\begin{array}{c}
y \\
x \\
w \\
M
\end{array}
\left[
\begin{array}{cccccc|c}
0 & 1 & \frac{7}{27} & -\frac{1}{9} & 0 & 0 & 8 + 18\left(-\frac{1}{9}\right) \\
1 & 0 & -\frac{5}{27} & \frac{2}{9} & 0 & 0 & 14 + 18\left(\frac{2}{9}\right) \\
0 & 0 & -\frac{1}{27} & -\frac{5}{9} & 1 & 0 & 40 + 18\left(-\frac{5}{9}\right) \\
\hline
0 & 0 & \frac{20}{27} & \frac{1}{9} & 0 & 1 & 82 + 18\left(\frac{1}{9}\right)
\end{array}
\right].
$$

The company should make 4 more paring knives $[18(\frac{2}{9}) = 4]$ and 2 fewer pocket knives $[18(-\frac{1}{9}) = -2]$. Doing so will increase the profits by \$2 $[18(\frac{1}{9}) = 2]$.

(c) With h additional units of steel available, the right-hand column of the final tableau will be similar to the preceding tableau but with 18 replaced by h:

		(Labor)	(Steel)	(Wood)			
	x	y	u	v	w	M	
y	0	1	$\frac{7}{27}$	$-\frac{1}{9}$	0	0	$8 + h\left(-\frac{1}{9}\right)$
x	1	0	$-\frac{5}{27}$	$\frac{2}{9}$	0	0	$14 + h\left(\frac{2}{9}\right)$
w	0	0	$-\frac{1}{27}$	$-\frac{5}{9}$	1	0	$40 + h\left(-\frac{5}{9}\right)$
M	0	0	$\frac{20}{27}$	$\frac{1}{9}$	0	1	$82 + h\left(\frac{1}{9}\right)$

Therefore, the number of paring knives made should be $14 + h\left(\frac{2}{9}\right)$, and the number of pocket knives made should be $8 + h\left(-\frac{1}{9}\right)$. The new profit will be $82 + h\left(\frac{1}{9}\right)$ dollars. This analysis is valid provided that each of the entries in the upper part of the right-hand column of the tableau is nonnegative. The restrictions on h given by each of these entries are as follows:

Entry	Restriction
$8 + h\left(-\frac{1}{9}\right)$	$h \le 72$
$14 + h\left(\frac{2}{9}\right)$	$h \ge -63$
$40 + h\left(-\frac{5}{9}\right)$	$h \le 72$

All of these restrictions will be satisfied if h is between -63 and 72.

Now Try Exercise 1

■

Matrix Formulations of Linear Programming Problems Linear programming problems can be neatly stated in terms of matrices. Such formulations provide a convenient way to define the *dual* of a linear programming problem, an important concept that is studied in Section 4.5. To introduce the matrix formulation of a linear programming problem, we first need the concept of inequality for matrices.

Let A and B be two matrices of the same size. We say that A is less than or equal to B (denoted $A \le B$) if each entry of A is less than or equal to the corresponding entry of B. For instance, we have the following matrix inequalities:

$$\begin{bmatrix} 2 & -3 \\ \frac{1}{2} & 0 \end{bmatrix} \le \begin{bmatrix} 5 & -1 \\ 1 & 0 \end{bmatrix} \quad \text{and} \quad \begin{bmatrix} 5 \\ 6 \end{bmatrix} \le \begin{bmatrix} 8 \\ 9 \end{bmatrix}.$$

The symbol $\ge$ has an analogous meaning for matrices.

EXAMPLE 2 **Matrix formulation of a linear programming problem** Let

$$A = \begin{bmatrix} 6 & 3 \\ 1 & 1 \\ 2 & 6 \end{bmatrix}, \quad B = \begin{bmatrix} 96 \\ 18 \\ 72 \end{bmatrix}, \quad C = \begin{bmatrix} 80 & 70 \end{bmatrix}, \quad X = \begin{bmatrix} x \\ y \end{bmatrix}.$$

Carry out the indicated matrix multiplications in the following statement: Maximize CX subject to the constraints $AX \le B$, $X \ge \mathbf{0}$. ($\mathbf{0}$ is the matrix of all zeros.)

Solution $CX = \begin{bmatrix} 80 & 70 \end{bmatrix} \begin{bmatrix} x \\ y \end{bmatrix} = \begin{bmatrix} 80x + 70y \end{bmatrix}$

$$AX = \begin{bmatrix} 6 & 3 \\ 1 & 1 \\ 2 & 6 \end{bmatrix} \begin{bmatrix} x \\ y \end{bmatrix} = \begin{bmatrix} 6x + 3y \\ x + y \\ 2x + 6y \end{bmatrix}$$

$$AX \leq B \text{ means } \begin{bmatrix} 6x + 3y \\ x + y \\ 2x + 6y \end{bmatrix} \leq \begin{bmatrix} 96 \\ 18 \\ 72 \end{bmatrix} \quad \text{or} \quad \begin{cases} 6x + 3y \leq 96 \\ x + y \leq 18 \\ 2x + 6y \leq 72. \end{cases}$$

$$X \geq \mathbf{0} \text{ means } \begin{bmatrix} x \\ y \end{bmatrix} \geq \begin{bmatrix} 0 \\ 0 \end{bmatrix} \quad \text{or} \quad \begin{cases} x \geq 0 \\ y \geq 0. \end{cases}$$

Hence, the statement "Maximize CX subject to the constraints $AX \leq B$, $X \geq \mathbf{0}$" is a matrix formulation of the furniture manufacturing problem.

Now Try Exercise 13 ■

Another concept that is needed for the definition of the dual of a linear programming problem is the *transpose of a matrix*. If A is an $m \times n$ matrix, then the matrix A^T (read "A transpose") is the $n \times m$ matrix whose ijth entry is the jith entry of A. The rows of A^T are the columns of A, and vice versa.

EXAMPLE 3 **Finding the transpose of a matrix** Find the transpose of

(a) $\begin{bmatrix} 3 & -2 & 4 \\ 6 & 5 & 0 \end{bmatrix}$ (b) $\begin{bmatrix} 5 \\ 2 \\ 1 \end{bmatrix}$

Solution (a) Since the given matrix has two rows and three columns, its transpose will have three rows and two columns. The entries of the first row of the transpose will be the entries in the first column of the original matrix:

$$\begin{bmatrix} 3 & 6 \\ & \end{bmatrix}.$$

The entries of the second and third rows are obtained in a similar manner from the second and third columns of the original matrix. Therefore,

$$\begin{bmatrix} 3 & -2 & 4 \\ 6 & 5 & 0 \end{bmatrix}^T = \begin{bmatrix} 3 & 6 \\ -2 & 5 \\ 4 & 0 \end{bmatrix}.$$

(b) Since the given matrix has three rows and one column, its transpose has one row and three columns. That row consists of the single column of the original matrix:

$$\begin{bmatrix} 5 \\ 2 \\ 1 \end{bmatrix}^T = \begin{bmatrix} 5 & 2 & 1 \end{bmatrix}.$$

Now Try Exercise 7 ■

ES In Section 3.3, we showed how to generate a sensitivity report with Solver. Table 1 shows the second table in the sensitivity sheet report for the furniture manufacturing problem. The cells $B$1, $B$2, and $B$3 represent carpentry, finishing, and upholstery, respectively. For each of the three factors of production, the table gives its shadow price along with the amounts by which the right side of its constraint can be increased or decreased without affecting the shadow price. (*Note:* The spreadsheet uses the expression 1E+30 to represent infinity.) For instance, the right side of the carpentry constraint can be any number from 81 to 108 without changing the shadow price for carpentry. That is, the interval $[81, 108]$ is the range of feasibility for the carpentry constraint.

TABLE 1 The second table of the sensitivity sheet

Constraints

Cell	Name	Final Value	Shadow Price	Constraint R.H. Side	Allowable Increase	Allowable Decrease
$B$1		96	3.333333333	96	12	15
$B$2		18	60	18	2	2
$B$3		52	0	72	1E+30	20
$B$4		14	0	0	14	1E+30
$B$5		4	0	0	4	1E+30

PRACTICE PROBLEMS 4.4

Consider the furniture manufacturing problem, whose final simplex tableau appears in Section 4.2.

1. Suppose that the number of labor-hours for finishing that are available each day is decreased by 2. What will be the effect on the optimal number of chairs and sofas produced and on the profit?

2. For what range of values of h will a sensitivity analysis on the effect of a change of h labor-hours for finishing be valid?

EXERCISES 4.4

Exercises 1 and 2 refer to the Cutting Edge Knife Company problem of Example 1.

1. Suppose that the number of labor-hours that are available each day is increased by 54. Use sensitivity analysis to determine the effect on the optimal number of knives produced and on the profit.

2. For what range of values of h will a sensitivity analysis on the effect of a change of h labor-hours be valid?

Exercises 3 and 4 refer to the transportation problem of Example 3 in Section 4.3.

3. Suppose that the number of TV sets stocked in the College Park warehouse is increased to 50. What will be the effect on the optimal numbers of TV sets shipped from each warehouse to each store and what will be the change in the cost?

4. For what range of values of h will a sensitivity analysis on the effect of a change of h in the number of TV sets

stocked in the College Park store be valid?

5. Refer to Exercise 23 of Section 4.2. For what values of h will a sensitivity analysis on the effect of a change of h pounds of peat be valid?

6. Consider the nutrition problem in Example 1 of Section 3.2. Solve the problem by the simplex method and then determine the optimal quantities of soybeans and rice in the diet, and the new cost, if the daily requirement for calories is increased to 1700. For what range of values of h will a sensitivity analysis on the effects of a change of h calories be valid?

In Exercises 7–10, find the transpose of the given matrix.

7. $\begin{bmatrix} 9 & 4 \\ 1 & 8 \\ 1 & -3 \end{bmatrix}$

8. $\begin{bmatrix} 4 \\ 0 \\ 6 \end{bmatrix}$

9. $\begin{bmatrix} 7 & 6 & 5 & 1 \end{bmatrix}$

10. $\begin{bmatrix} 5 & 2 \\ 3 & -1 \end{bmatrix}$

11. Is it true that the transpose of the transpose of a matrix is the original matrix?

12. Give an example of a matrix that is its own transpose.

In Exercises 13 and 14, give the matrix formulation of the linear programming problem.

13. Minimize $7x + 5y + 4z$ subject to

$$\begin{cases} 3x + 8y + 9z \geq 75 \\ x + 2y + 5z \geq 80 \\ 4x + y + 7z \geq 67 \\ x \geq 0, \quad y \geq 0, \quad z \geq 0. \end{cases}$$

14. Maximize $20x + 30y$ subject to

$$\begin{cases} 7x + 8y \leq 55 \\ x + 2y \leq 78 \\ x \leq 25 \\ x \geq 0, \quad y \geq 0. \end{cases}$$

15. Give a matrix formulation of the Cutting Edge Knife Company problem of Example 1.

16. Does every linear programming problem have a matrix formulation? If not, under what conditions will a linear programming problem have a matrix formulation?

In Exercises 17 and 18, let

$$C = \begin{bmatrix} 2 & 3 \end{bmatrix}, \quad X = \begin{bmatrix} x \\ y \end{bmatrix}, \quad A = \begin{bmatrix} 7 & 4 \\ 5 & 8 \\ 1 & 3 \end{bmatrix},$$

$$B = \begin{bmatrix} 33 \\ 44 \\ 55 \end{bmatrix}, \quad and \quad U = \begin{bmatrix} u \\ v \\ w \end{bmatrix}.$$

17. Give the linear programming problem whose matrix formulation is "Minimize CX subject to the constraints $AX \geq B$, $X \geq \mathbf{0}$."

18. Give the linear programming problem whose matrix formulation is "Maximize $B^T U$ subject to the constraints $A^T U \leq C^T$, $U \geq \mathbf{0}$."

Exercises 19 and 20 require a spreadsheet.

19. Use an electronic spreadsheet such as Excel to create a sensitivity report for the transportation problem of Example 2 of Section 3.3. Use the report to determine the shadow price and range of feasibility for the numbers of sets shipped from College Park.

20. Use an electronic spreadsheet such as Excel to create a sensitivity report for the nutrition problem of Example 1 of Section 3.2. Use the report to determine the shadow prices and ranges of feasibility for each of the three nutritional factors.

SOLUTIONS TO PRACTICE PROBLEMS 4.4

1. Since the finishing column in the original final tableau is

$$\begin{array}{c} -1 \\ 2 \\ -10 \\ \hline 60, \end{array}$$

the right-hand column in the new tableau is

$$\begin{array}{c} 14 + (-2)(-1) \\ 4 + (-2)(2) \\ 20 + (-2)(-10) \\ \hline 1400 + (-2)(60). \end{array}$$

Therefore, the new values of x, y, and M are 16, 0, and 1280.

2. Using h instead of -2 in the marginal analysis, we find that the right-hand column of the new final tableau is

$$\begin{array}{c} 14 + h(-1) \\ 4 + h(2) \\ 20 + h(-10) \\ \hline 1400 + h(60). \end{array}$$

Of course, this analysis is valid only if the three numbers above the line are not negative. That is, $14 + h(-1) \geq 0$, $4 + h(2) \geq 0$, and $20 + h(-10) \geq 0$. These three inequalities can be simplified to $h \leq 14$, $h \geq -2$, and $h \leq 2$. Therefore, in order to satisfy all three inequalities, h must be in the range $-2 \leq h \leq 2$.

4.5 Duality

Each linear programming problem may be converted into a related linear programming problem called its *dual*. The dual problem is sometimes easier to solve than the original problem and, moreover, it has the same optimum value. Furthermore, the solution of the dual problem often can provide valuable insights into the original problem. To understand the relationship between a linear programming problem and its dual, it is best to begin with a concrete example.

Problem A   Maximize the objective function $6x + 5y$ subject to the constraints

$$\begin{cases} 4x + 8y \leq 32 \\ 3x + 2y \leq 12 \\ x \geq 0, \quad y \geq 0. \end{cases}$$

The dual of Problem A is the following problem.

Problem B   Minimize the objective function $32u + 12v$ subject to the constraints

$$\begin{cases} 4u + 3v \geq 6 \\ 8u + 2v \geq 5 \\ u \geq 0, \quad v \geq 0. \end{cases}$$

The relationship between the two problems is easiest to see if we write them in their matrix formulations. Problem A is

$$\text{Maximize } \begin{bmatrix} 6 & 5 \end{bmatrix} \begin{bmatrix} x \\ y \end{bmatrix} \text{ subject to the constraints}$$

$$\begin{bmatrix} 4 & 8 \\ 3 & 2 \end{bmatrix} \begin{bmatrix} x \\ y \end{bmatrix} \leq \begin{bmatrix} 32 \\ 12 \end{bmatrix} \quad \text{and} \quad \begin{bmatrix} x \\ y \end{bmatrix} \geq \begin{bmatrix} 0 \\ 0 \end{bmatrix}.$$

Problem B is

$$\text{Minimize } \begin{bmatrix} 32 & 12 \end{bmatrix} \begin{bmatrix} u \\ v \end{bmatrix} \text{ subject to the constraints}$$

$$\begin{bmatrix} 4 & 3 \\ 8 & 2 \end{bmatrix} \begin{bmatrix} u \\ v \end{bmatrix} \geq \begin{bmatrix} 6 \\ 5 \end{bmatrix} \quad \text{and} \quad \begin{bmatrix} u \\ v \end{bmatrix} \geq \begin{bmatrix} 0 \\ 0 \end{bmatrix}.$$

Each of the numeric matrices in Problem B is the transpose of one of the matrices in Problem A. Let

$$C = \begin{bmatrix} 6 & 5 \end{bmatrix}, \quad X = \begin{bmatrix} x \\ y \end{bmatrix}, \quad A = \begin{bmatrix} 4 & 8 \\ 3 & 2 \end{bmatrix}, \quad B = \begin{bmatrix} 32 \\ 12 \end{bmatrix}, \quad U = \begin{bmatrix} u \\ v \end{bmatrix}, \quad \mathbf{0} = \begin{bmatrix} 0 \\ 0 \end{bmatrix}.$$

Problem A is

"Maximize CX subject to the constraints $AX \leq B$, $X \geq \mathbf{0}$."

Problem B is

"Minimize $B^T U$ subject to the constraints $A^T U \geq C^T$, $U \geq \mathbf{0}$."

Problem A is referred to as the *primal* problem and Problem B as its *dual*.

> **The Dual of a Linear Programming Problem**
>
> **1.** If the original (primal) problem has the form
>
> Maximize CX subject to the constraints $AX \leq B$, $X \geq \mathbf{0}$,
>
> then the dual problem is
>
> Minimize $B^T U$ subject to the constraints $A^T U \geq C^T$, $U \geq \mathbf{0}$.
>
> **2.** If the original (primal) problem has the form
>
> Minimize CX subject to the constraints $AX \geq B$, $X \geq \mathbf{0}$,
>
> then the dual problem is
>
> Maximize $B^T U$ subject to the constraints $A^T U \leq C^T$, $U \geq \mathbf{0}$.

Note that Problem A is a standard maximization problem. That is, all of the inequalities involve $\leq$, except for $x \geq 0$ and $y \geq 0$. Problem B is a standard minimization problem in that all of its inequalities are $\geq$. The coefficients of the objective function for Problem A are the numbers on the right-hand side of the inequalities of Problem B, and vice versa. The coefficient matrices for the left-hand sides of the inequalities are transposes of one another. In an analogous way, we can start with a standard minimization problem and define its dual to be a standard maximization problem. Any linear programming problem can be put into one of these two standard forms. (If an inequality points in the wrong direction, we need only multiply it by -1.) Therefore, every linear programming problem has a dual.

EXAMPLE 1 **Finding the dual in standard form** Determine the dual of the following linear programming problem. Minimize $18x + 20y + 2z$ subject to the constraints

$$\begin{cases} 3x - 5y - 2z \leq 4 \\ 6x \quad\quad - 8z \geq 9 \\ x \geq 0, \quad y \geq 0, \quad z \geq 0. \end{cases}$$

Solution We first put the problem into standard form. Since the primal problem is a minimization problem, we must write all constraints with the inequality sign $\geq$. To put the first inequality in this form, we multiply by -1 to obtain

$$-3x + 5y + 2z \geq -4.$$

We now write the problem in matrix form:

Minimize $\begin{bmatrix} 18 & 20 & 2 \end{bmatrix} \begin{bmatrix} x \\ y \\ z \end{bmatrix}$ subject to the constraints

$$\begin{bmatrix} -3 & 5 & 2 \\ 6 & 0 & -8 \end{bmatrix} \begin{bmatrix} x \\ y \\ z \end{bmatrix} \geq \begin{bmatrix} -4 \\ 9 \end{bmatrix} \quad \text{and} \quad \begin{bmatrix} x \\ y \\ z \end{bmatrix} \geq \begin{bmatrix} 0 \\ 0 \\ 0 \end{bmatrix}.$$

The dual is

Maximize $\begin{bmatrix} -4 & 9 \end{bmatrix} \begin{bmatrix} u \\ v \end{bmatrix}$ subject to the constraints

$$\begin{bmatrix} -3 & 6 \\ 5 & 0 \\ 2 & -8 \end{bmatrix} \begin{bmatrix} u \\ v \end{bmatrix} \leq \begin{bmatrix} 18 \\ 20 \\ 2 \end{bmatrix} \quad \text{and} \quad \begin{bmatrix} u \\ v \end{bmatrix} \geq \begin{bmatrix} 0 \\ 0 \end{bmatrix}.$$

Multiplying the matrices, we obtain the following:
Maximize $-4u + 9v$ subject to

$$\begin{cases} -3u + 6v \le 18 \\ 5u \qquad \le 20 \\ 2u - 8v \le \;\; 2 \\ u \ge 0, \quad v \ge 0. \end{cases}$$

Now Try Exercise 1

■

Let us now return to Problems A and B to examine the connection between the solutions of a linear programming problem and its dual problem. Problems A and B both involve two variables and hence can be solved by the geometric method of Chapter 3. Figure 1 shows their respective feasible sets and the vertices that yield the optimum values of the objective functions. The feasible sets do not look alike, and the optimal vertices are different. However, both problems have the same optimum value, 27. The relationship between the two problems is brought into even sharper focus by looking at the final tableaux that arise when the two problems are solved by the simplex method. (*Note*: In Problem B, the original variables are u and v and the slack variables have been named x and y.)

FINAL TABLEAUX

Problem A

	x	y	u	v	M	
y	0	1	$\frac{3}{16}$	$-\frac{1}{4}$	0	3
x	1	0	$-\frac{1}{8}$	$\frac{1}{2}$	0	2
M	0	0	$\frac{3}{16}$	$\frac{7}{4}$	1	27

Problem B

	u	v	x	y	M	
v	0	1	$-\frac{1}{2}$	$\frac{1}{4}$	0	$\frac{7}{4}$
u	1	0	$\frac{1}{8}$	$-\frac{3}{16}$	0	$\frac{3}{16}$
M	0	0	2	3	1	-27

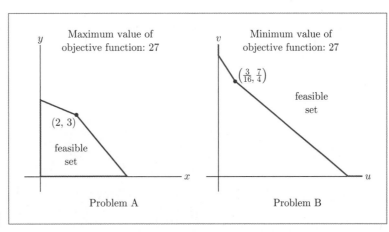

Problem A — Maximum value of objective function: 27; feasible set with vertex (2, 3).
Problem B — Minimum value of objective function: 27; feasible set with vertex $\left(\frac{3}{16}, \frac{7}{4}\right)$.

Figure 1

The final tableau for Problem A contains the solution to Problem B $(u = \frac{3}{16}$, $v = \frac{7}{4})$ in the final entries of the u and v columns. Similarly, the final tableau for Problem B gives the solution to Problem A $(x = 2, y = 3)$ in the final entries of its x and y columns. This situation always occurs. The solutions to a linear programming problem and its dual problem may be obtained simultaneously by solving just one of the problems using the simplex method and applying the following theorem.

Fundamental Theorem of Duality

1. If either the primal problem or the dual problem has an optimal solution, then they both have an optimal solution and their objective functions have the same values at these optimal points.

2. If both the primal and the dual problems have a feasible solution, then they both have optimal solutions and their objective functions have the same value at these optimal points.

3. The solution of one of these problems by the simplex method yields the solution of the other problem as the final entries in the columns associated with the slack variables.

EXAMPLE 2   **Solving a minimization problem using its dual**   Solve the linear programming problem of Example 1 by applying the simplex method to its dual problem.

Solution   In the solution to Example 1, the dual problem is as follows: Maximize $-4u + 9v$ subject to the constraints

$$\begin{cases} -3u + 6v \le 18 \\ 5u \qquad \le 20 \\ 2u - 8v \le \ 2 \\ u \ge 0, \quad v \ge 0. \end{cases}$$

Since there are three nontrivial inequalities, the simplex method calls for three slack variables. Denote the slack variables by x, y, and z. Let $M = -4u + 9v$, and apply the simplex method.

$$
\begin{array}{c}
\begin{array}{ccccccc}
\quad u & v & x & y & z & M &
\end{array} \\
\begin{array}{c} x \\ y \\ z \\ M \end{array}
\left[
\begin{array}{cccccc|c}
-3 & 6 & 1 & 0 & 0 & 0 & 18 \\
5 & 0 & 0 & 1 & 0 & 0 & 20 \\
2 & -8 & 0 & 0 & 1 & 0 & 2 \\
4 & -9 & 0 & 0 & 0 & 1 & 0
\end{array}
\right]
\end{array}
$$

$$
\begin{array}{c}
\begin{array}{ccccccc}
\quad u & v & x & y & z & M &
\end{array} \\
\begin{array}{c} v \\ y \\ z \\ M \end{array}
\left[
\begin{array}{cccccc|c}
-\frac{1}{2} & 1 & \frac{1}{6} & 0 & 0 & 0 & 3 \\
5 & 0 & 0 & 1 & 0 & 0 & 20 \\
-2 & 0 & \frac{4}{3} & 0 & 1 & 0 & 26 \\
-\frac{1}{2} & 0 & \frac{3}{2} & 0 & 0 & 1 & 27
\end{array}
\right]
\end{array}
$$

$$
\begin{array}{c}
\begin{array}{cccccc}
u & v & x & y & z & M
\end{array} \\
\begin{array}{c}
v \\ u \\ z \\ M
\end{array}
\left[
\begin{array}{cccccc|c}
0 & 1 & \frac{1}{6} & \frac{1}{10} & 0 & 0 & 5 \\
1 & 0 & 0 & \frac{1}{5} & 0 & 0 & 4 \\
0 & 0 & \frac{4}{3} & \frac{2}{5} & 1 & 0 & 34 \\
0 & 0 & \frac{3}{2} & \frac{1}{10} & 0 & 1 & 29
\end{array}
\right]
\end{array}
$$

Since the maximum value of the dual problem is 29, we know that the minimum value of the original problem is also 29. Looking at the last row of the final tableau, we conclude that this minimum value is assumed when $x = \frac{3}{2}$, $y = \frac{1}{10}$, and $z = 0$.

Now Try Exercise 11   ∎

In Example 2 the dual problem was easier to solve than the original problem given in Example 1. Thus, we see how consideration of the dual problem may simplify the solution of linear programming problems in some cases.

An Economic Interpretation of the Dual Problem   To illustrate the interpretation of the dual problem, let's reconsider the furniture manufacturing problem. Recall that this problem asked us to maximize the profit from the sale of x chairs and y sofas subject to limitations on the amount of labor available for carpentry, finishing, and upholstery. In mathematical terms, the problem required us to maximize $80x + 70y$ subject to the constraints

$$
\begin{cases}
6x + 3y \le 96 \\
x + y \le 18 \\
2x + 6y \le 72 \\
x \ge 0, \quad y \ge 0.
\end{cases}
$$

Its dual problem is to minimize $96u + 18v + 72w$ subject to the constraints

$$
\begin{cases}
6u + v + 2w \ge 80 \\
3u + v + 6w \ge 70 \\
u \ge 0, \quad v \ge 0, \quad w \ge 0.
\end{cases}
$$

The variables u, v, and w can be assigned a meaning so that the dual problem has a significant interpretation in terms of the original problem.

First, recall the following table of data (labor-hours except as noted):

	Chair	Sofa	Available labor-hours
Carpentry	6	3	96
Finishing	1	1	18
Upholstery	2	6	72
Profit	$80	$70	

Suppose that we have an opportunity to hire out all our workers. Suppose that hiring out carpenters will yield a profit of u dollars per hour, the finishers v dollars per hour, and the upholsterers w dollars per hour. Of course, u, v, and w all must be ≥ 0. However, there are other constraints that we should reasonably impose. Any scheme for hiring out the workers should generate at least as much profit as is currently being generated in the construction of chairs and sofas. In terms of the potential profits from hiring the workers out, the labor involved in constructing a chair will generate

$$6u + v + 2w$$

dollars of profit. And this amount should be at least equal to the $80 profit that could be earned by using the labor to construct a chair. That is, we have the constraint

$$6u + v + 2w \geq 80.$$

Similarly, considering the labor involved in building a sofa, we derive the constraint

$$3u + v + 6w \geq 70.$$

Since there are available 96 hours of carpentry, 18 hours of finishing, and 72 hours of upholstery, the total profit from hiring out the workers would be

$$96u + 18v + 72w.$$

Thus, the problem of determining the least acceptable profit from hiring out the workers is equivalent to the following: Minimize $96u + 18v + 72w$ subject to the constraints

$$\begin{cases} 6u + v + 2w \geq 80 \\ 3u + v + 6w \geq 70 \\ u \geq 0, \quad v \geq 0, \quad w \geq 0. \end{cases}$$

This is just the dual of the furniture manufacturing problem. The values u, v, and w are measures of the value of an hour's labor by each type of worker. Earlier we referred to them as *shadow prices*. The fundamental theorem of duality asserts that the minimum acceptable profit that can be achieved by hiring the workers out is equal to the maximum profit that can be generated if they make furniture.

As we saw in the furniture manufacturing problem, the maximum profit, $M = 1400$, is achieved when

$$x = 14, \qquad y = 4, \qquad u = 0, \qquad v = 0, \qquad w = 20.$$

For the dual problem, we can read from the final tableau of the primal problem that the minimum acceptable profit, $M = 1400$, is achieved when

$$u = \tfrac{10}{3}, \qquad v = 60, \qquad w = 0, \qquad x = 0, \qquad y = 0.$$

The solution of the primal problem gives the activity level that meets the constraints imposed by the resources and provides the maximum profit. Such a model is called an *allocation problem*. On the other hand, the solution of the dual problem assigns values to each of the resources of production. The solution of the dual problem might be used by an insurance salesperson or by an accountant to impute a value to each resource. It is referred to as a *valuation problem*.

We summarize the situation briefly as follows:

If the original problem is a maximization problem,

Maximize CX subject to $AX \le B$ and $X \ge \mathbf{0}$,

then we interpret the solution matrix X as the *activity matrix* in which each entry gives the optimal level of each activity. The matrix B is the *capacity* or *resources matrix*, where each entry represents the available amount of a (scarce) resource. The matrix C is the *profit matrix* whose entries are the unit profits for each activity represented in X. The dual solution matrix U is the *imputed value matrix*, which gives the imputed value of each of the resources in the production process.

If the original problem is a minimization problem,

Minimize CX subject to $AX \ge B$ and $X \ge \mathbf{0}$,

then we interpret X as the *activity matrix* in which each entry gives the optimal level of an activity. B is the *requirements matrix* in which each entry is a minimum required level of production for some commodity. C is the *cost matrix*, where each entry is the unit cost of the corresponding activity in X. The dual solution matrix U is the *imputed cost matrix* whose entries are the costs imputed to the required commodities.

We see that $x = 14 > 0$ in the solution to the primal furniture manufacturing problem, and $x = 0$ in the dual. Also, $u = \frac{10}{3} > 0$ in the dual and $u = 0$ in the primal problem. In general, this is the complementary nature of the solutions to the primal and dual problems. Each variable that has a positive value in the solution of the primal problem has the value 0 in the solution of the dual. Similarly, if a variable has a positive value in the solution to the dual problem, then it has the value 0 in the solution of the primal problem. This result is called the *principle of complementary slackness*.

The economic interpretation of complementary slackness is this: If a slack variable from the primal problem is positive, then having available more of the corresponding resource cannot improve the value of the objective function. For instance, since $w > 0$ in the solution of the primal problem and $2x + 6y + w = 72$, not all of the labor available for upholstery will be used in the optimal production schedule. Therefore, the shadow price of the resource is zero. As we saw in Section 4.4, the shadow price of a resource is the final value in the column of its slack variable in the primal problem. But this number is the value of the main variable in the solution of the dual problem.

When we gave an economic interpretation to the dual of the furniture manufacturing problem, we first had to assign units to each of the variables of the dual problem. For instance, u was in units of profit per labor-hour of carpentry, or

$$\frac{\text{profit}}{\text{labor-hours of carpentry}}.$$

The following systematic procedure can often be used to obtain the units for the variables of the dual problem from units appearing in the primal problem. Assume that the primal problem is stated in one of the two forms given previously.

1. Replace each entry of the matrix A by its units, written in fraction form. Label each column and row with the corresponding variable.
2. Replace each entry of the matrix C by its units, written in fraction form.
3. To find the units for a variable of the dual problem, select any entry in its row in A, divide the corresponding entry in C by the entry chosen in A, and simplify the fraction.

For the furniture manufacturing problem, the matrices are

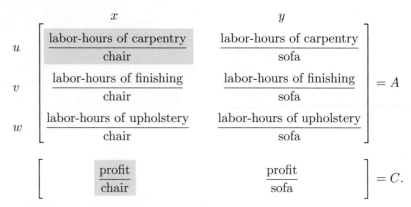

Using the first entry in the row labeled u, the units for u in the dual problem are

$$\frac{\text{profit}}{\text{chair}} \div \frac{\text{labor-hours of carpentry}}{\text{chair}} = \frac{\text{profit}}{\cancel{\text{chair}}} \cdot \frac{\cancel{\text{chair}}}{\text{labor-hours of carpentry}}$$

$$= \frac{\text{profit}}{\text{labor-hours of carpentry}}.$$

Here, u is measured in dollars per labor-hour of carpentry.

EXAMPLE 3   **Optimizing dietary allotments**   A rancher needs to provide a daily dietary supplement to the minks on his ranch. He needs 6 units of protein and 5 units of carbohydrates to add to their regular feed each day. Bran X costs 32 cents per ounce and supplies 4 units of protein and 8 units of carbohydrates per ounce. Wheatchips cost 12 cents per ounce, and each ounce supplies 3 units of protein and 2 units of carbohydrates.

(a) Determine the mixture of Bran X and Wheatchips that will meet the daily requirements at minimum cost.

(b) Solve and interpret the dual problem.

Solution (a) Let x be the number of ounces of Bran X and y the number of ounces of Wheatchips to be added to the feed. Then we must find the values of x and y that minimize $32x + 12y$ subject to the constraints

$$\begin{cases} 4x + 3y \geq 6 \\ 8x + 2y \geq 5 \\ x \geq 0, \quad y \geq 0. \end{cases}$$

For simplicity, we solve the dual problem. Let u and v be the dual variables. We must find the values of u and v that maximize $6u + 5v$ subject to the constraints

$$\begin{cases} 4u + 8v \leq 32 \\ 3u + 2v \leq 12 \\ u \geq 0, \quad v \geq 0. \end{cases}$$

The final tableau for the dual problem (with slack variables x and y) is

$$
\begin{array}{c|ccccc|c}
 & u & v & x & y & M & \\
\hline
v & 0 & 1 & \frac{3}{16} & -\frac{1}{4} & 0 & 3 \\
u & 1 & 0 & -\frac{1}{8} & \frac{1}{2} & 0 & 2 \\
\hline
M & 0 & 0 & \frac{3}{16} & \frac{7}{4} & 1 & 27
\end{array}.
$$

The solution to the rancher's problem is to add $x = \frac{3}{16}$ ounce of Bran X and $y = \frac{7}{4} = 1\frac{3}{4}$ ounces of Wheatchips to the daily feed at a minimum cost of 27 cents per day.

(b) The solution of the dual problem can be read from the preceding tableau: $u = 2$ and $v = 3$. To find the units of u and v, we construct a chart from the original problem:

$$
\begin{array}{c}
\qquad\qquad x \qquad\qquad\qquad\qquad\qquad y \\[4pt]
\begin{array}{c}
u \\[24pt] v
\end{array}
\left[
\begin{array}{cc}
\dfrac{\text{units of protein}}{\text{ounce Bran } X} & \dfrac{\text{units of protein}}{\text{ounce Wheatchips}} \\[14pt]
\dfrac{\text{units of carbohydrates}}{\text{ounce Bran } X} & \dfrac{\text{units of carbohydrates}}{\text{ounce Wheatchips}}
\end{array}
\right]
\end{array}
$$

$$
\left[
\begin{array}{cc}
\dfrac{\text{cents}}{\text{ounce Bran } X} & \dfrac{\text{cents}}{\text{ounce Wheatchips}}
\end{array}
\right].
$$

The units for u in the dual problem are

$$
\frac{\text{cents}}{\text{ounce Bran } X} \div \frac{\text{units of protein}}{\text{ounce Bran } X} = \frac{\text{cents}}{\text{ounce Bran } X} \cdot \frac{\text{ounce Bran } X}{\text{units of protein}}
$$

$$
= \frac{\text{cents}}{\text{units of protein}}.
$$

We can interpret the dual this way: If someone were to provide the perfect daily supplement for the minks with 6 units of protein and 5 units of carbohydrates, the rancher would expect to pay 27 cents. He certainly would pay no more, since he could mix his own supplement for that price per day. The value of the protein is 2 cents per unit and the value of the carbohydrates is 3 cents per unit. ∎

A Useful Application of the Dual   One type of decision that businesses must make is whether or not to introduce new products. The following example uses a linear programming problem and its dual to determine the proper course of action.

EXAMPLE 4   **Introducing new products**   Consider the furniture manufacturing problem once again. Suppose that the manufacturer has the same resources but is considering adding a new product to his line, love seats. The manufacture of a love seat requires 3 hours of carpentry, 2 hours of finishing, and 4 hours of upholstery. What profit must he gain per love seat in order to justify adding love seats to his product line?

Solution Let z denote the number of love seats to be produced each day, and let p be the profit per love seat. The new furniture manufacturing problem is as follows:
Maximize $80x + 70y + pz$ subject to

$$\begin{cases} 6x + 3y + 3z \leq 96 \\ x + y + 2z \leq 18 \\ 2x + 6y + 4z \leq 72 \\ x \geq 0, \quad y \geq 0, \quad z \geq 0. \end{cases}$$

What is the dual of this problem? All inequalities are $\leq$ in the original maximization problem so we may proceed directly to write the dual.
Minimize $96u + 18v + 72w$ subject to

$$\begin{cases} 6u + v + 2w \geq 80 \\ 3u + v + 6w \geq 70 \\ 3u + 2v + 4w \geq p \\ u \geq 0, \quad v \geq 0, \quad w \geq 0. \end{cases}$$

Whereas the new furniture manufacturing problem has one more variable than the original problem, its dual has one more constraint than the dual of the original problem,

$$3u + 2v + 4w \geq p.$$

If the optimal solution to the original problem (just making chairs and sofas) were to remain optimal in the new problem, the variable z would not enter the set of group II variables. If that were the case, the solution to the dual would also remain optimal. But that solution was

$$u = \tfrac{10}{3}, \qquad v = 60, \qquad w = 0.$$

And the new dual would require that $3u + 2v + 4w \geq p$. Since

$$3\left(\tfrac{10}{3}\right) + 2(60) + 4(0) = 130,$$

if the profit per love seat is at most \$130, the previous solution will remain optimal. That is, the manufacturer should make 14 chairs and 4 sofas and 0 love seats. However, if the profit per love seat exceeds \$130, the variable z will enter the set of group II variables, and we will find an optimal production schedule in which $z > 0$.

Now Try Exercise 19   ∎

PRACTICE PROBLEMS 4.5

A linear programming problem involving three variables and four nontrivial inequalities has the number 52 as the maximum value of its objective function.

1. How many variables and nontrivial inequalities will the dual problem have?

2. What is the optimum value for the objective function of the dual problem?

EXERCISES 4.5

In Exercises 1–6, determine the dual problem of the given linear programming problem.

1. Maximize $4x + 2y$ subject to the constraints

$$\begin{cases} 5x + y \leq 80 \\ 3x + 2y \leq 76 \\ x \geq 0, \quad y \geq 0. \end{cases}$$

2. Minimize $30x + 60y + 50z$ subject to the constraints

$$\begin{cases} 5x + 3y + z \geq 2 \\ x + 2y + z \geq 3 \\ x \geq 0, \quad y \geq 0, \quad z \geq 0. \end{cases}$$

3. Minimize $10x + 12y$ subject to the constraints

$$\begin{cases} x + 2y \geq 1 \\ -x + y \geq 2 \\ 2x + 3y \geq 1 \\ x \geq 0, \quad y \geq 0. \end{cases}$$

4. Maximize $80x + 70y + 120z$ subject to the constraints

$$\begin{cases} 6x + 3y + 8z \leq 768 \\ x + y + 2z \leq 144 \\ 2x + 5y \leq 216 \\ x \geq 0, \quad y \geq 0, \quad z \geq 0. \end{cases}$$

5. Minimize $3x + 5y + z$ subject to the constraints

$$\begin{cases} 2x - 4y - 6z \leq 7 \\ y \geq 10 - 8x - 9z \\ x \geq 0, \quad y \geq 0, \quad z \geq 0. \end{cases}$$

6. Maximize $2x - 3y + 4z - 5w$ subject to the constraints

$$\begin{cases} x + y + z + w - 6 \leq 10 \\ 7x + 9y - 4z - 3w \geq 5 \\ x \geq 0, \quad y \geq 0, \quad z \geq 0, \quad w \geq 0. \end{cases}$$

7. The final simplex tableau for the linear programming problem of Exercise 1 is as follows. Give the solution to the problem and to its dual.

	x	y	u	v	M	
x	1	0	$\frac{2}{7}$	$-\frac{1}{7}$	0	12
y	0	1	$-\frac{3}{7}$	$\frac{5}{7}$	0	20
M	0	0	$\frac{2}{7}$	$\frac{6}{7}$	1	88

8. The final simplex tableau for the *dual* of the linear programming problem of Exercise 2 is as follows. Give the solution to the problem and to its dual.

	u	v	x	y	z	M	
v	5	1	1	0	0	0	30
y	-7	0	-2	1	0	0	0
z	-4	0	-1	0	1	0	20
M	13	0	3	0	0	1	90

9. The final simplex tableau for the *dual* of the linear programming problem of Exercise 3 is as follows. Give the solution to the problem and to its dual.

	u	v	w	x	y	M	
x	3	0	5	1	1	0	22
v	2	1	3	0	1	0	12
M	3	0	5	0	2	1	24

10. The final simplex tableau for the linear programming problem of Exercise 4 is as follows. Give the solution to the problem and to its dual.

	x	y	z	u	v	w	M	
x	1	0	0	$\frac{5}{12}$	$-\frac{5}{3}$	$\frac{1}{12}$	0	98
z	0	0	1	$-\frac{1}{8}$	1	$-\frac{1}{8}$	0	21
y	0	1	0	$-\frac{1}{6}$	$\frac{2}{3}$	$\frac{1}{6}$	0	4
M	0	0	0	$\frac{20}{3}$	$\frac{100}{3}$	$\frac{10}{3}$	1	10,640

In Exercises 11–14, determine the dual problem. Solve either the original problem or its dual by the simplex method and then give the solutions to both.

11. Minimize $3x + y$ subject to the constraints

$$\begin{cases} x + y \geq 3 \\ 2x \geq 5 \\ x \geq 0, \quad y \geq 0. \end{cases}$$

12. Minimize $3x + 5y + z$ subject to the constraints

$$\begin{cases} x + y + z \geq 20 \\ y + 2z \geq 0 \\ x \geq 0, \quad y \geq 0, \quad z \geq 0. \end{cases}$$

13. Maximize $10x + 12y + 10z$ subject to the constraints

$$\begin{cases} x - 2y \leq 6 \\ 3x + z \leq 9 \\ y + 3z \leq 12 \\ x \geq 0, \quad y \geq 0, \quad z \geq 0. \end{cases}$$

14. Maximize $x + 3y$ subject to the constraints

$$\begin{cases} x + y \le 7 \\ x + 2y \le 10 \\ x \ge 0, \quad y \ge 0. \end{cases}$$

15. (*Cutting Edge Knife Co.*) Give an economic interpretation to the dual of the Cutting Edge Knife Company problem of Example 1 of Section 4.4.

16. (*Electronics Manufacture*) Give an economic interpretation to the dual of Exercise 10 of Section 4.3.

17. (*Mining*) Give an economic interpretation to the dual of the mining problem of Exercise 6 of Section 3.1 and Exercise 14 of Section 3.2.

18. (*Nutrition*) Give an economic interpretation to the dual of the nutrition problem of Example 1 of Section 3.2.

19. (*Cutting Edge Knife Co.*) Consider the Cutting Edge Knife Company problem of Example 1 of Section 4.4. Suppose that the company is thinking of also making

table knives. If each table knife requires 4 labor-hours, 6 units of steel, and 2 units of wood, what profit must be realized per knife to justify adding this product?

In Exercises 20 and 21, use a graphing calculator or mathematical software to solve the linear programming problems by applying the simplex method to the dual of the problem.

20. Minimize $3x + y$ subject to the constraints

$$\begin{cases} x + y \ge 3 \\ 2x \quad \ge 5 \\ x \ge 0, \quad y \ge 0. \end{cases}$$

21. Minimize $16x + 42y$ subject to the constraints

$$\begin{cases} x + 3y \ge 5 \\ 2x + 4y \ge 8 \\ x \ge 0, \quad y \ge 0. \end{cases}$$

SOLUTIONS TO PRACTICE PROBLEMS 4.5

1. It will have four variables and three nontrivial inequalities. The number of variables in the dual problem is always the same as the number of nontrivial inequalities in the original problem. The number of nontrivial inequalities in the dual problem is the same as the number of variables in the original problem.

2. The optimum is a minimum value of 52. The original problem and the dual problem always have the same optimum values. However, if this value is a maximum for one of the problems, it will be a minimum for the other.

CHAPTER 4 SUMMARY

1. A linear programming problem is in *standard form* if the linear objective function is to be maximized, every variable is constrained to be nonnegative, and all other constraints are of the form [linear polynomial] $\le$ [nonnegative constant].

2. To form the *initial simplex tableau* corresponding to a linear programming problem in standard form:

   Step 1 For each constraint of the form [linear polynomial] $\le$ [nonnegative constant], introduce a *slack variable* and write the constraint as an equation.

   Step 2 Introduce a variable M to represent the quantity to be maximized, and form the equation $-$[objective function] $+ M = 0$.

   Step 3 Form the augmented matrix corresponding to the system of linear equations from Steps

1 and 2, with the equation from Step 2 at the bottom. This matrix is the initial simplex tableau.

3. The simplex method entails pivoting around entries in the simplex tableau until the bottom row contains no negative entries except perhaps the entry in the last column. The solution can be read off the final tableau by letting the variables heading columns with 0 entries in every row but the ith row take on the value in the ith row of the right-most column, and letting the other variables equal 0.

4. The simplex method can be used to solve a linear programming problem in *nonstandard* form as follows:

   (a) If the problem is a minimization problem, convert it to a maximization problem by multiplying the objective function by -1.

(b) Form the initial tableau, and eliminate any negative entries in the upper part of the last column by pivoting.

(c) Apply the pivoting process for problems in standard form to the resulting tableau.

5. The constraints in a linear programming problem usually represent limitations in the availability of resources. *Sensitivity analysis* is the analysis of the effects of small changes in the constraints. The *shadow price* of a resource is the change in the optimal value of the objective function resulting from a unit increase in the availability of that resource. In the final simplex tableau, the bottom entry in each slack variable's column gives the shadow price of the resource associated with that variable.

6. A linear programming problem in standard form can be expressed in terms of matrices: Maximize CX subject to constraints $AX \leq B$ and $X \geq \mathbf{0}$, where A is the matrix of coefficients of the linear polynomials, B is the column vector of the right-hand constants, and C is a row vector of the coefficients in the objective function.

7. Given a linear programming problem in standard form, its *dual* problem is "Minimize $B^T U$ subject to $A^T U \geq C^T$, $U \geq \mathbf{0}$," where U is the column matrix consisting of the slack variables of the original problem. The original problem is called the *primal problem*.

8. The *fundamental theorem of duality* states that if either the primal problem or the dual problem has an optimal feasible solution, then they both have optimal feasible solutions and their objective functions have the same optimal value. Furthermore, the final simplex tableau for either of these problems yields the solution of the other as the final entries in the columns associated with the slack variables.

REVIEW OF FUNDAMENTAL CONCEPTS

1. What is the standard form of a linear programming problem?

2. What is a slack variable? a group I variable? a group II variable?

3. Explain how to construct a simplex tableau corresponding to a linear programming problem in standard form.

4. Give the steps for carrying out the simplex method for linear programming problems in standard form.

5. Explain how to convert a minimization problem to a maximization problem.

6. Give the steps for carrying out the simplex method for problems in nonstandard form.

7. Describe how to obtain the dual of a linear programming problem.

8. State the fundamental theorem of duality.

9. Explain how to obtain the matrix formulation of a linear programming problem.

10. What is meant by "sensitivity analysis"?

11. Explain how the dual can be used to decide whether to introduce a new product.

SUPPLEMENTARY EXERCISES

In Exercises 1–10, use the simplex method to solve the linear programming problems.

1. Maximize $3x + 4y$ subject to the constraints

$$\begin{cases} 2x + y \leq 7 \\ -x + y \leq 1 \\ x \geq 0, \quad y \geq 0. \end{cases}$$

2. Maximize $2x + 5y$ subject to the constraints

$$\begin{cases} x + y \leq 7 \\ 4x + 3y \leq 24 \\ x \geq 0, \quad y \geq 0. \end{cases}$$

3. Maximize $2x + 3y$ subject to the constraints

$$\begin{cases} x + 2y \leq 14 \\ x + y \leq 9 \\ 3x + 2y \leq 24 \\ x \geq 0, \quad y \geq 0. \end{cases}$$

4. Maximize $3x + 7y$ subject to the constraints

$$\begin{cases} x + 2y \leq 10 \\ 4x + 3y \leq 30 \\ -2x + y \leq 0 \\ x \geq 0, \quad y \geq 0. \end{cases}$$

5. Minimize $x + y$ subject to the constraints

$$\begin{cases} 7x + 5y \ge 40 \\ x + 4y \ge 9 \\ x \ge 0, \quad y \ge 0. \end{cases}$$

6. Minimize $3x + 2y$ subject to the constraints

$$\begin{cases} x + y \ge 6 \\ x + 2y \ge 0 \\ x \ge 0, \quad y \ge 0. \end{cases}$$

7. Minimize $20x + 30y$ subject to the constraints

$$\begin{cases} x + 4y \ge 8 \\ x + y \ge 5 \\ 2x + y \ge 7 \\ x \ge 0, \quad y \ge 0. \end{cases}$$

8. Minimize $5x + 7y$ subject to the constraints

$$\begin{cases} 2x + y \ge 10 \\ 3x + 2y \ge 18 \\ x + 2y \ge 10 \\ x \ge 0, \quad y \ge 0. \end{cases}$$

9. Maximize $36x + 48y + 70z$ subject to the constraints

$$\begin{cases} x \le 4 \\ y \le 6 \\ z \le 8 \\ 4x + 3y + 2z \le 38 \\ x \ge 0, \quad y \ge 0, \quad z \ge 0. \end{cases}$$

10. Maximize $3x + 4y + 5z + 4w$ subject to the constraints

$$\begin{cases} 6x + 9y + 12z + 15w \le 672 \\ x - y + 2z + 2w \le 92 \\ 5x + 10y - 5z + 4w \le 280 \\ x \ge 0, \quad y \ge 0, \quad z \ge 0, \quad w \ge 0. \end{cases}$$

11. Determine the dual problem of the linear programming problem in Exercise 3.

12. Determine the dual problem of the linear programming problem in Exercise 7.

13. The final simplex tableau for the linear programming problem of Exercise 3 is as follows. Give the solution to the problem and to its dual.

	x	y	u	v	w	M	
y	0	1	1	-1	0	0	5
x	1	0	-1	2	0	0	4
w	0	0	1	-4	1	0	2
M	0	0	1	1	0	1	23

14. The final simplex tableau for the *dual* of the linear programming problem of Exercise 7 is as follows. Give the solution to the problem and to its dual.

	u	v	w	x	y	M	
v	0	1	$\frac{7}{3}$	$\frac{4}{3}$	$-\frac{1}{3}$	0	$\frac{50}{3}$
u	1	0	$-\frac{1}{3}$	$-\frac{1}{3}$	$\frac{1}{3}$	0	$\frac{10}{3}$
M	0	0	2	4	1	1	110

15, 16. For each of the linear programming problems in Exercises 3 and 7, identify the matrices A, B, C, X, and U and state the problem and its dual in terms of matrices.

17. (*Manufacturing—Resource Allocation*) Consider Exercise 34 of Section 3.2.

(a) Solve the problem by the simplex method.

(b) The Bluejay Lacrosse Stick Company is considering diversifying by also making tennis rackets. A tennis racket requires 1 labor-hour for cutting, 4 labor-hours for stringing, and 2 labor-hours for finishing. How much profit must the company be able to make on each tennis racket in order to justify the diversification?

18. (*Stereo Store*) Consider the stereo store of Exercise 18 in Section 4.2. A fourth brand of stereo system has appeared on the market. Brand D takes up 3 cubic feet of storage space and generates a sales commission of \$30. What profit would the store have to realize on the sale of each brand D stereo set in order to justify carrying it?

CHAPTER TEST

1. Form the initial simplex tableau corresponding to the following linear programming problem. Maximize $2x + y - 3z$ subject to the constraints

$$\begin{cases} x + y - 2z \le 10 \\ 2x - y + 3z \le 18 \\ x + 3y + z \le 21 \\ x \ge 0, \quad y \ge 0, \quad z \ge 0. \end{cases}$$

2. Give the solution of the linear programming problem corresponding to the final tableau

$$\begin{bmatrix} x & y & z & u & v & w & M & \\ 1 & 10 & 0 & 3 & -7 & 0 & 0 & 35 \\ 0 & -7 & 0 & 8 & 14 & 1 & 0 & 42 \\ 0 & 6 & 1 & -1 & 8 & 0 & 0 & 30 \\ \hline 0 & 14 & 0 & 0 & \frac{7}{2} & 0 & 1 & 560 \end{bmatrix}.$$

Also, give the solution to the dual problem.

3. State the maximization problem corresponding to the following tableau, and solve it using the simplex method.

$$\begin{bmatrix} x & y & u & v & M & \\ 6 & 7 & 1 & 0 & 0 & 120 \\ 15 & 5 & 0 & 1 & 0 & 195 \\ \hline -3 & 4 & 0 & 0 & 1 & 0 \end{bmatrix}$$

4. Use the simplex method to solve the following linear programming problem. Minimize $12x + 5y$ subject to the constraints

$$\begin{cases} y \ge -\frac{1}{2}x + 5 \\ y \le -x + 10 \\ y \ge -3x + 10 \\ x \ge 0, \quad y \ge 0. \end{cases}$$

5. State the dual of the following linear programming problem. Minimize $3x + 2y$ subject to the constraints

$$\begin{cases} x + y \ge 6 \\ 2x - y \ge 3 \\ x \ge 0, \quad y \ge 0. \end{cases}$$

6. (*Newspaper Production*) Consider the following linear programming problem, but do *not* solve it.

A newspaper publisher puts out a morning paper and an evening paper. He can sell all the papers he prints, but paper and labor are in short supply. Each morning paper requires 2 pounds of paper and 2 minutes of labor to print, while each evening paper requires 1 pound of paper and 3 minutes of labor to print. Each day, 6000 pounds of paper and 9600 minutes of labor are available. Morning papers sell for $.50 each, and evening papers sell for $.35 each. How many morning and evening papers should be printed in order to maximize revenue?

(a) Give the matrix formulation of the problem.

(b) Give the matrix formulation of the dual of the problem.

(c) State the dual problem without matrices. What do the variables represent?

(d) Give an economic interpretation to the dual of the problem.

CHAPTER 4 PROJECT

Shadow Prices

Jason's House of Cheese and Nuts offers two cheese assortments for holiday gift giving. In his supply refrigerator Jason has 3600 ounces of Cheddar, 1498 ounces of Brie, and 2396 ounces of Stilton. The St. Nick assortment contains 10 ounces of Cheddar, 5 ounces of Brie, and 6 ounces of Stilton. The Holly assortment contains 8 ounces of Cheddar, 3 ounces of Brie, and 8 ounces of Stilton. Each St. Nick assortment sells for $4.00 and each Holly assortment sells for $3.50. How many of each assortment should be produced and sold in order to maximize Jason's revenue?

1. Solve the problem geometrically.

2. By looking at your graph from part 1, can you determine the shadow price of Cheddar?

3. Solve the problem using the simplex method. The solution should be the same as in part 1. Verify your answer to part 2 by looking at your final tableau.

4. What are the shadow prices for Brie and Stilton?

5. What would the maximum revenue be if there were 3620 ounces of Cheddar, 1500 ounces of Brie, and 2400 ounces of Stilton?

6. Go back to the original problem and state its dual problem. What information do the original slack variables u, v, and w give us about the dual problem? Determine the solution to the dual problem from your final tableau in part 3, and give an economic interpretation.

CHAPTER 5

Sets and Counting

IN this chapter we introduce some ideas useful in the study of probability (Chapter 6). Our first topic, the theory of sets, will provide a convenient language and notation in which to discuss probability. Using set theory, we develop a number of counting principles that can also be applied to computing probabilities.

5.1 Sets

In many applied problems one must consider collections of various sorts of objects. For example, a survey of unemployment might consider the collection of all U.S. cities with current unemployment greater than 7%. A study of birthrates might consider the collection of countries with a current birthrate less than 20 per 1000 population. Such collections are examples of sets. A *set* is any collection of objects.

The objects, which may be countries, cities, years, numbers, letters, or anything else, are called the *elements* of the set. A set is often specified by a listing of its elements inside a pair of braces. For example, the set whose elements are the first six letters of the alphabet is written

$$\{a, b, c, d, e, f\}.$$

Similarly, the set whose elements are the even numbers between 1 and 11 is written

$$\{2, 4, 6, 8, 10\}.$$

We can also specify a set by giving a description of its elements (without actually listing the elements). For example, the set $\{a, b, c, d, e, f\}$ can also be written

{the first six letters of the alphabet},

and the set $\{2, 4, 6, 8, 10\}$ can be written

{all even numbers between 1 and 11}.

For convenience, we usually denote sets by capital letters, A, B, C, and so on. The great diversity of sets is illustrated by the following examples.

1. Let C = {possible sequences of outcomes of tossing a coin three times}. If we let H denote "heads" and T denote "tails," the various sequences can be easily described:

$$C = \{HHH, THH, HTH, HHT, TTH, THT, HTT, TTT\},$$

where, for instance, THH means "first toss tails, second toss heads, third toss heads."

2. Let B = {license plate numbers consisting of three letters followed by three digits}. Some typical elements of B are

SBG 602, GXZ 179, YHJ 006.

The number of elements in B is sufficiently large so that listing all of them is impractical. However, in this chapter we develop a technique that allows us to calculate the number of elements of B.

3. The graph of the equation $y = x^2$ is the set of all points (a, b) in the plane for which $b = a^2$. This set has infinitely many elements.

Sets arise in many practical contexts, as the next example shows.

EXAMPLE 1 **Listing the elements of a set** Table 1 gives the rate of inflation, as measured by the percentage change in the consumer price index, for the years from 1985 to 2004. Let

$$A = \{\text{years from 1985 to 2004 in which inflation was above 4\%}\}$$
$$B = \{\text{years from 1985 to 2004 in which inflation was below 3\%}\}.$$

Determine the elements of A and B.

Solution By reading Table 1, we see that

$$A = \{1988, 1989, 1990, 1991\}$$
$$B = \{1986, 1994, 1995, 1997, 1998, 1999, 2001, 2002, 2003, 2004\}.$$

Now Try Exercises 9(a) and (b) ∎

TABLE 1			
Year	Inflation (%)	Year	Inflation (%)
1985	3.6	1995	2.8
1986	1.9	1996	3.0
1987	3.6	1997	2.3
1988	4.1	1998	1.6
1989	4.8	1999	2.2
1990	5.4	2000	3.4
1991	4.2	2001	2.8
1992	3.0	2002	1.6
1993	3.0	2003	2.3
1994	2.6	2004	2.7

Suppose that we are given two sets, A and B. Then it is possible to form new sets from A and B as follows: The *union* of A and B, denoted $A \cup B$, is the set consisting of all those elements that belong to either A *or* B (or both). The *intersection* of A and B, denoted $A \cap B$, is the set consisting of those elements that belong to both A *and* B. For example, let

$$A = \{1, 2, 3, 4\} \qquad B = \{1, 3, 5, 7, 11\}.$$

Then, since $A \cup B$ consists of those elements belonging to either A or B (or both), we have

$$A \cup B = \{1, 2, 3, 4, 5, 7, 11\}.$$

Moreover, since $A \cap B$ consists of those elements that belong to both A and B, we have

$$A \cap B = \{1, 3\}.$$

EXAMPLE 2   **The intersection and union of sets**   Table 2 on the next page gives the rates of unemployment and inflation for the years from 1991 to 2004. Let

$A = \{$years from 1991 to 2004 in which unemployment is at least 6%$\}$

$B = \{$years from 1991 to 2004 in which the inflation rate is at least 3%$\}$.

(a) Describe the sets $A \cap B$ and $A \cup B$.

(b) Determine the elements of A, B, $A \cap B$, and $A \cup B$.

Solution   (a)   From the descriptions of A and B, we have

$$A \cap B = \{\text{years from 1991 to 2004 in which unemployment is}$$
$$\text{at least 6\% and inflation is at least 3\%}\}$$
$$A \cup B = \{\text{years from 1991 to 2004 in which either unemployment}$$
$$\text{is at least 6\% or inflation is at least 3\% (or both)}\}.$$

(b)   From the table we see that

$$A = \{1991, 1992, 1993, 1994, 2003\}$$

$$B = \{1991, 1992, 1993, 1996, 2000\}$$

$$A \cap B = \{1991, 1992, 1993\}$$

$$A \cup B = \{1991, 1992, 1993, 1994, 1996, 2000, 2003\}.$$

TABLE 2

Year	Unemployment (%)	Inflation (%)
1991	6.8	4.2
1992	7.5	3.0
1993	6.9	3.0
1994	6.1	2.6
1995	5.6	2.8
1996	5.4	3.0
1997	4.9	2.3
1998	4.5	1.6
1999	4.2	2.2
2000	4.0	3.4
2001	4.8	2.8
2002	5.8	1.6
2003	6.0	2.3
2004	5.3	2.7

Now Try Exercises 9(c) and (d)

We have defined the union and the intersection of two sets. In a similar manner, we can define the union and intersection of any number of sets. For example, if A, B, and C are three sets, then their union, denoted $A \cup B \cup C$, is the set whose elements are precisely those that belong to at least one of the sets A, B, and C. Similarly, the intersection of A, B, and C, denoted $A \cap B \cap C$, is the set consisting of those elements that belong to all the sets A, B, and C. In a similar way, we may define the union and intersection of more than three sets.

Suppose that we are given a set A. We may form new sets by selecting elements from A. Sets formed in this way are called *subsets* of A. More precisely, a set B is called a *subset* of A provided that every element of B is also an element of A. For example, let $A = \{1, 2, 3\}$, $B = \{1, 3\}$. Since the elements of B (1 and 3) are also elements of A, B is a subset of A.

One set that is considered very often is the set that contains no elements at all. This set is called the *empty set* (or *null set*) and is written $\emptyset$. The empty set is a subset of every set.[1]

EXAMPLE 3 **Listing the subsets of a set** Let $A = \{a, b, c\}$. Find all subsets of A.

Solution Since A contains three elements, every subset of A has at most three elements. We look for subsets according to the number of elements:

---

[1] Here is why: Let A be any set. Every element of $\emptyset$ also belongs to A. If you do not agree, then you must produce an element of $\emptyset$ that does not belong to A. But you cannot, since $\emptyset$ has no elements. So $\emptyset$ is a subset of A.

Number of elements in subset	Possible subsets
0	$\emptyset$
1	$\{a\}, \{b\}, \{c\}$
2	$\{a, b\}, \{a, c\}, \{b, c\}$
3	$\{a, b, c\}$

Thus we see that A has eight subsets, namely those listed on the right. (Note that we count A as a subset of itself.)

Now Try Exercise 5

It is usually convenient to regard all sets involved in a particular discussion as subsets of a single larger set. Thus, for example, if a problem involves the sets $\{a, b, c\}, \{e, f\}, \{g\}, \{b, x, y\}$, then we can regard all of these as subsets of the set

$$U = \{\text{all letters of the alphabet}\}.$$

Since U contains all sets being discussed, it is called a *universal set* (for the particular problem). In this book we shall specify the particular universal set we have in mind or it will be clearly defined by the context.

Suppose that U is a universal set and A is a subset of U. The set of elements of U that are not in A is called the *complement* of A, denoted A'. For example, if

$$U = \{1, 2, 3, 4, 5, 6, 7, 8, 9\} \quad \text{and} \quad A = \{2, 4, 6, 8\},$$

then

$$A' = \{1, 3, 5, 7, 9\}.$$

EXAMPLE 4 **Finding the complement of a set** Let $U = \{a, b, c, d, e, f, g\}$, $S = \{a, b, c\}$, and $T = \{a, c, d\}$. List the elements of the following sets.

(a) S' (b) T' (c) $(S \cap T)'$ (d) $S' \cap T'$

Solution (a) S' consists of those elements of U that are not in S, so $S' = \{d, e, f, g\}$.

(b) Similarly, $T' = \{b, e, f, g\}$.

(c) To determine $(S \cap T)'$, we must first determine $S \cap T$:

$$S \cap T = \{a, c\}.$$

Then we determine the complement of this set:

$$(S \cap T)' = \{b, d, e, f, g\}.$$

(d) We determined S' and T' in parts (a) and (b). The set $S' \cap T'$ consists of the elements that belong to both S' and T'. Therefore, referring to parts (a) and (b), we have

$$S' \cap T' = \{e, f, g\}.$$

Now Try Exercises 9(e) and (f)

PRACTICE PROBLEMS 5.1

1. Let $U = \{a, b, c, d, e, f, g\}$, $R = \{a, b, c, d\}$, $S = \{c, d, e\}$, and $T = \{c, e, g\}$. List the elements of the following sets.

   (a) R' (b) $R \cap S$

   (c) $(R \cap S) \cap T$ (d) $R \cap (S \cap T)$

2. Let $U = \{$all Nobel Prize winners$\}$, $W = \{$women who have won Nobel Prizes$\}$, $A = \{$Americans who have won Nobel Prizes$\}$, $L = \{$winners of the Nobel Prize in literature$\}$. Describe the following sets.

   (a) W' (b) $A \cap L'$ (c) $W \cap A \cap L'$

3. Refer to Problem 2. Use set-theoretic notation to describe $\{$Nobel Prize winners who are American men or recipients of the Nobel Prize in literature$\}$.

EXERCISES 5.1

1. Let $U = \{1, 2, 3, 4, 5, 6, 7\}$, $S = \{1, 2, 3, 4\}$, and $T = \{1, 3, 5, 7\}$. List the elements of the following sets.
   (a) S' (b) $S \cup T$ (c) $S \cap T$ (d) $S' \cap T$

2. Let $U = \{1, 2, 3, 4, 5\}$, $S = \{1, 2, 3\}$, and $T = \{5\}$. List the elements of the following sets.
   (a) S' (b) $S \cup T$ (c) $S \cap T$ (d) $S' \cap T$

3. Let $U = \{$all letters of the alphabet$\}$, $R = \{a, b, c\}$, $S = \{c, d, e, f\}$, and $T = \{x, y, z\}$. List the elements of the following sets.
   (a) $R \cup S$ (b) $R \cap S$ (c) $S \cap T$

4. Let $U = \{a, b, c, d, e, f, g\}$, $R = \{a\}$, $S = \{a, b\}$, and $T = \{b, d, e, f, g\}$. List the elements of the following sets.
   (a) $R \cup S$ (b) $R \cap S$ (c) T' (d) $T' \cup S$

5. List all subsets of the set $\{1, 2\}$.

6. List all subsets of the set $\{1\}$.

7. (*College Students*) Let $U = \{$all college students$\}$, $M = \{$all male college students$\}$, and $F = \{$all college students who like football$\}$. Describe the elements of the following sets.
   (a) $M \cap F$ (b) M' (c) $M' \cap F'$ (d) $M \cup F$

8. (*Corporations*) Let $U = \{$all corporations$\}$, $S = \{$all corporations with headquarters in New York City$\}$, and $T = \{$all privately owned corporations$\}$. Describe the elements of the following sets.
   (a) S' (b) T' (c) $S \cap T$ (d) $S \cap T'$

9. (*S&P Index*) The Standard and Poor's Index measures the price of a certain collection of 500 stocks. Table 3 on the next page compares the percentage change in the index during the first 5 days of certain years with the percentage change for the entire year. Let $U = \{$all years from 1977 to 2004$\}$, $S = \{$all years during which the index increased by 2% or more during the first 5 days$\}$, and $T = \{$all years for which the index increased by 16% or more during the entire year$\}$. List the elements of the following sets.
   (a) S (b) T (c) $S \cap T$
   (d) $S \cup T$ (e) $S' \cap T$ (f) $S \cap T'$

10. (*S&P Index*) Refer to Table 3. Let $U = \{$all years from 1977 to 2004$\}$, $A = \{$all years during which the index declined during the first 5 days$\}$, and $B = \{$all years during which the index declined for the entire year$\}$. List the elements of the following sets.
    (a) A (b) B (c) $A \cap B$
    (d) $A' \cap B$ (e) $A \cap B'$

11. (*S&P Index*) Refer to Exercise 9. Describe verbally the fact that $S \cap T'$ has three elements.

12. (*S&P Index*) Refer to Exercise 10. Describe verbally the fact that $A \cap B'$ has eight elements.

13. Let $U = \{a, b, c, d, e, f\}$, $R = \{a, b, c\}$, $S = \{a, c, e\}$, and $T = \{e, f\}$. List the elements of the following sets.
    (a) $(R \cup S)'$ (b) $R \cup S \cup T$
    (c) $R \cap S \cap T$ (d) $R \cap S \cap T'$
    (e) $R' \cap S \cap T$ (f) $S \cup T$
    (g) $(R \cup S) \cap (R \cup T)$
    (h) $(R \cap S) \cup (R \cap T)$
    (i) $R' \cap T'$

14. Let $U = \{1, 2, 3, 4, 5\}$, $R = \{1, 3, 5\}$, $S = \{3, 4, 5\}$, and $T = \{2, 4\}$. List the elements of the following sets.
    (a) $R \cap S \cap T$ (b) $R \cap S \cap T'$ (c) $R \cap S' \cap T$
    (d) $R' \cap T$ (e) $R \cup S$ (f) $R' \cup R$
    (g) $(S \cap T)'$ (h) $S' \cup T'$

In Exercises 15–20, simplify the given expression.

15. $(S')'$ 16. $S \cap S'$ 17. $S \cup S'$
18. $S \cap \varnothing$ 19. $T \cap S \cap T'$ 20. $S \cup \varnothing$

(Corporation) A large corporation classifies its many divisions by their performance in the preceding year. Let $P = \{$divisions that made a profit$\}$, $L = \{$divisions that had an increase in labor costs$\}$, and $T = \{$divisions whose total revenue increased$\}$. Describe the sets in Exercises 21–26 using set-theoretic notation.

21. $\{$divisions that had increases in labor costs or total revenue$\}$

22. $\{$divisions that did not make a profit$\}$

	Percent change	Percent change		Percent change	Percent change
Year	for first 5 days	for year	Year	for first 5 days	for year
2004	1.8	9.0	1990	0.1	−6.6
2003	3.4	26.4	1989	1.2	27.3
2002	1.1	−23.4	1988	−1.5	12.4
2001	−1.9	−13.0	1987	6.2	2.0
2000	−1.9	−10.1	1986	−1.6	14.6
1999	3.7	19.5	1985	−1.9	26.3
1998	−1.5	26.7	1984	2.4	1.4
1997	1.0	31.0	1983	3.3	17.3
1996	0.4	20.3	1982	−2.4	14.8
1995	0.3	34.1	1981	−2.0	−9.7
1994	0.7	−1.5	1980	0.9	25.8
1993	−1.5	7.1	1979	2.8	12.3
1992	0.2	4.5	1978	−4.7	1.1
1991	−4.6	26.3	1977	−2.3	−11.5

TABLE 3 Percentage change in the Standard and Poor's index

23. {divisions that made a profit despite an increase in labor costs}

24. {divisions that had an increase in labor costs and either were unprofitable or did not increase their total revenue}

25. {profitable divisions with increases in labor costs and total revenue}

26. {divisions that were unprofitable or did not have increases in either labor costs or total revenue}

(Automobile Insurance) An automobile insurance company classifies applicants by their driving records for the previous three years. Let $S = \{$applicants who have received speeding tickets$\}$, $A = \{$applicants who have caused accidents$\}$, and $D = \{$applicants who have been arrested for driving while intoxicated$\}$. Describe the sets in Exercises 27–32 using set-theoretic notation.

27. {applicants who have not received speeding tickets}

28. {applicants who have caused accidents and been arrested for drunk driving}

29. {applicants who have received speeding tickets, caused accidents, or been arrested for drunk driving}

30. {applicants who have not been arrested for drunk driving but have received speeding tickets or have caused accidents}

31. {applicants who have not both caused accidents and received speeding tickets but who have been arrested for drunk driving}

32. {applicants who have not caused accidents or have not been arrested for drunk driving}

(College Teachers and Students) Let $U = \{$people at Mount College$\}$, $A = \{$students at Mount College$\}$, $B = \{$teachers at Mount College$\}$, $C = \{$females at Mount College$\}$, and $D = \{$males at Mount College$\}$. Describe verbally the sets in Exercises 33–40.

33. $A \cap D$ 34. $B \cap C$ 35. $A \cap B$ 36. $B \cup C$

37. $A \cup C'$ 38. $(A \cap D)'$ 39. D' 40. $D \cap U$

(Ice Cream Preferences) Let $U = \{$all people$\}$, $S = \{$people who like strawberry ice cream$\}$, $V = \{$people who like vanilla ice cream$\}$, and $C = \{$people who like chocolate ice cream$\}$. Describe the sets in Exercises 41–46 using set-theoretic notation.

41. {people who don't like strawberry ice cream}

42. {people who like vanilla but not chocolate ice cream}

43. {people who like vanilla or chocolate but not strawberry ice cream}

44. {people who don't like any of the three flavors of ice cream}

45. {people who like neither chocolate nor vanilla ice cream}

46. {people who like only strawberry ice cream}

47. Let U be the set of vertices in Fig. 1 on the next page. Let $R = \{$vertices (x, y) with $x > 0\}$, $S = \{$vertices (x, y) with $y > 0\}$, and $T = \{$vertices (x, y) with $x \leq y\}$. List the elements of the following sets.

(a) R (b) S (c) T

(d) $R' \cup S$ (e) $R' \cap T$ (f) $R \cap S \cap T$

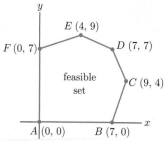

Figure 1

48. (*Topping Choices*) Sam ordered a baked potato at a restaurant. The waitress offered him butter, cheese, and chives as toppings. How many different ways could he have his potato? List them.

49. Let $S = \{1, 3, 5, 7\}$ and $T = \{2, 5, 7\}$. Give an example of a subset of T that is not a subset of S.

50. Suppose that S and T are subsets of the set U. Under what circumstance will $S \cap T = T$?

51. Suppose that S and T are subsets of the set U. Under what circumstance will $S \cup T = T$?

52. Find three subsets of the set of integers from 1 through 10, R, S, and T, such that $R \cup (S \cap T)$ is different from $(R \cup S) \cap T$.

SOLUTIONS TO PRACTICE PROBLEMS 5.1

1. (a) $\{e, f, g\}$

(b) $\{c, d\}$

(c) $\{c\}$. This problem asks for the intersection of two sets. The first set is $R \cap S = \{c, d\}$ and the second set is $T = \{c, e, g\}$. The intersection of these sets is $\{c\}$.

(d) $\{c\}$. Here again the problem asks for the intersection of two sets. However, now the first set is $R = \{a, b, c, d\}$ and the second set is $S \cap T = \{c, e\}$. The intersection of these sets is $\{c\}$.
[*Note*: It should be expected that the set $(R \cap S) \cap T$ is the same as the set $R \cap (S \cap T)$, for each set consists of those elements that are in all three sets. Therefore, each of these sets equals the set $R \cap S \cap T$.]

2. (a) $W' = \{$men who have won Nobel Prizes$\}$. This is so since W' consists of those elements of U that are not in W—that is, those Nobel Prize winners who are not women.

(b) $A \cap L' = \{$Americans who have received Nobel Prizes in fields other than literature$\}$.

(c) $W \cap A \cap L' = \{$American women who have received Nobel Prizes in fields other than literature$\}$. This is so since to qualify for $W \cap A \cap L'$, a Nobel Prize winner must simultaneously be in W, in A, and in L'—that is, a woman, an American, and not a winner of the Nobel Prize in literature.

3. $(A \cap W') \cup L$

5.2 A Fundamental Principle of Counting

A counting problem is one that requires us to determine the number of elements in a set S. Counting problems arise in many applications of mathematics and comprise the mathematical field of *combinatorics*. We shall study a number of different sorts of counting problems in the remainder of this chapter.

If S is any set, we will denote the number of elements in S by $n(S)$. For example, if $S = \{1, 7, 11\}$, then $n(S) = 3$, and if $S = \{a, b, c, d, e, f, g, h, i\}$, then $n(S) = 9$. Of course, if $S = \emptyset$, the empty set, then $n(S) = 0$. (The empty set contains no elements.)

Let us begin by stating one of the fundamental principles of counting, the *inclusion–exclusion principle*.

> **Inclusion–Exclusion Principle** Let S and T be sets. Then
> $$n(S \cup T) = n(S) + n(T) - n(S \cap T). \tag{1}$$

Notice that formula (1) connects the four quantities $n(S \cup T)$, $n(S)$, $n(T)$, and $n(S \cap T)$. Given any three, the remaining quantity can be determined by using this formula.

To test the plausibility of the inclusion–exclusion principle, consider this example. Let $S = \{a, b, c, d, e\}$ and $T = \{a, c, g, h\}$. Then

$$S \cup T = \{a, b, c, d, e, g, h\} \qquad n(S \cup T) = 7$$
$$S \cap T = \{a, c\} \qquad n(S \cap T) = 2.$$

In this case the inclusion–exclusion principle reads

$$n(S \cup T) = n(S) + n(T) - n(S \cap T)$$
$$7 \quad = \quad 5 \; + \; 4 \; - \quad 2,$$

which is correct.

Here is the reason for the validity of the inclusion–exclusion principle: The left side of formula (1) is $n(S \cup T)$, the number of elements in either S or T (or both). As a first approximation to this number, add the number of elements in S to the number of elements in T, obtaining $n(S) + n(T)$. However, if an element lies in both S and T, it is counted twice—once in $n(S)$ and again in $n(T)$. To make up for this double counting we must subtract the number of elements counted twice, namely $n(S \cap T)$. So doing gives us $n(S) + n(T) - n(S \cap T)$ as the number of elements in $S \cup T$.

The next example illustrates a typical use of the inclusion–exclusion principle in an applied problem.

EXAMPLE 1 **Using the inclusion–exclusion principle** In the year 2004, *Executive* magazine surveyed the presidents of the 500 largest corporations in the United States. Of these 500 people, 310 had degrees (of any sort) in business, 238 had undergraduate degrees in business, and 184 had graduate degrees in business. How many presidents had both undergraduate and graduate degrees in business?

Solution Let

$$S = \{\text{presidents with an undergraduate degree in business}\}$$
$$T = \{\text{presidents with a graduate degree in business}\}.$$

Then

$$S \cup T = \{\text{presidents with at least one degree in business}\}$$
$$S \cap T = \{\text{presidents with both undergraduate and graduate degrees in business}\}.$$

From the data given we have

$$n(S) = 238 \qquad n(T) = 184 \qquad n(S \cup T) = 310.$$

The problem asks for $n(S \cap T)$. By the inclusion–exclusion principle we have

$$n(S \cup T) = n(S) + n(T) - n(S \cap T)$$
$$310 = 238 + 184 - n(S \cap T)$$
$$n(S \cap T) = 112.$$

That is, exactly 112 of the presidents had both undergraduate and graduate degrees in business.

Now Try Exercise 9

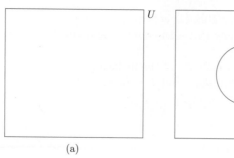
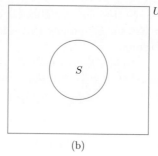
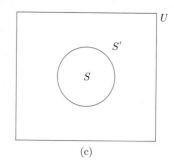

(a)   (b)   (c)

Figure 1

It is possible to visualize sets geometrically by means of drawings known as *Venn diagrams*. Such graphical representations of sets are very useful tools in solving counting problems. In order to describe Venn diagrams, let us begin with a single set S contained in a universal set U. Draw a rectangle and view its points as the elements of U [Fig. 1(a)]. To show that S is a subset of U, we draw a circle inside the rectangle and view S as the set of points in the circle [Fig. 1(b)]. The resulting diagram is called a Venn diagram of S. It illustrates the proper relationship between S and U. Since S' consists of those elements of U that are not in S, we may view the portion of the rectangle that is outside of the circle as representing S' [Fig. 1(c)].

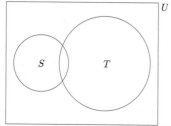

Venn diagrams are particularly useful for visualizing the relationship between two or more sets. Suppose that we are given two sets S and T in a universal set U. As before, we represent each of the sets by means of a circle inside the rectangle (Fig. 2).

Figure 2

We can now illustrate a number of sets by shading in appropriate regions of the rectangle. For instance, in Fig. 3 we have shaded the regions corresponding to T, $S \cup T$, and $S \cap T$, respectively.

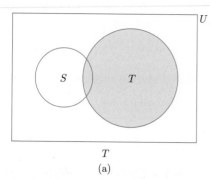
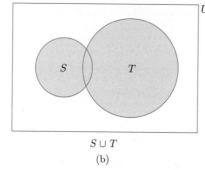
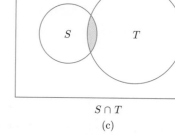

T $S \cup T$ $S \cap T$

(a)   (b)   (c)

Figure 3

EXAMPLE 2   **Shading portions of a Venn diagram**   Shade the portions of the rectangle corresponding to the sets

(a) $S \cap T'$ (b) $(S \cap T')'$

Solution (a) $S \cap T'$ consists of the points in S and in T', that is, the points in S and not in T. So we shade the points that are in the circle S but are not in the circle T [Fig. 4(a)].

(b) $(S \cap T')'$ is the complement of the set $S \cap T'$. Therefore, it consists of exactly those points not shaded in Fig. 4(a). [See Fig. 4(b).]

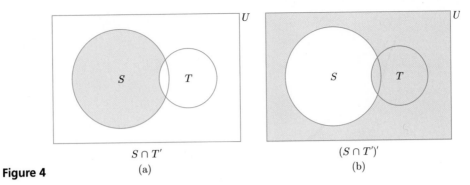

Figure 4

$S \cap T'$
(a)

$(S \cap T')'$
(b)

Now Try Exercises 15 and 19

In a similar manner, Venn diagrams can illustrate intersections and unions of three sets. Some representative regions are shaded in Fig. 5.

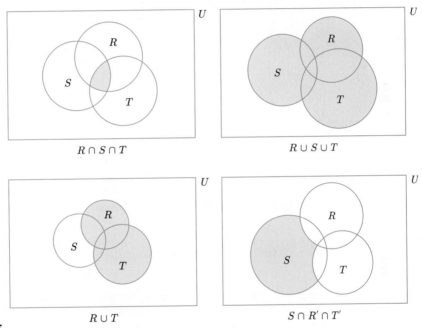

$R \cap S \cap T$

$R \cup S \cup T$

$R \cup T$

$S \cap R' \cap T'$

Figure 5

There are many formulas expressing relationships between intersections and unions of sets. Possibly the most fundamental are the two formulas known as De Morgan's laws.

De Morgan's Laws Let S and T be sets. Then
$$(S \cup T)' = S' \cap T' \quad \text{and} \quad (S \cap T)' = S' \cup T'. \tag{2}$$

In other words, De Morgan's laws state that to form the complement of a union (or intersection), form the complements of the individual sets and change unions to intersections (or intersections to unions).

Verification of De Morgan's Laws Let us utilize Venn diagrams to describe $(S \cup T)'$. In Fig. 6(a) on the next page we have shaded the region corresponding

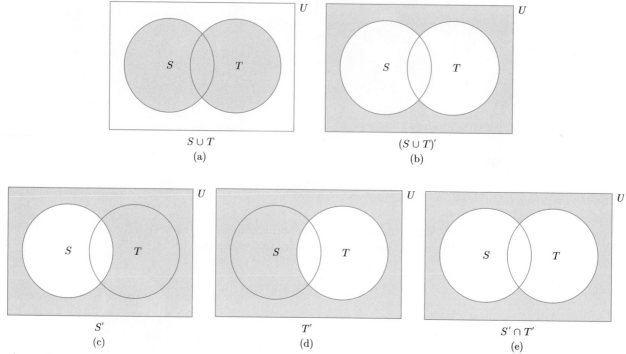

Figure 6

to $S \cup T$. In Fig. 6(b) we have shaded the region corresponding to $(S \cup T)'$. On the other hand, in Fig. 6(c) we have shaded the region corresponding to S' and in Fig. 6(d) the region corresponding to T'. By considering the common shaded regions of Fig. 6(c) and (d), we arrive at the shaded region corresponding to $S' \cap T'$ [Fig. 6(e)]. Note that this is the same region as shaded in Fig. 6(b). Therefore,

$$(S \cup T)' = S' \cap T'.$$

This verifies the first of De Morgan's laws. The proof of the second law is similar. ∎

PRACTICE PROBLEMS 5.2

1. Draw a two-circle Venn diagram and shade the portion corresponding to the set $(S \cap T') \cup (S \cap T)$.

2. Suppose that $n(S) + n(T) = n(S \cup T)$. What can you conclude about S and T?

EXERCISES 5.2

1. Find $n(S \cup T)$, given that $n(S) = 5$, $n(T) = 4$, and $n(S \cap T) = 2$.

2. Find $n(S \cup T)$, given that $n(S) = 17$, $n(T) = 13$, and $n(S \cap T) = 9$.

3. Find $n(S \cap T)$, given that $n(S) = 7$, $n(T) = 8$, and $n(S \cup T) = 15$.

4. Find $n(S \cap T)$, given that $n(S) = 4$, $n(T) = 12$, and $n(S \cup T) = 15$.

5. Find $n(S)$, given that $n(T) = 7$, $n(S \cap T) = 5$, and $n(S \cup T) = 13$.

6. Find $n(T)$, given that $n(S) = 14$, $n(S \cap T) = 6$, and $n(S \cup T) = 14$.

7. If $n(S) = n(S \cap T)$, what can you conclude about S and T?

8. If $n(S) = n(S \cup T)$, what can you conclude about S and T?

9. (*Languages*) Suppose that each of the 219 million adults in South America is fluent in Portuguese or Spanish. If 120 million are fluent in Portuguese and

116 million are fluent in Spanish, how many are fluent in both languages?

10. (*Course Enrollments*) Suppose that all of the 1000 first-year students at a certain college are enrolled in a math or an English course. Suppose that 400 are taking both math and English and 600 are taking English. How many are taking a math course?

11. (*Symmetry of Letters*) Of the 26 capital letters of the alphabet, 11 have vertical symmetry (for instance, A, M, and T), 9 have horizontal symmetry (such as B, C, and D), and 4 have both (H, I, O, X). How many letters have no symmetry?

12. (*Magazine Subscriptions*) A survey of employees in a certain company revealed that 300 people subscribe to *Newsweek*, 200 subscribe to *Time*, and 50 subscribe to both. How many people subscribe to at least one of these magazines?

13. (*Automobile Options*) Motors Inc. manufactured 325 cars with automatic transmissions, 216 with power steering, and 89 with both of these options. How many cars were manufactured with at least one of the two options?

14. (*Investments*) A survey of 100 investors in stocks and bonds revealed that 80 investors owned stocks and 70 owned bonds. How many investors owned both stocks and bonds?

In Exercises 15–26, draw a two-circle Venn diagram and shade the portion corresponding to the set.

15. $S' \cap T$

16. $S' \cap T'$

17. $S \cup T'$

18. $S' \cup T'$

19. $(S' \cap T)'$

20. $(S \cap T)'$

21. $(S \cap T') \cup (S' \cap T)$

22. $(S \cap T) \cup (S' \cap T')$

23. $S \cup (S \cap T)$

24. $S \cup (T' \cup S)$

25. $S \cup S'$

26. $S \cap S'$

In Exercises 27–38, draw a three-circle Venn diagram and shade the portion corresponding to the set.

27. $R \cap S \cap T'$

28. $R' \cap S' \cap T$

29. $R \cup (S \cap T)$

30. $R \cap (S \cup T)$

31. $R \cap (S' \cup T)$

32. $R' \cup (S \cap T')$

33. $R \cap T$

34. $S \cap T'$

35. $R' \cap S' \cap T'$

36. $(R \cup S \cup T)'$

37. $(R \cap T) \cup (S \cap T')$

38. $(R \cup S') \cap (R \cup T')$

In Exercises 39–44, use De Morgan's laws to simplify the given expression.

39. $S' \cup (S \cap T)'$

40. $T \cap (S \cup T)'$

41. $(S' \cup T)'$

42. $(S' \cap T')'$

43. $T \cup (S \cap T)'$

44. $(S' \cap T)' \cup S$

In Exercises 45–50, give a set-theoretic expression that describes the shaded portion of the Venn diagram.

45.

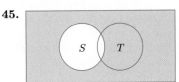

46.

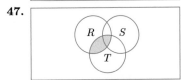

47.

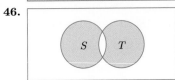

48.

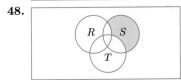

49.

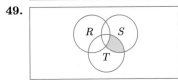

50.

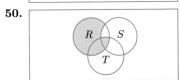

By drawing a Venn diagram, replace each of the expressions in Exercises 51–53 with one involving at most one union and the complement symbol applied only to R, S, and T.

51. $(T \cap S) \cup (T \cap R) \cup (R \cap S') \cup (T \cap R' \cap S')$

52. $(R \cap S) \cup (S \cap T) \cup (R \cap S' \cap T')$

53. $((R \cap S') \cup (S \cap T') \cup (T \cap R'))'$

(*Citizenship*) Assume the universal set U is the set of all people living in the United States. Let A be the set of all citizens, B be the set of all legal aliens, and C be the set of all illegal aliens. Let D be the set of all children under 5 years of age, E be the set of children from 5 to 18 years old, and F be the set of everyone over the age of 18. Let G be the set of all people who are employed. Describe in words the sets in Exercises 54–59.

54. $E \cap G \cap B$

55. $C' \cup (G \cap F)$

56. $A \cap (F \cup G)$

57. $F \cap G'$

58. $A \cap B$

59. $(A' \cup B) \cap G'$

SOLUTIONS TO PRACTICE PROBLEMS 5.2

1. $(S \cap T') \cup (S \cap T)$ is given as a union of two sets, $S \cap T'$ and $S \cap T$. The Venn diagrams for these two sets are given in Fig. 7(a) and (b). The desired set consists of the elements that are in one or the other (or both) of the two sets. Therefore, its Venn diagram is obtained by shading everything that is shaded in either Fig. 7(a) or (b) [see Fig. 7(c)]. [*Note:* Looking at Fig. 7(c) reveals that $(S \cap T') \cup (S \cap T)$ and S are

the same set. Often Venn diagrams can be used to simplify complicated set-theoretic expressions.]

2. From the inclusion–exclusion principle, we obtain

$$n(S \cap T) = 0 \text{—that is, } S \cap T = \varnothing.$$

We conclude that S and T have no elements in common.

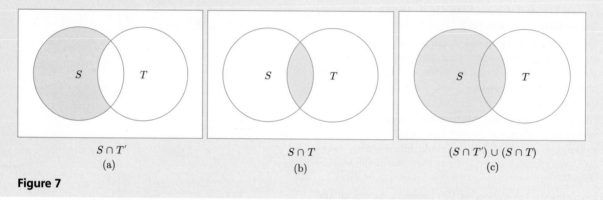

$$S \cap T'$$
(a)

$$S \cap T$$
(b)

$$(S \cap T') \cup (S \cap T)$$
(c)

Figure 7

5.3 Venn Diagrams and Counting

In this section we discuss the use of Venn diagrams in solving counting problems. The techniques developed are especially useful in analyzing survey data.

Each Venn diagram divides the universal set U into a certain number of regions. For example, the Venn diagram for a single set divides U into two regions—the inside and outside of the circle [Fig. 1(a)]. The Venn diagram for two sets divides U into four regions [Fig. 1(b)]. And the Venn diagram for three sets divides U into eight regions [Fig. 1(c)]. Each of the regions is called a *basic region* for the Venn diagram. Knowing the number of elements in each basic region is of great use in many applied problems. As an illustration consider the following example.

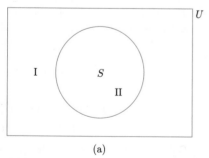

(a)

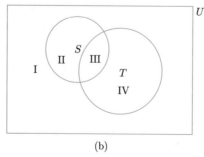

(b)

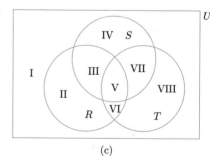

(c)

Figure 1

EXAMPLE 1 **Nobel prize winners** Let

$$U = \{\text{winners of the Nobel Prize during the period 1901–2004}\}$$
$$A = \{\text{American winners of the Nobel Prize during the period 1901–2004}\}$$
$$C = \{\text{winners of the Nobel Prize in chemistry during the period 1901–2004}\}$$
$$P = \{\text{winners of the Nobel Peace Prize during the period 1901–2004}\}.$$

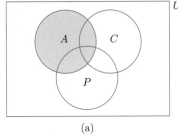

Figure 2

These sets are illustrated in the Venn diagram of Fig. 2 in which each basic region has been labeled with the number of elements in it.

(a) How many Americans received a Nobel Prize during the period 1901–2004?

(b) How many Americans received Nobel Prizes in fields other than chemistry and peace during this period?

(c) How many Americans received the Nobel Peace Prize during this period?

(d) How many people received Nobel Prizes during this period?

Solution (a) The number of Americans who received a Nobel Prize is the total contained in the circle A, which is

$$209 + 24 + 1 + 53 = 287 \quad [\text{Fig. 3(a)}].$$

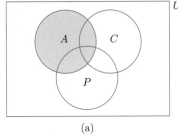
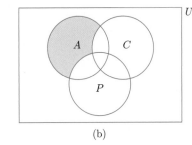
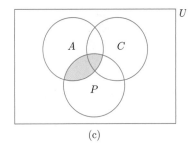

(a)   (b)   (c)

Figure 3

(b) The question asks for the number of Nobel laureates in A but not in C and not in P. So start with the A circle and eliminate those basic regions belonging to C or P [Fig. 3(b)]. There remains a single basic region with 209 Nobel laureates. Note that this region corresponds to $A \cap C' \cap P'$.

(c) The question asks for the number of elements in both A and P—that is, $n(A \cap P)$. But $A \cap P$ comprises two basic regions [Fig. 3(c)]. Thus, to compute $n(A \cap P)$ we add the numbers in these basic regions to obtain $24 + 1 = 25$ Americans who have won the Nobel Peace Prize.

(d) The number of recipients is just $n(U)$, and we obtain it by adding together the numbers corresponding to the basic regions. We obtain

$$301 + 209 + 53 + 1 + 24 + 91 + 0 + 84 = 763. \quad \blacksquare$$

One need not always be given the number of elements in each of the basic regions of a Venn diagram. Very often these data can be deduced from given information.

EXAMPLE 2 **Corporate presidents** Consider the set of 500 corporate presidents of Example 1, Section 5.2.

(a) Draw a Venn diagram displaying the given data, and determine the number of elements in each basic region.

(b) Determine the number of presidents having exactly one degree (graduate or undergraduate) in business.

Solution (a) Recall that we defined the following sets:

$$S = \{\text{presidents with an undergraduate degree in business}\}$$
$$T = \{\text{presidents with a graduate degree in business}\}.$$

We were given the following data:

$$n(S) = 238 \qquad n(T) = 184 \qquad n(S \cup T) = 310.$$

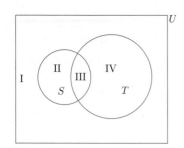

Figure 4

We draw a Venn diagram corresponding to S and T (Fig. 4). Notice that none of the given information corresponds to a basic region of the Venn diagram. So we must use our wits to determine the number of presidents in each of the regions I–IV. Region I is the complement of $S \cup T$, so it contains

$$n(U) - n(S \cup T) = 500 - 310 = 190$$

presidents. Region III is just $S \cap T$. By using the inclusion–exclusion principle, in Example 1, Section 5.2, we determined that $n(S \cap T) = 112$. Now the total number of presidents in II and III combined equals $n(S)$, or 238. Therefore, the number of presidents in II is

$$238 - 112 = 126.$$

Similarly, the number of presidents in IV is

$$184 - 112 = 72.$$

Thus we may fill in the data to obtain a completed Venn diagram (Fig. 5).

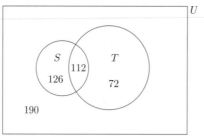

Figure 5

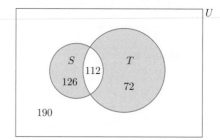

Figure 6

(b) The number of people with exactly one business degree corresponds to the shaded region in Fig. 6. Adding together the number of presidents in each of these regions gives $126 + 72 = 198$ presidents with exactly one business degree.

Now Try Exercise 13

∎

Here is another example illustrating the procedure for determining the number of elements in each of the basic regions of a Venn diagram.

EXAMPLE 3 **Advertising media** An advertising agency finds that of its 170 clients, 115 use television (T), 100 use radio (R), 130 use magazines (M), 75 use television and radio, 95 use radio and magazines, 85 use television and magazines, and 70 use all three. Use these data to complete a Venn diagram displaying the use of mass media (Fig. 7).

Solution Of the various data given, only the last item corresponds to one of the eight basic regions of the Venn diagram, namely the "70" corresponding to the use of all three media. So we begin by entering this number in the diagram [Fig. 8(a)]. We can fill in the rest of the Venn diagram by working with the remaining information one piece at a time in the reverse order that it is given. Since 85 clients advertise in television and magazines, $85 - 70 = 15$ advertise in television and magazines but not on radio. The appropriate region is labeled in Fig. 8(b). In Fig. 8(c) the next two pieces of information have been used in the same way to fill in two more basic regions. In Fig. 8(c) we observe that three of the four basic regions comprising M have been filled in. Since $n(M) = 130$, we deduce that the number of clients advertising only in magazines is $130 - (15 + 70 + 25) = 130 - 110 = 20$ [Fig. 9(a)]. By similar reasoning the number of clients using only radio advertising and the number using only television advertising can be determined [Fig. 9(b)]. Adding together the numbers in the three circles gives the number of clients utilizing television, radio, or magazines as $25 + 5 + 0 + 15 + 70 + 25 + 20 = 160$. Since there were 170 clients in total, the remainder—or $170 - 160 = 10$ clients—use none of these media. Figure 9(c) gives a complete display of the data.

$n(U) = 170$

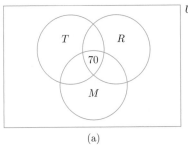

Figure 7

Now Try Exercise 15(a)

■

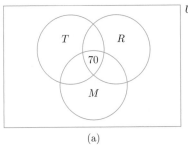

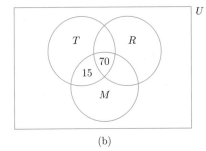

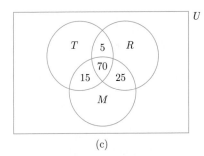

(a)   (b)   (c)

Figure 8

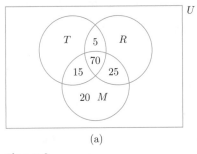

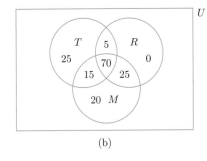

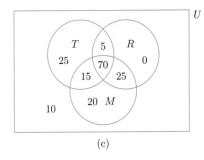

(a)   (b)   (c)

Figure 9

PRACTICE PROBLEMS 5.3

1. Of the 1000 first-year students at a certain college, 700 take mathematics courses, 300 take mathematics and economics courses, and 200 do not take any mathematics or economics courses. Represent these data in a Venn diagram.

2. Refer to the Venn diagram from Problem 1.

   (a) How many of the first-year students take an economics course?

   (b) How many take an economics course but not a mathematics course?

EXERCISES 5.3

In Exercises 1–12, let R, S, and T be subsets of the universal set U. Draw an appropriate Venn diagram and use the given data to determine the number of elements in each basic region.

1. $n(U) = 14$, $n(S) = 5$, $n(T) = 6$, $n(S \cap T) = 2$.
2. $n(U) = 20$, $n(S) = 11$, $n(T) = 7$, $n(S \cap T) = 7$.
3. $n(U) = 20$, $n(S) = 12$, $n(T) = 14$, $n(S \cup T) = 18$.
4. $n(S') = 6$, $n(S \cup T) = 10$, $n(S \cap T) = 5$, $n(T) = 7$.
5. $n(U) = 75$, $n(S) = 15$, $n(T) = 25$, $n(S' \cap T') = 40$.
6. $n(S) = 9$, $n(T) = 11$, $n(S \cap T) = 5$, $n(S') = 13$.
7. $n(S) = 3$, $n(S \cup T) = 6$, $n(T) = 4$, $n(S' \cup T') = 9$.
8. $n(U) = 15$, $n(S) = 8$, $n(T) = 9$, $n(S \cup T) = 14$.
9. $n(U) = 44$, $n(R) = 17$, $n(S) = 17$, $n(T) = 17$, $n(R \cap S) = 7$, $n(R \cap T) = 6$, $n(S \cap T) = 5$, $n(R \cap S \cap T) = 2$.
10. $n(U) = 29$, $n(R) = 10$, $n(S) = 12$, $n(T) = 10$, $n(R \cap S) = 1$, $n(R \cap T) = 5$, $n(S \cap T) = 4$, $n(R \cap S \cap T) = 1$.
11. $n(R') = 22$, $n(R \cup S) = 21$, $n(S) = 14$, $n(T) = 22$, $n(R \cap S) = 7$, $n(S \cap T) = 9$, $n(R \cap T) = 11$, $n(R \cap S \cap T) = 5$.
12. $n(U) = 64$, $n(R \cup S \cup T) = 45$, $n(R) = 22$, $n(T) = 26$, $n(R \cap S) = 4$, $n(S \cap T) = 6$, $n(R \cap T) = 8$, $n(R \cap S \cap T) = 1$.

13. (*Music Preferences*) A survey of 70 high school students revealed that 35 like folk music, 15 like classical music, and 5 like both. How many of the students surveyed do not like either folk or classical music?

14. (*Nobel Prize Winners*) A total of 763 Nobel Prizes had been awarded by 2004. Fifteen of the 101 prizes in literature were awarded to Scandinavians. Scandinavians received a total of 55 awards. How many Nobel Prizes outside of literature have been awarded to non-Scandinavians?

15. (*Class Enrollment*) Out of 35 students in a finite math class, 22 are male, 19 are business majors, 27 are first-year students, 14 are male business students, 17 are male first-year students, 15 are first-year students who are business majors, and 11 are male first-year business majors.

    (a) Use these data to complete a Venn diagram displaying the characteristics of the students.

    (b) How many upperclass female nonbusiness majors are in the class?

    (c) How many female business majors are in the class?

16. (*Exercise Preferences*) A survey of 100 college faculty who exercise regularly found that 45 jog, 30 swim, 20 cycle, 6 jog and swim, 1 jogs and cycles, 5 swim and cycle, and 1 does all three. How many of the faculty members do not do any of these three activities? How many just jog?

17. (*Voting Preferences*) One hundred college students were surveyed after voting in an election involving a Democrat and a Republican. Fifty were first-year students, 55 voted Democratic, and 25 were non–first-year students who voted Republican. How many first-year students voted Democratic?

18. (*Union Membership and Education Status*) A group of 100 workers were asked if they were college graduates and if they belonged to a union. Sixty were not college graduates, 20 were nonunion college graduates, and 30 were union members. How many of the workers were neither college graduates nor union members?

19. (*Diagnostic Test Results*) A class of 30 students was given a diagnostic test on the first day of a mathematics course. At the end of the semester, only 2 of the 21 students who had passed the diagnostic test failed the course. A total of 23 students passed the course. How many students managed to pass the course even though they had failed the diagnostic test?

20. (*Air-Traffic Controllers*) A group of applicants for training as air-traffic controllers consisted of 35 pilots, 20 veterans, 30 pilots who were not veterans, and 50 people who were neither veterans nor pilots. How large was the group?

21. (*Analysis of Sonnet*) One of Shakespeare's sonnets has a verb in 11 of its 14 lines, an adjective in 9 lines, and both in 7 lines. How many lines have a verb but no adjective? An adjective and no verb? Neither an adjective nor a verb?

(Exam Performance) Of the 130 students who took a mathematics exam, 90 correctly answered the first question, 62 correctly answered the second question, and 50 correctly answered both questions. Exercises 22–26 refer to these students.

22. How many students correctly answered either the first or second question?

23. How many students did not answer either of the two questions correctly?

24. How many students answered either the first or the second question correctly, but not both?

25. How many students answered the second question correctly, but not the first?

26. How many students missed the second question?

(Football Cards) A collector of football cards has 2200 cards. He has 1500 players from the National Football League (NFL), 900 who played defense, and 400 who played defense for the NFL. Exercises 27–32 refer to these football players.

27. How many players either were in the NFL or played defense?

28. How many players played defense but were not in the NFL?

29. How many players played offense but were not in the NFL?

30. How many players played offense in the NFL?

31. How many players either played defense or were in the NFL, but not both?

32. How many players did not play defense for the NFL?

(Music Preferences) A campus radio station surveyed 190 students to determine the types of music they liked. The survey revealed that 114 liked rock, 50 liked country, and 41 liked classical music. Moreover, 14 liked rock and country, 15 liked rock and classical, 11 liked classical and country, and 5 liked all three types of music. Exercises 33–40 refer to the students in this survey.

33. How many students like rock only?

34. How many students like country but not rock?

35. How many students like classical and country but not rock?

36. How many students like classical or country but not rock?

37. How many students like exactly one of the three types of music?

38. How many students do not like any of the three types of music?

39. How many students like at least two of the three types of music?

40. How many students do not like either rock or country?

(News Dissemination) A merchant surveyed 400 people to determine the way they learned about an upcoming sale. The survey showed that 180 learned about the sale from the radio, 190 from television, 190 from the newspaper, 80 from radio and television, 90 from radio and newspapers, 50 from television and newspapers, and 30 from all three sources. Exercises 41–46 refer to the people in this survey.

41. How many people learned of the sale from newspapers or radio, but not both?

42. How many people learned of the sale only from newspapers?

43. How many people learned of the sale from radio or television but not the newspaper?

44. How many people learned of the sale from at least two of the three media?

45. How many people learned of the sale from exactly one of the three media?

46. How many people learned of the sale from radio and television but not the newspaper?

47. *(Magazine Preferences)* One hundred and eighty business executives were surveyed to determine if they regularly read *Fortune*, *Time*, or *Money* magazines. Seventy-five read *Fortune*, 70 read *Time*, 55 read *Money*, 45 read exactly two of the three magazines, 25 read *Fortune* and *Time*, 25 read *Time* and *Money*, and 5 read all three magazines. How many read none of the three magazines?

48. *(Small Businesses)* A survey of the characteristics of 100 small businesses that had failed revealed that 95 of them either were undercapitalized, had inexperienced management, or had a poor location. Four of the businesses had all three of these characteristics. Forty businesses were undercapitalized but had experienced management and good location. Fifteen businesses had inexperienced management but sufficient capitalization and good location. Seven were undercapitalized and had inexperienced management. Nine were undercapitalized and had poor location. Ten had inexperienced management and poor location. How many of the businesses had poor location? Which of the three characteristics was most prevalent in the failed businesses?

49. *(Course Enrollments)* **(PE)** Table 1 shows the number of students enrolled in each of three science courses at Gotham College. Although no students are enrolled in all three courses, 15 are enrolled in both Chemistry and Physics, 10 are enrolled in both Physics and Biology, and 5 are enrolled in both Biology and Chemistry. How many different students are enrolled in at least one of these science courses?

   (a) 75   **(b)** 80   **(c)** 90   **(d)** 100   **(e)** 130

TABLE 1	
Course	**Enrollment**
Chemistry	60
Physics	40
Biology	30

(Foreign Language Courses) Use a Venn diagram to find the number of people in the sets given in Exercises 50–54. A survey in a local high school shows that of the 4000 students in the school, 2000 take French (F), 3000 take Spanish (S), and 500 take Latin (L). The survey shows that 1500 take both French and Spanish, 300 take both French and Latin, and 200 take Spanish and Latin. Fifty students take all three languages.

50. $L \cap (F \cup S)$

51. $(L \cup F \cup S)'$

52. $L \cup S \cup F'$

53. L'

54. $F \cap S' \cap L'$

SOLUTIONS TO PRACTICE PROBLEMS 5.3

1. Draw a Venn diagram with two circles, one for mathematics (M) and one for economics (E) [Fig. 10(a)]. This Venn diagram has four basic regions, and our goal is to label each basic region with the proper number of students. The numbers for two of the basic regions are given directly. Since "300 take mathematics and economics," $n(M \cap E) = 300$. Since "200 do not take any mathematics or economics courses," $n((M \cup E)') = 200$ [Fig. 10(b)]. Now "700 take mathematics courses." Since M is made up of two basic regions and one region has 300 elements, the other basic region of M must contain 400 elements [Fig. 10(c)]. At this point all but one of the basic regions have been labeled and $400 + 300 + 200 = 900$ students have

been accounted for. Since there is a total of 1000 students, the remaining basic region has 100 students [Fig. 10(d)].

2. (a) 400. "Economics" refers to the entire circle E, which is made up of two basic regions, one having 300 elements and the other 100. (A common error is to interpret the question as asking for the number of first-year students who take economics exclusively and therefore give the answer 100. To say that a person takes an economics course does not imply anything about the person's enrollment in mathematics courses.)

   (b) 100

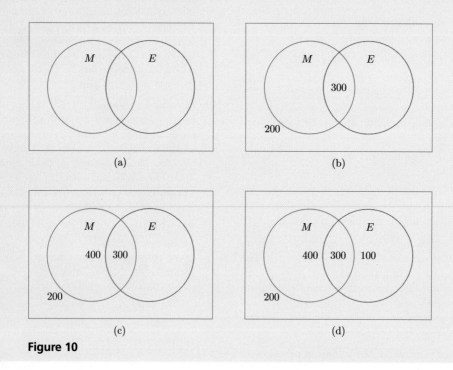

(a)

(b)

(c)

(d)

Figure 10

5.4 The Multiplication Principle

In this section we introduce a second fundamental principle of counting, the *multiplication principle*. By way of motivation, consider the following example.

EXAMPLE 1 **Counting paths through a maze** A medical researcher wishes to test the effect of a drug on a rat's perception by studying the rat's ability to run a maze while under the influence of the drug. The maze is constructed so that to arrive at the exit point C, the rat must pass through a central point B. There are five paths from the entry point A to B, and three paths from B to C. In how many different ways can the rat run the maze from A to C? (See Fig. 1.)

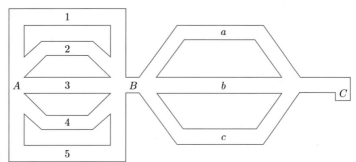

Figure 1

Solution The paths from A to B have been labeled 1 through 5, and the paths from B to C have been labeled a through c. The various paths through the maze can be schematically represented as in Fig. 2. The diagram shows that there are five ways to go from A to B. For each of these five ways, there are three ways to go from B to C. So there are five groups of three paths each, and therefore $5 \cdot 3 = 15$ possible paths from A to C. (A diagram such as Fig. 2, called a *tree diagram*, is useful in enumerating the various possibilities in counting problems.) ■

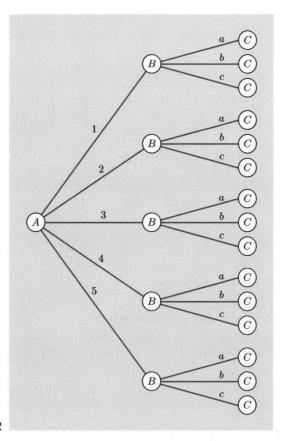

Figure 2

In the preceding problem, choosing a path is a task that can be broken up into two consecutive operations.

Choose path from A to B	Choose path from B to C
Operation 1	Operation 2

The first operation can be performed in five ways and, after the first operation has been carried out, the second can be performed in three ways. And we determined that the entire task can be performed in $5 \cdot 3 = 15$ ways. The same reasoning as just used yields the following useful counting principle:

> **Multiplication Principle**   Suppose that a task is composed of two consecutive operations. If operation 1 can be performed in m ways and, for each of these, operation 2 can be performed in n ways, then the complete task can be performed in $m \cdot n$ ways.

EXAMPLE 2   **Counting routes for a trip**   An airline passenger must fly from New York to Frankfurt via London. There are 8 flights leaving New York for London. All of these provide connections on any one of 19 flights from London to Frankfurt. In how many different ways can the passenger book reservations?

Solution   The task "fly from New York to Frankfurt" is composed of two consecutive operations:

Fly from New York to London	Fly from London to Frankfurt
Operation 1	Operation 2

From the data given, the multiplication principle implies that the task can be accomplished in $8 \cdot 19 = 152$ ways.

Now Try Exercise 1   ■

It is possible to generalize the multiplication principle to tasks consisting of more than two operations.

> **Generalized Multiplication Principle**   Suppose that a task consists of t operations performed consecutively. Suppose that operation 1 can be performed in m_1 ways; for each of these, operation 2 in m_2 ways; for each of these, operation 3 in m_3 ways; and so forth. Then the task can be performed in
> $$m_1 \cdot m_2 \cdot m_3 \cdot \ \cdots \ \cdot m_t \quad \text{ways.}$$

EXAMPLE 3   **Officers for a board of directors**   A corporation has a board of directors consisting of 10 members. The board must select from among its members a chairperson, vice chairperson, and secretary. In how many ways can this be done?

Solution The task "select the three officers" can be divided into three consecutive operations:

Select chairperson	Select vice chairperson	Select secretary

Since there are 10 directors, operation 1 can be performed in 10 ways. After the chairperson has been selected, there are 9 directors left as possible candidates for vice chairperson, so that for each way of performing operation 1, operation 2 can be performed in 9 ways. After this has been done, there are 8 directors who are possible candidates for secretary, so operation 3 can be performed in 8 ways. By the generalized multiplication principle, the number of possible ways to perform the sequence of three operations equals $10 \cdot 9 \cdot 8$, or 720. So the officers of the board can be selected in 720 ways.

Now Try Exercise 13   ■

In Example 3 we made important use of the phrase "for each of these" in the generalized multiplication principle. The operation "select a vice chairperson" can be performed in 10 ways, since any member of the board is eligible. However, when we view the selection process as a sequence of operations of which "select a vice chairperson" is the second operation, the situation has changed. *For each way* that the first operation is performed, one person will have been used up; hence there will be only 9 possibilities for choosing the vice chairperson.

EXAMPLE 4 **Posing for a group picture** In how many ways can a baseball team of nine members arrange themselves in a line for a group picture?

Solution Choose the players by their place in the picture, say from left to right. The first can be chosen in nine ways; for each of these choices the second can be chosen in eight ways; for each of these choices the third can be chosen in seven ways; and so forth. So the number of possible arrangements is

$$9 \cdot 8 \cdot 7 \cdot 6 \cdot 5 \cdot 4 \cdot 3 \cdot 2 \cdot 1 = 362{,}880.$$

Now Try Exercise 24   ■

EXAMPLE 5 **License plates** A certain state uses automobile license plates that consist of three letters followed by three digits. How many such license plates are there?

Solution The task in this case, "form a license plate," consists of a sequence of six operations: three for choosing letters and three for choosing digits. Each letter can be chosen in 26 ways and each digit in 10 ways. So the number of license plates is

$$26 \cdot 26 \cdot 26 \cdot 10 \cdot 10 \cdot 10 = 17{,}576{,}000.$$

Now Try Exercise 33   ■

PRACTICE PROBLEMS 5.4

1. There are six seats available in a sedan. In how many ways can six people be seated if only three can drive?

2. A multiple-choice exam contains 10 questions, each having 3 possible answers. How many different ways are there of completing the exam?

EXERCISES 5.4

1. (*Routes*) If there are three routes from College Park to Baltimore and five routes from Baltimore to New York, how many routes are there from College Park to New York via Baltimore?

2. (*Choosing an Outfit*) How many different outfits consisting of a coat and a hat can be chosen from two coats and three hats?

3. (*Two-Letter Words*) How many different two-letter words (including nonsense words) can be formed when repetition of letters is allowed?

4. (*Two-Letter Words*) How many different two-letter words (including nonsense words) can be formed such that the two letters are distinct?

5. (*Railroad Tickets*) A railway has 20 stations. If the names of the point of departure and the destination are printed on each ticket, how many different kinds of single tickets must be printed?

6. (*Railroad Tickets*) Refer to Exercise 5. How many different kinds of tickets are needed if each ticket may be used in either direction between two stations?

7. (*Gloves*) A man has five different pairs of gloves. In how many ways can he select a right-hand glove and a left-hand glove that do not match?

8. (*License Plates*) How many license plates consisting of two letters followed by four digits are possible?

9. (*Group Picture*) How many ways can five people be arranged in a line for a group picture?

10. (*Arranging Books*) In how many different ways can four books be arranged on a bookshelf?

11. (*Coin Tosses*) Toss a coin six times and observe the sequence of heads or tails that results. How many different sequences are possible?

12. (*Coin Tosses*) Refer to Exercise 11. In how many of the sequences are the first and last tosses identical?

13. (*Winners*) Twenty athletes enter an Olympic event. How many different possibilities are there for winning the Gold Medal, Silver Medal, and Bronze Medal?

14. (*Ranking Teams*) A sportswriter is asked to rank eight teams. How many different orderings are possible?

15. (*Electing Captains*) In how many different ways can a 30-member football team select a captain and an assistant captain?

16. (*Choosing an Outfit*) How many different outfits can be selected from two coats, three hats, and two scarves?

17. (*Rearranging Letters*) How many different words (including nonsense words) can be formed using the four letters of the word "MATH"?

18. (*Travel Options*) If you can travel from Frederick, Maryland, to Baltimore, Maryland, by car, bus, or train and from Baltimore to London by airplane or ship, how many different ways are there to go from Frederick to London?

19. (*Exam Questions*) An exam contains five "true or false" questions. In how many different ways can the exam be completed?

20. (*Pairs of Initials*) A company has 700 employees. Explain why there must be two people with the same pair of initials.

21. (*Serial Numbers*) A computer manufacturer assigns serial numbers to its computers. The first symbol of a serial number is either A, B, or C, indicating the manufacturing plant. The second and third symbols taken together are one of the numbers $01, 02, \ldots, 12$, indicating the month of manufacture. The final four symbols are digits. How many possible serial numbers are there?

22. (*Forming Words*) How many four-letter words (including nonsense words) can be made from the letters of "statistics," assuming that each word may not have repeated letters?

23. (*Forming Words*) How many four-letter words (including nonsense words) can be made from the letters h, o, t, s, m, x, and e for each of the following conditions?

   (a) Letters can be repeated.

   (b) Letters cannot be repeated.

   (c) Words must begin with an h, and repetitions are allowed.

   (d) Words must end with a vowel, and repetitions are not allowed.

24. (*Group Picture*) A group of five boys and three girls is to be photographed.

   (a) How many ways can they be arranged in one row?

   (b) How many ways can they be arranged with the girls in the front row and the boys in the back row?

25. (*Batting Orders*) The manager of a Little League baseball team has picked the nine starting players for a game. How many different batting orders are possible under each of the following conditions?

   (a) There are no restrictions.

   (b) The pitcher must bat last.

   (c) The pitcher must bat last, the catcher eighth, and the shortstop first.

26. (*Shading of Venn Diagrams*) How many different ways can a Venn diagram with two circles be shaded?

27. (*Shading of Venn Diagrams*) How many different ways can a Venn diagram with three circles be shaded?

28. (*Club Officers*) A club can elect a member as president and a different member as treasurer in 506 different ways. How many members does the club have?

29. (*Exam Questions*) An exam contains six true or false statements. In how many ways can the exam be completed if leaving the answer blank is also an option?

30. (*Test Volunteers*) A physiologist wants to test the effects of exercise and meditation on blood pressure. She devises four different exercise programs and three different meditation programs. If she wants 10 subjects for each combination of exercise and meditation program, how many volunteers must she recruit?

31. (*Menu Selections*) A college student eats all his meals at a restaurant offering six breakfast specials, seven lunch specials, and four dinner specials. How many days can he go without repeating an entire day's menu selection?

32. (*Entrances*) A classroom building has 7 different doors. In how many ways can a student enter by one door and exit by a different door?

33. (*License Plates*) A California license plate consists of a digit followed by three letters, and then three digits. How many such license plates are there?

34. (*Sweaters*) A clothing store offers three styles of sweaters with each sweater available in six colors. How many different sweaters are there?

35. (*Area Codes*) Before 1995, three-digit area codes for the United States had the following restrictions:

  (a) Neither 0 nor 1 could be used as the first digit.

  (b) 0 or 1 had to be used for the second digit.

  (c) There were no restrictions on the third digit.
  How many different area codes were possible?

36. (*Area Codes*) Refer to Exercise 35. Beginning in 1995, restriction (b) was lifted and any digit could be used in the second position. How many different area codes are possible?

37. (*Handshakes*) Two ten-member basketball teams play a game. After the game, each of the members of the winning team shakes hands once with each member of both teams. How many handshakes take place?

38. (*Chair Varieties*) A furniture manufacturer makes three types of upholstered chairs and offers a choice of 20 fabrics. How many different chairs are available?

39. (*Colored Houses*) Six houses in a row are each to be painted with one of the colors red, blue, green, and yellow. In how many different ways can the houses be painted so that no two adjacent houses are of the same color?

40. (*Numbers*) How many three-digit odd numbers can be formed using the digits 1, 2, 3, 4, 5, 6, and 7?

41. (*Exam Questions*) Each of the 10 questions on a multiple-choice exam has four possible answers. How many different ways are there for a student to answer the questions? Assume that every question must be answered.

42. (*Mismatched Shoes*) Fred has 10 different pairs of shoes. In how many ways can he put on a pair of shoes that do not match?

43. (*Transportation Options*) Suppose that Jack wants to go from Florida to Maine via New York and can travel each leg of the journey by bus, car, train, or airplane. How many different ways can Jack make the trip?

44. (*Menu Selections*) A restaurant menu lists 6 appetizers, 10 entrées, and 5 desserts. How many ways can a diner select a three-course meal?

45. (*Band Selections*) (**PE**) In how many arrangements can a band play three waltzes and three tangos in a row without repeating any song, such that the first, third, and fifth songs are waltzes?

  (a) 6   **(b)** 12   **(c)** 16   **(d)** 36   **(e)** 720

46. (*Milk Delivery*) How many ways can a milkman deliver 10 distinguishable bottles of milk on a street containing five houses?

47. (*Computer Options*) A computer manufacturer offers a computer with a choice of four types of monitors, two types of keyboards, and three types of hard drives. How many different computers are offered?

48. (*Ballots*) Seven candidates for mayor, 4 candidates for city council president, and 12 propositions are being put before the electorate. How many different ballots could be cast, assuming that every voter votes on each of the items? If voters can choose to leave any item blank, how many different ballots are possible?

49. (*Gift Wrapping*) The gift-wrap desk at a large department store offers 5 box sizes, 10 wrapping papers, 7 colors of ribbon in two widths, and 9 special items to be added on the bow. How many different ways are there to gift-wrap a package assuming that the customer must choose at least a box but need not choose any of the other offerings?

50. (*Selecting Fruit*) José was told to get a dozen oranges, eight apples, and a half-pound of grapes. When he gets to the store he finds five varieties of oranges, five varieties of apples, and two varieties of grapes. Assuming that he buys only one variety of each type of fruit, how many different bags of fruit could he bring home?

51. (*College Applications*) Allison is preparing her applications for college. She will apply to three community colleges and has to fill out five parts to each of those applications. She will apply to three four-year schools, each of which has a six-part application. How many application segments must she complete?

52. (*Car Options*) An automobile dealer is offering five models of a particular car. On each model the customer may choose cloth or leather seats, each available in three colors, automatic or manual transmission, and a CD player or a cassette player. The car can be ordered in any one of eight exterior colors. How many different cars can be ordered?

53. (*Game of Clue*) In the game of *Clue*, there are six suspects in a murder that was committed in any one of nine rooms, using any one of six weapons. How many scenarios are possible?

54. (*Numbers*) James chooses a number from 1 to 100 (inclusive), Janet chooses a number between 1 and 100 that is divisible by 7, and Sandy chooses a number between 1 and 100 that is divisible by 21. How many triples of numbers are possible?

55. (*Numbers*) How many four-digit numbers can be formed using the digits $\{1, 2, 3, 4, 5, 6, 7\}$ if adjacent digits must be different?

56. (*Fruit Baskets*) A grocery store makes up fruit baskets using as many as four apples, three pears, and four oranges. A basket must contain at least one piece of fruit. How many different fruit selections are possible? (From *The Mathematics Teacher*, December 1991)

57. (*Social Security Numbers*) How many Social Security numbers are available if the only restriction is that the number 000-00-0000 cannot be assigned?

58. (*Displaying Paintings*) An art gallery has two paintings by each of four artists. In how many ways can the eight paintings be displayed in a row if paintings by the same artist must be adjacent to each other?

59. (*Paths to Texas*) Consider the following triangular display of letters. Start with the letter T at the top and move down the triangle to a letter S at the bottom. From any given letter move only to one of the letters directly below it on the left or right. How many different paths spell *TEXAS*?

$$T$$
$$E \quad E$$
$$X \quad X \quad X$$
$$A \quad A \quad A \quad A$$
$$S \quad S \quad S \quad S \quad S$$

60. (*Choosing an Outfit*) A girl dresses in the morning in a blouse, a skirt, and shoes. She always wears standard white socks. She wants to wear a different combination on every day of the year. If she has the same number of blouses, skirts, and pairs of shoes, how many of each article would she need to have a different combination every day? (From *The Mathematics Teacher*, February 1996)

61. (*Seating Arrangements*) Three couples go on a movie date. In how many ways can they be seated in a row of six seats so that each couple is seated together?

62. (*Three-Letter Words*) How many different three-letter words (including nonsense words) are there in which successive letters are different?

63. (*Seating Arrangements*) A panel of eight experts is to present opposing arguments in a discussion of legalizing marijuana for medical purposes. The four proponents arrive first and decide to sit in such a way that the panelists will alternate pro and con. In how many ways can the four proponents be seated? In how many ways can all eight panelists be seated if they adhere to the same seating rules?

SOLUTIONS TO PRACTICE PROBLEMS 5.4

1. 360. Pretend that you are given the task of seating the six people. This task consists of six operations performed consecutively, as shown in Table 1. After you have performed operation 1, five people will remain, and any one of these five can be seated in the middle front seat. After operation 2, four people remain, and so on. By the generalized multiplication principle, the task can be performed in $3 \cdot 5 \cdot 4 \cdot 3 \cdot 2 \cdot 1 = 360$ ways.

TABLE 1

Operation	Number of ways operation can be performed
1: Select person to drive	3
2: Select person for middle front seat	5
3: Select person for right front seat	4
4: Select person for left rear seat	3
5: Select person for middle rear seat	2
6: Select person for right rear seat	1

2. 3^{10}. The task of answering the questions consists of 10 consecutive operations, each of which can be performed in three ways. Therefore, by the generalized multiplication principle, the task can be performed in

$$\underbrace{3 \cdot 3 \cdot 3 \cdot \cdots \cdot 3}_{10 \text{ terms}} \text{ ways.}$$

(*Note*: The answer can be left as 3^{10} or can be multiplied out to 59,049.)

5.5 Permutations and Combinations

In preceding sections we have solved a variety of counting problems using Venn diagrams and the generalized multiplication principle. Let us now turn our attention to two types of counting problems that occur very frequently and that can be solved using formulas derived from the generalized multiplication principle. These problems involve what are called permutations and combinations, which are particular types of arrangements of elements of a set. The sorts of arrangements we have in mind are illustrated in two problems:

Problem A How many words (by which we mean strings of letters) of two distinct letters can be formed from the letters $\{a, b, c\}$?

Problem B A construction crew has three members. A team of two must be chosen for a particular job. In how many ways can the team be chosen?

Each of the two problems can be solved by enumerating all possibilities.

Solution of Problem A There are six possible words, namely

$$ab \quad ac \quad ba \quad bc \quad ca \quad cb.$$

Solution of Problem B Designate the three crew members by a, b, and c. Then there are three possible two-person teams, namely

$$ab \quad ac \quad bc.$$

(Note that ba, the team consisting of b and a, is the same as the team ab.)

We deliberately set up both problems using the same letters in order to facilitate comparison. Both problems are concerned with counting the numbers of arrangements of the elements of the set $\{a, b, c\}$, taken two at a time, without allowing repetition (for example, aa was not allowed). However, in Problem A the order of the arrangement mattered, whereas in Problem B it did not. Arrangements of the sort considered in Problem A are called *permutations*, whereas those in Problem B are called *combinations*.

More precisely, suppose that we are given a set of n objects.[1] Then a *permutation of n objects taken r at a time* is an arrangement of r of the n objects in a specific order. So, for example, Problem A was concerned with permutations of the three objects, a, b, c ($n = 3$), taken two at a time ($r = 2$). A *combination of n objects taken r at a time* is a selection of r objects from among the n, with order disregarded. Thus, for example, in Problem B we considered combinations of the three objects a, b, c ($n = 3$), taken two at a time ($r = 2$).

---

[1] All are assumed to be different.

It is convenient to introduce the following notation for counting permutations and combinations. Let

$P(n, r)$ = the number of permutations of n objects taken r at a time

$C(n, r)$ = the number of combinations of n objects taken r at a time.

Thus, for example, from our solutions to Problems A and B, we have

$$P(3, 2) = 6 \qquad C(3, 2) = 3.$$

Very simple formulas for $P(n, r)$ and $C(n, r)$ allow us to calculate these quantities for any n and r. Let us begin by stating the formula for $P(n, r)$. For $r = 1, 2, 3$ we have, respectively,

$$P(n, 1) = n$$
$$P(n, 2) = n(n - 1) \qquad \text{(two factors),}$$
$$P(n, 3) = n(n - 1)(n - 2) \qquad \text{(three factors),}$$

and, in general,

$$P(n, r) = n(n - 1)(n - 2) \cdot \cdots \cdot (n - r + 1) \qquad (r \text{ factors}). \qquad (1)$$

This formula is verified at the end of this section.

EXAMPLE 1 **Applying the permutation formula** Compute the following numbers.

(a) $P(100, 2)$ (b) $P(6, 4)$ (c) $P(5, 5)$

Solution (a) Here $n = 100$, $r = 2$. So we take the product of two factors, beginning with 100:

$$P(100, 2) = 100 \cdot 99 = 9900.$$

(b) $P(6, 4) = 6 \cdot 5 \cdot 4 \cdot 3 = 360$

(c) $P(5, 5) = 5 \cdot 4 \cdot 3 \cdot 2 \cdot 1 = 120$

Now Try Exercise 1   ■

In order to state the formula for $C(n, r)$, we must introduce some further notation. Suppose that r is any positive integer. We denote by $r!$ (read "r factorial") the product of all positive integers from r down to 1:

$$r! = r \cdot (r - 1) \cdot \cdots \cdot 2 \cdot 1.$$

For instance,

$$1! = 1$$
$$2! = 2 \cdot 1 = 2$$
$$3! = 3 \cdot 2 \cdot 1 = 6$$
$$4! = 4 \cdot 3 \cdot 2 \cdot 1 = 24$$
$$5! = 5 \cdot 4 \cdot 3 \cdot 2 \cdot 1 = 120.$$

In terms of this notation we can state a very simple formula for $C(n, r)$, the number of combinations of n things taken r at a time.

$$C(n,r) = \frac{P(n,r)}{r!} = \frac{n(n-1) \cdot \cdots \cdot (n-r+1)}{r(r-1) \cdot \cdots \cdot 1} \qquad (2)$$

This formula is verified at the end of this section.

EXAMPLE 2 **Applying the combination formula** Compute the following numbers.

(a) $C(100, 2)$ (b) $C(6, 4)$ (c) $C(5, 5)$

Solution (a) $C(100, 2) = \dfrac{P(100, 2)}{2!} = \dfrac{100 \cdot 99}{2 \cdot 1} = 4950$

(b) $C(6, 4) = \dfrac{P(6, 4)}{4!} = \dfrac{6 \cdot 5 \cdot 4 \cdot 3}{4 \cdot 3 \cdot 2 \cdot 1} = 15$

(c) $C(5, 5) = \dfrac{P(5, 5)}{5!} = \dfrac{5 \cdot 4 \cdot 3 \cdot 2 \cdot 1}{5 \cdot 4 \cdot 3 \cdot 2 \cdot 1} = 1$

Now Try Exercise 5   ■

EXAMPLE 3 **Applying the permutation and combination formulas** Solve Problems A and B using formulas (1) and (2).

Solution The number of two-letter words that can be formed from the three letters a, b, and c is equal to $P(3, 2) = 3 \cdot 2 = 6$, in agreement with our previous solution.

The number of two-worker teams that can be formed from three individuals is equal to $C(3, 2)$, and

$$C(3, 2) = \frac{P(3, 2)}{2!} = \frac{3 \cdot 2}{2 \cdot 1} = 3,$$

in agreement with our previous result.   ■

EXAMPLE 4 **Selecting a committee** The board of directors of a corporation has 10 members. In how many ways can they choose a committee of 3 board members to negotiate a merger?

Solution Since the committee of three involves no ordering of its members, we are concerned here with combinations. The number of combinations of 10 people taken 3 at a time is $C(10, 3)$, which is

$$C(10, 3) = \frac{10 \cdot 9 \cdot 8}{3 \cdot 2 \cdot 1} = 120.$$

Thus there are 120 choices for the committee.

Now Try Exercise 51   ■

EXAMPLE 5 **Outcomes of a horse race** Eight horses are entered in a race in which a first, second, and third prize will be awarded. Assuming no ties, how many different outcomes are possible?

Solution In this example we are considering ordered arrangements of three horses, so we are dealing with permutations. The number of permutations of eight horses taken three at a time is

$$P(8,3) = 8 \cdot 7 \cdot 6 = 336,$$

so the number of possible outcomes of the race is 336.

Now Try Exercise 44   ■

EXAMPLE 6 **Polling sample** A political pollster wishes to survey 1500 individuals chosen from a sample of 5,000,000 adults. In how many ways can the 1500 individuals be chosen?

Solution No ordering of the 1500 individuals is involved, so we are dealing with combinations. So the number in question is $C(5{,}000{,}000,\ 1500)$, a number too large to be written down in digit form. (It has several thousand digits!) But it could be calculated with the aid of a computer.

Now Try Exercise 45   ■

EXAMPLE 7 **Selecting club officers** A club has 10 members. In how many ways can they choose a slate of four officers, consisting of a president, vice president, secretary, and treasurer?

Solution In this problem we are dealing with an ordering of four members. (The first is the president, the second the vice president, and so on.) So we are dealing with permutations, and the number of ways of choosing the officers is

$$P(10,4) = 10 \cdot 9 \cdot 8 \cdot 7 = 5040.$$

Now Try Exercise 25   ■

Verification of the Formulas for P(n,r) and C(n,r) Let us first derive the formula for $P(n,r)$, the number of permutations of n objects taken r at a time. The task of choosing r objects (in a given order) consists of r consecutive operations (Fig. 1). The first operation can be performed in n ways. For each way that the first operation is performed, one object will have been used up and so we can perform the second operation in $n-1$ ways, and so on. For each way of performing the sequence of operations $1, 2, 3, \ldots, r-1$, the rth operation can be performed in $n - (r-1) = n - r + 1$ ways. By the generalized multiplication principle, the task of choosing the r objects from among the n can be performed in

$$n(n-1) \cdot \cdots \cdot (n-r+1) \quad \text{ways.}$$

That is,

$$P(n,r) = n(n-1) \cdot \cdots \cdot (n-r+1),$$

which is formula (1).

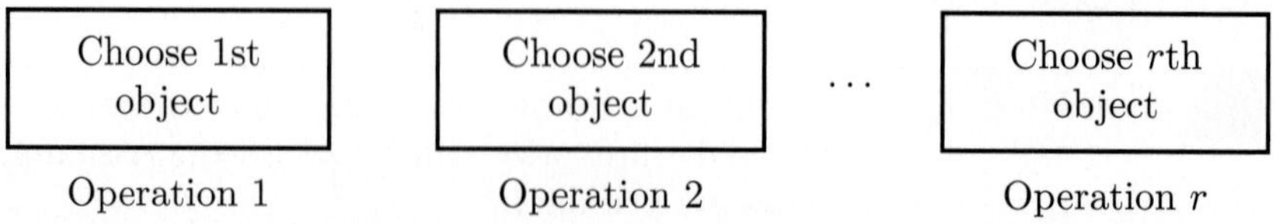

Choose 1st object		Choose 2nd object		...		Choose rth object

Figure 1   Operation 1   Operation 2   Operation r

Let us now verify the formula for $C(n,r)$, the number of combinations of n objects taken r at a time. Each such combination is a set of r objects and therefore can be ordered in

$$P(r,r) = r(r-1) \cdot \cdots \cdot 2 \cdot 1 = r!$$

ways by formula (1). In other words, each different combination of r objects gives rise to $r!$ permutations of the same r objects. On the other hand, each permutation of n objects taken r at a time gives rise to a combination of n objects taken r at a time, by simply ignoring the order of the permutation. Thus, if we start with the $P(n,r)$ permutations, we will have all the combinations of n objects taken r at a time, with each combination repeated $r!$ times. Thus

$$P(n,r) = r!\,C(n,r).$$

On dividing both sides of the equation by $r!$, we obtain formula (2).

GC Most graphing calculators have commands to compute $P(n,r)$, $C(n,r)$, and $n!$. For instance, the MATH key on the TI-83 leads to the PRB menu of Fig. 2, which contains the commands **nPr**, **nCr**, and **!**. Figure 3 shows how these commands are used. *Note*: The number **8.799226775E15** represents $8.799226775 \times 10^{15}$. On the TI-89, the first three items on the MATH/Probability menu are **!**, **nPr**, and **nCr**. Figure 4 shows how these commands are used.

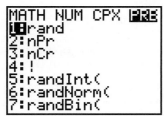
Figure 2

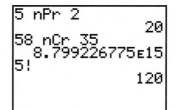
Figure 3

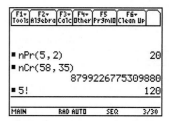
Figure 4

ES The values of $n!$, $P(n,r)$, and $C(n,r)$ are calculated in an Excel spreadsheet with the functions $\text{FACT}(n)$, $\text{PERMUT}(n,r)$, and $\text{COMBIN}(n,r)$.

PRACTICE PROBLEMS 5.5

1. Calculate the following values.

    (a) 3!   **(b)** 7!   **(c)** $P(7,3)$ **(d)** $C(7,3)$

2. A newborn child is to be given a first name and a middle name from a selection of 10 names. How many different possibilities are there?

EXERCISES 5.5

For Exercises 1–20, calculate the values.

1. $P(4,2)$ **2.** $P(5,1)$ **3.** $P(6,3)$

4. $P(5,4)$ **5.** $C(10,3)$ **6.** $C(12,2)$

7. $C(5,4)$ **8.** $C(6,3)$ **9.** $P(5,1)$

10. $P(5,5)$ **11.** $P(n,1)$ **12.** $P(n,2)$

13. $C(4,4)$ **14.** $C(n,2)$ **15.** $C(n,n-2)$

16. $C(n,1)$ **17.** 6!   **18.** $\dfrac{10!}{4!}$

19. $\dfrac{9!}{7!}$ **20.** 7!

21. (*Group Picture*) In how many ways can four people line up in a row for a group picture?

22. (*Contest Winners*) How many different outcomes of "winner" and "runner-up" are possible if there are six contestants in a pie-eating contest?

23. (*Book Selection*) How many different selections of two books can be made from a set of nine books?

24. (*Pizza Varieties*) A pizza parlor offers five toppings for the plain cheese base of the pizzas. How many different pizzas are possible that use three of the toppings?

25. (*Contest Winners*) How many ways are there to choose first, second, and third prizes in an art contest with 15 entrants?

26. (*Waiting in Line*) In how many ways can six people line up at a single counter to order food at McDonald's?

27. (*Banana Split Options*) A deluxe chocolate banana split is made with three scoops of chocolate ice cream, one banana, and a choice of 4 out of 10 possible toppings. How many different deluxe chocolate banana splits can be ordered?

28. (*Selecting Sweaters*) Suppose that you own 10 sweaters. How many ways can you select four of them to take on a trip?

29. (*Selecting Sweaters*) Suppose that you own 10 sweaters and are going on a trip. How many ways can you select six of them to leave at home?

30. (*Selecting Sweaters*) Why do Exercises 28 and 29 have the same answer?

31. (*Player Introductions*) The five starting players of a basketball team are introduced one at a time. In how many different ways can they be introduced?

32. (*Supreme Court Decisions*) In how many different ways can the nine members of the Supreme Court reach a six-to-three decision?

33. (*Conference Games*) In an eight-team football conference, each team plays every other team exactly once. How many games must be played?

34. (*Choosing Exam Questions*) A student is required to work exactly five problems from an eight-problem exam. In how many ways can the problems be chosen?

35. (*Arranging Books*) How many ways can you arrange 5 of 10 books on a shelf?

36. (*Selecting Books*) How many ways can you choose 5 of your 10 books to put in your backpack?

37. (*Guest Lists*) How many ways can you choose 5 out of 10 friends to invite to a dinner party?

38. (*Giving Gifts*) How many ways can you distribute $1, $2, $5, $10, $20 to 5 of your 10 friends? Assume that no one gets more than one bill.

39. (*CD Changer*) Suppose that you have 35 CDs and your CD player has five slots numbered 1 through 5. How many ways can you fill your CD player?

40. (*Job Interviews*) Of the 20 applicants for a job, 4 will be selected for intensive interviews. In how many ways can the selection be made?

41. (*Diskettes*) In a batch of 100 computer diskettes, 7 are defective. A sample of three diskettes is to be selected from the batch. How many samples are possible? How many of the samples consist of all defective diskettes?

42. (*Course Selection*) A student must choose five courses out of seven that he would like to take. How many possibilities are there?

43. (*Three-Letter Words*) How many different three-letter words are there having no repetition of letters?

44. (*Ranking Teams*) A sportswriter makes a preseason guess of the top 5 football teams (in order) from among 40 major teams. How many different possibilities are there?

45. (*Senate Committees*) In how many different ways can a committee of 5 senators be selected from the 100 members of the U.S. Senate?

46. (*Race Winners*) Theoretically, how many possibilities are there for first, second, and third places in a marathon race with 1000 entries?

(*Poker Hands*) Exercises 47–50 refer to poker hands. A poker hand consists of 5 cards selected from a deck of 52 cards.

47. How many different poker hands are there?

48. How many different poker hands consist entirely of aces and kings?

49. How many different poker hands consist entirely of clubs?

50. How many different poker hands consist entirely of red cards?

51. (*Choosing a Board of Directors*) A fraternity has 20 members. In how many ways can it choose a three-person board of directors?

52. (*Choosing Appetizers*) A restaurant offers an "appetizer-plate special" consisting of five selections from its list of appetizers. If there are more than 700 different possible appetizer-plate specials, what is the least possible number of appetizers?

53. (*Distributing Sandwiches*) Five students order different sandwiches at a campus eatery. The waiter forgets who ordered what and gives out the sandwiches at random. In how many different ways can the sandwiches be distributed?

54. (*Nautical Signals*) A nautical signal consists of three flags arranged vertically on a flagpole. If a sailor has flags of six different colors, how many different signals are possible?

55. (*Selecting Colleges*) A high school student decides to apply to four of the eight Ivy League colleges. In how many possible ways can the four colleges be selected?

56. (*Numbers*) In how many six digit numbers are the digits strictly decreasing when read from left to right? *Hint*: Start with the number 9876543210 and remove four digits.

57. (*Choosing Candy*) Two children, Moe and Joe, are allowed to select candy from a plate of nine pieces of candy. Moe, being younger, is allowed to choose first

but can only take two candies. Joe is then allowed to take three of the remaining candies. Joe complains that he has fewer choices than Moe. Is Joe correct? How many choices will each child have?

58. (*Franchises*) In how many ways can three new basketball franchises be distributed to the five cities that have applied for them?

59. (*Picture Arrangements*) A nursery school teacher has collected a picture from each of the 12 children in the class. She wants to hang the pictures in four rows of three. How many different arrangements are possible?

60. (*Racetrack Betting*) Most racetracks have "compound" bets on two or more horses. An *exacta* is a bet in which the first and second finishers in a race are specified in order. A *quinella* is a bet on the first two finishers in a race, with order not specified. With a field of nine horses, how many different exacta bets can be placed? Quinella bets?

61. (*Arranging Books*) In how many ways can five mathematics books and four novels be placed on a bookshelf if the mathematics books must be together?

62. (*Arranging Books*) George has three books by each of his four favorite authors. In how many ways can the books be placed on a shelf if books by the same author must be together?

63. (*Arranging Paintings*) An art gallery has seven paintings by a new artist. The director wants to place four of them in a row on one wall of the gallery. How many different arrangements are possible?

64. (*Baseball Lineup*) On a children's baseball team, there are four players who can play at any of the following infield positions: catcher, first base, third base, and shortstop. There are five possible pitchers, none of whom plays any other position. And there are four players that can play any of the three outfield positions (right, left, center) or second base. In how many ways can the coach assign players to positions?

65. (*Batting Orders*) How many batting orders are possible for a team of nine baseball players in which the pitcher always bats last and the first baseman bats in either the third or fourth spot?

66. (*Selecting Performers*) A cabaret show is put together by a manager who wishes to present two comedians and three singers. He can choose from 14 comedians and 20 singers. In how many ways can he select the acts for the show?

67. (*Performance Order*) The cabaret show manager of Exercise 66 also controls the order of the five acts he selects. In how many ways can he organize the program once he has decided on which acts to present?

68. (*MRI Processing*) Seven patients are waiting to have an MRI scan, but there is time to process only three of them before the office closes for the day. In how many ways can the three patients be chosen?

69. (*Poker Hands*) How many five-card combinations from a standard playing-card deck have cards from exactly two suits?

70. (*Ten-Letter Words*) A 10-letter word consists of 4 *A*'s and 6 *B*'s. How many different words are possible if no two *A*'s can be next to each other? *Hint*: Start with *BBBBBB* and decide where to insert the *A*'s.

71. (*ZIP Codes*) How many five-digit ZIP codes are possible in which the product of the digits is even? (From *The Mathematics Teacher*, February 1997)

72. (*League Games*) In a six-team softball league, each team plays every other team three times during the season. How many games must be scheduled?

73. (*Handshakes*) At a party, everyone shakes hands with everyone else. If 45 handshakes take place, how many people are at the party?

74. (*Distributing Jerseys*) (**PE**) Each of the 25 contestants in a race wears a jersey in either a single color or a pair of colors. What is the minimum number of colors so that no two contestants have the same color jersey?

(a) 5 **(b)** 6 **(c)** 7 **(d)** 25 **(e)** 50

75. (*Side Dishes*) A restaurant offers its customers a choice of 3 side dishes with each meal. The side dishes can be chosen from a list of fifteen possibilities with duplications allowed. For instance, a customer can order two sides of mashed potatoes and one side of string beans. Show that there are 680 possible choices for the three side dishes.

76. (*Ice Cream Specials*) An ice cream parlor offers a special consisting of three scoops of ice cream chosen from 16 different flavors. Duplication of flavors is allowed. For instance, one possibility is two scoops of chocolate and one scoop of vanilla. Show that there are 816 different possible choices for the special.

In Exercises 77–79, use a graphing calculator, a spreadsheet, or mathematical software to calculate the answers.

77. (*Course Options*) Students attending a college that operates on the semester system choose five courses each semester from a catalog containing 752 courses. The students who attend a college on the trimester system choose three courses each term from 937 courses in the catalog. Assume all courses are taught every term. In the standard four-year undergraduate program, which students have a greater number of different programs?

78. (*Arranging Microchips*) A computer manufacturer has 50 distinct microchips to place into a rectangular array that is 5 units wide by 10 units long.

(a) In how many ways can the chips be arranged?

(b) Ten of the chips control special functions. How many arrangements are possible if these must occupy the first column?

(c) Find the number of arrangements having no special-function chips in the first column.

79. (*Lottery*)

(a) Calculate the number of possible lottery tickets if the player must choose five distinct numbers from 0 to 44, inclusive, where the order does not matter. The winner must match all five.

(b) Calculate the number of lottery tickets if the player must choose four distinct numbers from 0

to 99, inclusive, where the order does not matter. The winner must match all four.

(c) In which lottery does the player have a better chance of choosing the randomly selected winning numbers?

(d) Find the answer to (c) if the order in which the numbers appear on the ticket must match the order on the winning ticket.

SOLUTIONS TO PRACTICE PROBLEMS 5.5

1. **(a)** $3! = 3 \cdot 2 \cdot 1 = 6$

(b) $7! = 7 \cdot 6 \cdot 5 \cdot 4 \cdot 3 \cdot 2 \cdot 1 = 5040$

(c) $P(7,3) = \underbrace{7 \cdot 6 \cdot 5}_{3 \text{ factors}} = 210$

[In general, $P(n, r)$ is the product of the first r factors in the descending expansion of $n!$.]

(d) $C(7,3) = \dfrac{7 \cdot 6 \cdot 5}{3 \cdot 2 \cdot 1} = \dfrac{7 \cdot \cancel{6} \cdot 5}{\cancel{3} \cdot \cancel{2} \cdot 1} = 35$

[A convenient procedure to follow when calculating $C(n, r)$ is first to write the product expansion of $r!$ in the denominator and then to write in the nu-

merator an integer from the descending expansion of $n!$ above each integer in the denominator.]

2. 90. The first question to be asked here is whether permutations or combinations are involved. Two names are to be selected, and the order of the names is important. (The name Amanda Beth is different from the name Beth Amanda.) Since the problem asks for arrangements of 10 names taken 2 at a time in a *specific order*, the number of arrangements is $P(10, 2) = 10 \cdot 9 = 90$. In general, order is important if a different outcome results when two items in the selection are interchanged.

5.6 Further Counting Problems

In Section 5.5 we introduced permutations and combinations and developed formulas for counting all permutations (or combinations) of a given type. Many counting problems can be formulated in terms of permutations or combinations. But to use the formulas of Section 5.5 successfully, we must be able to recognize these problems when they occur and to translate them into a form in which the formulas may be applied. In this section we practice doing that. We consider five typical applications giving rise to permutations or combinations. At first glance, the first two applications may seem to have little practical significance. However, they suggest a common way to "model" outcomes of real-life situations having two equally likely results.

As our first application, consider a coin-tossing experiment in which we toss a coin a fixed number of times. We can describe the outcome of the experiment as a sequence of "heads" and "tails." For instance, if a coin is tossed three times, then one possible outcome is "heads on the first toss, tails on the second toss, and tails on third toss." This outcome can be abbreviated as HTT. We can use the methods of the preceding section to count the number of possible outcomes having various prescribed properties.

EXAMPLE 1 **Tossing a coin ten times** Suppose that an experiment consists of tossing a coin 10 times and observing the sequence of heads and tails.

(a) How many different outcomes are possible?

(b) How many different outcomes have exactly four heads?

Solution (a) Visualize each outcome of the experiment as a sequence of 10 boxes, where each box contains one letter, H or T, with the first box recording the result of the first toss, the second box recording the result of the second toss, and so forth.

$$\boxed{H}\boxed{T}\boxed{H}\boxed{T}\boxed{T}\boxed{T}\boxed{H}\boxed{T}\boxed{H}\boxed{T}$$
$$1 \quad 2 \quad 3 \quad 4 \quad 5 \quad 6 \quad 7 \quad 8 \quad 9 \quad 10$$

Each box can be filled in two ways. So by the generalized multiplication principle, the sequence of 10 boxes can be filled in

$$\underbrace{2 \cdot 2 \cdot \; \cdots \; \cdot 2}_{\text{10 factors}} = 2^{10}$$

ways. So there are $2^{10} = 1024$ different possible outcomes.

(b) An outcome with 4 heads corresponds to filling the boxes with 4 H's and 6 T's. A particular outcome is determined as soon as we decide where to place the H's. The 4 boxes to receive H's can be selected from the 10 boxes in $C(10, 4)$ ways. So the number of outcomes with 4 heads is

$$C(10, 4) = \frac{10 \cdot 9 \cdot 8 \cdot 7}{4 \cdot 3 \cdot 2 \cdot 1} = 210.$$

Now Try Exercises 1(a) and (b)   ■

Ideas similar to those applied in Example 1 are useful in counting even more complicated sets of outcomes of coin-tossing experiments. The second part of our next example highlights a trick that can often save time and effort.

EXAMPLE 2   **Tossing a coin ten times**   Consider the coin-tossing experiment of Example 1.

(a) How many different outcomes have at most two heads?

(b) How many different outcomes have at least three heads?

Solution (a) The outcomes with at most two heads are those having 0, 1, or 2 heads. Let us count the number of these outcomes separately:

0 heads: There is 1 outcome, namely T T T T T T T T T T.

1 head: To determine such an outcome, we just select the box in which to put the single H. And this can be done in $C(10, 1) = 10$ ways.

2 heads: To determine such an outcome, we just select the boxes in which to put the two H's. And this can be done in $C(10, 2) = (10 \cdot 9)/(2 \cdot 1) = 45$ ways.

Adding up all the possible outcomes, we see that the number of outcomes with at most two heads is equal to $1 + 10 + 45 = 56$.

(b) "At least three heads" refers to an outcome with either 3, 4, 5, 6, 7, 8, 9, or 10 heads. The total number of such outcomes is

$$C(10, 3) + C(10, 4) + \cdots + C(10, 10).$$

This sum can, of course, be calculated, but there is a less tedious way to solve the problem. Just start with all outcomes [1024 of them by Example 1(a)] and subtract those with at most two heads [56 of them by part (a)]. So the number of outcomes with at least three heads is $1024 - 56 = 968$.

Now Try Exercises 1(c) and (d)   ■

Let us now turn to a different sort of counting problem, namely one that involves counting the number of paths between two points.

EXAMPLE 3   **Routes through a city**   In Fig. 1 we have drawn a partial map of the streets in a certain city. A tourist wishes to walk from point A to point B. We have drawn two possible routes from A to B. What is the total number of routes (with no backtracking) from A to B?

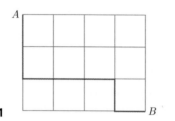 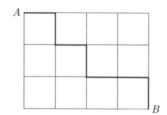

Figure 1

Solution   Any particular route can be described by giving the directions of each block walked in the appropriate order. For instance, the route on the left of Fig. 1 is described as "a block south, a block south, a block east, a block east, a block east, a block south, a block east." Using S for south and E for east, this route can be designated by the string of letters SSEEESE. Similarly, the route on the right is ESESEES. Note that each route is then described by a string of seven letters, of which three are S's (we must go three blocks south) and four are E's (we must go four blocks east). Selecting a route is thus the same as placing three S's in a string of seven boxes:

$$\boxed{}\ \boxed{S}\ \boxed{}\ \boxed{S}\ \boxed{}\ \boxed{}\ \boxed{S}$$

The three boxes to receive S's can be selected in $C(7,3) = 35$ ways. So the number of paths from A to B is 35.

Now Try Exercise 3   ■

Let us now move on to a third type of counting problem. Suppose that we have an urn in which there are a certain number of red balls and a certain number of white balls. We perform an experiment that consists of selecting a number of balls from the urn and observing the color distribution of the sample selected. (This model may be used, for example, to describe the process of selecting people to be polled in a survey. The different colors correspond to different opinions.) By using familiar counting techniques we can calculate the number of possible samples having a given color distribution. The next example illustrates a typical computation.

EXAMPLE 4   **Selecting balls from an urn**   An urn contains 25 numbered balls, of which 15 are red and 10 are white. A sample of 5 balls is to be selected.

(a) How many different samples are possible?

(b) How many samples contain all red balls?

(c) How many samples contain 3 red balls and 2 white balls?

(d) How many samples contain at least 4 red balls?

Solution (a) A sample is just an unordered selection of 5 balls out of 25. There are $C(25, 5)$ such samples. Numerically, we have

$$C(25, 5) = \frac{25 \cdot 24 \cdot 23 \cdot 22 \cdot 21}{5 \cdot 4 \cdot 3 \cdot 2 \cdot 1} = 53{,}130$$

samples.

(b) To form a sample of all red balls we must select 5 balls from the 15 red ones. This can be done in $C(15, 5)$ ways—that is, in

$$C(15, 5) = \frac{15 \cdot 14 \cdot 13 \cdot 12 \cdot 11}{5 \cdot 4 \cdot 3 \cdot 2 \cdot 1} = 3003$$

ways.

(c) To answer this question we use both the multiplication principle and the formula for $C(n, r)$. We form a sample of 3 red balls and 2 white balls using a sequence of two operations:

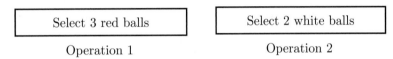

Select 3 red balls	Select 2 white balls
Operation 1	Operation 2

The first operation can be performed in $C(15, 3)$ ways and the second in $C(10, 2)$ ways. Thus the total number of samples having 3 red and 2 white balls is $C(15, 3) \cdot C(10, 2)$. That is,

$$C(15, 3) = \frac{15 \cdot 14 \cdot 13}{3 \cdot 2 \cdot 1} = 455$$

$$C(10, 2) = \frac{10 \cdot 9}{2 \cdot 1} = 45$$

$$C(15, 3) \cdot C(10, 2) = 455 \cdot 45 = 20{,}475.$$

So the number of possible samples is 20,475.

(d) A sample with at least 4 red balls has either 4 or 5 red balls. By part (b) the number of samples with 5 red balls is 3003. Using the same reasoning as in part (c), the number of samples with 4 red balls is $C(15, 4) \cdot C(10, 1) = 1365 \cdot 10 = 13{,}650$. Thus the total number of samples having at least 4 red balls is $13{,}650 + 3003 = 16{,}653$.

Now Try Exercise 5

■

PRACTICE PROBLEMS 5.6

1. (*School Board*) A newspaper reporter wants an indication of how the 15 members of the school board feel about a certain proposal. She decides to question a sample of 6 of the board members.

   (a) How many different samples are possible?

   (b) Suppose that 10 of the board members support the proposal and 5 oppose it. How many of the samples reflect the distribution of the board? That is, in how many of the samples do 4 people support the proposal and 2 oppose it?

2. (*Free Throws*) A basketball player shoots eight free throws and lists the sequence of results of each trial in order. Let S represent "success" and F represent "failure." Then, for instance, FFSSSSSS represents the outcome of missing the first two shots and hitting the rest.

   (a) How many different outcomes are possible?

   (b) How many of the outcomes have six successes?

EXERCISES 5.6

1. (*Heads and Tails*) An experiment consists of tossing a coin six times and observing the sequence of heads and tails.

   (a) How many different outcomes are possible?

   (b) How many different outcomes have exactly three heads?

   (c) How many different outcomes have more heads than tails?

   (d) How many different outcomes have at least two heads?

2. (*Routes through City Streets*) Refer to the map in Fig. 2. How many routes are there from A to B?

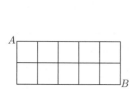
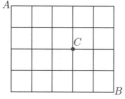

Figure 2   **Figure 3**

3. (*Routes through City Streets*) Refer to the map in Fig. 3. How many routes are there from A to B?

4. (*Selecting Balls from an Urn*) An urn contains 12 numbered balls, of which 8 are red and 4 are white. A sample of 4 balls is to be selected.

   (a) How many different samples are possible?

   (b) How many samples contain all red balls?

   (c) How many samples contain 2 red balls and 2 white balls?

   (d) How many samples contain at least 3 red balls?

   (e) How many samples contain a different number of red balls than white balls?

5. (*Selecting Apples*) A bag of 10 apples contains 2 rotten apples and 8 good apples. A shopper selects a sample of 3 apples from the bag.

   (a) How many different samples are possible?

   (b) How many samples contain all good apples?

   (c) How many samples contain at least 1 rotten apple?

6. (*Heads and Tails*) An experiment consists of tossing a coin 8 times and observing the sequence of heads and tails.

   (a) How many different outcomes are possible?

   (b) How many different outcomes have exactly 3 heads?

   (c) How many different outcomes have at least 2 heads?

   (d) How many different outcomes have 4 heads or 5 heads?

7. (*Dorm Assignments*) How many ways can a group of 100 students be assigned to dorms A, B, and C, with 25 assigned to dorm A, 40 to dorm B, and 35 to dorm C?

8. (*World Series*) In the World Series the American League team ("A") and the National League team ("N") play until one team wins four games. If the sequence of winners is designated by letters (NAAAA means the National League won the first game and lost the next four), how many different sequences are possible?

9. (*Routes through City Streets*) Refer to the map in Fig. 3. How many of the routes from A to B pass through the point C?

10. (*Selecting Fuses*) A package contains 100 fuses, of which 10 are defective. A sample of 5 fuses is selected at random.

    (a) How many different samples are there?

    (b) How many of the samples contain 2 defective fuses?

    (c) How many of the samples contain at least 1 defective fuse?

11. (*Senate Committees*) In how many ways can a committee of 5 senators be selected from the 100 members of the U.S. Senate so that no two committee members are from the same state?

12. (*Exam Questions*) An exam contains five "true or false" questions. How many of the 32 different ways of answering these questions contain 3 or more correct answers?

13. (*Exam Questions*) A student is required to work exactly 6 problems from a 10-problem exam and must work exactly 3 of the first 4 problems. In how many ways can the six problems be chosen?

14. (*Arranging Books*) The new book shelf at a library contains two novels, six biographies, and four how-to books. In how many ways can the books be arranged on the shelf if the books for each category are placed together?

15. (*Jury Selection*) In how many ways can 12 jurors and 2 alternates be chosen from a group of 20 prospective jurors?

16. (*Subcommittee Selection*) A committee has four male and five female members. In how many ways can a subcommittee consisting of two males and two females be selected?

17. (*Investment Portfolio*) In how many ways can an investor put together a portfolio of five stocks and six bonds selected from her favorite nine stocks and eight bonds?

18. (*Game Outcomes*) A football team plays 10 games. In how many ways can these games result in five wins, four losses, and one tie?

19. (*Marketing Calls*) A telemarketer makes 15 phone calls in 1 hour. In how many ways can the outcomes of the calls be three sales, eight no-sales, and four no-answers?

20. (*Free-Throws*) During practice, a basketball player shoots 10 free-throws. In how many ways can the outcomes result in seven hits and three misses?

21. (*License Plates*) A license plate contains six letters with no repetitions allowed. How many different license plates are possible?

22. (*Technology Options*) A college mathematics department has 15 calculus classes. Five classes will use graphing calculators and four classes will use computer software. In how many different ways can the classes be selected?

23. (*Three-Letter Words*) How many different three-letter words (i.e., sequences of letters) can be formed from the letters of the word JUPITER?

24. (*Seven-Letter Words*) The shortest word containing all five vowels is SEQUOIA. How many seven-letter sequences of letters, with no repeated letters, contain all five vowels?

25. (*Arranging Letters*) In how many arrangements of the letters in the word ABSTEMIOUS are the vowels in alphabetical order?

26. (*Teaching Assignments*) Out of a group of 20 senior education majors at the University of Maryland, five will be selected to student-teach in Montgomery County, four will be selected to student-teach in Prince Georges County, and three will be selected to student-teach in Howard County. In how many ways can this be done?

27. (*Group Picture*) The student council at Gotham College is made up of four freshmen, five sophomores, six juniors, and seven seniors. A yearbook photographer would like to line up three council members from each class for a picture. How many different pictures are possible if each group of classmates stands together?

28. (*Delivery Schedule*) A truck driver has to deliver bread to five grocery stores. In how many different ways can he schedule the order of his stops?

29. (*Delivery Schedule*) Suppose that the stores in Exercise 28 are located in two towns far enough apart that the driver wants to make all stops in one town before going on to the next. In how many different ways can he schedule the order of his stops if the larger town has three stores?

30. (*Job Allocation*) Of the 14 new programmers hired by a major software company, 3 will be selected to work on programming languages, 4 will be selected to work on word processing software, and 5 will be selected to work on spreadsheet software. In how many ways can this be done?

31. (*Poker*) How many poker hands consist of 3 aces and 2 kings?

32. (*Poker*) How many poker hands consist of 2 aces, 2 cards of another denomination, and 1 card of a third denomination?

33. (*Poker*) How many poker hands consist of 3 cards of one denomination and 2 cards of another denomination? (Such a poker hand is called a "full house.")

34. (*Poker*) How many poker hands consist of 2 cards of one denomination, 2 cards of another (different) denomination, and 1 card of a third denomination? (Such a poker hand is called "two pairs.")

35. (*Detour-Prone ZIP Codes*) A five-digit ZIP code is said to be detour-prone if it looks like a valid and different ZIP code when read upside down (Fig. 4). For instance, 68901 and 88111 are detour-prone, whereas 32145 and 10801 are not. How many of the 10^5 possible ZIP code numbers are detour-prone?

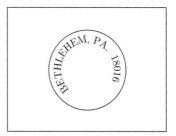

Figure 4

36. (*Dorm Assignments*) In how many ways can a residence director assign six students to four dormitory rooms if two rooms are doubles, two rooms are singles, and two of the students cannot be placed together?

37. (*Committee Selection*) A class has 12 students, of which 3 are seniors. How many committees of size 4 can be selected if at least one member of each committee must be a senior?

38. (*Dance Routines*) A dance team knows 15 routines of which 8 are tap, 5 are ballet, and 2 are modern. The program can consist of any five routines. In how many ways can the manager choose which routines to present if there are no restrictions?

39. (*Dance Routines*) Refer to Exercise 38. In how many ways can the manager choose the routines if at least one routine of each type must be included?

40. (*Seating Arrangements*) In how many ways can 6 married couples sit in a row if men and women alternate?

41. (*Seating Arrangements*) In how many ways can 6 married couples sit in a row if no 2 women sit next to each other?

42. (*Family Dinner*) A family has 6 members. In how many different ways can exactly three of them come to dinner? In how many different ways can no more than three of them come to dinner?

43. (*Family Dinner*) A family has 6 members. How many different family groups can come to dinner?

44. (*Placemat Arrangement*) A family has 6 members. Each member has a placemat with his or her name on it. In how many ways can the placemats be placed around a round table?

45. (*Selecting Marbles*) There are 8 marbles in a bag, numbered 1 through 8. Three (1, 2, 3) are blue and five (4, 5, 6, 7, 8) are white. Choose three marbles from the bag. In how many samples will the number of blue marbles chosen exceed the number of white marbles chosen?

In Exercises 46–49, use a graphing calculator, a spreadsheet, or mathematical software to calculate the answers.

46. (*Class Formation*) The dean at a small college wishes to form an experimental section of General Psychology with 22 students chosen from 20 women and 18 men interested in social science.

(**a**) How many different such classes are possible?

(**b**) The dean decides the class should have 12 women and 10 men. Compare the number of possible classes to the solution to (a).

47. (*Bridge*) A bridge hand contains 13 cards. What percentage of bridge hands contains all four aces?

48. (*Bridge*) Which is more likely—a bridge hand with four aces or one with exactly two kings and two queens?

49. (*Bridge*) Which is more likely—a bridge hand with four aces or one with the two red kings, the two red queens, and no other kings or queens?

SOLUTIONS TO PRACTICE PROBLEMS 5.6

1. (a) $C(15, 6)$. Each sample is an unordered selection of 15 objects taken 6 at a time.

(b) $C(10, 4) \cdot C(5, 2)$. Asking for the number of samples of a certain type is the same as asking for the number of ways that the task of forming such a sample can be performed. This task is composed of two consecutive operations. Operation 1, selecting 4 people from among the 10 that support the proposal, can be performed in $C(10, 4)$ ways. Operation 2, selecting 2 people from among the 5 people that oppose the proposal, can be performed in $C(5, 2)$ ways. Therefore, by the multiplication principle, the complete task can be performed in $C(10, 4) \cdot C(5, 2)$ ways.
[*Note:* $C(15, 6) = 5005$ and $C(10, 4) \cdot C(5, 2) = 2100$. Therefore, less than half of the possible samples reflect the true distribution of the school board.]

2. (a) 2^8 or 256. Apply the generalized multiplication principle.

(b) $C(8, 6)$ or 28. Each outcome having 6 successes corresponds to a sequence of 8 letters of which 6 are S's and 2 are F's. Such an outcome is specified by selecting the 6 locations for the S's from among the 8 locations, and this has $C(8, 6)$ possibilities.

5.7 The Binomial Theorem

In Sections 5.5 and 5.6 we dealt with permutations and combinations and, in particular, derived a formula for $C(n, r)$, the number of combinations of n objects taken r at a time. Namely, we have

$$C(n, r) = \frac{P(n, r)}{r!} = \frac{n(n - 1) \cdot \, \cdots \, \cdot (n - r + 1)}{r!}. \tag{1}$$

Actually, formula (1) was verified in case both n and r are positive integers. But it is useful to consider $C(n, r)$ also in case $r = 0$. In this case we are considering the number of combinations of n things taken 0 at a time. There is clearly only one such combination: the one containing no elements. Therefore,

$$C(n, 0) = 1. \tag{2}$$

Here is another convenient formula for $C(n, r)$:

$$C(n, r) = \frac{n!}{r! \, (n - r)!}. \tag{3}$$

For instance, according to formula (3),

$$C(8,3) = \frac{8!}{3!\,(8-3)!} = \frac{8!}{3!\,5!} = \frac{8 \cdot 7 \cdot 6 \cdot \cancel{5} \cdot \cancel{4} \cdot \cancel{3} \cdot \cancel{2} \cdot \cancel{1}}{3 \cdot 2 \cdot 1 \cdot \cancel{5} \cdot \cancel{4} \cdot \cancel{3} \cdot \cancel{2} \cdot \cancel{1}} = \frac{8 \cdot 7 \cdot 6}{3 \cdot 2 \cdot 1},$$

which agrees with the result given by formula (1).

Verification of Formula (3)   Note that

$$n(n-1) \cdot \cdots \cdot (n-r+1) =$$

$$\frac{n(n-1) \cdot \cdots \cdot (n-r+1)\cancel{(n-r)}\cancel{(n-r-1)} \cdot \cdots \cdot \cancel{2} \cdot \cancel{1}}{\cancel{(n-r)}\cancel{(n-r-1)} \cdot \cdots \cdot \cancel{2} \cdot \cancel{1}} = \frac{n!}{(n-r)!}.$$

Then, by formula (1), we have

$$C(n,r) = \frac{n(n-1) \cdot \cdots \cdot (n-r+1)}{r!} = \frac{\dfrac{n!}{(n-r)!}}{r!} = \frac{n!}{r!\,(n-r)!},$$

which is formula (3).   ■

Note that for $r = 0$, formula (3) reads

$$C(n,0) = \frac{n!}{0!\,(n-0)!} = \frac{n!}{0!\,n!} = \frac{1}{0!}.$$

Let us agree that the value of $0!$ is 1. Then the right-hand side of the equation above is 1, so that formula (3) also holds for $r = 0$.

Formula (3) can be used to prove many facts about $C(n,r)$. For example, the following formula is useful in calculating $C(n,r)$ for large values of r:

$$C(n,r) = C(n, n-r). \tag{4}$$

Suppose that we wish to calculate $C(100, 98)$. If we apply formula (4), we have

$$C(100, 98) = C(100, 100 - 98) = C(100, 2) = \frac{100 \cdot 99}{2 \cdot 1} = 4950.$$

Verification of Formula (4)   Apply formula (3) to evaluate $C(n, n-r)$:

$$C(n, n-r) = \frac{n!}{(n-r)!\,(n-(n-r))!} = \frac{n!}{(n-r)!\,r!}$$

$$= C(n,r) \quad \text{[by formula (3) again]}.$$

The formula is intuitively reasonable since each time we select a subset of r elements we are excluding a subset of $n - r$ elements. Thus there are as many subsets of $n - r$ elements as there are subsets of r elements.   ■

An alternative notation for $C(n,r)$ is $\binom{n}{r}$. Thus, for example,

$$\binom{5}{2} = C(5, 2) = \frac{5 \cdot 4}{2 \cdot 1} = 10.$$

The symbol $\binom{n}{r}$ is called a *binomial coefficient*. To discover why, let us tabulate the values of $\binom{n}{r}$ for some small values of n and r.

$$n = 2: \quad \binom{2}{0} = 1 \quad \binom{2}{1} = 2 \quad \binom{2}{2} = 1$$

$$n = 3: \quad \binom{3}{0} = 1 \quad \binom{3}{1} = 3 \quad \binom{3}{2} = 3 \quad \binom{3}{3} = 1$$

$$n = 4: \quad \binom{4}{0} = 1 \quad \binom{4}{1} = 4 \quad \binom{4}{2} = 6 \quad \binom{4}{3} = 4 \quad \binom{4}{4} = 1$$

$$n = 5: \quad \binom{5}{0} = 1 \quad \binom{5}{1} = 5 \quad \binom{5}{2} = 10 \quad \binom{5}{3} = 10 \quad \binom{5}{4} = 5 \quad \binom{5}{5} = 1$$

Each row consists of the coefficients that arise in expanding $(x + y)^n$. To see this, inspect the results of expanding $(x + y)^n$ for $n = 2$, 3, 4, and 5:

$$(x + y)^2 = x^2 + 2xy + y^2$$
$$(x + y)^3 = x^3 + 3x^2y + 3xy^2 + y^3$$
$$(x + y)^4 = x^4 + 4x^3y + 6x^2y^2 + 4xy^3 + y^4$$
$$(x + y)^5 = x^5 + 5x^4y + 10x^3y^2 + 10x^2y^3 + 5xy^4 + y^5.$$

Compare the coefficients in any row with the values in the corresponding row of binomial coefficients. Note that they are the same. Thus we see that the binomial coefficients arise as coefficients in multiplying out powers of the binomial $x + y$; hence the name *binomial coefficient*.

What we observed for the exponents $n = 2$, 3, 4, and 5 holds true for any positive integer n. We have the following result, a proof of which is given at the end of this section.

Binomial Theorem

$$(x + y)^n = \binom{n}{0}x^n + \binom{n}{1}x^{n-1}y + \binom{n}{2}x^{n-2}y^2 + \cdots + \binom{n}{n-1}xy^{n-1} + \binom{n}{n}y^n$$

EXAMPLE 1 **Applying the binomial theorem** Expand $(x + y)^6$.

Solution By the binomial theorem,

$$(x + y)^6 = \binom{6}{0}x^6 + \binom{6}{1}x^5y + \binom{6}{2}x^4y^2 + \binom{6}{3}x^3y^3$$
$$+ \binom{6}{4}x^2y^4 + \binom{6}{5}xy^5 + \binom{6}{6}y^6.$$

Furthermore,

$$\binom{6}{0} = 1 \quad \binom{6}{1} = \frac{6}{1} = 6 \quad \binom{6}{2} = \frac{6 \cdot 5}{2 \cdot 1} = 15$$

$$\binom{6}{3} = \frac{6 \cdot 5 \cdot 4}{3 \cdot 2 \cdot 1} = 20 \quad \binom{6}{4} = \frac{6 \cdot 5 \cdot 4 \cdot 3}{4 \cdot 3 \cdot 2 \cdot 1} = 15$$

$$\binom{6}{5} = \frac{6 \cdot 5 \cdot 4 \cdot 3 \cdot 2}{5 \cdot 4 \cdot 3 \cdot 2 \cdot 1} = 6 \quad \binom{6}{6} = \frac{6 \cdot 5 \cdot 4 \cdot 3 \cdot 2 \cdot 1}{6 \cdot 5 \cdot 4 \cdot 3 \cdot 2 \cdot 1} = 1.$$

Thus

$$(x + y)^6 = x^6 + 6x^5y + 15x^4y^2 + 20x^3y^3 + 15x^2y^4 + 6xy^5 + y^6.$$

Now Try Exercise 19   ■

The binomial theorem can be used to count the number of subsets of a set, as shown in the next example.

EXAMPLE 2   **Counting the number of subsets of a set**   Determine the number of subsets of a set with five elements.

Solution   Let us count the number of subsets of each possible size. A subset of r elements can be chosen in $\binom{5}{r}$ ways, since $C(5, r) = \binom{5}{r}$. So the set has $\binom{5}{0}$ subsets with 0 elements, $\binom{5}{1}$ subsets with 1 element, $\binom{5}{2}$ subsets with 2 elements, and so on. Therefore, the total number of subsets is

$$\binom{5}{0} + \binom{5}{1} + \binom{5}{2} + \binom{5}{3} + \binom{5}{4} + \binom{5}{5}.$$

On the other hand, the binomial theorem for $n = 5$ gives

$$(x + y)^5 = \binom{5}{0}x^5 + \binom{5}{1}x^4y + \binom{5}{2}x^3y^2 + \binom{5}{3}x^2y^3 + \binom{5}{4}xy^4 + \binom{5}{5}y^5.$$

Set $x = 1$ and $y = 1$ in this formula.

$$(1 + 1)^5 = \binom{5}{0}1^5 + \binom{5}{1}1^4 \cdot 1 + \binom{5}{2}1^3 \cdot 1^2 + \binom{5}{3}1^2 \cdot 1^3 + \binom{5}{4}1 \cdot 1^4 + \binom{5}{5}1^5$$

$$2^5 = \binom{5}{0} + \binom{5}{1} + \binom{5}{2} + \binom{5}{3} + \binom{5}{4} + \binom{5}{5}$$

Thus the total number of subsets of a set with five elements (the right side) equals $2^5 = 32$.

Now Try Exercise 27   ■

There is nothing special about the number 5 in the preceding example. An analogous argument gives the following result:

> A set of n elements has 2^n subsets.

EXAMPLE 3   **Counting pizza options**   A pizza parlor offers a plain cheese pizza to which any number of six possible toppings can be added. How many different pizzas can be ordered?

Solution   Ordering a pizza requires selecting a subset of the six possible toppings. Since the set of six toppings has 2^6 different subsets, there are 2^6 or 64 different pizzas. (Note that the plain cheese pizza corresponds to selecting the empty subset of toppings.)

Now Try Exercise 31   ■

Proof of the Binomial Theorem   Note that

$$(x+y)^n = \underbrace{(x+y)(x+y) \cdot \cdots \cdot (x+y)}_{n \text{ factors}}.$$

Multiplying out these factors involves forming all products, where one term is selected from each factor, and then combining like products. For instance,

$$(x+y)(x+y)(x+y) = x \cdot x \cdot x + x \cdot x \cdot y + x \cdot y \cdot x + y \cdot x \cdot x$$
$$+ x \cdot y \cdot y + y \cdot x \cdot y + y \cdot y \cdot x + y \cdot y \cdot y.$$

The first product on the right, $x \cdot x \cdot x$, is obtained by selecting the x-term from each of the three factors. The next term, $x \cdot x \cdot y$, is obtained by selecting the x-terms from the first two factors and the y-term from the third. The next product, $x \cdot y \cdot x$, is obtained by selecting the x-terms from the first and third factors and the y-term from the second. And so on. There are as many products containing two x's and one y as there are ways of selecting the factor from which to pick the y-term—namely $\binom{3}{1}$.

In general, when multiplying the n factors $(x+y)(x+y)\cdots(x+y)$, the number of products having k y's (and therefore $(n-k)$ x's) is equal to the number of different ways of selecting the k factors from which to take the y-term—that is, $\binom{n}{k}$. Therefore, the coefficient of $x^{n-k}y^k$ is $\binom{n}{k}$. This proves the binomial theorem. ■

PRACTICE PROBLEMS 5.7

1. Calculate $\binom{12}{8}$.

2. An ice cream parlor offers 10 flavors of ice cream and 5 toppings. How many different servings are possible if each choice consists of one flavor of ice cream and as many toppings as desired?

EXERCISES 5.7

Calculate the value for each of Exercises 1–18.

1. $\binom{6}{2}$
2. $\binom{7}{3}$
3. $\binom{8}{1}$
4. $\binom{9}{9}$

5. $\binom{18}{16}$
6. $\binom{25}{24}$
7. $\binom{7}{0}$
8. $\binom{6}{1}$

9. $\binom{8}{8}$
10. $\binom{9}{0}$
11. $\binom{n}{n-1}$

12. $\binom{n}{n}$
13. $0!$
14. $1!$

15. $n \cdot (n-1)!$
16. $\dfrac{n!}{n}$

17. $\binom{6}{0} + \binom{6}{1} + \binom{6}{2} + \binom{6}{3} + \binom{6}{4} + \binom{6}{5} + \binom{6}{6}$

18. $\binom{7}{0} + \binom{7}{1} + \binom{7}{2} + \binom{7}{3} + \binom{7}{4} + \binom{7}{5} + \binom{7}{6} + \binom{7}{7}$

19. Determine the first three terms in the binomial expansion of $(x+y)^{10}$.

20. Determine the first three terms in the binomial expansion of $(x+y)^{20}$.

21. Determine the last three terms in the binomial expansion of $(x+y)^{15}$.

22. Determine the last three terms in the binomial expansion of $(x+y)^{12}$.

23. Determine the middle term in the binomial expansion of $(x+y)^{20}$.

24. Determine the middle term in the binomial expansion of $(x+y)^{10}$.

25. Determine the coefficient of x^4y^7 in the binomial expansion of $(x+y)^{11}$.

26. Determine the coefficient of x^8y^5 in the binomial expansion of $(x+y)^{13}$.

27. How many different subsets can be chosen from a set of six elements?

28. How many different subsets can be chosen from a set of 100 elements?

29. (*Restaurant Tip*) How many different tips could you leave in a restaurant if you had a nickel, a dime, a quarter, and a half-dollar?

30. (*Pizza Options*) A pizza parlor offers mushrooms, green peppers, onions, and sausage as toppings for the plain cheese base. How many different types of pizzas can be made?

31. (*Cable TV Options*) A cable TV franchise offers 20 basic channels plus a selection (at an extra cost per channel) from a collection of 5 premium channels. How many different options are available to the subscriber?

32. (*Salad Options*) A salad bar offers a base of lettuce to which tomatoes, chickpeas, beets, pinto beans, olives, and green peppers can be added. Five salad dressings are available. How many different salads are possible? (Assume that each salad contains at least lettuce and at most one salad dressing.)

33. (*Book Selection*) In how many ways can a selection of at least one book be made from a set of eight books?

34. (*Dessert Choices*) In how many ways can a selection of at most five desserts be made from a dessert trolley containing six desserts?

35. (*Pizza Options*) Armand's Chicago Pizzeria offers thin-crust and deep-dish pizzas in 9-, 12-, and 14-inch sizes, with 15 possible toppings. How many different types of pizzas can be ordered?

36. (*CD Selection*) In how many ways can a selection of at least two CDs be made from a set of seven CDs?

37. (*Appetizers Selection*) In how many ways can a selection of at most five appetizers be made from a menu containing seven appetizers?

38. (*Ice Cream Sundaes*) An ice cream parlor offers four flavors of ice cream, three sauces, and two types of nuts. How many different sundaes consisting of a single flavor of ice cream are possible?

39. How many subsets of the set $\{a, b, c, d, e\}$ do not contain the letter c?

40. How many subsets of the set $\{1, 2, 3, 4, 5\}$ do not contain an even digit?

41. (*Lab Projects*) Students in a physics class are required to complete at least two out of a collection of eight lab projects. In how many ways can a student satisfy the requirement?

42. Suppose that a set has an odd number of elements. Explain why half of the subsets will have an odd number of elements.

43. (*Seminar Attendance*) How many different groups of students can show up for a seminar with an enrollment of 12?

44. (*Menu Planning*) In planning a menu, a dietitian must choose 2 carbohydrates from the 7 available ones, 3 vegetables from the 10 available ones, and 1 protein from the 5 available ones. How many menus are possible?

45. Show that in a Venn diagram containing three sets, there are eight basic regions. Use binomial coefficients to explain why.

46. (*Car Showroom*) A car dealership has 20 different models that could be placed on display in the showroom, but it has room for at most 5 of these. In how many ways can the manager choose the "show cars"? (Assume that at least 1 car will be shown.)

47. (*Car Showroom*) A car dealership has 20 different models, of which 8 are two-door and 12 are four-door vehicles. In how many ways can the manager choose three of the two-door and two of the four-door models for a showroom display?

48. (*Car Sales*) In how many ways can a car dealership with 20 cars available on a particular day make at least one sale?

49. What is the coefficient of y^4 in the expansion of $(x - 3y)^7$?

50. (a) Use equation (4) to show that

$$\binom{5}{0} - \binom{5}{1} + \binom{5}{2} - \binom{5}{3} + \binom{5}{4} - \binom{5}{5} = 0.$$

(b) For what values of n can equation (4) be used to show that

$$\binom{n}{0} - \binom{n}{1} + \binom{n}{2} - \binom{n}{3} + \cdots \pm \binom{n}{n} = 0?$$

(c) Use the binomial theorem to prove the result in part (a). *Hint*: Apply the binomial theorem to $(x + y)^5$ with $x = 1$ and $y = -1$.

(d) For what values of n can the binomial theorem be used to prove the result in part (b)?

51. (*Club Contingent*) The Scrabble club at Gotham College agrees to be represented at a national intercollegiate Scrabble tournament. The contingent can consist of as many members of the club as desired, but must consist of at least one person. If the club has 15 members, how many different contingents are possible?

SOLUTIONS TO PRACTICE PROBLEMS 5.7

1. 495. $\binom{12}{8}$ is the same as $C(12,8)$, which equals $C(12, 12 - 8)$ or $C(12, 4)$.

$$C(12, 4) = \frac{12 \cdot 11 \cdot 10 \cdot 9}{4 \cdot 3 \cdot 2 \cdot 1} = \frac{\cancel{12} \cdot 11 \cdot \overset{5}{\cancel{10}} \cdot 9}{\cancel{4} \cdot \cancel{3} \cdot \cancel{2} \cdot 1} = 495$$

2. 320. The task of deciding what sort of serving to have consists of two operations. The first operation, selecting the flavor of ice cream, can be performed in 10 ways. The second operation, selecting the toppings, can be performed in 2^5 or 32 ways, since selecting the toppings amounts to selecting a subset from the set of 5 toppings, and a set of 5 elements has 2^5 subsets. (Notice that selecting the empty subset corresponds to ordering a plain dish of ice cream.) By the multiplication principle, the task can be performed in $10 \cdot 32 = 320$ ways.

5.8 Multinomial Coefficients and Partitions

Permutation and combination problems are only two of the many types of counting problems. By appropriately generalizing the binomial coefficients to the *multinomial coefficients*, we can consider certain generalizations of combinations, namely *partitions*. To introduce the notion of a partition, let us return to combinations and look at them from another viewpoint. Suppose that we consider combinations of n objects taken r at a time. View the n objects as the elements of a set S. Then each combination determines an ordered division of S into two subsets, S_1 and S_2, the first containing the r elements selected and the second containing the $n - r$ elements remaining (Fig. 1). We see that

$$S = S_1 \cup S_2 \quad \text{and} \quad n(S_1) + n(S_2) = n.$$

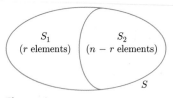

S_1 (r elements) S_2 ($n - r$ elements) S

Figure 1

This ordered division is called an *ordered partition of type* $(r, n - r)$. We know that the number of such partitions is just the number of ways of selecting the first subset, that is, $n!/[r!(n - r)!]$. If we let $n_1 = n(S_1) = r$ and $n_2 = n(S_2) = n - r$, then we find that the number of ordered partitions of type (n_1, n_2) is $n!/n_1! n_2!$.

We may generalize the aforementioned situation as follows: Let S be a set of n elements. An *ordered partition of S of type* $(n_1, n_2, \ldots, n_m)$ is a decomposition of S into m subsets (given in a specific order) $S_1, S_2, \ldots, S_m$, where no two of these intersect and where

$$n(S_1) = n_1, \qquad n(S_2) = n_2, \quad \ldots \quad , n(S_m) = n_m$$

(Fig. 2). Since S has n elements, we clearly must have $n = n_1 + n_2 + \cdots + n_m$.

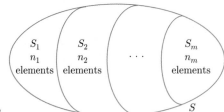

S_1 n_1 elements $\quad$ S_2 n_2 elements $\quad \cdots \quad$ S_m n_m elements $\quad$ S

Figure 2

EXAMPLE 1 **Ordered partitions of a set** List all ordered partitions of $S = \{a, b, c, d\}$ of type $(1, 1, 2)$.

Solution

$$(\{a\}, \{b\}, \{c, d\}) \qquad (\{c\}, \{a\}, \{b, d\})$$
$$(\{a\}, \{c\}, \{b, d\}) \qquad (\{c\}, \{b\}, \{a, d\})$$
$$(\{a\}, \{d\}, \{b, c\}) \qquad (\{c\}, \{d\}, \{a, b\})$$
$$(\{b\}, \{a\}, \{c, d\}) \qquad (\{d\}, \{a\}, \{b, c\})$$
$$(\{b\}, \{c\}, \{a, d\}) \qquad (\{d\}, \{b\}, \{a, c\})$$
$$(\{b\}, \{d\}, \{a, c\}) \qquad (\{d\}, \{c\}, \{a, b\})$$

Note that the ordered partition $(\{a\}, \{b\}, \{c, d\})$ is different from the ordered partition $(\{b\}, \{a\}, \{c, d\})$, since in the first S_1 is $\{a\}$, whereas in the second S_1 is $\{b\}$. The order in which the subsets are given is significant. ∎

We saw earlier that the number of ordered partitions of type (n_1, n_2) for a set of n elements is $n!/n_1! n_2!$. This result generalizes.

> **Number of Ordered Partitions of Type $(n_1, n_2, \ldots, n_m)$** Let S be a set of n elements. Then the number of ordered partitions of S of type $(n_1, n_2, \ldots, n_m)$ is
>
> $$\frac{n!}{n_1! n_2! \cdots n_m!}. \tag{1}$$

The number of ordered partitions of type $(n_1, n_2, \ldots, n_m)$ for a set of n elements is often denoted

$$\binom{n}{n_1, n_2, \ldots, n_m}.$$

Using the preceding notation, result (1) says that

$$\binom{n}{n_1, n_2, \ldots, n_m} = \frac{n!}{n_1! n_2! \cdots n_m!}.$$

The binomial coefficient $\binom{n}{n_1}$ can also be written $\binom{n}{n_1, n_2}$. The number

$$\binom{n}{n_1, n_2, \ldots, n_m}$$

is known as a *multinomial coefficient* since it appears as the coefficient of $x_1^{n_1} x_2^{n_2} \cdots x_m^{n_m}$ in the expansion of $(x_1 + x_2 + \cdots + x_m)^n$.

EXAMPLE 2 **Ordered partitions of a set** Let S be a set of four elements. Use the formula in (1) to determine the number of ordered partitions of S of type $(1, 1, 2)$.

Solution Here $n = 4$, $n_1 = 1$, $n_2 = 1$, and $n_3 = 2$. Therefore, the number of ordered partitions of type $(1, 1, 2)$ is

$$\binom{4}{1, 1, 2} = \frac{4!}{1! \, 1! \, 2!} = \frac{4 \cdot 3 \cdot 2 \cdot 1}{1 \cdot 1 \cdot 2 \cdot 1} = 12.$$

This result is the same as obtained in Example 1 by enumeration. ∎

Now Try Exercise 1

EXAMPLE 3 **Assigning tasks to construction workers** A work crew consists of 12 construction workers, all having the same skills. A construction job requires four welders, three concrete workers, three heavy equipment operators, and two bricklayers. In how many ways can the 12 workers be assigned to the required tasks?

Solution Each assignment of jobs corresponds to an ordered partition of the type $(4, 3, 3, 2)$. The number of such ordered partitions is

$$\binom{12}{4, 3, 3, 2} = \frac{12!}{4!\, 3!\, 3!\, 2!} = 277{,}200.$$

Now Try Exercise 15 ■

Sometimes all of the m subsets of an ordered partition are required to have the same number of elements. If the set has n elements and each of the m subsets has r elements, then the number of ordered partitions of type

$$(\underbrace{r, r, \dots, r}_{m})$$

is

$$\binom{n}{r, r, \dots, r} = \frac{n!}{r!\, r! \cdots r!} = \frac{n!}{(r!)^m}. \tag{2}$$

EXAMPLE 4 **Counting the number of bridge hands** In the game of bridge, four players seated in a specific order are each dealt 13 cards. How many different possibilities are there for the hands dealt to the players?

Solution Each deal results in an ordered partition of the 52 cards of type $(13, 13, 13, 13)$. The number of such partitions is

$$\binom{52}{13, 13, 13, 13} = \frac{52!}{(13!)^4}.$$

This number is approximately 5.36×10^{28}. ■

Unordered Partitions Determining the number of unordered partitions of a certain type is a complex matter. We will restrict our attention to the special case in which each subset is of the same size.

EXAMPLE 5 **Unordered partitions of a set** List all unordered partitions of $S = \{a, b, c, d\}$ of type $(2, 2)$.

Solution
$$(\{a, b\}, \{c, d\})$$
$$(\{a, c\}, \{b, d\})$$
$$(\{a, d\}, \{b, c\})$$
■

Note that the partition $(\{c, d\}, \{a, b\})$ is the same as the partition $(\{a, b\}, \{c, d\})$ when order is not taken into account.

Number of Unordered Partitions of Type (*r, r, . . . , r*)   Let S be a set of n elements where $n = m \cdot r$. Then the number of unordered partitions of S of type $(r, r, \ldots, r)$ is

$$\frac{1}{m!} \cdot \frac{n!}{(r!)^m}. \tag{3}$$

Formula (3) follows from the fact that each unordered partition of the m subsets gives rise to $m!$ ordered partitions. Therefore,

$$(m!) \, [\text{number of unordered partitions}] = [\text{number of ordered partitions}]$$

or

$$[\text{number of unordered partitions}] = \frac{1}{m!} \cdot [\text{number of ordered partitions}]$$

$$= \frac{1}{m!} \cdot \frac{n!}{(r!)^m} \quad [\text{by formula (2)}].$$

EXAMPLE 6   **Unordered partitions of a set**   Let S be a set of four elements. Use formula (3) to determine the number of unordered partitions of S of type $(2, 2)$.

Solution   Here $n = 4$, $r = 2$, and $m = 2$. Therefore, the number of unordered partitions of type $(2, 2)$ is

$$\frac{1}{2!} \cdot \frac{4!}{(2!)^2} = \frac{1}{2} \cdot \frac{4 \cdot 3 \cdot 2 \cdot 1}{(2 \cdot 1)^2} = 3.$$

This result is the same as that obtained in Example 5 by enumeration.

Now Try Exercise 11   ■

EXAMPLE 7   **Grouping construction workers**   A construction crew contains 12 workers, all having similar skills. In how many ways can the workers be divided into four groups of three?

Solution   The order of the four groups is not relevant. (It does not matter which is labeled S_1 and which S_2, and so on. Only the composition of the groups is important.) Applying formula (3) with $n = 12$, $r = 3$, and $m = 4$, we see that the number of ways is

$$\frac{1}{m!} \cdot \frac{n!}{(r!)^m} = \frac{1}{4!} \cdot \frac{12!}{(3!)^4}$$

$$= \frac{1}{4 \cdot 3 \cdot 2 \cdot 1} \cdot \frac{12 \cdot 11 \cdot 10 \cdot 9 \cdot 8 \cdot 7 \cdot 6 \cdot 5 \cdot 4 \cdot 3 \cdot 2 \cdot 1}{6^4}$$

$$= \frac{\overset{2}{\cancel{12}} \cdot 11 \cdot 10 \cdot \cancel{9} \cdot \overset{2}{\cancel{8}} \cdot 7 \cdot \cancel{6} \cdot 5}{\cancel{6} \cdot \cancel{6} \cdot \cancel{6} \cdot \cancel{6}} = 15{,}400.$$

Now Try Exercise 19   ■

PRACTICE PROBLEMS 5.8

1. A foundation wishes to award one grant of $100,000, two grants of $10,000, five grants of $5000, and five grants of $2000. Its list of potential grant recipients has been narrowed to 13 possibilities. In how many ways can the awards be made?

2. In how many different ways can six medical interns be put into three groups of two and assigned to

   (a) The radiology, neurology, and surgery departments?

   (b) Share living quarters?

EXERCISES 5.8

Let S be a set of n elements. Determine the number of ordered partitions of the types in Exercises 1–10.

1. $n = 5$; $(3, 1, 1)$
2. $n = 5$; $(2, 1, 2)$
3. $n = 6$; $(2, 1, 2, 1)$
4. $n = 6$; $(3, 3)$
5. $n = 7$; $(3, 2, 2)$
6. $n = 7$; $(4, 1, 2)$
7. $n = 12$; $(4, 4, 4)$
8. $n = 8$; $(3, 3, 2)$
9. $n = 12$; $(5, 3, 2, 2)$
10. $n = 8$; $(2, 2, 2, 2)$

Let S be a set of n elements. Determine the number of unordered partitions of the types in Exercises 11–14.

11. $n = 15$; $(3, 3, 3, 3, 3)$
12. $n = 10$; $(5, 5)$
13. $n = 18$; $(6, 6, 6)$
14. $n = 12$; $(4, 4, 4)$

15. (*Stock Reports*) A brokerage house regularly reports the behavior of a group of 20 stocks, each stock being reported as "up," "down," or "unchanged." How many different reports can show seven stocks up, five stocks down, and eight stocks unchanged?

16. (*Investment Ratings*) An investment advisory service rates investments as A, AA, and AAA. On a certain week, it rates 15 investments. In how many ways can it rate five investments in each of the categories?

17. (*Observation Groups*) A psychology experiment observes groups of four individuals. In how many ways can an experimenter choose 5 groups of 4 from among 20 subjects?

18. (*Weather*) In a certain month (of 30 days) it rains 10 days, snows 2 days, and is clear 18 days. In how many ways can such weather be distributed over the month?

19. (*Orientation Groups*) During orientation, new students are divided into groups of five people. In how many ways can 4 groups be chosen from among 20 people?

20. (*Awarding Prizes*) Of the nine contestants in a contest, three will receive cars, three will receive TV sets, and three will receive radios. In how many different ways can the prizes be awarded?

21. (*Job Promotions*) A corporation has four employees that it wants to place in high executive positions. One will become president, one will become vice president, and two will be appointed to the board of directors. In how many different ways can this be accomplished?

22. (*Forming Committees*) The 10 members of a city council decide to form two committees of six to study zoning ordinances and street-repair schedules, with an overlap of two committee members. In how many ways can the committees be formed? (*Hint*: Specify three groups, not two.)

23. (*Field Trip*) In how many ways can the 14 children in a third-grade class be paired up for a trip to a museum?

24. (*Work Schedule*) A sales representative must travel to three cities twice each in the next 10 days. Her non-travel days are spent in the office. In how many different ways can she schedule her travel, assuming that she does not want to spend four consecutive days in the office?

25. (*Basketball Teams*) Ten students in a physical education class are to be divided into five-member teams for a basketball game. In how many ways can the two teams be selected?

26. Derive formula (1) using the generalized multiplication principle and the formula for $\binom{n}{r}$. (*Hint*: First select the elements of S_1, then the elements of S_2, and so on.)

In Exercises 27 and 28, use a graphing calculator, a spreadsheet, or mathematical software to calculate the answers.

27. (*Assignments to Seminars*) Calculate the number of ways that 38 students can be assigned to four seminars of size 10, 12, 10, and 6, respectively.

28. (*Campaign Tasks*) Calculate the number of ways that 65 phone numbers can be distributed to 5 campaign workers if each worker gets the same number of names.

SOLUTIONS TO PRACTICE PROBLEMS 5.8

1. Each choice of recipients is an ordered partition of the 13 finalists into a first subset of one ($100,000 award), a second subset of two ($10,000 award), a third subset of five ($5000 award), and a fourth subset of five ($2000 award). The number of ways to choose the recipients is thus

$$\binom{13}{1,2,5,5} = \frac{13!}{1!\,2!\,5!\,5!}$$

$$= \frac{13 \cdot 12 \cdot 11 \cdot \cancel{10} \cdot \cancel{9} \cdot \cancel{8} \cdot 7 \cdot 6 \cdot \cancel{5} \cdot \cancel{4} \cdot \cancel{3} \cdot \cancel{2} \cdot \cancel{1}}{1 \cdot \cancel{2} \cdot 1 \cdot \cancel{5} \cdot \cancel{4} \cdot \cancel{3} \cdot \cancel{2} \cdot 1 \cdot 5 \cdot 4 \cdot 3 \cdot 2 \cdot 1}$$

$$= 13 \cdot 12 \cdot 11 \cdot 3 \cdot 7 \cdot 6 = 216{,}216.$$

2. Each partition is of the type $(2, 2, 2)$. In part (a) the order of the subsets is important, whereas in part (b) the order is irrelevant. Consider the partitions

$$(\{\text{Dr. A, Dr. B}\}, \{\text{Dr. C, Dr. D}\}, \{\text{Dr. E, Dr. F}\})$$

and

$$(\{\text{Dr. C, Dr. D}\}, \{\text{Dr. A, Dr. B}\}, \{\text{Dr. E, Dr. F}\}).$$

With respect to part (a) these two partitions are different since in one Drs. A and B are assigned to the radiology department and in the other they are assigned to the neurology department. With respect to part (b), these two partitions are the same since, for instance, Drs. A and B are roommates in both partitions. Therefore, the answers are

(a) $\dbinom{6}{2,2,2} = \dfrac{6!}{(2!)^3} = \dfrac{6 \cdot 5 \cdot \cancel{4} \cdot 3 \cdot \cancel{2} \cdot 1}{\cancel{2} \cdot \cancel{2} \cdot \cancel{2}} = 90.$

(b) $\dfrac{1}{3!} \cdot \dfrac{6!}{(2!)^3} = \dfrac{1}{\cancel{6}} \cdot \dfrac{\cancel{6} \cdot 5 \cdot \cancel{4} \cdot 3 \cdot \cancel{2} \cdot 1}{\cancel{2} \cdot \cancel{2} \cdot \cancel{2}} = 15.$

CHAPTER 5 SUMMARY

1. A *set* is a collection of objects. Each object is called an *element* of the set. The *empty set* is the set containing no objects.

2. The *union* of two sets is the set consisting of all elements that belong to **at least one** of the sets. The *intersection* of two sets is the set consisting of all elements that belong to **both** of the sets.

3. Set A is a *subset* of set B if every element of set A is also an element of set B. In each situation or problem, all sets are considered to be subsets of a *universal set*. The set of all elements in the universal set that do not belong to the set A is called the *complement of A*, denoted A'.

4. The inclusion-exclusion principle says that the number of elements in the union of two sets is the sum of the number of elements in each set minus the number of elements in their intersection.

5. A Venn diagram consists of a rectangle containing overlapping circles and is used to depict relationships among sets. The rectangle represents the universal set and the circles represent subsets of the universal set.

6. De Morgan's laws state that the complement of the union (intersection) of two sets is the intersection (union) of their complements.

7. The multiplication principle states that the number of ways a sequence of several independent operations can be performed is the product of the number of ways each individual operation can be performed.

8. The number of ordered arrangements, each called a *permutation*, of n objects taken r at a time is

$$P(n, r) = n(n-1)(n-2) \cdots (n-r+1).$$

9. The number of unordered arrangements, each called a *combination*, of n objects taken r at a time is

$$C(n, r) = \frac{P(n, r)}{r!} = \frac{n(n-1)(n-2) \cdots (n-r+1)}{r(r-1) \cdots 1}.$$

$C(n, r)$ is also denoted $\binom{n}{r}$.

10. The formula $C(n, r) = C(n, n-r)$ simplifies the computation of $C(n, r)$ when r is greater than $\frac{n}{2}$.

11. The binomial theorem states that

$$(x + y)^n = \binom{n}{0}x^n + \binom{n}{1}x^{n-1}y + \binom{n}{2}x^{n-2}y^2$$

$$+ \cdots + \binom{n}{n-1}xy^{n-1} + \binom{n}{n}y^n.$$

12. A set of n elements has 2^n subsets.

13. Let S be a set of n elements, and suppose that $n = n_1 + n_2 + \cdots + n_m$ where each number in the sum is a positive integer. Then the number of ordered partitions of S into subsets of sizes $n_1, n_2, \ldots, n_m$ is

$$\frac{n!}{n_1!\,n_2! \cdots n_m!}.$$

This number is also denoted

$$\binom{n}{n_1, n_2, \ldots, n_m}.$$

14. Let S be a set of n elements, where $n = m \cdot r$. Then the number of unordered partitions of S into m subsets of size r is $\dfrac{1}{m!} \cdot \dfrac{n!}{(r!)^m}$.

REVIEW OF FUNDAMENTAL CONCEPTS

1. What is a set?

2. What is a subset of a set?

3. What is an element of a set?

4. Define a universal set.

5. Define the empty set.

6. Use a Venn diagram to draw the complement of a set A.

7. Use a Venn diagram to draw the sets $A \cap B$ and $A \cup B$.

8. Use a Venn diagram to draw the sets $A \cap (B \cup C)$ and $A \cup (B \cap C)$.

9. State the generalized multiplication principle for counting.

10. What is meant by a permutation of n items taken r at a time?

11. How would you calculate the number of permutations of n items taken r at a time?

12. What is the difference between a permutation and a combination?

13. How would you calculate the number of combinations of n items taken r at a time?

14. Give a formula that can be used to calculate each of the following:

$$n! \qquad \binom{n}{r} \qquad C(n,r) \qquad P(n,r).$$

15. State the binomial theorem.

16. If a set contains n elements, how many subsets does it have?

17. Explain what is meant by an ordered partition of a set.

18. Explain how to calculate the number of ordered partitions of a set.

19. Give a formula that can be used to calculate

$$\binom{n}{n_1, n_2, \ldots, n_m}.$$

KEY FORMULAS

Inclusion–Exclusion Principle:

$$n(S \cup T) = n(S) + n(T) - n(S \cap T)$$

DeMorgan's Law:

$$(S \cup T)' = S' \cap T' \quad \text{and} \quad (S \cap T)' = S' \cup T'$$

Permutations:

$$P(n, r) = n(n-1)(n-2) \cdots (n-r+1) \quad (r \text{ factors})$$

Combinations:

$$C(n, r) = \frac{P(n, r)}{r!} = \frac{n(n-1) \cdots (n-r+1)}{r(r-1) \cdots 1}$$

$$= \frac{n!}{r!(n-r)!}$$

A Property of Binomial Coefficients: $C(n, r) = C(n, n-r)$

Binomial Theorem:

$$(x+y)^n = \binom{n}{0}x^n + \binom{n}{1}x^{n-1}y + \binom{n}{2}x^{n-2}y^2$$

$$+ \cdots + \binom{n}{n-1}xy^{n-1} + \binom{n}{n}y^n$$

Number of subsets in a set of n elements: 2^n

Number of Ordered Partitions of Type $(n_1, n_2, \ldots, n_m)$, where $n = n_1 + \cdots + n_m$:

$$\frac{n!}{n_1! \, n_2! \cdots n_m!}$$

Number of Unordered Partitions of Type $(r, r, \ldots, r)$, where there are m r's and $n = m \cdot r$:

$$\frac{1}{m!} \cdot \frac{n!}{(r!)^m}$$

SUPPLEMENTARY EXERCISES

1. List all subsets of the set $\{a, b\}$.

2. Draw a two-circle Venn diagram and shade the portion corresponding to the set $(S \cup T')'$.

3. (*Tennis Finalists*) There are 16 contestants in a tennis tournament. How many different possibilities are there for the two people who will play in the final round?

4. (*Team Picture*) In how many ways can a coach and five basketball players line up in a row for a picture if the coach insists on standing at one of the ends of the row?

5. Draw a three-circle Venn diagram and shade the portion corresponding to the set $R' \cap (S \cup T)$.

6. Calculate the first three terms in the binomial expansion of $(x + y)^{12}$.

7. (*Balls in an Urn*) An urn contains 14 numbered balls, of which 8 are red and 6 are green. How many different possibilities are there for selecting a sample of 5 balls in which 3 are red and 2 are green?

8. (*Testing a Drug*) Sixty people with a certain medical condition were given pills. Fifteen of these people received placebos. Forty people showed improvement, and 30 of these people received an actual drug. How many of the people who received the drug showed no improvement?

9. (*Appliance Purchase*) An appliance store carries seven different types of washing machines and five different types of dryers. How many different combinations are possible for a customer who wants to purchase a washing machine and a dryer?

10. (*Contest Prizes*) There are 12 contestants in a contest. Two will receive trips around the world, four will receive cars, and six will receive TV sets. In how many different ways can the prizes be awarded?

11. (*Languages*) Out of a group of 115 applicants for jobs at the World Bank, 70 speak French, 65 speak Spanish, 65 speak German, 45 speak French and Spanish, 35 speak Spanish and German, 40 speak French and German, and 35 speak all three languages. How many of the people speak none of the three languages?

12. Calculate $\binom{17}{15}$.

(Environmental Poll) The 100 members of the Earth Club were asked what they felt the club's priorities should be in the coming year: clean water, clean air, or recycling. The responses were 45 for clean water, 30 for clean air, 42 for recycling, 13 for both clean air and clean water, 20 for clean air and recycling, 16 for clean water and recycling, and 9 for all three. Exercises 13–20 refer to this poll.

13. How many members thought the priority should be clean air only?

14. How many members thought the priority should be clean water or clean air, but not both?

15. How many members thought the priority should be clean water or recycling but not clean air?

16. How many members thought the priority should be clean air and recycling but not clean water?

17. How many members thought the priority should be exactly one of the three issues?

18. How many members thought recycling should not be a priority?

19. How many members thought the priority should be recycling but not clean air?

20. How many members thought the priority should be something other than one of these three issues?

21. (*Nine-Letter Words*) How many different nine-letter words (i.e., sequences of letters) can be made using four S's and five T's?

22. (*Passing an Exam*) Forty people take an exam. How many different possibilities are there for the set of people who pass the exam?

23. (*Winter Sports*) A survey at a small New England college showed that 400 students skied, 300 played ice hockey, and 150 did both. How many students participated in at least one of these sports?

24. (*Meal Choices*) How many different meals can be chosen if there are 6 appetizers, 10 main dishes, and 8 desserts, assuming a meal consists of one item from each category?

25. (*Test Scoring*) On an essay test there are 5 questions worth 20 points each. In how many ways can a student get 10 points on one question, 15 points on each of three questions, and 20 points on another question?

26. (*Seven-Digit Numbers*) How many seven-digit numbers are even and have a 3 in the hundreds place?

27. (*Telephone Numbers*) How many telephone numbers are theoretically possible if all numbers are of the form *abc-def-ghij* and neither of the first two leading digits (*a* and *d*) is zero?

28. Calculate the number of different strings of letters of length at least 1 but less than 11.

29. How many strings of length 8 can be formed from the symbols *a*, *b*, *c*, *d*, and *e*? How many of the strings have at least one *e*?

30. (*Basketball Teams*) How many different 5-person basketball teams can be formed from a pool of 12 players?

31. (*Selecting Students*) Fourteen students in the 100-student eighth grade are to be chosen to tour the United Nations. How many different groups of 14 are possible?

32. (*Journal Subscriptions*) In one ZIP code, there are 40,000 households. Of them, 4000 households get *Fancy Diet Magazine*, 10,000 households get *Clean Living Journal*, and 1500 households get both publications. How many households get neither?

33. (*Computer Program*) At each stage in a decision process, a computer program has three branches. There are 10 stages at which these branches appear. How many different paths could the process follow?

34. (*Filling Jobs*) Sixty people apply for 10 job openings. In how many ways can all the jobs be filled?

35. (*Generating Tests*) A computerized test generator can generate any one of 5 problems for each of the 10 areas being tested. How many different tests can be generated?

36. If a string of six letters cannot contain any vowels (a, e, i, o, u), how many strings are possible?

37. (*Choosing Delegations*) How many different 4-person delegations can be chosen from 10 ambassadors?

38. (*Subdividing a Class*) In how many ways can a teacher divide a class of 21 students into groups of 7 students each?

39. (*Distributing Candy*) In how many ways can 14 different candies be distributed to 14 scouts?

40. (*Senate Subdivision*) In how many ways can 100 senators be divided into groups of 20 each?

41. (*Senate Subdivision*) In how many ways can 100 senators be assigned to groups of 20 each if the 2 senators from the state of New York cannot be in the same group?

42. (*Palindromes*) Determine the number of palindromes of length 3 that can be formed using the letters in the word PALINDROME.

43. (*Palindromes*) Determine the number of palindromes of length 4 that can be formed using the letters in the word PALINDROME.

44. (*Palindromes*) Determine the number of palindromes of length 5 that can be formed using the letters in the word PALINDROME.

45. (*Hat Displays*) A designer of a window display wants to form a pyramid with 15 hats. She wants to place the five men's hats in the bottom row, the four women's hats in the next row, next the three baseball caps, then the two berets, and a clown's hat at the top. All the hats are different. How many displays are possible?

46. (*Seat Assignments*) In how many ways can 5 people be assigned to seats in a 12-seat room?

47. (*Poker*) A poker hand consists of five cards. How many different poker hands contain all cards of the same suit? (Such a hand is called a "flush.")

48. How many three-digit numbers are there in which no two digits are alike?

49. How many three-digit numbers are there in which exactly two digits are alike?

50. (*Poker*) How many hands of five cards contain exactly three aces?

51. (*Greek-Letter Societies*) Fraternity and sorority names consist of two or three letters from the Greek alphabet. How many different names are there in which no letter appears more than once? (The Greek alphabet contains 24 letters.)

(Party Guests) In Exercises 52–54 suppose there are three boys and three girls at a party.

52. How many different pairings of the six into three boy-girl pairs can be formed?

53. In how many ways can they be seated in a row such that no person is seated next to someone of the same sex?

54. In how many ways can they be seated at a round table such that no person is seated next to someone of the same sex?

55. If 10 lines are drawn in the plane so that none of them are parallel and no three lines intersect at the same point, how many points of intersection are there? If each of these 10 lines forms the boundary of the bounded feasible set of a system of linear inequalities, how many of the intersections occur outside the feasible set?

56. (*Splitting Classes*) Two elementary school teachers have 24 students each. The first teacher splits his students into four groups of six. The second teacher splits her students into six groups of four. Which teacher has more options?

57. (*Consultation Schedule*) A consulting engineer agrees to spend three days at Widgets International, four days at Gadgets Unlimited, and three days at Doodads Incorporated in the next two workweeks. In how many different ways can she schedule her consultations?

58. (*Arranging Books*) A set of books can be arranged on a bookshelf in 120 different ways. How many books are in the set?

59. (*Batting Order*) How many batting orders are possible in a nine-member baseball team if the catcher must bat fourth and the pitcher last?

60. (*Choosing Committees*) In how many ways can a committee of 5 people be chosen from 12 married couples if

  (a) The committee must consist of two men and three women?

  (b) A husband and wife cannot both serve on the committee?

61. (*Voting*) Suppose that you are voting in an election for state delegate. Two state delegates are to be elected from among seven candidates. In how many different ways can you cast your ballot? (*Note:* You may vote for two candidates. However, some people "single-shoot," and others don't pull any levers.)

62. (*Spelling Algebra*) The object is to start with the letter A on top and to move down the diagram to the C at the bottom. From any given letter, move only to one of the letters directly below it on the left or right. If these rules are followed, how many different paths spell *ALGEBRAIC*?

$$A$$
$$L \quad L$$
$$G \quad G \quad G$$
$$E \quad E \quad E \quad E$$
$$B \quad B \quad B \quad B \quad B$$
$$R \quad R \quad R \quad R$$
$$A \quad A \quad A$$
$$I \quad I$$
$$C$$

63. (*Call Letters*) The call letters of radio stations in the U.S. consist of either three or four letters, where the first letter is K or W. How many different call letters are possible?

In Exercises 64–66, use a graphing calculator, a spreadsheet, or mathematical software to calculate the answer.

64. (*Arranging Students*) A group of students can be arranged in a row of seats in 479,001,600 ways. How many students are there?

65. (*Assigning Jobs*) There are 25 people in a department who must be deployed to work on three projects requiring 10, 9, and 6 people. All the people are eligible for all jobs. Calculate the number of ways this can be done.

66. (*License Plates*)

    (a) Calculate the number of license plates that can be formed using three distinct letters and three distinct numbers in any order.

    (b) Compare the solution to (a) with the number of license plates consisting of three distinct letters followed by three distinct numbers. (See Section 4, Example 5.)

Conceptual Exercises

67. What is the relationship between the two sets A and B if $A \cap B = \varnothing$?

68. What is the relationship between the two sets A and B if $n(A \cup B) = n(A) + n(B)$?

69. Explain why $n \cdot (n - 1)! = n!$.

70. Use the result in Exercise 69 to explain why 0! is defined to be 1.

71. Express in your own words the difference between a permutation and a combination.

72. Consider a group of 10 people. Without doing any computation, explain why the number of committees of size 6 is equal to the number of committees of 4 people.

73. Without doing any computation, explain why $C(10, 3) = C(10, 7)$.

74. Without doing any computation, explain why $C(10, 4) + C(10, 5) = C(11, 5)$. *Hint*: Suppose a committee of size 5 is to be chosen from a pool of 11 people, and John Doe is one of the people. How many committees are there which include John? How many committees are there that don't include John?

CHAPTER TEST

1. Calculate each of the following:

    (a) 4!   **(b)** $P(7, 3)$ **(c)** $C(18, 16)$

    (d) $\binom{6}{0}$ **(e)** $\binom{5}{2, 1, 2}$

2. True or false?

    (a) $\{a, b\} = \{b, a, b\}$.

    (b) If $A \cup B = A$, then $A \cap B = B$.

    (c) $(A \cap B)' = A' \cap B'$.

3. Let $U = \{a, b, c, d, e\}$, $S = \{b, c, d\}$, and $T = \{a, c, e\}$. List the elements of the following sets.

    (a) $S' \cap T$ **(b)** $(S \cup T)'$

4. Let

$$U = \{\text{all people}\}$$
$$C = \{\text{certified public accountants}\}$$

and

$$E = \{\text{self-employed people}\}.$$

Describe verbally the sets $C \cap E$ and $C \cup E'$.

5. Draw a three-circle Venn diagram and shade the portion corresponding to the set $A \cap (B' \cup C)$.

6. (*Group Membership*) The Choral Society and the Drama Club at State University hold a joint party. Of the 60 people attending, 40 are members of the Choral Society and 30 are members of the Drama Club. How many are members of both groups?

7. (*Freshmen Survey*) Out of a group of 100 college freshmen, 9 are education majors, 46 describe themselves as middle-of-the-road politically, and 82 did volunteer work during the past year. Of the education majors, 5 describe themselves as middle-of-the-road politically and 7 did volunteer work during the past year. Thirty-two students did volunteer work during the past year and describe themselves as middle-of-the-road politically. Three freshmen are education majors who describe themselves as middle-of-the-road politically and who did volunteer work during the past year. How many of the freshmen did not meet any of the three criteria? (*Note*: The data for this problem are based on a survey conducted in the fall semester of 2004 by the American Council of Education.)

8. (*Executive Positions*) An Internet company is considering three candidates for CEO, five candidates for CFO, and four candidates for marketing director. In how many different ways can these positions be filled?

9. (*Exam Questions*)

   (a) In choosing problems for an exam consisting of 10 true/false questions, a teacher uses a data bank of 20 easy and 30 hard questions. How many ways can he select the questions for an exam consisting of 5 questions of each type?

   (b) A student who is totally unprepared for the exam decides to answer six of the questions true and the remainder false without even reading the questions. How many different ways can he answer the questions?

10. How many three-digit numbers can be formed from the numbers $\{1, 2, 3, 4, 5, 6, 7\}$ if no digit is repeated?

11. (*Family Picture*) A family consisting of two parents and four children is to be seated in a row for a picture. How many different arrangements are possible in which the children are seated together?

12. (*Selecting Students*) The 30 students in a math class consist of 10 science majors and 20 humanities majors. In how many ways can a group of 6 students be selected so that 3 students from each type of major are in the group?

13. Determine the coefficient of $x^7 y^5$ in the binomial expansion of $(x + y)^{12}$.

14. (*Hot Dogs*) The most popular condiments[2] for hot dogs are mustard, ketchup, onions, relish, chili, sauerkraut, and mayonnaise. How many types of hot dogs can be ordered?

15. (*Assigning Volunteers*) In how many ways can eight volunteers be assigned to four pairs for visiting the sick?

---

[2]*Source*: Mingo, J., and E. Barrett, *Just Curious, Jeeves*, Ask Jeeves, Inc., 2000.

CHAPTER 5 PROJECT

Pascal's Triangle

In the following triangular table, known as Pascal's triangle, the entries in the nth row are the binomial coefficients $\binom{n}{0}, \binom{n}{1}, \binom{n}{2}, \ldots, \binom{n}{n}$.

				1				0th row
			1		1			1st row
		1		2		1		2nd row
	1		3		3		1	3rd row
1		4		6		4		1 4th row

$$
\begin{array}{ccccccccccccc}
 & & & & & 1 & & & & & & & \text{0th row}\\
 & & & & 1 & & 1 & & & & & & \text{1st row}\\
 & & & 1 & & 2 & & 1 & & & & & \text{2nd row}\\
 & & 1 & & 3 & & 3 & & 1 & & & & \text{3rd row}\\
 & 1 & & 4 & & 6 & & 4 & & 1 & & & \text{4th row}\\
1 & & 5 & & 10 & & 10 & & 5 & & 1 & & \text{5th row}\\
\end{array}
$$

1   6   15   20   15   6   1   6th row

1   7   21   35   35   21   7   1   7th row

Observe that each number (other than the ones) is the sum of the two numbers directly above it. For example, in the 5th row the number 5 is the sum of the numbers 1 and 4 from the 4th row, and the number 10 is the sum of the numbers 4 and 6 from the 4th row. This fact is known as *Pascal's formula*. Namely, the formula says that

$$
\binom{n}{r} = \binom{n-1}{r-1} + \binom{n-1}{r}.
$$

1. For what values of n and r does Pascal's formula say that the number 10 in the triangle above is the sum of the numbers 4 and 6?

2. Derive Pascal's formula from the fact that $C(n,r) = \dfrac{n!}{r!\,(n-r)!}$.

3. Derive Pascal's formula from the fact that $C(n,r)$ is the number of ways of selecting r objects from a set of n objects. (*Hint*: Let x denote the nth object of the set. Count the number of ways a subset of r objects containing x can be selected and then count the number of ways a subset of r objects not containing x can be selected.)

4. Use Pascal's formula to extend Pascal's triangle to the 12th row. Determine the values of $\binom{12}{5}$ and $\binom{12}{6}$ from the extended triangle.

5. (a) Show that for any positive integer n,

$$
\binom{n}{0} + \binom{n}{1} + \binom{n}{2} + \binom{n}{3} + \cdots + \binom{n}{n} = 2^n.
$$

   Hint: Apply the binomial theorem to $(x+y)^n$ with $x = 1$, $y = 1$.

   (b) Show that for any positive integer n,

$$
\binom{n}{0} - \binom{n}{1} + \binom{n}{2} - \binom{n}{3} + \cdots \pm \binom{n}{n} = 0.
$$

   Hint: Apply the binomial theorem to $(x+y)^n$ with $x = 1$, $y = -1$.

   (c) Show that for the 7th row of Pascal's triangle, the sum of the even-numbered elements equals the sum of the odd-numbered elements; that is,

$$
\binom{7}{0} + \binom{7}{2} + \binom{7}{4} + \binom{7}{6} = \binom{7}{1} + \binom{7}{3} + \binom{7}{5} + \binom{7}{7}.
$$

   (d) Use the result of parts (a) and (b) to show that for any row of Pascal's triangle, the sum of the even-numbered elements equals the sum of the odd-numbered elements, and give that common sum for the nth row in terms of n. [*Hint*: Add the two equations in parts (a) and (b).]

(e) Suppose S is a set of n elements. Use the result of part (d) to determine the number of subsets of S that have an even number of elements.

6. **(a)** Show that for any positive integer n,

$$1 + 2 + 4 + 8 + \cdots + 2^n = 2^{n+1} - 1.$$

(*Hint*: Let $S = 1 + 2 + 4 + 8 + \cdots + 2^n$, multiply both sides of the equation by 2, and subtract the first equation from the second.)

(b) Show that the sum of the elements of any row of Pascal's triangle equals one more than the sum of the elements of all previous rows.

7. **(a)** Consider the 7th row of Pascal's triangle. Observe that each interior number (that is, a number other than 1) is divisible by 7. For what values of n, for $1 \le n \le 12$, are the interior numbers of the nth row divisible by n?

(b) Confirm that each of the values of n from part (a) is a prime number. (*Note*: A number p is a *prime* number if the only positive integers that divide it are p and 1.) Prove that if p is a prime number, then each interior number of the pth row of Pascal's triangle is divisible by p. [*Hint*: Use the fact that $\binom{p}{r} = \frac{p(p-1)(p-2)\cdots(p-r+1)}{1\cdot 2\cdot 3\cdots\cdots r}$.]

(c) Show that for any prime number p, the sum of the interior numbers of the pth row is $2^p - 2$.

(d) Calculate $2^p - 2$ for $p = 7$ and show that it is a multiple of 7.

(e) Use the results of parts (b) and (c) to show that for any prime number p, $2^p - 2$ is a multiple of p. (*Note*: This is a special case of Fermat's theorem, which states that for any prime number p and any integer a, $a^p - a$ is a multiple of p.)

8. There are 4 odd numbers in the 6th row of Pascal's triangle $(1, 15, 15, 1)$ and $4 = 2^2$ is a power of 2. For the 0th through 12th rows of Pascal's triangle, show that the number of odd numbers in each row is a power of 2.

In Fig. 1(a) each odd number in the first eight rows of Pascal's triangle has been replaced by a dot and each even number has been replaced by a space. In Fig. 1(b) the pattern for the first four rows is shown in blue. Notice that this pattern appears twice in the next four rows, with the two appearances separated by an inverted triangle. Figure 1(c) shows the locations of the odd numbers in the first 16 rows of Pascal's triangle, and Fig. 1(d) shows that the pattern for the first eight rows appears twice in the next eight rows separated by an inverted triangle. Figure 2 demonstrates that the first 32 rows of Pascal's triangle have the same property.

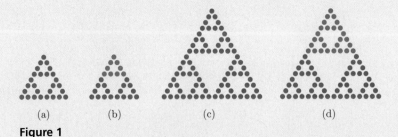

(a)   (b)   (c)   (d)

Figure 1

9. Assume that the property shown in Figs. 1 and 2 continues to hold for subsequent rows of Pascal's triangle. Use this result to explain why the number of odd numbers in each row of Pascal's triangle is a power of 2.

(a)

(b)

Figure 2

Probability

IN this chapter we discuss probability, the mathematics of chance. We consider the basic concepts that allow us to associate realistic probabilities to random events that we see in both our personal and professional lives.

6.1 Introduction

Uncertainty faces us every day. We wake up in the morning and check the weather report (60% chance of rain). We decide on our breakfast cereal (oatmeal might reduce cholesterol). We choose the route to school or work (average delay on Route 450 is 20 minutes, average delay on the alternate is 15 minutes). We call our stockbroker and check the bank rates to decide how to handle our paycheck. We get a flat tire on the way to an important date (what's the chance of that?). And so on.

Many events in the world around us exhibit a random character. Yet, by repeated observations of such events we can often discern long-term patterns that persist despite random, short-term fluctuations. Probability is the branch of mathematics devoted to the study of such events.

Human beings have always been interested in games of chance and gambling. We have evidence that games such as dice have been in existence since 3000 B.C. But the mathematical treatment of such games did not begin until the fifteenth century in Italy. The French contributed to the literature in the seventeenth century in an attempt to calculate probabilities and develop the theory. The foundations of modern probability theory are generally credited to Kolmogorov, the Russian mathematician, who in 1933 proposed the axioms on which the present subject of probability rests. Even as we enter the twenty-first century, contemporary mathematicians continue to develop new ideas that are applied to such varied areas as investment and risk analysis, card shuffling, diagnostic medical procedures, cryptography, and the efficacy of treatment modalities in medical and social programs. To obtain an idea of the sorts of events that are considered, let us consider a concrete example from the field of medicine.

Suppose that we wish to analyze the reliability of a skin test for active pulmonary tuberculosis. Unfortunately, such a test is not completely reliable. On one hand, the test may be negative even for a person with tuberculosis. On the other hand, the test may be positive for a person who does not have tuberculosis. For the moment, let us concentrate on errors of the first sort and consider only individuals actually having tuberculosis. Suppose that by observing the results of the test on increasingly large populations of tuberculosis patients we accumulate the data shown in Table 1. Note that out of each group of tuberculosis patients the test fails to identify a certain number. However, the data do exhibit a pattern. It appears that out of a very large population of tuberculosis patients the skin test will successfully identify about 98% of them. In fact, it appears that as the size of the population is increased, the relative frequency m/N more and more closely approximates the number .98. In a situation like this, we say that the skin test detects tuberculosis with a 98% likelihood or that the *probability* that the test detects tuberculosis (when present) is .98.

TABLE 1

Number of tuberculosis patients N	Number of positive test results m	Relative frequency of positive test results m/N
100	97	.97
500	494	.988
1000	981	.981
10,000	9806	.9806
50,000	49,008	.98016
100,000	98,005	.98005

More generally, the *probability of an event* is a number that expresses the long-run likelihood that the event will occur. Such numbers are always chosen to lie between 0 and 1. The smaller the probability, the less likely the event is to occur. So, for example, an event having probability .1 is rather unlikely to occur; an event with probability .9 is very likely to occur; and an event with probability .5 is just as likely to occur as not.

We assign probabilities to events on the following intuitive basis: The probability of an event should represent the long-run proportion of the time that the event can be expected to occur. For example, an event with probability .9 can be expected to occur 90% of the time, and an event with probability .1 can be expected to occur 10% of the time.

As we shall see, many real-life problems require us to calculate probabilities from known data. Here is one example that arises in connection with the skin test for tuberculosis.

> **Medical Diagnosis** A clinic tests for active pulmonary tuberculosis. If a person has tuberculosis, the probability of a positive test result is .98. If a person does not have tuberculosis, the probability of a negative test result is .99. The incidence of tuberculosis in a certain city is 2 cases per 10,000 population. Suppose that an individual is tested and a positive result is noted. What is the probability that the individual actually has active pulmonary tuberculosis?

Before we can solve this problem, it will be necessary to do considerable preliminary work. We begin this work in Section 6.2, where we introduce a convenient language for discussing events and the process of observing them. In Section 6.3 we introduce probabilities of events, and in Sections 6.4, 6.5, 6.6, and 6.7 we develop methods for calculating probabilities of various sorts of events. The solution of the medical diagnosis problem is presented in Section 6.6. Section 6.8 is devoted to the simulation of simple experiments.

6.2 Experiments, Outcomes, and Events

The events whose probabilities we wish to compute all arise as outcomes of various experiments. So as our first step in developing probability theory, let us describe, in mathematical terms, the notions of experiment, outcome, and event.

For our purposes an *experiment* is an activity with an observable outcome. Here are some typical examples of experiments.

Experiment 1 Flip a coin and observe the side that faces upward.

Experiment 2 Allow a conditioned rat to run a maze, and observe which one of the three possible paths it takes.

Experiment 3 Choose a year and tabulate the amount of rainfall in New York City during that year.

We think of an experiment as being performed repeatedly. Each repetition of the experiment is called a *trial*. In each trial we observe the *outcome* of the experiment. For example, a possible outcome of Experiment 1 is "heads"; a possible outcome of Experiment 2 is "path 3"; and a possible outcome of Experiment 3 is "37.23 inches."

To describe an experiment in mathematical language, we construct a model of the experiment. It is most convenient to form the set consisting of all possible outcomes of the experiment. This set is called the *sample space* of the experiment. For example, if S_1, S_2, S_3 are the sample spaces for Experiments 1, 2, 3, respectively, then we immediately see that

$$S_1 = \{\text{heads, tails}\}$$
$$S_2 = \{\text{path 1, path 2, path 3}\}.$$

Moreover, since any nonnegative number is a candidate for the amount of rainfall, we have

$$S_3 = \{\text{all numbers} \geq 0\}.$$

We can describe an experiment in terms of the sample space as follows:

> Suppose that an experiment has a sample space S. Then each trial has as its outcome one of the elements of S.

Thus, for example, each trial for Experiment 1 has as its outcome one of the elements of the set

$$S_1 = \{\text{heads, tails}\}.$$

Henceforth, we shall always describe experiments in terms of their respective sample spaces. So it is important to be able to recognize the appropriate sample space in each instance. The next few examples should help you obtain the necessary facility in doing this.

EXAMPLE 1   **Sample space for rolling a die**   An experiment consists of tossing a die and observing the number on the uppermost face. Describe the sample space S for this experiment.

Solution   There are six outcomes of the experiment, corresponding to the six possible numbers on the uppermost face. Therefore,

$$S = \{1, 2, 3, 4, 5, 6\}.$$ ■

EXAMPLE 2   **Sample space for the number of people in a queue**   Once an hour a supermarket manager observes the number of people standing in checkout lines. The store has space for at most 30 customers to wait in line. What is the sample space S for this experiment?

Solution   The outcome of the experiment is the number of people standing in checkout lines. This number may be $0, 1, 2, \ldots,$ or 30. Therefore,

$$S = \{0, 1, 2, \ldots, 30\}.$$ ■

EXAMPLE 3   **Sample space for rolling a pair of dice**   An experiment consists of throwing two dice, one red and one green, and observing the uppermost face on each. What is the associated sample space S?

Solution   Each outcome of the experiment can be regarded as an ordered pair of numbers, the first representing the number on the red die and the second the number on the green die. Thus, for example, the pair of numbers $(3, 5)$ represents the outcome "3 on the red die, 5 on the green die." The sample space consists of all possible pairs of numbers (r, g), where r and g are each one of the numbers 1, 2, 3, 4, 5, 6. This sample space has 36 elements:

$$
\begin{aligned}
S = \{ & (1,1), \ (1,2), \ (1,3), \ (1,4), \ (1,5), \ (1,6), \\
& (2,1), \ (2,2), \ (2,3), \ (2,4), \ (2,5), \ (2,6), \\
& (3,1), \ (3,2), \ (3,3), \ (3,4), \ (3,5), \ (3,6), \\
& (4,1), \ (4,2), \ (4,3), \ (4,4), \ (4,5), \ (4,6), \\
& (5,1), \ (5,2), \ (5,3), \ (5,4), \ (5,5), \ (5,6), \\
& (6,1), \ (6,2), \ (6,3), \ (6,4), \ (6,5), \ (6,6) \}.
\end{aligned}
$$

Note that if we were interested in the *sum* of the numbers on the uppermost faces, the sample space would be quite different:

$$S = \{2, 3, 4, 5, \ldots, 12\}.$$

The preceding array helps demonstrate what the elements of this second sample space should be. Having the results of an experiment enumerated in a simple way often helps to elucidate the more complicated multistep experiment.

Now Try Exercise 3(a)

■

The sample space for an experiment should be chosen so that every outcome is included and there is no overlap. Since we will be assigning probabilities to the elements of the sample space, we choose a sample space for its utility and not for its simplicity. Personal choices might differ; there is no one right answer.

Consider the experiment of observing the number of heads on three tosses of a fair coin. One possible sample space is $\{0, 1, 2, 3\}$. But we will see later that for practical purposes we might prefer

$$S = \{\text{TTT, TTH, THT, HTT, THH, HTH, HHT, HHH}\},$$

which gives the results (heads or tails) on each of the three tosses.

EXAMPLE 4 **Sample space for pollutant levels** The Environmental Protection Agency orders Middle States Edison Corporation to install "scrubbers" to remove the pollutants from its smokestacks. To monitor the effectiveness of the scrubbers, the corporation installs monitoring devices to record the levels of sulfur dioxide, particulate matter, and oxides of nitrogen in the smokestack emissions. Consider the monitoring operation as an experiment. Describe the associated sample space.

Solution Each reading of the instruments consists of an ordered triple of numbers (x, y, z), where x = level of sulfur dioxide, y = level of particulate matter, and z = level of oxides of nitrogen. The sample space thus consists of all possible triples (x, y, z), where $x \geq 0$, $y \geq 0$, and $z \geq 0$.

Now Try Exercise 7(a)

■

The sample spaces in Examples 1, 2, and 3 are *finite*. That is, the associated experiments have only a finite number of possible outcomes. However, the sample space of Example 4 is *infinite*, since there are infinitely many triples (x, y, z), where $x \geq 0$, $y \geq 0$, and $z \geq 0$.

Now that we have discussed experiments and their outcomes, let us turn our attention to the notion of "event." In connection with our preceding discussion, we can define many events whose probabilities we might wish to know. For example, in connection with Experiment 2, we can consider the event

"A conditioned rat chooses either path 2 or path 3."

Here are two events associated with Experiment 3:

"The annual rainfall in New York City exceeds 50 inches."

"The annual rainfall in New York City is less than 35 inches."

It is easy to describe events in terms of the sample space. For example, let us consider the die-tossing experiment of Example 1 and the following events.

I. An even number occurs.

II. A number greater than 2 occurs.

We saw previously that the sample space S for this experiment is

$$S = \{1, 2, 3, 4, 5, 6\}.$$

Assume that the experiment is performed. Then event I occurs precisely when the outcome of the experiment is 2, 4, or 6. That is, event I occurs precisely when the outcome belongs to the set

$$E_{\text{I}} = \{2, 4, 6\}.$$

Note that this set is a subset of the sample space S. Similarly, we can describe event II by the set

$$E_{\text{II}} = \{3, 4, 5, 6\}.$$

Event II occurs precisely when the outcome of the experiment is an element of E_{II}.

The sets E_{I} and E_{II} contain all the information we need in order to completely describe events I and II. This observation suggests the following definition of an event in terms of the sample space.

> An *event* E is a subset of the sample space. We say that the event *occurs* when the outcome of the experiment is an element of E.

The next few examples provide some practice in describing events as subsets of the sample space.

EXAMPLE 5   **Events associated with a queue**   Consider the supermarket of Example 2. Describe the following events as subsets of the sample space.

(a) Fewer than 5 people are waiting in line.

(b) More than 23 people are waiting in line.

(c) No people are waiting in line.

Solution   We saw that the sample space for this experiment is given by

$$S = \{0, 1, 2, \ldots, 30\}.$$

(a) If fewer than 5 people are waiting in line, then the number of people waiting is 0, 1, 2, 3, or 4. So the subset of S corresponding to event (a) is

$$\{0, 1, 2, 3, 4\}.$$

(b) In this case the number waiting must be 24, 25, 26, 27, 28, 29, or 30. So the event is just the subset

$$\{24, 25, 26, 27, 28, 29, 30\}.$$

(c) $\{0\}$ ∎

EXAMPLE 6   **Tossing a coin three times**   Suppose that an experiment consists of tossing a coin three times and observing the sequence of heads and tails. (Order counts.)

(a) Determine the sample space S.

(b) Determine the event $E =$ "exactly two heads."

Solution (a) Denote "heads" by H and "tails" by T. Then a typical outcome of the experiment is a sequence of H's and T's. So, for instance, the sequence HTT would stand for a head followed by two tails. We exhibit all such sequences and arrive at the sample space S:

$$S = \{\text{HHH, HHT, HTH, THH, HTT, THT, TTH, TTT}\}.$$

(b) Here are the outcomes in which exactly two heads occur: HHT, HTH, THH. Therefore, event E is

$$E = \{\text{HHT, HTH, THH}\}.$$

Now Try Exercise 3(b)

■

EXAMPLE 7   **Political poll**   A political poll surveys a group of people to determine their income levels and political affiliations. People are classified as either low-, middle-, or upper-level income and as either Democrat, Republican, or Independent.

(a) Find the sample space corresponding to the poll.

(b) Determine the event $E_1 =$ "Independent."

(c) Determine the event $E_2 =$ "low income and not Independent."

(d) Determine the event $E_3 =$ "neither upper income nor Independent."

Solution (a) Let us abbreviate low, middle, and upper income, respectively, by the letters L, M, and U, respectively. And let us abbreviate Democrat, Republican, and Independent by the letters D, R, and I, respectively. Then a response to the poll can be represented as a pair of letters. For example, the pair (L, D) refers to a low-income-level Democrat. The sample space S is then given by

$$S = \{(\text{L, D}), (\text{L, R}), (\text{L, I}), (\text{M, D}), (\text{M, R}), (\text{M, I}), (\text{U, D}), (\text{U, R}), (\text{U, I})\}.$$

(b) For event E_1 the income level may be anything, but the political affiliation is Independent. Thus

$$E_1 = \{(\text{L, I}), (\text{M, I}), (\text{U, I})\}.$$

(c) For event E_2 the income level is low and the political affiliation may be either Democrat or Republican, so that

$$E_2 = \{(\text{L, D}), (\text{L, R})\}.$$

(d) For event E_3 the income level may be either low or middle and the political affiliation may be Democrat or Republican. Thus

$$E_3 = \{(\text{L, D}), (\text{L, R}), (\text{M, D}), (\text{M, R})\}.$$

Now Try Exercise 9

■

The *New York Times* of November 21, 1996 reported that in the New York City public schools with total enrollment of 1.06 million students, 88.1% of the students are in regular classrooms, 4.5% get part-time special education, and 7.4% get full-time special education. Students in New York City attend schools in one of the five boroughs (Manhattan, Bronx, Brooklyn, Queens, Staten Island). If we choose a student at random, we can determine whether the student is in a regular classroom (C), gets part-time special education (P), or gets full-time special education (F).

Likewise, for each student we can determine in what borough (M, Bx, Bk, Q, SI) the student goes to school. The sample space is

$$S = \{(C, M), (C, Bx), (C, Bk), (C, Q), (C, SI),$$
$$(P, M), (P, Bx), (P, Bk), (P, Q), (P, SI),$$
$$(F, M), (F, Bx), (F, Bk), (F, Q), (F, SI)\}.$$

A student corresponding to the sample point (F, Bk) is getting full-time special education in Brooklyn.

As we have seen, an event is a subset of the sample space. Two events are worthy of special mention. The first is the event corresponding to the empty set, ∅. This is called the *impossible event*, since it can never occur. The second special event is the set S itself. Every outcome is an element of S, so S always occurs. For this reason S is called the *certain event*.

One particular advantage of defining experiments and events in terms of sets is that it allows us to define new events from given ones by applying the operations of set theory. When so doing we always let the sample space S play the role of universal set. (All outcomes belong to the universal set.)

If E and F are events, then so are $E \cup F$, $E \cap F$, and E'. For example, consider the die-tossing experiment of Example 1. Then

$$S = \{1, 2, 3, 4, 5, 6\}.$$

Let E and F be the events given by

$$E = \{3, 4, 5, 6\} \qquad F = \{1, 4, 6\}.$$

Then we have

$$E \cup F = \{1, 3, 4, 5, 6\}$$
$$E \cap F = \{4, 6\}$$
$$E' = \{1, 2\}.$$

$E \cup F$

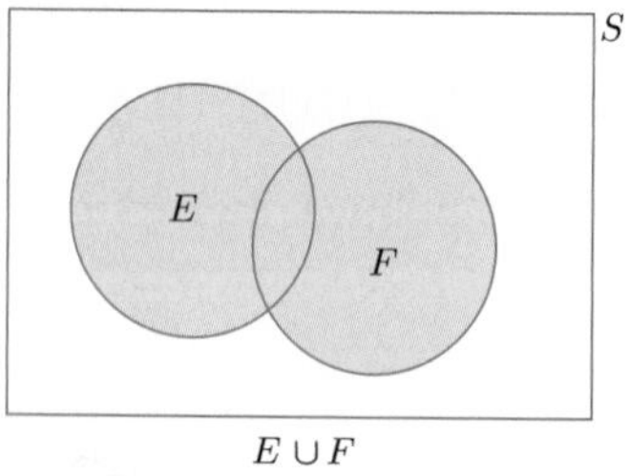

Figure 1

Let us interpret the events $E \cup F$, $E \cap F$, and E' using Venn diagrams. In Fig. 1 we have drawn a Venn diagram for $E \cup F$. Note that $E \cup F$ occurs precisely when the experimental outcome belongs to the shaded region—that is, to either E or F. Thus we have the following result.

> The event $E \cup F$ occurs precisely when either E or F (or both) occurs.

Similarly, we can interpret the event $E \cap F$. This event occurs when the experimental outcome belongs to the shaded region of Fig. 2—that is, to both E and F. Thus we have an interpretation for $E \cap F$:

> The event $E \cap F$ occurs precisely when both E and F occur.

Finally, the event E' consists of all those outcomes not in E (Fig. 3). Therefore, we have

> The event E' occurs precisely when E does not occur.

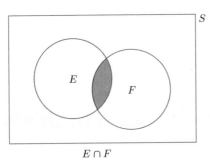

$E \cap F$

Figure 2

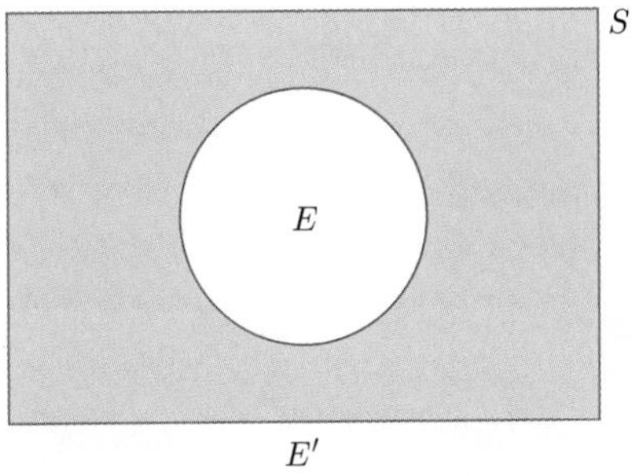

E'

Figure 3

EXAMPLE 8 **Events related to pollution control** Consider the pollution monitoring described in Example 4. Let E, F, and G be the events

$$E = \text{``level of sulfur dioxide} \geq 100\text{''}$$

$$F = \text{``level of particulate matter} \leq 50\text{''}$$

$$G = \text{``level of oxides of nitrogen} \leq 30.\text{''}$$

Describe the following events.

(a) $E \cap F$ (b) E' (c) $E \cup G$ (d) $E' \cap F \cap G$

Solution (a) $E \cap F = $ "level of sulfur dioxide ≥ 100 *and* level of particulate matter ≤ 50."

(b) $E' = $ "level of sulfur dioxide < 100."

(c) $E \cup G = $ "level of sulfur dioxide ≥ 100 *or* level of oxides of nitrogen ≤ 30."

(d) $E' \cap F \cap G = $ "level of sulfur dioxide < 100 *and* level of particulate matter ≤ 50 *and* level of oxides of nitrogen ≤ 30."

Now Try Exercise 7(b)

Suppose that E and F are events in a sample space S. We say that E and F are *mutually exclusive* (or *disjoint*) provided that $E \cap F = \emptyset$. In terms of Venn diagrams, we may represent a pair of mutually exclusive events as a pair of circles with no points in common (Fig. 4). If the events E and F are mutually exclusive, then E and F cannot simultaneously occur; if E occurs, then F does not; and if F occurs, then E does not.

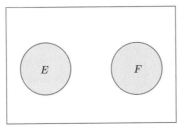

Figure 4 E and F are mutually exclusive.

EXAMPLE 9 **Determining whether events are mutually exclusive** Let $S = \{a, b, c, d, e, f, g\}$ be a sample space, and let $E = \{a, b, c\}$, $F = \{e, f, g\}$, and $G = \{c, d, f\}$.

(a) Are E and F mutually exclusive?

(b) Are F and G mutually exclusive?

Solution (a) $E \cap F = \emptyset$, and so E and F are mutually exclusive.

(b) $F \cap G = \{f\}$, and so F and G are *not* mutually exclusive.

Now Try Exercise 11 ■

PRACTICE PROBLEMS 6.2

1. (*Light Bulbs*) A machine produces light bulbs. As part of a quality control procedure, a sample of five light bulbs is collected each hour and the number of defective light bulbs among these is observed.

   (a) What is the sample space for this experiment?

   (b) Describe the event "there are at most two defective light bulbs" as a subset of the sample space.

2. (*Citrus Fruit*) Suppose that there are two crates of

citrus fruit and each crate contains oranges, grapefruit, and tangelos. An experiment consists of selecting a crate and then selecting a piece of fruit from that crate. Both the crate and the type of fruit are noted. Refer to the crates as crate I and crate II.

   (a) What is the sample space for this experiment?

   (b) Describe the event "a tangelo is selected" as a subset of the sample space.

EXERCISES 6.2

1. (*Committee Selection*) A committee of two people is to be selected from five people, R, S, T, U, and V.

   (a) What is the sample space for this experiment?

   (b) Describe the event "R is on the committee" as a subset of the sample space.

   (c) Describe the event "neither R nor S is on the committee" as a subset of the sample space.

2. A letter is selected at random from the word "MISSISSIPPI."

   (a) What is the sample space for this experiment?

   (b) Describe the event "the letter chosen is a vowel" as a subset of the sample space.

3. (*Heads and Tails*) An experiment consists of tossing a coin two times and observing the sequence of heads and tails.

   (a) What is the sample space of this experiment?

   (b) Describe the event "the first toss is a head" as a subset of the sample space.

4. (*Four-Sided Dice*) A pair of four-sided dice, each with the numbers from 1 to 4 on their sides, are rolled and the numbers facing down are observed.

   (a) List the sample space.

   (b) Describe each of the following events as a subset of the sample space.

   (i) Both numbers are even.
   (ii) At least one number is even.
   (iii) Neither number is less than or equal to 2.
   (iv) The sum of the numbers is 6.
   (v) The sum of the numbers is greater than or equal to 5.
   (vi) The numbers are the same.

   (vii) A 2 or 3 occurs, but not both 2 and 3.
   (viii) No 4 appears.

5. (*Selecting from Urns*) Suppose that we have two urns—call them urn I and urn II—each containing red balls and white balls. An experiment consists of selecting an urn and then selecting a ball from that urn and noting its color.

   (a) What is a suitable sample space for this experiment?

   (b) Describe the event "urn I is selected" as a subset of the sample space.

6. (*Coin Tosses*) An experiment consists of tossing a coin four times and observing the sequence of heads and tails.

   (a) What is the sample space of this experiment?

   (b) Determine the event $E_1 =$ "more heads than tails occur."

   (c) Determine the event $E_2 =$ "the first toss is a head."

   (d) Determine the event $E_1 \cap E_2$.

7. (*Efficiency Studies*) A corporation efficiency expert records the time it takes an assembly line worker to perform a particular task. Let E be the event "more than 5 minutes," F the event "less than 8 minutes," and G the event "less than 4 minutes."

   (a) Describe the sample space for this experiment.

   (b) Describe the events $E \cap F$, $E \cap G$, E', F', $E' \cap F$, $E' \cap F \cap G$, and $E \cup F$.

8. (*Product Reliability*) A manufacturer of kitchen appliances tests the reliability of its refrigerators by recording in a laboratory test the elapsed time between consecutive failures. Let E be the event "more than nine

months" and F the event "less than two years." Describe the events $E \cap F$, $E \cup F$, E', F', and $(E \cup F)'$.

9. (*Student Poll*) A campus survey is taken to correlate the number of years that students have been on campus with their political leanings. Students are classified as first-year, sophomore, junior, or senior and as conservative or liberal.

   (a) Find the sample space corresponding to the poll.

   (b) Determine the event $E_1 = $ "conservative."

   (c) Determine the event $E_2 = $ "junior and liberal."

   (d) Determine the event $E_3 = $ "neither first-year nor conservative."

10. (*Automobiles*) An experiment consists of selecting a car at random from a college parking lot and observing the color and make. Let E be the event "the car is red," F be the event "the car is a Chevrolet," G be the event "the car is a green Ford," and H be the event "the car is black or a Chrysler."

    (a) Which of the following pairs of events are mutually exclusive?

(i) E and F	(ii) E and G
(iii) F and G	(iv) E and H
(v) F and H	(vi) G and H
(vii) E' and G	(viii) F' and H'

    (b) Describe each of the following events:

(i) $E \cap F$	(ii) $E \cup F$
(iii) E'	(iv) F'
(v) G'	(vi) H'
(vii) $E \cup G$	(viii) $E \cap G$
(ix) $E \cap H$	(x) $E \cup H$
(xi) $G \cap H$	(xii) $E' \cap F'$
(xiii) $E' \cup G'$	

11. Let $S = \{1, 2, 3, 4, 5, 6\}$ be a sample space,

    $$E = \{1, 2\} \qquad F = \{2, 3\} \qquad G = \{1, 5, 6\}.$$

    (a) Are E and F mutually exclusive?

    (b) Are F and G mutually exclusive?

12. Show that if E is any event, and E' its complement, then E and E' are mutually exclusive.

13. Let $S = \{a, b, c\}$ be a sample space. Determine all possible events associated with S.

14. Let S be a sample space with n outcomes. How many events are associated with S?

15. Let $S = \{1, 2, 3, 4\}$ be a sample space, $E = \{1\}$, and $F = \{2, 3\}$. Are the events $E \cup F$ and $E' \cap F'$ mutually exclusive?

16. Let S be any sample space, and E, F any events associated with S. Are the events $E \cup F$ and $E' \cap F'$ mutually exclusive? (*Hint:* Apply De Morgan's laws.)

17. (*Coin Tosses*) Suppose that 10 coins are tossed and the number of heads observed.

    (a) Describe the sample space for this experiment.

18. (b) Describe the event "more heads than tails" in terms of the sample space.

18. (*Coin Tosses*) Suppose that five nickels and five dimes are tossed and the numbers of heads from each group recorded.

    (a) Describe the sample space for this experiment.

    (b) Describe the event "more heads on the nickels than on the dimes" in terms of the sample space.

19. (*Genetic Traits*) An experiment consists of observing the eye color and sex of the students at a certain school. Let E be the event "blue eyes," F the event "male," and G the event "brown eyes and female."

    (a) Are E and F mutually exclusive?

    (b) Are E and G mutually exclusive?

    (c) Are F and G mutually exclusive?

20. (*Genetic Traits*) Consider the experiment and events of Exercise 19. Describe the following events.

(a) $E \cup F$	(b) $E \cap G$	(c) E'
(d) F'	(e) $(G \cup F) \cap E$	(f) $G' \cap E$

21. (*Restaurant Queue*) Suppose that you observe the length of the line at a fast-food restaurant. Describe the sample space.

22. (*Restaurant Service Times*) Suppose that you observe the time (in minutes) to be served at a fast-food restaurant. Describe the sample space.

23. (*Restaurant Arrival Times*) Suppose that you observe the time (in minutes) between customer arrivals at a fast-food restaurant. Describe the sample space.

24. (*Restaurant Queue and Service Times*) Suppose that you observe the length of the line when a customer arrives and the length of time (in minutes) it takes for him or her to be served in a fast-food restaurant. Describe the sample space.

25. (*Lottery Numbers*) New York, New Jersey, and Connecticut each have lotteries of various kinds; in particular, a three-digit number is chosen at random each week in each of the three states. Each week the New York papers publish all three states' winning numbers. What would the sample space look like? Let E be the event that all the numbers are even. Let F be the event that all the numbers are more than 699. Determine the events E' and $E \cap F$.

26. (*Gender of Children*) In the December 1, 1996 issue of *Parade* magazine, Marilyn Vos Savant discusses the following situation: "A woman and a man (who are unrelated) each has two children. At least one of the woman's children is a boy, and the man's older child is a boy."

    (a) Describe the sample space of possible families for the man and the sample space of possible families for the woman.

(b) In the sample space describing the woman's family, let E = the event that there are two boys in the woman's family. Show the elements of E.

(c) In the sample space describing the man's family, let F = the event that there are two boys in the man's family. Show the elements of F.

27. (*The Game of Clue*) Anthony E. Pratt, the inventor of the game *Clue*, died in 1996. *Clue* is a board game in which players are given the opportunity to solve a murder that has six suspects, six possible weapons, and nine possible rooms where the murder may have oc-

curred. The six suspects are Colonel Mustard, Miss Scarlet, Professor Plum, Mrs. White, Mr. Green, and Mrs. Peacock. Determine a sample space for the choice of murderer. Discuss how to form a sample space with the entire solution to the murder, giving murderer, weapon, and site.

(a) How many outcomes would the sample space have?

Let E be the event that the murder occurred in the library. Let F be the event that the weapon was a gun.

(b) Describe $E \cap F$.   **(c)** Describe $E \cup F$.

SOLUTIONS TO PRACTICE PROBLEMS 6.2

1. (a) $\{0, 1, 2, 3, 4, 5\}$. The sample space is the set of all outcomes of the experiment. At first glance it might seem that each outcome is a set of five light bulbs. What is observed, however, is not the specific sample but rather the number of defective bulbs in the sample. Therefore, the outcome must be a number.

(b) $\{0, 1, 2\}$. "At most 2" means "2 or less."

2. (a) $\{$(crate I, orange), (crate I, grapefruit), (crate I, tangelo), (crate II, orange), (crate II, grapefruit), (crate II, tangelo)$\}$. Two selections are being made and both should be recorded.

(b) $\{$(crate I, tangelo), (crate II, tangelo)$\}$. This set consists of those outcomes in which a tangelo is selected.

6.3   Assignment of Probabilities

In Section 6.2 we introduced the sample space of an experiment and used it to describe events. We complete our description of experiments by introducing probabilities associated to events. For the remainder of this chapter, let us limit our discussion to experiments with only a finite number of outcomes.[1]

What do we mean when we say that the probability of getting a "head" on a toss of a fair coin is 50%? What is meant by the statement that the chance was 1 in 100 of choosing a particular number from the numbers $0, 1, 2, 3, \ldots, 97, 98, 99$? Similar notions are used every day. Probability occurs in the news, in strategies used in games and in war, and in decisions that involve medical treatments, insurance policies, and all kinds of risk.

Suppose that you took a coin and tossed it 152 times and kept track of the results on a chart recording the number of heads and tails as you tossed. You might try this with a real coin, or as we will see later, with a simulation device of some kind. Assume that the tally looked like this:

	Number	Relative frequency
Heads	67	$67/152 = 44\%$
Tails	85	$85/152 = 56\%$
Total	152	1 or 100%

How "fair" does the coin seem to be? We have computed the actual relative frequency of the two outcomes and noted that heads occurs on about 44% of the tosses, and tails on about 56% of the tosses. For this experiment, the *experimental probability*, or *relative frequency*, of heads is 44%.

---

[1] This restriction will remain in effect until our discussion of the normal distribution in Section 7.6.

This experiment is just that—an experiment—and repeating it would generally yield different tallies and different experimental probabilities. But if the coin is fair, the probability assigned to these two outcomes is the fixed theoretical value of 50% for heads and 50% for tails. What this tells us is that in theory, if a fair coin is tossed many, many times, heads will occur about 50% of the time. The experimental probability (relative frequency) of an event may be quite different from the probability.

Suppose that an experiment has a sample space S consisting of a finite number of outcomes $s_1, s_2, \ldots, s_N$. To each outcome we associate a number, called the *probability of the outcome*, which represents the relative likelihood that the outcome will occur. Suppose that to the outcome s_1 we associate the probability p_1, to the outcome s_2 the probability p_2, and so forth. We can summarize these data in a chart of the following sort:

Outcome	Probability
s_1	p_1
s_2	p_2
$\vdots$	$\vdots$
s_N	p_N

Such a chart is called the *probability distribution* for the experiment. The numbers $p_1, p_2, \ldots, p_N$ are chosen so that each probability represents the long-run proportion of trials in which the associated outcome can be expected to occur. Central to our assignment of probabilities is the requirement that the total probability is 1. That is,

$$p_1 + p_2 + \cdots + p_N = 1.$$

The next three examples illustrate some methods for determining probability distributions.

EXAMPLE 1   **Probability distribution for coin toss**   Toss an unbiased coin and observe the side that faces upward. Determine the probability distribution for this experiment.

Solution   Since the coin is unbiased, we expect each of the outcomes "heads" and "tails" to be equally likely. We assign the two outcomes equal probabilities, namely $\frac{1}{2}$. The probability distribution is

Outcome	Probability
Heads	$\frac{1}{2}$
Tails	$\frac{1}{2}$

◼

EXAMPLE 2   **Probability distribution for roll of a die**   Toss a die and observe the side that faces upward. Determine the probability distribution for this experiment.

Solution   There are six possible outcomes, namely 1, 2, 3, 4, 5, 6. Assuming that the die is unbiased, these outcomes are equally likely. So we assign to each outcome the probability $\frac{1}{6}$. Here is the probability distribution for the experiment:

Outcome	Probability	Outcome	Probability
1	$\frac{1}{6}$	4	$\frac{1}{6}$
2	$\frac{1}{6}$	5	$\frac{1}{6}$
3	$\frac{1}{6}$	6	$\frac{1}{6}$

Probabilities may be assigned to the elements of a sample space using common sense about the physical nature of the experiment. The fair coin has two sides, both equally likely to be face up. The balanced die has six equally probable faces. However, it may not be possible to use intuition alone to decide on a realistic probability to assign to individual sample elements. Sometimes it is necessary to conduct an experiment and use the data to shed light on the long-run relative frequency with which events occur. The following example demonstrates this technique.

EXAMPLE 3   **Probability distribution for traffic volume**   Traffic engineers measure the volume of traffic on a major highway during the rush hour from 5 to 6 P.M. By observing the number of cars that pass a fixed point for 300 consecutive weekdays, they collect the following data:

Number of cars observed	Frequency observed
≤ 1000	30
1001–3000	45
3001–5000	135
5001–7000	75
> 7000	15

(a) Describe the sample space associated to this experiment.

(b) Assign a probability distribution to this experiment.

Solution (a) The experiment consists of counting the number of cars during rush hour on 300 weekdays and assigning each day to a category depending on the number of cars observed on that day. So, for instance, if we observe 6241 cars on a particular day, we add one to the tally for the category 5001–7000. We let

$$s_1 = \text{``}\leq 1000 \text{ cars''}$$

$$s_2 = \text{``1001–3000 cars''}$$

$$s_3 = \text{``3001–5000 cars''}$$

$$s_4 = \text{``5001–7000 cars''}$$

$$s_5 = \text{``}>7000 \text{ cars.''}$$

The sample space is

$$S = \{s_1, s_2, s_3, s_4, s_5\}.$$

(b) For each outcome we use the available data to compute its relative frequency. For example, on 30 days of the 300 observed, the number of cars passing the fixed point is ≤ 1000. So the outcome s_1 occurred in $30/300 = 10\%$ of the observations. If we assume that the 300 consecutive observations are representative of rush hours in general, it seems reasonable to assign to the outcome s_1 the probability .10. Similarly, we can assign probabilities to the other outcomes based on the percentages of observations in which they occur.

Outcome	Probability
s_1	$\frac{30}{300} = .10$
s_2	$\frac{45}{300} = .15$
s_3	$\frac{135}{300} = .45$
s_4	$\frac{75}{300} = .25$
s_5	$\frac{15}{300} = .05$

This method of assigning probabilities to outcomes is valid only insofar as the observed trials are "representative." If such probability models are to be used for planning roadways or traffic patterns, they would have to be tested extensively to be sure that the probabilities are realistic representations of the long-run frequency of events.

In fact, in many instances we cannot rely on intuition or perform such experiments to help us in assigning probabilities; instead, we must use our knowledge of sets and counting to construct a theoretical model of the experiment along with associated probabilities. We need to keep in mind that certain fundamental properties must hold when the set of outcomes (the sample space) is a finite set. These can be observed in the following probability distribution for the number of heads in four tosses of a fair coin:

Events	Probability
0 heads	$\frac{1}{16}$
1 head	$\frac{4}{16} = \frac{1}{4}$
2 heads	$\frac{6}{16} = \frac{3}{8}$
3 heads	$\frac{4}{16} = \frac{1}{4}$
4 heads	$\frac{1}{16}$
Total	1

Note that the column labeled "Events" contains all possible outcomes in the sample space. Also, all of the entries in the probability column are nonnegative numbers between 0 and 1. Furthermore, the sum of the probabilities is 1. These properties hold for every probability distribution.

Let an experiment have outcomes $s_1, s_2, \ldots, s_N$ with respective probabilities $p_1, p_2, p_3, \ldots, p_N$. Then the numbers $p_1, p_2, p_3, \ldots, p_N$ must satisfy two basic properties

Fundamental Property 1   Each of the numbers $p_1, p_2, \ldots, p_N$ is between 0 and 1.

Fundamental Property 2 $p_1 + p_2 + \cdots + p_N = 1$.

Roughly speaking, Fundamental Property 1 says that the likelihood of each outcome lies between 0% and 100%, whereas Fundamental Property 2 says that there is a 100% likelihood that one of the outcomes $s_1, s_2, \ldots, s_N$ will occur. The two fundamental properties may be easily verified for the probability distributions of Examples 1, 2, and 3.

EXAMPLE 3 (Continued)   Verify that the probabilities assigned to the outcomes of Example 3 satisfy Fundamental Properties 1 and 2.

Solution   The probabilities are .10, .15, .45, .25, and .05. Clearly, each is between 0 and 1, so Property 1 is satisfied. Adding the probabilities shows that their sum is 1, satisfying Property 2. ∎

Suppose that we are given an experiment with a finite number of outcomes. Let us now assign to each event E a probability, which we denote by $\Pr(E)$. If E consists of a single outcome, say $E = \{s\}$, then E is called an *elementary event*. In this case we associate to E the probability of the outcome s. If E consists of more than one outcome, we may compute $\Pr(E)$ via the *addition principle*.

Addition Principle   Suppose that an event E consists of the finite number of outcomes $s, t, u, \ldots, z$. That is,

$$E = \{s, t, u, \ldots, z\}.$$

Then

$$\Pr(E) = \Pr(s) + \Pr(t) + \Pr(u) + \cdots + \Pr(z).$$

We supplement the addition principle with the convention that the probability of the impossible event ∅ is 0. This is certainly reasonable, since the impossible event never occurs.

EXAMPLE 4   **Probability distribution for the number of boys in two-child families**   Observe two-child families. Describe the sample space for counting the number of boys, and assign probabilities to each outcome.

Solution   The sample space $S = \{\text{GG, GB, BG, BB}\}$ describes the sex and birth order in two-child families. If we assume that each of the four outcomes in S is equally likely to occur, we should assign probability $\frac{1}{4}$ to each outcome. Then

$$\Pr(\text{no boys}) = \Pr(\text{GG}) = \tfrac{1}{4}$$
$$\Pr(\text{one boy}) = \Pr(\text{GB}) + \Pr(\text{BG}) = \tfrac{1}{4} + \tfrac{1}{4} = \tfrac{1}{2}$$
$$\Pr(\text{two boys}) = \Pr(\text{BB}) = \tfrac{1}{4}.$$

We are making a reasonable assignment of probabilities here, although U.S. statistics show that 51% of all live births are boys and 49% are girls. If we wish to take many years of census data into account for a more accurate model, we would not assign equal probabilities to each outcome in S. ∎

EXAMPLE 5   **Probability associated with a die**   Suppose that we toss a die and observe the side that faces upward. What is the probability that an odd number will occur?

Solution The event "odd number occurs" corresponds to the subset of the sample space given by

$$E = \{1, 3, 5\}.$$

That is, the event occurs if a 1, 3, or 5 appears on the side that faces upward. By the addition principle,

$$\Pr(E) = \Pr(1) + \Pr(3) + \Pr(5).$$

As we observed in Example 2, each of the outcomes in the die-tossing experiment has probability $\frac{1}{6}$. Therefore,

$$\Pr(E) = \tfrac{1}{6} + \tfrac{1}{6} + \tfrac{1}{6} = \tfrac{1}{2}.$$

So we expect an odd number to occur approximately half of the time.

Now Try Exercise 5

■

EXAMPLE 6   **Probability associated with traffic volume**   Consider the traffic study of Example 3. What is the probability that at most 5000 cars will use the highway during rush hour?

Solution The event

"at most 5000 cars"

is the same as

$$\{s_1, s_2, s_3\},$$

where we use the same notation for the outcomes as we used in Example 3. Thus the probability of the event is

$$\Pr(s_1) + \Pr(s_2) + \Pr(s_3) = .10 + .15 + .45 = .70.$$

Therefore, we expect that traffic will involve ≤ 5000 cars in approximately 70% of the rush hours.

Now Try Exercise 23(a)

■

EXAMPLE 7   **Probability associated with the roll of a pair of dice**   Suppose that we toss a red die and a green die and observe the numbers on the sides that face upward.

(a) Calculate the probabilities of the elementary events.

(b) Calculate the probability that the two dice show the same number.

Solution (a) As shown in Example 3 of Section 6.2, the sample space consists of 36 pairs of numbers:

$$S = \{(1, 1), (1, 2), \ldots, (6, 5), (6, 6)\}.$$

Each of these pairs is equally likely to occur. (How could the dice show favoritism to a particular pair?) Therefore, each outcome is expected to occur about $\frac{1}{36}$ of the time, and the probability of each elementary event is $\frac{1}{36}$.

(b) The event

$$E = \text{"both dice show the same number"}$$

consists of six outcomes:

$$E = \{(1, 1), (2, 2), (3, 3), (4, 4), (5, 5), (6, 6)\}.$$

Thus, by the addition principle,

$$\Pr(E) = \tfrac{1}{36} + \tfrac{1}{36} + \tfrac{1}{36} + \tfrac{1}{36} + \tfrac{1}{36} + \tfrac{1}{36} = \tfrac{6}{36} = \tfrac{1}{6}.$$

Now Try Exercise 3

∎

EXAMPLE 8 **Probability associated with a small lottery** A person playing a certain lottery can win $100, $10, or $1, can break even, or can lose $10. These five outcomes with their corresponding probabilities are given by the probability distribution in Table 1.

(a) Which outcome has the greatest probability?

(b) Which outcome has the least probability?

(c) What is the probability that the person will win some money?

Solution (a) Table 1 reveals that the outcome -10 has the greatest probability, .50. (A person playing the lottery repeatedly can expect to lose $10 about 50% of the time.) This outcome is just as likely to occur as not.

TABLE 1

Winnings	Probability
100	.02
10	.05
1	.40
0	.03
-10	.50

(b) The outcome 100 has the least probability, .02. A person playing the lottery can expect to win $100 about 2% of the time. (This outcome is quite unlikely to occur.)

(c) We are asked to determine the probability that the event E occurs, where $E = \{100, 10, 1\}$. By the addition principle,

$$\Pr(E) = \Pr(100) + \Pr(10) + \Pr(1)$$
$$= \quad .02 \quad + \quad .05 \quad + \quad .40$$
$$= \quad .47.$$

∎

Here is a useful formula that relates $\Pr(E \cup F)$ to $\Pr(E \cap F)$:

Inclusion–Exclusion Principle Let E and F be any events. Then

$$\Pr(E \cup F) = \Pr(E) + \Pr(F) - \Pr(E \cap F).$$

In particular, if E and F are mutually exclusive, then

$$\Pr(E \cup F) = \Pr(E) + \Pr(F).$$

Note the similarity of this principle to the principle of the same name that was used in Section 5.2 to count the elements in a set.

EXAMPLE 9 **Probability associated with rolling a die** In tossing a fair die, we observe the uppermost face. What is the probability that the result is odd or greater than 4?

Solution The event "odd number occurs" corresponds to the set

$$E = \{1, 3, 5\}.$$

The event "number greater than 4 occurs" corresponds to the set

$$F = \{5, 6\}.$$

From Example 5 we see that $\Pr(E) = \frac{1}{2}$. We use the addition principle to find $\Pr(F)$:

$$\Pr(F) = \Pr(5) + \Pr(6) = \frac{1}{6} + \frac{1}{6} = \frac{1}{3}.$$

The event "odd number or number greater than 4 occurs" is $E \cup F$. By the inclusion–exclusion principle, and the fact that $E \cap F = \{5\}$,

$$\Pr(E \cup F) = \Pr(E) + \Pr(F) - \Pr(E \cap F)$$
$$= \frac{1}{2} + \frac{1}{3} - \frac{1}{6} = \frac{4}{6} = \frac{2}{3}.$$

Of course, we could have simply found $E \cup F = \{1, 3, 5, 6\}$ and used the addition principle to get the same answer. ∎

EXAMPLE 10 **Resource availability** A factory needs two raw materials. The probability of not having an adequate supply of material A is .05, whereas the probability of not having an adequate supply of material B is .03. A study determines that the probability of a shortage of both A and B is .01. What proportion of the time can the factory operate?

Solution Let E be the event "shortage of A" and F the event "shortage of B." We are given that

$$\Pr(E) = .05 \qquad \Pr(F) = .03 \qquad \Pr(E \cap F) = .01.$$

The factory can operate only if it has both raw materials. Therefore, we must calculate the proportion of the time in which there is no shortage of material A or material B. A shortage of A or B is the event $E \cup F$. By the inclusion–exclusion principle,

$$\begin{aligned}
\Pr(E \cup F) &= \Pr(E) + \Pr(F) - \Pr(E \cap F) \\
&= .05 \; + \; .03 \; - \; .01 \\
&= .07.
\end{aligned}$$

Thus the factory is likely to be short of one raw material or the other 7% of the time. Therefore, the factory can expect to operate 93% of the time.

Now Try Exercise 21

∎

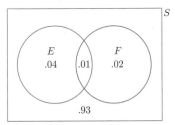

Figure 1

Probabilities involving unions and intersections of events are often conveniently displayed in a Venn diagram. Figure 1 displays the probabilities from Example 10.

Odds Frequently in applications we meet statements like these:

The odds of a Republican victory are 3 to 2.

The odds of a recession next year are 1 to 3.

Such statements may be readily translated into the language of probability. For example, consider the first statement. It means that if the election were repeated often (a theoretical possibility), then for every two Democratic wins there would be three Republican wins. That is, the Republicans would win $\frac{3}{5}$ (or 60%) of the elections. In terms of probability, this means that the probability of a Republican win is .6. In a similar way, we can translate the second statement into probabilistic terms. The statement says that if we consider a large number of years experiencing conditions identical to this year's, then for every one that is followed by a recession, three years are not. That is, the probability of having a recession next year is $\frac{1}{4}$.

We may generalize our reasoning to obtain the following result:

> If the odds in favor of an event are a to b, the probability of the event is $\dfrac{a}{a+b}$.

In the lottery of Example 8, the odds of winning \$100 are 1 to 49, since the probability of winning \$100 is $.02 = \frac{2}{100} = \frac{1}{50}$.

EXAMPLE 11 **Chance of rain** Suppose that the odds of rain tomorrow are 5 to 3. What is the probability that rain will occur?

Solution The probability that rain will occur is

$$\frac{5}{5+3} = \frac{5}{8}.$$

Now Try Exercise 11

There is a systematic way to find the odds in favor of the event E whose probability is known. Consider the fraction

$$\frac{\Pr(E)}{1 - \Pr(E)}$$

and write it in the form $\frac{a}{b}$, where a and b are integers having no common divisor. Then the odds in favor of E are a to b.

EXAMPLE 12 **The game of Clue** In the game of *Clue*, there are 6 equally likely suspects (Colonel Mustard, Mrs. Peacock, Professor Plum, Mr. Green, Miss Scarlet, Mrs. White), one of whom committed a murder that players are trying to solve. What are the odds that Colonel Mustard killed the victim?

Solution Let E be the event that Colonel Mustard committed the crime. $\Pr(E)$ is $\frac{1}{6}$.

$$\frac{\Pr(E)}{1 - \Pr(E)} = \frac{\frac{1}{6}}{\frac{5}{6}} = \frac{1}{5}.$$

Therefore, the odds that Colonel Mustard committed the crime are 1 to 5.

Now Try Exercise 13

▶ *Note* We usually state odds with the larger number first, so in Example 12 we would more naturally claim that the odds are 5 to 1 that Colonel Mustard did *not* kill the victim. ◀

PRACTICE PROBLEMS 6.3

1. (*T-Maze*) A mouse is put into a T-maze (a maze shaped like a "T") (Fig. 2). If he turns to the left he receives cheese, and if he turns to the right he receives a mild shock. This trial is done twice with the same mouse and the directions of the turns recorded.

   (a) What is the sample space for this experiment?

   (b) Why would it not be reasonable to assign each outcome the same probability?

2. (a) What are the odds in favor of an event that is just as likely to occur as not?

   (b) Is there any difference between the odds 6 to 4 and the odds 3 to 2?

3. Suppose that E and F are any events. Show that

   $$\Pr(E) = \Pr(E \cap F) + \Pr(E \cap F').$$

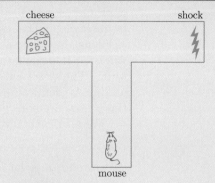

cheese   shock

mouse

Figure 2

Exercises 6.3

1. (*Word Frequencies*) There are 774,746 words in the Bible. The word "and" occurs 46,277 times and the word "Lord" occurs 1855 times.[2] Suppose that a word is selected at random from the Bible.

   (a) What is the probability that the word is "and"?

   (b) What is the probability that the word is "and" or "Lord"?

   (c) What is the probability that the word is neither "and" nor "Lord"?

2. (*Heads and Tails*) An experiment consists of tossing a coin two times and observing the sequence of heads and tails. Each of the four outcomes has the same probability of occurring.

   (a) What is the probability that "HH" is the outcome?

   (b) What is the probability of the event "at least one head"?

3. (*Dice Toss*) Suppose that a red die and a green die are tossed and the numbers on the sides that face upward are observed. (See Example 7.)

   (a) What is the probability that the numbers add up to 8?

   (b) What is the probability that the sum of the numbers is less than 5?

4. (*U.S. States*) A state is selected at random from the 50 states of the United States. What is the probability that it is one of the six New England states?

5. (*Roulette*) The modern American roulette wheel has 38 slots, which are labeled with 36 numbers evenly divided between red and black, plus two green numbers 0 and 00. What is the probability that the ball lands on a green number?

6. (*Selecting a Number*) An experiment consists of selecting a number at random from the set of numbers $\{1, 2, 3, 4, 5, 6, 7, 8, 9\}$. Find the probability that the number selected is

   (a) Less than 4. (b) Odd.

   (c) Less than 4 or odd.

7. (*Horse Race*) Three horses, call them A, B, and C, are going to race against each other. The probability that A will win is $\frac{1}{3}$ and the probability that B will win is $\frac{1}{2}$.

   (a) What is the probability that C will win? (Assume that there are no ties.)

   (b) What are the odds that C will win?

8. Which of the following probabilities are feasible for an experiment having sample space $\{s_1, s_2, s_3\}$?

   (a) $\Pr(s_1) = .4$, $\Pr(s_2) = .4$, $\Pr(s_3) = .4$

   (b) $\Pr(s_1) = .5$, $\Pr(s_2) = .7$, $\Pr(s_3) = -.2$

   (c) $\Pr(s_1) = 2$, $\Pr(s_2) = 1$, $\Pr(s_3) = \frac{1}{2}$

   (d) $\Pr(s_1) = \frac{1}{4}$, $\Pr(s_2) = \frac{1}{2}$, $\Pr(s_3) = \frac{1}{4}$

9. An experiment with outcomes s_1, s_2, s_3, s_4 is described by the following probability table:

Outcome	Probability
s_1	.1
s_2	.6
s_3	.2
s_4	.1

   (a) What is $\Pr(\{s_1, s_2\})$?

   (b) What is $\Pr(\{s_2, s_4\})$?

10. An experiment with outcomes $s_1, s_2, s_3, s_4, s_5, s_6$ is described by the following probability table:

Outcome	Probability
s_1	.05
s_2	.25
s_3	.05
s_4	.01
s_5	.63
s_6	.01

    Let $E = \{s_1, s_2\}$ and $F = \{s_3, s_5, s_6\}$.

    (a) Determine $\Pr(E)$ and $\Pr(F)$.

    (b) Determine $\Pr(E')$.

    (c) Determine $\Pr(E \cap F)$.

    (d) Determine $\Pr(E \cup F)$.

11. Convert the following odds to probabilities.

    (a) 10 to 1   (b) 1 to 2   (c) 4 to 5

12. (*Poker*) In poker, the probability of being dealt a hand containing a pair of jacks or better is about $\frac{1}{6}$. What are the corresponding odds?

13. (*Candy Bars*) Nine percent of all candy bars sold in the United States are Snickers™ bars. What are the odds that a randomly selected candy bar purchased is a Snickers™ bar?

14. (*Birth States*) The odds of Americans living in the state where they were born is 16 to 9. What is the probability that an American selected at random lives in his or her birth state?

15. (*Horse Race*) If the odds for Secretariat to win a horse race are 11:7, what is the probability that Secretariat wins? Loses?

---

[2] According to *The People's Almanac* by Wallechinsky and Wallace (New York: Doubleday, 1975).

16. (*Class Election*) Four people are running for class president, Liz, Sam, Sue, and Tom. The probabilities of Sam, Sue, and Tom winning are .18, .23, and .31, respectively.

    (a) What is the probability of Liz winning?

    (b) What is the probability that a boy wins?

    (c) What is the probability that Tom loses?

    (d) What are the odds that Sue loses?

    (e) What are the odds that a girl wins?

    (f) What are the odds that Sam wins?

17. Let E and F be events for which $\Pr(E) = .6$, $\Pr(F) = .5$, and $\Pr(E \cap F) = .4$. Find

    (a) $\Pr(E \cup F)$

    (b) $\Pr(E \cap F')$ (*Hint*: Make a Venn diagram similar to Fig. 1.)

18. Let E and F be events for which $\Pr(E) = .4$, $\Pr(F) = .5$, and $\Pr(E \cap F') = .3$. Find

    (a) $\Pr(E \cap F)$ (b) $\Pr(E \cup F)$

19. Suppose $\Pr(E) = .4$ and $\Pr(F) = .5$, where E and F are mutually exclusive. Find $\Pr(E \cup F)$.

20. (*Pair of Dice*) Suppose a pair of dice is rolled. The probability is $\frac{1}{6}$ that the sum of the numbers on the uppermost faces is 7, and the probability is $\frac{1}{18}$ that the sum is 11. Find the probability that the sum is 7 or 11.

21. (*Grades*) Joe feels that the probability of getting an A in history is .7, the probability of getting an A in psychology is .8, and the probability of getting an A in history or psychology is .9. What is the probability that he will get an A in both subjects?

22. Suppose $\Pr(E) = .5$, $\Pr(F) = .6$, and $\Pr(E \cup F) = .9$. Find $\Pr(E \cap F)$.

23. (*Supermarket Queue*) A statistical analysis of the wait (in minutes) at the checkout line of a certain supermarket yields the following probability distribution:

Wait (in minutes)	Probability
At most 3	.10
More than 3 and at most 5	.20
More than 5 and at most 10	.25
More than 10 and at most 15	.25
More than 15	.20

    (a) What is the probability of waiting more than 3 minutes but at most 15?

    (b) If you observed the waiting times of a representative sample of 10,000 supermarket customers, approximately how many would you expect to wait for more than 3 minutes but at most 15?

24. What is the probability of the certain event?

25. (*College Applications*) The following table was derived from a survey of college freshmen attending baccalaureate colleges and universities.[3] Each probability is the likelihood that a randomly selected freshman applied to the specified number of colleges. For instance, 19% of the freshmen applied to just one college and therefore the probability that a student selected at random applied to just one college is .19. Convert these data into a probability distribution.

Number of Colleges Applied to	Probability
1	.19
2 or less	.32
3 or less	.47
4 or less	.64
20 or less	1

26. (*College Applications*) Refer to the data of Exercise 25. What is the probability that a student applied to three or more colleges?

27. (*Employees' Ages*) In a study of the ages of its employees, over a period of several years a university finds the following:

Age (years)	Probability
20–34	.15
20–49	.70
20–64	.90
20–79	1.00

    (a) Find the probability associated with each of the events: 20–34 years, 35–49 years, 50–64 years, and 65–79 years.

    (b) Find the probability that an employee selected at random is at least 50 years old.

28. (*House Sales*) A realtor analyzes the office's sales over the past two years. She sorts the sales by the age of the house at the time of sale.

Age	Number of Houses Sold
1–2	1200
3–4	1570
5–6	1600
7–8	1520
9–10	1480

    (a) Determine the probability that a sale chosen at random is a house between five and six years old.

---

[3] *The American Freshman*: *National Norms for Fall 2004*, Los Angeles, Calif.: American Council on Education, 2005.

(b) What are the odds that the house is less than seven years old?

29. (*Computer Usage*) Computer usage in a community is segmented by the kind of use:

Kind of use	Frequency
School only	20%
Work only	22%
Home computer only	20%
No computer anywhere	17%

(a) Explain why the percentages listed above do not add up to 100%.

(b) What is the probability that someone in the community uses a computer?

30. (*Odds of an Earthquake*) The probability that there will be a major earthquake in the San Francisco area during the next 30 years is .7. What are the corresponding odds?

31. (*Bookies*) Gamblers usually give odds *against* an event happening. For instance, if a bookie gives the odds 4 to 1 that the Yankees win the next World Series, he is stating that the probability that the Yankees will win are $\frac{1}{5}$ or .2. Also, if a bettor bets $1 that the Yankees will win and the Yankees do win, then the bettor will receive $5 (his original bet + a profit of $4). The following are typical odds set by a bookie for the eventual winner in a fictitious four-team league: Sparks (5 to 3), Meteors (3 to 1), Asteroids (3 to 2), Suns (4 to 1).

(a) Convert the odds to probabilities of winning for each team.

(b) Add the four probabilities.

(c) Explain why the answer to (b) makes sense.

SOLUTIONS TO PRACTICE PROBLEMS 6.3

1. (a) {LL, LR, RR, RL}. Here LL means that the mouse turned left both times, LR means that the mouse turned left the first time and right the second, and so on.

(b) The mouse will learn something from the first trial. If he turned left the first time and got rewarded, then he is more likely to turn left again on the second trial. Hence LL should have a greater probability than LR. Similarly, RL should be more likely than RR.

2. (a) 1 to 1. An event that is just as likely to occur as not has probability $\frac{1}{2}$. So if the odds are a to b,

then we may set $a = 1$, $a + b = 2$. Thus $a = 1$, $b = 1$. (The odds could also be given as 2 to 2, 3 to 3, etc.)

(b) No. Odds of 6 to 4 correspond to a probability of $6/(6 + 4) = \frac{6}{10} = \frac{3}{5}$. Odds of 3 to 2 correspond to a probability of $3/(3 + 2) = \frac{3}{5}$. (There are always many different ways to express the same odds.)

3. The sets $(E \cap F)$ and $(E \cap F')$ have no elements in common and so are mutually exclusive. Since $E = (E \cap F) \cup (E \cap F')$, the result follows from the inclusion–exclusion principle.

6.4 Calculating Probabilities of Events

As mentioned in Section 6.3, there are several ways to assign probabilities to the events of a sample space. One way is to perform the experiment many times and assign probabilities based on empirical data. Sometimes our intuition about situations suffices, as in simple coin-tossing experiments. But we frequently are faced with forming a model of an experiment to assign probabilities consistent with Fundamental Properties 1 and 2. This section shows how the counting techniques of Chapter 5 may be used to extend these ideas to more complex situations. In addition, the exceptionally wide range of applications that make use of probability theory is illustrated.

Experiments with Equally Likely Outcomes In the experiments associated to many common applications, all outcomes are equally likely—that is, they all have the same probability. This is the case, for example, if we toss an unbiased coin or select a person at random from a population. If a sample space has N equally likely outcomes, then the probability of each outcome is $1/N$ (since the probabilities must add up to 1). If we use this fact, the probability of any event is easy to compute. Namely, suppose that E is an event consisting of M outcomes. Then, by the addition principle,

$$\text{Pr}(E) = \underbrace{\frac{1}{N} + \frac{1}{N} + \cdots + \frac{1}{N}}_{M \text{ times}} = \frac{M}{N}.$$

We can restate this fundamental result as follows:

Let S be a sample space consisting of N equally likely outcomes. Let E be any event. Then

$$\text{Pr}(E) = \frac{[\text{number of outcomes in } E]}{N}. \tag{1}$$

In order to apply formula (1) in particular examples, it is necessary to compute N, the number of equally likely outcomes in the sample space, and [number of outcomes in E]. Often these quantities can be determined using the counting techniques of Chapter 5. Some illustrative computations are provided in Examples 1 through 7 here.

We should mention that, although the urn and dice problems considered in this section and the next might seem artificial and removed from applications, many applied problems can be described in mathematical terms as urn or dice-tossing experiments. We begin our discussion with two examples involving abstract urn problems. Then in two more examples we show the utility of urn models by applying them to quality control and medical screening problems (Examples 3 and 5, respectively).

EXAMPLE 1 **Selecting balls from an urn** An urn contains eight white balls and two green balls. A sample of three balls is selected at random. What is the probability of selecting only white balls?

Solution The experiment consists of selecting 3 balls from the 10. Since the order in which the 3 balls are selected is immaterial, the samples are combinations of 10 balls taken 3 at a time. The total number of samples is therefore $\binom{10}{3}$, and this is N, the number of elements in the sample space. Since the selection of the sample is random, all samples are equally likely, and thus we can use formula (1) to compute the probability of any event. The problem asks us to compute the probability of the event $E =$ "all three balls selected are white." Since there are 8 white balls, the number of different samples in which all are white is $\binom{8}{3}$. Thus

$$\text{Pr}(E) = \frac{[\text{number of outcomes in } E]}{N} = \frac{\binom{8}{3}}{\binom{10}{3}} = \frac{56}{120} = \frac{7}{15}. \qquad \blacksquare$$

EXAMPLE 2 **Selecting balls from an urn** An urn contains eight white balls and two green balls. A sample of three balls is selected at random. What is the probability that the sample contains at least one green ball?

Solution As in Example 1, there are $N = \binom{10}{3}$ equally likely outcomes. Let F be the event "at least one green ball is selected." Let us determine the number of different outcomes in F. These outcomes contain either one or two green balls. There are $\binom{2}{1}$ ways to select one green ball from two; and for each of these, there are $\binom{8}{2}$ ways to select two white balls from eight. By the multiplication principle, the number of samples containing one green ball equals $\binom{2}{1}\binom{8}{2}$. Similarly, the number of samples containing two green balls equals $\binom{2}{2}\binom{8}{1}$. Note that although the sample size is 3, there are only two green balls in the urn. Since the sampling is done without replacement, no sample can have more than two greens. Therefore, the number of outcomes in F—namely the number of samples having at least one green ball—equals

$$\binom{2}{1}\binom{8}{2} + \binom{2}{2}\binom{8}{1} = 2 \cdot 28 + 1 \cdot 8 = 64,$$

so

$$\Pr(F) = \frac{[\text{number of outcomes in } F]}{N} = \frac{64}{\binom{10}{3}} = \frac{64}{120} = \frac{8}{15}.$$

Now Try Exercise 9   ■

EXAMPLE 3   **Quality control**   A toy manufacturer inspects boxes of toys before shipment. Each box contains 10 toys. The inspection procedure consists of randomly selecting three toys from the box. If any are defective, the box is not shipped. Suppose that a given box has two defective toys. What is the probability that it will be shipped?

Solution   This problem is not really new! We solved it in disguise as Example 1. The urn can be regarded as a box of toys, and the balls as individual toys. The white balls are nondefective toys and the green balls defective toys. The random selection of three balls from the urn is just the inspection procedure. And the event "all three balls selected are white" corresponds to the box being shipped. As we calculated previously, the probability of this event is $\frac{7}{15}$. (Since $\frac{7}{15} \approx .47$, there is approximately a 47% chance of shipping a box with two defective toys. This inspection procedure is not particularly effective!)

Now Try Exercise 7   ■

EXAMPLE 4   **Selecting students**   A professor is randomly choosing a group of three students to do an oral presentation. In her class of 10 students, 2 are on the debate team. What is the chance that the professor chooses at least one of the debaters for the group?

Solution   This is the same as Example 2. Just think of the debate team members as the green balls and the others in the class as white balls. The chance of getting at least one debate team member in the group is $\frac{8}{15} \approx .53$.

Now Try Exercise 29   ■

EXAMPLE 5   **Medical screening**   Suppose that a cruise ship returns to the United States from the Far East. Unknown to anyone, 4 of its 600 passengers have contracted a rare disease. Suppose that the Public Health Service screens 20 passengers, selected at random, to see whether the disease is present aboard ship. What is the probability that the presence of the disease will escape detection?

Solution The sample space consists of samples of 20 drawn from among the 600 passengers. There are $\binom{600}{20}$ such samples. The number of samples containing none of the sick passengers is $\binom{596}{20}$. Therefore, the probability of not detecting the disease is

$$\frac{\binom{596}{20}}{\binom{600}{20}} = \frac{\frac{596!}{20!\,576!}}{\frac{600!}{20!\,580!}} = \frac{596!}{600!} \cdot \frac{580!}{576!}$$

$$= \frac{596!}{600 \cdot 599 \cdot 598 \cdot 597 \cdot 596!} \cdot \frac{580 \cdot 579 \cdot 578 \cdot 577 \cdot 576!}{576!}$$

$$= \frac{580 \cdot 579 \cdot 578 \cdot 577}{600 \cdot 599 \cdot 598 \cdot 597} \approx .87.$$

So there is approximately an 87% chance that the disease will escape detection.

Now Try Exercise 13   ∎

The Complement Rule The *complement rule* relates the probability of an event E to the probability of its complement E'. When applied together with counting techniques, it often simplifies computation of probabilities.

> **Complement Rule**   Let E be any event, E' its complement. Then
> $$\Pr(E) = 1 - \Pr(E').$$

For example, recall Example 1. We determined the probability of the event

$$E = \text{``all three balls selected are white''}$$

associated to the experiment of selecting three balls from an urn containing eight white balls and two green balls. We found that $\Pr(E) = \frac{7}{15}$. On the other hand, in Example 2 we determined the probability of the event

$$F = \text{``at least one green ball is selected.''}$$

The event E is the complement of F:

$$E = F'.$$

So, by the complement rule,

$$\Pr(F) = 1 - \Pr(F') = 1 - \Pr(E) = 1 - \tfrac{7}{15} = \tfrac{8}{15},$$

in agreement with the calculations of Example 2.

The complement rule is especially useful in situations where $\Pr(E')$ is easier to compute than $\Pr(E)$. One of these situations arises in the celebrated *birthday problem*.

EXAMPLE 6 **The famous "birthday problem"**   A group of five people is to be selected at random. What is the probability that two or more of them have the same birthday?

Solution For simplicity we ignore February 29. Furthermore, we assume that each of the 365 days in a year is an equally likely birthday (not an unreasonable assumption). The experiment we have in mind is this. Pick out five people and observe their birthdays. The outcomes of this experiment are strings of five dates, corresponding to the birthdays. For example, one outcome of the experiment is

(June 2, April 6, Dec. 20, Feb. 12, Aug. 5).

Each birth date has 365 different possibilities. So, by the generalized multiplication principle, the total number N of possible outcomes of the experiment is

$$N = 365 \cdot 365 \cdot 365 \cdot 365 \cdot 365 = 365^5.$$

Let E be the event "at least two people have the same birthday." It is very difficult to calculate directly the number of outcomes in E. However, it is comparatively simple to compute the number of outcomes in E' and hence to compute $\Pr(E')$. This is because E' is the event "all five birthdays are different." An outcome in E' can be selected in a sequence of five steps:

Select a day	Select a different day	Select yet a different day	Select yet a different day	Select yet a different day

These five steps will result in a sequence of five different birthdays. The first step can be performed in 365 ways; for each of these, the next step in 364; for each of these, the next step in 363; for each of these, the next step in 362; and for each of these, the last step in 361 ways. Therefore, E' contains $365 \cdot 364 \cdot 363 \cdot 362 \cdot 361$ [or $P(365, 5)$] outcomes, and

$$\Pr(E') = \frac{365 \cdot 364 \cdot 363 \cdot 362 \cdot 361}{365^5} \approx .973.$$

By the complement rule,

$$\Pr(E) = 1 - \Pr(E') \approx 1 - .973 = .027.$$

So the likelihood is about 2.7% that two or more of the five people will have the same birthday.

Now Try Exercise 15   ■

The experiment of Example 6 can be repeated using samples of 8, 10, 20, or any number of people. As before, let E be the event "at least two people have the same birthday," so that $E' = $ "all the birthdays are different." If a sample of r people is used, then the same reasoning as used previously yields

$$\Pr(E') = \frac{365 \cdot 364 \cdot \cdots \cdot (365 - r + 1)}{365^r}.$$

Table 1 gives the values of $\Pr(E) = 1 - \Pr(E')$ for various values of r. You may be surprised by the numbers in the table. Even with as few as 23 people it is more likely than not that at least two people have the same birthday. With a sample of 50 people we are almost certain to have two with the same birthday. (Try this experiment in your dormitory or class.)

EXAMPLE 7 **Rolling a die five times** A die is tossed five times. What is the probability of obtaining exactly three 4's?

TABLE 1	Probability that, in a randomly selected group of r people, at least two will have the same birthday									
r	5	10	15	20	22	23	25	30	40	50
$\Pr(E)$	.027	.117	.253	.411	.476	.507	.569	.706	.891	.970

Solution There are $6^5 = 7776$ possible outcomes when a die is tossed five times. These are equally likely to occur. The event we seek is $E =$ "three 4's." How many outcomes are in E? There are $1 \cdot 1 \cdot 1 \cdot 5 \cdot 5 = 25$ ways to get three 4's followed by two tosses yielding anything except 4. But the three 4's need not appear in the first three tosses. What if the 4's appear as 4xx44? The number of possibilities is $1 \cdot 5 \cdot 5 \cdot 1 \cdot 1 = 25$. Thus we need to determine in how many positions these three 4's can appear in the sequence of five tosses. That is just $C(5, 3) = 10$. Therefore, E contains $(10)(25)$ elements, and

$$\Pr(\text{exactly three 4's in five tosses of a die}) = \frac{(10)(25)}{7776} \approx .03.$$

Thus there is about a 3% chance that in five tosses of a fair die we obtain three 4's.

Now Try Exercise 31 ■

Verification of the Complement Rule If S is the sample space, then $\Pr(S) = 1$, $E \cup E' = S$, and $E \cap E' = \emptyset$. Therefore, by the inclusion–exclusion principle,

$$\Pr(S) = \Pr(E \cup E') = \Pr(E) + \Pr(E').$$

So we have

$$1 = \Pr(E) + \Pr(E') \quad \text{and} \quad \Pr(E) = 1 - \Pr(E'). \qquad ■$$

PRACTICE PROBLEMS 6.4

1. (*Children*) A couple decides to have four children. What is the probability that among the children there will be at least one boy and at least one girl?

2. (a) Find the probability that all the numbers are different in three spins of a roulette wheel. [*Note:* A roulette wheel has 38 numbers.]

   (b) Guess how many spins are required in order that the probability that all the numbers are different will be less than .5.

EXERCISES 6.4

1. (*Course Selection*) Each of three people randomly chooses one of three calculus sections to take (A, B, or C).

   (a) What is the probability that they all choose the same one?

   (b) What is the probability that they each choose a different section?

2. (*Track Positions*) Michael and Christopher are among seven contestants in a race to be run on a seven-lane track. If the runners are assigned to the lanes at random, what is the probability that Michael will be assigned to the inside lane and Christopher will be assigned to the outside lane?

3. Suppose you are asked to choose a whole number between 1 and 13, inclusive.

   (a) What is the probability that it is odd?

   (b) What is the probability that it is even?

   (c) What is the probability that it is a multiple of 3?

(d) What is the probability that it is odd or a multiple of 3?

4. Five numbers are chosen at random from the whole numbers between 1 and 13, inclusive, with replacement.

   (a) What is the probability that all the numbers are even?

   (b) What is the probability that all the numbers are odd?

   (c) What is the probability that at least one of the numbers is odd?

5. Five numbers are chosen at random from the whole numbers between 1 and 13, inclusive, without replacement.

   (a) What is the probability that all the numbers are even?

   (b) What is the probability that all the numbers are odd?

   (c) What is the probability that at least one of the numbers is odd?

6. (*Balls in an Urn*) An urn contains 40 balls, some red and some white. If the probability of selecting a red ball is .45, how many red balls are in the urn?

7. (*Quality Control*) A factory produces fuses, which are packaged in boxes of 10. Three fuses are selected at random from each box for inspection. The box is rejected if at least one of these three fuses is defective. What is the probability that a box containing five defective fuses will be rejected?

8. (*Balls in an Urn*) An urn contains six white balls and five red balls. A sample of four balls is selected at random from the urn. What is the probability that the sample contains two white balls and two red balls?

9. (*Balls in an Urn*) An urn contains five red balls and four white balls. A sample of two balls is selected at random from the urn. What is the probability that at least one of the balls is red?

10. (*Opinion Polling*) Of the nine members of the board of trustees of a college, five agree with the president on a certain issue. The president selects three trustees at random and asks for their opinions. What is the probability that at least two of them will agree with him?

(Selecting Students) Exercises 11–14 refer to a classroom of children (12 boys and 10 girls) in which seven students are chosen to go to the blackboard.

11. What is the probability that at least two girls are chosen?

12. What is the probability that more boys than girls are chosen?

13. What is the probability that no boys are chosen?

14. What is the probability that the first three children chosen are boys?

15. (*Date Conflict*) Without consultation, each of four organizations announces a one-day convention to be held during June. Find the probability that at least two organizations specify the same day for their convention.

16. Five letters are selected from the alphabet, one at a time with replacement, to form a five-letter "word." What is the probability that the "word" has five different letters?

17. (*Track Position*) Michael is one of seven contestants entered in two races to be run on a seven-lane track. If in each race the runners are assigned to the lanes at random, what is the probability that Michael will be assigned to the inside lane at least once?

18. (*Hotel Choices*) An airport limousine has four passengers and stops at six different hotels. What is the probability that two or more people will be staying at the same hotel? (Assume that each person is just as likely to stay in one hotel as another.)

19. (*Weather Conditions*) In a certain agricultural region the probability of a drought during the growing season is .2, the probability of a severe cold spell is .15, and the probability of both is .1. Find the probability of

   (a) either a drought or a severe cold spell.

   (b) neither a drought nor a severe cold spell.

   (c) not having a drought.

20. (*Coin Tosses*) A coin is to be tossed seven times. What is the probability of obtaining four heads and three tails?

21. Let E and F be events such that

$$\Pr(E) = .3 \quad \Pr(F') = .6 \quad \text{and} \quad \Pr(E \cup F) = .7.$$

What is $\Pr(E \cap F)$?

22. (*Manufacturing Defects*) In a certain manufacturing process the probability of a type I defect is .12, the probability of a type II defect is .22, and the probability of having both types of defects is .02. Find the probability of having neither type of defect.

23. (*Socks*) A man has six different pairs of socks, from which he selects two socks at random. What is the probability that the selected socks will match?

24. (*Opinion Polling*) Of the 15 members on a Senate committee, 10 plan to vote "yes" and 5 plan to vote "no" on an important issue. A reporter attempts to predict the outcome of the vote by questioning six of the senators. Find the probability that this sample is precisely representative of the final vote. That is, find the probability that four of the six senators questioned plan to vote "yes."

25. (*Group Picture*) A man, a woman, and their three children randomly stand in a row for a family picture. What is the probability that the parents will be standing next to each other?

26. (*Poker Hands*) In poker the probabilities of being dealt a flush, a straight, and a straight flush are .0019654, .0039246, and .0000154, respectively. What is the probability of being dealt a straight or a flush?

27. (*Street Routes*) Figure 1 shows a partial map of the streets in a certain city. A tourist starts at point A and selects at random a path to point B. (We shall assume that he walks only south and east.) Find the probability that

(**a**) he passes through point C.

(**b**) he passes through point D.

(**c**) he passes through point C and point D.

(**d**) he passes through point C or point D.

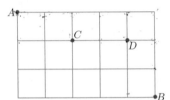

Figure 1

28. (*Boys and Girls*) A couple decides to have four children. What is the probability that they will have more girls than boys?

29. (*Committee Selection*) A law firm has six senior and four junior partners. A committee of three partners is selected at random to represent the firm at a conference. What is the probability that at least one of the junior partners is on the committee?

30. (*Rotten Tomato*) A bag contains nine tomatoes, of which one is rotten. A sample of three tomatoes is selected at random. What is the probability that the sample contains the rotten tomato?

31. (*Coin Tosses*) A coin is to be tossed six times. What is the probability of obtaining exactly three heads?

32. (*Book Selection*) A vacationer has brought along four novels and four nonfiction books. One day the person selects two at random to take to the beach. What is the probability that both are novels?

(*Illinois Lotto*) *Exercises 33 and 34*[1] *refer to the Illinois Lottery Lotto game. In this game, the player chooses six different integers from 1 to 40. If the six match (in any order) the six different integers drawn by the lottery, the player wins the grand prize jackpot, which starts at $1 million and grows weekly until won. Multiple winners split the pot equally. For each $1 bet, the player must pick two (presumably different) sets of six integers.*

33. What is the probability of winning the Illinois Lottery Lotto with a $1 bet?

34. In the game week ending June 18, 1983, 2 million people bought $1 tickets and 78 people matched all six winning integers and split the jackpot. If all numbers were selected randomly, the likelihood of having so many joint winners would be about 10^{-115}. Can you think of any reason that such an unlikely event occurred? (*Note:* The winning numbers were 7, 13, 14, 21, 28, and 35.) What would be the best strategy in selecting the numbers to ensure that in the event you won, you would probably not have to share the jackpot with too many people?

35. (*Senate Committee*) The U.S. Senate consists of two senators from each of the 50 states. Five senators are to be selected at random to form a committee. What is the probability that no two members of the committee are from the same state?

36. (*Subcommittees*) A politician knows that a committee vote is stacked against her, 6 to 3. However, she has the option of letting a randomly selected subcommittee decide the issue. Show that the smaller the subcommittee is, the better her chances of winning the vote. (Consider only subcommittees with an odd number of members, so that tie votes are precluded.)

37. (*Birthdays*) What is the probability that in a group of 25 people at least one person has a birthday on June 13? Why is your answer different from the probability displayed in Table 1 for $r = 25$?

38. (*Letter Positions*) What is the probability that a random arrangement of the letters in the word GEESE has all the E's adjacent to one another?

39. (*Guest List*) Fred is having a dinner party and is limited to 10 guests. He has 16 men friends and 12 women friends, including Laura and Mary.

(**a**) If Fred chooses his guests at random, what is the probability that Mary and Laura are invited?

(**b**) If Fred decides to invite 5 men and 5 women, what is the probability that Mary and Laura are invited?

40. (*Baseball Predictions*) In the American League, the East, Central, and West divisions consist of 5, 5, and 4 baseball teams, respectively. A sportswriter predicts the winner of each of the three divisions by choosing a team completely at random in each division. What is the probability that the sportswriter will predict at least one winner correctly?

41. (*Baseball Predictions*) Suppose that the sportswriter in Exercise 40 eliminates from each division one team that clearly has no chance of winning and predicts a winner at random from the remaining teams. Assuming the eliminated teams don't end up surprising anyone, what is the writer's chance of predicting at least one winner?

---

[1]The data for these exercises were taken from Allan J. Gottlieb's "Puzzle Corner" in *Technology Review*, February/March 1985.

42. (*Baseball Predictions*) Suppose that the sportswriter in Exercise 40 simply puts the 14 team names in a hat and draws 3 completely at random. Does this increase or decrease the writer's chance of picking at least one winner?

43. (*Powerball Lottery*) The winner of the Powerball[2] lottery must correctly pick a set of 5 numbers from 1 through 49 and then correctly pick one number (called the Powerball) from 1 through 42.

    (a) What are the odds of winning the Powerball lottery?

    (b) What is the probability of winning the Powerball lottery?

44. (*License Plate Game*) Johnny and Doyle are driving along on a lightly traveled road. Johnny proposes the following game. They will look at the license plates of oncoming cars and focus on the last two digits. For instance, the license plates ABC512 or 7412BG would yield 12, and the license plate XY406T would yield the number 6. Johnny bets Doyle that at least two of the next fifteen cars will yield the same number. What is the probability that Johnny wins? Assume that each of the 100 possible numbers is equally likely to occur.

45. (*Numbers*) Two distinct numbers are selected at random from the set $\{1, 2, 3, 4, 5, 6, 7\}$. What is the probability that their product is an even number? *Hint*: First find the probability that the product is odd.

46. (*Matching Socks*) Ten white socks and six red socks are in a drawer. If two socks are selected at random, what is the probability that both socks are the same color?

47. (*British Lottery*) In the British lottery, a player pays 1 euro for a ticket and selects 6 numbers from the numbers 1 through 49. If he matches exactly three of the six numbers drawn, he receives 10 euros. What is the probability of selecting exactly three of the six numbers drawn?

48. (*Swiss Lottery*) The winning combination in the Swiss lottery consists of six numbers drawn from the numbers 1 through 45. What is the probability that a single ticket has no matches?

49. (*Place Settings*) Fred has five place settings consisting of a dinner plate, a salad plate, and a bowl. Each

setting is a different color. If Fred randomly selects a dinner plate, a salad plate, and a bowl, what is the probability that they all have different colors.

50. (*Balls in an Urn*) An urn contains red balls and white balls. The probability of removing two red balls, without replacement, from the urn is $\frac{2}{5}$. The probability of removing three red balls, without replacement, from the urn is $\frac{1}{5}$. Explain why there must be two more red balls than white balls. How many red balls were in the urn? *Hint*: Let r be the number of red balls in the urn, and let w be the number of white balls in the urn. Show that $r = w + 2$.

In Exercises 51–55, use a graphing calculator, a spreadsheet, or mathematical software to calculate the probabilities.

51. (*Balls in an Urn*) Two balls are to be selected at random without replacement from an urn that contains 40 balls, some red and some white. If the probability of selecting two red balls is .1, how many red balls are in the urn?

52. (*Pick a Card*) Find the probability that at least two people in a group of size $n = 5$ select the same card when drawing from a 52-card deck with replacement. Determine for what size group the probability of such a match first exceeds .5.

53. (*Rolling a Die*) A die is tossed 24 times.

    (a) How many 3's would you expect?

    (b) Calculate the probability of getting exactly four 3's.

54. (*Presidential Choices*) There were 16 presidents of the Continental Congresses from 1774 to 1788. Each of the five students in a seminar in American history chooses one of these on which to do a report. If all presidents are equally likely to be chosen, calculate the probability that at least two students choose the same president.

55. (*Term Papers*) A political science class has 20 students, each of whom chooses a topic from a list for a term paper. How big a pool of topics is necessary for the probability of at least one duplicate to drop below 50%?

---

[2]Powerball drawings are held twice a week. If you buy 50 tickets for each drawing, you can expect to win once every 5000 years.

SOLUTIONS TO PRACTICE PROBLEMS 6.4

1. Each possible outcome is a string of four letters composed of B's and G's. By the generalized multiplication principle, there are 2^4 or 16 possible outcomes. Let E be the event "children of both sexes." Then $E' = \{BBBB, GGGG\}$, and

$$Pr(E') = \frac{[\text{number of outcomes in } E']}{[\text{total number of outcomes}]} = \frac{2}{16} = \frac{1}{8}.$$

Therefore,

$$Pr(E) = 1 - Pr(E') = 1 - \frac{1}{8} = \frac{7}{8}.$$

So the probability is 87.5% that they will have children of both sexes.

2. (a) Each sequence of three numbers is just as likely to occur as any other. Therefore,

Pr(numbers different)

$$= \frac{[\text{number of outcomes with numbers different}]}{[\text{number of possible outcomes}]}$$

$$= \frac{38 \cdot 37 \cdot 36}{38^3} \approx .92.$$

(b) 8

6.5 Conditional Probability and Independence

The probability of an event depends, often in a critical way, on the sample space in question. In this section we explore this dependence in some detail by introducing what are called *conditional probabilities*.

To illustrate the dependence of probabilities on the sample space, consider the following example.

EXAMPLE 1 **College students** Suppose that a certain mathematics class contains 26 students. Of these, 14 are economics majors, 15 are first-year students, and 7 are neither. Suppose that a person is selected at random from the class.

(a) What is the probability that the person is both an economics major and a first-year student?

(b) Suppose we are given the additional information that the person selected is a first-year student. What is the probability that he or she is also an economics major?

Solution Let E denote the set of economics majors and F the set of first-year students. A complete Venn diagram of the class can be obtained with the techniques of Section 5.3. See Fig. 1.

(a) In selecting a student from the class, the sample space consists of all 26 students. Since the choice is random, all students are equally likely to be selected. The event "economics major and first-year student" corresponds to the set $E \cap F$ of the Venn diagram. Therefore,

$$Pr(E \cap F) = \frac{[\text{number of outcomes in } E \cap F]}{[\text{number of possible outcomes}]} = \frac{10}{26} = \frac{5}{13}.$$

So the probability of selecting a first-year economics major is $\frac{5}{13}$.

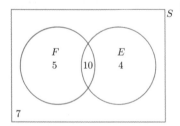

Figure 1

(b) If we know that the student selected is a first-year student, then the possible outcomes of the experiment are restricted. They must belong to F. In other words, given the additional information, we must alter the sample space from "all students" to "first-year students." Since each of the 15 first-year students is equally likely to be selected, and since 10 of the 15 first-year students are economics majors, the probability of choosing an economics major under these circumstances is equal to $\frac{10}{15} = \frac{2}{3}$. ■

Let us consider Example 1 more carefully. In part (a) the sample space is the set of all students in the mathematics class, E is the event "student is an economics major" and F is the event "student is a first-year student." On the other hand, part (b) poses a condition, "student is a first-year student." The condition is satisfied by every element of F. We are being asked to find the conditional probability of E given F, written $\Pr(E|F)$. To do this we shall restrict our attention to the new (restricted) sample space, now just the elements of F. Thus, we consider only first-year students and ask, of these, what is the probability of choosing an economics major. We assign a value to $\Pr(E|F)$ via the following formula:

Conditional Probability

$$\Pr(E|F) = \frac{\Pr(E \cap F)}{\Pr(F)}, \tag{1}$$

provided that $\Pr(F) \neq 0$.

We will provide an intuitive justification of this formula shortly. However, we first give two applications.

EXAMPLE 2 **Probablities associated with two-children families** Consider all families with two children (not twins). Assume that all elements of the sample space {BB, BG, GB, GG} are equally likely. (Here, for instance, BG denotes the birth sequence "boy girl.") Let E be the event {BB} and F the event "at least one boy." Calculate $\Pr(E|F)$.

Solution Each of the four birth sequences is equally likely and has probability $\frac{1}{4}$. In particular, since $F = \{$BB, BG, GB$\}$, $E \cap F = \{$BB$\}$, and we have $\Pr(E \cap F) = \frac{1}{4}$. Also, $\Pr(F) = \frac{1}{4} + \frac{1}{4} + \frac{1}{4} = \frac{3}{4}$. The conditional probability $\Pr(E|F)$ equals the probability that a two-child family has two boys given that it has at least one. This conditional probability has the value

$$\Pr(E|F) = \frac{\Pr(E \cap F)}{\Pr(F)} = \frac{\frac{1}{4}}{\frac{3}{4}} = \frac{1}{3}.$$

[This result is contrary to many people's intuition. They reason that among families with at least one boy, the other child is just as likely to be a boy as a girl. Therefore, they conclude incorrectly that the probability $\Pr(E|F)$ is $\frac{1}{2}$.]

Now Try Exercise 49 ■

EXAMPLE 3 **Earnings and education** Twenty percent of the employees of Acme Steel Company are college graduates. Of all its employees, 25% earn more than $50,000 per year, and 15% are college graduates earning more than $50,000. What is the probability that an employee selected at random earns more than $50,000 per year, given that he or she is a college graduate?

Solution Let H and C be the events

$$H = \text{"earns more than \$50,000 per year"}$$
$$C = \text{"college graduate."}$$

We are asked to calculate $\Pr(H|C)$. The given data are

$$\Pr(H) = .25 \qquad \Pr(C) = .20 \qquad \Pr(H \cap C) = .15.$$

By formula (1), we have

$$\Pr(H|C) = \frac{\Pr(H \cap C)}{\Pr(C)} = \frac{.15}{.20} = \frac{3}{4}.$$

Thus $\frac{3}{4}$ of all college graduates at Acme Steel earn more than \$50,000 per year.

Now Try Exercise 13(b)   ■

Suppose that an experiment has N equally likely outcomes. Then we may apply the following formula to calculate $\Pr(E|F)$.

> **Conditional Probability in Case of Equally Likely Outcomes**
>
> $$\Pr(E|F) = \frac{[\text{number of outcomes in } E \cap F]}{[\text{number of outcomes in } F]}, \qquad (2)$$
>
> provided that $[\text{number of outcomes in } F] \neq 0$.

For instance, in Example 2, the different birth sequences were all equally likely. Moreover, the number of outcomes in $E \cap F$ (two boys) is 1, whereas the number of outcomes in F (at least one boy) is 3. Then formula (2) gives that $\Pr(E|F) = \frac{1}{3}$, in agreement with the calculation of Example 2.

Let us now justify formulas (1) and (2).

Justification of Formula (1) Formula (1) is a definition of conditional probability and as such does not really need any justification. (We can make whatever definitions we choose!) However, let us proceed intuitively and show that the definition is reasonable, in the sense that formula (1) gives the expected long-run proportion of occurrences of E given that F occurs. Assume that our experiment is performed repeatedly, say for 10,000 trials. We would expect F to occur in approximately $10,000 \Pr(F)$ trials. Among these, the trials for which E also occurs are exactly those for which both E and F occur. In other words, the trials for which E also occurs are exactly those for which the event $E \cap F$ occurs; and this event has probability $\Pr(E \cap F)$. Therefore, out of the original 10,000 trials, there should be approximately $10,000 \Pr(E \cap F)$ in which E and F both occur. Thus, considering only those trials in which F occurs, the proportion in which E also occurs is

$$\frac{10,000 \Pr(E \cap F)}{10,000 \Pr(F)} = \frac{\Pr(E \cap F)}{\Pr(F)}.$$

Thus, at least intuitively, it seems reasonable to define $\Pr(E|F)$ by formula (1). ■

Justification of Formula (2) Suppose that the number of outcomes of the experiment is N. Then

$$\Pr(F) = \frac{[\text{number of outcomes in } F]}{N}$$

$$\Pr(E \cap F) = \frac{[\text{number of outcomes in } E \cap F]}{N}.$$

Therefore, using formula (1), we have

$$\Pr(E|F) = \frac{\Pr(E \cap F)}{\Pr(F)}$$

$$= \frac{\dfrac{[\text{number of outcomes in } E \cap F]}{N}}{\dfrac{[\text{number of outcomes in } F]}{N}}$$

$$= \frac{[\text{number of outcomes in } E \cap F]}{[\text{number of outcomes in } F]}.$$ ■

From formula (1), multiplying both sides of the equation by $\Pr(F)$, we can deduce the following useful fact.

Product Rule   If $\Pr(F) \neq 0$,

$$\Pr(E \cap F) = \Pr(F) \cdot \Pr(E|F). \tag{3}$$

The next example illustrates the use of this rule.

EXAMPLE 4   **Color-blind males**   Assume that a certain school contains an equal number of female and male students and that 5% of the male population is color-blind. Find the probability that a randomly selected student is a color-blind male.

Solution   Let $M =$ "male" and $B =$ "color-blind." We wish to calculate $\Pr(B \cap M)$. From the given data,

$$\Pr(M) = .5 \quad \text{and} \quad \Pr(B|M) = .05.$$

Therefore, by the product rule,

$$\Pr(B \cap M) = \Pr(M) \cdot \Pr(B|M) = (.5)(.05) = .025.$$ ■

Often an event G can be described as a sequence of two other events E and F. That is, G occurs if F occurs and then E occurs. The product rule allows us to compute the probability of G as the probability of F times the conditional probability $\Pr(E|F)$. The next example illustrates this point.

EXAMPLE 5   **Cards**   A sequence of two playing cards is drawn at random (without replacement) from a standard deck of 52 cards. What is the probability that the first card is red and the second is black?

Solution   The event in question is a sequence of two events, namely,

$$F = \text{"the first card is red"}$$
$$E = \text{"the second card is black."}$$

Since half the deck consists of red cards, $\Pr(F) = \frac{1}{2}$. If we are given that F occurs, then there are only 51 cards left in the deck, of which 26 are black, so

$$\Pr(E|F) = \frac{26}{51}.$$

By the product rule

$$\Pr(E \cap F) = \Pr(F) \cdot \Pr(E|F) = \frac{1}{2} \cdot \frac{26}{51} = \frac{13}{51}.$$ ■

The product rule may be generalized to sequences of three events E_1, E_2, and E_3:

$$\Pr(E_1 \cap E_2 \cap E_3) = \Pr(E_1) \cdot \Pr(E_2|E_1) \cdot \Pr(E_3|E_1 \cap E_2).$$

Similar formulas hold for sequences of four or more events.

One of the most important applications of conditional probability is in the discussion of independent events. Intuitively, two events are independent of each other if the occurrence of one has no effect on the likelihood that the other will occur. For example, suppose that we toss a die twice. Let the events E and F be

$$F = \text{``first throw is a 6''}$$
$$E = \text{``second throw is a 3.''}$$

Then intuitively these events are independent of one another. Throwing a 6 on the first throw has no effect whatsoever on the outcome of the second throw. On the other hand, suppose that we draw a sequence of two cards at random (without replacement) from a deck. Then the events

$$F = \text{``first card is red''}$$
$$E = \text{``second card is black''}$$

are not independent of one another, at least intuitively. Indeed, whether or not we draw a red on the first card affects the likelihood of drawing a black on the second.

The notion of independence of events is easily formulated. If E and F are events in a sample space and $\Pr(F) \neq 0$, then the product rule states that $\Pr(E \cap F) = \Pr(E|F) \cdot \Pr(F)$. However, if the occurrence of event F does not affect the likelihood of the occurrence of event E, we would expect that $\Pr(E|F) = \Pr(E)$. Substitution then shows that $\Pr(E \cap F) = \Pr(E) \cdot \Pr(F)$.

Let E and F be events. We say that E and F are *independent* provided that
$$\Pr(E \cap F) = \Pr(E) \cdot \Pr(F).$$

If $\Pr(E) \neq 0$ and $\Pr(F) \neq 0$, then our definition is equivalent to the intuitive statement of independence stated in terms of conditional probability. The two may be used interchangeably.

Let E and F be events with nonzero probability. E and F are *independent* provided that
$$\Pr(E|F) = \Pr(E) \quad \text{and} \quad \Pr(F|E) = \Pr(F).$$

EXAMPLE 6 **Two throws of a die** An experiment consists of observing the outcome of two consecutive throws of a die. Let E and F be the events

$$E = \text{``the first throw is a 3''}$$
$$F = \text{``the second throw is a 6.''}$$

Show that these events are independent.

Solution Clearly, $\Pr(E) = \Pr(F) = \frac{1}{6}$. To compute $\Pr(E|F)$, assume that F occurs. Then there are six possible outcomes:

$$F = \{(1,6),(2,6),(3,6),(4,6),(5,6),(6,6)\},$$

and all outcomes are equally likely. Moreover,

$$E \cap F = \{(3,6)\},$$

so that

$$\Pr(E|F) = \frac{[\text{number of outcomes in } E \cap F]}{[\text{number of outcomes in } F]} = \frac{1}{6} = \Pr(E).$$

Similarly, $\Pr(F|E) = \Pr(F)$. So E and F are independent events, in agreement with our intuition. ∎

EXAMPLE 7 **Cards** Suppose that an experiment consists of observing the results of drawing two consecutive cards from a 52-card deck. Let E and F be the events

$$E = \text{"second card is black"}$$
$$F = \text{"first card is red."}$$

Are these events independent?

Solution There are the same number of outcomes with the second card red as with the second card black, so $\Pr(E) = \frac{1}{2}$. To compute $\Pr(E|F)$, note that if F occurs, then there are 51 equally likely choices for the second card, of which 26 are black, so that $\Pr(E|F) = \frac{26}{51}$. Note that $\Pr(E|F) \neq \Pr(E)$, so E and F are not independent, in agreement with our intuition.

Now Try Exercise 5 ∎

EXAMPLE 8 **Heads and tails** Suppose that we toss a coin three times and record the sequence of heads and tails. Let E be the event "at most one head occurs" and F the event "both heads and tails occur." Are E and F independent?

Solution Using the abbreviations H for "heads" and T for "tails," we have

$$E = \{\text{TTT, HTT, THT, TTH}\}$$
$$F = \{\text{HTT, HTH, HHT, THH, THT, TTH}\}$$
$$E \cap F = \{\text{HTT, THT, TTH}\}.$$

The sample space contains eight equally likely outcomes, so that

$$\Pr(E) = \frac{1}{2} \qquad \Pr(F) = \frac{3}{4} \qquad \Pr(E \cap F) = \frac{3}{8}.$$

Moreover,

$$\Pr(E) \cdot \Pr(F) = \frac{1}{2} \cdot \frac{3}{4} = \frac{3}{8},$$

which equals $\Pr(E \cap F)$. So E and F are independent.

Now Try Exercise 19 ∎

EXAMPLE 9 **Probabilities associated with four-children families** Suppose that a family has four children. Let E be the event "at most one boy" and F the event "at least one child of each sex." Are E and F independent?

Solution Let B stand for "boy" and G for "girl." Then

$$E = \{GGGG, GGGB, GGBG, GBGG, BGGG\}$$
$$F = \{GGGB, GGBG, GBGG, BGGG, BBBG, BBGB, BGBB,$$
$$GBBB, BBGG, BGBG, BGGB, GBBG, GBGB, GGBB\},$$

and the sample space consists of 16 equally likely outcomes. Furthermore,

$$E \cap F = \{GGGB, GGBG, GBGG, BGGG\}.$$

Therefore,

$$\Pr(E) = \tfrac{5}{16} \qquad \Pr(F) = \tfrac{7}{8} \qquad \Pr(E \cap F) = \tfrac{1}{4}.$$

In this example

$$\Pr(E) \cdot \Pr(F) = \tfrac{5}{16} \cdot \tfrac{7}{8} \neq \Pr(E \cap F).$$

So E and F are *not* independent events.

Now Try Exercise 51   ■

Examples 8 and 9 are similar, yet the events they describe are independent in one case and not the other. Although intuition is frequently a big help, in complex problems we shall need to use the definition of independence to verify that our intuition is correct.

EXAMPLE 10 **Reliability of a calculator** A new hand calculator is designed to be ultrareliable by having two independent calculating units. The probability that a given calculating unit fails within the first 1000 hours of operation is .001. What is the probability that at least one calculating unit will operate without failure for the first 1000 hours of operation?

Solution Let

$$E = \text{``calculating unit 1 fails in first 1000 hours''}$$
$$F = \text{``calculating unit 2 fails in first 1000 hours.''}$$

Then E and F are independent events, since the calculating units are independent of one another. Therefore,

$$\Pr(E \cap F) = \Pr(E) \cdot \Pr(F) = (.001)^2 = .000001$$
$$\Pr((E \cap F)') = 1 - .000001 = .999999.$$

Since $(E \cap F)' = $ "not both calculating units fail in first 1000 hours," the desired probability is .999999.

Now Try Exercise 17   ■

The concept of independent events can be extended to more than two events:

A set of events is said to be *independent* if, for each collection of events chosen from them, say $E_1, E_2, \ldots, E_n$, we have

$$\Pr(E_1 \cap E_2 \cap \cdots \cap E_n) = \Pr(E_1) \cdot \Pr(E_2) \cdot \cdots \cdot \Pr(E_n).$$

EXAMPLE 11 **Probabilities associated with independent events** Three events A, B, and C are independent; $\Pr(A) = .5$, $\Pr(B) = .3$, and $\Pr(C) = .2$.

(a) Calculate $\Pr(A \cap B \cap C)$.   (b) Calculate $\Pr(A \cap C)$.

Solution (a) $\Pr(A \cap B \cap C) = \Pr(A) \cdot \Pr(B) \cdot \Pr(C) = (.5)(.3)(.2) = .03$.

(b) $\Pr(A \cap C) = \Pr(A) \cdot \Pr(C) = (.5)(.2) = .1$. ∎

We shall leave as an exercise the intuitively reasonable result that if E and F are independent events, so are E and F', E' and F, and E' and F'. This result also generalizes to any collection of independent events.

EXAMPLE 12 **Quality control** A company manufactures stereo components. Experience shows that defects in manufacture are independent of one another. Quality control studies reveal that

2% of CD players are defective,

3% of amplifiers are defective,

7% of speakers are defective.

A system consists of a CD player, an amplifier, and two speakers. What is the probability that the system is not defective?

Solution Let C, A, S_1, and S_2 be events corresponding to defective CD player, amplifier, speaker 1, and speaker 2, respectively. Then

$$\Pr(C) = .02 \qquad \Pr(A) = .03 \qquad \Pr(S_1) = \Pr(S_2) = .07.$$

We wish to calculate $\Pr(C' \cap A' \cap S_1' \cap S_2')$. By the complement rule we have

$$\Pr(C') = .98 \qquad \Pr(A') = .97 \qquad \Pr(S_1') = \Pr(S_2') = .93.$$

Since we have assumed that C, A, S_1, and S_2 are independent, so are C', A', S_1', and S_2'. Therefore,

$$\Pr(C' \cap A' \cap S_1' \cap S_2') = \Pr(C') \cdot \Pr(A') \cdot \Pr(S_1') \cdot \Pr(S_2')$$
$$= (.98)(.97)(.93)^2 \approx .822.$$

Thus there is an 82.2% chance that the system is not defective.

Now Try Exercise 9 ∎

PRACTICE PROBLEMS 6.5

1. (*Cards*) Suppose there are three cards: one red on both sides; one white on both sides; and one having a side of each color. A card is selected at random and placed on a table. If the up side is red, what is the probability that the down side is red? (Try guessing at the answer before working it using the formula for conditional probability.)

2. Show that if events E and F are independent of each other, then so are E and F'. [*Hint*: Since $E \cap F$ and $E \cap F'$ are mutually exclusive, we have

$$\Pr(E) = \Pr(E \cap F) + \Pr(E \cap F').]$$

EXERCISES 6.5

In Exercises 1–4, let S be a sample space and E and F events associated with S. Suppose that $\Pr(E) = .5$, $\Pr(F) = .3$, *and* $\Pr(E \cap F) = .1$.

1. Calculate $\Pr(E|F)$ and $\Pr(F|E)$.

2. Are E and F independent events? Explain.

3. Calculate $\Pr(E|F')$.

4. Calculate $\Pr(E'|F')$.

5. (*Epidemiology*) A doctor studies the known cancer patients in a certain town. The probability that a randomly chosen resident has cancer is found to be .001. It is found that 30% of the town works for Ajax Chemical Company. The probability that an employee of Ajax has cancer is equal to .003. Are the events "has cancer" and "works for Ajax" independent of one another?

6. (*Value of College*) The proportion of individuals in a certain city earning more than $35,000 per year is .25. The proportion of individuals earning more than $35,000 and having a college degree is .10. Suppose that a person is randomly chosen and he turns out to be earning more than $35,000. What is the probability that he is a college graduate?

7. (*Fitness Tests*) A medical screening program administers three independent fitness tests. Of the persons taking the tests, 80% pass test I, 75% pass test II, and 60% pass test III. A participant is chosen at random.

   (a) What is the probability that she will pass all three tests?

   (b) What is the probability that she will pass at least two of the three tests?

8. (*System Reliability*) A stereo system contains 50 transistors. The probability that a given transistor will fail in 100,000 hours of use is .0005. Assume that the failures of the various transistors are independent of one another. What is the probability that no transistor will fail during the first 100,000 hours of use?

9. (*System Reliability*) A TV set contains five circuit boards of type A, five of type B, and three of type C. The probability of failing in its first 5000 hours of use is .01 for a type A circuit board, .02 for a type B circuit board, and .025 for a type C circuit board. Assuming that the failures of the various circuit boards are independent of one another, compute the probability that no circuit board fails in the first 5000 hours of use.

10. (*Light Bulbs*) A certain brand of a long-life bulb has probability .01 of burning out in less than 1000 hours. Suppose that we wish to light a corridor with a number of independent bulbs in such a way that at least one of the bulbs remains lit for 1000 consecutive hours. What is the minimum number of bulbs needed to ensure that the probability of success is at least .99999?

11. Let E and F be events with $\Pr(E) = \frac{1}{2}$, $\Pr(F) = \frac{1}{3}$, and $\Pr(E \cap F) = \frac{1}{4}$. Compute $\Pr(E|F)$ and $\Pr(F|E)$.

12. Let E and F be events with $\Pr(E) = .3$, $\Pr(F) = .6$, and $\Pr(E \cup F) = .7$. Find

    (a) $\Pr(E \cap F)$ (b) $\Pr(E|F)$ (c) $\Pr(F|E)$

    (d) $\Pr(E' \cap F)$ (e) $\Pr(E'|F)$

13. (*Opinion Polling*) Of the registered voters in a certain town, 50% are Democrats, 40% favor a school loan, and 30% are Democrats who favor a school loan. Suppose that a registered voter is selected at random from the town.

    (a) What is the probability that the person is not a Democrat and opposes the school loan?

    (b) What is the conditional probability that the person favors the school loan given that he or she is a Democrat?

    (c) What is the conditional probability that the person is a Democrat given that he or she favors the school loan?

14. (*Game Attendance*) Of the students at a certain college, 50% regularly attend the football games, 30% are first-year students, and 40% are upper-class students who do not regularly attend football games. Suppose that a student is selected at random.

    (a) What is the probability that the person both is a first-year student and regularly attends football games?

    (b) What is the conditional probability that the person regularly attends football games given that he is a first-year student?

    (c) What is the conditional probability that the person is a first-year student given that he regularly attends football games?

15. (*Coin Tosses*) A coin is tossed three times. What is the conditional probability that the outcome is HHH given that at least two heads occur?

16. (*Balls in an Urn*) Two balls are selected at random from an urn containing two white balls and three red balls. What is the conditional probability that both balls are white given that at least one of them is white?

17. (*Life Expectancies*) The probabilities that a person A and a person B will live an additional 15 years are .8 and .7, respectively. Assuming that their lifespans are independent, what is the probability that A or B will live an additional 15 years?

18. (*Spread of a Rumor*) Jane has two friends who do not know each other. Each of them has heard the same rumor. The probability that each will tell Jane is 60%. What is the probability that Jane does not hear the rumor from either of these friends?

19. (*Balls in an Urn*) A sample of two balls is drawn from an urn containing two white balls and three red balls. Are the events "the sample contains at least one white ball" and "the sample contains balls of both colors" independent?

20. (*Fishing*) The probability that a fisherman catches a tuna in any one excursion is .15. What is the probability that he catches a tuna on each of three excursions? on at least one of three excursions?

21. (*Prizes*) The probability that a prize appears in a box of breakfast cereal is .005. What is the probability that two boxes of cereal contain at least one prize?

22. (*Guessing on an Exam*) A "true–false" exam has 10 questions. Assuming that the questions are independent and that a student is guessing, find the probability that she gets 100%.

23. (*CD Players*) Suppose that in Sleepy Valley only 30% of those over 50 years old own CD players. Find the probability that among four randomly chosen people in that age group none owns a CD player.

24. Assume that A and B are events in a sample space and that $\Pr(A) = .40$ and $\Pr(B|A) = .25$. Find $\Pr(A \cap B)$. With the further assumption that $\Pr(B) = .30$, find $\Pr(A \cup B)$, $\Pr(A' \cap B)$, and $\Pr(A|B)$.

25. Show that if events E and F are independent of each other, then so are E' and F'.

26. Show that if E and F are independent events, then

$$\Pr(E \cup F) = 1 - \Pr(E') \cdot \Pr(F').$$

27. (*Free-Throws*) A basketball player with a free-throw shooting average of .6 is on the line for a one-and-one free throw. (That is, a second throw is allowed only if the first is successful.) What is the probability that the player will score 0 points? 1 point? 2 points? Assume that the two throws are independent.

28. Let $\Pr(F) > 0$.
   (a) Show that $\Pr(E'|F) = 1 - \Pr(E|F)$.
   (b) Find an example for which
   $$\Pr(E|F') \neq 1 - \Pr(E|F).$$

29. Use the inclusion–exclusion principle for (nonconditional) probabilities to show that if E, F, and G are events in S, then
   $$\Pr(E \cup F|G) = \Pr(E|G) + \Pr(F|G) - \Pr(E \cap F|G).$$

30. (*Marriage Statistics*) The percentage of the U.S. male population that has ever been married is given in Table 1. Explain how the table relates to the idea of conditional probability by giving a precise definition of "%

ever married among 25- to 29-year-olds."

TABLE 1	
Ages	Percent ever married
18–19	1.7
20–24	14.0
25–29	45.4
30–34	66.9
35–44	80.5
45–54	89.2

31. (*Mortality Rates*) Two communities are being compared as to their annual death rates. Community A is in the Sunbelt and community B is in Alaska.

   Community A: Population: 120,000
   Number of deaths: 12,000
   Community B: Population: 90,000
   Number of deaths: 4500

   (a) Find the death rate in community A. Note that this is a conditional probability, namely
   $$\Pr(\text{death}|A).$$

   (b) Express the probability you found in part (a) as a death rate per 1000 in the population of community A.

   (c) Find $\Pr(\text{death}|B)$ and compare to part (b).

32. (*Blood Tests*) A hospital uses two tests to classify blood. Every blood sample is subjected to both tests. The first test correctly identifies blood type with probability .7, and the second test correctly identifies blood type with probability .8. The probability that at least one of the tests correctly identifies the blood type is .9.

   (a) Find the probability that both tests correctly identify the blood type.

   (b) Determine the probability that the second test is correct given that the first test is correct.

   (c) Determine the probability that the first test is correct given that the second test is correct.

   (d) Are the events "test I correctly identifies the blood type" and "test II correctly identifies the blood type" independent?

33. (*E.R. Waits*) Sixty-five percent of the patients in the emergency room of a hospital are seen by a physician immediately. The remainder are kept waiting in the waiting room. Eighty percent of those kept waiting are seen within 2 hours. Seventy-five percent of those seen immediately are admitted to the hospital. Forty percent of those seen within 2 hours (but not immediately) are admitted to the hospital, and 10% of those who wait more than 2 hours are admitted to the hospital. Let A be the event "the patient is seen immediately." Let B be the event "the patient is not seen immediately but is seen within 2 hours." Let C be the event "the patient waits more than 2 hours." Let H be the event "the patient is admitted to the hospital."

(a) Find $\Pr(B)$. (b) Find $\Pr(C)$. (c) Find $\Pr(H)$.

34. (*Course Choices*) Out of 250 students interviewed at a community college, 90 were taking mathematics but not computer science, 160 were taking mathematics, and 50 were taking neither mathematics nor computer science. Find the probability that a student chosen at random was

(a) taking just computer science.

(b) taking mathematics or computer science, but not both.

(c) taking computer science.

(d) not taking mathematics.

(e) taking mathematics, given that the student was taking computer science.

(f) taking computer science, given that the student was taking mathematics.

(g) taking mathematics, given that the student was taking computer science or mathematics.

(h) taking computer science, given that the student was not taking mathematics.

(i) not taking mathematics, given that the student was not taking computer science.

35. (*Sports Choices*) Out of 250 third-grade boys, 120 played baseball, 140 played soccer, and 50 played both. Find the probability that a boy chosen at random

(a) did not play either sport.

(b) played exactly one sport.

(c) played soccer but not baseball.

(d) played soccer, given that he played baseball.

(e) played baseball, given that he did not play soccer.

(f) did not play soccer, given that he did not play baseball.

36. (*Student Employment*) Table 2 provides some information about students at a certain college. Find the probability that a student selected at random is

(a) a senior.

(b) working full time.

(c) working part time, given that the student is a first-year student.

(d) a first-year student, given that the student does not work.

(e) a junior or senior, given that the student does not work.

(f) working part time or full time, given that the student is a sophomore or a junior.

TABLE 2			
	Works full time	Works part time	Not working
First-year	130	460	210
Sophomore	100	500	150
Junior	80	420	100
Senior	200	300	50

37. (*Party Affiliation*) Table 3 describes the voters in a certain district. Find the probability that a voter chosen at random is a

(a) Democrat.

(b) male.

(c) Independent, given that the voter is female.

(d) male, given that the voter is a Republican.

(e) female, given that the voter is not an Independent.

(f) Democrat, given that the voter is not a Republican.

TABLE 3			
	Democrat	Republican	Independent
Male	400	700	300
Female	600	300	200

38. (*Epidemiology*) Table 4 describes the incidence of a disease called the X virus.

TABLE 4			
	White	African American	Other
Have X virus	1,000	1,200	50
Disease free	100,000	35,000	10,000

(a) Find the probability that someone in this population is disease free.

(b) Find the probability that a person has the X virus.

(c) If a person is an African American, what is the probability that he or she has the X virus?

(d) Given a person who is not African American, what is the probability that he or she has the X virus?

(e) Would you say that race and incidence of the disease are independent? Explain.

39. (*Batting Average*) One interpretation of a baseball player's batting average is as the probability of getting a hit each time the player goes to bat. For instance, a player with a .300 average has probability .3 of getting a hit.

(a) If a player with .3 probability of getting a hit bats four times in a game and each at-bat is an independent event, what is the probability of the player getting at least one hit in the game?

(b) What is the probability of the player in part (a) starting off the season with at least one hit in each of the first 10 games?

(c) If there are 20 players with a .300 average, what is the probability that at least one of them will start the season with a 10-game hitting streak?

40. (*Military Personnel*) Table 5 shows the numbers of officers and enlisted persons on active military duty on May 31, 2005. Let E be the event that a person selected at random from the active-duty military personnel is enlisted, and let N be the event that a person selected at random from the active-duty military personnel is in the Navy.

TABLE 5

	Army	Navy	Marine Corps	Air Force
Officer	81,621	54,149	19,089	72,720
Enlisted	404,027	305,505	158,694	281,321

(a) Find each of the following probabilities: $\Pr(E)$, $\Pr(N)$, $\Pr(E \cap N)$, $\Pr(E|N)$, $\Pr(N|E)$.

(b) Are the events E and N independent?

41. (*Balls in an Urn*) An urn contains 10 red balls and 15 white balls. A ball is selected at random from the urn and not replaced. Then, a second ball is selected. Show that the probability that the second ball is red is the same as the probability that the first ball is red.

42. (*Choosing Numbers*) Juan randomly chooses a positive integer that is divisible by 7 and is less than 70. Chita randomly selects a positive integer less than 70 that is divisible by 11. What is the probability that they choose the same number?

43. (*Bag of Marbles*) A bag contains four blue marbles, three white marbles, and six red marbles. Three marbles are selected from the bag. What is the probability they are all red? What is the probability that they are all the same color?

44. (*Spinners*) A spinner on a wheel with the numbers 2, 5, 6, and 9 in segments of equal area is spun at random. Another spinner on a wheel with equal segments marked with numbers 3, 7, and 8 is also spun. What is the probability that the sum of the two numbers chosen by the spinners is odd?

45. (*Sandwiches*) Horatio goes to the deli for sandwiches and orders two roast beef sandwiches and two ham sandwiches. They are wrapped and put into a paper bag. What is the probability that two sandwiches chosen at random will both be roast beef? What is the probability that the two sandwiches chosen at random will be alike?

46. (*Numbers*) Ten slips of paper have numbers 1 through 10. Sue chooses two slips at random. What is the probability that the sum of the numbers chosen is even?

47. (*Bag of Marbles*) A bag contains five red marbles and eight white marbles. If a sample of four marbles contains one white marble, what is the probability that all the marbles in the sample are white?

48. (*Cards*) Eight people each have a deck of well-shuffled cards. Each chooses a card at random from his or her own deck. What is the probability that there is at least one match?

49. (*Families*) The following problem created considerable controversy when its answer appeared in the column "Ask Marilyn" in *Parade* magazine. "A woman and a man (unrelated) each have two children. At least one of the woman's children is a boy, and the man's older child is a boy. Do the chances that the woman has two boys equal the chances that the man has two boys?" Determine the two probabilities.

50. (*Conditional Probabilities*) Suppose a male is selected at random from the adult male population. Let F be the event "he played varsity sports in college." Which of the following two probabilities is greater?

(a) $\Pr($"he works in a bank and is an avid sports fan"$|F)$

(b) $\Pr($"he works in a bank"$|F)$

51. (*Roll a Die*) Roll a die and consider the following two events: $E = \{2, 4, 6\}$, $F = \{3, 6\}$. Are the events E and F independent?

52. (*Balls in an Urn*) An urn contains both red and white balls. A ball is selected at random from the urn and not replaced. Then, a second ball is selected. Show that the sequence of colors "red, white" has the same probability of occurring as the sequence "white, red."

53. (*Children*) A newly married couple decides to keep having children until they have a girl. What is the probability that the couple will have at least five children? *Note*: Assume no multiple births.

54. (*Tossing Dice*) What is the probability that at least one of the dice shows a 1 or a 2 when four dice are tossed?

55. (*Lottery*) Suppose you decide to bet $1000 in a weekly lottery for which the probability of winning the jackpot is p, where p is very small. Are your chances of winning the jackpot better if you bet 1000 different combinations in a single lottery or bet a single combination in 1000 successive weekly lotteries? To answer this question, determine the following:

(a) the probability of winning the jackpot when you bet 1000 different combinations in a single lottery.

(b) the probability of winning the jackpot at least once when you bet a single combination in 1000 successive weekly lotteries.

Note: $(1 - p)^{1000} \approx 1 - 1000p + 499{,}500p^2$.

56. (*Roulette*) If you bet on the number 7 in roulette, the probability of winning on a single spin of the wheel is $\frac{1}{38}$. Suppose you bet on "7" for 38 consecutive spins.

   (a) Which of the following numbers do you think is closest to the probability of winning at least once: 1, .64, or .5?

   (b) Calculate the probability of winning at least once.

In Exercises 57–59, use a graphing calculator, a spreadsheet, or mathematical software to calculate the probabilities.

57. (*Gender and Unemployment*) Consider Table 6, with figures in thousands, pertaining to the 2004 American civilian labor force. Determine if gender and unemployment status are independent by finding the probabilities $\Pr(W)$, $\Pr(U)$, and $\Pr(W \cap U)$.

TABLE 6

	Employed (E)	Unemployed (U)	Totals
Men (*M*)	74,524	4456	78,980
Women (*W*)	64,728	3694	68,422
Totals	139,252	8150	147,402

Source: U.S. Bureau of Labor Statistics.

58. (*Medical Diagnosis*) A diagnostic test is given to a large population to determine its efficacy. Of the 313 people for whom the test is positive, 260 have the condition. The rest are free of the condition. Of the 8249 people for whom the test is negative, 14 have the condition

and 8235 do not.

   (a) Determine the probability of a false positive.

   (b) Determine the probability that a person with a positive test result actually has the disease.

   (c) What is the prevalence of the disease in this group?

59. (*Cards*)

   (a) Find the probability of drawing four aces from a deck of 52 cards in four repeated draws without replacement.

   (b) Calculate the probability of drawing two red queens followed by two red kings in four draws from the deck without replacement.

   (c) Which of the two events in (a) or (b) is more likely?

60. (*Bridge*) A bridge hand consists of 13 cards selected from a deck of 52 cards.

   (a) How many possible bridge hands contain at least one ace?

   (b) How many possible bridge hands contain the ace of spades?

   (c) Let $E =$ "the hand contains at least two aces." Let $F =$ "the hand contains at least one ace." Let $G =$ "the hand contains the ace of spades." Which is the greater probability: $\Pr(E|F)$ or $\Pr(E|G)$?

61. (*Roulette*) Find that value of N for which the probability of winning in Exercise 56 at least once in N trials is about .5.

SOLUTIONS TO PRACTICE PROBLEMS 6.5

1. $\frac{2}{3}$. Let F be the event that the up side is red and E the event that the down side is red. $\Pr(F) = \frac{1}{2}$ since half the faces are red. $F \cap E$ is the event that both sides of the card are red—that is, that the card that is red on both sides was selected, an event with probability $\frac{1}{3}$. By (2),

$$\Pr(E|F) = \frac{\Pr(E \cap F)}{\Pr(F)} = \frac{\frac{1}{3}}{\frac{1}{2}} = \frac{2}{3}.$$

(This result may seem more intuitively evident when you realize that two-thirds of the time the card will have the same color on the bottom as on the top.)

2. By the hint,

$$\Pr(E \cap F') = \Pr(E) - \Pr(E \cap F)$$
$$= \Pr(E) - \Pr(E) \cdot \Pr(F)$$
$$\text{(since } E \text{ and } F \text{ are independent)}$$
$$= \Pr(E)[1 - \Pr(F)]$$
$$= \Pr(E) \cdot \Pr(F')$$
$$\text{(by the complement rule)}.$$

Therefore, E and F' are independent events.

6.6 Tree Diagrams

In solving many probability problems, it is helpful to represent the various events and their associated probabilities by a *tree diagram*. To explain this useful notion, suppose that we wish to compute the probability of an event that results from

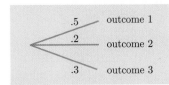

Figure 1

performing a sequence of experiments. The various outcomes of each experiment are represented as branches emanating from a point. For example, Fig. 1 represents an experiment with three outcomes. Notice that each branch has been labeled with the probability of the associated outcome. For example, the probability of outcome 1 is .5.

We represent experiments performed one after another by stringing together diagrams of the sort shown in Fig. 1, proceeding from left to right. For example, the diagram in Fig. 2 indicates that first we perform experiment A, having three outcomes, labeled 1–3. If the outcome is 1 or 2, we perform experiment B. If the outcome is 3, we perform experiment C. The probabilities on the right are conditional probabilities. For example, the top probability is the probability of outcome a (of B) given outcome 1 (of A). The probability of a sequence of outcomes may then be computed by multiplying the probabilities along a path. For example, to calculate the probability of outcome 2 followed by outcome b, we must calculate $\Pr(2 \text{ and } b) = \Pr(2) \cdot \Pr(b|2)$. To carry out this calculation, trace out the sequence of outcomes. Multiplying the probabilities along the path gives $(.2)(.6) = .12$—the probability of outcome 2 followed by outcome b.

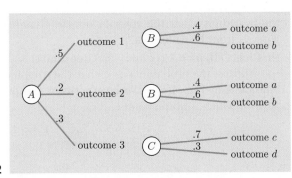

Figure 2

The next example illustrates the use of tree diagrams in calculating probabilities.

EXAMPLE 1   **Political polling** A pollster is hired by a presidential candidate to determine his support among the voters of Pennsylvania's two big cities: Philadelphia and Pittsburgh. The pollster designs the following sampling technique: Select one of the cities at random and then poll a voter selected at random from that city. Suppose that in Philadelphia two-fifths of the voters favor the Republican candidate and three-fifths favor the Democratic candidate. Suppose that in Pittsburgh two-thirds of the voters favor the Republican candidate and one-third favor the Democratic candidate.

(a) Draw a tree diagram describing the survey.

(b) Find the probability that the voter polled is from Philadelphia and favors the Republican candidate.

(c) Find the probability that the voter favors the Republican candidate.

(d) Find the probability that the voter is from Philadelphia, given that he or she favors the Republican candidate.

Solution (a) The survey proceeds in two steps: First, select a city, and second, poll a voter. Figure 3(a) shows the possible outcomes of the first step and the associated probabilities. For each outcome of the first step there are two possibilities for the second step: The person selected could favor the Republican or the Demo-

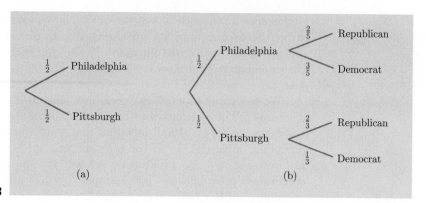

Figure 3

crat. In Fig. 3(b) we have represented these possibilities by drawing branches emanating from each of the outcomes of the first step. The probabilities on the new branches are actually conditional probabilities. For instance,

$$\tfrac{2}{5} = \Pr(\text{Rep}|\text{Phila}),$$

the probability that the voter favors the Republican candidate, given that the voter is from Philadelphia.

(b) $\Pr(\text{Phila} \cap \text{Rep}) = \Pr(\text{Phila}) \cdot \Pr(\text{Rep}|\text{Phila}) = \tfrac{1}{2} \cdot \tfrac{2}{5} = \tfrac{1}{5}.$

That is, the probability is $\tfrac{1}{5}$ that the combined outcome corresponds to the blue path in Fig. 4(a). We have written the probability $\tfrac{1}{5}$ at the end of the path to which it corresponds.

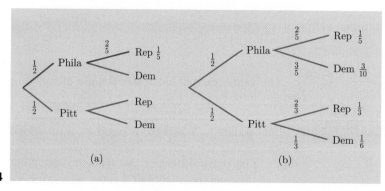

Figure 4

(c) In Fig. 4(b) we have computed the probabilities for each path of the tree as in part (b). Namely, the probability for a given path is the product of the probabilities for each of its segments. We are asked for $\Pr(\text{Rep})$. There are two paths through the tree leading to Republican, namely

Philadelphia $\cap$ Republican   or   Pittsburgh $\cap$ Republican.

The probabilities of these two paths are $\tfrac{1}{5}$ and $\tfrac{1}{3}$, respectively. So the probability that the Republican is favored equals $\tfrac{1}{5} + \tfrac{1}{3} = \tfrac{8}{15}.$

(d) Here we are asked for $\Pr(\text{Phila}|\text{Rep})$. By the definition of conditional probability,

$$\Pr(\text{Phila}|\text{Rep}) = \frac{\Pr(\text{Phila} \cap \text{Rep})}{\Pr(\text{Rep})} = \frac{\tfrac{1}{5}}{\tfrac{8}{15}} = \frac{3}{8}.$$

Now Try Exercise 13   ■

Note that from part (c) we might be led to conclude that the Republican candidate is leading, with $\frac{8}{15}$ of the vote. However, we must always be careful when interpreting surveys. The results depend heavily on the survey design. For example, the survey drew half of its sample from each of the cities. However, Philadelphia is a much larger city and is leaning toward the Democratic candidate—so much so, in fact, that in terms of popular vote the Democratic candidate would win, contrary to our expectations drawn from (c). A pollster must be very careful in designing the procedure for selecting people.

We now finally solve the medical diagnosis problem introduced in Section 6.1.

EXAMPLE 2   **Medical screening** Suppose that the reliability of a skin test for active pulmonary tuberculosis (TB) is specified as follows: Of people with TB, 98% have a positive reaction and 2% have a negative reaction; of people free of TB, 99% have a negative reaction and 1% have a positive reaction.[1] From a large population of which 2 per 10,000 persons have TB, a person is selected at random and given a skin test, which turns out to be positive. What is the probability that the person has active pulmonary tuberculosis?

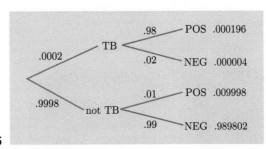

Figure 5

Solution The given data are organized in Fig. 5. The procedure called for is as follows: First select a person at random from the population. There are two possible outcomes: The person has TB,

$$\Pr(\text{TB}) = \frac{2}{10,000} = .0002,$$

or the person does not have TB,

$$\Pr(\text{not TB}) = 1 - .0002 = .9998.$$

For each of these two possibilities, the possible test results and conditional probabilities are given. Multiplying the probabilities along each of the paths through the tree gives the probabilities of the different outcomes. The resulting probabilities are written on the right in Fig. 5. The problem asks for the conditional probability that a person has TB, given that the test is positive. By definition,

$$\Pr(\text{TB}|\text{POS}) = \frac{\Pr(\text{TB} \cap \text{POS})}{\Pr(\text{POS})} = \frac{.000196}{.000196 + .009998} = \frac{.000196}{.010194} \approx .02.$$

Therefore, the probability is .02 that a person with a positive skin test has TB.[2] In other words, although the skin test is quite reliable, only about 2% of those with

---

[1]The probability that a person with the disease has a positive test is called the *sensitivity* of the test. The probability that a disease-free person has a negative test is called the *specificity* of the test. In this case the sensitivity and specificity of the skin test are .98 and .99, respectively.

[2]The probability that a person who tests positive actually has the disease is called the *predictive value of the positive test.*

a positive test turn out to have active TB. This result must be taken into account when large-scale medical diagnostic tests are planned. Because the group of people without TB is so much larger than the group with TB, the small error in the former group is magnified to the point where it dominates the calculation.

Now Try Exercise 31 ■

▶ *Note* The numerical data presented in Example 2 are only approximate. Variations in air quality for different localities within the United States cause variations in the incidence of TB and the reliability of skin tests. ◀

Tree diagrams come in all shapes and sizes. Three or more branches might emanate from a single point, for example, and some trees may not have the symmetry of those in Examples 1 and 2. Tree diagrams arise whenever an activity can be thought of as a sequence of simpler activities.

EXAMPLE 3 **Quality control** A box contains five good light bulbs and two defective ones. Bulbs are selected one at a time (without replacement) until a good bulb is found. Find the probability that the number of bulbs selected is (i) one, (ii) two, (iii) three.

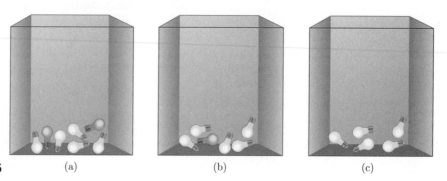

Figure 6   (a)   (b)   (c)

Solution The initial situation in the box is shown in Fig. 6(a). A bulb selected at random will be good (G) with probability $\frac{5}{7}$ and defective (D) with probability $\frac{2}{7}$. If a good bulb is selected, the activity stops. Otherwise, the situation is as shown in Fig. 6(b), and a bulb selected at random has probability $\frac{5}{6}$ of being good and probability $\frac{1}{6}$ of being defective. If the second bulb is good, the activity stops. If the second bulb is defective, then the situation is as shown in Fig. 6(c). At this point a bulb has probability 1 of being good.

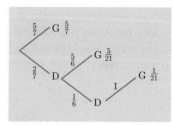

Figure 7

The tree diagram corresponding to the sequence of activities is given in Fig. 7. Each of the three paths has a different length. The probability associated with the length of each path has been computed by multiplying the probabilities for its branches. The first path corresponds to the situation where only one bulb is selected, the second path corresponds to two bulbs, and the third path to three bulbs. Therefore,

$$\text{(i) } \Pr(1) = \tfrac{5}{7} \qquad \text{(ii) } \Pr(2) = \tfrac{5}{21} \qquad \text{(iii) } \Pr(3) = \tfrac{1}{21}.$$

Now Try Exercise 11 ■

PRACTICE PROBLEMS 6.6

Fifty percent of the students enrolled in a business statistics course had previously taken a finite mathematics course. Thirty percent of these students received an A for the statistics course, whereas 20% of the other students received an A for the statistics course.

1. Draw a tree diagram and label it with the appropriate probabilities.

2. What is the probability that a student selected at random previously took a finite mathematics course and did not receive an A in the statistics course?

3. What is the probability that a student selected at random received an A in the statistics course?

4. What is the conditional probability that a student previously took a finite mathematics course, given that he or she received an A in the statistics course?

EXERCISES 6.6

In Exercises 1–4, draw trees representing the sequence of experiments.

1. Experiment I is performed. Outcome *a* occurs with probability .3, and outcome *b* occurs with probability .7. Then experiment II is performed. Its outcome *c* occurs with probability .6, and its outcome *d* occurs with probability .4.

2. Experiment I is performed twice. The three outcomes of experiment I are equally likely.

3. (*Quality Control*) A stereo repair shop uses a two-step diagnostic procedure to repair amplifiers. Step I locates the problem in an amplifier with probability .8. Step II (which is executed only if step I fails to locate the problem) locates the problem with probability .6.

4. (*Personnel Categories*) A training program is used by a corporation to direct hirees to appropriate jobs. The program consists of two steps. Step I identifies 30% as management trainees, 60% as nonmanagerial workers, and 10% to be fired. In step II, 75% of the management trainees are assigned to managerial positions, 20% are assigned to nonmanagerial positions, and 5% are fired. In step II, 60% of the nonmanagerial workers are kept in the same category, 10% are assigned to management positions, and 30% are fired.

5. (*Quality Control*) Refer to Exercise 3. What is the probability that the procedure will fail to locate the problem?

6. (*Personnel Categories*) Refer to Exercise 4. What is the probability that a randomly chosen hiree will be assigned to a management position at the end of the training period?

7. (*Personnel Categories*) Refer to Exercise 4. What is the probability that a randomly chosen hiree will be fired by the end of the training period?

8. (*Personnel Categories*) Refer to Exercise 4. What is the probability that a randomly chosen hiree will be designated a management trainee but *not* be appointed to a management position?

9. (*Selecting from Urns*) Suppose that we have a white urn containing two white balls and one red ball and we have a red urn containing one white ball and three red balls. An experiment consists of selecting at random a ball from the white urn and then (without replacing the first ball) selecting at random a ball from the urn having the color of the first ball. Find the probability that the second ball is red.

10. (*Cards, Coins, Dice*) A card is drawn from a 52-card deck. If the card is a picture card, we toss a coin. If the card is not a picture card, we toss a die. Find the probability that we end the sequence with a "6" on the die. Find the probability that we end the sequence with a "head" on the coin.

11. (*Cards*) A card is drawn from a 52-card deck. We continue to draw until we have drawn a king or until we have drawn five cards, whichever comes first. Draw a tree diagram that illustrates the experiment. Put the appropriate probabilities on the tree. Find the probability that the drawing ends before the fourth draw.

12. (*Final Exams*) There are three sections of a mathematics course available at convenient times for a student. There is a 20% chance that Professor Jones gives a final exam, a 10% chance that Professor Cates gives a final exam, and a 5% chance that Professor Smithson gives a final. At other times there are two biology sections, and in those the probabilities of a final are 20% and 13%. Find the probability that a student who randomly chooses one mathematics course and one biology course has to take at least one final examination.

13. (*Quality Control*) Twenty percent of the library books in the fiction section are worn and need replacement. Ten percent of the nonfiction holdings are worn. The library's holdings are 40% fiction and 60% nonfiction. Draw a tree diagram to illustrate how to find the probability that a book chosen at random from this library is worn and needs replacement.

14. (*School Bussing*) Draw a tree diagram that illustrates the following. Three-fifths of kindergarten children are bussed to school, while two-fifths of the first to fifth graders are bussed. The school has grades K through

5, and 17.5% of the students are in kindergarten. Determine the probability that a child chosen at random from the school is bussed to school.

15. (*Color Blindness*) Color blindness is a sex-linked, inherited condition that is much more common among men than women. Suppose that 5% of all men and .4% of all women are color-blind. A person is chosen at random and found to be color-blind. What is the probability that the person is male? (You may assume that 50% of the population are men and 50% are women.)

16. (*T-maze*) A mouse is put into a T-maze (a maze shaped like a "T"). In this maze he has the choice of turning to the left and being rewarded with cheese or going to the right and receiving a mild shock. Before any conditioning takes place (i.e., on trial 1), the mouse is equally likely to go to the left or to the right. After the first trial his decision is influenced by what happened on the previous trial. If he receives cheese on any trial, the probabilities of his going to the left or right become .9 and .1, respectively, on the following trial. If he receives the electric shock on any trial, the probabilities of his going to the left or right on the next trial become .7 and .3, respectively. What is the probability that the mouse will turn left on the second trial?

17. (*T-maze*) Refer to Exercise 16. What is the probability that the mouse will turn left on the third trial?

18. (*Manufacturing*) A factory has two machines that produce bolts. Machine I produces 60% of the daily output of bolts, and 3% of its bolts are defective. Machine II produces 40% of the daily output, and 2% of its bolts are defective.

   (a) What is the probability that a bolt selected at random will be defective?

   (b) If a bolt is selected at random and found to be defective, what is the probability that it was produced by machine I?

19. (*Heads or Tails*) Three ordinary quarters and a fake quarter with two heads are placed in a hat. One quarter is selected at random and tossed twice. If the outcome is "HH," what is the probability that the fake quarter was selected?

20. (*Medical Diagnosis*) Suppose that the reliability of a test for hepatitis is specified as follows: Of people with hepatitis, 95% have a positive reaction and 5% have a negative reaction; of people free of hepatitis, 90% have a negative reaction and 10% have a positive reaction. From a large population of which .05% of the people have hepatitis, a person is selected at random and given the test. If the test is positive, what is the probability that the person actually has hepatitis?

21. (*Tennis*) Kim has a strong first serve; whenever it is good (that is, in) she wins the point 75% of the time. Whenever her second serve is good, she wins the point

50% of the time. Sixty percent of her first serves and 75% of her second serves are good.

   (a) What is the probability that Kim wins the point when she serves?

   (b) If Kim wins a service point, what is the probability that her first serve was good?

22. (*Accidental Nuclear War*) Suppose that during any year the probability of an accidental nuclear war is .0001 (provided, of course, that there hasn't been one in a previous year). Draw a tree diagram representing the possibilities for the next three years. What is the probability that there will be an accidental nuclear war during the next three years?

23. (*Accidental Nuclear War*) Refer to Exercise 22. What is the probability that there will be an accidental nuclear war during the next n years?

24. (*Coin Tosses*) A coin is to be tossed at most five times. The tosser wins as soon as the number of heads exceeds the number of tails and loses as soon as three tails have been tossed. Draw a tree diagram for this game and calculate the probability of winning.

25. (*Cards*) Suppose that instead of tossing a coin, the player in Exercise 24 draws up to five cards from a deck consisting only of three red and three black cards. The drawer wins as soon as the number of red cards exceeds the number of black cards and loses as soon as three black cards have been drawn. Does the tree diagram for the card game have the same shape as the tree diagram for the coin game? Is there any difference in the probability of winning? If so, which game has the greater probability of winning?

26. (*Cards*) A man has been guessing the colors of cards drawn from a standard deck. During the first 50 draws he kept track of the number of cards of each color. What is the probability of guessing the color of the fifty-first card?

27. (*Genetics*) Traits passed from generation to generation are carried by genes. For a certain type of pea plant, the color of the flower produced by the plant (either red or white) is determined by a pair of genes. Each gene is of one of the types C (dominant gene) or c (recessive gene). Plants for which both genes are of type c (said to have genotype cc) produce white flowers. All other plants—that is, plants of genotypes CC and Cc—produce red flowers. When two plants are crossed, the offspring receives one gene from each parent.

   (a) Suppose you cross two pea plants of genotype Cc. What is the probability that the offspring produces white flowers? red flowers?

   (b) Suppose you have a batch of red-flowering pea plants, of which 60% have genotype Cc and 40% have genotype CC. If you select one of these plants at random and cross it with a white-flowering pea plant, what is the probability that the offspring will produce red flowers?

28. (*Genetics*) Refer to Exercise 27. Suppose a batch of 99 pea plants contains 33 plants of each of the three genotypes.

(a) If you select one of these plants at random and cross it with a white-flowering pea plant, what is the probability that the offspring will produce white flowers?

(b) If you select one of the 99 pea plants at random, cross it with a white-flowering pea plant, and the offspring produces red flowers, what is the probability that the selected plant had genotype Cc?

29. (*College Faculty*) At a local college, four sections of economics are taught during the day and two sections are taught at night. Seventy-five percent of the day sections are taught by full-time faculty. Forty percent of the evening sections are taught by full-time faculty. If Jane has a part-time teacher for her economics course, what is the probability that she is taking a night class?

30. (*U.S. Car Production*) Car production in the United States in 2005 was distributed among car manufacturers as follows.[3]

U.S. car production	Type	Percentage of type by brand	
60%	Domestic	Chrysler	23%
		Ford	31%
		General Motors	46%
40%	Foreign	Honda	20%
		Toyota	32%
		Other	48%

This means that 60% of the cars produced in the United States were manufactured by domestic companies; of them, 23% were Chryslers, 31% were Fords, and 46% were General Motors products.

(a) A 2005 automobile is chosen at random. What is the probability that it is a General Motors car?

(b) What is the probability that a randomly selected 2005 automobile is a Ford or a Toyota?

(*Medical Screening*) *Exercises 31 and 32 refer to Example 2.*

31. Find the predictive value of the negative skin test; that is, the probability that a person who tests negative is free of tuberculosis.

32. The predictive values of a diagnostic test do not depend entirely on the sensitivity and specificity of the test. They also depend on the prevalence of the disease being tested for. Find the predictive values of the positive and negative skin tests if 2 out of 100 people in the population have tuberculosis.

33. (*Balls in an Urn*) Urn I contains 5 red balls and 5 white balls. Urn II contains 12 white balls. A ball is selected at random from urn I and placed in urn II. Then a ball is selected at random from urn II. What is the probability that the second ball is white?

34. (*Balls in an Urn*) An urn contains five red balls and three green balls. One ball is selected at random and then replaced by a ball of the other color. Then a second ball is selected at random. What is the probability that the second ball is green?

35. (*Coin Tosses*) Two people toss two coins each. What is the probability that they get the same number of heads?

36. (*Quality Control*) A light bulb maufacturer knows that .05% of all bulbs manufactured are defective. A testing machine is 99% effective. If a randomly selected light bulb is tested and found to be defective, what is the probability that it actually is defective?

37. (*Tennis*) When a tennis player hits his first serve as hard as possible (called a *blast*), he gets the ball in (that is; within bounds) 60% of the time. When the blast first serve is in, he wins the point 80% of the time. When the first serve is out, his gentler second serve wins the point 45% of the time. Draw a tree diagram representing the probabilities of winning the point for the first two serves. Use the tree diagram to determine the probability that the server eventually wins the point when his first serve is a blast.

38. (*Golf*) Bud is a very consistent golfer. On par three holes, he always scores a 4. Lou, on the other hand, is quite erratic. Seventy percent of the time Lou scores a 3 and thirty percent of the time he scores a 6.

(a) If Bud and Lou play a single par three hole together, who is more likely to win; that is, to have the lowest score?

(b) If Bud and Lou play two consecutive par three holes, who is more likely to have the lowest total score?

---

[3] U.S. Light Vehicle Production by Manufacturer (*Ward's AutoInfo Bank*).

SOLUTIONS TO PRACTICE PROBLEMS 6.6

1.

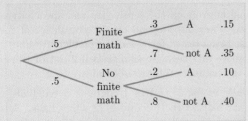

2. The event "finite math and not A" corresponds to the second path of the tree diagram, which has probability .35.

3. This event is satisfied by the first or third paths and therefore has probability $.15 + .10 = .25$.

4. $\Pr(\text{finite math}|A) = \dfrac{\Pr(\text{finite math and } A)}{\Pr(A)} = \dfrac{.15}{.25} = .6.$

6.7 Bayes' Theorem

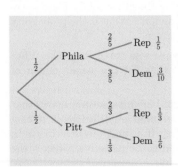

Figure 1

Let us reconsider the polling survey in Example 1(d) of Section 6.6 (see Fig. 1). Given that the person chosen at random favors the Republican candidate, what is the probability that the respondent is from Philadelphia? We found this probability, $\Pr(\text{Phila}|\text{Rep})$, by finding

$$\frac{\Pr(\text{Phila} \cap \text{Rep})}{\Pr(\text{Rep})}.$$

Let us analyze the components of this calculation. First, recall that

$$\Pr(\text{Phila} \cap \text{Rep}) = \Pr(\text{Phila}) \cdot \Pr(\text{Rep}|\text{Phila}).$$

Second,

$$\Pr(\text{Rep}) = \Pr(\text{Phila} \cap \text{Rep}) + \Pr(\text{Pitt} \cap \text{Rep})$$

$$= \Pr(\text{Phila}) \cdot \Pr(\text{Rep}|\text{Phila}) + \Pr(\text{Pitt}) \cdot \Pr(\text{Rep}|\text{Pitt}),$$

by using the tree diagram. Denote the events "Phila," "Pitt," "Rep," and "Dem" by the letters A, B, R, and D, respectively. Then

$$\Pr(\text{Phila}|\text{Rep}) = \Pr(A|R)$$

$$= \frac{\Pr(A \cap R)}{\Pr(R)}$$

$$= \frac{\Pr(A) \cdot \Pr(R|A)}{\Pr(A) \cdot \Pr(R|A) + \Pr(B) \cdot \Pr(R|B)}.$$

This is a special case of Bayes' theorem.

We summarize a simple form of Bayes' theorem. Suppose that A is an event in S, and B_1 and B_2 are mutually exclusive events such that $B_1 \cup B_2 = S$. Then

$$\Pr(B_1|A) = \frac{\Pr(B_1) \cdot \Pr(A|B_1)}{\Pr(B_1) \cdot \Pr(A|B_1) + \Pr(B_2) \cdot \Pr(A|B_2)}.$$

We have the same type of result for the situation in which we have three mutually exclusive sets B_1, B_2, and B_3 whose union is all of S. We state Bayes' theorem for that case and leave the general case for n mutually exclusive sets for the end of the section.

Bayes' Theorem ($n = 3$)   If B_1, B_2, and B_3 are mutually exclusive events, and $B_1 \cup B_2 \cup B_3 = S$, then for any event A in S,

$$\Pr(B_1|A) = \frac{\Pr(B_1) \cdot \Pr(A|B_1)}{\Pr(B_1) \cdot \Pr(A|B_1) + \Pr(B_2) \cdot \Pr(A|B_2) + \Pr(B_3) \cdot \Pr(A|B_3)}.$$

See Fig. 2.

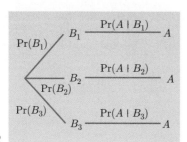

Figure 2

EXAMPLE 1   **Medical screening**   Solve the tuberculosis skin test problem of Example 2 of Section 6.6 by using Bayes' theorem.

Solution   The observed event A is "positive skin test result." There are two possible events leading to A—namely,

$$B_1 = \text{"person has tuberculosis"}$$

$$B_2 = \text{"person does not have tuberculosis."}$$

We wish to calculate $\Pr(B_1|A)$. From the data given we have

$$\Pr(B_1) = \frac{2}{10{,}000} = .0002$$

$$\Pr(B_2) = .9998$$

$$\Pr(A|B_1) = \Pr(\text{POS}|\text{TB}) = .98$$

$$\Pr(A|B_2) = \Pr(\text{POS}|\text{not TB}) = .01.$$

Therefore, by Bayes' theorem,

$$\Pr(B_1|A) = \frac{\Pr(B_1)\Pr(A|B_1)}{\Pr(B_1)\Pr(A|B_1) + \Pr(B_2)\Pr(A|B_2)}$$

$$= \frac{(.0002)(.98)}{(.0002)(.98) + (.9998)(.01)} \approx .02,$$

in agreement with our calculation of Example 2 of Section 6.6.

Now Try Exercise 11   ■

The advantages of Bayes' theorem over the use of tree diagrams are that (1) we do not need to draw the tree diagram to calculate the desired probability, and (2) we need not compute extraneous probabilities. These advantages become significant in dealing with experiments having many outcomes.

EXAMPLE 2 **Quality control** A printer has seven book-binding machines. For each machine, Table 1 gives the proportion of the total book production that it binds and the probability that the machine produces a defective binding. For instance, machine 1 binds 10% of the books and produces a defective binding with probability .03. Suppose that a book is selected at random and found to have a defective binding. What is the probability that it was bound by machine 1?

TABLE 1

Machine	Proportion of books bound	Probability of defective binding
1	.10	.03
2	.05	.03
3	.20	.02
4	.15	.02
5	.25	.01
6	.15	.02
7	.10	.03

Solution In this example we have seven mutually exclusive events whose union is the entire sample space (the book was bound by one, and only one, of the seven machines). Bayes' theorem can be extended to any finite number of B_i's.

Let B_i $(i = 1, 2, \ldots, 7)$ be the event that the book was bound by machine i, and let A be the event that the book has a defective binding. Then, for example,

$$\Pr(B_1) = .10 \quad \text{and} \quad \Pr(A|B_1) = .03.$$

The problem asks for the reversed conditional probability, $\Pr(B_1|A)$. By Bayes' theorem,

$$\Pr(B_1|A) = \frac{\Pr(B_1)\Pr(A|B_1)}{\Pr(B_1)\Pr(A|B_1) + \Pr(B_2)\Pr(A|B_2) + \cdots + \Pr(B_7)\Pr(A|B_7)}$$

$$= \frac{(.10)(.03)}{(.10)(.03) + (.05)(.03) + (.20)(.02) + (.15)(.02) + (.25)(.01) + (.15)(.02) + (.10)(.03)}$$

$$= \frac{.003}{.02} = .15.$$

Now Try Exercise 1 ■

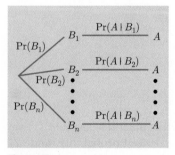

Figure 3

Derivation of Bayes' Theorem To derive Bayes' theorem in general, we consider a two-stage tree. Suppose that at the first stage there are the events $B_1, B_2, \ldots, B_n$, which are mutually exclusive and exhaust all possibilities. Let us examine only the paths of the tree leading to event A at the second stage of the experiment (see Fig. 3). Suppose we are given that the event A occurs. What is $\Pr(B_1|A)$? First, consider $\Pr(B_1 \cap A)$. This can be seen from Fig. 3 to be $\Pr(B_1) \cdot \Pr(A|B_1)$. Next we calculate $\Pr(A)$. Recall that A occurs at stage 2, preceded at stage 1 by either event $B_1, B_2, \ldots,$ or B_n. Since $B_1, B_2, \ldots, B_n$ are mutually exclusive,

$$\Pr(A) = \Pr(B_1 \cap A) + \Pr(B_2 \cap A) + \cdots + \Pr(B_n \cap A).$$

Each of the elements in the sum can be calculated by the product rule, or directly from Fig. 3:

$$\Pr(B_1 \cap A) = \Pr(B_1) \cdot \Pr(A|B_1)$$
$$\Pr(B_2 \cap A) = \Pr(B_2) \cdot \Pr(A|B_2)$$
$$\vdots$$
$$\Pr(B_n \cap A) = \Pr(B_n) \cdot \Pr(A|B_n).$$

The result is the following:

> **Bayes' Theorem**   If $B_1, B_2, \ldots, B_n$ are mutually exclusive events, and if $B_1 \cup B_2 \cup \cdots \cup B_n = S$, then for any event A in S,
>
> $$\Pr(B_1|A) = \frac{\Pr(B_1) \cdot \Pr(A|B_1)}{\Pr(B_1) \cdot \Pr(A|B_1) + \Pr(B_2) \cdot \Pr(A|B_2) + \cdots + \Pr(B_n) \cdot \Pr(A|B_n)}$$
>
> $$\Pr(B_2|A) = \frac{\Pr(B_2) \cdot \Pr(A|B_2)}{\Pr(B_1) \cdot \Pr(A|B_1) + \Pr(B_2) \cdot \Pr(A|B_2) + \cdots + \Pr(B_n) \cdot \Pr(A|B_n)},$$
>
> and so forth.

PRACTICE PROBLEMS 6.7

(*Quality Control*) *Refer to Example 2. Suppose that a book is selected at random and found to have a defective binding.*

1. What is the probability that the book was bound by machine 2?

2. By what machine is the book most likely to have been bound?

EXERCISES 6.7

1. (*Accident Rates*) An automobile insurance company has determined the accident rate (probability of having at least one accident during a year) for various age groups (see Table 2). Suppose that a policyholder calls in to report an accident. What is the probability that he or she is over 60?

TABLE 2

Age group	Proportion of total insured	Accident rate
Under 21	.05	.06
21–30	.10	.04
31–40	.25	.02
41–50	.20	.015
51–60	.30	.025
Over 60	.10	.04

2. (*Quality Control*) An electronic device has six different types of transistors. For each type of transistor, Table 3 gives the proportion of the total number of transistors of that type and the failure rate (probability of failing within one year). If a transistor fails, what is the probability that it is type 1?

TABLE 3

Type	Proportion of total	Failure rate
1	.30	.0002
2	.25	.0004
3	.20	.0005
4	.10	.001
5	.05	.002
6	.10	.004

3. (*Student Performance*) The enrollment in a certain course is 10% first-year students, 30% sophomores, 40% juniors, and 20% seniors. Past experience has shown that the likelihood of receiving an A in the course is .2 for first-year students, .4 for sophomores, .3 for juniors, and .1 for seniors. Find the probability that a student who receives an A is a sophomore.

4. (*Larceny Rates*) A metropolitan police department maintains statistics of larcenies reported in the various precincts of the city. It records the proportion of the city population in each precinct and the precinct larceny rate (= the proportion of the precinct population reporting a larceny within the past year). These statistics are summarized in Table 4. A larceny victim is randomly chosen from the city population. What is the probability he or she comes from Precinct 3?

TABLE 4

Precinct	Proportion of population	Larceny rate
1	.20	.01
2	.10	.02
3	.40	.05
4	.30	.04

5. (*Cars and Income*) Table 5 gives the distribution of incomes and shows the proportion of two-car families by income level for a certain suburban county. Suppose that a randomly chosen family has two or more cars. What is the probability that its income is at least $25,000 per year?

TABLE 5

Annual family income	Proportion of people	Proportion having two or more cars
< $10,000	.10	.2
$10,000–$14,999	.20	.5
$15,000–$19,999	.35	.6
$20,000–$24,999	.30	.75
≥ $25,000	.05	.9

6. (*Voter Turnout*) Table 6 gives the distribution of voter registration and voter turnouts for a certain city. A randomly chosen person is questioned at the polls. What is the probability that the person is an Independent?

TABLE 6

	Proportion registered	Proportion turnout
Democrat	.50	.4
Republican	.20	.5
Independent	.30	.7

7. (*Mathematics Exam*) In a calculus course, the instructor gave an algebra exam on the first day of class to

help students determine whether or not they had enrolled in the appropriate course. Eighty percent of the students in the class passed the exam. Forty percent of those who passed the exam on the first day of class earned an A in the course, whereas only twenty percent of those who failed the exam earned an A in the course. What is the probability that a student selected at random passed the exam on the first day of class, given that he or she earned an A in the course?

8. (*Demographics*) Table 7 shows the percentages of various portions of the U.S. population in 2004 based on age and sex. Suppose that a person is chosen at random from the entire population.

TABLE 7

Age Group	U.S. Population % of population	% male
Under 5 yrs	7	51
5–19 yrs	21	51
20–44 yrs	36	51
45–64 yrs	24	49
Over 64 yrs	12	42

(a) What is the probability that the person chosen is male?

(b) Given that the person chosen is male, find the probability that he is between 5 and 19 years old.

9. (*Bilingual Employees*) A multinational company has five divisions: A, B, C, D, and E. The percentage of employees from each division who speak at least two languages fluently is shown in Table 8.

TABLE 8

Division	Number of employees	Percentage of employees that are bilingual
A	20,000	20
B	15,000	15
C	25,000	12
D	30,000	10
E	10,000	10
Total	100,000	

(a) Find the probability that an employee selected at random is bilingual.

(b) Find the probability that a bilingual employee selected at random works for division C.

10. (*Customized Dice*) A specially made pair of dice has only one- and two-spots on the faces. One of the dice has three faces with a one-spot and three faces with a two-spot. The other die has two faces with a one-spot and four faces with a two-spot. One of the dice

is selected at random and then rolled six times. If a two-spot shows up only once, what is the probability that it is the die with four two-spots?

11. (*Mammogram Accuracy*) The *New York Times* of January 24, 1997, discusses the recommendation of a special panel concerning mammograms for women in their 40s. About 2% of women aged 40 to 49 years old develop breast cancer in their 40s. But the mammogram used for women in that age group has a high rate of false positives and false negatives; the false positive rate is .30, and the false negative rate is .25. If a woman in her 40s has a positive mammogram, what is the probability that she actually has breast cancer?

12. (*Drug Testing*) A *false negative* in a diagnostic test is a test result that is negative even though the patient has the condition. A *false positive*, on the other hand, is a test result that is positive although the patient does not have the condition. A drug-testing laboratory produces false negative results 2% of the time and false positive results 5% of the time. Suppose that the laboratory has been hired by a company in which 10% of the employees use drugs.

   (a) If an employee tests positive for drug use, what is the probability that he or she actually uses drugs?

   (b) What is the probability that a nondrug user will test positive for drug use twice in a row?

   (c) What is the probability that someone who tests positive twice in a row is not a drug user?

13. (*Cards*) Thirteen cards are dealt from a deck of 52 cards.

   (a) What is the probability that the ace of spades is one of the 13 cards?

   (b) Suppose one of the 13 cards is chosen at random and found *not* to be the ace of spades. What is the probability that *none* of the 13 cards is the ace of spades?

   (c) Suppose the experiment in part (b) is repeated a total of 10 times (replacing the card looked at each time), and the ace of spades is not seen. What is the probability that the ace of spades actually *is* one of the 13 cards?

14. (*Cookie Jars*) There are two cookie jars on the shelf in the kitchen. The red one has 10 chocolate-chip cookies and 15 gingersnaps. The blue jar has 20 chocolate-chip cookies and 10 gingersnaps. James goes down in the middle of the night and without turning on the light chooses a jar at random and then chooses a cookie at random. If the cookie is chocolate chip, what is the probability that he got the cookie from the blue jar?

15. (*Scholarship Winners*) Twenty percent of the contestants in a scholarship competition come from Pylesville

High School, 40% come from Millerville High School, and the remainder come from Lakeside High School. Two percent of the Pylesville students are among the scholarship winners; 3% of the Millerville contestants and 5% of the Lakeside contestants win.

   (a) If a winner is chosen at random, what is the probability that he or she is from Lakeside?

   (b) What percentage of the winners are from Pylesville?

16. (*College Majors*) There are three sections of English 101. In Section I there are 25 students, of whom 5 are mathematics majors. In Section II there are 20 students, of whom 6 are mathematics majors. In Section III there are 35 students, of whom 5 are mathematics majors. A student in English 101 is chosen at random. Find the probability that the student is from Section I, given that he or she is a mathematics major.

17. (*Pregnancy Test*) An over-the-counter pregnancy test claims to be 99% accurate. Actually, what the insert says is that if the test is performed properly, it is 99% sure to detect a pregnancy.

   (a) What is the probability of a false negative?

   (b) Let us assume that the probability is 98% that the test result is negative for a woman who is not pregnant. If the woman estimates that her chances of being pregnant are about 40% and the test result is positive, what is the probability that she is actually pregnant?

18. (*Medical Testing*) A test for a condition is very sensitive and has a high probability of false positives, say 20%. Its rate of false negatives is 10%. The condition is estimated to exist in 65% of all patients sent for screening. If the test is positive, what is the chance the patient has the condition? Suppose that the condition is much more rare in the population—say Pr(condition) = .30. Given the same testing situation, what is Pr(condition|pos)?

19. (*Steroid Testing*) It is estimated that 10% of Olympic athletes use steroids. The test currently being used to detect steroids is said to be 93% effective in correctly detecting steroids in users. It yields false positives in only 2% of the tests. A country's best weightlifter tests positive. What is the probability that he actually takes steroids?

20. (*Manufacturing Reliability*) Ten percent of the pens made by Apex are defective. Only 5% of the pens made by its competitor, B-ink, are defective. Since Apex pens are cheaper than B-ink pens, an office orders 70% of its stock from Apex and 30% from B-ink. A pen is chosen at random and found to be defective. What is the probability that it was produced by Apex?

SOLUTIONS TO PRACTICE PROBLEMS 6.7

1. The problem asks for $\Pr(B_2|A)$. Bayes' theorem gives this probability as a quotient with numerator $\Pr(B_2)\Pr(A|B_2)$ and the same denominator as in the solution to Example 2. Therefore,

$$\Pr(B_2|A) = \frac{\Pr(B_2)\Pr(A|B_2)}{.02} = \frac{(.05)(.03)}{.02} = .075.$$

2. To solve this problem we must compute the seven conditional probabilities

$$\Pr(B_1|A), \ \Pr(B_2|A), \ \ldots, \ \Pr(B_7|A)$$

and see which one is the largest. The first two have already been computed. Using the method of the preceding problem we find that

$$\Pr(B_3|A) = .20, \ \Pr(B_4|A) = .15, \ \Pr(B_5|A) = .125,$$
$$\Pr(B_6|A) = .15, \ \text{and} \ \Pr(B_7|A) = .15.$$

Therefore, the book was most likely bound by machine 3.

6.8 Simulation

Simulation is a method of imitating an experiment by using an artificial device to substitute for the real thing. The technique is used often in industrial and scientific applications. For example, in the 1970s Deaconess Hospital in St. Louis was planning to add an extension to the hospital with 144 medical-surgical beds. The planners knew that an increase in the number of beds would require additional operating rooms and recovery room beds. By studying the pattern of patients already being treated in the existing hospital, Homer H. Schmitz and N. K. Kwak[1] constructed a mathematical model to imitate the rate of flow of patients through the planned hospital (with the additional beds) to see what the typical operating suite schedule would look like. But they did not use patients! They used random numbers and a computer in their simulation model. They built a model in which they could vary the number of operating rooms and adjust the schedule. By repeating the experiment many times, they found the optimal number of new operating rooms and recovery room beds to complement the added beds.

Calculators have a command, usually called **rand** or **randInt**, that can be used to select a number at random from a specified set of numbers. Each number in the set is just as likely to be selected as any other. Although random number tables are generally available and many of the techniques discussed here can be used with such tables, we recommend using a graphing calculator or a computer if at all possible when doing simulations. On the TI-83, the command **randInt(1,n)** generates a random integer from 1 through n, and **randInt(1,n,r)** generates a list of r random numbers. (The command **randInt** is the fifth item on the MATH/PRB menu.) On the TI-89, the command **rand(n)** generates a random integer from 1 through n, and **seq(rand(n),x,1,r)** generates a list of r random numbers. (The command **rand** is the fourth item on the MATH/Probability menu and the command **seq** is the first element on the MATH/List menu.) On either calculator, after the list of random numbers has been assigned to a list variable, the command **SortA($list$)** on the TI-83 or **SortA $list$** on the TI-89 sorts the list in ascending order. (**SortA** is found on the LIST/OPS menu of the TI-83, and on the MATH/List menu of the TI-89.) Some examples of the use of these commands follows:

1. *Simulate the toss of a single die*:

    randInt(1,6)   TI-83
    rand(6)   TI-89

---

[1] Homer H. Schmitz and N. K. Kwak, "Monte Carlo simulation of operating-room and recovery-room usage," *Operations Research* **20**, 1972, pp. 1171–1180.

2. *Simulate the toss of a pair of dice*:

> `randInt(1,6) + randInt(1,6)`   TI-83
> `rand(6) + rand(6)`   TI-89

3. *Simulate ten tosses of a pair of dice*:

> `randInt(1,6,10) + randInt(1,6,10)`   TI-83
> `seq(rand(6) + rand(6),x,1,10)`   TI-89

4. *Simulate the selection of a ball from an urn containing 7 white balls and 3 red balls*: Think of the white balls as numbered from 1 to 7 and the red balls as numbered from 8 to 10.

> `randInt(1,10)`   TI-83
> `rand(10)`   TI-89

5. *Simulate the outcome of a free throw by Michael Jordan, who had an 83% free-throw average*: Consider each shot as a random whole number from 1 to 100, where a number from 1 to 83 represents a successful free throw and a number from 84 to 100 represents a miss.

> `randInt(1,100)`   TI-83
> `rand(100)`   TI-89

There are several ways to generate random numbers on an Excel spreadsheet. The formula =**RAND()** produces a randomly selected number from 0 to 1, excluding 1. The formula =**RANDBETWEEN(m,n)** produces a randomly selected whole number from m to n. The Random Number Generation routine that is part of the Data Analysis tool generates a random sample from a probability distribution. (See Appendix C for details.) *Note*: The RANDBETWEEN function is only available if the Analysis Toolpack has been installed.

EXAMPLE 1   **Heads and tails**   Simulate 7 tosses of a fair coin. Then count the number of heads and tails.

Solution   (*Graphing Calculator*)   We will generate and sort seven numbers that are each either 1 or 2. The number 1 will be interpreted as a Heads and the number 2 will be interpreted as a Tails. See Figures 1 and 2. This simulation of 7 coins tosses yields 4 Heads and 3 Tails on the TI-83, and 5 Heads and 2 Tails on the TI-89.

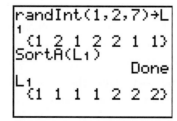

Figure 1

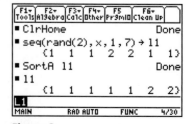

Figure 2

(*Electronic Spreadsheet*)   In Fig. 3, each cell in the range A1:E5 was generated with the equation =**RANDBETWEEN(0,1)**. If we interpret **0** as a tail and **1** as a head, then 9 heads (and therefore 11 tails) were produced. The COUNTIF function counts the number of ones appearing in the range of cells.   ∎

F5	▼		=	=COUNTIF(A1:E5,"=1")		
	A	B	C	D	E	F
1	1	1	1	0	1	
2	0	1	0	1	1	
3	0	0	0	0	1	
4	0	0	0	1	0	
5	0	0	0	0	0	9

Figure 3. Example 1 solved with a spreadsheet.

EXAMPLE 2 **Seventy-two rolls of a die** Simulate 72 tosses of a fair die, and tabulate the results. Compare your results with the theoretical probabilities.

Solution (*TI-83 Graphing Calculator*) Generate the 72 random whole numbers from 1 to 6 and store them in L_1. On the TI-83, you may tally the results by setting the STATPLOT as shown in Fig. 4 and setting the WINDOW to $[0, 8]$ by $[-10, 30]$ to allow the maximum frequency to show in the bar graph. Such a result is shown in Fig. 5. The GRAPH key will sketch a special type of bar graph indicating the number of tosses resulting in each of the possible outcomes, 1, 2, 3, 4, 5, and 6. Use the TRACE command and the cursor to determine the number of items in each of the bars. In Fig. 5, for example, we can read from the graph that there are $n = 9$ tosses that are greater than or equal to 2 and less than 3 (for a die, this means exactly 2).

Figure 4

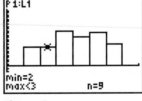

Figure 5

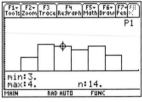

Figure 6

(*TI-89 Graphing Calculator*) The following steps produce the histogram shown in Figure 6.

1. In the home screen, execute **seq(rand(6),x,1,72)** → **L1**
2. In the **Y=** editor, highlight Plot1, and press ENTER to open the Plot1 dialog box.
3. Set **Plot Type** to **Histogram**, set **x** to **L1**, set **Hist. Bucket Width** to 1, and then press ENTER.
4. Uncheck any functions that are defined in the **Y=** editor.
5. Press ◆ [WINDOW] and set the window to $[0, 8]$ by $[-10, 30]$.
6. Press ◆ [GRAPH] to display the histogram.
7. Press F3 and then move the cursor to the right to see the heights of the rectangles. Figure 6 shows that outcome 3 occurred 14 times.

Since the probability of each outcome is $\frac{1}{6}$ and the number of tosses is 72, we would expect that each of the outcomes would occur $(\frac{1}{6})(72) = 12$ times. Observations of the graph or a survey of the generated list will give us the actual number of occurrences in each simulation of 72 tosses.

(*Electronic Spreadsheet*) In Fig. 7, each cell in column B has the content **=1/6**. The Random Number Generation routine from the Data Analysis tool used the probability distribution in A1:B6 to generate the random sample in D1:L8. Then

	A	B	C	D	E	F	G	H	I	J	K	L	M	N	O
1	1	0.166667		5	5	1	6	2	6	2	5	5		Bin	Frequency
2	2	0.166667		4	1	5	5	2	5	2	1	3		1	13
3	3	0.166667		4	6	4	4	5	5	4	3	1		2	14
4	4	0.166667		3	2	4	2	6	6	5	3	5		3	10
5	5	0.166667		4	6	2	4	3	1	4	1	1		4	15
6	6	0.166667		4	4	4	6	3	4	1	2	2		5	12
7	Prob. Dist.			2	3	1	5	1	3	3	2	4		6	8
8				2	4	1	2	1	2	6	3	1			

Figure 7. Example 2 solved with a spreadsheet.

the Histogram routine from the Data Analysis tool used the random sample and the numbers in column A to create the frequency distribution table in N1:O7. (See Appendix C for details.)

Now Try Exercise 1   ∎

EXAMPLE 3 **Simulation of recovery room** Patients having surgery fall into three categories. Sixty percent of them require two hours in the recovery room, 30% require one hour in the recovery room, and the remainder require a half-hour. Simulate the number of hours of recovery room time required by 25 patients.

Solution (*Graphing Calculator*) One technique is to generate 25 random whole numbers from 1 to 10. Numbers 1 to 6 represent a patient who requires two hours in the recovery room, numbers 7 to 9 represent a patient who requires one hour in the recovery room, and the number 10 represents a patient needing a half-hour in the recovery room. After the 25 random numbers are generated, they can be sorted and then displayed. The right-arrow key can be used to scroll along the list so that we can count all those random entries from 1 to 6, those from 7 to 9, and the 10s. Here is one possible outcome:

$$1, 1, 2, 2, 2, 4, 4, 4, 4, 5, 5, 5, 5, 5, 6, \quad 7, 7, 8, 8, 8, 9, \quad 10, 10, 10, 10,$$

giving 15 patients who require two hours, 6 patients who require one hour, and 4 patients who require a half-hour of recovery time. The total amount of time in the recovery room needed by these patients is $15(2) + 6(1) + 4(\frac{1}{2}) = 38$ hours.

(*Electronic Spreadsheet*) In Fig. 8, the Random Number Generation routine from the Data Analysis tool used the probability distribution in A1:B3 to generate the random sample in D1:H5. Then, the Histogram routine from the Data Analysis tool used the random sample and the numbers in column A to create the frequency distribution table in J1:K4. The total amount of time in the recovery room needed by these patients is $4(\frac{1}{2}) + 7(1) + 14(2) = 37$ hours.

Now Try Exercise 5   ∎

	A	B	C	D	E	F	G	H	I	J	K
1	1/2	0.1		1	1	2	2	2		Bin	Frequency
2	1	0.3		1	1	2	2	2		1/2	4
3	2	0.6		0.5	2	2	0.5	0.5		1	7
4	Prob. Dist.			2	2	0.5	2	1		2	14
5				2	2	2	1	1			

Figure 8. Example 3 solved with a spreadsheet.

EXAMPLE 4   **Simulation of a queue**   Customers steadily arrive at a bank during the hour from 9 A.M. to 10 A.M. so that the line of customers is never empty. There are three tellers, and each customer requires a varying amount of time with a teller. For simplicity, we assume that 40% of the customers need 3 minutes, 50% need 5 minutes, and 10% need 8 minutes. Each customer enters the queue at the end and goes to the first available teller when reaching the front of the queue. Simulate the service process.

(a) Show how many of the first 20 customers each of the tellers is able to service on a random day and in that hour.

(b) If all 20 customers were at the bank when it opened at 9 A.M., what was the average time spent at the bank once it opened?

Solution   (a) (*Graphing Calculator*)   We generate 20 random whole numbers from 1 to 10. We consider numbers 1 to 4 as representing customers requiring 3 minutes, numbers 5 to 9 as customers requiring 5 minutes, and 10 as customers needing 8 minutes. We do not sort them, because we want to preserve the randomness of their arrival. Here is a typical list:

Cust. #:   1, 2, 3, 4, 5, 6, 7, 8, 9, 10, 11, 12, 13, 14, 15, 16, 17, 18, 19, 20
Random #:   10, 7, 2, 3, 1, 6, 4, 3, 1, 10, 6, 5, 2, 3, 7, 7, 10, 10, 5, 5.

To determine the schedule, let the tellers be A, B, and C. Then the first customer goes to teller A. Since her random number is 10, she requires 8 minutes, occupying teller A until 9:08. Meanwhile, the second customer, with random number 7, goes to teller B, where he needs 5 minutes. He occupies teller B until 9:05. Then customer #3, with random number 2, goes to teller C until 9:03. Since teller C finishes first, customer #4 steps up to C at 9:03. That customer (with random number 3) requires 3 minutes and leaves teller C at 9:06. In case two tellers are free at the same moment, let us use the convention that the tellers are chosen in alphabetical order, with A first. From Table 1, we see that teller A served 6 of the first 20 customers, teller B served 6, and C served 8 customers. They completed the first 20 transactions at 9:34 A.M.

(*Electronic Spreadsheet*)   In Fig. 9, the Random Number Generation routine from the Data Analysis tool could have used the probability distribution in A1:B3 to generate the random sample in E1:X2, the same sequence of random numbers presented in the third column of Table 1. Table 1 can be used as before (with the second column removed) to obtain the same answer.

Figure 9. The random sample in Example 4(a) generated with a spreadsheet.

	A	B	C	D	E	F	G	H	I	J	K	L	M	N	O	P	Q	R	S	T	U	V	W	X
1	3	0.4		Cust. #:	1	2	3	4	5	6	7	8	9	10	11	12	13	14	15	16	17	18	19	20
2	5	0.5		Random #:	8	5	3	3	3	5	3	3	3	8	5	5	3	3	5	5	8	8	5	5
3	8	0.1																						
4	Prob. Dist.																							

(b) We can find the average amount of time spent after the bank opened by totaling the time spent by all customers (use the number of minutes after 9 A.M. in the "End time" column) and dividing by 20, the number of customers. This gives $(8 + 5 + 3 + 6 + 8 + 11 + 11 + 11 + 14 + 19 + 16 + 19 + 19 + 22 + 24 + 24 + 30 + 32 + 29 + 34)/20 = 17.25$ minutes. Thus, on the average, a person who was at the bank at 9 A.M. required 17.25 minutes after the bank opened to be served and to complete his transaction.

Now Try Exercise 7

TABLE 1

Customer #	Random #	Time req.	Teller	Start time	End time
1	10	8	A	9:00	9:08
2	7	5	B	9:00	9:05
3	2	3	C	9:00	9:03
4	3	3	C	9:03	9:06
5	1	3	B	9:05	9:08
6	6	5	C	9:06	9:11
7	4	3	A	9:08	9:11
8	3	3	B	9:08	9:11
9	1	3	A	9:11	9:14
10	10	8	B	9:11	9:19
11	6	5	C	9:11	9:16
12	5	5	A	9:14	9:19
13	2	3	C	9:16	9:19
14	3	3	A	9:19	9:22
15	7	5	B	9:19	9:24
16	7	5	C	9:19	9:24
17	10	8	A	9:22	9:30
18	10	8	B	9:24	9:32
19	5	5	C	9:24	9:29
20	5	5	C	9:29	9:34

The simulation should be repeated many times to determine the typical outcome. So you might simulate 50 days from 9 A.M. to 10 A.M. to see how long it takes to service the first 20 customers and, on the average, how long a customer spends in the bank if he or she is one of the first 20 people there when the bank opens. It might be useful to see the effect of using four tellers or see what happens when the probabilities of the service times are defined differently.

In practice, the arrival times of the customers are also random and can be built into the simulation. A time-and-motion study would be used to determine the appropriate probabilistic model of the arrival process.

EXERCISES 6.8

1. (*Rolling a Die*) Simulate 36 tosses of a fair die. Give the relative frequency and the corresponding theoretical probability of each of the outcomes and compare them.

2. (*Rolling Dice*) Simulate 96 tosses of a pair of dice where the sum is observed. Give the relative frequency and the corresponding theoretical probability of each of the outcomes. Make a table showing your results. Repeat the experiment 6 times and consider the total as if 576 tosses were simulated. [*Note:* (6)(96) = 576.]

3. (*Free-Throws*) Simulate 10 free-throws for Michael Jordan, whose free throw average was 83%. How many of the shots were successful?

4. (*Baseball*) A baseball player has a batting average of .331. Simulate 10 at-bats for this player and tell how many hits he gets. (A batting average of .331 means that 33.1% of at-bats result in a hit.)

5. (*Test Taking*) A student who has not studied for a 10-question multiple-choice test, with 4 choices among the answers (a, b, c, d) for each question, decides to simulate such a test and answer the questions according to a simulation in which each choice of answer has the same probability. Use a calculator to generate a simulated answer sheet. Assume the correct answers are a, b, b, c, d, d, a, c, b, a. What is the student's score?

6. (*Balls in an Urn*) In sampling 4 balls at random from an urn containing 30 balls, *without replacement* after each draw, we consider the balls as numbered 1 to 30. In selecting random whole numbers from 1 to 30, we ignore any number that has already been selected and continue the selection until we obtain a sample of size 4. Assume there are 20 red balls and 10 green balls in the urn. Draw 10 samples of size 4 and tabulate the number of red balls in each sample. Compare your results with the theoretical probability.

7. (*Registration Queue*) Students are queued up at the registrar's office when the registration windows open at 8 A.M. There are four open windows; students approach the first open window as they advance to the front of the queue. Assume that 10% of the students require 5 minutes of service time, 30% of the students require 7 minutes of service time, 40% require 10 minutes, and 20% require 15 minutes. Simulate the service of the first 20 students in a random queue. Show the schedule of service at the four windows (A, B, C, D),

determine how long it takes to process these students, and give the average time from 8 A.M. to leaving the service window.

8. (*Bank Queue*) Simulate the bank queue of Example 4 using four tellers. Give the time needed to process the first 20 customers and the average time spent by each customer in the bank after 9 A.M.

9. (*Gas Queue*) A gas station with four self-serve pumps has determined that 80% of all customers completely fill their gas tanks and the remaining 20% fill their tank with a fixed dollar amount's worth of fuel. Suppose that it takes an average of 5 minutes for a complete fill-up and 3 minutes for a partial fill-up. Suppose also that from 5 PM until 6 PM customers arrive steadily so that there is always a line and that the next customer in line proceeds to the next available pump. Simulate this process for 30 customers.

10. (*Rolling Three Dice*) Simulate 108 tosses of three dice, and show the frequency of each possible sum of the faces: $3, 4, \ldots, 18$.

CHAPTER 6 SUMMARY

1. The *sample space* of an experiment is the set of all possible outcomes of the experiment. Each subset of the sample space is called an *event*. We say that an event *occurs* when the outcome is an element of the event.

2. The event $E \cup F$ occurs when either E or F (or both) occurs. The event $E \cap F$ occurs when both E and F occur. The event E' occurs when E does not occur.

3. Two events are *mutually exclusive* if they cannot both occur at the same time.

4. A *probability distribution* for a finite sample space associates a probability with each outcome of the sample space. Each probability is a number between 0 and 1, and the sum of the probabilities is 1. The probability of an event is the sum of the probabilities of the outcomes in the event.

5. The inclusion-exclusion principle states that the probability of the union of two events is the sum of the probabilities of the events minus the probability of their intersection. If the two events are mutually exclusive, the probability of the union is just the sum of the probabilities of the events.

6. We say that the *odds* in favor of an event are a to b if the probability of the event is $a/(a + b)$. Intuitively, the event is expected to occur a times for every b times it does not occur.

7. For a sample space with a finite number of equally likely outcomes, the probability of an event is the number of elements in the event divided by the number of elements in the sample space.

8. The probability of the complement of an event is 1 minus the probability of the event.

9. $\Pr(E|F)$, the conditional probability that E occurs given that F occurs, is computed as $\Pr(E \cap F)/\Pr(F)$. For a sample space with a finite number of equally likely outcomes, it can be computed as $n(E \cap F)/n(F)$.

10. The product rule states that if $\Pr(F) \neq 0$, then $\Pr(E \cap F) = \Pr(F) \cdot \Pr(E|F)$.

11. E and F are *independent* events if $\Pr(E \cap F) = \Pr(E) \cdot \Pr(F)$. Equivalently, E and F [with $\Pr(F) \neq 0$] are independent events if $\Pr(E|F) = \Pr(E)$.

12. A collection of events is said to be *independent* if for each collection of events chosen from them, the probability that all the events occur equals the product of the probabilities that each occurs.

13. Tree diagrams provide a useful device for determining probabilities of combined outcomes in a sequence of experiments.

14. Bayes' theorem states that if $B_1, B_2, \ldots, B_n$ are mutually exclusive events whose union is the entire sample space and A is an event, then for each event B_i,

$$\Pr(B_i|A) = \frac{\Pr(B_i)\Pr(A|B_i)}{\Pr(B_1)\Pr(A|B_1) + \cdots + \Pr(B_n)\Pr(A|B_n)}.$$

15. The ability to generate random numbers allows us to simulate the outcomes of experiments.

REVIEW OF FUNDAMENTAL CONCEPTS

1. Describe how to form a sample space for an experiment.
2. Using the language of sets and assuming that A and B are events in a sample space S, write the following events in set notation: $(A$ or $B)$; $(A$ and $B)$; not A.
3. In a sample space, what is the probability of the empty set?
4. What subset in a sample space corresponds to the certain event?
5. Draw Venn diagrams for the events A or B, A and B, not A.
6. Write a formula for the probability of the event $A \cup B$ assuming that you know $\Pr(A)$, $\Pr(B)$, and $\Pr(A \cap B)$.
7. Explain the difference between mutually exclusive events and independent events.
8. State the addition principle.
9. Suppose the probability of an event is k/n. What are the odds that the event will occur?
10. Suppose the odds that an event occurs are a to b. What is the probability that the event will occur?
11. State the inclusion–exclusion principle for two events.
12. What is the definition of $\Pr(E|F)$?
13. What is Bayes' Theorem?
14. What is a tree diagram?

KEY FORMULAS

Addition Principle:
$$\Pr(E) = \Pr(s) + \Pr(t) + \Pr(u) + \cdots + \Pr(z),$$
where the event E consists of the finite number of outcomes $s, t, u, \ldots, z$

Inclusion–Exclusion Principle:
$$\Pr(E \cup F) = \Pr(E) + \Pr(F) - \Pr(E \cap F)$$

Mutually Exclusive Events: $\Pr(E \cup F) = \Pr(E) + \Pr(F)$, when E and F are mutually exclusive.

Converting Odds to a Probability: If the odds in favor of an event E are a to b, then $\Pr(E) = \dfrac{a}{a+b}$.

Conditional Probability for Equally Likely Outcomes:
$$\Pr(E|F) = \frac{n(E \cap F)}{n(F)}$$

Product Rule: If $\Pr(F) \neq 0$, then $\Pr(E \cap F) = \Pr(F) \cdot \Pr(E|F)$.

Independent Events: If E and F are independent events with nonzero probabilities, then $\Pr(E|F) = \Pr(E)$ and $\Pr(F|E) = \Pr(F)$.

Bayes' Theorem: If $B_1, B_2, \ldots, B_n$ are mutually exclusive events, and if $B_1 \cup B_2 \cup \cdots \cup B_n = S$, then for any event A in S,
$$\Pr(B_k|A) = \frac{\Pr(B_k) \cdot \Pr(A|B_k)}{\Pr(B_1) \cdot \Pr(A|B_1) + \cdots + \Pr(B_n) \cdot \Pr(A|B_n)}$$
for $k = 1, 2, \ldots, n$.

SUPPLEMENTARY EXERCISES

1. (*Coin Tosses*) A coin is to be tossed five times. What is the probability of obtaining at least one head?
2. (*Coin Tosses*) Suppose that we toss a coin three times and observe the sequence of heads and tails. Let E be the event that "the first toss lands heads" and F the event that "there are more heads than tails." Are E and F independent?
3. (*Prizes*) Each box of a certain brand of candy contains either a toy airplane or a toy gun. If one-third of the boxes contain an airplane and two-thirds contain a gun, what is the probability that a person who buys two boxes of candy will receive both an airplane and a gun?
4. (*Committee Seclection*) A committee consists of five men and five women. If three people are selected at random from the committee, what is the probability that they all will be men?
5. (*Public and Private Colleges*) Out of the 50 colleges in a certain state, 25 are private, 15 offer engineering majors, and 5 are private colleges offering engineering majors. If a college is selected at random, what is the conditional probability that it offers an engineering major given that it is a public college?
6. (*Tax Audits*) An auditing procedure for income tax returns has the following characteristics: If the return is incorrect, the probability is 90% that it will be rejected; if the return is correct, the probability is 95% that it will be accepted. Suppose that 80% of all income tax returns are correct. If a return is audited and rejected, what is the probability that the return was actually correct?

7. (*Numbers*) A number is chosen at random from the numbers 1 to 100. What is the probability that the number is divisible by 5?

8. (*Numbers*) A number is chosen at random from the numbers 1 to 10,000. What is the probability that the number is divisible by 5?

9. (*Numbers*) A number is chosen at random from the numbers 1 to 10,000. What is the probability that the number is divisible by 3 or 5?

10. (*Numbers*) A number is chosen at random from the numbers 1 to 10,000. What is the probability that the number is divisible by 3 or 12?

11. (*Seating Arrangements*) Jack and Hugo attend a party at which there are two tables of 8 for dinner. If guests are assigned to seats at random, what is the probability that Jack and Hugo will be seated at the same table?

12. (*Student Honors*) Five students are to receive special honors at commencement. Of the five, two are engineering majors.

    (a) What is the probability that the two engineering majors will be called up as the first two students to receive their awards?

    (b) What is the probability that the two engineering majors will be called up consecutively to receive their awards?

13. (*Exam Questions*) Prior to taking an essay examination, students are given 10 questions to prepare. Six of the 10 will appear on the exam. One student decides to prepare only eight of the questions. Assume that the questions are equally likely to be chosen by the professor.

    (a) What is the probability that she has prepared every question appearing on the test?

    (b) What is the probability that both questions she did not prepare appear on the test?

14. (*Code Words*) A collection of code words consists of all strings of seven characters, where each of the first three characters can be any letter or digit and each of the last four characters must be a digit. For example, 7A32765 is allowed but 7A3B765 is not.

    (a) What is the probability that a code word chosen at random begins with ABC?

    (b) What is the probability that a code word chosen at random ends with 6578?

    (c) What is the probability that a code word chosen at random ends with a four-digit number divisible by 3?

    (d) What is the probability that a code word chosen at random consists of three letters followed by four even digits?

15. (*Archery*) Two archers shoot at a moving target. One can hit the target with probability $\frac{1}{4}$ and the other with probability $\frac{1}{3}$. Assuming that their efforts are independent events, what is the probability that

    (a) Both will hit the target?

    (b) At least one will hit the target?

16. If the odds in favor of an event are 7 to 5, what is the probability that the event will occur?

17. (*Olympic Swimmers*) In an Olympic swimming event, two of the seven contestants are American. The contestants are randomly assigned to lanes 1 through 7. What is the probability that the Americans are assigned to the first two lanes?

18. (*Balls in an Urn*) An urn contains three balls numbered 1, 2, and 3. Balls are drawn one at a time without replacement until the sum of the numbers drawn is four or more. Find the probability of stopping after exactly two balls are drawn.

19. (*Dice*) A red die and a green die are tossed as a pair. Let E be the event that "the red die shows a 2" and let F be the event that "the sum of the numbers is 8." Are the events E and F independent?

20. Let E and F be events with $\Pr(E) = .4$, $\Pr(F) = .3$, and $\Pr(E \cup F) = .5$. Find $\Pr(E|F)$.

21. (*Weighing Produce*) A supermarket has three employees who package and weigh produce. Employee A records the correct weight 98% of the time. Employees B and C record the correct weight 97% and 95% of the time, respectively. Employees A, B, and C handle 40%, 40%, and 20% of the packaging, respectively. A customer complains about the incorrect weight recorded on a package he has purchased. What is the probability that the package was weighed by employee C?

22. (*Days of Week*) Three people are chosen at random. What is the probability that at least two of them were born on the same day of the week?

23. Let B and A be independent events for which the probability that at least one of them occurs is $\frac{1}{2}$ and the probability that B occurs but A does not occur is $\frac{1}{3}$. Find $\Pr(A)$.

24. (*Balls in an Urn*) An urn contains 10 balls numbered 1 through 10. Seven balls are drawn one at a time at random without replacement. Find the probability that exactly three odd-numbered balls are drawn and they occur on odd-numbered draws from the urn.

25. (*Bills in Envelopes*) Each of three sealed opaque envelopes contains two bills. One envelope contains two $1 bills, another contains two $5 bills, and the third contains a $1 bill and a $5 bill. An envelope is selected at random and a bill is taken from the envelope at random. If it is a $5 bill, what is the probability that the other bill in the envelope is also a $5 bill?

26. (*Carnival Game*) A carnival huckster has placed a coin under one of three cups and asks you to guess which cup contains the coin. After you select a cup, he removes one of the unselected cups, which he guarantees does

not contain the coin. You may now either stay with your original choice or switch to the other remaining cup. What decision will give you the greater probability of winning?

27. (*Commuting Time*) The odds of an American worker living within 20 minutes of work are 13 to 12. What is the probability that a worker selected at random lives within 20 minutes of work?

28. (*Demographics*) Twenty-six percent of all Americans are under 18 years old. What are the odds that a person selected at random is under 18?

29. (*Letters*) If the nine letters A, C, D, E, I, N, O, T, and U are arranged to form a word, what is the probability that it will be one of the meaningful words EDUCATION, AUCTIONED, or CAUTIONED?

30. (*Dragons*) An island contains an equal number of one-headed, two-headed, and three-headed dragons. If a dragon head is picked at random, what is the likelihood of its belonging to a one-headed dragon?

31. (*Coin Tosses*) Two players each toss a coin three times. What is the probability that they get the same number of tails?

32. (*Dice*) Suppose that a pair of dice is tossed. Given that a double does not occur, what is the probability that one die shows a three?

33. (*Left-Handedness*) According to *Bottom Line Personal* of March 30, 1991, the chances of having a left-handed child are 4 in 10 if both parents are left-handed, 2 in 10 if one parent is left-handed, and only 1 in 10 if neither parent is left-handed. Suppose a left-handed child is chosen at random from a population in which 25% of the adults are left-handed. What is the probability that the child's parents are both left-handed?

34. (*Languages*) Of the 120 students in a class, 30 speak Chinese, 50 speak Spanish, 75 speak French, 12 speak Spanish and Chinese, 30 speak Spanish and French, and 15 speak Chinese and French. Seven students speak all three languages. A student is chosen at random. What is the probability that he speaks none of these languages?

35. What is the probability that a whole number between 100 and 400 contains the digit 2?

36. (*Committee Composition*) (**PE**) A committee is composed of w women and m men. If three women and two men are added to the committee, and if one person is selected at random from the enlarged committee, then the probability that a woman is selected can be represented by

    (a) $\dfrac{w}{m}$ (b) $\dfrac{w}{w+m}$ (c) $\dfrac{w+3}{m+2}$

    (d) $\dfrac{w+3}{w+m+3}$ (e) $\dfrac{w+3}{w+m+5}$

37. (*Political Poll*) (**PE**) Of a group of people surveyed in a political poll, 60% said they would vote for candidate R. Of those who said they would vote for R,

90% actually voted for R, and of those who did not say they would vote for R, 5% actually voted for R. What percent of the group voted for R?

   (a) 56%   (b) 59%   (c) 62%   (d) 65%   (e) 74%

38. (*Coin Tosses*) When three coins are tossed, what is the probability of at least one tail appearing given that at least one head appeared?

39. (*Rolling a Die*) What is the probability of having each of the numbers one through six appear in six consecutive rolls of a die?

40. (*Balls in an Urn*) An urn contains 10 red balls and 20 green balls. If four balls are drawn one at a time without replacement, what is the probability that the sequence of colors will be red, green, green, red?

41. (*Drawing Cards*) A card is drawn at random from a deck of cards. Then the card is replaced and the deck is thoroughly shuffled. This process is repeated two more times.

    (a) What is the probability that all three cards are aces?

    (b) What is the probability that at least one of the cards is an ace?

In Exercises 42–45, use a graphing calculator, a spreadsheet, or mathematical software to calculate the probabilities.

42. (*Birthday Problem*) How many people are needed so that the probability of at least two matching birthdays is at least 80%?

43. (*Medical Screening*) A drug company wants to test a drug for a chronic disease. The company wants a sample of 500 people who have the disease. It tests 12,735 people for the disease, which generally affects 5% of the population. The test is known to have a 2% false positive rate and a 4% false negative rate. Six hundred and fifty people test positive for the disease. Based on the test results, how many of the total population actually can be expected to have the disease?

44. (*Rolling a Die*) Simulate an experiment in which either you toss a die until you get a 6 or you have tossed the die 15 times. Record the number of times you toss the die until the first 6. Find the relative frequency of each outcome in the set $\{1, 2, 3, \ldots, 15\}$. Perform the experiment at least 20 times.

45. (*Drawing Cards*) Simulate an experiment in which you draw 3 cards from a deck of 52 cards, with replacement after each draw. Determine the number of spades in the sample and repeat the experiment 20 times. Find the relative frequency of at least 2 spades in a sample of size 3. What is the theoretical probability of at least 2 spades in a sample of size 3?

Conceptual Exercises

46. Give an example of two events that are mutually exclusive. Give an example of two sets that are not mutually exclusive.

47. Give an example of two independent events.

48. Describe the difference between disjoint events and independent events in your own words.

49. Explain why two mutually exclusive events with

nonzero probabilities cannot be independent.

50. Give an example of two events that are mutually exclusive and not independent.

51. Explain why it is intuitively clear that if E and F are independent events, then so are E and F'.

52. What additional information would you need to know in order to compute $\Pr(E \cap F)$ if you already know $\Pr(E)$ and $\Pr(F)$?

CHAPTER TEST

1. (*Coins*) A box contains a penny, a nickel, a dime, a quarter, and a half dollar. You select two coins at random from the box.

  (a) Construct a sample space for this situation.

  (b) List the elements of the event E in which the total value of the coins you have selected is an even number of cents.

2. Convert the following odds to probabilities.

  (a) 3 to 5   **(b)** 1,000,000 to 1

3. (*Checkers*) When Tommy plays checkers against his father, he wins 40% of the time. When he plays against his mother, he wins 30% of the time. What are the odds of him beating his father? losing to his mother?

4. (*Candidates for Office*) Some of the candidates for president of the computer club at Riverdale High are seniors, and the rest are juniors. Let J be the event in which a junior is elected, and let F be the event in which a female is elected. Describe the following events:

  (a) $J \cap F'$ **(b)** $(J \cap F)'$ **(c)** $F' \cup J$

5. (*Heads and Tails*) Emilio tosses a fair coin five times. What is the probability that he gets at least one head and at least one tail?

6. (*Measles Vaccine*) Fifteen percent of children attending kindergarten have not had a measles vaccine. In a class of 20 students, what is the probability that at least one has not had a measles vaccine?

7. Suppose E and F are events in a sample space, with $\Pr(E) = \frac{1}{4}$, $\Pr(F') = \frac{3}{8}$, and $\Pr(E' \cap F) = \frac{1}{2}$. Determine the following:

  (a) $\Pr(F)$ **(b)** $\Pr(E \cup F)$ **(c)** $\Pr(F|E')$

  (d) Are E and F mutually exclusive?

  (e) Are E and F independent?

8. (*Quality Control*) Ten of a certain kind of gadget are tested. The probability that any given such gadget is defective is .1. What is the probability that at least one of the 10 gadgets is defective? What is the probability that exactly one of the 10 gadgets is defective?

9. (*Cards and Marbles*) Wanda has a deck of 52 cards, a box, and an urn. The box contains 5 red marbles and

10 green marbles, while the urn contains 12 red marbles and 8 green marbles. Wanda picks a card at random from the deck. If it is a face card (jack, queen, or king), she then picks a marble from the box. Otherwise, she picks a marble from the urn.

  (a) Determine the probability that she picks a red marble, given that she drew a face card from the deck.

  (b) What is the probability that she ends up with a red marble?

  (c) If E is the event in which Wanda picks a face card and F is the event in which she picks a red marble, are E and F independent? mutually exclusive?

  (d) If Wanda ends up with a green marble, what is the probability that she drew a face card?

10. (*Loaded Die*) The probability distribution for the result of rolling a loaded die is given in Table 1.

TABLE 1

Outcome	1	2	3	4	5	6
Probability	.15	.20	.10	.25	.25	.05

  (a) Determine the probability that the outcome is an odd number.

  (b) Given that the outcome is odd, what is the probability that the outcome is greater than 4?

11. (*Class Rank and Gender*) A finite mathematics class of 50 students is composed of freshmen and sophomores according to Table 2. A student from the class is selected at random.

TABLE 2

	Freshmen	Sophomores
Females	12	8
Males	14	16

  (a) What is the probability that the selected student is either female or a sophomore?

  (b) What is the probability that the selected student is male, given that the student is a freshman?

CHAPTER 6 PROJECT

Two Paradoxes

First Paradox: *Under certain circumstances you have your best chance of winning a tennis match if you play most of your games against the best possible opponent.*

Alice and her two sisters, Betty and Carol, are avid tennis players. Betty is the best of the three sisters, and Carol plays at the same level as Alice. Alice defeats Carol 50% of the time but only defeats Betty 40% of the time.

Alice's mother offers to give her $100 if she can win two consecutive games when playing three alternating games against her two sisters. Since the games will alternate, Alice has two choices for the sequence of opponents. One possibility is to play the first game against Betty, followed by a game with Carol, and then another game with Betty. We will refer to this sequence as BCB. The other possible sequence is CBC.

1. Make a guess of the best sequence for Alice to choose; the one having the majority of the games against the weaker opponent or the one having the majority of the games against the stronger opponent.

2. Calculate the probability of Alice getting the $100 reward if she chooses the sequence CBC.

3. Calculate the probability of Alice getting the $100 reward if she chooses the sequence BCB.

4. Which sequence should Alice choose?

5. How would you explain to someone who didn't know probability why the sequence you chose is best?

Second Paradox: *The probability of a male applicant being admitted to a graduate school can be higher than the probability for a female applicant even though for each department the probability of a female being admitted is higher.* (This apparent contradiction is known as Simpson's paradox.)

To simplify matters, consider a university with two professional graduate programs, medicine and law. Suppose that last year 1000 men and 1000 women applied and the outcome was as shown in Table 1.

TABLE 1

	Men			Women		
	Applied	Accepted	Rejected	Applied	Accepted	Rejected
Law	700	560	140	400	340	60
Medicine	300	40	260	600	160	440

6. What is the probability that a male applicant was accepted to a professional program? female?

7. Which gender does the university appear to be favoring?

8. What is the probability that a male applicant was accepted to law school? female?

9. What is the probability that a male applicant was accepted to medical school? female?

10. Which gender do the individual professional schools appear to be favoring?

11. Without using probability, justify the apparent contradiction between the answers for part 7 and part 10.

Probability and Statistics

STATISTICS is the branch of mathematics that deals with data: their collection, description, analysis, and use in prediction. In this chapter we present some topics in statistics that can be used as a springboard to further study. Since we are presenting a series of topics rather than a comprehensive survey, we will bypass large areas of statistics without saying anything about them. However, the discussion should give you some feeling for the subject. Section 7.1 shows how bar charts, pie charts, histograms, and box plots help us turn raw data into a visual form that often allows us to see patterns in the data quickly. Section 7.2 discusses the problem of describing data by means of a distribution and a histogram. It also introduces the concept of random variables. Section 7.3 presents the binomial distribution, one of the most commonly used distributions in statistical applications. Sections 7.4 and 7.5 introduce the mean and the standard deviation, the two most frequently used descriptive statistics, and illustrate how Chebychev's inequality can be used in making estimations. Sections 7.6 and 7.7 explore the binomial distribution further and introduce the normal distribution, which has special importance in statistical analysis.

7.1   Visual Representations of Data

Data can be presented in raw form or organized and displayed in tables or charts. In this section we use various types of charts to visualize and analyze data.

In the fall of 2004, freshmen at 440 baccalaureate colleges and universities in the United States answered an extensive questionnaire.[1] One question asked each student to give the highest degree they planned to pursue. See Table 1. (*Note*: The Medical category includes Dental and Veterinary degrees.) Such a table is often referred to as a *frequency table* since it presents the frequency with which each response occurs.

TABLE 1		
Highest Degree Planned	**Number**	**Percent**
Bachelor's	68,600	23.7
Master's	118,675	41.0
Doctorate	50,365	17.4
Medical	26,919	9.3
Law	14,183	4.9
Other	10,710	3.7
Total	289,452	100.0

Bar Charts   The numbers from the example in Table 1 are displayed in the *bar chart* of Fig. 1. This pictorial display gives a good feel for the relative number of students planning to earn each degree.

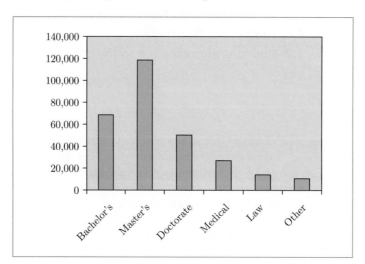

Figure 1.   Bar chart for highest degree planned.

The percentages in the right column of Table 1 give the percent of the students out of this group of 289,452 students who plan to pursue each degree. The bar chart for the percentages is shown in Fig. 2. It looks exactly like the bar chart in Fig. 1. The only difference is the labeling of the tick marks along the y-axis.

Pie Charts   Another popular type of chart that can be used to display data consisting of several categories is the *pie chart*. It consists of a circle subdivided into

[1]The detailed results of the questionnaire are given in *The American Freshman: National Norms for Fall 2004* (Los Angeles: American Council on Education, 2005, UCLA).

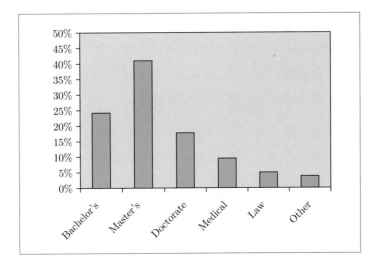

Figure 2. Bar chart for highest degree planned.

sectors (slices of pie), where each sector corresponds to a category. The area of each sector is proportional to the percentage of items in that category. This is accomplished by making the central angle of each sector equal to 360° times the percentage associated with the segment.

EXAMPLE 1 **Freshman aspirations** Create a pie chart for the "highest degree planned" data. Label each sector with its category and percentage.

Solution Step 1. Use the rightmost column of Table 1 to obtain the central angles for the sectors. See Table 2.

TABLE 2

Highest Degree Planned	Percent	360° × Percent
Bachelor's	23.7	85.3°
Master's	41.0	147.6°
Doctorate	17.4	62.6°
Medical	9.3	33.5°
Law	4.9	17.6°
Other	3.7	13.3°

Step 2. Draw a circle, draw a horizontal radius line extending from the center of the circle, and then use a protractor to draw an angle of measure approximately 85.3° with the radius line as the initial side of the angle. See Fig. 3(a).

Step 3. Draw an angle of approximately 147.6°, using the terminal side of the angle drawn in Step 2 as the initial side. See Fig. 3(b).

Step 4. Continue as in Step 3 to draw each sector, and then label the sectors with their categories and percentages.

Now Try Exercise 7 ■

Histograms In the "highest degree planned" example, the responses to the question were words. The data to be organized consisted of six different words, where each occurred with a high frequency. In many cases to be analyzed, the data are

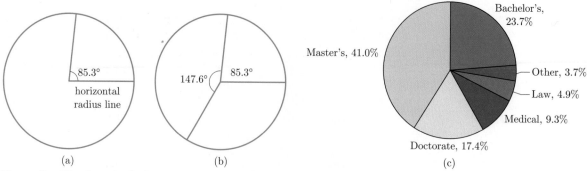

Figure 3. Pie chart for highest degree planned.

a collection of numbers. For instance, the data could consist of ages, weights, or test scores of individuals. In such cases, the x-axis in a bar chart is labeled with numbers, as in an ordinary x-y coordinate system, and the data are referred to as *numerical data*. The bars in this type of bar chart have no space between them, and the bar chart is called a *histogram*.

EXAMPLE 2 **Tabulating quiz scores** Figure 4 gives the quiz scores for a class of 25 students.

Figure 4 8 7 6 10 5 10 7 1 8 0 10 5 9 3 8 6 10 4 9 10 7 0 9 5 8

(a) Organize the data into a frequency table.

(b) Create a histogram for the data.

Solution (a) An easy way to count the number of exams for each score is to write down the numbers from 0 through 10, and considering the quiz papers one at a time, make a slash mark alongside the score for each paper. Such a tabulation produces Table 3(a). In Table 3(b) the slash marks have been totaled.

(b) The histogram (see Fig. 5) is drawn on an x-y coordinate system. Note that each bar is centered over its corresponding score. This is accomplished by having the base of each bar extend one-half unit on each side of its score. For instance, the base of the bar corresponding to a score of 10 extends from 9.5 to 10.5.

TABLE 3 Tabulation of quiz scores

10	⫽⫽⫽	10	5
9	///	9	3
8	////	8	4
7	///	7	3
6	//	6	2
5	///	5	3
4	/	4	1
3	/	3	1
2		2	0
1	/	1	1
0	//	0	2
(a)		(b)	

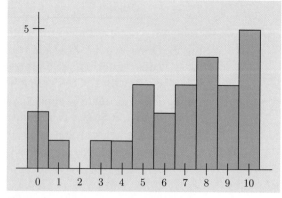

Figure 5. Histogram for quiz scores.

Now Try Exercise 13

(TI-83) The steps required to create and explore the histogram with a TI-83 graphing calculator are shown in Figures 6 through 9. (The details of carrying out the task are presented in Appendix B.) For Fig. 6, the stat list editor was invoked with $\boxed{\text{STAT}}$ **1**, the scores were placed in the L_1 column, and their frequencies were placed in the L_2 column. The screen in Fig. 7 was invoked with $\boxed{\text{2nd}}$ [STAT PLOT] **1**. "On" was selected in the On/Off row, the histogram icon was selected as the Type, Xlist was left at its default value "L_1", and Freq was set to L_2. In Fig. 8, Xmin was set to the left coordinate of the first rectangle, Xmax was set to the right coordinate of the last rectangle, Xscl was set to the length of the base of each rectangle, and Ymin and Ymax were set to be large enough to display the rectangles. Pressing $\boxed{\text{GRAPH}}$ produced the histogram in Fig. 9. Then, $\boxed{\text{TRACE}}$ was pressed and the cursor moved to the rectangle corresponding to a score of 8. The numbers at the bottom of the screen say that the rectangle extends from 7.5 to 8.5 on the x-axis and that $n = 4$ students had grades of 8 on the quiz.

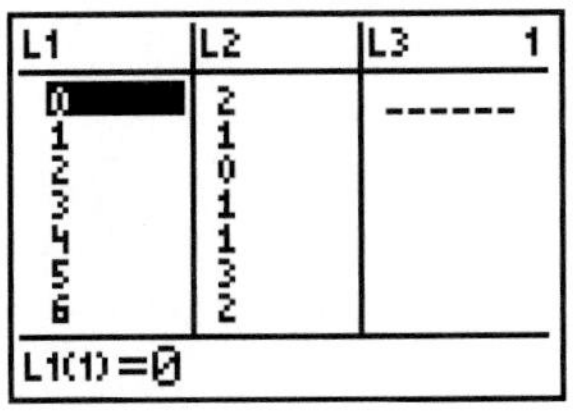

Figure 6

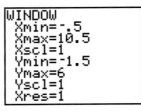

Figure 7

Figure 8

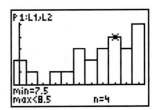

Figure 9

(TI-89) The following steps produce a display similar to the one in Figure 9, which shows the grade distribution from Example 2.

1. Invoke the Data/Matrix editor from the APPS menu.
2. Select 3 to set up a new data variable dialog box.
3. Set **Type** to **Data**, and give the variable a name such as *g* for grades.
4. Press $\boxed{\text{ENTER}}$ twice to invoke a spreadsheet.
5. Place the numbers 0 through 10 in the **c1** column and the frequency figures (2, 1, 0, 1, etc.) in the **c2** column.
6. In the **Y=** editor, highlight **Plot 1**, and press $\boxed{\text{ENTER}}$ to open the Plot 1 dialog box.
7. Set **Plot Type** to **Histogram**, set **x** to **c1**, set **Hist. Bucket Width** to 1, set **Use Freq and Categories** to **YES**, set **Freq** to **c2**.
8. Press $\boxed{\text{ENTER}}$ twice to return to the **Y=** editor. Uncheck any functions that are defined in the **Y=** editor.
9. Press $\boxed{\blacklozenge}$[WINDOW] and set the window to $[-.5, 10.5]$ *by* $[-2, 6.5]$, with **xscale** = 1. Thus, **xmin** is set to the left coordinate of the first rectangle, **xmax** is set to the right coordinate of the last rectangle, and **xscale** is set to the length of the base of each rectangle.
10. Press $\boxed{\blacklozenge}$[GRAPH] to display the histogram.
11. Press $\boxed{\text{F3}}$ and then move the cursor to the right to see the heights of the rectangles.

Median, Quartiles, and Box Plots When an instructor returns exam papers, he or she usually gives students some indication of how they did overall. Occasionally the instructor gives the average of the grades, and perhaps the standard deviation. (These two topics are discussed extensively in the remainder of this chapter.) However, most instructors state the *median* of the grade distribution. The median grade is the grade that divides the bottom 50% of the grades from the top 50%. To find the median of a set of N numbers, first arrange the numbers in increasing or decreasing order. The median is the middle number if N is odd and the average of the two middle numbers if N is even.

EXAMPLE 3 **Medians** Find the medians of the following two sets of data.

(a) Tiger Woods's scores on four rounds of golf in the 2005 U.S. Masters tournament: 74 66 65 71

(b) The 25 quiz scores discussed in Example 2

Solution (a) Here $N = 4$, an even number. Arranged in increasing order, the four scores are

$$65, \ 66, \ 71, \ 74.$$

The middle two scores are 66 and 71. The median is their average. Therefore,

$$\text{median} = \frac{66 + 71}{2} = \frac{137}{2} = 68.5.$$

(b) Here $N = 25$, an odd number. The position of the middle number is $\dfrac{25 + 1}{2} = 13$. The tabulation of quiz scores in Table 3 can be used instead of an ordering of the scores. The median will be the thirteenth highest score. Adding up the numbers of scores of 10's, 9's, and 8's gives 12 scores. Therefore, the thirteenth score must be a 7. That is, the median is 7. ■

Graphing calculators can display a picture, called a *box plot*, that analyzes a set of data and shows not only the median, but also the *quartiles*. The quartiles are the medians of the sets of data below and above the median. The median of the numbers less than the median is called the first quartile and is denoted Q_1. The median of the numbers greater than the median is called the third quartile and is denoted Q_3. A box plot also is useful in showing pictorially the spread of the data. Figure 10 gives the box plot on a TI-83 for the set of 25 quiz scores presented in Example 2. The steps for obtaining the box plot are the same as for the histogram as shown in Figs. 6 through 9, with the exception that in Fig. 7 the fifth icon is selected. Figure 11, which shows the median, appears when $\boxed{\text{TRACE}}$ is pressed. Pressing the arrow keys reveals the smallest quiz score (minX), the largest quiz score (maxX), and the first and third quartiles. See Figs. 12–15. Box plots similar to those in Figs. 10 through 15 can be displayed on the TI-89. The steps are the same as those for the histogram except that in Step 7, **Plot Type** should be set to **Box Plot**.

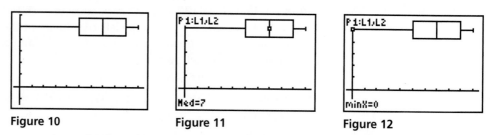

Figure 10   **Figure 11**   **Figure 12**

Figure 16 shows the five pieces of information given by a box plot. This information is referred to as the *five-number summary* of the data. The median is also called the second quartile and is denoted Q_2. Essentially, the three quartiles

P 1:L1,L2	P 1:L1,L2	P 1:L1,L2
maxX=10	q1=5	q3=9
Figure 13	**Figure 14**	**Figure 15**

divide the data into four approximately equal parts, each part consisting of roughly 25% of the numbers. The length of the rectangular part of the box plot, which is $Q_3 - Q_1$, is called the *interquartile range*. The quartiles provide information about the dispersion of the data. The interquartile range is the length of the interval in which approximately the middle 50% of the data lie.

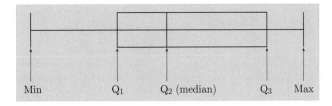

Figure 16. A general box plot.

Min Q_1 Q_2 (median) Q_3 Max

EXAMPLE 4   **Five-number summary and interquartile range**   Find the five-number summary and the interquartile range for the following set of numbers: 1   3   6   10   15   21   28   36   45   55.

Solution   The numbers are given in ascending order, and there are 10 numbers. Immediately we see that min = 1 and max = 55. The next number to be found is the median. Since 10 is an even number, the median is the average of the middle two numbers.

$$1 \quad 3 \quad 6 \quad 10 \quad \mathbf{15} \quad \mathbf{21} \quad 28 \quad 36 \quad 45 \quad 55$$

That is,

$$Q_2 = \text{median} = \frac{15 + 21}{2} = 18.$$

The numbers to the left of the median are 1   3   6   10   15, and the numbers to the right of the median are 21   28   36   45   55. These lists have medians 6 and 36, respectively. Therefore, $Q_1 = 6$, $Q_3 = 36$, and the interquartile range is $Q_3 - Q_1 = 30$.

Now Try Exercise 19   ■

PRACTICE PROBLEMS 7.1

Suppose a list consists of 17 numbers in increasing order. Clearly the first number is the min and the last number is the max.

1. Which number in the list is the median?

2. Which numbers in the list are used to obtain the first quartile?

3. Which numbers in the list are used to obtain the third quartile?

EXERCISES 7.1

In Exercises 1–4, display the data in a bar chart that shows frequencies on the y-axis.

1. 2005 School Enrollments (in millions)

Type	Enrollment
Elementary	38.4
Secondary	16.2
College	16.7

Source: U.S. Dept. of Education, National Center for Education Statistics, *Digest of Educational Statistics.*

2. 2005 U.S. Defense Employees

Branch	Officers and Enlistees
Army	485,648
Navy	359,654
Marine Corps	177,783
Air Force	354,041

Source: U.S. Dept. of Defense, *Defense.*

3. Areas of the Great Lakes

Lake	Area (sq mi)
Superior	31,700
Michigan	22,300
Huron	23,100
Erie	9,910
Ontario	7,550

Source: *Encyclopedia Britannica.*

4. Bachelor's Degrees Earned in 2002, by Field

Field of Study	Number of Degrees
Business	281,330
Social Sciences	132,874
Education	106,383
Psychology	76,671
Engineering	73,964
Health Sciences	70,517
Other	550,161

Source: U.S. National Center for Education Statistics, *Digest of Education Statistics*, annual.

5. (*School Enrollments*) Display the data from the table in Exercise 1 in a bar chart showing percentages on the y-axis.

6. (*U.S. Defense Employees*) Display the data from the table in Exercise 2 in a bar chart showing percentages on the y-axis.

7. (*Interest on Public Debt*) (**PE**) In 2005, the interest on the public debt accounted for about 7.5% of the federal budget. If the federal budget is displayed in a pie chart, what should be the size of the central angle of the sector corresponding to the interest on the public debt?

(**a**) $7.5°$ (**b**) $27°$ (**c**) $60°$ (**d**) $90°$ (**e**) $120°$

8. (*U.S. Defense Employees*) Display the data from the table in Exercise 2 in a pie chart.

9. (*Great Lakes*) Display the data from the table in Exercise 3 in a pie chart.

10. (*Bachelor's Degrees*) Display the data from the table in Exercise 4 in a pie chart.

(Freshman Aspirations) Exercises 11 and 12 refer to the pie chart in Fig. 3(c).

11. What is the probability that a freshman selected at random in the fall of 2004 planned to obtain a Master's or Doctorate degree?

12. What is the probability that a freshman selected at random in the fall of 2004 planned to obtain a medical or law degree?

13. (*Vice-Presidential Tie Breakers*) The number of tie-breaking votes cast by each of the 21 vice presidents of the United States who served during the twentieth century are shown. Draw a histogram for these data.

0, 0, 4, 10, 0, 2, 3, 3, 4, 1, 7,
8, 0, 4, 2, 0, 0, 1, 7, 0, 4

14. (*Presidential Ages*) The ages at inauguration of the 43 presidents from George Washington to George W. Bush are shown. Draw a histogram for the ages.

57, 61, 57, 57, 58, 57, 61, 54, 68, 51, 49, 64, 50,
48, 65, 52, 56, 46, 54, 49, 50, 47, 55, 55, 54, 42,
51, 56, 55, 51, 54, 51, 60, 62, 43, 55, 56, 61, 52,
69, 64, 46, 54

15. (**PE**) For what value of n will the median of the numbers $\{2, 3, 5, 20\}$ be the same as the median of the numbers $\{n, 1, 3, 5, 15\}$?

(**a**) 2 (**b**) 3 (**c**) 4 (**d**) 5 (**e**) 6

16. (**PE**) If n is a whole number between 0 and 7, then the median of the numbers $\{1, 3, 5, n, 2, 4, 6\}$ must be

(**a**) n (**b**) 3.5 (**c**) either 2 or 3
(**d**) either 3 or 4 (**e**) either 5 or 6

In Exercises 17 and 18, draw the box plot corresponding to the given five-number summary.

17. min $= 2$, $Q_1 = 5$, $Q_2 = 7$, $Q_3 = 10$, max $= 15$

18. min $= 10$, $Q_1 = 13$, $Q_2 = 17$, $Q_3 = 20$, max $= 25$

In Exercises 19–22, find the five-number summary and the interquartile range for the given set of numbers, and then draw the box plot.

19. 10, 11, 13, 14, 16, 17, 19, 20, 21, 23, 24

20. 3, 6, 8, 9, 11, 14, 18

21. 20, 25, 31, 38, 42, 47, 51, 54, 56

22. 7, 17, 26, 34, 41, 47, 52, 56, 59, 61

23. Match each of the histograms in column I with the corresponding box plot in column II.

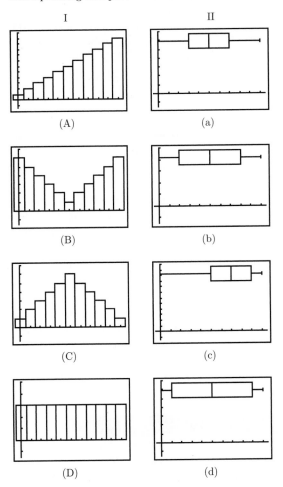

(A)   (a)

(B)   (b)

(C)   (c)

(D)   (d)

24. (*Food Cost*) The box plot for the price (in cents) of a can of tomato soup is shown.

20   35 45   65   129

   (a) Approximately what percentage of the soups are priced below 35 cents?

(b) Approximately what percentage of the soups are priced above 65 cents?

(c) Approximately what percentage of the soups are priced below 45 cents?

(d) Approximately what percentage of the soups are priced between 45 and 65 cents?

(e) What is the median price of a can of soup?

25. (*Test Scores*) Consider the following box plot of scores on a standardized test.

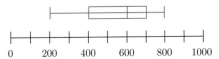

0   200   400   600   800   1000

(a) Give the five-number summary of the data.

(b) Approximately what percentage of the scores is below 400?

(c) Approximately what percentage of the scores is between 400 and 600?

(d) Approximately what percentage of the scores is higher than 600?

(e) Approximately what percentage of the scores is between 200 and 700?

26. (*Batting Averages*) The data below give the top 10 batting averages at the end of the 2004 season for the American League and the National League. Compare the data using a box plot and discuss which of the leagues seems to have the better players.

American League: .372, .340, .337, .334, .321, .318, .316, .314, .313, .311

National League: .362, .347, .335, .334, .331, .326, .324, .319, .318, .316

27. (*Batting Averages*) The Atlanta Braves played a game against the Milwaukee Brewers. The batting averages of each team's players (except for pitchers and designated hitters) on that day were as follows:

Braves: .229, .243, .317, .281, .296, .200, .345, .227, .350, .286

Brewers: .317, .150, .270, .333, .250, .200, .359, .280, .091

Draw a box plot for each team and comment on which team seems to be more likely to win. Explain your answer.

SOLUTIONS TO PRACTICE PROBLEMS 7.1

1. Since there are an odd number of numbers in the list, the middle number (that is, the ninth number) is the median. *Note*: When N, the number of numbers, is odd, the median is the $\frac{N+1}{2}$th number.

2. There are 8 numbers in the truncated list consisting of the numbers to the left of the median. Since 8 is even, the median of this truncated list is the average of the middle two numbers. That is, the first quartile is the average of the fourth and fifth numbers. *Note*: When N is even, the median of a list of N numbers

is the average of the $\frac{N}{2}$th number and the number following the $\frac{N}{2}$th number.

3. There are 8 numbers in the truncated list consisting of the numbers to the right of the median. Since 8 is even, the median of this truncated list is the average of the middle two numbers of the truncated list. So the third quartile is the average of the fourth and fifth numbers of the truncated list; that is, the average of the thirteenth and fourteenth numbers of the original list.

7.2 Frequency and Probability Distributions

Our goal in this section is to describe a given set of data in terms that allow for interpretation and comparison. As we shall see, both graphical and tabular displays of data can be useful for this purpose.

Our modern technological society has a fetish about gathering statistical data. It is hardly possible to glance at a newspaper or a magazine and not be confronted with massive arrays of statistics gathered from studies of schools, churches, the economy, and so forth. One of the chief tasks confronting us is to interpret in a meaningful way the data collected and to make decisions based on our interpretations. The mathematical tools for doing this belong to that part of mathematics called statistics.

To get an idea of the problems considered in statistics, let us consider a concrete example. Mr. Jones, a businessman, is interested in purchasing a car dealership. Two dealerships are for sale, and each dealer has provided him with data describing past sales. Dealership A provided 1 year's worth of data, dealership B, 2 years' worth. The data are summarized in Table 1. The problem confronting Mr. Jones is that of analyzing the data to determine which car dealership to buy.

TABLE 1		
	Number of occurrences	
Weekly sales	**Dealership A**	**Dealership B**
5	2	20
6	2	0
7	13	0
8	20	10
9	10	12
10	4	50
11	1	12

These data are presented in a form often used in statistical surveys. For each possible value of a statistical variable (in this case the number of cars sold weekly) we have tabulated the number of occurrences. Such a tabulation is called a *frequency distribution*. Although a frequency distribution is a very useful way of displaying and summarizing survey data, it is by no means the most efficient form in which to analyze such data. For example, it is difficult to compare dealership A with dealership B using only Table 1.

Comparisons are much more easily made if we use proportions rather than actual numbers of occurrences. For example, instead of recording that dealership A had weekly sales of 5 cars during 2 weeks of the year, let us record that the proportion of the observed weeks in which dealership A had weekly sales of 5 was $\frac{2}{52} \approx .04$. Similarly, by dividing each of the entries in the right column by 52, we obtain a new table describing the sales of dealership A (Table 2).[1] We similarly can construct a new table for dealership B (Table 3).

TABLE 2	
Weekly sales	**Proportion of occurrences, dealership A**
5	$\frac{2}{52} \approx .04$
6	$\frac{2}{52} \approx .04$
7	$\frac{13}{52} = .25$
8	$\frac{20}{52} \approx .38$
9	$\frac{10}{52} \approx .19$
10	$\frac{4}{52} \approx .08$
11	$\frac{1}{52} \approx .02$

TABLE 3	
Weekly sales	**Proportion of occurrences, dealership B**
5	$\frac{20}{104} \approx .19$
6	0
7	0
8	$\frac{10}{104} \approx .10$
9	$\frac{12}{104} \approx .12$
10	$\frac{50}{104} \approx .48$
11	$\frac{12}{104} \approx .12$

These tables are called *relative frequency distributions*. In general, consider an experiment with the numerical outcomes $x_1, x_2, \ldots, x_r$. Suppose that the number of occurrences of x_1 is f_1, the number of occurrences of x_2 is f_2, and so forth (Table 4). The frequency distribution lists all the outcomes of the experiment and the number of times each occurred. (For the sake of simplicity, we usually arrange $x_1, x_2, \ldots, x_r$ in increasing order.)

TABLE 4		
Outcome	**Frequency**	**Relative frequency**
x_1	f_1	f_1/n
x_2	f_2	f_2/n
$\vdots$	$\vdots$	$\vdots$
x_r	f_r	f_r/n
Total	n	1

Suppose that the total number of occurrences is n. Then the relative frequency of outcome x_1 is f_1/n, the relative frequency of outcome x_2 is f_2/n, and so forth. The relative frequency distribution pairs each outcome with its relative frequency. The sum of the frequencies in a frequency distribution is n. The sum of the relative frequencies in a relative frequency distribution is 1.

The frequency or relative frequency distribution is obtained directly from the performance of an experiment and the collection of data observed at each trial of the experiment. For example, we might imagine a coin-tossing experiment in which a coin is tossed five times and the number of occurrences of heads is observed. On

[1]For simplification we shall round off the data of this example to two decimal places.

each performance of the experiment we might observe 0, 1, 2, 3, 4, or 5 heads. We could repeat the experiment, say 90 times, and record the outcomes. We have collected data like that in Table 5. While doing the experiment we would record the frequencies $f_0, f_1, f_2, \ldots, f_5$ of the various outcomes and then divide by 90 to obtain the relative frequencies $f_0/90, f_1/90, f_2/90, \ldots, f_5/90$. The sum of the frequency column is the number of trials of the experiment (here, 90), and the sum of the relative frequency column is 1.

TABLE 5		
Number of heads	**Frequency**	**Relative frequency**
0	3	$\frac{3}{90} \approx .03$
1	14	$\frac{14}{90} \approx .16$
2	23	$\frac{23}{90} \approx .26$
3	27	$\frac{27}{90} = .30$
4	17	$\frac{17}{90} \approx .19$
5	6	$\frac{6}{90} \approx .07$
Total	90	$\frac{90}{90} = 1.00$

It is often possible to gain useful insight into an experiment by representing its relative frequency distribution in graphical form. For instance, let us graph the relative frequency distribution for car dealership A. Begin by drawing a number line (Fig. 1).

Figure 1

The numbers that represent possible outcomes of the experiment (weekly car sales) are 5, 6, 7, 8, 9, 10, 11. Locate each of these numbers on the number line. Above each number erect a rectangle whose base is one unit wide and whose height is the relative frequency of that number. Above the number 5, for example, we draw a rectangle of height .04. The completed graph is shown in Fig. 2. For the sake of comparison we have also drawn a graph of the relative frequency distribution for dealership B.

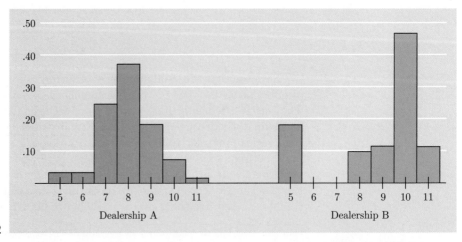

Figure 2

As we mentioned earlier, graphs of the type just drawn are called histograms. They vividly illustrate the data being considered. For example, comparison of the histograms of Fig. 2 reveals significant differences between the two dealerships. On the one hand, dealership A is very consistent. Most weeks its sales are in the middle range of 7, 8, or 9. On the other hand, dealership B can often achieve very high sales (it had sales of 10 in 48% of the weeks) at the expense of a significant number of weeks of low sales (sales of 5 in 19% of the weeks). The histogram for the coin-tossing experiment recorded previously appears in Fig. 3.

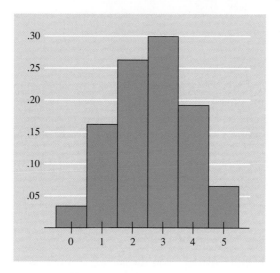

Figure 3. Relative frequency distribution for number of heads in 5 tosses—experimental results.

It is possible to use histograms to represent the relative frequencies of events as areas. To illustrate the procedure, consider the histogram corresponding to dealership A. Each rectangle has width 1 and height equal to the relative frequency of a particular outcome of the experiment. For instance, the highest rectangle is centered over the number 8 and has height .38, the relative frequency of the outcome 8. Note that the area of the rectangle is

$$\text{area} = (\text{height})(\text{width}) = (.38)(1) = .38.$$

In other words, the area of the rectangle equals the relative frequency of the corresponding outcome. We have verified this in the case of the outcome 8, but it is true in general. In a similar fashion, we may represent the relative frequency of more complicated events as areas. Consider, for example, the event $E =$ "sales between 7 and 10, inclusive." This event consists of the set of outcomes $\{7, 8, 9, 10\}$, and so its relative frequency is the sum of the respective relative frequencies of the outcomes 7, 8, 9, and 10. Therefore, the relative frequency of the event E is the area of the blue region of the histogram in Fig. 4 on the next page.

An important thing to notice is that so far we have made the tables and the histograms for actual, as opposed to theoretical, experiments. That is, the tables and the histograms were produced from collections of sample data that were obtained by actually recording the outcomes of experiments. We shall now look at theoretical experiments and continue to explore the important notion of *probability distribution*. In many cases, data of an actual experiment are best interpreted when we can construct a theoretical model for the experiment.

Let us reconsider the coin-tossing experiment in which a coin was tossed five times and the number of occurrences of heads recorded. If the coins are fair coins, then we can set up a model for the experiment by noting once again that the possible outcomes are 0, 1, 2, 3, 4, or 5 heads. We will construct the probability

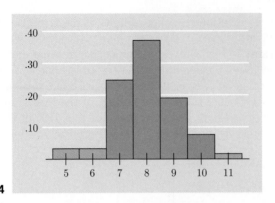

Figure 4

TABLE 6	
Number of heads	**Probability**
0	$\dfrac{\binom{5}{0}}{2^5} = \dfrac{1}{32}$
1	$\dfrac{\binom{5}{1}}{2^5} = \dfrac{5}{32}$
2	$\dfrac{\binom{5}{2}}{2^5} = \dfrac{10}{32}$
3	$\dfrac{\binom{5}{3}}{2^5} = \dfrac{10}{32}$
4	$\dfrac{\binom{5}{4}}{2^5} = \dfrac{5}{32}$
5	$\dfrac{\binom{5}{5}}{2^5} = \dfrac{1}{32}$

distribution for this experiment by listing the outcomes in the sample space with their probabilities. The probabilities of 0, 1, 2, 3, 4, 5 heads can be obtained with the methods of Chapter 6. The number of distinct sequences of 5 tosses is 2^5, or 32. The number of sequences having k heads (and $5 - k$ tails) is

$$C(5, k) = \binom{5}{k}.$$

Thus

$$\Pr(k \text{ heads}) = \frac{\binom{5}{k}}{2^5} \qquad (k = 0, 1, 2, 3, 4, 5).$$

The probability distribution for the experiment is shown in Table 6.

The histogram for a probability distribution is constructed in the same way as the histogram for a relative frequency distribution. Each outcome is represented on the number line, and above each outcome we erect a rectangle of width 1 and of height equal to the probability corresponding to that outcome (see Fig. 5).

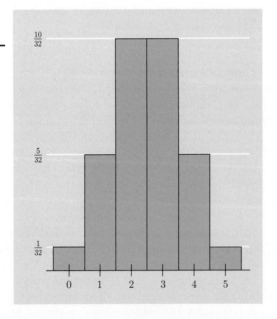

Figure 5. Probability distribution for number of heads in 5 tosses—theoretical results.

We note that the histogram in Fig. 5 is based on a theoretical model of coin tossing, whereas the histogram in Fig. 3 was drawn from experimental results only available after the experiment was actually performed and observed.

Just as we used the histogram for the relative frequency distribution to picture the relative frequency of an event, we may also use the histogram of a probability distribution to picture the probability of an event. For instance, to find the probability of at least 3 heads on 5 tosses of a fair coin, we need only add the probabilities of the outcomes: 3 heads, 4 heads, 5 heads. That is,

$$\text{Pr(at least 3 heads)} = \text{Pr(3 heads)} + \text{Pr(4 heads)} + \text{Pr(5 heads)}.$$

Since each of these probabilities is equal to the area of a rectangle in the histogram of Fig. 5, the area of the blue region in Fig. 6 equals the probability of the event. This result is a special case of the following fact:

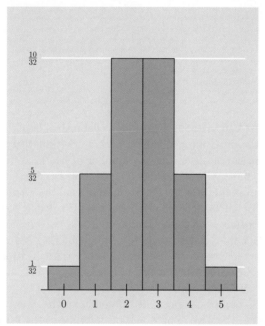

Figure 6

In a histogram of a probability distribution, the probability of an event E is the sum of the areas of the rectangles corresponding to the outcomes in E.

EXAMPLE 1 **Probabilities and histograms** The histogram of a probability distribution is as given in Fig. 7. Indicate the portion of the histogram whose area is the probability of the event "more than 50."

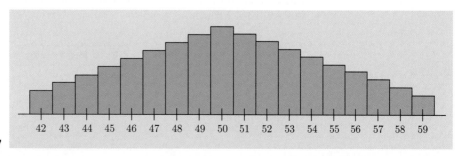

Figure 7

Solution The event "more than 50" is the set of outcomes $\{51, 52, 53, 54, 55, 56, 57, 58, 59\}$. So we indicate in blue the portion of the histogram corresponding to these outcomes (Fig. 8). The area of the blue region is the probability of the event "more than 50."

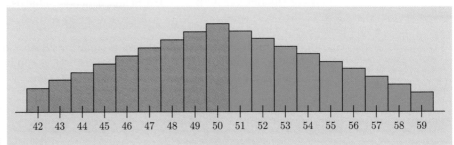

Figure 8

Now Try Exercise 11

Random Variables Consider a theoretical experiment with numerical outcomes. Denote the outcome of the experiment by the letter X. For example, if the experiment consists of observing the number of heads in five tosses of a fair coin, then X assumes one of the six values 0, 1, 2, 3, 4, 5. Since the values of X are determined by the unpredictable random outcomes of the experiment, X is called a *random variable* or, more specifically, the *random variable associated with the experiment*.

The random variable notation is often convenient and is commonly used in probability and statistics texts. In considering several different experiments, it is sometimes necessary to use letters other than X to stand for random variables. It is customary, however, to use only capital letters, such as X, Y, Z, W, U, V, for random variables.

If k is one of the possible outcomes of the experiment with associated random variable X, then we denote the probability of the outcome k by

$$\Pr(X = k).$$

For example, in the coin-tossing experiment described previously, if X is the number of heads in the five tosses, then

$$\Pr(X = 3)$$

denotes the probability of getting 3 heads. The probability distribution of the random variable X is shown in Table 7.

Rather than speak of the probability distribution associated with the model of an experiment, we can speak of the *probability distribution associated with the corresponding random variable*. Such a probability distribution is a table listing the various values of X (i.e., outcomes of the experiment) and their associated probabilities with $p_1 + p_2 + \cdots + p_r = 1$:

TABLE 7	
k	$\Pr(X = k)$
0	$\frac{1}{32}$
1	$\frac{5}{32}$
2	$\frac{10}{32}$
3	$\frac{10}{32}$
4	$\frac{5}{32}$
5	$\frac{1}{32}$

k	$\Pr(X = k)$
x_1	p_1
x_2	p_2
$\vdots$	$\vdots$
x_r	p_r

EXAMPLE 2   **Selecting balls from an urn**   Consider the urn of Example 2 of Section 6.4, in which there are eight white balls and two green balls. A sample of three balls is chosen at random from the urn. Let X denote the number of green balls in the sample. Find the probability distribution of X.

Solution   As in Example 2 of Section 6.4, we note that there are $N = \binom{10}{3} = 120$ equally likely outcomes. X can be 0, 1, or 2. From that example we conclude that

$$\Pr(X = 1) = \frac{56}{120} = \frac{7}{15} \quad \text{and} \quad \Pr(X = 2) = \frac{8}{120} = \frac{1}{15}.$$

$\Pr(X = 0)$ can be found by noting that there are

$$\binom{2}{0}\binom{8}{3} = (1)(56)$$

elements in the event "no greens." Hence,

$$\Pr(X = 0) = \frac{56}{120} = \frac{7}{15}.$$

The probability distribution for X is given by the following table.

k	$\Pr(X = k)$
0	$\frac{7}{15}$
1	$\frac{7}{15}$
2	$\frac{1}{15}$

Now Try Exercise 7   ■

EXAMPLE 3   **Tossing a pair of dice**   Let X denote the random variable defined as the sum of the upper faces appearing when two dice are thrown. Determine the probability distribution of X and draw its histogram.

Solution   The experiment of throwing two dice leads to 36 possibilities, each having probability $\frac{1}{36}$.

$$
\begin{array}{cccccc}
(1,1) & (1,2) & (1,3) & (1,4) & (1,5) & (1,6) \\
(2,1) & (2,2) & (2,3) & (2,4) & (2,5) & (2,6) \\
(3,1) & (3,2) & (3,3) & (3,4) & (3,5) & (3,6) \\
(4,1) & (4,2) & (4,3) & (4,4) & (4,5) & (4,6) \\
(5,1) & (5,2) & (5,3) & (5,4) & (5,5) & (5,6) \\
(6,1) & (6,2) & (6,3) & (6,4) & (6,5) & (6,6)
\end{array}
$$

The sum of the numbers in each pair gives the value of X. For example, the pair $(3, 1)$ corresponds to $X = 4$. Note that the pairs corresponding to a given value of X lie on a diagonal, as shown in the preceding chart, where we have indicated all pairs corresponding to $X = 4$. It is now easy to calculate the number of pairs corresponding to a given value k of X and from it the probability $\Pr(X = k)$. For example, there are three pairs adding to 4, so

$$\Pr(X = 4) = \tfrac{3}{36} = \tfrac{1}{12}.$$

Performing this calculation for all k from 2 to 12 gives the following probability distribution and the histogram shown in Fig. 9.

k	$\Pr(X = k)$	k	$\Pr(X = k)$
2	$\frac{1}{36}$	8	$\frac{5}{36}$
3	$\frac{1}{18}$	9	$\frac{1}{9}$
4	$\frac{1}{12}$	10	$\frac{1}{12}$
5	$\frac{1}{9}$	11	$\frac{1}{18}$
6	$\frac{5}{36}$	12	$\frac{1}{36}$
7	$\frac{1}{6}$		

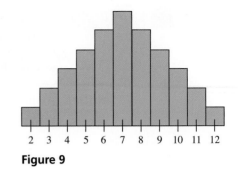

Figure 9

Now Try Exercise 5

The advantage of the random variable notation is that the variable X can be treated algebraically and we can consider expressions such as X^2. This is just the random variable corresponding to the experiment whose outcomes are the squares of the outcomes of the original experiment. Similarly, we can consider random variables such as $X + 3$ and $(X - 2)^2$. One important example of algebraic manipulation of random variables appears in Section 7.5. There are many others.

EXAMPLE 4 **Probability distribution of X^2** Suppose that a random variable X has probability distribution given by the following table:

k	$\Pr(X = k)$
-1	.2
0	.3
1	.1
2	.4

Determine the probability distribution of the random variable X^2.

Solution The outcomes of X^2 are the squares of the outcomes of X. The probabilities of the outcomes of X^2 are determined by the probabilities of the outcomes of X. The possible outcomes of X^2 are $k = 1$ (which results from the case $X = -1$ or from the case $X = 1$), $k = 0$, and $k = 4$. Since

$$\Pr(X^2 = 1) = \Pr(X = -1) + \Pr(X = 1),$$

the probability distribution of X^2 is as follows.

k	$\Pr(X^2 = k)$
0	.3
1	.3
4	.4

Now Try Exercise 13

GC Appendices B and D show how to draw histograms with a graphing calculator.

ES Appendix C shows how to use Excel's Chart Wizard to create histograms.

PRACTICE PROBLEMS 7.2

1. (*Carnival Game*) In a certain carnival game a wheel is divided into five equal parts, of which two are red and three are white. The player spins the wheel until the marker lands on "red" or until three spins have occurred. The number of spins is observed. Determine the probability distribution for this experiment.

2. (*Carnival Game*) Refer to the carnival game of Problem 1. Suppose that the player pays $1 to play this game and receives 50 cents for each spin. Determine the probability distribution for the experiment of playing the game and observing the player's earnings.

EXERCISES 7.2

1. (*Final Grades*) Table 8 gives the frequency distribution for the final grades in a course. (Here A = 4, B = 3, C = 2, D = 1, F = 0.) Determine the relative frequency distribution associated with these data and draw the associated histogram.

	TABLE 8	
Grade	Number of occurrences	
0	2	
1	3	
2	10	
3	6	
4	4	

2. (*Gas Queue*) The number of cars waiting to be served at a gas station was counted at the beginning of every minute during the morning rush hour. The frequency distribution is given in Table 9. Determine the relative frequency distribution associated with these data and draw the histogram.

TABLE 9	
Number of cars waiting	Number of occurrences
0	0
1	9
2	21
3	15
4	12
5	3

3. (*Weather Requests*) The telephone company counted the number of people dialing the weather each minute on a rainy morning from 5 A.M. to 6 A.M. The frequency distribution is given in Table 10. Determine the relative frequency distribution associated with these data.

TABLE 10	
Number of calls during minute	Number of occurrences
20	3
21	3
22	0
23	6
24	18
25	12
26	0
27	9
28	6
29	3

4. (*Production Level*) A production manager counted the number of items produced each hour during a 40-hour workweek. The frequency distribution is given in Table 11. Determine the relative frequency distribution associated with these data.

TABLE 11	
Number produced during hour	Number of occurrences
50	2
51	0
52	4
53	6
54	14
55	8
56	4
57	0
58	0
59	2

5. (*Coin Tosses*) A fair coin is tossed three times and the number of heads is observed. Determine the probability distribution for this experiment and draw its histogram.

6. (*Archery*) An archer can hit the bull's-eye of the target with probability $\frac{1}{3}$. She shoots until she hits the bull's-eye or until four shots have been taken. The number of shots is observed. Determine the probability distribution for this experiment.

7. (*Selecting Balls from an Urn*) An urn contains three red balls and four white balls. A sample of three balls is selected at random and the number of red balls observed. Determine the probability distribution for this experiment and draw its histogram.

8. (*Die Toss*) A die is rolled and the number on the top face is observed. Determine the probability distribution for this experiment and draw its histogram.

9. (*Carnival Game*) In a certain carnival game the player selects two balls at random from an urn containing two red balls and four white balls. The player receives $5 if he draws two red balls and $1 if he draws one red ball. He loses $1 if no red balls are in the sample. Determine the probability distribution for the experiment of playing the game and observing the player's earnings.

10. (*Carnival Game*) In a certain carnival game a player pays $1 and then tosses a fair coin until either a "head" occurs or he has tossed the coin four times. He receives 50 cents for each toss. Determine the probability distribution for the experiment of playing the game and observing the player's earnings.

11. Figure 10 is the histogram for a probability distribution. What is the probability that the outcome is between 5 and 7, inclusive?

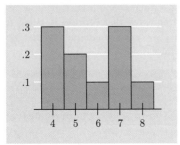

Figure 10

12. Figure 11 is the histogram for a probability distribution. To what event do the blue rectangles correspond?

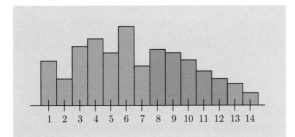

Figure 11

Let the random variables X and Y have the probability distributions listed in Table 12. Determine the probability distributions of the random variables in Exercises 13–20.

TABLE 12			
k	$\Pr(X = k)$	k	$\Pr(Y = k)$
0	.1	5	.3
1	.2	10	.4
2	.3	15	.1
3	.2	20	.1
4	.2	25	.1

13. X^2 **14.** Y^2

15. $X - 1$ **16.** $Y - 15$

17. $\frac{1}{5}Y$ **18.** $2X^2$

19. $(X + 1)^2$ **20.** $\left(\frac{1}{5}Y + 1\right)^2$

21. (*Grade Distributions*) Two classes take the same examination and the grades are recorded in Table 13. We assigned the integers 0 through 4 to the grades F through A, respectively. Table 13 gives the frequency distribution of grades in each class. Find the relative frequency distribution and the histogram for each class. Describe the difference in the grade distributions for the two classes.

TABLE 13			
		Number of students	
Grade		**9 A.M. class**	**10 A.M. class**
F	(0)	10	16
D	(1)	15	23
C	(2)	20	15
B	(3)	10	21
A	(4)	5	25

22. (*Grade Distributions*) Using Table 13, answer the following questions.

(a) What percentage of the students in the 9 A.M. class have grades of C or less?

(b) What percentage of the 10 A.M. class have grades of C or less?

(c) What percentage of the 9 A.M. class have grades of D or F?

(d) What percentage of the students in both classes combined have grades of C or more?

Exercises 23–27 refer to the tables associated with Exercises 1–4.

23. (*Grade Distribution*) For the data in Table 8, determine the percentage of students who get C or higher.

24. (*Gas Queue*) The data in Table 9 have been tabulated for every minute of the (60-minute) rush hour. What percentage of the time is the waiting line 4 or more cars?

25. (*Weather Requests*) Use the data in Table 10.

(a) For what percentage of the 60 minutes from 5 A.M. to 6 A.M. are there either less than 22 or more than 27 calls to the weather number?

(b) For what percentage of the hour are there between 23 and 25 calls (inclusive)?

(c) Draw the histogram.

(d) What would be your estimate of the average number of calls coming in during a minute of the hour for which the data have been tabulated? Explain.

26. (*Production Level*) Draw a histogram for the relative frequency distribution of Table 11.

27. (*Production Level*) For the data in Table 11, answer the following questions.

(a) What is the highest number of items produced in any one hour?

(b) What percentage of the time is that maximum number of items produced?

(c) What number of items is produced with the highest frequency?

(d) In what percentage of the 40 hours do production levels exceed 54 items?

(e) Estimate the average number of items produced per hour in this week. Explain.

28. Assume that X and Y are random variables with the given probability distributions.

k	$\Pr(X = k)$	$\Pr(Y = k)$
1	.30	.20
2	.40	.20
3	.20	.20
4	.10	.40

(a) Draw the histogram for X.

(b) Draw the histogram for Y.

(c) Find $\Pr(X = 2 \text{ or } 3)$.

(d) Find $\Pr(Y = 2 \text{ or } 3)$.

(e) Find the probability that X is at least 2.

(f) Find the probability that $X + 3$ is at least 5.

(g) Find the probability that Y^2 is at most 9.

(h) Find the probability that Y is at most 10.

(i) Find the probability distribution of $2X$.

(j) Find the probability distribution of $(Y + 2)^2$.

(k) Which of X or Y has the higher average value? Why would you think so?

29. Here is the probability distribution of the random variable U.

k	$\Pr(U = k)$
0	$\frac{3}{15}$
1	$\frac{2}{15}$
2	$\frac{4}{15}$
3	$\frac{5}{15}$
4	?

(a) Determine the probability that $U = 4$.

(b) Find $\Pr(U \geq 2)$.

(c) Find the probability that U is at most 3.

(d) Find the probability that $U + 2$ is less than 4.

(e) Draw the histogram of the distribution of U.

SOLUTIONS TO PRACTICE PROBLEMS 7.2

1. Since the outcomes are the numbers of spins, there are three possible outcomes: one, two, and three spins. The probabilities for each of these outcomes can be computed from a tree diagram (Fig. 12). For instance, the outcome two (spins) occurs if the first spin lands on white and the second spin on red. The probability of this outcome is $\frac{3}{5} \cdot \frac{2}{5} = \frac{6}{25}$.

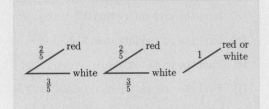

Figure 12

k	$\Pr(X = k)$
1	$\frac{2}{5}$
2	$\frac{6}{25}$
3	$\frac{9}{25}$

2. The same game is being played as in Problem 1, except that now the outcome we are concentrating on is the player's financial situation at the end of the game. The player's earnings depend on the number of spins as follows: one spin results in −$.50 earnings (i.e., a loss of 50 cents); two spins result in $0 earnings (i.e., breaking even); and three spins result in $.50 earnings (i.e., the player ends up ahead by 50 cents). The probabilities for these three situations are the same as before.

Earnings	Probability
−$.50	$\frac{2}{5}$
0	$\frac{6}{25}$
$.50	$\frac{9}{25}$

7.3 Binomial Trials

In this section we fix our attention on the simplest experiments: those with just two outcomes. These experiments, called *binomial trials* (or *Bernoulli trials*), occur in many applications. Here are some examples of binomial trials.

1. Toss a coin and observe the outcome, heads or tails.

2. Administer a drug to a sick individual and classify the reaction as "effective" or "ineffective."

3. Manufacture a light bulb and classify it as "nondefective" or "defective."

The outcomes of a binomial trial are usually called "success" and "failure." Of course, the labels "success" and "failure" need have no connection with the usual meanings of these words. For example, in experiment 2 we might label the outcome "ineffective" as "success" and "effective" as "failure." Throughout our discussion of binomial trials we will always denote the probability of "success" by p and probability of "failure" by q. Since a binomial trial has only two outcomes, we have $p + q = 1$, or

$$q = 1 - p.$$

Consider a particular binomial trial and the following experiment. Repeat the binomial trial n times and observe the number of successes that occur. Assume that the n successive trials are independent of one another. The fundamental problem of the theory of binomial trials is to calculate the probabilities of the outcomes of this experiment.

Let X be a random variable associated with the experiment. X is the number of "successes" in the n trials of the experiment. For example, if we toss a coin 20 times and assume that "heads" is a "success," then $X = 3$ means that the experiment resulted in 3 "heads" and 17 "tails." In an experiment of n trials, the number of "successes" can be any one of the numbers $0, 1, 2, \ldots, n$. These are the possible values of X.

We write $\Pr(X = k)$ to denote the probability that $X = k$, namely, the probability that k of the n trials result in success. As we saw in the coin-tossing experiment in Section 7.2, we can find the probability distribution of X using the methods of counting and basic probability principles developed earlier in the book.

> If X is the number of "successes" in n independent trials, where in each trial the probability of a "success" is p, then (with $q = 1 - p$)
>
> $$\Pr(X = k) = \binom{n}{k} p^k q^{n-k} \tag{1}$$
>
> for $k = 0, 1, 2, \ldots, n$.

Note that the right side of (1) is one of the terms in the binomial expansion of $(p + q)^n$ (see Section 5.7). We say that X is a *binomial random variable* with parameters n and p. The derivation of (1) is given at the end of this section.

Let X be the number of heads in five tosses of a fair coin. Then X is a binomial random variable with parameters $p = \frac{1}{2}$ and $n = 5$. The calculation for this particular variable appears in Section 7.2. The probability distribution is

k	$\Pr(X = k)$
0	$\frac{1}{32}$
1	$\frac{5}{32}$
2	$\frac{10}{32}$
3	$\frac{10}{32}$
4	$\frac{5}{32}$
5	$\frac{1}{32}$

By (1)

$$\Pr(X = k) = \binom{5}{k} \left(\frac{1}{2}\right)^k \left(\frac{1}{2}\right)^{5-k}.$$

Substitution of the values of k (0, 1, 2, 3, 4, 5) in (1) gives the probabilities in the table.

EXAMPLE 1 **Baseball** Each time at bat the probability that a baseball player gets a hit is .300. He comes up to bat four times in a game. Assume that his times at bat are independent trials. Find the probability that he gets (a) exactly two hits and (b) at least two hits.

Solution Each at-bat is considered an independent binomial trial. A "success" is a hit. So $p = .300$, $q = 1 - p = .700$, and $n = 4$. Therefore, X is the number of hits in four at-bats or the number of "successes" in 4 trials.

(a) We need to determine $\Pr(X = 2)$. From formula (1) with $k = 2$, we have

$$\Pr(X = 2) = \binom{4}{2}(.300)^2(.700)^{4-2} = 6(.09)(.49) = .2646.$$

(b) "At least two hits" means $X \geq 2$. Applying formula (1) with $k = 2, 3$, and 4, we have

$$\Pr(X \geq 2) = \Pr(X = 2) + \Pr(X = 3) + \Pr(X = 4)$$

$$= \binom{4}{2}(.300)^2(.700)^2 + \binom{4}{3}(.300)^3(.700)^1 + \binom{4}{4}(.300)^4(.700)^0$$

$$= 6(.09)(.49) + 4(.027)(.700) + 1(.0081)(1)$$

$$= .2646 + .0756 + .0081 = .3483.$$

So the batter can be expected to get at least two hits out of four at-bats in about 35% of the games.

Now Try Exercise 1 ■

EXAMPLE 2 **Women in the labor force** Statistics[1] show that 61% of all married women in the United States are in the labor force. Five married U.S. women are randomly selected. Find the probability that at least one of them is in the labor force. Assume that each selection is an independent binomial trial.

Solution Let "success" be "in the labor force." Then

$$p = .61 \qquad q = 1 - p = .39 \quad \text{and} \quad n = 5.$$

Therefore, X is the number of women (out of the five selected) that are in the labor force. Then

$$\Pr(X \geq 1) = 1 - \Pr(X = 0) = 1 - \binom{5}{0}(.61)^0(.39)^5$$

$$\approx 1 - .009 = .991.$$

Thus, in a group of five randomly selected married U.S. women, there is about a 99.1% chance that at least one of them is in the labor force.

Now Try Exercise 17 ■

EXAMPLE 3 **Quality control** A plumbing-supplies manufacturer produces faucet washers, which are packaged in boxes of 300. Quality control studies have shown that 2% of the washers are defective. What is the probability that a box of washers contains exactly 9 defective washers?

Solution Deciding whether a single washer is or is not defective is a binomial trial. Since we wish to consider the number of defective washers in a box, let "success" be the outcome "defective." Then

$$p = .02 \qquad q = 1 - .02 = .98 \qquad n = 300.$$

The probability that 9 out of 300 washers are defective equals

$$\Pr(X = 9) = \binom{300}{9}(.02)^9(.98)^{291} \approx .07.$$

---

[1]Statistical Abstract of the United States; U.S. Census Bureau, 1999.

Now Try Exercise 7

EXAMPLE 4   **Veterinary medicine**   The recovery rate for a certain cattle disease is 25%. If 40 cattle are afflicted with the disease, what is the probability that exactly 10 will recover?

Solution   In this example the binomial trial consists of observing a single cow, with recovery as "success." Then

$$p = .25 \qquad q = 1 - .25 = .75 \qquad n = 40.$$

The probability of 10 successes is

$$\Pr(X = 10) = \binom{40}{10}(.25)^{10}(.75)^{30} \approx .14.$$

Now Try Exercise 11

The numerical work in Examples 3 and 4 required a calculator [to compute, for example, $(.98)^{291}$]. In other problems the calculations can be stickier. Suppose, for example, that we return to Example 4. What is the probability that 16 or more cattle recover? Using formula (1) to compute the probabilities that $16, 17, \ldots, 40$ cattle recover, the desired probability is

$$\Pr(X = 16) + \Pr(X = 17) + \cdots + \Pr(X = 40)$$
$$= \binom{40}{16}(.25)^{16}(.75)^{24} + \binom{40}{17}(.25)^{17}(.75)^{23} + \cdots + \binom{40}{40}(.25)^{40}(.75)^{0}.$$

Each of the terms in this sum is difficult to compute. The thought of computing all of them should be sufficient motivation to seek an alternative approach. Fortunately, there is a reasonably simple method of approximating a sum of this type. We will illustrate the technique in Section 7.7.

Verification of Formula (1)   We first consider the case where $n = 3$ and then generalize. Consider a three-trial binomial experiment with two possible results (S or F) on each trial. Assume that the trials are independent and the probability of "S" on each trial is p. It follows that the probability of "F" on each trial is $1 - p$, which we denote by q. The tree in Fig. 1 represents the possible outcomes of the experiment.

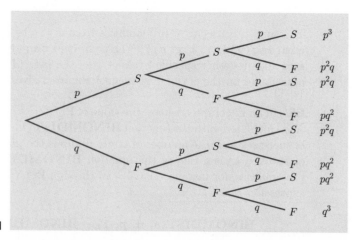

Figure 1

The probability of each individual branch is obtained by multiplying the probabilities along the branch. Then, the probability of two successes in three trials, for example, is the sum of the probabilities of each branch representing two S's and one F. There are three such branches, each with probability $p^2 q$. For the general case, suppose we want the probability of k successes in n trials. Each branch having k S's and $(n-k)$ F's has the associated probability $p^k q^{n-k}$. How many such branches are there in the tree? We can use the methods of Chapter 5 to count the number of branches with k S's and $(n-k)$ F's. What we want to know is the number of ways in which we can arrange k S's and $(n-k)$ F's. This is just $\binom{n}{k}$. So

$$\Pr(X = k) = \binom{n}{k} p^k q^{n-k}. \qquad \blacksquare$$

GC Binomial probabilities are easily calculated on most graphing calculators. Let's solve Example 1, first on the TI-83 and then on the TI-89.

(TI-83) The value of **binompdf(n,p,x)** is the probability of x successes in n trials, where the probability of success is p. (To display **binompdf(**, press $\boxed{\text{2nd}}$ [DISTR] **0**.) The sum of consecutive probabilities from $x = r$ to $x = s$ successes is the value of **sum(seq(binompdf(n,p,X),X,r,s))**. (To display the word **sum**, press $\boxed{\text{2nd}}$[LIST], move the cursor to MATH, and press **5**. To display the word **seq(**, press $\boxed{\text{2nd}}$[LIST], move the cursor to OPS, and press **5**.) Figure 2 uses these two commands to solve Example 1.

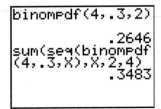

Figure 2

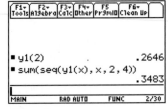

Figure 3

(TI-89) A convenient way to calculate the binomial probabilities for Example 1, is to invoke the **Y=** editor and then assign to **y1** the expression

$$\texttt{nCr(4,x)*.3\^{}x*.7\^{}(4-x)}.$$

The sum of consecutive probabilities from $x = r$ to $x = s$ successes is the value of **sum(seq(y1(x),x,r,s))**. The command **seq** creates a list, and the command **sum** adds together specified values from the list. (The words **sum(** and **seq(** are both found on the MATH/List menu.) Figure 3 gives the solution to Example 1.

ES In an Excel spreadsheet, the value of $\Pr(X = k)$ for a binomial random variable X is calculated with the function **BINOMDIST(k, n, p, 0)** where k is the number of successes, n is the number of trials, and p is the probability of success. The value of $\Pr(X \le k)$ is given by the function **BINOMDIST(k, n, p, 1)**. Consecutive probabilities, for instance $\Pr(X = r)$ through $\Pr(X = s)$, can be summed with the expression

$$\textbf{BINOMDIST}(s, n, p, 1) - \textbf{BINOMDIST}(r - 1, n, p, 1).$$

PRACTICE PROBLEMS 7.3

1. A number is selected at random from the numbers 0 through 9999. What is the probability that the number is a multiple of 5?

2. If the experiment in Problem 1 is repeated 20 times, with replacement, what is the probability of getting four numbers that are multiples of 5?

EXERCISES 7.3

1. (*Die Tosses*) A single die is tossed four times. Find the probability that exactly two of the tosses show a "one."

2. (*Coin Tosses*) Find the probability of obtaining exactly three heads when tossing a fair coin six times.

3. (*Sales*) A salesperson determines that the probability of making a sale to a customer is $\frac{1}{4}$. What is the probability of making a sale to three of the next four customers?

4. (*Basketball*) A basketball player makes free throws with probability .7. What is the probability of making exactly two out of five free throws?

5. (*Voter Preferences*) Suppose that 60% of the voters in a state intend to vote for a certain candidate. What is the probability that a survey polling five people reveals that two or fewer intend to vote for that candidate?

6. (*Exam Questions*) An exam consists of six "true or false" questions. What is the probability that a person can get five or more correct by just guessing?

7. (*Chemistry Majors*) Ten percent of all undergraduates at a university are chemistry majors. In a random sample of eight students, find the probability that exactly two are chemistry majors.

8. (*Committee Selection*) Sixty percent of all students at a university are female. A committee of five students is selected at random. Only one is a woman. Find the probability that no more than one woman is selected. What might be your conclusion about the way the committee was chosen?

9. (*Commuter Stickers*) Thirty percent of all cars crossing a toll bridge have a commuter sticker. What is the probability that among 10 randomly selected cars waiting to cross the bridge at least 2 have commuter stickers?

10. (*Die Tosses*) A die is tossed 12 times. What is the probability of at least two 5's?

11. (*Silver Cars*) Forty percent of a particular model of car are silver. What is the probability that in the next 10 observations of this model you observe 5 silver cars?

12. (*Remedial English*) Fifteen percent of the students who take a screening test are assigned to a remedial English class. In a group of eight students, what is the probability that three will be placed in the remedial class?

13. (*Consumer Preferences*) Nine customers at a supermarket are asked independently if they use brand X laundry soap. In general, 20% of the population use this brand. What is the probability that among the nine, more than two people use brand X?

14. Write the probability distribution for a binomial random variable with parameters $n = 6$, $p = \frac{1}{4}$.

15. Calculate the probability distribution for a binomial random variable with $n = 8$, $p = .40$.

16. (*Darts*) The probability is .64 that a dartist scores a bull's-eye on a single toss of a dart. What is the most probable number of bull's-eyes for him to score in his next ten tosses? (*Note*: First make a guess in order to test your intuition. Then calculate the probabilities of 6 successes and 7 successes.)

17. (*Coin Tosses*) (**PE**) A coin is tossed 4 times. What is the probability that at least one head appears?

(a) $\frac{1}{4}$ (b) $\frac{1}{2}$ (c) $\frac{7}{8}$ (d) $\frac{15}{16}$ (e) $\frac{31}{32}$

18. (a) Explain why

$$\binom{n}{k} = \binom{n}{n-k}$$

for $k = 0, 1, 2, \ldots, n$.

(b) Let X be the random variable associated with binomial trials with $n = 10$ and $p = \frac{1}{2}$. Show that

$$\Pr(X = k) = \Pr(X = 10 - k)$$

for $k = 0, 1, 2, \ldots, n$.

19. (*Jury Verdict*) A jury has 12 jurors. A vote of at least 10 of 12 for "guilty" is necessary for a defendant to be convicted of a crime. Assume that each juror acts independently of the others and that the probability that any one juror makes the correct decision on a defendant is .80. If the defendant is guilty, what is the probability that the jury makes the correct decision?

20. (*Genetics*) Every offspring inherits a gene for hair color from each parent. We denote the dominant gene by A and the recessive gene by a. If a person has AA or Aa, then the person exhibits the dominant characteristic. We call a person with the genes Aa a hybrid. A person with aa exhibits the recessive characteristic. Two hybrid parents have three children. Find the probability that at least one child exhibits the recessive characteristic.

21. (*Coin Tosses*) A coin is tossed until four heads occur. What is the probability that the fourth head occurs on the tenth toss?

22. (*Die Tosses*) A single die is rolled 10 times and the number of sixes is observed. What is the probability that a six appears 9 times, given that it appears at least 9 times?

23. (*Selecting Teams*) The students in a small class are divided into three teams—*A*, *B*, and *C*. Each week two of the teams are selected at random to participate in a competition. What is the probability that team *C* is selected at least twice during the next three weeks?

24. (*Rolling Dice*) When six dice are rolled, what is the probability that exactly five of the dice show a 5 or a 6?

25. (*Free Throws*) A basketball player makes 82% of his free throws. What is the most likely number of free throws for him to make in his next 10 tries? (*Note:* First make a guess in order to test your intuition. Then calculate the probabilities of 8 free throws and 9 free throws.)

26. (*Multiple-Choice Exam*) An exam consists of ten multiple-choice questions where each question has four choices. If you guess the answers completely at random, what is the probability that you answer at least two questions correctly?

27. (*Tennis*) Suppose that in a single game of tennis, the server wins each point with a probability of .6. The game ends when one player accumulates at least four points and is ahead by at least two points. For instance, denoting the players by A and B, player A would win with any of the following outcomes; ABAAA, AABB-BAAA, ABAABA. What is the probability that the server wins and the game ends after 6 points?

In Exercises 28–31, use a graphing calculator, a spreadsheet, or mathematical software to calculate the probabilities.

28. (*Quality Control*) Consider Example 3.

   (a) Calculate the probability that a box of washers contains exactly 12 defective washers.

   (b) What is the probability that at most 9 washers will be defective?

29. (*Veterinary Medicine*) Consider Example 4.

   (a) Calculate the probability that exactly 20 cattle will recover.

   (b) Calculate the sum $\Pr(X = 16) + \Pr(X = 17) + \cdots + \Pr(X = 40)$ discussed in the paragraph following Example 4.

30. (*Jury Duty*) Suppose there is a .40% chance of being selected for jury duty in September. A school system has 900 teachers.

   (a) What is the probability that 8 teachers will be chosen for jury duty in September?

   (b) Create a table showing the probabilities of having 0 through 6 teachers chosen.

   (c) What is the probability that at most 9 teachers will be chosen?

31. (*Dice Tosses*) When a pair of dice is tossed, the probability of obtaining seven is $\frac{1}{6}$. Suppose a pair of dice is tossed 25 times.

   (a) Calculate the probability that 8 sevens occur.

   (b) Create a table showing the probabilities of having 0 through 6 sevens occur.

   (c) What is the probability that 10 or more sevens will occur?

SOLUTIONS TO PRACTICE PROBLEMS 7.3

1. The number of multiples of 5 that occur in the numbers from 0 to 9999 is 2000. Since each of the 10,000 choices is as likely as any other, the probability is $2000/10,000 = .2$.

2. Let "success" be "the selected number is a multiple of 5." Then the selection of a number at random is a binomial trial. So $p = .2$ and $n = 20$.

$$\Pr(X = 4) = \binom{20}{4}(.2)^4(.8)^{16} \approx .2182.$$

7.4   The Mean

It is important at this point to recognize the difference between a *population* and a *sample*. A population is a set of all elements about which information is desired. A sample is a subset of a population that is analyzed in an attempt to estimate certain properties of the entire population. For instance, suppose that we are interested in finding out some characteristics of the automobile industry but do not have access to the entire population of dealerships in the United States. We must choose a random and representative sample of these dealerships and concentrate on gathering information from it.

A numerical descriptive measurement made on a sample is called a *statistic*. Such a measurement made on a population is called a *parameter* of the population. Usually, since we cannot have access to entire populations, we rely on our experimental results to obtain statistics, and we attempt to use the statistics to estimate the parameters of the population. One of the measurements we all have used is the arithmetic average. For instance, we could choose a representative random sample of $n = 200$ car dealerships across the country, determine the average weekly sales for these dealerships, and then use this value to estimate the average weekly sales of all car dealerships in the United States.

It is a familiar notion to find the *average* (also called the *mean*) of a set of numbers. For example, to obtain the average of 11, 17, 18, and 10, we add these numbers and divide by 4:

$$[\text{average}] = \frac{11 + 17 + 18 + 10}{4} = 14.$$

If we have gathered a sample of n numbers $x_1, x_2, \ldots, x_n$, the *sample mean* is

$$\overline{x} = \frac{x_1 + x_2 + \cdots + x_n}{n}.$$

EXAMPLE 1 **Average weekly car sales**   Compute the sample mean of the weekly sales of dealership A of Section 7.2.

Solution Recall that the weekly sales of dealership A are given in Table 1. Thus we may form the sample mean of the weekly sales figures as follows:

$$\tfrac{1}{52}[(5 + 5) + (6 + 6) + \underbrace{(7 + \cdots + 7)}_{13 \text{ times}} + \underbrace{(8 + \cdots + 8)}_{20 \text{ times}}$$

$$+ \underbrace{(9 + \cdots + 9)}_{10 \text{ times}} + \underbrace{(10 + \cdots + 10)}_{4 \text{ times}} + 11]$$

$$= \frac{5 \cdot 2 + 6 \cdot 2 + 7 \cdot 13 + 8 \cdot 20 + 9 \cdot 10 + 10 \cdot 4 + 11 \cdot 1}{52} \qquad (1)$$

$$= 5 \cdot \tfrac{2}{52} + 6 \cdot \tfrac{2}{52} + 7 \cdot \tfrac{13}{52} + 8 \cdot \tfrac{20}{52} + 9 \cdot \tfrac{10}{52} + 10 \cdot \tfrac{4}{52} + 11 \cdot \tfrac{1}{52}$$

$$\approx 7.96.$$

Thus the sample mean of the weekly sales is approximately 7.96 cars. ∎

TABLE 1

Weekly sales	Number of occurrences
5	2
6	2
7	13
8	20
9	10
10	4
11	1

▶ *Note* We considered the data from dealership A to be a sample since it was only one year's data and we are interested in making comparisons and using them to predict the future sales of the two dealerships. ◀

Let us reexamine the calculation above. The sample mean is given by expression (1). We did this calculation by adding together the observed sales for the 52 weeks of the year and then dividing by 52. To make the calculation more efficient, we noticed that we could group all of the weeks in which we observed five sales, all of the weeks in which there were six sales, and so on. Instead of adding 52 numbers, we simply multiplied each observed value of "sales" by its frequency of occurrence and then divided by 52. Another alternative is to multiply each observed value of "sales" by the relative frequency with which it occurred (e.g., $5 \times \frac{2}{52}$), and then add. That is, to compute the mean weekly sales we need only add up the products [number of sales] · [relative frequency of occurrence] over all possible sales figures. If the data for any sample are displayed in a frequency table or a relative frequency table, the sample mean can be calculated in a similar fashion. Since the calculation of $\bar{x}$ depends on the sample values, $\bar{x}$ is a *statistic*.

Sample Mean   Suppose that an experiment has as outcomes the numbers $x_1, x_2, \ldots, x_r$. Suppose the frequency of x_1 is f_1, the frequency of x_2 is f_2, and so forth, and that

$$f_1 + f_2 + \cdots + f_r = n.$$

Then

$$\bar{x} = \frac{x_1 f_1 + x_2 f_2 + \cdots + x_r f_r}{n},$$

or

$$\bar{x} = x_1 \left(\frac{f_1}{n} \right) + x_2 \left(\frac{f_2}{n} \right) + \cdots + x_r \left(\frac{f_r}{n} \right).$$

Similar calculations are made when we work with entire populations. Assume the population size is N and that $x_1, x_2, \ldots, x_N$ are the population values. The *population mean*, denoted by the Greek letter μ (mu), is given by the formula

$$\mu = \frac{x_1 + x_2 + \cdots + x_N}{N}.$$

If the data have been grouped into a frequency or relative frequency table, a formula analogous to the one for samples is used:

$$\mu = x_1 \left(\frac{f_1}{N} \right) + x_2 \left(\frac{f_2}{N} \right) + \cdots + x_r \left(\frac{f_r}{N} \right).$$

To help distinguish the parameter from the statistic, all parameters are denoted by Greek letters and all statistics by English letters. We use the common convention of denoting a sample size by lowercase n and a population size by uppercase N.

EXAMPLE 2 **Average life expectancy of deer**   An ecologist observes the life expectancy of a certain species of deer held in captivity. Based on a population of 1000 deer, he observes the data shown in Table 2. What is the mean life expectancy of this population of deer?

Solution We convert the given data into a relative frequency distribution by replacing observed frequencies by relative frequencies [= (observed frequency)/1000] (Table 3). The mean of these data is

$$\mu = 1 \cdot 0 + 2 \cdot (.06) + 3 \cdot (.18) + 4 \cdot (.25) + 5 \cdot (.20) + 6 \cdot (.12)$$
$$+ 7 \cdot (.05) + 8 \cdot (.12) + 9 \cdot (.02) = 4.87.$$

So the mean life expectancy of this population of deer is 4.87 years.

Now Try Exercise 3 ∎

TABLE 2	
Age at death (years)	Number observed
1	0
2	60
3	180
4	250
5	200
6	120
7	50
8	120
9	20

TABLE 3	
Age at death (years)	Relative frequency
1	0
2	.06
3	.18
4	.25
5	.20
6	.12
7	.05
8	.12
9	.02

EXAMPLE 3 **Average weekly car sales** Which car dealership should Mr. Jones buy if he wants the one that will, on the average, sell more cars?

Solution We have seen in Example 1 that the mean of the sample data for dealership A is $\overline{x}_A \approx 7.96$. On the other hand, associated to the data for dealership B (Table 4), we find the sample mean:

$$\overline{x}_B = 5 \cdot \left(\tfrac{20}{104}\right) + 6 \cdot 0 + 7 \cdot 0 + 8 \cdot \left(\tfrac{10}{104}\right) + 9 \cdot \left(\tfrac{12}{104}\right) + 10 \cdot \left(\tfrac{50}{104}\right) + 11 \cdot \left(\tfrac{12}{104}\right)$$
$$\approx 8.85.$$

Thus the average sales of dealership A are 7.96 cars per week, whereas those of dealership B are 8.85 cars per week. If we make the assumption that past sales history predicts future sales, Mr. Jones should buy dealership B.

Now Try Exercise 5

■

TABLE 4	
Weekly sales	Relative frequency
5	$\frac{20}{104}$
6	0
7	0
8	$\frac{10}{104}$
9	$\frac{12}{104}$
10	$\frac{50}{104}$
11	$\frac{12}{104}$

Expected Value The sample and population mean have analogs in the theoretical setting of random variables. Suppose that X is a random variable with the following probability distribution:

x_i	$\Pr\left(X = x_i\right)$
x_1	p_1
x_2	p_2
$\vdots$	$\vdots$
x_N	p_N

Then the values of X (namely $x_1, x_2, \ldots, x_N$) are the possible outcomes of an experiment. The expected value of X, denoted $E(X)$, is defined as follows:

The expected value of the random variable X:
$$E(X) = x_1 p_1 + x_2 p_2 + \cdots + x_N p_N.$$

Since the value of p_i represents the theoretical relative frequency of the outcome x_i, the formula for $E(X)$ is similar to the formula for the mean of a relative frequency distribution. Thus the expected value of the random variable X is also called the *mean* of the probability distribution of X and may be denoted either by $E(X)$ or by μ_X. Frequently, $E(X)$ is used interchangeably with the Greek letter μ when the context is clear.

The expected value of a random variable is the center of the probability distribution in the sense that it is the balance point of the histogram. For example, let X = the number of heads in 5 tosses of a fair coin. The probability distribution appears in Table 5 and the histogram in Fig. 1. We can calculate the mean μ of X:

$$\mu_X = 0\left(\tfrac{1}{32}\right) + 1\left(\tfrac{5}{32}\right) + 2\left(\tfrac{10}{32}\right) + 3\left(\tfrac{10}{32}\right) + 4\left(\tfrac{5}{32}\right) + 5\left(\tfrac{1}{32}\right) = \tfrac{80}{32} = 2.5.$$

TABLE 5	X = number of heads in 5 tosses of a fair coin	
k	$\Pr(X = k)$	$k \cdot \Pr(X = k)$
0	$\frac{1}{32}$	0
1	$\frac{5}{32}$	$\frac{5}{32}$
2	$\frac{10}{32}$	$\frac{20}{32}$
3	$\frac{10}{32}$	$\frac{30}{32}$
4	$\frac{5}{32}$	$\frac{20}{32}$
5	$\frac{1}{32}$	$\frac{5}{32}$
Totals 1		$\mu = \frac{80}{32} = 2.5$

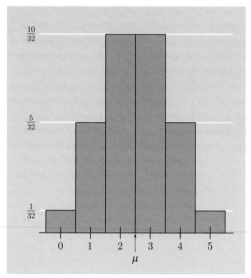

Figure 1. Probability distribution of number of heads.

The mean is shown at the bottom of the histogram (Fig. 1). In contrast, we note that the sample mean, $\overline{x}$, for the coin-tossing experiment tabulated in Section 7.2 was 2.66 (see Table 6). We rarely find that the sample mean $\overline{x}$ is exactly the theoretical value μ_X.

TABLE 6	Observed frequency of heads		
Number of heads x_i	Frequency f_i	Relative frequency (f_i/n)	$x_i \cdot (f_i/n)$
0	3	$\frac{3}{90}$	0
1	14	$\frac{14}{90}$	.16
2	23	$\frac{23}{90}$	.51
3	27	$\frac{27}{90}$	.90
4	17	$\frac{17}{90}$	.76
5	6	$\frac{6}{90}$	.33
Totals $n = 90$		1	$2.66 = \overline{x}$

We use binomial random variables so often that it is helpful to know that we have an easy formula for $E(X)$ in such cases.

If X is a binomial random variable with parameters n and p, then

$$\text{E}(X) = np. \tag{2}$$

If X is the number of heads in five tosses of a fair coin, then using the formula, we see that $\text{E}(X) = 5\left(\frac{1}{2}\right) = 2.5$, which is consistent with our previous calculation. Formula (2) will be verified at the end of this section.

EXAMPLE 4   **Quality control**   Consider the plumbing-supplies manufacturer of Example 3 of Section 7.3. Find the average number of defective washers per box.

Solution   The number of defective washers in each box is a binomial random variable with $n = 300$ and $p = .02$. The average number of defective washers per box is the expected value of X. $\text{E}(X) = np = 300 \cdot (.02) = 6$.

Now Try Exercise 15   ■

EXAMPLE 5   **Average result from tossing a pair of dice**   Let the random variable X denote the sum of the faces appearing after tossing two dice. Determine $\text{E}(X)$.

TABLE 7

k	$\Pr(X = k)$	k	$\Pr(X = k)$
2	$\frac{1}{36}$	8	$\frac{5}{36}$
3	$\frac{1}{18}$	9	$\frac{1}{9}$
4	$\frac{1}{12}$	10	$\frac{1}{12}$
5	$\frac{1}{9}$	11	$\frac{1}{18}$
6	$\frac{5}{36}$	12	$\frac{1}{36}$
7	$\frac{1}{6}$		

Solution   We determined the probability distribution of X in Example 3 of Section 7.2 (Table 7). Therefore,

$$\text{E}(X) = 2 \cdot \tfrac{1}{36} + 3 \cdot \tfrac{1}{18} + 4 \cdot \tfrac{1}{12} + 5 \cdot \tfrac{1}{9} + 6 \cdot \tfrac{5}{36} + 7 \cdot \tfrac{1}{6}$$
$$+ 8 \cdot \tfrac{5}{36} + 9 \cdot \tfrac{1}{9} + 10 \cdot \tfrac{1}{12} + 11 \cdot \tfrac{1}{18} + 12 \cdot \tfrac{1}{36} = 7.$$

Clearly, 7 is the balance point of the histogram shown in Section 7.2.

Now Try Exercise 13   ■

The expected value of a random variable may be used to analyze games of chance, as the next two examples show.

EXAMPLE 6   **Expected winnings from a dice game**   Two people play a dice game. A single die is thrown. If the outcome is 1 or 2, then A pays B $2. If the outcome is 3, 4, 5, or 6, then B pays A $4. What are the long-run expected winnings for A?

Solution Let X be the random variable representing the payoff to A. Then X assumes the possible values -2 and 4. Moreover, since the probability of 1 or 2 on the die is $\frac{1}{3}$, we have

$$\Pr(X = -2) = \tfrac{1}{3}.$$

Similarly,

$$\Pr(X = 4) = \tfrac{2}{3}.$$

Therefore,

$$\mathrm{E}(X) = (-2) \cdot \tfrac{1}{3} + 4 \cdot \tfrac{2}{3} = 2.$$

In other words, the expected payoff to A is \$2 per play. If the game is repeated a large number of times, then on the average A should profit \$2 per play. For example, in 1000 games, we expect that A will profit \$2000.

Now Try Exercise 9

In evaluating a game of chance we use the expected value of the winnings to determine how fair the game is. The expected value of a completely fair game is zero. Let us compute the expected value of the winnings for two variations of the game roulette. American and European roulette games differ in both the nature of the wheel and the rules for playing.

American roulette wheels have 38 numbers (1 through 36 plus 0 and 00), of which 18 are red, 18 are black, and 2 are green. Many different types of bets are possible. We shall consider the "red" bet. When you bet \$1 on red, you win \$1 if a red number appears and you lose \$1 otherwise.

European roulette wheels have 37 numbers (1 through 36 plus 0). The rules of European roulette differ from the American rules. One variation is as follows: When you bet \$1 on "red," you win \$1 if the ball lands on a red number and lose \$1 if the ball lands on a black number. However, if the ball lands on the green number (0), then your bet stays on the table (the bet is said to be "imprisoned") and the payoff is determined by the result of the next spin. If a red number appears, you receive your \$1 bet back, and if a black number appears, you lose your \$1 bet. However, if the green number (0) appears, you get back half of your bet, \$.50. The tree diagrams for American and European roulette are given in Fig. 2.

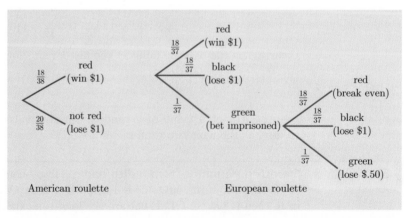

Figure 2

EXAMPLE 7   **Expected winnings from roulette**

(a) Set up the probability distribution tables for the earnings in American roulette and European roulette for the $1 bet on red.

(b) Compute the expected values for the probability distributions in part (a).

Solution   (a) For American roulette there are only two possibilities: earnings of $1 or of −$1. These occur with probabilities $\frac{18}{38}$ and $\frac{20}{38}$, respectively.

For European roulette the possible earnings are $1, $0, −$.50, or −$1. There are two ways in which to lose $1, one with probability $\frac{18}{37}$ and the other with probability $\frac{1}{37} \cdot \frac{18}{37} = \frac{18}{1369}$. Therefore,

$$\Pr(\text{lose } \$1) = \tfrac{18}{37} + \tfrac{18}{1369} = \tfrac{684}{1369}.$$

These probability distributions are tabulated in Table 8.

TABLE 8

American roulette		European roulette	
Earnings	Probability	Earnings	Probability
1	$\frac{18}{38}$	1	$\frac{18}{37}$
−1	$\frac{20}{38}$	0	$\frac{18}{1369}$
		$-\frac{1}{2}$	$\frac{1}{1369}$
		−1	$\frac{684}{1369}$

(b) American roulette:

$$\mu = 1 \cdot \tfrac{18}{38} + (-1) \cdot \tfrac{20}{38} = -\tfrac{2}{38} \approx -.0526.$$

European roulette:

$$\mu = 1 \cdot \tfrac{18}{37} + 0 \cdot \tfrac{18}{1369} + \left(-\tfrac{1}{2}\right)\tfrac{1}{1369} + (-1)\tfrac{684}{1369}$$

$$= -\tfrac{1}{74} \approx -.0135.$$

Now Try Exercise 7   ■

The secrets of Nicholas Dandolas, one of the famous gamblers of the twentieth century, are revealed by Ted Thackrey, Jr., in *Gambling Secrets of Nick the Greek* (Rand-McNally, 1968). The chapter entitled "Roulette" is subtitled "For Europeans Only." Looking at the probabilities of winning does not reveal any significant advantage of European roulette over American roulette: $\frac{18}{37}$ is not much bigger than $\frac{18}{38}$. Also, the chance in European roulette to break even is very small. The real difference between the two games is revealed by the expected values. Someone playing American roulette will lose, on the average, about $5\frac{1}{4}$ cents per $1 bet, whereas for European roulette the average loss is about $1\frac{1}{3}$ cents. In both cases you expect to lose money in the long run, but in American roulette you lose nearly four times as much.

In summary, there are three things to which the mean applies: samples, populations, and probability distributions.

Verification of Formula (2)   Note that

$$\mu = 0 \cdot \Pr(X = 0) + 1 \cdot \Pr(X = 1) + \cdots + n \cdot \Pr(X = n)$$

$$= 0 \cdot \binom{n}{0} p^0 q^{n-0} + 1 \cdot \binom{n}{1} p^1 q^{n-1} + \cdots + n \cdot \binom{n}{n} p^n q^{n-n}.$$

Moreover, note that

$$k \binom{n}{k} = k \cdot \frac{n(n-1) \cdot \ \cdots \ \cdot (n-k+1)}{k(k-1) \cdot \ \cdots \ \cdot 2 \cdot 1} = n \cdot \frac{(n-1) \cdot \ \cdots \ \cdot (n-k+1)}{(k-1) \cdot \ \cdots \ \cdot 2 \cdot 1}$$

$$= n \binom{n-1}{k-1} \qquad (k = 1, 2, 3, \ldots, n).$$

Therefore,

$$\mu = 1 \cdot \binom{n}{1} p^1 q^{n-1} + 2 \cdot \binom{n}{2} p^2 q^{n-2} + \cdots + n \cdot \binom{n}{n} p^n q^0$$

$$= n \binom{n-1}{0} p^1 q^{n-1} + n \binom{n-1}{1} p^2 q^{n-2} + \cdots + n \binom{n-1}{n-1} p^n q^0$$

$$= np \left[\binom{n-1}{0} p^0 q^{n-1} + \binom{n-1}{1} p^1 q^{n-2} + \cdots + \binom{n-1}{n-1} p^{n-1} q^0 \right]$$

$$= np(p+q)^{n-1} \qquad \text{(by the binomial theorem)}$$

$$= np \qquad \text{(since } p + q = 1\text{).} \qquad \blacksquare$$

PRACTICE PROBLEMS 7.4

1. (*Life Insurance*) A 74-year-old man pays $100 for a one-year life insurance policy, which pays $2000 in the event that he dies during the next year. According to life insurance tables, the probability of a 74-year-old man living one additional year is .95. Write down the probability distribution for the possible financial outcome and determine its expected value.

2. (*Life Insurance*) According to life insurance tables, the probability that a 74-year-old man will live an additional five years is .7. How much should a 74-year-old be willing to pay for a policy that pays $2000 in the event of death any time within the next 5 years?

EXERCISES 7.4

1. Find the expected value for the probability distribution in Table 9.

TABLE 9

Value	Probability
0	.15
1	.2
2	.1
3	.25
4	.3

2. Find the expected value for the probability distribution in Table 10.

TABLE 10

Value	Probability
-1	.1
$-\frac{1}{2}$	.4
0	.25
$\frac{1}{2}$	.2
1	.05

3. (*Grades*) A college student received the following course grades for 10 (three-credit) courses during his freshman year: 4, 4, 4, 3, 3, 3, 3, 2, 2, 1.

(a) Find his grade point average by adding the grades and dividing by 10.

(b) Write down the relative frequency table.

(c) Find the mean of the relative frequency distribution in part (b).

4. (*Gymnastic Scores*) An Olympic gymnast received the following scores from six judges: 9.8, 9.8, 9.4, 9.2, 9.2, 9.0.

(a) Find the average score by adding the scores and dividing by 6.

(b) Write down the relative frequency table.

(c) Find the mean of the relative frequency distribution in part (b).

5. (*Comparing Toothpastes*) Table 11 gives the relative frequency of the number of cavities for two groups of children trying different brands of toothpaste. Calculate the sample means to determine which group had fewer cavities.

TABLE 11

Number of cavities	Relative frequency	
	Group A	Group B
0	.3	.2
1	.3	.3
2	.2	.3
3	.1	.1
4	0	.1
5	.1	0

6. (*Investments*) Table 12 gives the possible returns of two different investments and their probabilities. Calculate the means of the probability distributions to determine which investment has the greater expected return.

TABLE 12

Investment A	
Return	Probability
$1000	.2
$2000	.5
$3000	.3
Investment B	
Return	Probability
−$3000	.1
0	.3
$4000	.6

7. (*Roulette*) In American roulette, a bettor may place a $1 bet on any one of the 38 numbers on the roulette wheel. He wins $35 (plus the return of his bet) if the ball lands on his number; otherwise, he loses his bet. Write down the probability distribution for the earnings from this type of bet and find the expected value.

8. (*Roulette*) In American roulette, a dollar may be bet on a pair of numbers. The expected earnings for this type of bet is $-\$\frac{1}{19}$. How much money does the bettor receive if the ball lands on one of the two numbers?

9. (*Carnival Game*) In a carnival game, the player selects two coins from a bag containing two silver dollars and six slugs. Write down the probability distribution for the winnings and determine how much the player would have to pay so that he would break even, on the average, over many repetitions of the game.

10. (*Carnival Game*) In a carnival game, the player selects balls one at a time, without replacement, from an urn containing two red and four white balls. The game proceeds until a red ball is drawn. The player pays $1 to play the game and receives $\$\frac{1}{2}$ for each ball drawn. Write down the probability distribution for the player's earnings and find its expected value.

11. (*Life Insurance*) Using life insurance tables, a retired man determines that the probability of living 5 more years is .9. He decides to take out a life insurance policy that will pay $10,000 in the event that he dies during the next 5 years. How much should he be willing to pay for this policy? (Do not take account of interest rates or inflation.)

12. (*Life Insurance*) Using life insurance tables, a retired couple determines that the probability of living 5 more years is .9 for the man and .95 for the woman. They decide to take out a life insurance policy that will pay $10,000 if either one dies during the next 5 years and $15,000 if both die during that time. How much should they be willing to pay for this policy? (Assume that their life spans are independent events.)

13. (*Dice*) A pair of dice is tossed and the larger of the two numbers showing is recorded. Find the expected value of this experiment.

14. (*Baseball*) Ted is a consistent hitter with a .275 batting average. How many hits is he expected to have in his next 40 at-bats?

15. (*Rolling a Die*) A die is rolled 30 times. What is the expected number of times that a 5 or a 6 will appear?

16. What is the probability of success for a binomial random variable with 20 trials whose expected value is 3?

17. (*Basketball*) A basketball player makes 40% of his three-point shots and 60% of his free throws. If he is fouled while taking a three-point shot, he is given three free throws. Which is greater: the expected number of points from taking a three-point shot or the expected number of points from taking three free throws?

18. (*Exam Scores*) (PE) A student's exam scores are 95, 88, and 79. What score must the student earn on the fourth exam to have an average (arithmetic mean) score of 90?

    (a) 96 (b) 97 (c) 98 (d) 99 (e) 100

19. (PE) If 5, 6, and x have the same average (arithmetic mean) as 2, 7, and 9, then x equals

    (a) 4 (b) 5 (c) 6 (d) 7 (e) 8

20. (*Candle Sales*) (PE) A store sold an average (arithmetic mean) of x candles per day for k days, and then sold y candles on the next day. What is the average number of candles sold daily for the $(k+1)$-day period?

    (a) $x + \dfrac{y}{k}$ (b) $\dfrac{kx + y}{k + 1}$ (c) $\dfrac{k(x + y)}{k + 1}$

    (d) $\dfrac{x + ky}{k + 1}$ (e) $x + \dfrac{y}{k + 1}$

21. (*Ticket Sales*) (PE) Last weekend a movie theater sold x adult tickets at $7 each and y children's tickets at $4 each. The average (arithmetic mean) revenue per ticket was

    (a) $\dfrac{28xy}{x + y}$ (b) $\dfrac{7x + 4y}{x + y}$ (c) $\dfrac{7x + 4y}{11}$

    (d) $\dfrac{28xy}{11}$ (e) $\dfrac{7x + 4y}{xy}$

22. (PE) If three distinct positive integers have an average (arithmetic mean) of 70, and if the smallest of the three integers is 50, then the largest of the three integers can be at most

    (a) 70 (b) 99 (c) 100 (d) 109 (e) 159

23. (PE) If three distinct positive integers have an average (arithmetic mean) of 200, and if the smallest of the three integers is 120, then the largest of the three integers can be at most

    (a) 200 (b) 350 (c) 359 (d) 459 (e) 460

24. (*Weekly Revenue*) (PE) A small business had an average (arithmetic mean) weekly revenue of $14,000 over the past three weeks. The revenue for the first week was twice the revenue of the third week, and the revenue for the second week was half the revenue of the third week. What was the revenue for the first week?

    (a) $6000 (b) $12,000 (c) $18,000

    (d) $20,000 (e) $24,000

25. (*Baseball Cards*) (PE) Tom, Dick, and Harry have an average (arithmetic mean) of 120 baseball cards. Tom has $1\frac{1}{2}$ times as many cards as Dick, and Harry has $\frac{1}{2}$ as many cards as Dick. How many baseball cards does Tom have?

    (a) 60 (b) 90 (c) 120 (d) 180 (e) 240

26. (*Exam Scores*) (PE) Three members of a study group each earned a score of 91 on an exam, and the other five members of the group each earned a score of 87. The average (arithmetic mean) score earned by the members of this study group is

    (a) 88.0 (b) 88.5 (c) 89.0 (d) 89.5 (e) 90.0

27. (*Batting Averages*) (PE) Five members of a baseball team have batting averages of .300, and the other four members of the team have batting averages of .350. The overall batting average among members of the team is closest to

    (a) .311 (b) .322 (c) .325 (d) .330 (e) .333

28. (*Magazine Sales*) (PE) Half of the magazines at a newsstand sell for an average (arithmetic mean) of $2.00 and the other half sell for an average of $2.50. If the total retail value of the magazines is $135.00, how many magazines are there at the newsstand?

    (a) 56 (b) 57 (c) 58 (d) 59 (e) 60

29. (*Truck Capacity*) (PE) A truck can carry a maximum of 75,000 pounds of cargo. How many cases of cargo can it carry if half of the cases have an average (arithmetic mean) weight of 20 pounds and the other half have an average weight of 30 pounds?

    (a) 3000 (b) 3200 (c) 3500 (d) 3750 (e) 4000

30. (*Theft Insurance*) (PE) Bob wishes to insure a priceless family heirloom against theft. The annual premium for policy A is $150, and it will pay $75,000 if the heirloom is stolen. Policy B will pay $100,000, but the annual premium for policy B is $250. Bob estimates the probability that the heirloom will be stolen in any given year and remains undecided between the two policies. What is this estimated probability?

    (a) .001 (b) .002 (c) .003 (d) .004 (e) .005

31. (*Rain Insurance*) (PE) The promoter of a football game is concerned that it will rain. She has the option of spending $8000 on insurance that will pay $40,000 if it rains. She estimates that the revenue from the game will be $60,000 if it does not rain and $25,000 if it does rain. What must the chance of rain be if she is ambivalent about this insurance?

    (a) 20% (b) 25% (c) 30% (d) 35% (e) 40%

SOLUTIONS TO PRACTICE PROBLEMS 7.4

1. There are two possibilities. If the man lives until the end of the year, he loses $100. If he dies during the year, his estate gains $1900 (the $2000 settlement minus the $100 premium).

Outcome	Probability
−$100	.95
$1900	.05

$$\mu = (-100)(.95) + (1900)(.05) = 0$$

(Thus, if the insurance company insures a large number of people, it should break even. Its profits will result from the interest that it earns on the money being held.)

2. Let x denote the cost of the policy. The probability distribution is as follows.

Outcome	Probability
$-x$	.7
$2000 - x$	.3

$$\mu = (-x)(.7) + (2000 - x)(.3)$$
$$= -.7x + 600 - .3x$$
$$= 600 - x$$

The expected value will be zero if $x = 600$. Therefore, the man should be willing to pay up to $600 for his policy.

7.5 The Variance and Standard Deviation

In Section 7.4 we introduced three analogous concepts: the mean of a sample ($\bar{x}$); the mean of a population (μ); and the mean or expected value of a random variable [E(X)]. The mean is probably the single most important number that can be used to describe a sample, a population, or a probability distribution of a random variable. The next most important number is the *variance*.

Roughly speaking, the variance measures the dispersal or spread of a distribution about its mean. The more closely concentrated the distribution about its mean, the smaller the variance; the more spread out, the larger the variance. Thus, for example, the probability distribution whose histogram is drawn in Fig. 1(a) has a smaller variance than that in Fig. 1(b).

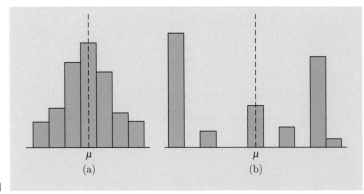

Figure 1

Let us now define the variance of a probability distribution of a random variable. Suppose X is a random variable with values $x_1, x_2, \ldots, x_N$ and respective probabilities $p_1, p_2, \ldots, p_N$. Suppose that the mean is μ. Then the deviations of the various outcomes from the mean are given by the N differences

$$x_1 - \mu, \quad x_2 - \mu, \quad \ldots, \quad x_N - \mu.$$

Since we want to give weight to the various deviations according to their likelihood of occurrence, it is tempting to multiply each deviation by its probability of occurrence.

However, this will not lead to a very satisfactory measure of deviation from the mean. This is because some of the differences will be positive and others negative. In the process of addition, deviations from the mean (both positive and negative deviations) will combine to yield a zero total deviation. To correct this, we consider instead the squares of the differences:

$$(x_1 - \mu)^2, \quad (x_2 - \mu)^2, \quad \ldots, \quad (x_N - \mu)^2,$$

which are all ≥ 0. To obtain a measure of deviation from the mean, we multiply each of these expressions by the probability of the corresponding outcome. The number thus obtained is called the *variance of the probability distribution* (or of the associated random variable). That is, the variance of a probability distribution is given by the formula

$$[\text{variance}] = (x_1 - \mu)^2 p_1 + (x_2 - \mu)^2 p_2 + \cdots + (x_N - \mu)^2 p_N.$$

EXAMPLE 1 **Computing variance** Compute the variance of the following probability distribution:

Outcome	Probability
0	.1
1	.3
2	.5
3	.1

Solution The mean is given by

$$\mu = 0 \cdot (.1) + 1 \cdot (.3) + 2 \cdot (.5) + 3 \cdot (.1) = 1.6.$$

In the notation of random variables, the calculations of the variance may be summarized as follows:

k	$\Pr(X = k)$	$k - \mu$	$(k - \mu)^2$	$(k - \mu)^2 \Pr(X = k)$
0	.1	$0 - 1.6 = -1.6$	2.56	.256
1	.3	$1 - 1.6 = -.6$	.36	.108
2	.5	$2 - 1.6 = .4$	.16	.080
3	.1	$3 - 1.6 = 1.4$	1.96	.196

$$[\text{variance}] = .256 + .108 + .080 + .196 = .640.$$

Now Try Exercise 1 ∎

Actually, a much more commonly used measure of dispersal about the mean is the *standard deviation*, which is just the square root of the variance:

$$[\text{standard deviation}] = \sqrt{[\text{variance}]}.$$

The most commonly used notation for standard deviation is the Greek letter σ (sigma). Thus, for example, for the probability distribution of Example 1 we have

$$\sigma = \sqrt{[\text{variance}]} = \sqrt{.64} = .8.$$

We denote the variance by σ^2 (sigma squared). The reason for using the standard deviation as opposed to the variance is that the former is expressed in the same units of measurement as X, whereas the latter is not.

In a similar way, we can define the variance and standard deviation for a collection of data—either for an entire population or for a sample drawn from a population. If we have collected data for an entire population under study, with values $x_1, x_2, \ldots, x_N$, then the variance can be found by first finding the mean, μ, and then finding σ^2:

$$\sigma^2 = \frac{1}{N} \left[(x_1 - \mu)^2 + (x_2 - \mu)^2 + \cdots + (x_N - \mu)^2 \right].$$

If the data have been grouped into a frequency table or relative frequency table, the appropriate formulas would be

$$\sigma^2 = \frac{1}{N} \left[(x_1 - \mu)^2 (f_1) + (x_2 - \mu)^2 (f_2) + \cdots + (x_r - \mu)^2 (f_r) \right]$$

or

$$\sigma^2 = (x_1 - \mu)^2 \left(\frac{f_1}{N} \right) + (x_2 - \mu)^2 \left(\frac{f_2}{N} \right) + \cdots + (x_r - \mu)^2 \left(\frac{f_r}{N} \right),$$

where the value x_1 occurs with frequency f_1, the value x_2 occurs with frequency f_2, and so on. Recall that N denotes the population size.

EXAMPLE 2   **Computing variance and standard deviation**   Compute the variance and the standard deviation for the population of scores on a five-question quiz as tabulated in Table 1.

Solution   We first find μ.

$$\mu = \tfrac{1}{60}[0(4) + 1(9) + 2(6) + 3(14) + 4(18) + 5(9)] = \tfrac{180}{60} = 3.$$

We find σ^2 by subtracting 3 from each of the test scores, squaring the differences, weighting each with its frequency, and dividing the resulting sum by $N = 60$. The computation is shown in Table 2. Therefore, $\sigma^2 = \tfrac{132}{60} = 2.2$. The standard deviation, which is found by taking the square root of the variance, is $\sigma \approx 1.48$. ∎

TABLE 1	
Score	Frequency
0	4
1	9
2	6
3	14
4	18
5	9
Total	60

TABLE 2				
x_i	f_i	$x_i - \mu$	$(x_i - \mu)^2$	$(x_i - \mu)^2 (f_i)$
0	4	−3	9	36
1	9	−2	4	36
2	6	−1	1	6
3	14	0	0	0
4	18	1	1	18
5	9	2	4	36
Totals	60			132

As we discussed in Section 7.4, we sometimes have only sample values at our disposal. If so, we must use sample statistics to estimate the population parameters. For example, we can use the sample mean $\overline{x}$ as an estimate of the population mean μ, and the sample variance as an estimate of the population variance, σ^2. In fact, if samples of size n were chosen repeatedly from a population and the sample mean were computed for each sample, then the average of these means should be close to the value of the population mean, μ. This is a desirable property for any statistic used to estimate a parameter—the averages of the statistic get arbitrarily close to the actual value of the parameter as the number of samples increases. Such an estimate is said to be *unbiased*. Thus $\bar{x}$ is an unbiased estimate of μ.

The situation with the variance is a little trickier. In order to have an unbiased estimate of the population variance σ^2, we must define the *sample variance*, s^2, in a slightly peculiar way. Suppose that $x_1, x_2, \ldots, x_n$ are the n sample values. Then the sample variance is

$$s^2 = \frac{1}{n-1}\left[(x_1 - \bar{x})^2 + (x_2 - \bar{x})^2 + \cdots + (x_n - \bar{x})^2\right],$$

and the sample standard deviation is $s = \sqrt{s^2}$. This is the usual definition and the way in which most statistical calculators do the computation of the sample variance. Note that the divisor is one less than the sample size. With this definition, s^2 is an unbiased estimate of σ^2. The formula for the sample standard deviation when the data are presented by a frequency table is analogous:

$$s^2 = \frac{1}{n-1}\left[(x_1 - \bar{x})^2(f_1) + (x_2 - \bar{x})^2(f_2) + \cdots + (x_r - \bar{x})^2(f_r)\right].$$

> The sample mean, $\bar{x}$, is an unbiased estimate of the population mean μ. The sample variance, s^2, is an unbiased estimate of the population variance σ^2.

EXAMPLE 3   **Computing sample standard deviation**   Compute the sample standard deviations for the frequency distributions of sales in car dealerships A and B.

Solution   The frequency distribution for dealership A is given by Table 3. In Example 1 of Section 7.4, we found the sample mean of the weekly sales to be $\bar{x}_A = 7.96$. Recall that we are treating the data collected from these dealerships as samples (one year's data from dealership A and two year's data from dealership B). Therefore, the sample variance for dealership A is given by

$$s_A^2 = \tfrac{1}{51}\big[(5 - 7.96)^2 \cdot 2 + (6 - 7.96)^2 \cdot 2 + (7 - 7.96)^2 \cdot 13$$
$$+ (8 - 7.96)^2 \cdot 20 + (9 - 7.96)^2 \cdot 10 + (10 - 7.96)^2 \cdot 4$$
$$+ (11 - 7.96)^2 \cdot 1\big] \approx 1.45.$$

The sample standard deviation, s_A, for dealership A is given by

$$s_A = \sqrt{s_A^2} = \sqrt{1.45} \approx 1.20 \text{ cars.}$$

In a similar way, we find that the sample standard deviation of dealership B is $s_B \approx 2.03$ cars. Since s_A is smaller than s_B, dealership B exhibited greater variation than dealership A during the time the sales were observed. On average, dealership B had higher weekly sales, but those of dealership A showed greater consistency. The sample statistics might help Mr. Jones decide which dealership to buy. He will have to decide if consistency is more important than the size of long-term average sales per week.

TABLE 3

Weekly sales	Frequency
5	2
6	2
7	13
8	20
9	10
10	4
11	1
Total	$n = 52$

Now Try Exercise 5   ■

There is an alternative formula for σ^2 which can simplify its calculation:

$$\sigma^2 = E(X^2) - \mu^2.$$

EXAMPLE 4 **Using the alternate formula for variance** Use the alternative formula for the variance to find σ^2 for X, the number of heads in five tosses of a coin.

Solution We tabulate the essential values (Table 4). From the table we can see that

$$E(X^2) = \tfrac{240}{32} = \tfrac{15}{2} \qquad \text{and} \qquad \mu = \tfrac{80}{32} = \tfrac{5}{2}.$$

Thus

$$\sigma^2 = E(X^2) - \mu^2 = \tfrac{15}{2} - \left(\tfrac{5}{2}\right)^2 = \tfrac{15}{2} - \tfrac{25}{4} = \tfrac{5}{4}.$$

TABLE 4

k	$\Pr(X = k)$	k^2	$k\Pr(X = k)$	$k^2 \Pr(X = k)$
0	$\frac{1}{32}$	0	0	0
1	$\frac{5}{32}$	1	$\frac{5}{32}$	$\frac{5}{32}$
2	$\frac{10}{32}$	4	$\frac{20}{32}$	$\frac{40}{32}$
3	$\frac{10}{32}$	9	$\frac{30}{32}$	$\frac{90}{32}$
4	$\frac{5}{32}$	16	$\frac{20}{32}$	$\frac{80}{32}$
5	$\frac{1}{32}$	25	$\frac{5}{32}$	$\frac{25}{32}$
			Totals $\frac{80}{32}$	$\frac{240}{32}$

Now Try Exercise 17

■

One of the big advantages of this alternative formula is that μ and σ^2 can be calculated at the same time; this means that we need only make one pass through the probability distribution. In addition, if the mean is an unwieldy number, we are spared the tedious task of subtracting it from the values of the variable, which sometimes produces even more unwieldy numbers that must be squared and summed.

Once again, binomial random variables are used so often that we find it convenient to have a formula for the variance that avoids messy calculations.

If X is a binomial random variable with parameters n and p, then

$$\sigma^2 = npq.$$

For instance, the variance in Example 4 can be calculated as

$$\sigma^2 = 5 \cdot \tfrac{1}{2} \cdot \tfrac{1}{2} = \tfrac{5}{4}.$$

The variance and standard deviation are used in many sophisticated statistical analyses, which are beyond the scope of this book. For example, we can use the value of the sample mean to estimate the population mean. The sample standard deviation helps us to determine the degree of accuracy of our estimate. If the sample standard deviation is small, indicating that the population is not widely dispersed about its mean, the estimated value of μ is likely to be close to the actual value of μ.

What does the standard deviation tell us about the dispersal of the data about the mean? *Chebychev's inequality* helps us to see that the larger the standard deviation, the more likely it is that we find extreme values in the data. The probability that an outcome falls more than c units away from the mean is at most σ^2/c^2.

> **Chebychev's Inequality** Suppose that a probability distribution with numerical outcomes has expected value μ and standard deviation σ. Then the probability that a randomly chosen outcome lies between $\mu - c$ and $\mu + c$ is at least $1 - (\sigma^2/c^2)$.

A verification of the Chebychev inequality can be found in most elementary statistics texts.

EXAMPLE 5 **Applying Chebychev's inequality** Suppose that a probability distribution has mean 5 and standard deviation 1. Use the Chebychev inequality to estimate the probability that an outcome lies between 3 and 7.

Solution Here $\mu = 5$, $\sigma = 1$. Since we wish to estimate the probability of an outcome lying between 3 and 7, we set $\mu - c = 3$ and $\mu + c = 7$. Thus $c = 2$. Then by the Chebychev inequality, the desired probability is at least

$$1 - \frac{\sigma^2}{c^2} = 1 - \frac{1}{4} = .75.$$

That is, if the experiment is repeated a large number of times, we expect at least 75% of the outcomes to be between 3 and 7. Also, we expect at most 25% of the outcomes to fall below 3 or above 7.

Now Try Exercise 9(a) ■

The Chebychev inequality has many practical applications, one of which is illustrated in the next example.

EXAMPLE 6 **Quality control** Apex Drug Supply Company sells bottles containing 100 capsules of penicillin. Due to the bottling procedure, not every bottle contains exactly 100 capsules. Assume that the average number of capsules in a bottle is indeed 100 ($\mu = 100$) and the standard deviation is 2 ($\sigma = 2$). If the company ships 5000 bottles, estimate the number having between 95 and 105 capsules inclusive.

Solution By Chebychev's inequality, the proportion of bottles having between $100 - 5$ and $100 + 5$ capsules should be at least

$$1 - (2^2/5^2) = \tfrac{21}{25} = .84.$$

That is, we expect at least 84% of the 5000 bottles, or 4200 bottles, to be in the desired range.

Now Try Exercise 11 ■

Note that the estimate provided by Chebychev's inequality is crude. In more advanced statistics books, you can find sharper estimates. Also, in the case of the normal distribution, we shall provide a much more precise way of estimating the probability of falling within c units of the mean, based on using a table of areas under the normal curve (see Section 7.6).

GC Sample means and standard deviations can be easily determined with a graphing calculator.

(TI-83) Consider the car dealership data from Table 3. In Fig. 2 the weekly sales are entered into list **L₁** and the frequencies into **L₂**. Figure 3 was invoked by entering

$$1-\textbf{Var Stats } \textbf{L}_1, \textbf{L}_2$$

from the home screen. (To display **1-Var Stats**, press $\boxed{\text{STAT}}$ $\boxed{\blacktriangleright}$ **1**.) The sample mean and the sample standard deviation are denoted by $\overline{\textbf{x}}$ and **Sx**.

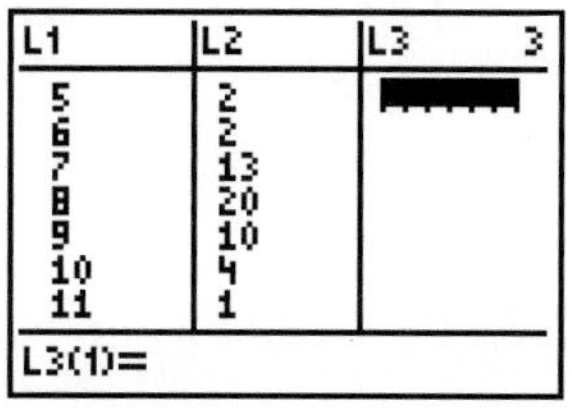

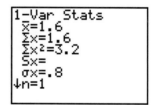

Figure 2 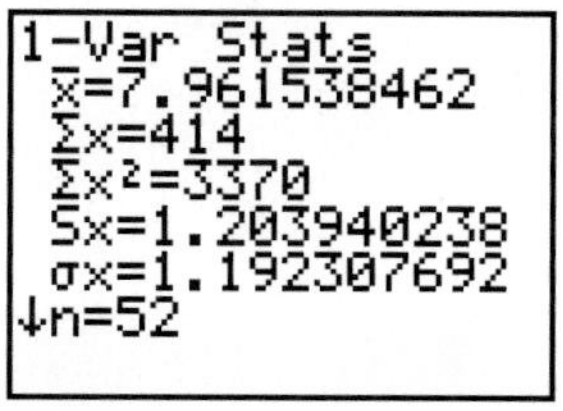   **Figure 3**

Means and standard deviations of probability distributions can be calculated in the same way by placing the probabilities in list **L₂**. See Fig. 4. Figure 5 shows the probability distribution from Example 1. The standard deviation is denoted **x**. Note that **n** is the sum of the entries in **L₂**.

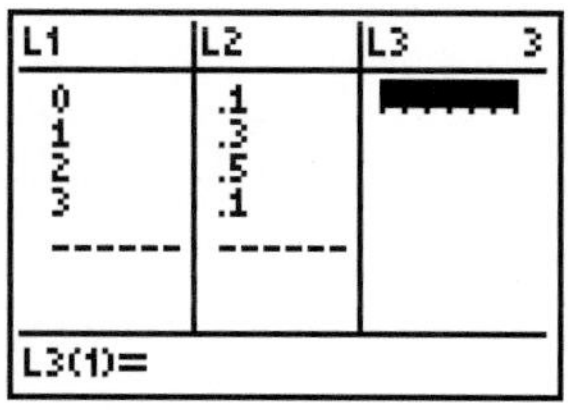

Figure 4   **Figure 5**

(TI-89) The following steps find the mean and standard deviation for the car dealership data in Table 3.

1. Invoke the Data/Matrix editor from the APPS menu.

2. Select 3 to set up a new data variable dialog box.

3. Set **Type** to **Data**, and give the variable a name such as **s** for sales.

4. Press $\boxed{\text{ENTER}}$ twice to invoke a spreadsheet.

5. Place the weekly sales figures in the c1 column and the frequency figures in the c2 column.

6. Press F5 to bring up the Calculate dialog box.

7. Set **Calculation Type** to **OneVar**, set **x** to **c1**, set **Use Freq and Categories** to **YES**, and set **Freq** to **c2**.

8. Press $\boxed{\text{ENTER}}$ twice to bring up a STAT VARS window. The sample mean and the sample deviation appear denoted by $\overline{\textbf{x}}$ and **Sx**.

ES　After the numbers in a sample have been placed into a range, the sample mean, variance, and standard deviation can be calculated by evaluating the functions AVERAGE, VAR, and STDEV of the range.

The formulas for the mean, variance, and standard deviation of a probability distribution are easily evaluated on a spreadsheet. Figure 6 shows the computations for the probability distribution of Example 1, and Fig. 7 shows the contents of the cells in Fig. 6.

	A	B	C	D	E
1	0	0.1	0	0.256	
2	1	0.3	0.3	0.108	
3	2	0.5	1	0.08	
4	3	0.1	0.3	0.196	
5			1.6	0.64	0.8
6			**Mean**	**Var.**	**St. Dev.**

Figure 6

	A	B	C	D	E
1	0	0.1	=A1*B1	=(A1-$C$5)^2*B1	
2	1	0.3	=A2*B2	=(A2-$C$5)^2*B2	
3	2	0.5	=A3*B3	=(A3-$C$5)^2*B3	
4	3	0.1	=A4*B4	=(A4-$C$5)^2*B4	
5			=SUM(C1:C4)	=SUM(D1:D4)	=SQRT(D5)
6			**Mean**	**Var.**	**St. Dev.**

Figure 7

PRACTICE PROBLEMS 7.5

1. (a) Compute the variance of the probability distribution in Table 5.

   (b) Using Table 5, find the probability that the outcome is between 22 and 24, inclusive.

2. Refer to the probability distribution of Problem 1. Use the Chebychev inequality to approximate the probability that the outcome is between 22 and 24.

TABLE 5

Outcome	Probability
21	$\frac{1}{16}$
22	$\frac{1}{8}$
23	$\frac{5}{8}$
24	$\frac{1}{8}$
25	$\frac{1}{16}$

EXERCISES 7.5

1. Compute the variance of the probability distribution in Table 6.

TABLE 6

Outcome	Probability
70	.5
71	.2
72	.1
73	.2

2. Compute the variance of the probability distribution in Table 7.

3. Determine by inspection which one of the probability distributions, A or B, in Fig. 8 has the greater variance.

TABLE 7

Outcome	Probability
-1	$\frac{1}{8}$
$-\frac{1}{2}$	$\frac{3}{8}$
0	$\frac{1}{8}$
$\frac{1}{2}$	$\frac{1}{8}$
1	$\frac{2}{8}$

4. Determine by inspection which one of the probability distributions, B or C, in Fig. 8 has the greater variance.

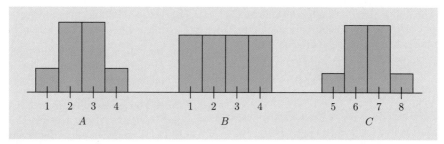

Figure 8

5. (*Investment Returns*) Table 8 gives the probability distribution for the possible returns from two different investments.

(a) Compute the mean and the variance for each investment.

(b) Which investment has the higher expected return (i.e., mean)?

(c) Which investment is less risky (i.e., has lesser variance)?

TABLE 8

Investment A

Return ($ millions)	Probability
−10	$\frac{1}{5}$
20	$\frac{3}{5}$
25	$\frac{1}{5}$

Investment B

Return ($ millions)	Probability
0	.3
10	.4
30	.3

6. (*Golf Scores*) Two golfers recorded their scores for 20 nine-hole rounds of golf. Golfer A's scores were

$$39, 39, 40, 40, 40, 40, 40, 40,$$
$$41, 41, 41, 41, 41, 41, 41, 42, 43, 43, 43, 44.$$

Golfer B's scores were

$$40, 40, 40, 41, 41, 41, 41, 42, 42, 42, 42, 42,$$
$$43, 43, 43, 43, 43, 43, 44, 44.$$

(a) Compute the sample mean and the variance of each golfer's scores.

(b) Who is the better golfer? (*Note*: The lower the score, the better.)

(c) Who is the more consistent golfer?

7. (*Weekly Sales*) Table 9 gives the relative frequency distribution for the weekly sales of two businesses.

(a) Compute the population mean and the variance for each business.

(b) Which business has the better sales record?

(c) Which business has the more consistent sales record?

TABLE 9

Sales	Relative frequency Business A	Business B
100	.1	0
101	.2	.2
102	.3	0
103	0	.2
104	0	.1
105	.2	.2
106	.2	.3

8. (*Course Grades*) Student A received the following course grades during her first year of college:

$$4, 4, 4, 4, 3, 3, 2, 2, 2, 0.$$

Student B received the following course grades during her first year:

$$4, 4, 4, 4, 4, 4, 3, 1, 1, 1.$$

(a) Write down the relative frequency distribution tables for each student, and compute the population means and variances.

(b) Which student had the better grade point average?

(c) Which student was more consistent?

9. Suppose that a probability distribution has mean 35 and standard deviation 5. Use the Chebychev inequality to estimate the probability that an outcome will lie between

(a) 25 and 45.   (b) 20 and 50.   (c) 29 and 41.

10. Suppose that a probability distribution has mean 8 and standard deviation .4. Use the Chebychev inequality to estimate the probability that an outcome will lie between
    (a) 6 and 10.   (b) 7.2 and 8.8.
    (c) 7.5 and 8.5.

11. (*Bulb Lifetimes*) For certain types of fluorescent lights the number of hours a bulb will burn before requiring replacement has a mean of 3000 hours and a standard deviation of 250 hours. Suppose that 5000 such bulbs are installed in an office building. Estimate the number that will require replacement between 2000 and 4000 hours from the time of installation.

12. (*Quality Control*) An electronics firm determines that the number of defective transistors in each batch averages 15 with standard deviation 10. Suppose that 100 batches are produced. Estimate the number of batches having between 0 and 30 defective transistors.

13. Suppose that a probability distribution has mean 75 and standard deviation 6. Use the Chebychev inequality to find the value of c for which the probability that the outcome lies between $75 - c$ and $75 + c$ is at least $\frac{7}{16}$.

14. Suppose that a probability distribution has mean 17 and standard deviation .2. Use the Chebychev inequality to find the value of c for which the probability that the outcome lies between $17 - c$ and $17 + c$ is at least $\frac{15}{16}$.

15. (*Dice*) The probability distribution for the sum of numbers obtained from tossing a pair of dice is given in Table 10.

TABLE 10	
Number	**Probability**
2	$\frac{1}{36}$
3	$\frac{2}{36}$
4	$\frac{3}{36}$
5	$\frac{4}{36}$
6	$\frac{5}{36}$
7	$\frac{6}{36}$
8	$\frac{5}{36}$
9	$\frac{4}{36}$
10	$\frac{3}{36}$
11	$\frac{2}{36}$
12	$\frac{1}{36}$

(a) Compute the mean and the variance of this probability distribution.

(b) Using the table, give the probability that the number is between 4 and 10, inclusive.

(c) Use the Chebychev inequality to estimate the probability that the number is between 4 and 10, inclusive.

16. (*Dice*) The probability distribution for the number of "ones" obtained from tossing 12 dice is given in Table 11. This probability distribution has mean 2 and standard deviation 1.291 ($\sigma^2 = \frac{5}{3}$).

TABLE 11	
Number of "ones"	**Probability**
0	.112
1	.269
2	.296
3	.197
4	.089
5	.028
6	.007
7	.001
8	.000
9	.000
10	.000
11	.000
12	.000

(a) Using the table, give the probability that the number of "ones" is between 0 and 4, inclusive.

(b) Use the Chebychev inequality to estimate the probability that the number of "ones" is between 0 and 4, inclusive.

17. Redo Example 1 using the alternate formula for the variance.

18. If X is a random variable, then the variance of X equals the variance of $X - a$ for any number a. Redo Exercise 1 using this result with $a = 70$.

19. If X is a random variable, then the variance of aX equals a^2 times the variance of X. Verify this result for the random variable in Exercise 2 with $a = 2$.

20. If X is a random variable, then

$$E(X - a) = E(X) - a \quad \text{and} \quad E(aX) = aE(X)$$

for any number a. Give intuitive justifications of these results.

TABLE 12 Ten largest U.S. universities

University	Enrollment
Miami-Dade Community College	51,717
University of Texas at Austin	50,616
Ohio State University, Main Campus	48,477
University of Minnesota, Twin Cities	46,597
University of Florida	46,515
Arizona State University, Main Campus	45,693
Texas A&M University	44,618
Michigan State University	44,227
University of Wisconsin, Madison	40,922
Pennsylvania State University, Main Campus	40,828

Source: U.S. Department of Education, National Center for Education Statistics, *Digest of Education Statistics 2003*.

TABLE 13 Ten largest U.S. public libraries

Library	Books
Boston Public Library	14.9
Chicago Public Library	10.7
Public Library of Cincinnati & Hamilton County	9.9
Queens Borough Public Library	9.7
County of Los Angeles Public Library	9.2
Detroit Public Library	7.3
New York Public Library	6.8
Free Library of Philadelphia	6.4
Dallas Public Library	5.9
Brooklyn Public Library	5.8

Source: American Library Association, *The Nation's Largest Libraries*.

In Exercises 21–24, use a graphing calculator, a spreadsheet, or mathematical software to calculate the answers.

21. (*University Enrollments*) Table 12 gives the fall 2001 enrollments of the 10 largest universities in the United States. Determine the mean and the standard deviation for these enrollments.

22. (*Public Libraries*) Table 13 gives the number of books (in millions) in the 10 largest public libraries in the United States. Determine the mean and the standard deviation for the number of books.

23. (*Ph.D. Degrees*) Table 14 summarizes the production of Ph.D. degrees in statistics at a certain university during the past 25 years. For instance, three Ph.D. degrees were awarded during 5 of the last 25 years. Determine the mean and the standard deviation for the number of degrees awarded each year.

24. (*Game of Chance*) Table 15 gives the probability distribution for the possible earnings from a certain game of chance. Determine the mean and the standard deviation for the earnings.

TABLE 14

Number of degrees	Number of years
3	5
4	7
5	8
6	2
7	1
8	2

TABLE 15

Earnings	Probability
−5	.23
−1	.32
1	.35
5	.07
10	.03

SOLUTIONS TO PRACTICE PROBLEMS 7.5

1. (a)

k	$\Pr(X = k)$	$k - \mu$	$(k-\mu)^2$	$(k-\mu)^2 \Pr(X = k)$
21	$\frac{1}{16}$	-2	4	$\frac{4}{16}$
22	$\frac{1}{8}$	-1	1	$\frac{1}{8}$
23	$\frac{5}{8}$	0	0	0
24	$\frac{1}{8}$	1	1	$\frac{1}{8}$
25	$\frac{1}{16}$	2	4	$\frac{4}{16}$

$$\mu = 21 \cdot \tfrac{1}{16} + 22 \cdot \tfrac{1}{8} + 23 \cdot \tfrac{5}{8} + 24 \cdot \tfrac{1}{8} + 25 \cdot \tfrac{1}{16}$$

$$= \tfrac{21}{16} + \tfrac{44}{16} + \tfrac{230}{16} + \tfrac{48}{16} + \tfrac{25}{16} = \tfrac{368}{16} = 23$$

$$[\text{variance}] = \tfrac{4}{16} + \tfrac{1}{8} + 0 + \tfrac{1}{8} + \tfrac{4}{16}$$

$$= \tfrac{2}{8} + \tfrac{1}{8} + 0 + \tfrac{1}{8} + \tfrac{2}{8} = \tfrac{6}{8} = \tfrac{3}{4}$$

(b) $\frac{7}{8}$. The probability that the outcome is between 22 and 24 is

$$\Pr(22) + \Pr(23) + \Pr(24) = \tfrac{1}{8} + \tfrac{5}{8} + \tfrac{1}{8} = \tfrac{7}{8}.$$

2. Probability $\geq \frac{1}{4}$. Here $\mu = 23$, $\sigma^2 = \frac{3}{4}$, and $c = 1$. By the Chebychev inequality, the probability that the outcome is between $23 - 1$ and $23 + 1$ is at least $1 - \left(\frac{3}{4}/1^2\right) = 1 - \frac{3}{4} = \frac{1}{4}$. [From 1(b) we obtained the actual probability of $\frac{7}{8}$, which is much greater than $\frac{1}{4}$. In the next section we will study a technique that gives better estimates. However, this technique holds only for a special type of probability distribution.]

7.6 The Normal Distribution

In this section we will see that the histogram for a binomial random variable can be approximated by a region under a smooth curve called a *normal curve*. Actually, these curves have a value in their own right in that they represent some random variables associated with experiments having infinitely many possible outcomes.

Toss a coin 20 times and observe the number of heads. By using formula (2) of Section 7.3 with $n = 20$ and $p = .5$, we can calculate the probability of k heads. The results are displayed in Table 1 on the next page. The data of Table 1 can be displayed in histogram form, as in Fig. 1 on the next page.

As we have seen, various probabilities may be interpreted as areas. For example, the probability that at most 9 heads occur is equal to the sum of the areas of all the rectangles to the left of the central one (Fig. 2). The shape of the histogram in Figs. 1 and 2 suggests that we might be able to approximate such areas by using a smooth bell-shaped curve. The curve shown in Fig. 3 is a good candidate. It is called a *normal curve* and plays an important role in statistics and probability. (Tables giving the areas under normal curves have been constructed and can be used to approximate binomial probabilities.) For instance, the area of the blue rectangles in Fig. 2 is approximately the same as the area under the normal curve to the left of 9.5 shown in Fig. 4 on page 394. As another example, the probability of obtaining exactly 10 heads in 20 tosses of a coin is the area of the blue rectangle in Fig. 5(a), which is approximated by the area under the normal curve shown in Fig. 5(b).

k	Probability of k heads	k	Probability of k heads
0	.0000	10	.1762
1	.0000	11	.1602
2	.0002	12	.1201
3	.0011	13	.0739
4	.0046	14	.0370
5	.0148	15	.0148
6	.0370	16	.0046
7	.0739	17	.0011
8	.1201	18	.0002
9	.1602	19	.0000
		20	.0000

TABLE 1 Probability of k heads (to four places)

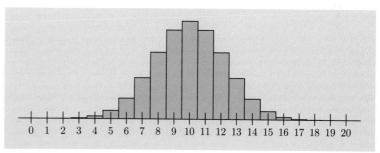

Figure 1

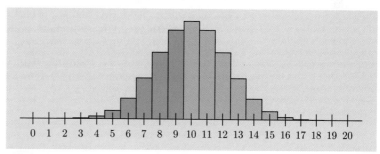

Figure 2

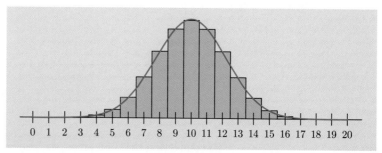

Figure 3

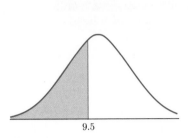

Figure 4

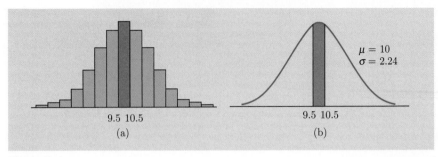

Figure 5

To be able to use normal curves in our computations, we need to study them more closely. Let us now take a glimpse into the realm of so-called continuous probability by studying appropriate experiments—namely, experiments with *normally distributed outcomes*. For such experiments, the probabilities of events are computed as areas under normal curves. It is no exaggeration to say that experiments with normally distributed outcomes are among the most significant in probability theory. Such experiments abound in the world around us. Here are a few examples:

1. Choose an individual at random and observe his or her IQ.

2. Choose a 1-day-old infant and observe his or her weight.

3. Choose an 8-year-old male at random and observe his height.

4. Choose a leaf at random from a particular tree and observe its length.

5. A lumber mill is cutting planks that are supposed to be 8 feet long; choose a plank at random and observe its actual length.

Associated to each of the foregoing experiments is a bell-shaped curve, as shown in Fig. 6. Such a curve is called a *normal curve*. The curve is symmetric about a vertical line drawn through its highest point. This line of symmetry indicates the mean value of the corresponding experiment. The mean value is denoted as usual by the Greek letter μ. For example, if in experiment 4 above the average length of the leaves on the tree is 5 inches, then $\mu = 5$ and the corresponding bell-shaped curve is symmetric about the line $x = 5$.

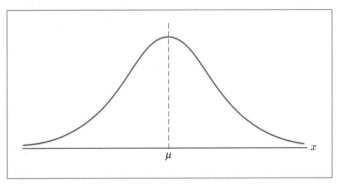

Figure 6

The connection between an experiment with normally distributed outcomes and its associated normal curve is as follows:

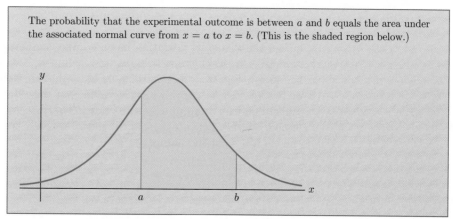

The probability that the experimental outcome is between a and b equals the area under the associated normal curve from $x = a$ to $x = b$. (This is the shaded region below.)

Figure 7

The total area under a normal curve is always 1. This is due to the fact that the probability that the variable X corresponding to the distribution takes on some numerical value on the x-axis is 1.

EXAMPLE 1 **Shading regions under the standard normal curve** A certain experiment has normally distributed outcomes with mean $\mu = 1$. Shade the region corresponding to the probabilities of the following outcomes.

(a) The outcome lies between 1 and 3.

(b) The outcome lies between 0 and 2.

(c) The outcome is less than .5.

(d) The outcome is greater than 2.

Solution The outcomes are plotted along the x-axis. We then shade the appropriate area under the curve.

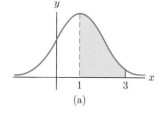
(a)

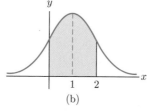
(b)

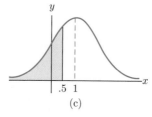
(c)

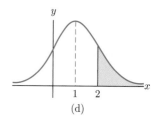
(d)

There are many different normal curves with the same mean. For instance, in Fig. 8 we have drawn three normal curves, all with $\mu = 0$. Roughly speaking, the difference between these normal curves is in the width of the center "hump."

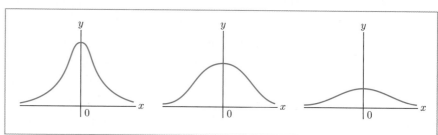

Figure 8

A sharper hump indicates that the outcomes are more likely to be close to the mean. A flatter hump indicates a greater likelihood for the outcomes to be spread out. As we have seen, the spread of the outcomes about the mean is described by the standard deviation, denoted by the Greek letter σ. In the case of a normal curve, the standard deviation has a simple geometric meaning: The normal curve "twists" (or, in calculus terminology, "inflects") at a distance σ on either side of the mean (Fig. 9). More specifically, a normal curve may be thought of as made up of two pieces: a "cap," which looks like an upside-down bowl; and a pair of legs, which curve in the opposite direction. The places at which the cap and legs are joined are at a distance σ from the mean. Thus it is clear that the size of σ controls the sharpness of the hump.

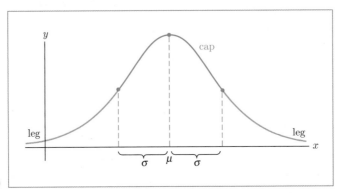

Figure 9

A normal curve is completely described by its mean μ and standard deviation σ. In fact, given μ and σ, we may write down the equation of the associated normal curve:

$$y = \frac{1}{\sigma\sqrt{2\pi}}\, e^{-\left(\frac{1}{2}\right)\left(\frac{x-\mu}{\sigma}\right)^2},$$

where $\pi \approx 3.1416$ and $e \approx 2.7183$. Fortunately, we will not need this rather complicated formula in what follows. But it is only fair to say that all theoretical work on the normal curve ultimately rests on this equation.

For our purposes we will compute areas under normal curves by consulting a table. One might expect that a separate table would be needed for each normal curve, but such is not the case. Only one table is needed: the table corresponding to the *standard normal curve*, which is the one for which $\mu = 0$ and $\sigma = 1$. So let us begin our discussion of areas under normal curves by considering the standard normal curve.

We usually use the letter Z to denote a random variable having the standard normal distribution. Let z be any number and let $A(z)$ denote the area under the standard normal curve to the left of z (Fig. 10). Table 2 gives $A(z)$ for various values of z, with the values of $A(z)$ rounded to four decimal places. Thus, $A(z) = \Pr(Z \leq z)$.[1] A more extensive table can be found in Table 1 of Appendix A. The efficient use of these tables depends on the following three facts:

1. The standard normal curve is symmetric about $z = 0$.
2. The total area under the standard normal curve is 1.
3. The probability that the standard normal variable Z lies to the left of the number z is the area $A(z)$ in Fig. 10.

These facts allow us to use the tables to find the areas of various types of regions.

---

[1] We could have said that $A(z) = \Pr(Z < z)$. However, since the region strictly to the left of z and that region with the line segment at z adjoined have the same area, $\Pr(Z < z)$ and $\Pr(Z \leq z)$ are the same. We always use the $\leq$ symbol.

TABLE 2

z	A(z)	z	A(z)	z	A(z)
−4.00	.0000	−1.25	.1056	1.50	.9332
−3.75	.0001	−1.00	.1587	1.75	.9599
−3.50	.0002	−.75	.2266	2.00	.9772
−3.25	.0006	−.50	.3085	2.25	.9878
−3.00	.0013	−.25	.4013	2.50	.9938
−2.75	.0030	0	.5000	2.75	.9970
−2.50	.0062	.25	.5987	3.00	.9987
−2.25	.0122	.50	.6915	3.25	.9994
−2.00	.0228	.75	.7734	3.50	.9998
−1.75	.0401	1.00	.8413	3.75	.9999
−1.50	.0668	1.25	.8944	4.00	1.0000

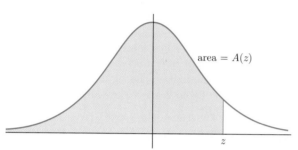

Figure 10

EXAMPLE 2 **Determining areas of regions under the standard normal curve** Use Table 2 to determine the areas of the regions under the standard normal curve pictured in Fig. 11.

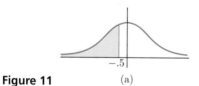 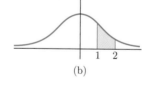 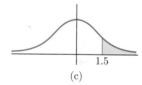

Figure 11   (a)   (b)   (c)

Solution (a) This region is just the portion of the curve to the left of −.5. So its area is $A(-.5)$. Looking down the middle pair of columns of the table, we find that $A(-.5) = .3085$. This means that

$$\Pr(Z \le -.5) = .3085.$$

(b) This region results from beginning with the region to the left of 2 and subtracting the region to the left of 1. We obtain an area of

$$A(2) - A(1) = .9772 - .8413 = .1359.$$

Thus

$$\Pr(1 \le Z \le 2) = .1359.$$

(c) This region can be thought of as the entire region under the curve, with the region to the left of 1.5 removed. Therefore, the area is

$$1 - A(1.5) = 1 - .9332 = .0668.$$

So

$$\Pr(Z \ge 1.5) = .0668.$$

Now Try Exercises 1, 3, and 5

EXAMPLE 3   **Finding a region under the standard normal curve** Find the value of z for which $\Pr(Z \geq z) = .1056$ (see Fig. 12).

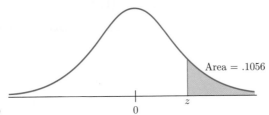

Area = .1056

Figure 12

Solution   Since the area under the standard normal curve is 1 and the curve is symmetric about $z = 0$, the area of the portion to the right of 0 must be .5. We draw a sketch of the standard normal curve, placing z on the axis to the right of 0. (This way, the area to the right of z will be less than .5.) Table 2 gives the values of $A(z)$, which are left tail areas. The area to the left of our z is

$$A(z) = 1 - .1056 = .8944.$$

From Table 2 we find that the value of z for which $A(z) = .8944$ is 1.25.

Now Try Exercise 9   ■

Percentiles   In large-scale testing, scores are frequently reported as percentiles rather than as raw scores. What does it mean to say that a score is "the 90th percentile"? It means that, roughly speaking, the score separates the bottom 90% of the scores from the top 10%.

> If a score S is the pth percentile of a normal distribution, then $p\%$ of all scores fall below S, and $(100 - p)\%$ of all scores fall above S.

EXAMPLE 4   **Determining a percentile of the standard normal distribution** What is the 50th percentile of the standard normal distribution?

Solution   The standard normal curve is symmetric about $z = 0$, and the total area under the curve is 1. Thus, 50% of the values of the standard normal variable fall below 0, and $(100 - 50)\% = 50\%$ of its values fall above 0. So 0 is the 50th percentile of the standard normal distribution.   ■

EXAMPLE 5   **Determining a percentile of the standard normal distribution** What is the 95th percentile of the standard normal distribution?

Solution   We shall call the value that we seek z_{95} to remind us that it is a score and that the probability that an outcome is to the left of it is 95% (see Fig. 13).
    Since Table 1 of Appendix A gives areas to the left of values of z, we should search the column marked $A(z)$ for the area we need—.95. We find that the closest value to .95 is .9505, and $A(1.65) = .9505$. Hence $z_{95} \approx 1.65$. This means that 95% of the time the standard normal variable falls below 1.65. Since $\mu = 0$ and $\sigma = 1$, another way of stating the result is that in the standard normal distribution, 95% of the values are less than 1.65 standard deviations above the mean.

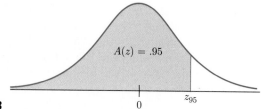

Figure 13

Now Try Exercise 13 ■

The problem of finding the area of a region under any normal curve can be reduced to finding the area of a region under the standard normal curve. To illustrate the computation procedure, let us consider a numerical example.

EXAMPLE 6   **Finding areas of regions under normal curves**   Find the area under the normal curve with $\mu = 3$, $\sigma = 2$ from $x = 1$ to $x = 5$. This represents $\Pr(1 \leq X \leq 5)$ for a random variable X having the given normal distribution.

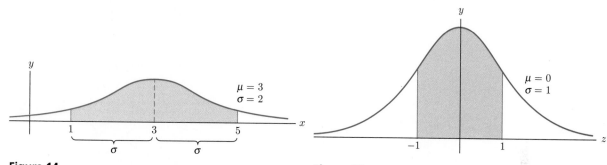

Figure 14   **Figure 15**

Solution   We have sketched the described region in Fig. 14. It extends from one standard deviation below the mean to one standard deviation above. Draw the corresponding region under the standard normal curve. That is, draw the region from one standard deviation below to one standard deviation above the mean (Fig. 15). It is a theorem that this new region has the same area as the original one. But the area in Fig. 15 may be computed from Table 2 as $A(1) - A(-1)$. So our desired area is

$$A(1) - A(-1) = .8413 - .1587 = .6826.$$

Now Try Exercise 25 ■

EXAMPLE 7   **Finding areas of regions under a normal curve**   Consider the normal curve with $\mu = 12$, $\sigma = 1.5$. Find the area of the region under the curve between $x = 11.25$ and $x = 15$. (Fig. 16).

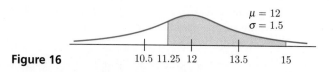

Figure 16

Solution Expressed as a probability, we want to find $\Pr(11.25 \leq X \leq 15)$ for a random variable X having a normal distribution with $\mu = 12$ and $\sigma = 1.5$. The number 11.25 is .75 below the mean 12. And .75 is $.75/1.5 = .5$ standard deviations. The number 15 is 3 above the mean. And 3 is $3/1.5 = 2$ standard deviations. Therefore, the region has the same area as the region under the standard normal curve from $-.5$ to 2, which is

$$A(2) - A(-.5) = .9772 - .3085 = .6687.$$

Now Try Exercise 23   ■

Suppose that a normal curve has mean μ and standard deviation σ. Then the area under the curve from $x = a$ to $x = b$ is

$$A\left(\frac{b - \mu}{\sigma}\right) - A\left(\frac{a - \mu}{\sigma}\right).$$

The numbers $b - \mu$ and $a - \mu$, respectively, measure the distances of b and a from the mean. The numbers $(b - \mu)/\sigma$ and $(a - \mu)/\sigma$ express these distances as multiples of the standard deviation σ. So the area under the normal curve from $x = a$ to $x = b$ is computed by expressing x in terms of standard deviations from the mean and then treating the curve as if it were the standard normal curve. We summarize the procedure.

If X is a random variable having a normal distribution with mean μ and standard deviation σ, then

$$\Pr(a \leq X \leq b) = \Pr\left(\frac{a - \mu}{\sigma} \leq Z \leq \frac{b - \mu}{\sigma}\right) = A\left(\frac{b - \mu}{\sigma}\right) - A\left(\frac{a - \mu}{\sigma}\right)$$

and

$$\Pr(X \leq x) = \Pr\left(Z \leq \frac{x - \mu}{\sigma}\right) = A\left(\frac{x - \mu}{\sigma}\right),$$

where Z has the standard normal distribution and $A(z)$ is the area under that distribution to the left of z.

Let us now use our knowledge of areas under normal curves to calculate probabilities arising in some applied problems.

EXAMPLE 8 **Birth weights of infants** Suppose that for a certain population the birth weights of infants in pounds are normally distributed with $\mu = 7.75$ and $\sigma = 1.25$. Find the probability that an infant's birth weight is more than 9 pounds, 10 ounces. (*Note*: 9 pounds, 10 ounces $= 9\frac{5}{8}$ pounds.)

Solution Let $X =$ infant's birth weight. Then X is a random variable having a normal distribution with $\mu = 7.75$ and $\sigma = 1.25$ pounds. $\Pr\left(X \geq 9\frac{5}{8}\right)$ is given by the area under the appropriate normal curve to the right of $9\frac{5}{8}$—that is, the area shaded in Fig. 17. Since $9\frac{5}{8} = 9.625$, the number $9\frac{5}{8}$ lies $9.625 - 7.75 = 1.875$ units above the mean. In turn, this is $1.875/1.25 = 1.5$ standard deviations. We can find the corresponding z-value in one calculation by finding

$$z = \frac{x - \mu}{\sigma} = \frac{9.625 - 7.75}{1.25} = 1.5.$$

Figure 17

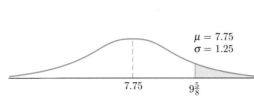

Figure 18

Thus 9.625 is 1.5 standard deviations above the mean. The area we seek is sketched under the standard normal curve in Fig. 18 and is

$$1 - A(1.5) = 1 - .9332 = .0668.$$

So the probability that an infant weighs more than 9 pounds, 10 ounces is .0668.

Now Try Exercise 27   ■

EXAMPLE 9 **Produce sales** A wholesale produce dealer finds that the number of boxes of bananas sold each day is normally distributed with $\mu = 1200$ and $\sigma = 100$. Find the probability that the number sold on a particular day is less than 1000.

Solution Let $X =$ the number of boxes of bananas sold each day. Since daily sales are normally distributed, the desired probability, $\Pr(X \leq 1000)$, is the area to the left of 1000 in the normal curve drawn in Fig. 19. The number 1000 is 2 standard deviations below the mean; that is, $x = 1000$ corresponds to a z-value of

$$z = \frac{x - \mu}{\sigma} = \frac{1000 - 1200}{100} = -2.$$

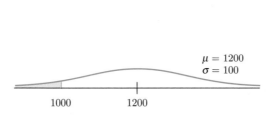

Figure 19

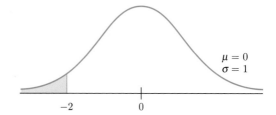

Figure 20

Therefore, the area we seek is $A(-2) = .0228$, shown in Fig. 20. The probability that less than 1000 boxes will be sold in a day is .0228.

Now Try Exercise 31   ■

EXAMPLE 10 **Birth weights of infants** Find the 95th percentile of infant birth weights if infant birth weights are normally distributed with $\mu = 7.75$ and $\sigma = 1.25$ pounds.

Solution Let us denote the 95th percentile of the infant birth weights by x_{95}. This means that the area to the left of x_{95} under the normal curve with $\mu = 7.75$ and $\sigma = 1.25$ is 95% (Fig. 21).

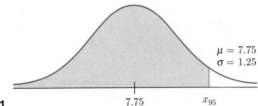

Figure 21

The corresponding value for the standard normal random variable would be z_{95}, which we already know to be 1.65. Thus x_{95} is 1.65 standard deviations above μ. That is,

$$x_{95} = 7.75 + (1.65)(1.25) \approx 9.81 \text{ pounds.}$$

Therefore, 95% of all infants have birth weights below 9.81 pounds (9 lb, 13 oz).

Now Try Exercise 33(c)

■

EXAMPLE 11 **Inventory size** The wholesale produce dealer of Example 9 wants to be 99% sure that she has enough boxes of bananas on hand each day to meet the demand. How many should she stock each day?

Solution Let x be the number of boxes of bananas that the produce dealer should stock. Since she wants to be 99% sure that the demand for boxes of bananas does not exceed x, we must find the 99th percentile of a normal distribution with $\mu = 1200$ and $\sigma = 100$. To help us remember what x really is, we will rename it x_{99}. The corresponding value for the standard normal random variable is z_{99}. Figures 22 and 23 show the appropriate areas—first under the given normal curve and then under the standard normal curve.

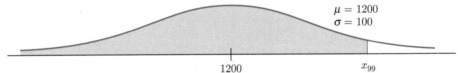

Figure 22

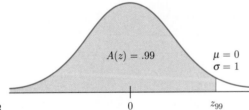

Figure 23

The area, $A(z)$, that we seek under the standard normal curve is .99. Referring to Table 1 of Appendix A, we get closest with $A(z) = .9906$, corresponding to $z_{99} = 2.35$. The value z_{99} is 2.35 standard deviations above the mean of its distribution. We conclude that x_{99} is also 2.35 standard deviations above its mean. Hence,

$$x_{99} = 1200 + (2.35)(100) = 1435 \text{ boxes.}$$

Therefore, we expect that on 99% of the days, 1435 boxes of bananas will meet the demand. More than 1435 boxes should be needed 1% of the time (one day out of 100). ■

We summarize the technique for finding percentiles of normal distributions.

If x_p is the pth percentile of a normal distribution with mean μ and standard deviation σ, then

$$x_p = \mu + z_p \cdot \sigma,$$

where z_p is the pth percentile of the standard normal distribution.

GC Normal curves can easily be graphed and normal probabilities calculated on most graphing calculators. Consider Example 6. Figure 24 contains the graph of the normal curve, and Fig. 25 gives the desired probability.

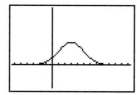

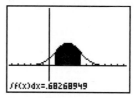

Figure 24
$[-6.4, 12.4]$ by $[-.2, .5].$

Figure 25

(TI-83) The graph of the normal curve in Fig. 24 was obtained by setting **Y₁=normalpdf(X,3,2)** in the **Y=** editor, and pressing GRAPH. (To display **normalpdf(**, press 2nd [DISTR] **1**. The general form of this function is **Y₁=normalpdf(X,μ,σ)**.) To obtain the result in Fig. 25, press 2nd [CALC] **7** and answer the questions by entering the appropriate numbers. Specifically, respond to **"Lower Limit?"** by typing in the number **1** and pressing ENTER, and respond to **"Upper Limit?"** by typing in the number **5** and pressing ENTER. To erase the shading under the curve, press 2nd [DRAW] **1**. To find the area of an infinite region, just give a large negative value for the lower bound or a large positive value for the upper bound.

The area under a normal curve also can be evaluated on the home screen with the command **fnInt(Y₁,X,a,b)** where Y₁ has been set to a normalpdf function. For instance, the area of the shaded region in Example 6 can be evaluated by typing **fnInt(Y₁,X,1,5)** and pressing ENTER. (From the home screen, **fnInt(** is displayed with MATH **9**.)

(TI-89) The graph of the normal curve similar to the one in Fig. 24 can be obtained by setting **y1=1/2($\sqrt{}$(2π))e^(-.5((x-3)/2)^2)** in the **Y=** editor, and pressing ♦[GRAPH]. To obtain a result similar to the one in Fig. 25, press F5 **7** and answer the questions by entering the appropriate numbers. Specifically, respond to **"Lower Limit?"** by typing in the number **1** and pressing ENTER, and respond to **"Upper Limit?"** by typing in the number **5** and pressing ENTER. To erase the shading under the curve, press 2nd [F6] **1**. To find the area of an infinite region, just give a large negative value for the lower bound or a large positive value for the upper bound.

ES If z and x are any numbers, then the value of $A(z)$ is given by the Excel function NORMSDIST(z), and the area to the left of x under the normal curve having mean μ and standard deviation σ is NORMDIST(x,μ,σ,TRUE). If y is a number between 0 and 1, then NORMSINV(y) is the value of z for which $A(z) = y$, and NORMINV(y,μ,σ) is the value of x for which the area to the left of x under the normal curve having mean μ and standard deviation σ is y. In Fig. 26, the answers to four examples from this section are computed in an Excel spreadsheet.

	A	B	C	D
1	**Example**	**Answer**		**Formula Used**
2	2 (b)	0.135905198		NORMSDIST(2) - NORMSDIST(1)
3	5	1.644853		NORMSINV(0.95)
4	9	0.022750062		NORMDIST(1000,1200,100,TRUE)
5	11	1432.634193		NORMINV(0.99,1200,100)

Figure 26

PRACTICE PROBLEMS 7.6

1. Refer to Fig. 27(a). Find the value of z for which the area of the shaded region is .0802.

2. Refer to the normal curve in Fig. 27(b). Express the following numbers in terms of standard deviations from the mean.

(a) 90   (b) 82   (c) 94   (d) 104

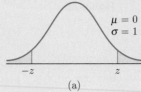
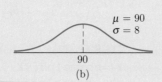

Figure 27

EXERCISES 7.6

In Exercises 1–8, use the table for $A(z)$ (Table 2) to find the areas of the shaded regions under the standard normal curve.

1.

2.

3.

4.

5.

6.

7.

8.

In Exercises 9–12, find the value of z for which the area of the shaded region under the standard normal curve is as specified.

9. Area is .0401.

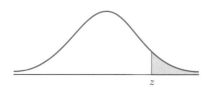

10. Area is .0456.

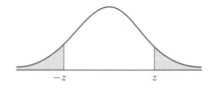

11. Area is .5468.

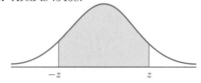

12. Area is .6915.

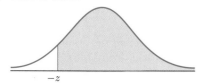

13. What is the 90th percentile of the standard normal distribution?

14. What is the 65th percentile of the standard normal distribution?

In Exercises 15–18, determine μ and σ by inspection.

15.

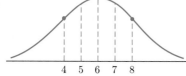

16.

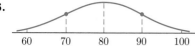

17.

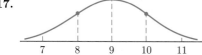

18.

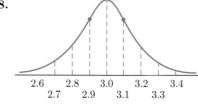

Exercises 19–22 refer to the normal curve with $\mu = 8$, $\sigma = \frac{3}{4}$.

19. Convert 6 into standard deviations from the mean.

20. Convert $9\frac{1}{4}$ into standard deviations from the mean.

21. **(PE)** What value is exactly 10 standard deviations above the mean?
(a) $15\frac{1}{2}$ (b) $\frac{1}{2}$ (c) 18 (d) $7\frac{1}{2}$ (e) $6\frac{3}{4}$

22. **(PE)** What value is exactly 2 standard deviations below the mean?
(a) $9\frac{1}{2}$ (b) 10 (c) 6 (d) $7\frac{1}{4}$ (e) $6\frac{1}{2}$

In Exercises 23–26, find the areas of the shaded regions under the given normal curves.

23.

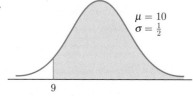

24.

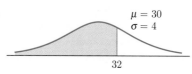

25.

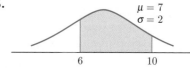

26.

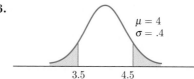

27. (*Elephant Heights*) Suppose that the height (at the shoulder) of adult African bull bush elephants is normally distributed with $\mu = 3.3$ meters and $\sigma = .2$ meter. The elephant on display at the Smithsonian Institution has height 4 meters and is the largest elephant on record. What is the probability that an adult African bull bush elephant has height 4 meters or more?

28. (*Bottling Reliability*) At a soft-drink bottling plant, the amount of cola put into the bottles is normally distributed with $\mu = 16\frac{3}{4}$ ounces and $\sigma = \frac{1}{2}$. What is the probability that a bottle will contain less than 16 ounces?

29. (*Manufacturing Reliability*) Bolts produced by a machine are acceptable provided that their length is within the range from 5.95 to 6.05 centimeters. Suppose that the lengths of the bolts produced are normally distributed with $\mu = 6$ centimeters and $\sigma = .02$. What is the probability that a bolt will be of an acceptable length?

30. (*IQ Scores*) Suppose that IQ scores are normally distributed with $\mu = 100$ and $\sigma = 10$.
(a) What percent of the population has an IQ score of 125 or more?
(b) Find the 90th percentile of IQ scores.

31. (*Gasoline Sales*) The amount of gas sold weekly by a certain gas station is normally distributed with $\mu = 30,000$ gallons and $\sigma = 4000$. If the station has 39,000 gallons on hand at the beginning of the week, what is the probability of its running out of gas before the end of the week?

32. (*Light Bulb Lifetimes*) Suppose that the lifetimes of a certain light bulb are normally distributed with $\mu = 1200$ hours and $\sigma = 160$. Find the probability that a light bulb will burn out in less than 1000 hours.

33. (*SAT Scores*) Assume that SAT verbal scores for a first-year class at a university are normally distributed with mean 520 and standard deviation 75.

    (a) The top 10% of the students are placed into the honors program for English. What is the lowest score for admittance into the honors program?

    (b) What is the range of the middle 90% of the SAT verbal scores at this university?

    (c) Find the 98th percentile of the SAT verbal scores.

34. (*Mailing Bags*) A mail-order house uses an average of 300 mailing bags per day. The number of bags needed each day is approximately normally distributed with $\sigma = 50$. How many bags must the company have on hand at the beginning of a day to be 99% certain that all orders can be filled?

35. (*Tire Lifetimes*) The lifetime of a certain brand of tires is normally distributed with mean $\mu = 30{,}000$ miles and standard deviation $\sigma = 5000$ miles. The company has decided to issue a warranty for the tires but does not want to replace more than 2% of the tires that it sells. At what mileage should the warranty expire?

36. Let X be a random variable with $\mu = 4$ and $\sigma = .5$.

    (a) Use the Chebychev inequality to estimate $\Pr(3 \leq X \leq 5)$.

    (b) If X were normally distributed, what would be the exact probability that X is between 3 and 5 inclusive?

    (c) Reconcile the difference between the answers to (a) and (b).

37. (*Soft-Drink Dispenser*) Let X be the amount of soda released by a soft-drink dispensing machine into a 6-ounce cup. Assume that X is normally distributed with $\sigma = .25$ ounces and that the average "fill" can be set by the vendor.

    (a) At what quantity should the average "fill" be set so that no more than .5% of the releases overflow the cup?

    (b) Using the average "fill" found in part (a), determine the minimal amount that will be dispensed in 99% of the cases.

In Exercises 38–41, use a graphing calculator, a spreadsheet, or mathematical software to obtain the information.

38. Draw the graph of the normal curve with $\mu = 10, \sigma = 2$ and the normal curve with $\mu = 10, \sigma = 3$ on the same coordinate system. Describe the effect on a normal curve of increasing σ.

39. Draw the graph of the normal curve with $\mu = 11, \sigma = 2$ and the normal curve with $\mu = 15, \sigma = 2$ on the same coordinate system. Describe the effect on a normal curve of increasing μ.

40. (*Heights*) In a certain population, heights (in inches) are normally distributed with $\mu = 67$ and $\sigma = 3$. Find the probability that a person selected at random has a height between 63 and 71 inches.

41. (*Manufacturing Reliability*) In a certain manufacturing process, lengths (in cm) of widgets are normally distributed with $\mu = 5.4$ and $\sigma = .6$. Find the probability that a widget selected at random has a length greater than 5.832 cm.

SOLUTIONS TO PRACTICE PROBLEMS 7.6

1. 1.75. Due to the symmetry of normal curves, each piece of the shaded region has area $\frac{1}{2}(.0802) = .0401$. Therefore, $A(-z) = .0401$ and so by Table 2, $-z = -1.75$. Thus $z = 1.75$.

2. (a) 0. Since 90 *is* the mean, it is 0 standard deviations from the mean.

    (b) −1. Since 82 is $90 - 8$, it is 8 units or 1 standard deviation below the mean.

    (c) .5. Here $94 = 90 + 4$ is 4 units or .5 standard deviations above the mean.

    (d) 1.75. Here $104 = 90 + 14$ is 14 units or $\frac{14}{8} = 1.75$ standard deviations above the mean.

7.7 Normal Approximation to the Binomial Distribution

In Section 7.3 we saw that complicated and tedious calculations can arise from the binomial probability distribution. For instance, determining the probability of getting at least 20 threes in 100 tosses of a die requires the computation

$$\binom{100}{20}\left(\frac{1}{6}\right)^{20}\left(\frac{5}{6}\right)^{80} + \binom{100}{21}\left(\frac{1}{6}\right)^{21}\left(\frac{5}{6}\right)^{79} + \cdots + \binom{100}{100}\left(\frac{1}{6}\right)^{100}\left(\frac{5}{6}\right)^{0}.$$

Mathematicians have shown that such probabilities can be closely approximated by using normal curves.

Consider the histograms of the binomial distributions in Figs. 1 and 2. The number of trials, n, increases from 5 to 40, with p fixed at .3. As n increases, the shape of the histogram more closely conforms to the shape of the region under a normal curve.

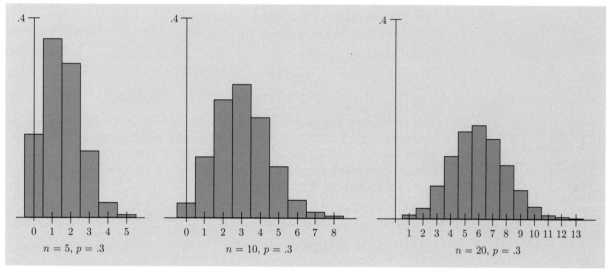

$n = 5, p = .3$ $n = 10, p = .3$ $n = 20, p = .3$

Figure 1. Binomial distribution.

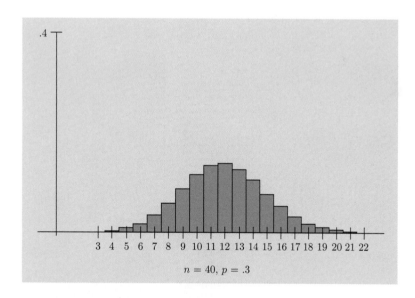

$n = 40, p = .3$

Figure 2. Binomial distribution.

We have the following result:

Suppose that we perform a sequence of n binomial trials with probability of success p and probability of failure q and observe the number of successes. Then the histogram for the resulting probability distribution may be approximated by the normal curve with $\mu = np$ and $\sigma = \sqrt{npq}$.

▶ *Note* This approximation is very accurate when both $np > 5$ and $nq > 5$. ◀

EXAMPLE 1 **Quality control** Refer to Example 3 of Section 7.3. A plumbing-supplies manufacturer produces faucet washers that are packaged in boxes of 300. Quality control studies have shown that 2% of the washers are defective. What is the probability that more than 10 of the washers in a single box are defective?

Solution Let $X =$ the number of defective washers in a box. Then X is a binomial random variable with $n = 300$ and $p = .02$. We will use the approximating normal curve with

$$\mu = np = 300(.02) = 6$$
$$\sigma = \sqrt{npq} = \sqrt{300(.02)(.98)} \approx 2.425.$$

The probability that more than 10 of the washers in a single box are defective is the sum of the areas of the blue rectangles centered at $11, 12, \ldots, 300$ in the histogram for the random variable X (see Fig. 3). The corresponding region under the approximating normal curve is shaded in Fig. 4. This is the area under the standard normal curve to the right of

$$z = \frac{10.5 - \mu}{\sigma} = \frac{10.5 - 6}{2.425} \approx 1.85.$$

The area of the region is $1 - A(1.85) = 1 - .9678 = .0322$. Therefore, approximately 3.22% of the boxes should contain more than 10 defective washers.

Now Try Exercise 1(c) ■

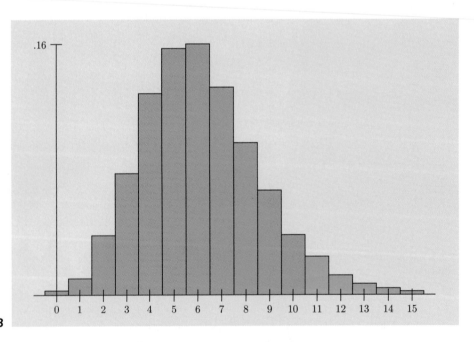

Figure 3

▶ **Note** In Fig. 4 we shaded the region to the right of 10.5 rather than to the right of 10. This gives a better approximation to the corresponding area under the histogram, since the rectangle corresponding to 11 "successes" has its left endpoint at 10.5. ◀

Let us consider an application to medical research.

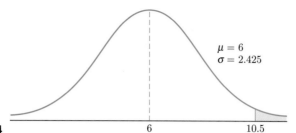

Figure 4

$\mu = 6$
$\sigma = 2.425$

6   10.5

EXAMPLE 2   **Veterinary medicine**   Consider the cattle disease of Example 4 of Section 7.3, from which 25% of the cattle recover. A veterinarian discovers a serum to combat the disease. In a test of the serum she observes that 16 of a herd of 40 recover. Suppose that the serum had not been used. What is the likelihood that at least 16 cattle would have recovered?

Solution   Let X be the number of cattle that recover. Then X is a binomial random variable with $n = 40$ independent trials. If the serum is not used, $p = .25$. The approximating normal curve has

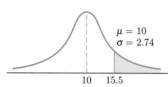

Figure 5

$\mu = 10$
$\sigma = 2.74$

10   15.5

$$\mu = np = 40(.25) = 10, \qquad \sigma = \sqrt{npq} = \sqrt{40(.25)(.75)} \approx 2.74.$$

The likelihood that at least 16 cattle would have recovered is $\Pr(X \geq 16)$. This corresponds to the area under the normal curve to the right of 15.5 (Fig. 5). The area to the right of 15.5 under a normal curve with $\mu = 10$ and $\sigma = 2.74$ is the same as the area under the standard normal curve to the right of

$$z = \frac{15.5 - \mu}{\sigma} = \frac{15.5 - 10}{2.74} \approx 2.01.$$

We find $1 - A(2.01) \approx 1 - A(2.00) = 1 - .9772 = .0228$. Thus, if the serum were not used, the veterinarian would expect about a 2% chance that 16 or more cattle recover. Thus the 16 observed recoveries probably did not occur by chance. The veterinarian can reasonably conclude that the serum is effective against the disease.

Now Try Exercise 3   ■

EXAMPLE 3   **Heads and tails**   Assume that a fair coin is tossed 100 times. Find the probability of observing exactly 50 heads.

Solution   Let X be the number of heads on $n = 100$ binomial trials with $p = .5$ probability of "success" on each trial. We are to find the area of the rectangle extending from 49.5 to 50.5 on the x-axis in the histogram for X. The actual probability is

$$\binom{100}{50}(.5)^{50}(.5)^{50} = \binom{100}{50}(.5)^{100}.$$

We approximate the probability with an area under the normal curve with $\mu = np = 100(.5) = 50$ and $\sigma = \sqrt{npq} = \sqrt{100(.5)(.5)} = 5$. The area we seek is sketched in Fig. 6. It is the area under the standard normal curve from

$$z = \frac{49.5 - 50}{5} = -.10 \quad \text{to} \quad z = \frac{50.5 - 50}{5} = .10.$$

Using Table 1 in Appendix A, we find that

$$A(.10) - A(-.10) = .5398 - .4602 = .0796.$$

So the likelihood of getting exactly 50 heads in 100 tosses of a fair coin is fairly small—about 8%.

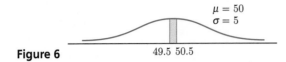

Figure 6   49.5 50.5

Now Try Exercise 1(a)

EXAMPLE 4   **Die tosses** Find the probability that in 100 tosses of a fair die we observe at least 20 threes.

Solution Let X be the number of threes observed in $n = 100$ trials, with $p = \frac{1}{6}$ the probability of a three on each trial. Then we need to find the area to the right of 19.5 under the normal curve with

$$\mu = np = 100\left(\tfrac{1}{6}\right) \approx 16.7 \quad \text{and} \quad \sigma = \sqrt{npq} = \sqrt{100\left(\tfrac{1}{6}\right)\left(\tfrac{5}{6}\right)} \approx 3.73.$$

This area is

$$1 - A\left(\frac{19.5 - 16.7}{3.73}\right) \approx 1 - A(.75) = 1 - .7734 = .2266.$$

Therefore, the probability of observing at least 20 threes is about .2266.

Now Try Exercise 7

PRACTICE PROBLEMS 7.7

1. (*Drug Testing*) A new drug is being tested on laboratory mice. The mice have been given a disease for which the recovery rate is $\frac{1}{2}$.

   (a) In the first experiment the drug is given to 5 of the mice and all 5 recover. Find the probability that the success of this experiment was due to luck. That is, find the probability that 5 out of 5 mice recover in the event that the drug has no effect on the illness.

   (b) In a second experiment the drug is given to 25 mice and 18 recover. Find the probability that 18 or more recover in the event that the drug has no effect on the illness.

2. (*Drug Testing*) What conclusions can be drawn from the results in Problem 1?

EXERCISES 7.7

In Exercises 1–14, use the normal curve to approximate the probability.

1. An experiment consists of 25 binomial trials, each having probability $\frac{1}{5}$ of success. Use an approximating normal curve to estimate the probability of

   (a) exactly 5 successes.

   (b) between 3 and 7 successes, inclusive.

   (c) less than 10 successes.

2. An experiment consists of 18 binomial trials, each having probability $\frac{2}{3}$ of success. Use an approximating normal curve to estimate the probability of

   (a) exactly 10 successes.

   (b) between 8 and 16 successes, inclusive.

   (c) more than 12 successes.

3. (*Drug Testing*) Laboratory mice are given an illness for which the usual recovery rate is $\frac{1}{6}$. A new drug is tested on 20 of the mice, and 8 of them recover. What is the probability that 8 or more would have recovered if the 20 mice had not been given the drug?

4. (*ESP*) A person claims to have ESP (extrasensory perception). A coin is tossed 16 times, and each time the person is asked to predict in advance whether the coin will land heads or tails. The person predicts correctly 75% of the time (i.e., on 12 tosses). What is the probability of being correct 12 or more times by pure guessing?

5. (*Roulette*) In American roulette, the probability of winning when betting "red" is $\frac{9}{19}$. What is the probability of being ahead after betting the same amount 90 times?

6. (*Wine Tasting*) A wine-taster claims that she can usually distinguish between domestic and imported wines. As a test, she is given 100 wines to test and correctly identifies 63 of them. What is the probability that she accomplished that good a record by pure guessing? That is, what is the probability of being correct 63 or more times out of 100 by pure guessing?

7. (*Basketball*) A basketball player makes each free throw with probability $\frac{3}{4}$. What is the probability of making 68 or more shots out of 75 trials?

8. (*Bookstore Customers*) A bookstore determines that two-fifths of the people who come into the store make a purchase. What is the probability that of the 54 people who come into the store during a certain hour, less than 14 make a purchase?

9. (*Baseball*) A baseball player has a lifetime batting average of .310. Find the probability that he gets at least 6 hits in 20 times at bat.

10. (*Advertising Campaign*) An advertising agency, which reached 25% of its target audience with its old campaign, has devised a new advertising campaign. In a sample of 1000 people, it finds that 290 have been reached by the new advertising campaign. What is the probability that at least 290 people would have been reached by the old campaign? Does the new campaign seem to be more effective?

11. (*Equipment Reliability*) A washing machine manufacturer finds that 2% of its washing machines break down within the first year. Find the probability that less than 15 out of a lot of 1000 washers break down within 1 year.

12. (*Color Blindness*) The incidence of color blindness among the men in a certain country is 20%. Find the expected number of color-blind men in a random sample of 70 men. What is the probability of finding exactly that number of color-blind men in a sample of size 70?

13. (*Product Reliability*) The probabilities of failure for each of three independent components in a device are .01, .02, and .01, respectively. The device fails only if all three components fail. Out of a lot of 1 million devices, how many would be expected to fail? Find the probability that more than three devices in the lot fail.

14. (*CD Players*) In a random sample of 250 college students, 50 of them own a compact-disc player. Estimate the probability that a college student chosen at random owns a compact-disc player. If actually 25% of all college students own compact-disc players, what is the probability that in a random survey of 250 students, at most 50 of them own the devices?

In Exercises 15–17, use a graphing calculator, a spreadsheet, or mathematical software to calculate the probabilities.

15. (*Coin Tosses*) In 100 tosses of a fair coin, let X be the number of heads. Find the exact value of $\Pr(49 \leq X \leq 51)$, and also find the normal approximation to that probability.

16. (*Die Tosses*) Let X be the number of 4's in 120 tosses of a fair die. Calculate the exact probability $\Pr(17 \leq X \leq 21)$, and also find the normal approximation to that probability.

17. (*Name Recognition*) Say that there is a 20% chance that a person chosen at random from the population has never heard of John Steinbeck. Find the exact probability that in 150 people we find exactly 30 people who have not heard of Steinbeck. Compare your result with the normal approximation to that probability.

18. (*Male Heights*) About 5% of American males are 6 feet 2 inches or taller. Calculate the exact probability that of 150 men at a business meeting, no more than five are 6 feet 2 inches or taller, and also find the normal approximation to that probability.

SOLUTIONS TO PRACTICE PROBLEMS 7.7

1. **(a)** Giving the drug to a single mouse is a binomial trial with "recovery" as "success" and "death" as "failure." If the drug has no effect, then the probability of success is $\frac{1}{2}$. The probability of five successes in five trials is given by formula (2) of Section 7.3, with $n = 5$, $p = \frac{1}{2}$, $q = \frac{1}{2}$, $k = 5$.

$$\Pr(X = 5) = \binom{5}{5} \left(\frac{1}{2}\right)^5 \left(\frac{1}{2}\right)^0$$

$$= \left(\frac{1}{2}\right)^5$$

$$= \frac{1}{32} = .03125$$

(b) As in part (a), this experiment is a binomial experiment with $p = \frac{1}{2}$. However, now $n = 25$. The probability that 18 or more mice recover is

$$\Pr(X = 18) + \Pr(X = 19) + \cdots + \Pr(X = 25).$$

This probability is the area of the blue portion of the histogram in Fig. 7(a). The histogram can be approximated by the normal curve with

$$\mu = 25 \cdot \frac{1}{2} = 12.5$$

and

$$\sigma = \sqrt{25 \cdot \frac{1}{2} \cdot \frac{1}{2}} = \sqrt{\frac{25}{4}} = \frac{5}{2} = 2.5$$

[Fig. 7(b)]. Since the blue portion of the histogram begins at the point 17.5, the desired probability is approximately the area of the shaded region under the normal curve. The number 17.5 is

$$\frac{17.5 - 12.5}{2.5} = \frac{5}{2.5} = 2$$

standard deviations to the right of the mean. Hence, the area under the curve is $1 - A(2) = 1 - .9772 = .0228$. Therefore, the probability that 18 or more mice recover is approximately .0228.

2. Both experiments offer convincing evidence that the drug is helpful in treating the illness. The likelihood of obtaining the results by pure chance is slim. The second experiment might be considered more conclusive than the first, since the result, if due to chance, has a lower probability.

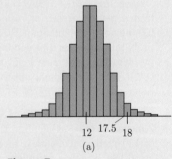

12 17.5 18

(a)

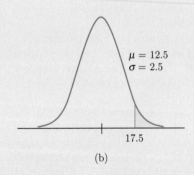

$\mu = 12.5$
$\sigma = 2.5$

17.5

(b)

Figure 7

Chapter 7 Summary

1. *Bar charts*, *pie charts*, *histograms*, and *box plots* help us turn raw data into visual forms that often allow us to see patterns in the data quickly.

2. The *median* of an ordered list of data is a number with the property that the same number of data items lie above it as below it. For an ordered list of N numbers, it is the middle number when N is odd, and the average of the two middle numbers when N is even.

3. For an ordered list of data, the *first quartile* Q_1 is the median of the list of data items below the median, and the *third quartile* Q_3 is the median of the list of data items above the median. The difference of the third and first quartiles is called the *interquartile range*. The sequence of numbers consisting of the lowest number, Q_1, the median, Q_3, and the highest number is called the *five-number summary*.

4. The probability distribution for a random variable can be displayed in a table or a histogram. With a histogram, the probability of an event is the sum of the areas of the rectangles corresponding to the outcomes in the event.

5. If the probability of success in each trial of a binomial experiment is p, then the probability of k successes in n trials is $\binom{n}{k}p^k(1-p)^{n-k}$.

6. The *sample mean* of a sample of n numbers is the sum of the numbers divided by n.

7. The *expected value* of a random variable is the sum of the products of each outcome and its probability.

8. The *variance* of a random variable is the sum of the products of the square of each outcome's distance from the expected value and the outcome's probability. The variance of the random variable X can also be computed as $E(X^2) - [E(X)]^2$.

9. A binomial random variable with parameters n and p has expected value np and variance $np(1-p)$.

10. The square root of the variance is called the *standard deviation*.

11. Chebychev's inequality states that the probability that an outcome of an experiment is within c units of the mean is at least $1 - (\sigma^2/c^2)$, where σ is the standard deviation.

12. A normal curve is identified by its mean (μ) and its standard deviation (σ). The standard normal curve has $\mu = 0$ and $\sigma = 1$. Areas of regions under the standard normal curve can be obtained with the aid of a table or graphing calculator.

13. A random variable is said to be normally distributed if the probability that an outcome lies between a and b is the area of the region under a normal curve from $x = a$ to $x = b$. After the numbers a and b are converted to standard deviations from the mean, the sought-after probability can be obtained as an area under the standard normal curve.

14. Probabilities associated with a binomial random variable with parameters n and p can be approximated with a normal curve having $\mu = np$ and $\sigma = \sqrt{np(1-p)}$.

REVIEW OF FUNDAMENTAL CONCEPTS

1. What is a bar chart? A pie chart? A histogram? A box plot?

2. What is the median of a list of numbers? The first quartile? The third quartile? The interquartile range? The five-number summary?

3. What is a frequency distribution? A relative frequency distribution? A probability distribution?

4. How is a histogram constructed from a distribution?

5. What is a random variable?

6. What is meant by the probability distribution of a discrete random variable?

7. What are the identifying features of a binomial random variable?

8. What is the formula for the probability of k successes in n independent binomial trials?

9. What is meant by the expectation (or expected value) of a random variable? Variance? Standard deviation?

10. What is the Chebychev inequality, and how is it used?

11. What is meant by a normal random variable?

12. What is meant by the pth percentile of a normal random variable?

13. How are binomial probabilities approximated with the normal distribution?

KEY FORMULAS

Binomial Probabilities: If X is the number of "successes" in n independent trials, where in each trial the probability of a "success" is p, then $\Pr(X = k) = \binom{n}{k}p^k q^{n-k}$, for $k = 0, 1, 2, \ldots, n$, where $q = 1 - p$.

Sample Mean: If $x_1, x_2, \ldots, x_n$ is a sample of n numbers, then

$$\overline{x} = \frac{x_1 + x_2 + \cdots + x_n}{n}.$$

Sample Mean: If $x_1, x_2, \ldots, x_r$ is a sample of n numbers where the frequency of x_1 is f_1, the frequency of x_2 is f_2,

and so forth, where $f_1 + f_2 + \cdots + f_r = n$, then

$$\overline{x} = \frac{x_1 f_1 + x_2 f_2 + \cdots + x_r f_r}{n}$$

$$= x_1\left(\frac{f_1}{n}\right) + x_2\left(\frac{f_2}{n}\right) + \cdots + x_r\left(\frac{f_r}{n}\right).$$

Population Variance: If $x_1, x_2, \ldots, x_r$ is a population of n numbers where the frequency of x_1 is f_1, the frequency of x_2 is f_2, and so forth, where $f_1 + f_2 + \cdots + f_r = n$, then

$$\sigma^2 = \frac{1}{N}\left[(x_1 - \mu)^2(f_1) + (x_2 - \mu)^2(f_2)\right.$$
$$\left. + \cdots + (x_r - \mu)^2(f_r)\right].$$

Sample Variance: If $x_1, x_2, \ldots, x_r$ is a sample of n numbers where the frequency of x_1 is f_1, the frequency of x_2 is f_2, and so forth, where $f_1 + f_2 + \cdots + f_r = n$, then

$$s^2 = \frac{1}{n-1}\left[(x_1 - \overline{x})^2(f_1) + (x_2 - \overline{x})^2(f_2)\right.$$
$$\left. + \cdots + (x_r - \overline{x})^2(f_r)\right].$$

Alternate Formula for Variance: $\sigma^2 = E(X^2) - \mu^2$

Mean and Variance of a Binomial Random Variable: $\mu = np$ and $\sigma^2 = npq$, where X has parameters n and p, where $q = 1 - p$.

Chebychev's Inequality: $\Pr(\mu - c \le X \le \mu + c) \ge 1 - \dfrac{\sigma^2}{c^2}$, where X has mean μ and standard deviation σ.

Normal Random Variable:

$$\Pr(a \le X \le b) = \Pr\left(\frac{a-\mu}{\sigma} \le Z \le \frac{b-\mu}{\sigma}\right)$$
$$= A\left(\frac{b-\mu}{\sigma}\right) - A\left(\frac{a-\mu}{\sigma}\right)$$

and

$$\Pr(X \le x) = \Pr\left(Z \le \frac{x-\mu}{\sigma}\right) = A\left(\frac{x-\mu}{\sigma}\right),$$

where X is a normal random variable with mean μ and standard deviation σ, Z has the standard normal distribution, and $A(z)$ is the area under that distribution to the left of z.

Percentiles for a Normal Distribution: $x_p = \mu + z_p \cdot \sigma$, where x_p is the pth percentile of a normal distribution with mean μ and standard deviation σ, and z_p is the pth percentile of the standard normal distribution.

Normal Approximation to the Binomial Distribution: The histogram for the binomial distribution with parameters n and p is approximated by the normal curve with $\mu = np$ and $\sigma = \sqrt{npq}$, where $q = 1 - p$.

SUPPLEMENTARY EXERCISES

1. (*U.S. Population*) Display the data from Table 1 in a bar chart that shows frequencies on the y-axis. Then display the data in a pie chart.

TABLE 1	U.S. population by region (in millions)
Region	**Population**
Northeast	54.6
Midwest	65.7
South	105.9
West	67.4

Source: U.S. Census Bureau, Census 2004.

2. Find the five-number summary and the interquartile range for the following set of numbers, and then draw the box plot.

$$1, 2, 3, 4, 5, 9, 14, 23$$

3. An experiment consists of three binomial trials, each having probability $\frac{1}{3}$ of success.

   (a) Determine the probability distribution table for the number of successes.

   (b) Use the table to compute the mean and the variance of the probability distribution.

4. Find the area of the shaded region under the standard normal curve shown in Fig. 1(a).

5. Find the area of the shaded region under the normal curve with $\mu = 5$, $\sigma = 3$ shown in Fig. 1(b).

6. (*Archery*) An archer has probability .3 of hitting a certain target. What is the probability of hitting the target exactly two times in four attempts?

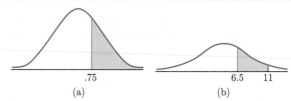

.75   6.5   11

(a)   (b)

Figure 1

7. Suppose that a probability distribution has mean 10 and standard deviation $\frac{1}{3}$. Use the Chebychev inequality to estimate the probability that an outcome will lie between 9 and 11.

8. Table 2 gives the probability distribution of the random variable X. Compute the mean and the variance of the random variable.

TABLE 2	
k	$\Pr(X = k)$
0	.2
1	.3
5	.1
10	.4

9. (*Heights of Adult Males*) The height of adult males in the United States is normally distributed with $\mu = 5.75$ feet and $\sigma = .2$ feet. What percent of the adult male population has height of 6 feet or greater?

10. (*Balls in an Urn*) An urn contains four red balls and four white balls. An experiment consists of selecting at random a sample of four balls and recording the number of red balls in the sample. Set up the probability distribution and compute its mean and variance.

11. (*Jury Selection*) In a certain city two-fifths of the registered voters are women. Out of a group of 54 voters allegedly selected at random for jury duty, 13 are women. A local civil liberties group has charged that the selection procedure discriminated against women. Use the normal curve to estimate the probability of 13 or fewer women being selected in a truly random selection process.

12. (*Quality Control*) In a complicated production process, $\frac{1}{4}$ of the items produced have to be readjusted. Use the normal curve to estimate the probability that out of a batch of 75 items, between 8 and 22 (inclusive) of the items require readjustment.

13. Figure 2(a) is a normal curve with $\mu = 80$ and $\sigma = 15$. Find the value of h for which the area of the shaded region is .8664.

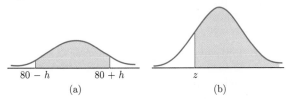

(a)　　　　　(b)

Figure 2

14. Figure 2(b) is a standard normal curve. Find the value of z for which the area of the shaded region is .7734.

15. (*Baseball*) The World Series in baseball consists of a sequence of games that terminates when one team (the winner) wins its fourth game. If the two teams are equally likely to win any one game, what is the probability that the series will last exactly four games? Five? Six? Seven?

16. (*Guessing on an Exam*) A true–false exam consists of ten 10-point questions. The instructor informs the students that six of the answers are *true* and four are

false. An unprepared student decides to guess the answer to each question with the use of the spinner in Fig. 3, which gives *true* 60% of the time. Determine the student's expected score. Can you think of a better strategy?

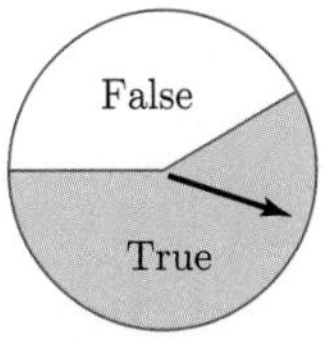

Figure 3

Conceptual Exercises

17. Give an example of a grade distribution for a class of students in which

    (a) scoring in the 3rd quartile is not very good.

    (b) scoring in the 3rd quartile corresponds to a perfect grade.

18. Give an example of a distribution of ten grades for which

    (a) the mean and median are equal.

    (b) the mean is less than the median.

    (c) the median is less than the mean.

19. What is the difference between a population mean and a sample mean?

20. Explain in your own words the meaning of "expected value."

21. Explain the type of probability situations for which the binomial distribution applies.

22. Give an example of a sequence of repeated trials that does not produce a binomial distribution.

CHAPTER TEST

1. (*Supermarket Queue*) The manager of a supermarket counts the number of customers waiting in the express checkout line at random times throughout the week. Her observations are summarized in the following frequency table.

Number Waiting in Line	Frequency
0	2
1	5
2	9
3	13
4	11
5	7
6	3

Construct the corresponding relative frequency table, and use it to estimate the probability that at most three customers are waiting in line.

2. Find the five-number summary and the interquartile range for the following set of numbers, and then draw the box plot.

$$20, 25, 26, 27, 29, 30, 33, 34, 37, 40, 42$$

3. (*Coin Tosses*) A fair coin is tossed twice. Let X be the number of heads.

    (a) Determine the probability distribution of X.

    (b) Determine the probability distribution of $2X + 5$.

4. (*Dice Tosses*) A pair of fair dice is rolled 12 times. *Note*: For each roll, the probability of getting a 7 is 1/6.

(a) What is Pr(the result is 7 exactly twice)?

(b) What is Pr(the result is 7 at least twice)?

(c) What is the expected number of times that the result is 7?

5. The probability distribution of a random variable X is given in the following table. Determine the mean, variance, and standard deviation of X.

k	$\Pr(X = k)$
-2	.3
0	.1
1	.4
3	.2

6. (*Dice Game*) Lucy and Ethel play a game of chance in which a pair of fair dice is rolled once. If the result is 7 or 11, then Lucy pays Ethel $10.00. Otherwise, Ethel pays Lucy $3.00. In the long run, which player comes out ahead, and by how much?

7. Let Z be the standard normal random variable. Determine the following.

(a) $\Pr(Z \le 1)$

(b) $\Pr(Z \ge -2.25)$

(c) $\Pr(-1 \le Z \le 1.15)$

8. (*Blood Sugar Levels for Diabetics*) Assume the fasting blood glucose level among diabetics is normally dis-tributed with mean 106 mg/100 ml and standard de-viation 10 mg/100 ml.

(a) What is the probability that a randomly selected diabetic will have a fasting blood glucose level of more than 116 mg/100 ml?

(b) What is the probability that the fasting blood glu-cose level of a randomly selected diabetic will be between 96 and 121 mg/100 ml?

(c) For what value of c will 24.2% of diabetics have a fasting blood glucose level of less than c mg/100 ml? more than c mg/100 ml?

9. (*Opinion Polling*) A new shopping mall is to be built, and 20% of the community's residents oppose the con-struction. A random sample of 25 residents is polled.

(a) Approximate the probability that more than 5 but at most 10 of the selected residents oppose the con-struction.

(b) Determine the exact probability that exactly 10 of the selected residents oppose the construction, and compare your answer with the normal approx-imation of that probability.

(c) Determine the exact probability that at least one of the selected residents opposes the construction, and compare your answer with the normal approx-imation of that probability.

CHAPTER 7 PROJECT

An Unexpected Expected Value[1]

An urn contains four red balls and six white balls. Suppose two balls are drawn at random from the urn, and let X be the number of red balls drawn. The probability of obtaining two red balls depends on whether the balls are drawn with or without replacement. The purpose of this project is to show that the expected number of red balls drawn is not affected by whether or not the first ball is replaced before the second ball is drawn.

1. Do you think the probability that both balls are red is higher if the first ball is replaced before the second ball is drawn, or if the first ball is not replaced?

2. Suppose the first ball is replaced before the second ball is drawn. Find the probability that both balls are red.

3. Suppose the first ball is not replaced before the second ball is drawn. Find the probability that both balls are red.

4. Was your intuitive guess in part 1 correct?

5. Do you think that the expected number of red balls drawn is higher if the balls are drawn with replacement or without replacement?

6. Suppose the first ball is replaced before the second ball is drawn. Find the expected number of red balls that will be drawn.

7. Suppose the first ball is not replaced before the second ball is drawn. Find the expected number of red balls that will be drawn.

8. Was your intuitive guess in part 5 correct?

9. Pretend that the balls are ping-pong balls that have been finely ground up, and that the red and white specks have been thoroughly mixed. Forty percent of the specks will be red and 60% will be white. Suppose you stir the specks and use a tablespoon to scoop out 10% of the specks. That is, suppose the tablespoon holds a quantity of specks corresponding to one ball.

   (a) What percentage of a red ball is contained in the spoon?

   (b) What percentage of the remaining specks in the urn are red?

   (c) If the spoonful of specks is replaced, does the percentage of red specks in the urn change?

   (d) Use the results from parts (b) and (c) to explain why the expected number of red balls as calculated in parts 6 and 7 is the same with and without replacement.

---

[1]The idea for this project was taken from the article "An Unexpected Expected Value," by Stephen Schwartzman, which appeared in the February 1993 issue of *The Mathematics Teacher*.

Markov Processes

SUPPOSE that we perform, one after the other, a sequence of experiments that have the same set of outcomes. The probabilities of the various outcomes of a particular experiment of the sequence may depend in some way on the outcomes of preceding experiments. The nature of such a dependency may be very complicated. In the extreme, the outcome of the current experiment may depend on the entire history of the outcomes of preceding experiments. However, there is a simple type of dependency that occurs frequently in applications and that we can analyze with fair ease. Namely, we suppose that the probabilities of the various outcomes of the current experiment depend (at most) on the outcome of the preceding experiment. In this case the sequence of experiments is called a *Markov process*. In this chapter we present some of the most elementary ideas concerning Markov processes and their applications.

8.1 The Transition Matrix

Here are some Markov processes that arise in applications.

EXAMPLE 1 **Investment** A particular utility stock is very stable and, in the short run, the probability that it increases or decreases in price depends only on the result of the preceding day's trading. The price of the stock is observed at 4 P.M. each day and is recorded as "increased," "decreased," or "unchanged." The sequence of observations forms a Markov process. ∎

EXAMPLE 2 **Medicine** A doctor tests the effect of a new drug on high blood pressure. Based on the effects of metabolism, a given dose is eliminated from the body in 24 hours. Blood pressure is measured once a day and is recorded as "high," "low," or "normal." The sequence of measurements forms a Markov process. ■

EXAMPLE 3 **Sociology** A sociologist postulates that the likelihood that, in certain countries, a woman will enter the labor force depends primarily on whether the woman's mother worked. He designs an experiment to test this hypothesis by viewing the sequence of career choices of a woman, her daughters, her granddaughters, her great-granddaughters, and so on as a Markov process. ■

Let us now introduce some vocabulary and mathematical machinery with which to study Markov processes. The experiments are performed at regular time intervals and have the same set of outcomes. These outcomes are called *states*, and the outcome of the current experiment is referred to as the *current state* of the process. After each time interval, the process may change its state. The transition from state to state can be described by tree diagrams, as is shown in the next example.

EXAMPLE 4 **Using a tree diagram to represent transitions** Refer to the utility stock of Example 1. Suppose that if the stock increases one day, the probability that on the next day it increases is .3, remains unchanged .2, decreases .5. On the other hand, if the stock is unchanged one day, the probability that on the next day it increases is .6, remains unchanged .1, decreases .3. If the stock decreases one day, the probability that it increases the next day is .3, is unchanged .4, decreases .3. Represent the possible transitions between states and their probabilities by tree diagrams.

Solution The Markov process has three states: "increases," "unchanged," and "decreases." The transitions from the first state ("increases") to the other states are

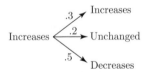

Note that each branch of the tree has been labeled with the probability of the corresponding transition. Similarly, the tree diagrams corresponding to the other two states are

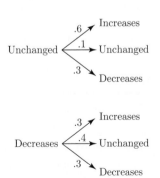

■

The three tree diagrams of Example 4 may be summarized in a single matrix. We insert the probabilities from a given tree down a column of the matrix, so that each column of the matrix records the information about transitions from one particular state. So the first column of the matrix is

$$\begin{bmatrix} .3 \\ .2 \\ .5 \end{bmatrix},$$

corresponding to transitions from the state "increases." The complete matrix is

		Current state		
		Increases	Unchanged	Decreases
	Increases	.3	.6	.3
Next state	Unchanged	.2	.1	.4
	Decreases	.5	.3	.3

This matrix, which records all data about transitions from one state to another, is called the *transition matrix of the Markov process.*

EXAMPLE 5 **Women in the labor force** Census studies from the 1960s reveal that in the United States 80% of the daughters of working women also work and that 30% of the daughters of nonworking women work. Assume that this trend remains unchanged from one generation to the next. Determine the corresponding transition matrix.

Solution There are two states, which we label "work" and "don't work." The first column corresponds to transitions from the first state—that is, from "work." The probability that the daughter of a working woman chooses *not* to work is $1 - .8 = .2$. Therefore, the first column is

$$\begin{bmatrix} .8 \\ .2 \end{bmatrix}.$$

In similar fashion, the second column is

$$\begin{bmatrix} .3 \\ .7 \end{bmatrix}.$$

The transition matrix is, therefore,

		Current generation	
		Work	Don't work
Next generation	Work	.8	.3
	Don't work	.2	.7

Now Try Exercise 13(a) ■

Here is the form of a general transition matrix for a Markov process:

		Current state				
		State 1	$\cdots$	State j	$\cdots$	State r
	State 1					
	$\vdots$					
Next state	State i			$\Pr(\text{next } i \mid \text{current } j)$		
	$\vdots$					
	State r					

Note that this matrix satisfies the following two properties:

1. All entries are greater than or equal to 0.

2. The sum of the entries in each column is 1.

Any square matrix satisfying properties 1 and 2 is called a *stochastic matrix*. (The word "stochastic" derives from the Greek word "stochastices," which means a person who predicts the future.)

Let us examine further the Markov process of Example 5. In 1960 about 40% of U.S. women worked and 60% did not. This distribution is described by the column matrix

$$\begin{bmatrix} .4 \\ .6 \end{bmatrix}_0,$$

which is called a *distribution matrix*. The subscript 0 is added to denote that this matrix describes generation 0. Their daughters constitute generation 1, and their granddaughters generation 2. There is a distribution matrix for each generation. The distribution matrix for generation n is

$$\begin{bmatrix} p_W \\ p_{DW} \end{bmatrix}_n,$$

where p_W is the percentage of women in generation n who work and p_{DW} is the percentage who don't work. Of course, the numbers p_W and p_{DW} are also probabilities. The number p_W is the probability that a woman selected at random from generation n works. Similarly, p_{DW} is the probability that a woman selected at random from generation n does not work. Shortly we shall give a method for calculating the distribution matrix for generation n.

In general, whenever a Markov process applies to a group with members in r possible states, a distribution matrix of the form

$$\begin{bmatrix} p_1 \\ p_2 \\ \vdots \\ p_r \end{bmatrix}_0$$

gives the initial percentages of members in each of the r states. Similarly, a matrix of the same type (with the subscript n) gives the percentages of members in each of the r states after n time periods. Note that each of the percentages in the distribution matrix is also a probability—the probability that a randomly selected member will be in the corresponding state.

EXAMPLE 6 **Projecting women in the labor force** In 1960, census figures showed that 40% of American women worked. Use the stochastic matrix of Example 5 to determine the percentage of working women in each of the next two generations.

Solution Let us denote by A the stochastic matrix of Example 5:

		Current generation	
		Work	Don't work
Next generation	Work	.8	.3
	Don't work	.2	.7

The initial distribution matrix is

$$\begin{bmatrix} .4 \\ .6 \end{bmatrix}_0 .$$

Our goal is to compute the distribution matrices for generations 1 and 2. With an eye toward performing similar calculations for other countries, let us use letters rather than specific numbers in these distribution matrices. Let

$$\begin{bmatrix} \ \\ \ \end{bmatrix}_0 = \begin{bmatrix} x_0 \\ y_0 \end{bmatrix}, \qquad \begin{bmatrix} \ \\ \ \end{bmatrix}_1 = \begin{bmatrix} x_1 \\ y_1 \end{bmatrix}, \qquad \begin{bmatrix} \ \\ \ \end{bmatrix}_2 = \begin{bmatrix} x_2 \\ y_2 \end{bmatrix}.$$

The state transitions from generation 0 to generation 1 may be displayed in a tree diagram.

Adding together the probabilities for the two paths leading to "work" in generation 1 gives

$$x_1 = .8x_0 + .3y_0. \tag{1}$$

Similarly, adding together the probabilities for the two paths leading to "don't work" in generation 1 gives

$$y_1 = .2x_0 + .7y_0. \tag{2}$$

Thus, (1) and (2) show that x_1 and y_1 can be computed from this pair of equations.

$$x_1 = .8x_0 + .3y_0$$
$$y_1 = .2x_0 + .7y_0$$

This system of equations is equivalent to the one matrix equation

$$\begin{bmatrix} .8 & .3 \\ .2 & .7 \end{bmatrix} \begin{bmatrix} x_0 \\ y_0 \end{bmatrix} = \begin{bmatrix} x_1 \\ y_1 \end{bmatrix}. \tag{3}$$

Or, to write equation (3) symbolically,

$$A \begin{bmatrix} \ \\ \ \end{bmatrix}_0 = \begin{bmatrix} \ \\ \ \end{bmatrix}_1 . \tag{4}$$

Now it is easy to do the arithmetic to compute the distribution matrix for generation 1 of American women:

$$A \begin{bmatrix} \ \\ \ \end{bmatrix}_0 = \begin{bmatrix} .8 & .3 \\ .2 & .7 \end{bmatrix} \begin{bmatrix} .4 \\ .6 \end{bmatrix}_0 = \begin{bmatrix} .5 \\ .5 \end{bmatrix}_1 .$$

That is, 50% of American women in generation 1 will work, and 50% will not.

To compute the distribution matrix for generation 2, we use the same reasoning as before, when we showed that to get

$$\begin{bmatrix} \ \\ \ \end{bmatrix}_1 \qquad \text{from} \qquad \begin{bmatrix} \ \\ \ \end{bmatrix}_0 ,$$

just multiply by A.

Similarly,

$$A \begin{bmatrix} \ \\ \ \end{bmatrix}_1 = \begin{bmatrix} \ \\ \ \end{bmatrix}_2 .$$

However, by (4), we have a formula for $\begin{bmatrix} \ \\ \ \end{bmatrix}_1$, which we can insert into the last equation, getting[1]

$$\begin{bmatrix} \ \\ \ \end{bmatrix}_2 = A \begin{bmatrix} \ \\ \ \end{bmatrix}_1 = A \left(A \begin{bmatrix} \ \\ \ \end{bmatrix}_0 \right) = A^2 \begin{bmatrix} \ \\ \ \end{bmatrix}_0 .$$

In other words,

$$A^2 \begin{bmatrix} \ \\ \ \end{bmatrix}_0 = \begin{bmatrix} \ \\ \ \end{bmatrix}_2 . \tag{5}$$

A simple calculation gives

$$A^2 = \begin{bmatrix} .70 & .45 \\ .30 & .55 \end{bmatrix},$$

so that we can now compute the distribution for generation 2 of American women:

$$\begin{bmatrix} .70 & .45 \\ .30 & .55 \end{bmatrix} \begin{bmatrix} .4 \\ .6 \end{bmatrix}_0 = \begin{bmatrix} .55 \\ .45 \end{bmatrix}_2 .$$

That is, after two generations 55% of American women work and 45% do not.

Now Try Exercise 7 ■

An argument similar to that used to derive equation (5) can be used to show that

$$A^3 \begin{bmatrix} \ \\ \ \end{bmatrix}_0 = \begin{bmatrix} \ \\ \ \end{bmatrix}_3 , \quad A^4 \begin{bmatrix} \ \\ \ \end{bmatrix}_0 = \begin{bmatrix} \ \\ \ \end{bmatrix}_4 , \quad A^5 \begin{bmatrix} \ \\ \ \end{bmatrix}_0 = \begin{bmatrix} \ \\ \ \end{bmatrix}_5 , \quad \dots .$$

A condensed notation is

$$A^n \begin{bmatrix} \ \\ \ \end{bmatrix}_0 = \begin{bmatrix} \ \\ \ \end{bmatrix}_n \qquad (n = 1, 2, 3, \dots). \tag{6}$$

That is, to compute the distribution matrix for generation n, merely compute the product of A^n times the distribution matrix for generation 0. Equation (6) can be used to predict the distribution matrix for any number of generations into the future, starting from any given distribution matrix. Let us now look at an entirely different type of situation that can be described by a Markov process.

EXAMPLE 7 **Taxi zones** Taxis pick up and deliver passengers in a city that is divided into three zones. Records kept by the drivers show that of the passengers picked up in zone I, 50% are taken to a destination in zone I, 40% to zone II, and 10% to zone III. Of the passengers picked up in zone II, 40% go to zone I, 30% to zone II, and 30% to zone III. Of the passengers picked up in zone III, 20% go to zone I, 60% to zone II, and 20% to zone III. Suppose that at the beginning of the day 60% of the taxis are in zone I, 10% in zone II, and 30% in zone III. What is the distribution of taxis in the various zones after all have had one rider? Two riders?

---

[1] Just as in elementary algebra, we define the powers $A^2, A^3, \dots$ of the matrix A by $A^2 = A \cdot A$, $A^3 = A \cdot A \cdot A, \dots$.

Solution This situation is an example of a Markov process. The states are the zones. The initial distribution of the taxis gives the zeroth distribution matrix:

$$\begin{bmatrix} .6 \\ .1 \\ .3 \end{bmatrix}_0 .$$

The stochastic matrix associated to the process is the one giving the probabilities of taxis starting in any one zone and ending up in any other. There is one column for each zone:

From zone:

		I	II	III
To zone:	I	.5	.4	.2
	II	.4	.3	.6
	III	.1	.3	.2

$= A.$

After all taxis have had one passenger, the distribution matrix is just

$$A\begin{bmatrix} .6 \\ .1 \\ .3 \end{bmatrix}_0 = \begin{bmatrix} .5 & .4 & .2 \\ .4 & .3 & .6 \\ .1 & .3 & .2 \end{bmatrix}\begin{bmatrix} .6 \\ .1 \\ .3 \end{bmatrix}_0 = \begin{bmatrix} .40 \\ .45 \\ .15 \end{bmatrix}_1 .$$

That is, 40% of the taxis are in zone I, 45% in zone II, and 15% in zone III. After all taxis have had two passengers, the distribution matrix is

$$A^2\begin{bmatrix} .6 \\ .1 \\ .3 \end{bmatrix}_0 ,$$

which, after some arithmetic, can be shown to be

$$\begin{bmatrix} .410 \\ .385 \\ .205 \end{bmatrix}_2 .$$

That is, after two passengers, 41% of the taxis are in zone I, 38.5% are in zone II, and 20.5% are in zone III.

Now Try Exercise 9   ■

The crucial formula used in both Examples 6 and 7 is

$$A\begin{bmatrix} \quad \\ \quad \end{bmatrix}_0 = \begin{bmatrix} \quad \\ \quad \end{bmatrix}_1 . \qquad (7)$$

From this one follows the more general formula

$$A^n\begin{bmatrix} \quad \\ \quad \end{bmatrix}_0 = \begin{bmatrix} \quad \\ \quad \end{bmatrix}_n .$$

We carefully proved (7) in the special case of Example 6. Let us do the same for Example 7. Suppose that the initial distribution of taxis is

$$\begin{bmatrix} x_0 \\ y_0 \\ z_0 \end{bmatrix} .$$

How many taxis end up in zone I after one passenger? The taxis in zone I come from three sources—zones I, II, and III. And the first row of the stochastic matrix

A gives the percentages of taxis starting out in each of the zones and ending up in zone I:

[percent of taxis going to zone I] $= [.5] \cdot$ [percent of taxis in zone I]

$+ [.4] \cdot$ [percent of taxis in zone II]

$+ [.2] \cdot$ [percent of taxis in zone III]

$= .5x_0 + .4y_0 + .2z_0.$

Indeed, the first entry, x_1, in the product

$$\begin{bmatrix} .5 & .4 & .2 \\ .4 & .3 & .6 \\ .1 & .3 & .2 \end{bmatrix} \begin{bmatrix} x_0 \\ y_0 \\ z_0 \end{bmatrix} = \begin{bmatrix} x_1 \\ \\ \end{bmatrix}$$

is just $.5x_0 + .4y_0 + .2z_0$. Similarly, the proportions of taxis in zones II and III coincide with the other two entries in the matrix product. So equation (7) really holds.

Interpretation of the Entries of A^n Label the columns and rows of the transition matrix A and A^n with the states as in the paragraph following Example 5. Then the entry in the ith row and jth column of A is the probability of the transition from state j to state i after one time period.

> The entry in the ith row and jth column of the matrix A^n is the probability of the transition from state j to state i after n time periods.

GC Let

$$A = \begin{bmatrix} .8 & .3 \\ .2 & .7 \end{bmatrix}$$

be the 2×2 matrix of Examples 5 and 6, and let

$$B = \begin{bmatrix} .4 \\ .6 \end{bmatrix}$$

be the initial distribution matrix. Successive distribution matrices are easy to obtain with a graphing calculator. The most recently displayed list, number, or matrix becomes the value of the variable **Ans** (on the TI-83) or the variable **ans(1)** (on the TI-89). (To obtain **Ans** or **ans(1)**, press 2nd[ANS].) In Figs. 1 and 3, after matrix B is displayed, it becomes the value of the variable **Ans** or **ans(1)**. Therefore, the value of **[A]*Ans** (or **a*ans(1)**) is the matrix product AB. Each time ENTER is pressed, the instruction **[A]*Ans** (or **a*ans(1)**) is repeated. That is, the preceding matrix is multiplied on the left by the matrix A. Figures 2 and 4, which result after ENTER is pressed two more times, show the additional matrices A^2B and A^3B.

```
[B]
            [[.4]
             [.6]]
[A]*Ans
            [[.5]
             [.5]]
```

Figure 1

```
[A]*Ans
            [[.5]
             [.5]]
            [[.55]
             [.45]]
            [[.575]
             [.425]]
```

Figure 2

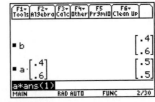

Figure 3

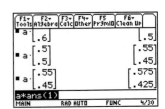

Figure 4

ES   Figure 5, which shows successive distribution matrices, can be created as follows:

1. Give the name **A** to the 2 × 2 matrix on the left.

2. Enter **.4** and **.6** into cells D2 and D3.

3. Select the range E2:E3, type in **=MMULT(A,D2:D3)**, and press Ctrl+Shift+Enter.

4. Select the range E2:E3, and drag its fill handle right to column I.

Figure 5

	A	B	C	D	E	F	G	H	I
1	**Matrix A**		**Generation**	**0**	**1**	**2**	**3**	**4**	**5**
2	0.8	0.3		0.4	0.5	0.55	0.575	0.5875	0.59375
3	0.2	0.7		0.6	0.5	0.45	0.425	0.4125	0.40625

PRACTICE PROBLEMS 8.1

1. Is $\begin{bmatrix} \frac{2}{5} & 1 \\ \frac{3}{5} & .2 \\ \frac{2}{5} & -.3 \end{bmatrix}$ a stochastic matrix?

2. (*Learning Process*) An elementary learning process consists of subjects participating in a sequence of events. Experiment shows that of the subjects not conditioned to make the correct response at the beginning of any event, 40% will be conditioned to make the correct response at the end of the event. Once a subject is conditioned to make the correct response, he stays conditioned.

(a) Set up the 2 × 2 stochastic matrix with columns and rows labeled N (not conditioned) and C (conditioned) that describes this situation.

(b) Compute A^3.

(c) If initially all the subjects are not conditioned, what percent of them will be conditioned after three events?

EXERCISES 8.1

In Exercises 1–6, determine whether or not the matrix is stochastic.

1. $\begin{bmatrix} 1 & .8 \\ 0 & .2 \end{bmatrix}$

2. $\begin{bmatrix} \frac{1}{3} & \frac{1}{3} \\ \frac{2}{3} & \frac{2}{3} \end{bmatrix}$

3. $\begin{bmatrix} .4 & .3 & .2 \\ .6 & .7 & .8 \end{bmatrix}$

4. $\begin{bmatrix} .4 & .5 & .1 \\ .3 & .4 & 0 \\ .3 & .2 & .9 \end{bmatrix}$

5. $\begin{bmatrix} \frac{1}{6} & \frac{5}{12} & 0 \\ \frac{1}{2} & \frac{1}{4} & .5 \\ \frac{1}{3} & \frac{1}{3} & .5 \end{bmatrix}$

6. $\begin{bmatrix} 1 & 0 & 0 \\ 0 & 1 & 0 \\ 0 & 0 & 1 \end{bmatrix}$

7. (*Women in the Labor Force*) Referring to Example 5, consider a typical group of French women, of whom 47% currently work. Assume that the same percentage of daughters follow in their mothers' footsteps as with the American women—that is, those given by the matrix

$$A = \begin{bmatrix} .8 & .3 \\ .2 & .7 \end{bmatrix}.$$

Use A and A^2 to determine the proportion of working women in the next two generations. (Round off to the nearest whole percent.)

8. (*Women in the Labor Force*) Repeat Exercise 7 for the women of Belgium, of whom 44% currently work. (Round off the percentage of women to the nearest whole percent.)

9. (*Taxi Zones*) Refer to Example 7 (taxi zones). If originally 40% of the taxis start in zone I, 40% in zone II, and 20% in zone III, how will the taxis be distributed after each has taken one passenger?

10. (*Fitness*) A group of physical fitness devotees works out in the gym every day. The workouts vary from strenuous to moderate to light. When their exercise routine was recorded, the following observation was made: Of the people who work out strenuously on a particular day, 40% will work out strenuously on the next day and 60% will work out moderately. Of the people who work out moderately on a particular day, 50% will work out strenuously and 50% will work out lightly on the next day. Of the people working out lightly on a particular day, 30% will work out strenuously on the next day, 20% moderately, and 50% lightly.

(a) Set up the 3×3 stochastic matrix with columns and rows labeled S, M, and L that describes these transitions.

(b) Suppose that on a particular Monday 80% have a strenuous, 10% a moderate, and 10% a light workout. What percent will have a strenuous workout on Wednesday?

11. (*T-Maze*) Each day mice are put into a T-maze (a maze shaped like a "T"; Fig. 6). In this maze they have the choice of turning to the left (rewarded with cheese) or to the right (receive cheese along with mild shock). After the first day their decision whether to turn left or right is influenced by what happened on the previous day. Of those that go to the left on a certain day, 90% go to the left on the next day and 10% go to the right. Of those that go to the right on a certain day, 70% go to the left on the next day and 30% go to the right.

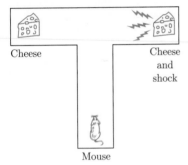

Figure 6

(a) Set up the 2×2 stochastic matrix with columns and rows labeled L and R that describes this situation.

(b) Compute the second power of the matrix in part (a).

(c) Suppose that on the first day (day 0) 50% go to the left and 50% go to the right. So, the initial distribution is given by the column matrix $\begin{bmatrix} .5 \\ .5 \end{bmatrix}_0$. Using the matrices in parts (a) and (b), find the distribution matrices for the next two days, days 1 and 2.

(d) Make a guess as to the percentage of mice that will go to the left after 50 days. (Do not compute.)

12. (*Bird Migrations*) The following matrix describes the migration patterns of a species of bird from year to year among three habitats, I, II, and III.

From habitat:

		I	II	III
To habitat:	I	.78	.07	.15
	II	.12	.85	.05
	III	.10	.08	.80

(a) What percentage of the birds that begin the year in habitat III migrate to habitat I during the year?

(b) Explain the meaning of the percentage .85 appearing in the center of the matrix.

(c) If there are 1000 birds in each habitat at the beginning of a year, how many will be in each habitat at the end of the year? At the end of two years?

13. (*Voter Patterns*) For a certain group of states, it was observed that 70% of the Democratic governors were succeeded by Democrats and 30% by Republicans. Also, 40% of the Republican governors were succeeded by Democrats and 60% by Republicans.

(a) Set up the 2×2 stochastic matrix with columns and rows labeled D and R that displays these transitions.

(b) Compute A^2 and A^3.

(c) Suppose that all the current governors are Democrats. Assuming that the current trend holds for three elections, what percent of the governors will then be Democrats?

14. (*Market Share*) A retailer stocks three brands of breakfast cereal. A survey is taken of 5000 people who purchase cereal weekly from this retailer. Each week Crispy Flakes loses 12% of its customers to Crunchy Nuggets and 19% to Toasty Cinnamon Twists. Crunchy Nuggets loses 16% of its customers to Crispy Flakes and 10% of its customers to Toasty Cinnamon Twists, and Toasty Cinnamon Twists loses 20% of its customers to Crispy Flakes and 14% to Crunchy Nuggets.

(a) Set up a stochastic matrix displaying these transitions.

(b) Suppose that this week 1500 people buy Crispy Flakes, 1500 buy Crunchy Nuggets, and 2000 people buy Toasty Cinnamon Twists. How many people will buy Crispy Flakes next week? How many will buy Toasty Cinnamon Twists in two weeks?

15. (*Population Movement*) A sociologist studying living patterns in a certain region determines that each year the population shifts between urban, suburban, and rural areas as shown in Fig. 7.

(a) Set up a stochastic matrix A that displays these transitions.

(b) What percentage of people who live in urban areas in 2006 will live in rural areas in 2008?

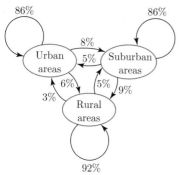

Figure 7

16. (*Analysis of a Poem*) In 1913, Markov analyzed a long poem written by a Russian author.[1] He found that vowels were followed by consonants 87.2% of the time (either in the same word or the next word), and that consonants were followed by vowels 66.3% of the time.

    (a) Set up the 2×2 stochastic matrix, with columns and rows labeled V and C, that describes this situation.

    (b) Find the probability that the second letter following a vowel is also a vowel.

17. Suppose that the matrices $\begin{bmatrix} .5 & .4 \\ .5 & .6 \end{bmatrix}$ and $\begin{bmatrix} .3 \\ .7 \end{bmatrix}_0$ are the stochastic and initial distribution matrices of a Markov process. Find the third and fourth distribution matrices. (Round entries to two decimal places.)

In Exercises 18–22, compute the first five powers of each matrix (round off to two decimal places).

18. $\begin{bmatrix} 1 & \frac{1}{2} \\ 0 & \frac{1}{2} \end{bmatrix}$

19. $\begin{bmatrix} \frac{1}{3} & \frac{1}{3} \\ \frac{2}{3} & \frac{2}{3} \end{bmatrix}$

20. $\begin{bmatrix} 0 & 1 \\ 1 & 0 \end{bmatrix}$

21. $\begin{bmatrix} .1 & .3 \\ .9 & .7 \end{bmatrix}$

22. $\begin{bmatrix} .2 & .2 & .2 \\ .3 & .3 & .3 \\ .5 & .5 & .5 \end{bmatrix}$

Stochastic matrices for which some power contains no zero entries are called regular *matrices. In Exercises 23 and 24, conjecture whether or not the given matrix is regular by looking at the first few powers of the matrix.*

23. $\begin{bmatrix} .5 & 0 \\ .5 & 1 \end{bmatrix}$

24. $\begin{bmatrix} .7 & 1 \\ .3 & 0 \end{bmatrix}$

In Exercises 25–30, use a graphing calculator, a spreadsheet, or mathematical software to calculate the answers.

25. Let A be the stochastic matrix $\begin{bmatrix} .1 & .6 \\ .9 & .4 \end{bmatrix}$, and let the initial distribution be $B = \begin{bmatrix} .5 \\ .5 \end{bmatrix}_0$.

    (a) Generate the next four distribution matrices.

    (b) Calculate $A^4 B$ and confirm that it is the same as the fourth distribution matrix.

26. Repeat Exercise 25 for the matrices $\begin{bmatrix} .4 & .2 \\ .6 & .8 \end{bmatrix}$ and $\begin{bmatrix} .7 \\ .3 \end{bmatrix}_0$.

27. Consider the matrices of Exercise 25. Beginning with the initial distribution matrix, generate 10 more distributions. Continue to generate 10 more. The matrices will get closer and closer to a certain 2×1 matrix. What is that matrix?

28. Repeat Exercise 27 for the matrices of Exercise 26.

29. Generate 35 successive powers of the matrix A from Exercise 25. (With a graphing calculator, the instruction **[A]*Ans** also can be used to compute successive powers of a square matrix.) The matrices will get closer and closer to a certain 2×2 matrix. What is that matrix? How is that matrix related to the 2×1 matrix found in Exercise 27?

30. Repeat Exercise 29 for the square matrix of Exercise 26.

SOLUTIONS TO PRACTICE PROBLEMS 8.1

1. No. It fails on all three conditions. The matrix is not square, the entry $-.3$ is not ≥ 0, and the sum of the entries in the first (and second) column is not equal to 1.

2. (a) $\begin{array}{c} \\ N \\ C \end{array} \begin{array}{c} N \quad C \\ \begin{bmatrix} .6 & 0 \\ .4 & 1 \end{bmatrix} \end{array}$. Since there are only two possibil-

ities and 40% of those not conditioned become conditioned, the remaining 60% stay not conditioned. After each event 100% of the conditioned stay conditioned, and therefore 0% become not conditioned.

---

[1]Markov, A.A.: An example of statistical analysis of the text of *Eugene Onegin. Bulletin de l'Académie Imperiale des Sciences de St Petersburg*, 7, series 6, pp. 153–62.

SOLUTIONS TO PRACTICE PROBLEMS 8.1

(b) $A^2 = \begin{bmatrix} .6 & 0 \\ .4 & 1 \end{bmatrix} \begin{bmatrix} .6 & 0 \\ .4 & 1 \end{bmatrix} = \begin{bmatrix} .36 & 0 \\ .64 & 1 \end{bmatrix}$,

$A^3 = A^2 \cdot A = \begin{bmatrix} .36 & 0 \\ .64 & 1 \end{bmatrix} \begin{bmatrix} .6 & 0 \\ .4 & 1 \end{bmatrix}$

$= \begin{bmatrix} .216 & 0 \\ .784 & 1 \end{bmatrix}$.

(c) Here $\begin{bmatrix} \ \end{bmatrix}_0 = \begin{bmatrix} 1 \\ 0 \end{bmatrix}$. Therefore,

$\begin{bmatrix} \ \end{bmatrix}_3 = A^3 \begin{bmatrix} 1 \\ 0 \end{bmatrix} = \begin{bmatrix} .216 & 0 \\ .784 & 1 \end{bmatrix} \begin{bmatrix} 1 \\ 0 \end{bmatrix} = \begin{bmatrix} .216 \\ .784 \end{bmatrix}$.

So 78.4% will be conditioned after three events.

8.2 Regular Stochastic Matrices

In the preceding section we studied the percentages of working women in various generations in America. We showed that if $\begin{bmatrix} \ \end{bmatrix}_0$ is the initial distribution matrix, then the distribution matrix $\begin{bmatrix} \ \end{bmatrix}_n$ for the nth generation is given by

$$\begin{bmatrix} \ \end{bmatrix}_n = A^n \begin{bmatrix} \ \end{bmatrix}_0, \tag{1}$$

where A is the stochastic matrix

$$A = \begin{bmatrix} .8 & .3 \\ .2 & .7 \end{bmatrix}.$$

In this section we are interested in determining long-term trends in Markov processes. To get an idea of what is meant, consider the example of Jordanian women in the labor force.

EXAMPLE 1 **Women in the labor force** In Jordan 25% of the women currently work. The effect of maternal influence of mothers on their daughters is given by the matrix

$$\begin{bmatrix} .6 & .2 \\ .4 & .8 \end{bmatrix}.$$

(a) How many women will work after $1, 2, 3, \ldots, 11$ generations?

(b) Estimate the long-term trend.

(c) Answer the same questions (a) and (b) for China, assuming that 48% of all Chinese women currently work and that the effect of maternal influence is the same as for Jordan.

Solution (a) The percentages of women working in generation n (for $n = 1, 2, 3, \ldots$) can be determined from equation (1):

$$\begin{bmatrix} \ \end{bmatrix}_n = \begin{bmatrix} .6 & .2 \\ .4 & .8 \end{bmatrix}^n \begin{bmatrix} .25 \\ .75 \end{bmatrix}_0.$$

After the mildly tedious job of raising the stochastic matrix to various powers, we obtain the results shown in Table 1.

(b) It appears from the accompanying table that the long-term trend is for one-third or $33\frac{1}{3}\%$ of all Jordanian women to work.

TABLE 1			
Generation	Percent of women working	Generation	Percent of women working
0	25	6	33.30
1	30	7	33.32
2	32	8	33.33
3	32.8	9	33.33
4	33.12	10	33.33
5	33.25	11	33.33

(c) The corresponding results for China can be computed by replacing the initial distribution matrix $\begin{bmatrix} .25 \\ .75 \end{bmatrix}_0$ by $\begin{bmatrix} .40 \\ .60 \end{bmatrix}_0$, reflecting that initially 40% of all Chinese women work. The results of the calculations are shown in Table 2. Again, the long-term trend is for one-third of the women to work. ∎

TABLE 2			
Generation	Percent of women working	Generation	Percent of women working
0	40	6	33.36
1	36	7	33.34
2	34.4	8	33.34
3	33.76	9	33.34
4	33.50	10	33.33
5	33.40	11	33.33

From Example 1 one might begin to suspect the following: The long-term trend is always for one-third of the women to work, independent of the initial distribution.

Verification To see why this rather surprising fact should hold, it is useful to examine the powers of A:

$$A^2 = \begin{bmatrix} .44 & .28 \\ .56 & .72 \end{bmatrix} \qquad A^3 = \begin{bmatrix} .376 & .312 \\ .624 & .688 \end{bmatrix} \qquad A^4 = \begin{bmatrix} .3504 & .3248 \\ .6496 & .6752 \end{bmatrix}$$

$$A^5 = \begin{bmatrix} .3402 & .3299 \\ .6598 & .6701 \end{bmatrix} \qquad A^6 = \begin{bmatrix} .3361 & .3320 \\ .6639 & .6680 \end{bmatrix} \qquad A^7 = \begin{bmatrix} .3344 & .3328 \\ .6656 & .6672 \end{bmatrix}$$

$$A^8 = \begin{bmatrix} .3338 & .3331 \\ .6662 & .6669 \end{bmatrix} \qquad A^9 = \begin{bmatrix} .3335 & .3332 \\ .6665 & .6668 \end{bmatrix} \qquad A^{10} = \begin{bmatrix} .3334 & .3333 \\ .6666 & .6667 \end{bmatrix}.$$

As A is raised to further powers, the matrices approach

$$\begin{bmatrix} \frac{1}{3} & \frac{1}{3} \\ \frac{2}{3} & \frac{2}{3} \end{bmatrix}. \tag{2}$$

Now suppose that initially the proportion of women working is x_0. That is, the initial distribution matrix is

$$\begin{bmatrix} x_0 \\ 1 - x_0 \end{bmatrix}.$$

Then, after n generations, the distribution matrix is

$$\left[\ \ \right]_n = A^n \left[\ \ \right]_0 = A^n \begin{bmatrix} x_0 \\ 1 - x_0 \end{bmatrix}.$$

But after many generations n is large, so that A^n is approximately the matrix (2). Thus,

$$\left[\ \ \right]_n \approx \begin{bmatrix} \frac{1}{3} & \frac{1}{3} \\ \frac{2}{3} & \frac{2}{3} \end{bmatrix} \begin{bmatrix} x_0 \\ 1 - x_0 \end{bmatrix} = \begin{bmatrix} \frac{1}{3}x_0 + \frac{1}{3}(1 - x_0) \\ \frac{2}{3}x_0 + \frac{2}{3}(1 - x_0) \end{bmatrix} = \begin{bmatrix} \frac{1}{3} \\ \frac{2}{3} \end{bmatrix}.$$

In other words, after n generations approximately one-third of the women work and two-thirds do not. ■

From the preceding calculations we see that the stochastic matrix A possesses a number of very special properties. First, as n gets large, A^n approaches the matrix

$$\begin{bmatrix} \frac{1}{3} & \frac{1}{3} \\ \frac{2}{3} & \frac{2}{3} \end{bmatrix}.$$

Second, any initial distribution approaches the distribution $\begin{bmatrix} \frac{1}{3} \\ \frac{2}{3} \end{bmatrix}$ after many generations. The limiting matrix

$$\begin{bmatrix} \frac{1}{3} & \frac{1}{3} \\ \frac{2}{3} & \frac{2}{3} \end{bmatrix}$$

is called the *stable matrix* of A, and the limiting distribution

$$\begin{bmatrix} \frac{1}{3} \\ \frac{2}{3} \end{bmatrix}$$

is called the *stable distribution* of A. Finally, note that all the columns of the stable matrix are the same and are equal to the stable distribution.

The matrices that share the aforementioned properties with A are very important. For these matrices one can predict a long-term trend, and this trend is independent of the initial distribution. An important class of matrices with these properties is the class of *regular stochastic matrices*.

A stochastic matrix is said to be *regular* if some power has all positive entries.

EXAMPLE 2 **Identifying regular stochastic matrices** Which of the following stochastic matrices are regular?

(a) $\begin{bmatrix} .6 & .2 \\ .4 & .8 \end{bmatrix}$ (b) $\begin{bmatrix} 0 & .5 \\ 1 & .5 \end{bmatrix}$ (c) $\begin{bmatrix} 0 & 1 \\ 1 & 0 \end{bmatrix}$

Solution (a) All entries are positive, so the matrix is regular.

(b) Here a zero occurs in the first power. However,

$$\begin{bmatrix} 0 & .5 \\ 1 & .5 \end{bmatrix}^2 = \begin{bmatrix} .5 & .25 \\ .5 & .75 \end{bmatrix},$$

which has all positive entries. So the original matrix is regular.

(c) Note that

$$\begin{bmatrix} 0 & 1 \\ 1 & 0 \end{bmatrix}^2 = \begin{bmatrix} 1 & 0 \\ 0 & 1 \end{bmatrix} \qquad \begin{bmatrix} 0 & 1 \\ 1 & 0 \end{bmatrix}^3 = \begin{bmatrix} 0 & 1 \\ 1 & 0 \end{bmatrix}$$

$$\begin{bmatrix} 0 & 1 \\ 1 & 0 \end{bmatrix}^4 = \begin{bmatrix} 1 & 0 \\ 0 & 1 \end{bmatrix} \qquad \begin{bmatrix} 0 & 1 \\ 1 & 0 \end{bmatrix}^5 = \begin{bmatrix} 0 & 1 \\ 1 & 0 \end{bmatrix}.$$

The even powers of the matrix are the 2×2 identity matrix, and the odd powers are the original matrix. Every power has a zero in it, so the matrix is not regular.

Now Try Exercises 1, 3, and 5 ∎

Regular matrices share all the properties observed in the special case. Moreover, there is a simple technique for computing the stable distribution (see property 4 in the following box) that spares us from having to multiply matrices.

Let A be a regular stochastic matrix.

1. The powers A^n approach a certain matrix as n gets large. This limiting matrix is called the *stable matrix* of A.

2. For any initial distribution $\begin{bmatrix} \\ \end{bmatrix}_0$, $A^n \begin{bmatrix} \\ \end{bmatrix}_0$ approaches a certain distribution $\begin{bmatrix} \\ \end{bmatrix}$. This limiting distribution is called the *stable distribution* of A.

3. All columns of the stable matrix are the same; they equal the stable distribution.

4. The stable distribution $X = [\ ]$ can be determined by solving the system of linear equations

$$\begin{cases} \text{sum of the entries of } X = 1 \\ AX = X. \end{cases}$$

EXAMPLE 3 **Finding the stable distribution of a regular stochastic matrix** Use property 4 to determine the stable distribution of the regular stochastic matrix

$$A = \begin{bmatrix} .6 & .2 \\ .4 & .8 \end{bmatrix}.$$

Solution Let $X = \begin{bmatrix} x \\ y \end{bmatrix}$ be the stable distribution. The condition "sum of the entries of $X = 1$" yields the equation

$$x + y = 1.$$

The condition $AX = X$ gives the equations

$$\begin{bmatrix} .6 & .2 \\ .4 & .8 \end{bmatrix} \begin{bmatrix} x \\ y \end{bmatrix} = \begin{bmatrix} x \\ y \end{bmatrix} \qquad \text{or} \qquad \begin{cases} .6x + .2y = x \\ .4x + .8y = y. \end{cases}$$

So we have the system

$$\begin{cases} x + y = 1 \\ .6x + .2y = x \\ .4x + .8y = y. \end{cases}$$

Combining terms in the second and third equations and eliminating the decimals by multiplying by 10, we have

$$\begin{cases} x + y = 1 \\ -4x + 2y = 0 \\ 4x - 2y = 0. \end{cases}$$

Note that the second and third equations are the same, except for a factor -1, so the last equation may be omitted. Now the system reads

$$\begin{cases} x + y = 1 \\ -4x + 2y = 0. \end{cases}$$

The diagonal form is obtained by the Gaussian elimination method:

$$\begin{bmatrix} 1 & 1 & | & 1 \\ -4 & 2 & | & 0 \end{bmatrix} \xrightarrow{[2]+4[1]} \begin{bmatrix} 1 & 1 & | & 1 \\ 0 & 6 & | & 4 \end{bmatrix}$$

$$\xrightarrow{\frac{1}{6}[2]} \begin{bmatrix} 1 & 1 & | & 1 \\ 0 & 1 & | & \frac{2}{3} \end{bmatrix}$$

$$\xrightarrow{[1]+(-1)[2]} \begin{bmatrix} 1 & 0 & | & \frac{1}{3} \\ 0 & 1 & | & \frac{2}{3} \end{bmatrix}$$

Thus

$$x = \tfrac{1}{3}, \qquad y = \tfrac{2}{3}.$$

So the stable distribution is

$$\begin{bmatrix} x \\ y \end{bmatrix} = \begin{bmatrix} \frac{1}{3} \\ \frac{2}{3} \end{bmatrix},$$

as we observed before. Note that once the stable distribution is determined, the stable matrix is easy to find. Just place the stable distribution in every column:

$$\begin{bmatrix} \frac{1}{3} & \frac{1}{3} \\ \frac{2}{3} & \frac{2}{3} \end{bmatrix}.$$

Now Try Exercise 7   ∎

EXAMPLE 4 **Finding the stable taxi distribution** In Section 8.1 we studied the distribution of taxis in three zones of a city. The movement of taxis from zone to zone was described by the regular stochastic matrix

$$A = \begin{bmatrix} .5 & .4 & .2 \\ .4 & .3 & .6 \\ .1 & .3 & .2 \end{bmatrix}.$$

In the long run, what percentage of taxis will be in each of the zones?

Solution Let $X = \begin{bmatrix} x \\ y \\ z \end{bmatrix}$ be the stable distribution of A. Then x is the long-term percentage of taxis in zone I, y the percentage in zone II, and z the percentage in zone III. X is determined by the equations

$$\begin{cases} x + y + z = 1 \\ \qquad\quad AX = X, \end{cases}$$

or, equivalently,

$$\begin{cases} x + \quad y + \quad z = 1 \\ .5x + .4y + .2z = x \\ .4x + .3y + .6z = y \\ .1x + .3y + .2z = z. \end{cases}$$

Rewriting the equations with all the terms involving the variables on the left, we get

$$\begin{cases} \quad\;\; x + \quad y + \quad z = 1 \\ -.5x + .4y + .2z = 0 \\ \quad .4x - .7y + .6z = 0 \\ \quad .1x + .3y - .8z = 0. \end{cases}$$

Applying the Gaussian elimination method to this system, we get the solution $x = .4$, $y = .4$, $z = .2$. Thus, after many trips, approximately 40% of the taxis are in zone I, 40% in zone II, and 20% in zone III.

Now Try Exercise 11

■

We have come full circle. We began Chapter 2 by solving systems of linear equations. Matrices were developed as a tool for solving such systems. Then we found matrices to be interesting in their own right. Finally, in the current chapter, we have used systems of linear equations to answer questions about matrices.

GC If A is a square matrix, then graphing calculators can compute A^n. (With the TI-83, n can be as large as 255. However, even higher powers can be obtained with instructions such as **Ans^255**.) For a regular stochastic matrix A, A^{255} should be an excellent approximation to the stable matrix of A. In Figures 1 and 3, A is the matrix of Example 3 and C is the matrix of Example 4. (*Note*: ▶ **Frac**, which stands for "display as a fraction," is obtained on the TI-83 by pressing MATH **1**.)

Calculators or computers also can be used to obtain a stable distribution by solving the system of linear equations or can be used to confirm a stable distribution by carrying out a matrix multiplication. The system of linear equations from Example 3 is solved in Figs. 2 and 4.

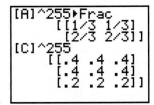

Figure 1

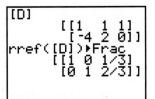

Figure 2

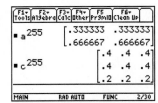

Figure 3

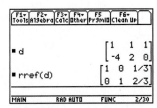

Figure 4

ES Successive powers of a matrix can be generated on a spreadsheet in much the same way that successive distribution matrices were generated in Fig. 5 of Section 8.1. The stable distribution can be found by using Solver to solve the appropriate system of linear equations.

APPENDIX Verification of Method for Obtaining the Stable Distribution

In this appendix we verify that the stable distribution X can be obtained by solving the system of equations

$$\begin{cases} \text{sum of the entries of } X = 1 \\ AX = X. \end{cases}$$

Let A be a regular stochastic matrix. Suppose that we take as the initial distribution its stable distribution X. Then the nth distribution matrix, or $A^n X$, approaches the stable distribution (by property 2), so that

$$X \approx A^n X \quad \text{for large } n.$$

Therefore,

$$AX \approx A \cdot A^n X = A^{n+1} X.$$

But $A^{n+1} X$ is the $(n+1)$st distribution matrix, which is also approximately X. Thus AX is approximately X. But the approximation can be made closer and closer by taking n large. Therefore, AX is arbitrarily close to X, or

$$AX = X. \tag{3}$$

This is a matrix equation in X. There is one other condition on X: X is a distribution matrix. Therefore,

$$\text{sum of the entries of } X = 1. \tag{4}$$

So, by (3) and (4), we find this system of linear equations for the entries of X:

$$\begin{cases} \text{sum of the entries of } X = 1 \\ AX = X. \end{cases}$$

PRACTICE PROBLEMS 8.2

1. Is $\begin{bmatrix} 0 & .2 & .5 \\ .5 & 0 & .5 \\ .5 & .8 & 0 \end{bmatrix}$ a regular stochastic matrix? Explain your answer.

2. Find the stable matrix for the regular stochastic matrix in Problem 1.

3. (*Cigarette Smokers*) In a study of cigarette smokers it was determined that of the people who smoked menthol cigarettes on a particular day, 10% smoked menthol the next day and 90% smoked nonmenthol. Of the people who smoked nonmenthol cigarettes on a particular day, the next day 30% smoked menthol and 70% smoked nonmenthol. In the long run, what percent of the people will be smoking nonmenthol cigarettes on a particular day?

EXERCISES 8.2

In Exercises 1–6, determine whether or not the matrix is a regular stochastic matrix.

1. $\begin{bmatrix} \frac{1}{4} & \frac{2}{7} \\ \frac{3}{4} & \frac{5}{7} \end{bmatrix}$ **2.** $\begin{bmatrix} .6 & 0 \\ .4 & 1 \end{bmatrix}$

3. $\begin{bmatrix} .3 & 1 \\ .7 & 0 \end{bmatrix}$ **4.** $\begin{bmatrix} 1 & 0 & .7 \\ 0 & 1 & .2 \\ 0 & 0 & .1 \end{bmatrix}$

5. $\begin{bmatrix} 0 & .8 & 0 \\ 1 & .1 & .5 \\ 0 & .1 & .5 \end{bmatrix}$ **6.** $\begin{bmatrix} .6 & .6 & .6 \\ .3 & .3 & .3 \\ .1 & .1 & .1 \end{bmatrix}$

In Exercises 7–11, find the stable distribution for the given regular stochastic matrix.

7. $\begin{bmatrix} .5 & .1 \\ .5 & .9 \end{bmatrix}$ **8.** $\begin{bmatrix} .4 & 1 \\ .6 & 0 \end{bmatrix}$

9. $\begin{bmatrix} .8 & .3 \\ .2 & .7 \end{bmatrix}$ **10.** $\begin{bmatrix} .3 & .1 & .2 \\ .4 & .8 & .6 \\ .3 & .1 & .2 \end{bmatrix}$

11. $\begin{bmatrix} .1 & .4 & .7 \\ .6 & .4 & .2 \\ .3 & .2 & .1 \end{bmatrix}$

12. (*Fitness*) Refer to Exercise 10 of Section 8.1. In the long run what percentage of the people will have a strenuous workout on a particular day?

13. (*T-Maze*) Refer to Exercise 11 of Section 8.1. What percentage of the mice will be going to the left after many days?

14. (*Voter Patterns*) Refer to Exercise 13 of Section 8.1. In the long run what proportion of the governors will be Democrats?

15. (*Genetics*) With respect to a certain gene, geneticists classify individuals as dominant, recessive, or hybrid. In an experiment, individuals are crossed with hybrids, then their offspring are crossed with hybrids, and so on. For dominant individuals, 50% of their offspring will be dominant and 50% will be a hybrid. For the recessive individuals 50% of their offspring will be recessive and 50% hybrid. For hybrid individuals (to be crossed with hybrids) their offspring will be 25% dominant, 25% recessive, and 50% hybrid. In the long run what percent of the individuals in a generation will be dominant?

16. (*Transportation Modes*) Commuters can get into town by car or bus. Surveys have shown that for those taking their car on a particular day, 20% take their car the next day and 80% take a bus. Also, for those taking a bus on a particular day, 50% take their car the next day and 50% take a bus. In the long run what percentage of the people take a bus on a particular day?

17. (*Weather Patterns*) The changes in weather from day to day on the planet Xantar form a regular Markov process. Each day is either rainy or sunny. If it rains one day, there is a 90% chance that it will be sunny the following day. If it is sunny one day, there is a 60% chance of rain the next day. In the long run, what is the daily likelihood of rain?

18. (*Computer Reliability*) A certain university has a computer room with 219 terminals. Each day there is a 3% chance that a given terminal will break and a 70% chance that a given broken terminal will be repaired. In the long run, about how many terminals in the room will be working?

19. As shown in Example 2, $\begin{bmatrix} 0 & 1 \\ 1 & 0 \end{bmatrix}$ is not a regular stochastic matrix. Show that the matrix has $\begin{bmatrix} .5 \\ .5 \end{bmatrix}$ as a stable distribution and explain why this fact does not contradict the main premise of this section.

20. Refer to the stochastic matrix in Example 6 of Section 8.1. In the long run, what percentage of American women will work?

In Exercises 21–24, use a graphing calculator, a spreadsheet, or mathematical software to calculate the answers.

21. Consider the stochastic matrix A, where

$$A = \begin{bmatrix} .85 & .35 \\ .15 & .65 \end{bmatrix}.$$

Approximate the stable matrix of A by raising A to a high power. Then find the exact stable distribution by solving an appropriate system of linear equations. Check your answer by forming the product of A and the stable distribution.

Repeat Exercise 21 for each of the matrices in Exercises 22–24.

22. $\begin{bmatrix} .3 & .1 \\ .7 & .9 \end{bmatrix}$ **23.** $\begin{bmatrix} .1 & .4 & .1 \\ .3 & .2 & .8 \\ .6 & .4 & .1 \end{bmatrix}$

24. $\begin{bmatrix} .4 & .2 & .4 \\ .1 & .3 & .1 \\ .5 & .5 & .5 \end{bmatrix}$

SOLUTIONS TO PRACTICE PROBLEMS 8.2

1. Yes. It is easily seen to be stochastic. Although it has some zero entries, there are no zero entries in

$$A^2 = \begin{bmatrix} .35 & .40 & .10 \\ .25 & .50 & .25 \\ .40 & .10 & .65 \end{bmatrix}.$$

2. $\begin{cases} x + y + z = 1 \\ \begin{bmatrix} 0 & .2 & .5 \\ .5 & 0 & .5 \\ .5 & .8 & 0 \end{bmatrix} \begin{bmatrix} x \\ y \\ z \end{bmatrix} = \begin{bmatrix} x \\ y \\ z \end{bmatrix} \end{cases}$

or $\begin{cases} x + y + z = 1 \\ .2y + .5z = x \\ .5x + .5z = y \\ .5x + .8y = z \end{cases}$

or $\begin{cases} x + y + z = 1 \\ -x + .2y + .5z = 0 \\ .5x - y + .5z = 0 \\ .5x + .8y - z = 0 \end{cases}$

To simplify the arithmetic, multiply each of the last three equations of the system by 10 to eliminate the decimals. Then apply the Gaussian elimination method.

$\begin{cases} x + y + z = 1 \\ -10x + 2y + 5z = 0 \\ 5x - 10y + 5z = 0 \\ 5x + 8y - 10z = 0 \end{cases}$ $\begin{cases} x + y + z = 1 \\ 12y + 15z = 10 \\ -15y = -5 \\ 3y - 15z = -5 \end{cases}$

Next interchange the second and third equations and pivot about $-15y$.

$\begin{cases} x + y + z = 1 \\ -15y = -5 \\ 12y + 15z = 10 \\ 3y - 15z = -5 \end{cases}$ $\begin{cases} x + z = \frac{2}{3} \\ y = \frac{1}{3} \\ 15z = 6 \\ -15z = -6 \end{cases}$

Pivoting about $15z$ yields $x = \frac{4}{15}$, $y = \frac{5}{15}$, $z = \frac{6}{15}$, so the stable matrix is

$$\begin{bmatrix} \frac{4}{15} & \frac{4}{15} & \frac{4}{15} \\ \frac{5}{15} & \frac{5}{15} & \frac{5}{15} \\ \frac{6}{15} & \frac{6}{15} & \frac{6}{15} \end{bmatrix}.$$

3. The regular stochastic matrix describing this daily transition is

$$\begin{array}{cc} & M \quad N \\ \begin{matrix} M \\ N \end{matrix} & \begin{bmatrix} .1 & .3 \\ .9 & .7 \end{bmatrix}. \end{array}$$

The stable distribution is found by solving

$$\begin{cases} x + y = 1 \\ \begin{bmatrix} .1 & .3 \\ .9 & .7 \end{bmatrix} \begin{bmatrix} x \\ y \end{bmatrix} = \begin{bmatrix} x \\ y \end{bmatrix} \end{cases}$$

or $\begin{cases} x + y = 1 \\ .1x + .3y = x \\ .9x + .7y = y \end{cases}$ or $\begin{cases} x + y = 1 \\ -.9x + .3y = 0 \\ .9x - .3y = 0. \end{cases}$

Since the last two equations are essentially the same, we need only solve the system consisting of the first two equations. Multiply the second equation by 10:

$\begin{cases} x + y = 1 \\ -9x + 3y = 0 \end{cases} \rightarrow \begin{cases} x + y = 1 \\ 12y = 9 \end{cases} \rightarrow \begin{cases} x = \frac{1}{4} \\ y = \frac{3}{4}. \end{cases}$

So the stable distribution is $\begin{bmatrix} \frac{1}{4} \\ \frac{3}{4} \end{bmatrix}$. The stable distribution tells us that in the long run, 25% smoke menthol and 75% smoke nonmenthol cigarettes on any particular day.

8.3 Absorbing Stochastic Matrices

In this section we study long-term trends for a certain class of matrices which are not regular—the absorbing stochastic matrices. By way of introduction, recall some general facts about stochastic matrices.

Stochastic matrices, such as

$$\begin{bmatrix} .3 & .5 & .1 & 0 \\ .2 & .2 & .8 & 0 \\ .1 & .3 & 0 & 0 \\ .4 & 0 & .1 & 1 \end{bmatrix},$$

describe state-to-state changes in certain processes. Each column of a stochastic matrix describes the transitions (or movements) from one specific state. For example, the first column of the preceding stochastic matrix indicates that at the end of one time period the probability is .3 that an object in state 1 stays in state 1, .2 that it goes to state 2, .1 that it goes to state 3, and .4 that it goes to state 4. Similarly, the second column indicates the probabilities for transitions from state 2, the third column from state 3, and the fourth from state 4.

If A is a stochastic matrix, then the columns of A^2 describe the transitions from the various states *after two time periods*. For instance, the third column of A^2 indicates the probabilities of an object starting out in state 3 and ending up in each of the states after two time periods. Similarly, the columns of A^n indicate the transitions from the various states after n periods.

Consider the stochastic matrix just described. Its fourth column illustrates a curious phenomenon. It indicates that if an object starts out in state 4, after one time period the probabilities of going to states 1, 2, or 3 are 0 and the probability of going to state 4 is 1. In other words, all of the objects in state 4 stay in state 4. A state with this property is called an *absorbing state*. More precisely, an absorbing state is a state that always leads back to itself.

EXAMPLE 1   **Finding absorbing states of a stochastic matrix**   Find all absorbing states of the stochastic matrix

$$\begin{bmatrix} 1 & 0 & .3 & 0 \\ 0 & 1 & .1 & 1 \\ 0 & 0 & .5 & 0 \\ 0 & 0 & .1 & 0 \end{bmatrix}.$$

Solution   To determine which states are absorbing, one must look at the columns. The first column describes the transitions from state 1. It says that state 1 leads to state 1 100% of the time and to the other states 0% of the time. So state 1 is absorbing. Column 2 describes transitions from state 2. It says that state 2 leads to state 2 100% of the time. So state 2 is absorbing. Clearly, the third column says that state 3 is not absorbing. For example, state 3 leads to state 1 with probability .3. At first glance, column 4 seems to say that state 4 is absorbing. But it is not, because column 4 says that state 4 leads to state 2 100% of the time. ∎

Based on Example 1, we can easily determine the absorbing states of any stochastic matrix: First, the corresponding column has a single 1 and the remaining entries 0. Second, the lone 1 must be located on the main diagonal of the matrix. That is, its row and column number must be the same. So, for example, state i is an absorbing state if and only if the ith entry in the ith column is 1 and all the remaining entries in that column are 0.

An *absorbing stochastic matrix* is a stochastic matrix in which (1) there is at least one absorbing state and (2) from any state it is possible to get to at least one absorbing state, either directly or through one or more intermediate states.

EXAMPLE 2   **Determining whether a stochastic matrix is absorbing**   Is the matrix

$$\begin{bmatrix} 1 & 0 & .3 & 0 \\ 0 & 1 & .1 & 1 \\ 0 & 0 & .5 & 0 \\ 0 & 0 & .1 & 0 \end{bmatrix}$$

an absorbing stochastic matrix?

Solution In Example 1 we showed that states 1 and 2 were absorbing. From column 3 we see that state 3 can lead to both state 1 and state 2: state 1 with probability .3 and state 2 with probability .1. From column 4 we see that state 4 does not lead to state 1, but it does lead to state 2. Thus states 3 and 4 both lead to absorbing states, and so the matrix is an absorbing stochastic matrix.

Now Try Exercises 1 and 3   ∎

In general, processes described by stochastic matrices can oscillate indefinitely from state to state in such a way that they exhibit no long-term trend. An example was given in Section 2 using the matrix

$$\begin{bmatrix} 0 & 1 \\ 1 & 0 \end{bmatrix}.$$

The idea of introducing absorbing states is to reduce the degree of oscillation. For when an absorbing state is reached, the process no longer changes. The main result of this section is that absorbing stochastic matrices exhibit a long-term trend. Further, we can determine this trend using a simple computational procedure.

When considering an absorbing stochastic matrix, we will always arrange the states so that the absorbing states come first, then the nonabsorbing states.

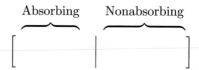

When the states are ordered in this manner, an absorbing stochastic matrix can be partitioned, or subdivided, into four submatrices.

$$\begin{array}{cc} \overbrace{\text{Absorbing}} & \overbrace{\text{Nonabsorbing}} \\ \left[\begin{array}{c|c} I & S \\ \hline 0 & R \end{array} \right] \end{array}$$

The matrix I is an identity matrix, and 0 denotes a matrix having all entries 0. The matrices S and R are the two pieces corresponding to the nonabsorbing states. For example, in the case of the absorbing stochastic matrix of Example 2, this partition is given by

$$\left[\begin{array}{cc|cc} 1 & 0 & .3 & 0 \\ 0 & 1 & .1 & 1 \\ \hline 0 & 0 & .5 & 0 \\ 0 & 0 & .1 & 0 \end{array} \right].$$

EXAMPLE 3 **Progression of a disease** The victims of a certain disease are classified into three states: cured, dead from the disease, or sick. Once a person is cured, he is permanently immune. Each year 70% of those sick are cured, 10% die from the disease, and 20% remain ill.

(a) Determine a stochastic matrix describing the progression of the disease.

(b) Determine the absorbing states.

Solution (a) There is one column for each state. Transferring the given data to the matrix gives the result:

$$
\begin{array}{c}
 \\
\text{Cured} \\
\text{Dead} \\
\text{Sick}
\end{array}
\begin{array}{ccc}
\text{Cured} & \text{Dead} & \text{Sick} \\
\end{array}
\left[\begin{array}{ccc}
1 & 0 & .7 \\
0 & 1 & .1 \\
0 & 0 & .2
\end{array}\right].
$$

(b) The absorbing states are "cured" and "dead," since these are the states that always lead back to themselves. The matrix is an absorbing stochastic matrix, since there is at least one absorbing state and the other state, "sick," can lead to an absorbing state. (In fact, in this case it leads to either of the absorbing states.)

Now Try Exercise 13(a)

■

EXAMPLE 4 **Long-term trend for an absorbing stochastic matrix** Find the long-term trend of the absorbing stochastic matrix

$$
A = \left[\begin{array}{cc|c}
1 & 0 & .2 \\
0 & 1 & .4 \\
\hline
0 & 0 & .4
\end{array}\right].
$$

Solution The long-term trend is found by raising A to various powers. Here are the results, accurate to three decimal places:

$$
A^2 = \left[\begin{array}{cc|c}
1 & 0 & .28 \\
0 & 1 & .56 \\
\hline
0 & 0 & .16
\end{array}\right]
\qquad
A^3 = \left[\begin{array}{cc|c}
1 & 0 & .312 \\
0 & 1 & .624 \\
\hline
0 & 0 & .064
\end{array}\right]
$$

$$
A^4 = \left[\begin{array}{cc|c}
1 & 0 & .325 \\
0 & 1 & .650 \\
\hline
0 & 0 & .026
\end{array}\right]
\qquad
A^5 = \left[\begin{array}{cc|c}
1 & 0 & .330 \\
0 & 1 & .660 \\
\hline
0 & 0 & .010
\end{array}\right]
$$

$$
A^6 = \left[\begin{array}{cc|c}
1 & 0 & .332 \\
0 & 1 & .664 \\
\hline
0 & 0 & .004
\end{array}\right]
\qquad
A^7 = \left[\begin{array}{cc|c}
1 & 0 & .333 \\
0 & 1 & .666 \\
\hline
0 & 0 & .001
\end{array}\right].
$$

Based on this numerical evidence, it appears as if the powers of A approach the matrix

$$
\left[\begin{array}{cc|c}
1 & 0 & \frac{1}{3} \\
0 & 1 & \frac{2}{3} \\
\hline
0 & 0 & 0
\end{array}\right].
$$

In other words, after a large number of time intervals, 100% of those originally in state 1 remain in state 1 (column 1), 100% of those originally in state 2 remain in state 2 (column 2), and those originally in state 3 end up in state 1 with probability one-third and in state 2 with probability two-thirds.

■

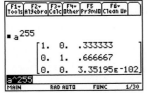

Example 4 exhibits three important features of the long-term behavior exhibited by all absorbing stochastic matrices. First, as in the case of regular stochastic matrices, the powers approach a particular matrix. This limiting matrix is called a *stable matrix*. Second, for absorbing stochastic matrices the long-term trend depends on the initial state. For example, the stable matrix just computed gives different results depending on the state in which you start. This is reflected by the fact that the three columns are different. (In the case of regular stochastic matrices, the long-term trend does not depend on the initial distribution. All columns of the stable matrix are the same. They all equal the stable distribution.) The third important point to notice is that, no matter what the initial state, all objects eventually go to absorbing states. The absorbing states act like magnets and attract all objects to themselves in the long run.

EXAMPLE 5   **Long-term trend for the progression of a disease**   Find the long-term trend of the disease described in Example 3.

Solution   The stochastic matrix here is

$$\begin{bmatrix} 1 & 0 & .7 \\ 0 & 1 & .1 \\ 0 & 0 & .2 \end{bmatrix}.$$

Its first few powers are given by

$$A^2 = \begin{bmatrix} 1 & 0 & .84 \\ 0 & 1 & .12 \\ 0 & 0 & .04 \end{bmatrix} \qquad A^3 = \begin{bmatrix} 1 & 0 & .868 \\ 0 & 1 & .124 \\ 0 & 0 & .008 \end{bmatrix}$$

$$A^4 \approx \begin{bmatrix} 1 & 0 & .874 \\ 0 & 1 & .125 \\ 0 & 0 & .002 \end{bmatrix} \qquad A^5 \approx \begin{bmatrix} 1 & 0 & .875 \\ 0 & 1 & .125 \\ 0 & 0 & .000 \end{bmatrix}.$$

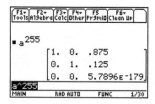

It appears that the powers approach the matrix

$$A = \begin{bmatrix} 1 & 0 & \frac{7}{8} \\ 0 & 1 & \frac{1}{8} \\ 0 & 0 & 0 \end{bmatrix}.$$

This is the stable matrix. In other words, reading from column 3, of those initially in state 3, "sick," the probability is seven-eighths of eventually being cured, and one-eighth of eventually dying of the disease.   ■

In Examples 4 and 5 we have computed the stable matrix by raising the given stochastic matrix to various powers. Actually, there is a formal computational procedure for determining the stable matrix. Suppose that we partition an absorbing stochastic matrix into submatrices

$$A = \begin{bmatrix} I & S \\ 0 & R \end{bmatrix}.$$

The stable matrix of A is

$$\left[\begin{array}{c|c} I & S(I-R)^{-1} \\ \hline 0 & 0 \end{array}\right].$$

[*Note*: The identity matrix I in $(I-R)^{-1}$ is chosen to be the same size as R in order to make the matrix subtraction permissible.]

EXAMPLE 6 **Determining stable matrices** Use the preceding formula to determine the stable matrices of

(a) $\left[\begin{array}{cc|c} 1 & 0 & .2 \\ 0 & 1 & .4 \\ \hline 0 & 0 & .4 \end{array}\right]$ (b) $\left[\begin{array}{cc|c} 1 & 0 & .7 \\ 0 & 1 & .1 \\ \hline 0 & 0 & .2 \end{array}\right].$

Solution (a) In this case $S = \begin{bmatrix} .2 \\ .4 \end{bmatrix}$, $R = [.4]$. The stable matrix is

$$\left[\begin{array}{c|c} I & S(I-R)^{-1} \\ \hline 0 & 0 \end{array}\right] = \left[\begin{array}{cc|c} 1 & 0 & S(I-R)^{-1} \\ 0 & 1 & \\ \hline 0 & 0 & 0 \end{array}\right].$$

Since R is 1×1, to compute $I - R$ we let I be the 1×1 identity matrix $[1]$. Then

$$I - R = [1] - [.4] = [1 - .4] = [.6].$$

Now $(I-R)^{-1}$ is a matrix that when multiplied by $I - R$ or $[.6]$, gives the 1×1 identity matrix $[1]$. Thus

$$(I-R)^{-1} = [.6]^{-1} = [1/.6]$$

and

$$S(I-R)^{-1} = \begin{bmatrix} .2 \\ .4 \end{bmatrix}[1/.6] = \begin{bmatrix} .2/.6 \\ .4/.6 \end{bmatrix} = \begin{bmatrix} \frac{1}{3} \\ \frac{2}{3} \end{bmatrix}.$$

Therefore, the stable matrix is given by

$$\left[\begin{array}{cc|c} 1 & 0 & \frac{1}{3} \\ 0 & 1 & \frac{2}{3} \\ \hline 0 & 0 & 0 \end{array}\right].$$

(b) In this case

$$S = \begin{bmatrix} .7 \\ .1 \end{bmatrix}, \quad R = [.2], \quad I - R = [1] - [.2] = [.8], \quad (I-R)^{-1} = [1/.8].$$

Therefore,

$$S(I-R)^{-1} = \begin{bmatrix} .7 \\ .1 \end{bmatrix}[1/.8] = \begin{bmatrix} .7/.8 \\ .1/.8 \end{bmatrix} = \begin{bmatrix} \frac{7}{8} \\ \frac{1}{8} \end{bmatrix}.$$

So the stable matrix is

$$\left[\begin{array}{c|c} I & S(I-R)^{-1} \\ \hline 0 & 0 \end{array}\right] = \left[\begin{array}{cc|c} 1 & 0 & \frac{7}{8} \\ 0 & 1 & \frac{1}{8} \\ \hline 0 & 0 & 0 \end{array}\right].$$

Now Try Exercise 5 (stable part only)

∎

EXAMPLE 7 **Gambler's ruin** Consider a game of chance with the following characteristics: A person repeatedly bets $1 each play. If he wins, he receives $1.[1] If he goes broke, he stops playing. Also, if he accumulates $3, he stops playing. On each play the probability of winning is .4 and of losing .6. What is the probability of eventually accumulating $3 if he starts with $1? $2?

Solution There are four states, corresponding to having $0, $3, $1, or $2. The first two are absorbing states. The stochastic matrix is

$$\begin{array}{c} \\ \$0 \\ \$3 \\ \$1 \\ \$2 \end{array} \begin{array}{cccc} \$0 & \$3 & \$1 & \$2 \end{array} \\ \left[\begin{array}{cc|cc} 1 & 0 & .6 & 0 \\ 0 & 1 & 0 & .4 \\ \hline 0 & 0 & 0 & .6 \\ 0 & 0 & .4 & 0 \end{array}\right].$$

The third column is derived in this way: If he has $1, there is a .6 probability of losing $1, which would mean going to state $0. Thus the first entry in the third column is .6. There is no way to get from $1 to $3 or from $1 to $1 after one play. So the second and third entries are 0. There is a .4 probability of winning $1—that is, of going from $1 to $2. So the last entry is .4. The fourth column is derived similarly.

In this example

$$S = \begin{bmatrix} .6 & 0 \\ 0 & .4 \end{bmatrix} \qquad R = \begin{bmatrix} 0 & .6 \\ .4 & 0 \end{bmatrix}.$$

To compute $S(I-R)^{-1}$, observe that

$$I - R = \begin{bmatrix} 1 & 0 \\ 0 & 1 \end{bmatrix} - \begin{bmatrix} 0 & .6 \\ .4 & 0 \end{bmatrix} = \begin{bmatrix} 1 & -.6 \\ -.4 & 1 \end{bmatrix}.$$

To compute $(I-R)^{-1}$, recall that

$$\begin{bmatrix} a & b \\ c & d \end{bmatrix}^{-1} = \begin{bmatrix} d/\Delta & -b/\Delta \\ -c/\Delta & a/\Delta \end{bmatrix}, \qquad \Delta = ad - bc \neq 0.$$

So, in this example, $\Delta = 1 \cdot 1 - (-.6)(-.4) = 1 - .24 = .76$ and

$$(I-R)^{-1} = \begin{bmatrix} 1 & -.6 \\ -.4 & 1 \end{bmatrix}^{-1} \approx \begin{bmatrix} 1.32 & .79 \\ .53 & 1.32 \end{bmatrix}$$

$$S(I-R)^{-1} \approx \begin{bmatrix} .6 & 0 \\ 0 & .4 \end{bmatrix} \begin{bmatrix} 1.32 & .79 \\ .53 & 1.32 \end{bmatrix} \approx \begin{bmatrix} .79 & .47 \\ .21 & .53 \end{bmatrix}.$$

---

[1] That is, he receives his bet of $1 plus winnings of $1.

Thus the stable matrix is

$$\left[\begin{array}{c|c} I & S(I-R)^{-1} \\ \hline 0 & 0 \end{array}\right] = \left[\begin{array}{cc|cc} 1 & 0 & .79 & .47 \\ 0 & 1 & .21 & .53 \\ 0 & 0 & 0 & 0 \\ 0 & 0 & 0 & 0 \end{array}\right].$$

We are interested in the probability that the gambler ends up with \$3. The percentage is different for each of the two starting amounts \$1 and \$2. Recall the meaning of the rows and columns:

$$\begin{array}{c} \\ \$0 \\ \$3 \\ \$1 \\ \$2 \end{array} \begin{array}{cccc} \$0 & \$3 & \$1 & \$2 \\ \left[\begin{array}{cc|cc} 1 & 0 & .79 & .47 \\ 0 & 1 & .21 & .53 \\ \hline 0 & 0 & 0 & 0 \\ 0 & 0 & 0 & 0 \end{array}\right]. \end{array}$$

Looking at the \$1 column, we see that if he starts with \$1, the probability is .21 that he ends up with \$3. Looking at the \$2 column, the probability is .53 that he ends up with \$3.

Now Try Exercise 17(a)

■

The Fundamental Matrix The matrix $(I-R)^{-1}$ used to compute the stable matrix is called the *fundamental matrix* and is denoted by the letter F. Its columns and rows should be labeled with the nonabsorbing states. The fundamental matrix directly provides certain useful probabilities for the process. Specifically, the ijth entry of F is the expected number of times the process will be in nonabsorbing state i if it starts in nonabsorbing state j. The sum of the entries of the jth column of F is the expected number of steps before absorption when the process begins in nonabsorbing state j.

The fundamental matrix for the gambler's ruin example is

$$\begin{array}{c} \\ F = \begin{array}{c} \$1 \\ \$2 \end{array} \end{array} \begin{array}{cc} \$1 & \$2 \\ \left[\begin{array}{cc} 1.32 & .79 \\ .53 & 1.32 \end{array}\right]. \end{array}$$

The first column of F indicates that when the gambler begins with \$1, he can expect to play $1.32 + .53 = 1.85$ times before quitting. He should have \$1 for an expected number of 1.32 plays and have \$2 for an expected number of .53 plays. The second column gives similar information when the gambler begins with \$2.

GC A graphing calculator can be used to obtain the stable matrix by raising the original stochastic matrix to a high power or by calculating $S(I-R)^{-1}$. In Figs. 1 and 2 on the next page, **[A]**, **[B]**, and **[C]** are the matrices A, S, and R of Example 7. In Fig. 2, the **MODE Float** option was set to 2. In Figs. 3 and 4, a, s, and r are the matrices A, S, and R of Example 7.

ES Successive powers of an absorbing stochastic matrix can be generated on a spreadsheet in much the same way that successive distribution matrices were generated in Fig. 5 of Section 8.1. The upper-right quadrant of the stable matrix can be obtained with the formula =MMULT(S,MINVERSE(I−R)).

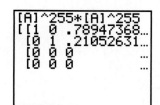

Figure 1   **Figure 2**   **Figure 3**   **Figure 4**

PRACTICE PROBLEMS 8.3

1. When an absorbing stochastic matrix is partitioned, the submatrix in the upper left is an identity matrix, denoted by I. Also, when finding the stable matrix, we subtract the submatrix R from an identity matrix, also denoted by I. Are these two identity matrices the same size?

2. Let A be an absorbing stochastic matrix. Interpret the entries of A^2.

3. Is $\begin{bmatrix} 1 & .4 & 0 \\ 0 & .2 & .1 \\ 0 & .4 & .9 \end{bmatrix}$ an absorbing stochastic matrix?

EXERCISES 8.3

In Exercises 1–4, determine whether the given matrix is an absorbing stochastic matrix.

1. $\begin{bmatrix} 1 & 0 & 0 & 0 \\ 0 & 1 & 0 & 0 \\ 0 & 0 & .8 & .1 \\ 0 & 0 & .2 & .9 \end{bmatrix}$ **2.** $\begin{bmatrix} 1 & 0 & 0 & .3 \\ 0 & 1 & 0 & .2 \\ 0 & 0 & 1 & .2 \\ 0 & 0 & 0 & .3 \end{bmatrix}$

3. $\begin{bmatrix} 1 & 0 & .4 \\ 0 & .5 & .3 \\ 0 & .5 & .3 \end{bmatrix}$ **4.** $\begin{bmatrix} 1 & 0 & 0 & 0 \\ 0 & 1 & 0 & 0 \\ 0 & 0 & 0 & 1 \\ 0 & 0 & 1 & 0 \end{bmatrix}$

The matrices in Exercises 5–10 are absorbing stochastic matrices. In each, identify R and S and compute the fundamental matrix and the stable matrix.

5. $\begin{bmatrix} 1 & 0 & .3 \\ 0 & 1 & .2 \\ 0 & 0 & .5 \end{bmatrix}$ **6.** $\begin{bmatrix} 1 & 0 & \frac{1}{2} \\ 0 & 1 & \frac{1}{6} \\ 0 & 0 & \frac{1}{3} \end{bmatrix}$

7. $\begin{bmatrix} 1 & 0 & .1 & 0 \\ 0 & 1 & .5 & .2 \\ 0 & 0 & .3 & .6 \\ 0 & 0 & .1 & .2 \end{bmatrix}$ **8.** $\begin{bmatrix} 1 & 0 & \frac{1}{4} & \frac{1}{6} \\ 0 & 1 & \frac{1}{6} & 0 \\ 0 & 0 & \frac{1}{4} & \frac{1}{2} \\ 0 & 0 & \frac{1}{3} & \frac{1}{3} \end{bmatrix}$

9. $\begin{bmatrix} 1 & 0 & 0 & .1 & .2 \\ 0 & 1 & 0 & .3 & 0 \\ 0 & 0 & 1 & 0 & .2 \\ 0 & 0 & 0 & .5 & 0 \\ 0 & 0 & 0 & .1 & .6 \end{bmatrix}$

10. $\begin{bmatrix} 1 & 0 & \frac{1}{4} & 0 & \frac{1}{3} \\ 0 & 1 & \frac{1}{4} & 0 & 0 \\ 0 & 0 & 0 & \frac{1}{3} & \frac{1}{6} \\ 0 & 0 & \frac{1}{2} & \frac{1}{3} & 0 \\ 0 & 0 & 0 & \frac{1}{3} & \frac{1}{2} \end{bmatrix}$

Note: $\begin{bmatrix} 1 & -\frac{1}{3} & -\frac{1}{6} \\ -\frac{1}{2} & \frac{2}{3} & 0 \\ 0 & -\frac{1}{3} & \frac{1}{2} \end{bmatrix}^{-1} = \begin{bmatrix} \frac{3}{2} & 1 & \frac{1}{2} \\ \frac{9}{8} & \frac{9}{4} & \frac{3}{8} \\ \frac{3}{4} & \frac{3}{2} & \frac{9}{4} \end{bmatrix}.$

(Gambler's Ruin) Exercises 11 and 12 refer to Example 7.

11. Interpret the entry .79 in the fundamental matrix.

12. If the gambler begins with \$2, what is the expected number of times that he will play before quitting?

13. *(Class Standings)* Suppose that the following data were obtained from the records of a certain two-year college. Of those who were freshmen (F) during a particular year, 80% became sophomores (S) the next year and 20% dropped out (D). Of those who were sophomores during a particular year, 90% graduated (G) by the next year and 10% dropped out.

(a) Set up the absorbing stochastic matrix with states D, G, F, S that describes this transition.

(b) Find the stable matrix.

(c) Determine the probability that an entering freshman will eventually graduate.

(d) Determine the expected number of years a student entering as a freshman will attend the college before either dropping out or graduating.

14. (*Progression of Mental Disease*) A research article[2] on the application of stochastic matrices to mental illness considers a person to be in one of four states: state I—chronically insane and hospitalized; state II—dead, with death occurring while unhospitalized; state III—sane; state IV—mildly insane and unhospitalized. States I and II are absorbing states. Suppose that of the people in state III, after one year 1.994% will be in state II, 98% in state III, and .006% in state IV. Also, suppose that of the people in state IV, after one year 2% will be in state I, 3% will be in state II, and 95% in state IV.

   (a) Set up the absorbing stochastic matrix that describes this transition.

   (b) Find the stable matrix.

   (c) Determine the probability that a person who is currently well will eventually be chronically insane.

   (d) For a person in state III, determine the expected number of years for which the person will be in state III.

15. (*Accounts Payable*) A retailer classifies accounts as having one of four possible states: "paid up," "overdue at most 30 days," "overdue less than 60 days but more than 30 days," and "bad." If no payment is made on an overdue account by the end of the month, the status moves to the next state. When a partial payment is made on an overdue account, its improved status depends on the size of the payment. Experience shows that the following matrix describes the changes in status of accounts.

	Paid	≤ 30	< 60	Bad
Paid	1	.4	.1	0
≤ 30	0	.4	.4	0
< 60	0	.2	.4	0
Bad	0	0	.1	1

   (a) What is the probability of an account eventually being paid off if it is currently overdue at most 30 days? Less than 60 days?

   (b) If an account is currently overdue at most 30 days, what is the expected number of months until it is either paid or bad?

   (c) If the company has $2000 in bills in the "≤ 30 day" category and $5000 in bills in the "< 60 day" category, how much can the retailer expect will eventually be paid up and how much will eventually be irretrievable bad debt?

16. (*Job Mobility*) The managers in a company are classified as top managers, middle managers, and first-line managers. Each year, 10% of top managers retire, 10% leave the company, 60% remain top managers, and 20% are demoted to middle managers. Each year, 5% of middle managers retire, 15% leave the company, 10% are promoted to top managers, 60% remain middle managers, and 10% are demoted to first-line managers. Each year, 5% of first-line managers retire, 25% leave the company, 10% are promoted to middle managers, and 60% remain first-line managers.

   (a) What is the probability of a top manager eventually retiring? A middle manager? A first-line manager?

   (b) If a person is currently a middle manager, what is the expected number of years that he or she will be with the company before either leaving or retiring?

   Note:

$$\begin{bmatrix} .4 & -.1 & 0 \\ -.2 & .4 & -.1 \\ 0 & -.1 & .4 \end{bmatrix}^{-1} = \begin{bmatrix} \frac{75}{26} & \frac{10}{13} & \frac{5}{26} \\ \frac{20}{13} & \frac{40}{13} & \frac{10}{13} \\ \frac{5}{13} & \frac{10}{13} & \frac{35}{13} \end{bmatrix}.$$

17. (*Gambler's Ruin*) As a variation of the gambler's ruin problem, suppose that on each play the probability of winning is $\frac{1}{2}$ and the gambler stops playing if he accumulates $4.

   (a) What is the probability of eventually going broke if he starts with $1? $2? $3?

   (b) If the gambler begins with $2, what is the expected number of times that he will play before quitting?

   Note:

$$\begin{bmatrix} 1 & -\frac{1}{2} & 0 \\ -\frac{1}{2} & 1 & -\frac{1}{2} \\ 0 & -\frac{1}{2} & 1 \end{bmatrix}^{-1} = \begin{bmatrix} \frac{3}{2} & 1 & \frac{1}{2} \\ 1 & 2 & 1 \\ \frac{1}{2} & 1 & \frac{3}{2} \end{bmatrix}.$$

In Exercises 18 and 19, use a graphing calculator, a spreadsheet, or mathematical software to carry out the matrix calculations.

18. Consider the absorbing stochastic matrix A, where

$$A = \begin{bmatrix} 1 & 0 & 0 & 0 & .6 \\ 0 & 1 & 0 & .5 & .1 \\ 0 & 0 & 1 & 0 & .1 \\ 0 & 0 & 0 & .2 & .2 \\ 0 & 0 & 0 & .3 & 0 \end{bmatrix}.$$

Approximate the stable matrix of A by raising A to a high power. Then find the exact stable distribution by calculating $S(I - R)^{-1}$.

[2]A.W. Marshall and H. Goldhamer, "An Application of Markov Processes to the Study of the Epidemiology of Mental Diseases," *American Statistical Association Journal*, March 1955, pp. 99–129.

19. (*Collecting Quotations*) A soft drink manufacturer puts one of three different quotations on the inside of each bottle cap. Assuming that each of the quotations is equally likely to appear, what is the probability of receiving all three after purchasing five of the soft drinks? Also, what is the expected number of soft drinks you have to purchase to have all three quotations?

20. Repeat Exercise 18 for the matrix

$$\begin{bmatrix} 1 & 0 & .5 & .4 & .2 \\ 0 & 1 & 0 & .3 & .7 \\ 0 & 0 & 0 & .2 & .1 \\ 0 & 0 & .3 & 0 & 0 \\ 0 & 0 & .2 & .1 & 0 \end{bmatrix}.$$

SOLUTIONS TO PRACTICE PROBLEMS 8.3

1. Sometimes they are, but in general they are not. The size of the first identity matrix equals the number of absorbing states, whereas the size of the second identity matrix is the same as that of the matrix R, which equals the number of nonabsorbing states.

2. Consider the entry in the ith row, jth column of A^2. Suppose that an object begins in state j. Then the entry gives the probability that it ends up in state i after two time periods.

3. Yes. (i) Its first state is absorbing. (ii) From each state it is possible to get to the first state. Note that it is possible to go directly from state 2 to state 1, since $a_{12} = .4 \neq 0$. Since $a_{13} = 0$, it is not possible to go directly from state 3 to state 1. However, this can be accomplished indirectly by going from state 3 to state 2 (possible since $a_{23} = .1 \neq 0$) and then from state 2 to state 1.

CHAPTER 8 SUMMARY

1. A Markov process is a sequence of experiments performed at regular time intervals involving *states*. As a result of each experiment, transitions between states occur with probabilities given by a matrix called the *transition matrix*. The ijth entry in the transition matrix is the conditional probability Pr(moving to state i|in state j).

2. A *stochastic matrix* is a square matrix for which every entry is greater than or equal to 0 and the sum of the entries in each column is 1. Every transition matrix is a stochastic matrix.

3. The nth distribution matrix gives the percentage of members in each state after n time periods.

4. A^n is obtained by multiplying together n copies of A. Its ijth entry is the conditional probability Pr(moving to state i after n time periods | in state j). Also, A^n times the initial distribution matrix gives the nth distribution matrix.

5. A stochastic matrix is called *regular* if some power of the matrix has only positive entries.

6. If A is a regular stochastic matrix, as n gets large the powers of the matrix, A^n, approach a certain matrix called the *stable matrix* of A and the distribution matrices approach a certain column matrix called the *stable distribution*. Each column of the stable matrix holds the stable distribution. The stable distribution can be found by solving $AX = X$, where the sum of the entries in X is equal to 1.

7. If the probability of moving from a state to itself is 1, we call that state an *absorbing state*. An *absorbing stochastic matrix* is a stochastic matrix with at least one absorbing state and in which from any state it is possible to eventually get to an absorbing state.

8. In an absorbing process, the transition matrix should be arranged so that absorbing states are listed before nonabsorbing states. The transition matrix will have the form

$$\left[\begin{array}{c|c} I & S \\ \hline 0 & R \end{array} \right],$$

where I is an identity matrix, 0 denotes a matrix of zeros, and S and R represent the transitions from nonabsorbing to absorbing states and from nonabsorbing to nonabsorbing states, respectively.

9. The *stable matrix* of the absorbing matrix in number 8 is

$$\left[\begin{array}{c|c} I & S(I-R)^{-1} \\ \hline 0 & 0 \end{array} \right].$$

10. The *fundamental matrix* of the absorbing matrix in number 8 is the matrix $(I-R)^{-1}$. When its columns and rows are labeled with the nonabsorbing states, its ijth entry is the expected number of times the process will be in nonabsorbing state i given that it started in nonabsorbing state j. The sum of the entries in the jth column is the expected number of steps before absorption when the process begins in state j.

REVIEW OF FUNDAMENTAL CONCEPTS

1. What is a Markov process?
2. What is a transition matrix? A stochastic matrix? A distribution matrix?
3. What is A^n? Give an interpretation of the entries of A^n.
4. How is the nth distribution matrix calculated from the initial distribution matrix?
5. Define *regular* stochastic matrix.
6. Define the stable matrix and the stable distribution of a regular stochastic matrix.

7. Explain how to find the stable distribution of a regular stochastic matrix.
8. What is meant by an absorbing state of a stochastic matrix?
9. What is an absorbing stochastic matrix?
10. Explain how to find the stable matrix of an absorbing stochastic matrix.
11. What is the fundamental matrix of an absorbing stochastic matrix and how is it used?

KEY FORMULAS

Distribution Matrices: $\begin{bmatrix} \ \\ \ \end{bmatrix}_n = A^n \begin{bmatrix} \ \\ \ \end{bmatrix}_0$

Absorbing Stochastic Matrix: The stable matrix of

$$\left[\begin{array}{c|c} I & S \\ \hline 0 & R \end{array}\right] \quad \text{is} \quad \left[\begin{array}{c|c} I & S(I-R)^{-1} \\ \hline 0 & 0 \end{array}\right].$$

$F = (I - R)^{-1}$ is the Fundamental Matrix, with its columns and rows labeled with the nonabsorbing states. The entry in row i, column j of F gives the expected number of times state i occurs when the process starts in state j. The sum of the entries in column j of F is the expected number of steps until absorption when the process begins in state j.

SUPPLEMENTARY EXERCISES

In Exercises 1–6, determine whether or not the given matrix is stochastic. If so, determine if it is regular, absorbing, or neither.

1. $\begin{bmatrix} 1 & .3 & 0 & 0 \\ 0 & .1 & 0 & 0 \\ 0 & .4 & .7 & .4 \\ 0 & .2 & .3 & .6 \end{bmatrix}$

2. $\begin{bmatrix} .1 & .1 & .1 & .1 \\ .2 & .2 & .2 & .2 \\ .3 & .3 & .3 & .3 \\ .4 & .4 & .4 & .4 \end{bmatrix}$

3. $\begin{bmatrix} 0 & .3 \\ 1 & .7 \end{bmatrix}$

4. $\begin{bmatrix} 1 & 0 & 0 \\ 0 & 1 & \frac{1}{3} \\ 0 & 0 & \frac{2}{3} \end{bmatrix}$

5. $\begin{bmatrix} 1 & \frac{1}{2} & 0 \\ 0 & \frac{1}{2} & 0 \\ 0 & \frac{1}{2} & 1 \end{bmatrix}$

6. $\begin{bmatrix} 1 & 0 & 0 & .3 \\ 0 & 1 & 0 & .3 \\ 0 & 0 & .5 & .3 \\ 0 & 0 & .5 & .1 \end{bmatrix}$

7. Find the stable distribution for the regular stochastic matrix $\begin{bmatrix} .6 & .5 \\ .4 & .5 \end{bmatrix}$.

8. Find the stable matrix for the absorbing stochastic matrix

$$\begin{bmatrix} 1 & 0 & 0 & \frac{1}{8} & \frac{1}{4} \\ 0 & 1 & 0 & \frac{1}{8} & 0 \\ 0 & 0 & 1 & 0 & \frac{1}{4} \\ 0 & 0 & 0 & \frac{1}{4} & \frac{1}{2} \\ 0 & 0 & 0 & \frac{1}{2} & 0 \end{bmatrix}.$$

9. (*Economic Mobility*) In a certain community currently 10% of the people are H (high income), 60% are M (medium income), and 30% are L (low income). Studies show that for the children of H parents, 50% also become H, 40% become M, and 10% become L. Of the children of M parents, 40% become H, 30% become M, and 30% become L. Of the children of L parents, 30% become H, 50% become M, and 20% become L.

   (a) Set up the 3×3 stochastic matrix that describes this situation.

   (b) What percent of the children of the current generation will have high incomes?

   (c) In the long run, what proportion of the population will have low incomes?

10. (*Quality Control*) In a certain factory some machines are properly adjusted and some need adjusting. Technicians randomly inspect machines and make adjustments. Suppose that of the machines that are properly adjusted on a particular day, 80% will also be properly adjusted the following day and 20% will need adjusting. Also, of the machines that need adjusting on a particular day, 30% will be properly adjusted the next day and 70% will still need adjusting.

   (a) Set up the 2×2 stochastic matrix with columns labeled P (properly adjusted) and N (need adjusting) that describes this situation.

(b) If initially all the machines are properly adjusted, what percent will need adjusting after 2 days?

(c) In the long run, what percent will be properly adjusted each day?

11. Find the stable matrix for the absorbing stochastic matrix

$$\begin{bmatrix} 1 & 0 & \frac{1}{6} & \frac{1}{2} & \frac{2}{5} \\ 0 & 1 & 0 & 0 & \frac{2}{5} \\ 0 & 0 & 0 & 0 & 0 \\ 0 & 0 & \frac{2}{3} & \frac{1}{2} & 0 \\ 0 & 0 & \frac{1}{6} & 0 & \frac{1}{5} \end{bmatrix}.$$

12. (*Mouse in a House*) Figure 1 gives the layout of a house with four rooms connected by doors. Room I contains a mousetrap and room II contains cheese. A mouse, after being placed in one of the rooms, will search for cheese; if unsuccessful after one minute, it will exit to another room by selecting one of the doors at random. (For instance, if the mouse is in room III, after one minute he will go to room II with probability $\frac{1}{3}$ and to room IV with probability $\frac{2}{3}$.) A mouse entering room I will be trapped and therefore no longer move. Also, a mouse entering room II will remain in that room.

II (cheese)	I (trap)
III	IV

Figure 1

(a) Set up the 4×4 absorbing stochastic matrix that describes this situation.

(b) If a mouse begins in room IV, what is the probability that he will find the cheese after 2 minutes?

(c) If a mouse begins in room IV, what is the probability that he will find the cheese in the long run?

(d) For a mouse beginning in room III, determine the expected number of minutes that will elapse before the mouse either finds the cheese or is trapped.

13. Which of the following is the stable distribution for the regular stochastic matrix $\begin{bmatrix} .4 & .4 & .2 \\ .1 & .1 & .3 \\ .5 & .5 & .5 \end{bmatrix}$?

(a) $\begin{bmatrix} .6 \\ .4 \\ 1 \end{bmatrix}$ **(b)** $\begin{bmatrix} .2 \\ .3 \\ .5 \end{bmatrix}$ **(c)** $\begin{bmatrix} .3 \\ .2 \\ .5 \end{bmatrix}$

14. (*Listening Preferences*) A city has two competing news stations. From a survey of regular listeners it was determined that of those who listen to station A on a particular day, 90% listen to station A the next day and 10% listen to station B. Of those who listen to

station B on a particular day, 20% listen to station A the next day and 80% listen to station B. If today 50% of the regular listeners listen to each station, what percentage of them would you expect to listen to station A 2 days from now?

15. (*Traffic Conditions*) Workday traffic conditions from 9 A.M. to 10 A.M. on the Baltimore Beltway can be characterized as Light, Moderate, and Heavy. The following stochastic matrix describes the day-to-day transitions.

	L	M	H
L	.70	.20	.10
M	.20	.75	.30
H	.10	.05	.60

(a) Interpret the numbers in the second column of the matrix.

(b) In the long run, what percent of the workdays fall into each category?

(c) Of 20 workdays in a month, how many are expected to have Heavy traffic on the Baltimore Beltway from 9 A.M. to 10 A.M.?

16. (*Mental Health*) A mental-health facility rates patients on their ability to live on their own. The state of a person's health is "able to work and considered cured (C)," or "long-term hospitalization or death (L)," or "group home (G)," or "short-term hospital care (S)." The following stochastic matrix describes the transitions from month to month. Use the fundamental matrix to determine the expected number of months spent in state G or S before being absorbed into state C or L.

	C	L	G	S
C	1	0	.60	.05
L	0	1	.10	.40
G	0	0	.20	.50
S	0	0	.10	.05

17. (*Reservoir Levels*) The contents of a reservoir depend on the available rainfall in the region and the demands on the water supply. Suppose a reservoir holds up to 4 units of water (a unit might be a million gallons) and policy for the use of the water for irrigation and drinking water never allows the contents of the reservoir to drop below 1 unit of water. The (rounded) amount of water in the reservoir from week to week seems to follow the transition matrix

	1	2	3	4
1	.20	.10	.05	.05
2	.30	.20	.20	.30
3	.40	.40	.50	.40
4	.10	.30	.25	.25

(a) Determine and interpret the stable distribution for the matrix.

(b) Suppose the weekly benefits to recreation in the area around the reservoir are estimated to be $4000 when there is 1 unit in the reservoir, $6000 when there are 2 units in the reservoir, $10,000 when there are 3 units, and $3000 when there are 4 units. Determine the average weekly benefits to be realized.

Conceptual Exercises

18. Explain why the entries in each column of a transition matrix must add up to 1.

19. Suppose A is a transition matrix and

$$A^4 = \begin{bmatrix} .74 & .18 \\ .26 & .82 \end{bmatrix}.$$

Interpret the number .26.

20. Explain why A^2 must be a stochastic matrix if A is a stochastic matrix.

CHAPTER TEST

1. Which of the following matrices are stochastic?

(a) $\begin{bmatrix} \frac{1}{2} & \frac{1}{3} \\ \frac{1}{2} & \frac{2}{3} \end{bmatrix}$ **(b)** $\begin{bmatrix} .1 & .9 \\ .5 & .5 \end{bmatrix}$

(c) $\begin{bmatrix} 1 & \frac{1}{3} \\ 0 & -\frac{2}{3} \end{bmatrix}$ **(d)** $\begin{bmatrix} \frac{1}{8} & \frac{1}{2} & 0 \\ \frac{3}{8} & 0 & 1 \\ \frac{1}{2} & \frac{1}{2} & 0 \end{bmatrix}$

2. Which of the following stochastic matrices are regular?

(a) $\begin{bmatrix} 1 & 0 \\ 0 & 1 \end{bmatrix}$ **(b)** $\begin{bmatrix} .1 & .5 \\ .9 & .5 \end{bmatrix}$

(c) $\begin{bmatrix} .4 & 1 \\ .6 & 0 \end{bmatrix}$

3. Which of the following matrices is the stable distribution for the regular stochastic matrix $\begin{bmatrix} .2 & .2 & .3 \\ .1 & .1 & .4 \\ .7 & .7 & .3 \end{bmatrix}$?

(a) $\begin{bmatrix} .3 \\ .2 \\ .5 \end{bmatrix}$ **(b)** $\begin{bmatrix} .25 \\ .25 \\ .50 \end{bmatrix}$

(c) $\begin{bmatrix} .35 \\ .35 \\ .30 \end{bmatrix}$ **(d)** $\begin{bmatrix} .2 \\ .1 \\ .7 \end{bmatrix}$

4. Find the stable distribution and the stable matrix for the regular stochastic matrix $\begin{bmatrix} \frac{1}{5} & \frac{3}{5} \\ \frac{4}{5} & \frac{2}{5} \end{bmatrix}$.

5. (*Browser Preferences*) Students at Gotham College regularly use the two browsers Internet Explorer® and Netscape®. Of the students who use Internet Explorer on a certain day, 70% use Internet Explorer the next day and 30% use Netscape. Of the students who use Netscape on a certain day, 60% use Internet Explorer the next day and 40% use Netscape. Suppose that during the first day of a semester 50% of the students use each browser.

(a) Set up the stochastic matrix displaying these transitions.

(b) What is the initial distribution matrix?

(c) What percent of the students will use Internet Explorer two days later?

(d) Show that $\begin{bmatrix} \frac{2}{3} \\ \frac{1}{3} \end{bmatrix}$ is the stable distribution for the stochastic matrix in part (a).

(e) Explain in a sentence the meaning of the number $\frac{2}{3}$ in the stable distribution from part (d).

6. Which of the following are absorbing stochastic matrices? For those that are not absorbing stochastic matrices, explain why.

(a) $\begin{bmatrix} 1 & 0 & .1 \\ 0 & 1 & .8 \\ 0 & 0 & .1 \end{bmatrix}$ **(b)** $\begin{bmatrix} 1 & 0 & 0 \\ 0 & .4 & .2 \\ 0 & .6 & .8 \end{bmatrix}$

(c) $\begin{bmatrix} 1 & .1 & 0 \\ 0 & .5 & 1 \\ 0 & .4 & 0 \end{bmatrix}$

7. (*Computer Preferences*) College math departments have been rapidly establishing their own computer labs. Of the departments with no labs, each year 10% set up labs using Apple® computers, 30% set up labs using IBM® compatible computers, and the remainder do not set up labs that year. Once a lab has been established with a certain type of computer, the lab is never abandoned and the brand of computer is never changed.

(a) Set up the absorbing stochastic matrix that describes these transitions.

(b) Find the stable matrix.

(c) In the long run, what percent of the math departments will have Apple computer labs?

(d) What is the expected number of years required for a math department to decide to set up its own computer lab if it currently does not have a lab?

CHAPTER 8 PROJECT

Doubly Stochastic Matrices

A square matrix is said to be *doubly stochastic* if the sum of the entries in each column is 1 and the sum of the entries in each row is 1. Some examples of doubly stochastic matrices are

$$\begin{bmatrix} .4 & .6 \\ .6 & .4 \end{bmatrix}, \quad \begin{bmatrix} .1 & .3 & .6 \\ .6 & .1 & .3 \\ .3 & .6 & .1 \end{bmatrix}, \quad \text{and} \quad \begin{bmatrix} .1 & .2 & .3 & .4 \\ .3 & .4 & .1 & .2 \\ .2 & .3 & .4 & .1 \\ .4 & .1 & .2 & .3 \end{bmatrix}.$$

1. Give another example of a 2×2 doubly stochastic matrix.

   (a) Is your matrix symmetric? That is, does it equal its own transpose?

   (b) Prove that every 2×2 doubly stochastic matrix is symmetric.

   (c) Show that the product of your matrix and the matrix

   $$\begin{bmatrix} .4 & .6 \\ .6 & .4 \end{bmatrix}$$

   is doubly stochastic.

   (d) Show that $\begin{bmatrix} \frac{1}{2} \\ \frac{1}{2} \end{bmatrix}$ is a stable distribution for your matrix.

2. Give another example of a 3×3 doubly stochastic matrix.

   (a) Is your matrix symmetric? Are all 3×3 doubly stochastic matrices symmetric?

   (b) Show that the product of your matrix and the matrix

   $$\begin{bmatrix} .1 & .3 & .6 \\ .6 & .1 & .3 \\ .3 & .6 & .1 \end{bmatrix}$$

   is doubly stochastic.

   (c) Show that $\begin{bmatrix} \frac{1}{3} \\ \frac{1}{3} \\ \frac{1}{3} \end{bmatrix}$ is a stable distribution for your matrix.

3. Give another example of a 4×4 doubly stochastic matrix.

   (a) Is your matrix symmetric? Are all 4×4 doubly stochastic matrices symmetric?

   (b) Show that the product of your matrix and the matrix

   $$\begin{bmatrix} .1 & .2 & .3 & .4 \\ .3 & .4 & .1 & .2 \\ .2 & .3 & .4 & .1 \\ .4 & .1 & .2 & .3 \end{bmatrix}$$

   is doubly stochastic.

   (c) Show that $\begin{bmatrix} \frac{1}{4} \\ \frac{1}{4} \\ \frac{1}{4} \\ \frac{1}{4} \end{bmatrix}$ is a stable distribution for your matrix.

We will now show that the product of any two $n \times n$ doubly stochastic matrices is doubly stochastic, and that any doubly stochastic $n \times n$ matrix has

$$\begin{bmatrix} \frac{1}{n} \\ \frac{1}{n} \\ \vdots \\ \frac{1}{n} \end{bmatrix}$$

as a stable distribution. Let E_n be the $n \times n$ matrix for which each entry is 1.

4. Show that A is a doubly stochastic $n \times n$ matrix if and only if $AE_n = E_n$ and $E_n A = E_n$.

5. Show that if A and B are doubly stochastic $n \times n$ matrices, then so is AB. *Hint*: Show that $(AB)E_n = E_n$ and $E_n(AB) = E_n$.

6. Show that if A is a doubly stochastic $n \times n$ matrix, then

$$\begin{bmatrix} \frac{1}{n} \\ \frac{1}{n} \\ \vdots \\ \frac{1}{n} \end{bmatrix}$$

is a stable distribution for A. *Hint*: First show that $A \begin{bmatrix} 1 \\ 1 \\ \vdots \\ 1 \end{bmatrix} = \begin{bmatrix} 1 \\ 1 \\ \vdots \\ 1 \end{bmatrix}$.

7. A collection of $n \times n$ matrices is said to be a *convex set*[1] if whenever A and B are in the set and t is a number between 0 and 1, then $tA + (1-t)B$ is also in the set. Show that the set of all $n \times n$ doubly stochastic matrices is a convex set.

---

[1] Convex sets are studied in advanced courses in applied matrix theory.

CHAPTER 9

The Theory of Games

ONE of the more interesting developments of twentieth-century mathematics has been the theory of games, a branch of mathematics used to analyze competitive phenomena. This theory has been applied extensively in many fields, including business, economics, psychology, and sociology. The 1994 and 2005 Nobel Prizes in Economics were awarded to economists for their groundbreaking work in integrating game theory into the study of economic behavior. Game theory has become one of the hottest areas of economics, with applications ranging from how the Federal Reserve sets interest rates, to how companies structure incentive pay for employees, to how companies bid on lucrative federal contracts. Mathematically, the theory of games blends the theory of matrices with probability theory. Although an extensive discussion is well beyond the scope of this book, we hope to give the flavor of the subject and some indication of the wide range of its applications.

9.1 Games and Strategies

Let us begin our study of game theory by analyzing a typical competitive situation. Suppose that in a certain town there are two furniture stores, Reliable Furniture Company and Cut-Rate Furniture Company, which compete for all furniture sales in the town. Each of the stores is planning a Labor Day sale and each has the option of marking its furniture down by 10% or 20%. The results of their decisions affect the total percentage of the market that each captures. On the basis of an analysis of past consumer tendencies, it is estimated that if Reliable chooses a 10% discount and so does Cut-Rate, then Reliable will capture 60% of the sales. If Reliable chooses a 10% discount but Cut-Rate chooses 20%, then Reliable will capture only 35% of the sales. On the other hand, if Reliable chooses a 20% discount but Cut-Rate chooses 10%, then Reliable will get 80% of the sales. If Reliable chooses a 20% discount and Cut-Rate also chooses 20%, then Reliable will get 50% of the sales. Each store is able to determine the other store's discount prior to the start of the sale and adjust its own discount accordingly. If you were a consultant to Reliable, what discount would you choose to obtain as large a share of the sales as possible?

To analyze the various possibilities, let us summarize the given data in a matrix. For the sake of brevity, denote Reliable by R and Cut-Rate by C. Then the data can be summarized in a matrix showing R's share in each case as follows:

$$\begin{array}{cc} & C \text{ discount} \\ & \begin{array}{cc} 10\% & 20\% \end{array} \\ R \text{ discount} \begin{array}{c} 10\% \\ 20\% \end{array} & \left[\begin{array}{cc} .6 & .35 \\ .8 & .5 \end{array} \right]. \end{array}$$

For example, the number in the second row, first column corresponds to an R discount of 20% and a C discount of 10%. In this case R will capture 80%, or .8, of the sales.

We may view R's choice of discount as choosing one of the rows of the matrix. Similarly, C's choice of discount amounts to choosing one of the columns of the matrix. Suppose that R and C both act rationally. What will be the result? Let us view things from R's perspective first. In scanning his options, he sees a .8 in the second row. So his first reaction might be to take a 20% discount and try for 80% of the sales. However, this route is very risky. As soon as C learns that R has chosen a 20% discount, C will set a 20% discount and lower R's share of the sales to 50%. So the result of choosing row 2 will be for R to capture only 50% of the sales. On the other hand, if R chooses row 1, then C will naturally choose a 20% discount to give R a 35% share of the sales. Of the options open to R, the 50% share is clearly the most desirable. So R will choose a 20% discount.

What about C? In setting his discount, C must choose a column of the matrix. Since the entries represent R's share of the sales, C wishes to make a choice resulting in as *small* a number as possible. If C chooses column 1, then R will immediately respond with a 20% discount in order to acquire 80% of the sales, a disaster for C. On the other hand, if C chooses column 2, then R will choose a 20% discount to obtain 50% of the sales. The best option open to C is to choose a 20% discount.

Thus we see that if both stores act rationally, they will each choose 20% discounts and each will capture 50% of the sales.

The preceding competitive situation is an example of a (mathematical) *game*. In such a game there are two or more players. In the example the players are R and C. Each player is allowed to make a move. In the example the moves are the choices of discount. As the result of a move by each player, there is a payoff to each player. The payoff to each player (store) is the percentage of total sales he captures. In our example, then, we have solved a problem that can be posed for any game:

Fundamental Problem of Game Theory How should each player decide his move in order to maximize his gain?

Indeed, in our example, R and C chose moves such that each maximized his own share of sales.

Throughout this chapter we consider only games with two players, whom we shall denote by R and C. (R and C stand for row and column, respectively.) Suppose that R can make moves $R_1, R_2, \ldots, R_m$ and that C can make moves $C_1, C_2, \ldots, C_n$. Further suppose that a move R_i by R and C_j by C results in a payoff of a_{ij} to R. Then the game can be represented by the following *payoff matrix*:

$$
R \text{ moves}
\begin{array}{c}
\\
R_1 \\
R_2 \\
\vdots \\
R_m
\end{array}
\overset{\begin{array}{cccc} C_1 & C_2 & \cdots & C_n \end{array}}{
\begin{bmatrix}
a_{11} & a_{12} & \cdots & a_{1n} \\
a_{21} & a_{22} & \cdots & a_{2n} \\
\vdots & \vdots & & \vdots \\
a_{m1} & a_{m2} & \cdots & a_{mn}
\end{bmatrix}
}.
$$

$$\overset{C \text{ moves}}{\rule{4cm}{0.4pt}}$$

Note that the payoff matrix is an $m \times n$ matrix with a_{ij} as the entry of the ith row, jth column. In our furniture store example the payoff matrix was just the matrix we used in our analysis. Note that a move by R corresponds to a choice of a *row* of the payoff matrix, whereas a move by C corresponds to a choice of a *column.*

Suppose that a given game is played repeatedly. The players can adopt various strategies to attempt to maximize their respective gains (or minimize their losses). In what follows we shall discuss the problem of determining strategies. The simplest type of strategy is one in which a player, on consecutive plays, consistently chooses the same row (or column). Such strategies are called *pure strategies* and are discussed in this section. Strategies involving varied moves are called *mixed strategies* and are discussed in Section 9.2.

In most examples of games, the payoffs to C are related in a simple way to the corresponding payoffs to R. For example, for the furniture stores the payoff to C is 100% minus the payoff to R. In another common type of game a payoff to R of a given amount results in a loss to C of the same amount, and vice versa. For such games the sum of the gains on each play is zero; hence they are called *zero-sum games.* An illustration of such a game is provided in the next example.

EXAMPLE 1 **Coin-matching game** Suppose that R and C play a coin-matching game. Each player can show either heads or tails. If R and C both show heads, then C pays R \$5. If R shows heads and C shows tails, then R pays C \$8. If R shows tails and C shows heads, then C pays R \$3. If R shows tails and so does C, then C pays R \$1.

(a) Determine the payoff matrix of this game.

(b) Suppose that R and C play the game repeatedly. Determine optimal pure strategies for R and C.

Solution (a) The payoff matrix is given by

$$
R
\begin{array}{c}
\\
\text{Heads} \\
\text{Tails}
\end{array}
\overset{\begin{array}{cc} \text{Heads} & \text{Tails} \end{array}}{
\begin{bmatrix}
5 & -8 \\
3 & 1
\end{bmatrix}
}.
$$

$$\overset{C}{}$$

Each entry specifies a payoff from C to R. The entry "-8" denotes a negative payoff to R, that is, a gain of \$8 to C.

(b) R would clearly like to choose heads so as to gain \$5. However, if R consistently chooses heads, C will retaliate by choosing tails, causing R to lose \$8. If R chooses tails, however, then the best C can do is choose tails to give a gain of \$1 to R. So R should clearly choose tails. Now we look at the game from C's point of view. Clearly, C's objective is to minimize the payment to R. If C consistently chooses heads, then R will notice the pattern and choose heads, at a cost to C of \$5. However, if C chooses tails, then the best that R can do is choose tails, thereby costing C \$1. So, clearly, the optimal move for C is to choose tails. ■

The reasoning just described is rather cumbersome. There is, however, an easy way to summarize what we have done. Let us first describe R's reasoning. R seeks to choose a row of the matrix that will maximize his payoff. However, once a row is chosen consistently, R can expect C to counter by choosing the least element of that row. Thus R should choose his move as follows:

Optimal Pure Strategy for R

1. For each row of the payoff matrix, determine the least element.

2. Choose the row for which this element is as large as possible.

For example, in the game of Example 1 we have circled the least element of each row:

$$\begin{bmatrix} 5 & \circledon{-8} \\ 3 & \circledon{1} \end{bmatrix}.$$

The largest circled element is 1. So R should choose the second row—that is, tails.

In a similar way we may describe the optimal strategy for C. C wishes to choose a column of the payoff matrix so as to minimize the payoff to R. However, C can expect R to adjust his choice to the maximum element of the column. Therefore, we can summarize the optimal strategy for C as follows:

Optimal Pure Strategy for C

1. For each column of the payoff matrix, determine the largest element.

2. Choose the column for which this element is as small as possible.

For example, in the game of Example 1 we have circled the largest element in each column:

$$\begin{bmatrix} \circledon{5} & -8 \\ 3 & \circledon{1} \end{bmatrix}.$$

The smallest circled element is 1, so C should choose the second column, or tails.

EXAMPLE 2 **Optimal pure strategy** Determine optimal pure strategies for R and C for the game whose payoff matrix is

$$\begin{bmatrix} -1 & 5 \\ 1 & 4 \\ 0 & -1 \end{bmatrix}.$$

Solution To determine the strategy for R, we first circle the smallest element in each row:

$$\begin{bmatrix} \boxed{-1} & 5 \\ \boxed{1} & 4 \\ 0 & \boxed{-1} \end{bmatrix}.$$

The largest of these is 1, so R should play the second row. To determine the strategy for C, we circle the largest element in each column:

$$\begin{bmatrix} -1 & \boxed{5} \\ \boxed{1} & 4 \\ 0 & -1 \end{bmatrix}.$$

The smallest of these is 1, so C should play the first column.

Now Try Exercise 5   ■

All the games considered so far have an important characteristic in common: There is an entry in the payoff matrix which is *simultaneously* the minimum element in its row and the maximum element in its column. Such an entry is called a *saddle point* for the game. As we have seen in the examples considered previously, if a game possesses a saddle point, then an optimal strategy is for R to choose the row containing the saddle point and for C to choose the column containing the saddle point.

A game need not have a saddle point. For example, the matrix

$$\begin{bmatrix} 2 & -2 \\ 0 & 1 \end{bmatrix}$$

is the payoff matrix of a game with no saddle point. The optimal pure strategy for R is to choose the row with the maximum of the circled elements in

$$\begin{bmatrix} 2 & \boxed{-2} \\ \boxed{0} & 1 \end{bmatrix}.$$

Thus, R chooses row 2. The optimal pure strategy for C is to choose the column with the minimum of the circled elements in

$$\begin{bmatrix} \boxed{2} & -2 \\ 0 & \boxed{1} \end{bmatrix}.$$

So C chooses column 2. No element is simultaneously the minimum element in its row and the maximum element in its column.

A game that has a saddle point is called a *strictly determined game*. If v is a saddle point for a strictly determined game, then if each player plays the optimal pure strategy, each repetition of the game will result in a payment of v to player R. The number v is called the *value* of the game.

EXAMPLE 3 **Value of a strictly determined game** Find the saddle point and the value of the strictly determined game given by the payoff matrix

$$\begin{bmatrix} -1 & -10 & 10 \\ 0 & 7 & 6 \\ 3 & 4 & 11 \\ 2 & 5 & 7 \end{bmatrix}.$$

Solution The least elements in the various rows are

$$\begin{bmatrix} -1 & \boxed{-10} & 10 \\ \boxed{0} & 7 & 6 \\ \boxed{3} & 4 & 11 \\ \boxed{2} & 5 & 7 \end{bmatrix}.$$

The maximum elements in the columns are

$$\begin{bmatrix} -1 & -10 & 10 \\ 0 & \boxed{7} & 6 \\ \boxed{3} & 4 & \boxed{11} \\ 2 & 5 & 7 \end{bmatrix}.$$

The element 3 in the third row, first column is a minimum in its row and a maximum in its column and so is a saddle point of the game. The value of the game is therefore 3: Each repetition of the game, assuming optimal strategies, results in a payoff of 3 to R.

Now Try Exercise 7 ■

EXAMPLE 4 **A child's game** R and C play a game in which they show 1 or 2 fingers simultaneously. It is agreed that C pays R an amount equal to the total number of fingers shown less 3 cents. Find the optimal strategy for each player and the value of the game.

Solution The payoff matrix is given by

$$\begin{array}{cc} & \begin{array}{cc} 1 & 2 \end{array} \\ \begin{array}{c} 1 \\ 2 \end{array} & \begin{bmatrix} 2-3 & 3-3 \\ 3-3 & 4-3 \end{bmatrix} \end{array} = \begin{array}{cc} & \begin{array}{cc} 1 & 2 \end{array} \\ \begin{array}{c} 1 \\ 2 \end{array} & \begin{bmatrix} -1 & 0 \\ 0 & 1 \end{bmatrix} \end{array}.$$

The saddle point is the element that is simultaneously the minimum of its row and the maximum of its column—so an optimal strategy is for R to show 2 fingers and for C to show 1 finger. The value of the game is 0.

Now Try Exercise 9 ■

A game may have more than one saddle point. Consider the game with payoff matrix

$$\begin{bmatrix} \boxed{1} & 2 & \boxed{1} \\ \boxed{1} & 5 & \boxed{1} \\ 0 & -7 & -1 \end{bmatrix}.$$

Each circled element is both the minimum element in its row and the maximum element in its column. There are four saddle points representing four optimal strategies. The value of the game, regardless of strategy, is 1. If a game has more than one saddle point, then the value of the game is the same at each of them.

PRACTICE PROBLEMS 9.1

Which of the following matrices are the payoff matrices of strictly determined games? For those that are, determine the saddle point and optimal pure strategy for each of the players.

1. $\begin{bmatrix} 1 & -1 & -3 \\ 0 & -2 & 3 \end{bmatrix}$ **2.** $\begin{bmatrix} 1 & -1 & 0 \\ 0 & -4 & 5 \end{bmatrix}$

3. $\begin{bmatrix} 1 & -2 & 1 \\ -2 & 1 & 1 \\ 1 & 1 & -2 \end{bmatrix}$

EXERCISES 9.1

Each of the following matrices is the payoff matrix for a strictly determined game. Determine optimal pure strategies for R and C.

1. $\begin{bmatrix} -1 & -2 \\ 0 & 3 \end{bmatrix}$ **2.** $\begin{bmatrix} -4 & 0 \\ 2 & 1 \end{bmatrix}$

3. $\begin{bmatrix} -2 & 4 & 1 \\ -1 & 3 & 5 \\ -3 & 5 & 2 \end{bmatrix}$ **4.** $\begin{bmatrix} 1 & -1 & 0 \\ 6 & 3 & 2 \\ 2 & -2 & 1 \end{bmatrix}$

5. $\begin{bmatrix} 0 & 3 \\ -1 & 1 \\ -2 & -4 \end{bmatrix}$ **6.** $\begin{bmatrix} 0 & -4 & 0 \\ -1 & -2 & 1 \end{bmatrix}$

Each of the following matrices is the payoff matrix for a strictly determined game. (a) Find a saddle point. (b) Determine the value of the game.

7. $\begin{bmatrix} 1 & 0 \\ 0 & -1 \end{bmatrix}$ **8.** $\begin{bmatrix} 2 & 3 \\ 4 & 5 \end{bmatrix}$

For each of the following games, give the payoff matrix and decide if the game is strictly determined. If so, determine the optimal strategies for R and C.

9. (*Matching Coins*) Suppose that R and C play a game by matching coins. On each play, C pays R the number of heads shown (0, 1, or 2) minus twice the number of tails shown.

10. (*Scissors, Paper, Stone*) In the child's game "scissors, paper, stone," each of two children calls out one of the three words. If they both call out the same word, then the game is a tie. Otherwise, "scissors" beats "paper" (since scissors can cut paper), "paper" beats "stone" (since paper can cover stone), and "stone" beats "scissors" (since stone can break scissors). Suppose that

the loser pays a penny to the winner.

11. (*Political Action*) Two candidates for political office must decide to be for, against, or neutral on a certain referendum. Pollsters have determined that if candidate R comes out for the referendum, then he will gain 8000 votes if candidate C also comes out for the referendum, will lose 1000 votes if candidate C comes out against, and will gain 1000 votes if candidate C comes out neutral. If candidate R comes out against, then he will lose 7000 votes (respectively gain 4000 votes, lose 2000 votes) if candidate C comes out for (respectively against, neutral on) the referendum. If candidate R is neutral, then he will gain 3000 votes if C is for or against and will gain 2000 votes if C is neutral.

12. (*Program Scheduling*) TV stations R and C each have a quiz show and a situation comedy to schedule for their 1 o'clock and 2 o'clock time slots. If they both schedule their quiz shows at 1 o'clock, then station R will take \$3000 in advertising revenue away from station C. If they both schedule their quiz shows at 2 o'clock, then station C will take \$2000 in advertising revenue from R. If they choose different hours for the quiz show, then R will take \$5000 in advertising from C by scheduling it at 2 o'clock, and \$2000 by scheduling it at 1 o'clock.

13. (*Card Game*) Player R has two cards: a red 5 and a black 10. Player C has three cards: a red 6, a black 7, and a black 8. They each place one of their cards on the table. If the cards are the same color, R receives the difference of the two numbers. If the cards are of different colors, C receives the minimum of the two numbers.

SOLUTIONS TO PRACTICE PROBLEMS 9.1

1. Not strictly determined. The minimum elements of the rows are

$$\begin{bmatrix} 1 & -1 & \boxed{-3} \\ 0 & \boxed{-2} & 3 \end{bmatrix};$$

the maximum elements of the columns are

$$\begin{bmatrix} \boxed{1} & \boxed{-1} & -3 \\ 0 & -2 & \boxed{3} \end{bmatrix}.$$

No element is simultaneously the minimum in its row and the maximum in its column.

2. Strictly determined. The least elements of the rows are

$$\begin{bmatrix} 1 & \boxed{-1} & 0 \\ 0 & \boxed{-4} & 5 \end{bmatrix}.$$

The largest elements of the columns are

$$\begin{bmatrix} \boxed{1} & \boxed{-1} & 0 \\ 0 & -4 & \boxed{5} \end{bmatrix}.$$

Thus -1 is a saddle point. The optimal strategy for R is to choose row 1; the optimal strategy for C is to choose column 2.

3. Not strictly determined. The minimum elements of the rows are

$$\begin{bmatrix} 1 & \boxed{-2} & 1 \\ \boxed{-2} & 1 & 1 \\ 1 & 1 & \boxed{-2} \end{bmatrix}.$$

The largest element for each column is 1. (Note that there are two choices for each largest element.) But none of the largest column elements is a least row element.

9.2 Mixed Strategies

In Section 9.1 we introduced strictly determined games and gave a method for determining optimal strategies for each player. However, not all games are strictly determined. For example, consider the game with payoff matrix

$$\begin{bmatrix} -1 & 5 \\ 2 & -3 \end{bmatrix}.$$

The minimum entries of the rows are

$$\begin{bmatrix} \boxed{-1} & 5 \\ 2 & \boxed{-3} \end{bmatrix},$$

whereas the maximum entries of the columns are

$$\begin{bmatrix} -1 & \boxed{5} \\ \boxed{2} & -3 \end{bmatrix}.$$

Note that no matrix entry is simultaneously the minimum in its row and the maximum in its column. Note also that no simple strategy of the type considered in Section 9.1 is sufficient to both maximize R's winnings and minimize C's losses. To see this, consider the game from R's point of view. Suppose that R repeatedly plays the strategy "first row," thereby attempting to win 5. After a few plays C will catch on to R's strategy and choose column 1, giving R a loss of 1. Similarly, if R consistently plays the strategy "second row," attempting to win 2, then C can thwart R by choosing the second column, to give R a loss of 3. It is clear that, in order to maximize his payoff, R should sometimes choose row 1 and sometimes row 2. One might expect that by choosing the rows on a probabilistic basis R can prevent C from anticipating his moves and amass enough positive payoffs to counteract the occasional negative ones. Thus we should investigate strategies of the type

$$A: \begin{cases} \text{Choose row 1 with probability .5} \\ \text{Choose row 2 with probability .5} \end{cases}$$

and

$$B: \begin{cases} \text{Choose row 1 with probability .9} \\ \text{Choose row 2 with probability .1.} \end{cases}$$

Such strategies are called *mixed strategies*. Either of the players can pursue such a strategy.

One way that R can carry out strategy A is to alternate between row 1 and row 2 on successive plays of the game. However, if C is at all clever, he will recognize this pattern and determine his play accordingly. What R should do is toss a coin and play row 1 whenever it lands heads and row 2 whenever it lands tails. Then there is no way that C can anticipate R's choice.

To carry out strategy B, R might use a card with a spinner attached at the center of a circle that is 90% red and 10% white. R would then determine his play by spinning the spinner and choosing row 1 if the spinner landed on the red part of the circle and row 2 if it landed on the white part. (See Fig. 1.)

It will be convenient to write mixed strategies in matrix form. Mixed strategies for R will be row matrices, and mixed strategies for C will be column matrices. Thus, for example, mixed strategy A above (for R) corresponds to the matrix

$$A: \begin{bmatrix} .5 & .5 \end{bmatrix},$$

whereas mixed strategy B (for R) corresponds to

$$B: \begin{bmatrix} .9 & .1 \end{bmatrix}.$$

The mixed strategy in which C chooses column 1 with probability .6 and column 2 with probability .4 corresponds to the column matrix

$$\begin{bmatrix} .6 \\ .4 \end{bmatrix}.$$

In comparing different mixed strategies, we use a number called their *expected value*. This number is just the average amount per game paid to R if the players pursue the given mixed strategies. The next example illustrates the computation of the expected value in a special case.

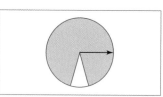

Figure 1

EXAMPLE 1   **Expected value of a mixed strategy game**   Suppose that a game has payoff matrix

$$\begin{bmatrix} -1 & 5 \\ 2 & -3 \end{bmatrix}.$$

Further suppose that R pursues the mixed strategy $\begin{bmatrix} .9 & .1 \end{bmatrix}$ and that C pursues the mixed strategy $\begin{bmatrix} .6 \\ .4 \end{bmatrix}$. Calculate the expected value of this game.

Solution   Let us view each repetition of the game as an experiment. There are four possible outcomes:

$$\text{(row 1, column 1)} \qquad \text{(row 1, column 2)}$$
$$\text{(row 2, column 1)} \qquad \text{(row 2, column 2).}$$

Let us compute the probability (relative frequency) with which each of the outcomes occurs. For example, consider the outcome (row 1, column 1). According to our assumptions about the strategies of R and C, R chooses row 1 with probability .9 and C chooses column 1 with probability .6. Since R and C make their respective choices independently of one another, the events "row 1" and "column 1" are independent. Therefore, we can compute the probability of the outcome (row 1, column 1) as follows:

$$\Pr(\text{row 1, column 1}) = \Pr(\text{row 1}) \cdot \Pr(\text{column 1}) = (.9)(.6) = .54.$$

That is, the outcome "row 1, column 1" will occur with probability .54. Similarly, we compute the probabilities of the other three outcomes. Table 1 gives the probability and the amount won by R for each outcome. The average amount that R wins per play is then given by

$$(-1)(.54) + (5)(.36) + (2)(.06) + (-3)(.04) = 1.26.$$

Hence the expected value of the given strategies is 1.26.

TABLE 1

Outcome	R wins	Probability
Row 1, column 1	-1	$(.9)(.6) = .54$
Row 1, column 2	5	$(.9)(.4) = .36$
Row 2, column 1	2	$(.1)(.6) = .06$
Row 2, column 2	-3	$(.1)(.4) = .04$

Now Try Exercise 1(a) ■

We may generalize the preceding computations. To see the pattern, let us first concentrate on 2×2 games. Suppose that a game has payoff matrix

$$\begin{bmatrix} a_{11} & a_{12} \\ a_{21} & a_{22} \end{bmatrix}.$$

Further suppose that R pursues a strategy $\begin{bmatrix} r_1 & r_2 \end{bmatrix}$. That is, R randomly chooses row 1 with probability r_1 and row 2 with probability r_2. Similarly, suppose that C pursues a strategy $\begin{bmatrix} c_1 \\ c_2 \end{bmatrix}$. Then the probabilities of the various outcomes can be tabulated as shown in Table 2. Thus, by following the reasoning used in the special case above, we see that the expected value of the strategies is the sum of the products of the payoffs times the corresponding probabilities:

$$a_{11}(r_1 c_1) + a_{12}(r_1 c_2) + a_{21}(r_2 c_1) + a_{22}(r_2 c_2).$$

TABLE 2

Outcome	Payoff to R	Probability
Row 1, column 1	a_{11}	$r_1 c_1$
Row 1, column 2	a_{12}	$r_1 c_2$
Row 2, column 1	a_{21}	$r_2 c_1$
Row 2, column 2	a_{22}	$r_2 c_2$

On the average, R gains this amount for each play. A somewhat tedious (but easy) calculation shows that

$$\begin{bmatrix} r_1 & r_2 \end{bmatrix} \begin{bmatrix} a_{11} & a_{12} \\ a_{21} & a_{22} \end{bmatrix} \begin{bmatrix} c_1 \\ c_2 \end{bmatrix} = \begin{bmatrix} r_1 & r_2 \end{bmatrix} \begin{bmatrix} a_{11}c_1 + a_{12}c_2 \\ a_{21}c_1 + a_{22}c_2 \end{bmatrix}$$

$$= \begin{bmatrix} a_{11}(r_1 c_1) + a_{12}(r_1 c_2) + a_{21}(r_2 c_1) + a_{22}(r_2 c_2) \end{bmatrix}. \quad (1)$$

Formula (1) is a special case of the following general fact:

Expected Value of a Pair of Strategies   Suppose that a game has payoff matrix

$$\begin{bmatrix} a_{11} & a_{12} & \cdots & a_{1n} \\ a_{21} & a_{22} & \cdots & a_{2n} \\ \vdots & \vdots & & \vdots \\ a_{m1} & a_{m2} & \cdots & a_{mn} \end{bmatrix}.$$

Suppose that R plays the strategy $\begin{bmatrix} r_1 & r_2 & \cdots & r_m \end{bmatrix}$ and that C plays the strategy

$$\begin{bmatrix} c_1 \\ c_2 \\ \vdots \\ c_n \end{bmatrix}.$$

Let e be the expected value of the pair of strategies. Then

$$\begin{bmatrix} r_1 & r_2 & \cdots & r_m \end{bmatrix} \begin{bmatrix} a_{11} & a_{12} & \cdots & a_{1n} \\ a_{21} & a_{22} & \cdots & a_{2n} \\ \vdots & \vdots & & \vdots \\ a_{m1} & a_{m2} & \cdots & a_{mn} \end{bmatrix} \begin{bmatrix} c_1 \\ c_2 \\ \vdots \\ c_n \end{bmatrix} = \begin{bmatrix} e \end{bmatrix}.$$

EXAMPLE 2   **Choosing the advantageous strategy**   Suppose that a game has payoff matrix

$$\begin{bmatrix} 2 & 0 & -1 \\ -1 & 3 & 4 \end{bmatrix}$$

and that R plays the strategy $\begin{bmatrix} .5 & .5 \end{bmatrix}$. Which of the following strategies is more advantageous for C?

$$A = \begin{bmatrix} .6 \\ .3 \\ .1 \end{bmatrix} \quad \text{or} \quad B = \begin{bmatrix} .3 \\ .3 \\ .4 \end{bmatrix}$$

Solution   We compare the expected value with C using strategy A to that with C using strategy B. With strategy A we have

$$\begin{bmatrix} .5 & .5 \end{bmatrix} \begin{bmatrix} 2 & 0 & -1 \\ -1 & 3 & 4 \end{bmatrix} \begin{bmatrix} .6 \\ .3 \\ .1 \end{bmatrix} = \begin{bmatrix} .90 \end{bmatrix}.$$

Using strategy B, we have

$$\begin{bmatrix} .5 & .5 \end{bmatrix} \begin{bmatrix} 2 & 0 & -1 \\ -1 & 3 & 4 \end{bmatrix} \begin{bmatrix} .3 \\ .3 \\ .4 \end{bmatrix} = \begin{bmatrix} 1.20 \end{bmatrix}.$$

Thus strategy A will yield an average payment per play of .90 to R, whereas strategy B will yield an average payment per play of 1.20. Clearly, it is to C's advantage to choose strategy A.   ■

EXAMPLE 3   **Chemical company's dumping strategy**   The Acme Chemical Corporation has two plants, each situated on the banks of the Blue River, 10 miles from one another.

A single inspector is assigned to check that the plants do not dump waste into the river. If he discovers plant A dumping waste, Acme is fined \$20,000. If he discovers plant B dumping waste, Acme is fined \$50,000. Suppose that the inspector visits one of the plants each day and that he chooses, on a random basis, to visit plant B 60% of the time. Acme schedules dumping from its two plants on a random basis, one plant per day, with plant B dumping waste on 70% of the days. How much is Acme's average fine per day?

Solution The competition between Acme and the inspector can be viewed as a non-strictly determined game, whose matrix is

$$
\begin{array}{cc}
 & \text{Inspect } A \quad \text{Inspect } B \\
\begin{matrix} \text{Plant } A \text{ dumps} \\ \text{Plant } B \text{ dumps} \end{matrix} &
\begin{bmatrix} -20{,}000 & 0 \\ 0 & -50{,}000 \end{bmatrix}.
\end{array}
$$

The strategy of Acme is given by the row matrix $\begin{bmatrix} .3 & .7 \end{bmatrix}$. The strategy of the inspector is given by the column matrix $\begin{bmatrix} .4 \\ .6 \end{bmatrix}$. The expected value of the strategies is the matrix product

$$
\begin{bmatrix} .3 & .7 \end{bmatrix}\begin{bmatrix} -20{,}000 & 0 \\ 0 & -50{,}000 \end{bmatrix}\begin{bmatrix} .4 \\ .6 \end{bmatrix} = \begin{bmatrix} -6000 & -35{,}000 \end{bmatrix}\begin{bmatrix} .4 \\ .6 \end{bmatrix}
$$

$$
= \begin{bmatrix} -23{,}400 \end{bmatrix}.
$$

In other words, Acme will be fined an average of \$23,400 per day for polluting the river.

Now Try Exercise 3 ■

In Section 9.3 we will alter payoff matrices by adding a fixed constant to each entry so that all entries become positive numbers. This does not alter the essential character of the game, in that good strategies for the original matrix will also be good strategies for the new matrix. The only difference is that the expected value is increased by the constant added. This procedure enables us to apply the methods of linear programming to the determination of optimal mixed strategies for zero-sum games that are not strictly determined.

EXAMPLE 4 **Payoffs with constant increments** In Example 1 we saw that for the game with payoff matrix

$$
\begin{bmatrix} -1 & 5 \\ 2 & -3 \end{bmatrix}
$$

and strategies $\begin{bmatrix} .9 & .1 \end{bmatrix}$, $\begin{bmatrix} .6 \\ .4 \end{bmatrix}$, the expected value was 1.26. Compute the expected value of those strategies for the matrix obtained by adding 4 to each entry.

Solution The new matrix is

$$
\begin{bmatrix} -1+4 & 5+4 \\ 2+4 & -3+4 \end{bmatrix} \quad \text{or} \quad \begin{bmatrix} 3 & 9 \\ 6 & 1 \end{bmatrix}.
$$

The expected value of the strategies for the new matrix is

$$\begin{bmatrix} .9 & .1 \end{bmatrix} \begin{bmatrix} 3 & 9 \\ 6 & 1 \end{bmatrix} \begin{bmatrix} .6 \\ .4 \end{bmatrix} = \begin{bmatrix} 3.3 & 8.2 \end{bmatrix} \begin{bmatrix} .6 \\ .4 \end{bmatrix} = \begin{bmatrix} 5.26 \end{bmatrix} .$$

As it should, the expected value has also increased by 4.   ∎

Suppose that the payoff matrix is

$$\begin{bmatrix} -\frac{1}{2} & \frac{5}{2} \\ 1 & -\frac{3}{2} \end{bmatrix} .$$

What is the expected value of the strategies $\begin{bmatrix} .9 & .1 \end{bmatrix}$ and $\begin{bmatrix} .6 \\ .4 \end{bmatrix}$? We see that

$$\begin{bmatrix} .9 & .1 \end{bmatrix} \begin{bmatrix} -\frac{1}{2} & \frac{5}{2} \\ 1 & -\frac{3}{2} \end{bmatrix} \begin{bmatrix} .6 \\ .4 \end{bmatrix} = \begin{bmatrix} .63 \end{bmatrix} .$$

We note that multiplying each element in the payoff matrix by 2 and adding 4 gives the matrix

$$\begin{bmatrix} 3 & 9 \\ 6 & 1 \end{bmatrix} ,$$

which with the preceding strategies gives the expected value 5.26 (see the solution to Example 4). Multiplying each element of the payoff matrix by 2 and adding 4 to each element produces the same effect on the expected value:

$$5.26 = 2(.63) + 4.$$

PRACTICE PROBLEMS 9.2

1. Suppose that the payoff matrix of a game is

$$\begin{bmatrix} 4 & -2 \\ -3 & 1 \end{bmatrix} .$$

Suppose that R plays the strategy $\begin{bmatrix} .6 & .4 \end{bmatrix}$. Which of the two strategies $\begin{bmatrix} .5 \\ .5 \end{bmatrix}$ or $\begin{bmatrix} .7 \\ .3 \end{bmatrix}$ is better for C?

2. Answer the question in Problem 1 for the game whose payoff matrix is

$$\begin{bmatrix} 9 & 3 \\ 2 & 6 \end{bmatrix} .$$

(This matrix is obtained by adding 5 to each entry of the matrix in Problem 1.)

EXERCISES 9.2

1. Suppose that a game has payoff matrix

$$\begin{bmatrix} 3 & -1 \\ -7 & 5 \end{bmatrix} .$$

Calculate the expected values for the following strategies and determine which of the following situations is most advantageous to R.

(a) R plays $\begin{bmatrix} .5 & .5 \end{bmatrix}$, C plays $\begin{bmatrix} .5 \\ .5 \end{bmatrix}$.

(b) R plays $\begin{bmatrix} 1 & 0 \end{bmatrix}$, C plays $\begin{bmatrix} .5 \\ .5 \end{bmatrix}$.

(c) R plays $\begin{bmatrix} .3 & .7 \end{bmatrix}$, C plays $\begin{bmatrix} .6 \\ .4 \end{bmatrix}$.

(d) R plays $\begin{bmatrix} .75 & .25 \end{bmatrix}$, C plays $\begin{bmatrix} .2 \\ .8 \end{bmatrix}$.

2. Suppose that a game has payoff matrix

$$\begin{bmatrix} 1 & 0 & 2 \\ -1 & 2 & 0 \\ 0 & -1 & -1 \end{bmatrix}.$$

Calculate the expected values for the following strategies and determine which of the following situations is most advantageous to C.

(a) R plays $\begin{bmatrix} 1 & 0 & 0 \end{bmatrix}$, C plays $\begin{bmatrix} .5 \\ .4 \\ .1 \end{bmatrix}$.

(b) R plays $\begin{bmatrix} .3 & .3 & .4 \end{bmatrix}$, C plays $\begin{bmatrix} .4 \\ .4 \\ .2 \end{bmatrix}$.

(c) R plays $\begin{bmatrix} 0 & .5 & .5 \end{bmatrix}$, C plays $\begin{bmatrix} .4 \\ 0 \\ .6 \end{bmatrix}$.

(d) R plays $\begin{bmatrix} .1 & .1 & .8 \end{bmatrix}$, C plays $\begin{bmatrix} .2 \\ .2 \\ .6 \end{bmatrix}$.

3. (*Inspector's Strategy*) Refer to Example 3. Suppose that the inspector changes his strategy and visits plant B 80% of the time. How much is Acme's average fine per day?

4. (*Inspector's Strategy*) Refer to Example 3. Suppose that the inspector visits plant B 30% of the time. How much is Acme's average fine per day?

5. (*A Letter Game*) Suppose that two players, R and C, write down letters of the alphabet. If both write vowels or both write consonants, then there is no payment to either player. If R writes a vowel and C writes a consonant, then C pays R \$2. If R writes a consonant and C writes a vowel, then R pays C \$1. Suppose that R chooses a consonant 75% of the plays and C chooses a vowel 40% of the plays. What is the average loss (or gain) of R per play?

6. (*Flood Insurance*) A small business owner must decide whether to carry flood insurance. She may insure her business for \$2 million for \$100,000, \$1 million for \$50,000, or \$.5 million for \$30,000. Her business is worth \$2 million. There is a flood serious enough to destroy her business an average of once every 10 years. In order to save insurance premiums, she decides each year on a probabilistic basis how much insurance to carry. She chooses \$2 million 20% of the time, \$1 million 20% of the time, \$.5 million 20% of the time, and no insurance 40% of the time. What is her average annual loss?

7. Two players, Robert and Carol, play a game with payoff matrix (to Robert)

$$\begin{bmatrix} 3 & -1 \\ -2 & 2 \end{bmatrix}.$$

Is the game strictly determined? Why?

Suppose Robert has strategy $\begin{bmatrix} .3 & .7 \end{bmatrix}$. The opponents agree that the game should be fair. Is it possible for this to be a fair game? What would Carol's strategy have to be?

8. Two players, Robert and Carol, play a game with payoff matrix (to Robert)

$$\begin{bmatrix} 3 & -1 \\ -2 & 2 \end{bmatrix}.$$

Is the game strictly determined? Why?

Suppose Robert has strategy $\begin{bmatrix} .7 & .3 \end{bmatrix}$. The opponents agree that the game should be fair. Is it possible for this to be a fair game? What would Carol's strategy have to be?

9. Two players, Robert and Carol, play a game with payoff matrix (to Robert)

$$\begin{bmatrix} 5 & -1 \\ -2 & 2 \end{bmatrix}.$$

Is the game strictly determined? Why?

Suppose Robert has strategy $\begin{bmatrix} .7 & .3 \end{bmatrix}$. The opponents agree that the game should be fair. Is it possible for this to be a fair game? What would Carol's strategy have to be?

10. Two players, Robert and Carol, play a game with payoff matrix (to Robert)

$$\begin{bmatrix} 5 & -1 \\ -2 & 2 \end{bmatrix}.$$

Is the game strictly determined? Why?

Suppose Robert has strategy $\begin{bmatrix} .3 & .7 \end{bmatrix}$. The opponents agree that the game should be fair. Is it possible for this to be a fair game? What would Carol's strategy have to be?

11. Assume two players, Renée and Carlos, play a game with the following payoff matrix (to Renée):

$$\begin{bmatrix} 1 & 2 & 4 \\ 1 & 0 & 5 \\ 0 & 1 & -1 \end{bmatrix}.$$

Is the game strictly determined? Determine the strategy for each player. What is the value of the game? Is the game fair?

12. The two players of Exercise 11, Renée and Carlos, play the game again, but this time the payoff matrix (to Renée) is

$$\begin{bmatrix} -3 & -2 & 6 \\ 2 & 0 & 2 \\ 5 & -2 & -4 \end{bmatrix}.$$

(a) Is the game strictly determined? Determine the strategy for each player.

(b) Is the game fair? Why?

1. If C plays $\begin{bmatrix} .5 \\ .5 \end{bmatrix}$, the expected value (to R) is

$$\begin{bmatrix} .6 & .4 \end{bmatrix} \begin{bmatrix} 4 & -2 \\ -3 & 1 \end{bmatrix} \begin{bmatrix} .5 \\ .5 \end{bmatrix} = \begin{bmatrix} 1.2 & -.8 \end{bmatrix} \begin{bmatrix} .5 \\ .5 \end{bmatrix} = \begin{bmatrix} .2 \end{bmatrix}.$$

If C plays $\begin{bmatrix} .7 \\ .3 \end{bmatrix}$, the expected value (to R) is

$$\begin{bmatrix} .6 & .4 \end{bmatrix} \begin{bmatrix} 4 & -2 \\ -3 & 1 \end{bmatrix} \begin{bmatrix} .7 \\ .3 \end{bmatrix} = \begin{bmatrix} 1.2 & -.8 \end{bmatrix} \begin{bmatrix} .7 \\ .3 \end{bmatrix} = \begin{bmatrix} .6 \end{bmatrix}.$$

Thus in the first case R gains an average of .2 per play, whereas in the second R gains .6. Since C wishes to minimize R's winnings, C should clearly play the first strategy.

2. The answer is the same as in Problem 1, since the new expected values will be 5 more than the original expected values and the first strategy will be better for C.

9.3 Determining Optimal Mixed Strategies

As we have seen, each choice of strategies by R and C results in an expected value, representing the average payoff to R per play. In this section we shall give a method for choosing the best strategies. Let us begin by clarifying our notion of optimality.

Optimal Strategy for *R*   To every choice of a strategy for R there is a best counterstrategy—that is, a strategy for C that results in the least expected value e. An *optimal mixed strategy for* R is one for which the expected value against C's best counterstrategy is as large as possible.

In a similar way we can define the optimal strategy for C.

Optimal Strategy for C   To every choice of a strategy for C there is a best counterstrategy—that is, a strategy for R that results in the largest expected value e. An *optimal mixed strategy for* C is one for which the expected value against R's best counterstrategy is as small as possible.

It is most surprising that the optimal strategies for R and C may be determined using linear programming. To see how this is done, let us consider a particular problem.

EXAMPLE 1   **Games as linear programs**   Suppose that a game has payoff matrix

$$\begin{bmatrix} 5 & 3 \\ 1 & 4 \end{bmatrix}.$$

Reduce the determination of an optimal strategy for R to a linear programming problem. *Note*: This game is not strictly determined. (Why?)

Solution   Suppose that R plays the strategy $\begin{bmatrix} r_1 & r_2 \end{bmatrix}$. What is C's best counterstrategy? If C plays $\begin{bmatrix} c_1 \\ c_2 \end{bmatrix}$, then the expected value of the game is

$$\begin{bmatrix} r_1 & r_2 \end{bmatrix} \begin{bmatrix} 5 & 3 \\ 1 & 4 \end{bmatrix} \begin{bmatrix} c_1 \\ c_2 \end{bmatrix} = \begin{bmatrix} 5r_1 + r_2 & 3r_1 + 4r_2 \end{bmatrix} \begin{bmatrix} c_1 \\ c_2 \end{bmatrix}$$

$$= \begin{bmatrix} (5r_1 + r_2)c_1 + (3r_1 + 4r_2)c_2 \end{bmatrix}.$$

If C pursues his best counterstrategy, then he will try to minimize the expected value of the game. That is, C will try to minimize

$$(5r_1 + r_2)c_1 + (3r_1 + 4r_2)c_2.$$

Since $c_1 \geq 0$, $c_2 \geq 0$, $c_1 + c_2 = 1$, this expression has as its minimum value the smaller of the terms $5r_1 + r_2$ or $3r_1 + 4r_2$. That is, if $5r_1 + r_2$ is the smaller, then C should choose the strategy $c_1 = 1$, $c_2 = 0$; whereas if $3r_1 + 4r_2$ is the smaller, then C should choose the strategy $c_1 = 0$, $c_2 = 1$. In any case the expected value of the game if C adopts his best counterstrategy is the smaller of $5r_1 + r_2$ and $3r_1 + 4r_2$. The goal of R is to maximize this expected value. In other words, the mathematical problem R faces is this:

Maximize the minimum of $5r_1 + r_2$ and $3r_1 + 4r_2$,

where $r_1 \geq 0$, $r_2 \geq 0$, $r_1 + r_2 = 1$.

Let v denote the minimum of $5r_1 + r_2$ and $3r_1 + 4r_2$. Clearly, $v > 0$. Then

$$\begin{aligned} 5r_1 + r_2 &\geq v \\ 3r_1 + 4r_2 &\geq v. \end{aligned} \tag{1}$$

Maximizing v is the same as minimizing $1/v$. Moreover, the inequalities (1) may be rewritten in the form

$$5\frac{r_1}{v} + \frac{r_2}{v} \geq 1$$

$$3\frac{r_1}{v} + 4\frac{r_2}{v} \geq 1. \tag{2}$$

Moreover, since $r_1 \geq 0$, $r_2 \geq 0$, and $r_1 + r_2 = 1$, we see that

$$\frac{r_1}{v} \geq 0, \qquad \frac{r_2}{v} \geq 0, \qquad \frac{r_1}{v} + \frac{r_2}{v} = \frac{1}{v}. \tag{3}$$

This suggests that we introduce new variables:

$$y_1 = \frac{r_1}{v}, \qquad y_2 = \frac{r_2}{v}.$$

Then (3) and (2) may be rewritten as

$$y_1 + y_2 = \frac{1}{v}$$

$$5y_1 + y_2 \geq 1$$

$$3y_1 + 4y_2 \geq 1$$

$$y_1 \geq 0, \quad y_2 \geq 0.$$

We wish to minimize $1/v$, so we may finally state our original question in terms of a linear programming problem: Minimize $y_1 + y_2$ subject to the constraints

$$\begin{cases} 5y_1 + y_2 \geq 1 \\ 3y_1 + 4y_2 \geq 1 \\ y_1 \geq 0, \quad y_2 \geq 0. \end{cases}$$

Now Try Exercise 1

In terms of the solution to this linear programming problem we may calculate R's optimal strategy as follows:

$$r_1 = vy_1 \qquad r_2 = vy_2, \qquad \text{where} \quad v = \frac{1}{y_1 + y_2}.$$

In the preceding derivation it was essential that the entries of the matrix were positive numbers, for this is how we derived that $v > 0$. The same reasoning used in Example 1 can be used in general to convert the determination of R's optimal strategy to a linear programming problem, *provided that the payoff matrix has positive entries*. If the payoff matrix does not have positive entries, then just add a large positive constant to each of the entries so as to give a matrix with positive entries. The new matrix will have the same optimal strategy as the original one. However, since all its entries are positive, we may use the previous reasoning to reduce determination of the optimal strategy to a linear programming problem.

Optimal Strategy for R   Let the payoff matrix of a game be

$$\begin{bmatrix} a_{11} & a_{12} & \cdots & a_{1n} \\ a_{21} & a_{22} & \cdots & a_{2n} \\ \vdots & \vdots & & \vdots \\ a_{m1} & a_{m2} & \cdots & a_{mn} \end{bmatrix},$$

where all entries of the matrix are positive numbers. Let $y_1, y_2, \ldots, y_m$ be chosen so as to minimize

$$y_1 + y_2 + \cdots + y_m$$

subject to the constraints

$$\begin{cases} y_1 \geq 0,\ y_2 \geq 0,\ \ldots,\ y_m \geq 0 \\ a_{11}y_1 + a_{21}y_2 + \cdots + a_{m1}y_m \geq 1 \\ a_{12}y_1 + a_{22}y_2 + \cdots + a_{m2}y_m \geq 1 \\ \qquad\qquad \vdots \\ a_{1n}y_1 + a_{2n}y_2 + \cdots + a_{mn}y_m \geq 1. \end{cases}$$

Let

$$v = \frac{1}{y_1 + y_2 + \cdots + y_m}.$$

Then an optimal strategy for R is $\begin{bmatrix} r_1 & r_2 & \cdots & r_m \end{bmatrix}$, where

$$r_1 = vy_1, \quad r_2 = vy_2, \quad \ldots, \quad r_m = vy_m.$$

Furthermore, if C adopts the best counterstrategy, then the expected value is v.

Note that the determination of $y_1, y_2, \ldots, y_m$ is a linear programming problem whose solution can be obtained using either the method of Chapter 3 (if $m = 2$) or the simplex method of Chapter 4 (any m). The next example illustrates the preceding result.

EXAMPLE 2   **Finding the optimal strategy for R**   Suppose that a game has payoff matrix

$$\begin{bmatrix} 5 & 3 \\ 1 & 4 \end{bmatrix}.$$

(a) Determine an optimal strategy for R.

(b) Determine the expected payoff to R if C uses the best counterstrategy.

Solution (a) The associated linear programming problem asks us to minimize $y_1 + y_2$ subject to the constraints

$$\begin{cases} y_1 \geq 0, \quad y_2 \geq 0 \\ 5y_1 + \ y_2 \geq 1 \\ 3y_1 + 4y_2 \geq 1. \end{cases}$$

In Fig. 1 we have sketched the feasible set for this problem and evaluated the objective function at each vertex.

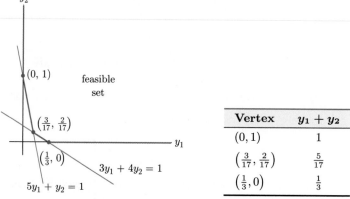

Vertex	$y_1 + y_2$
$(0, 1)$	1
$\left(\frac{3}{17}, \frac{2}{17}\right)$	$\frac{5}{17}$
$\left(\frac{1}{3}, 0\right)$	$\frac{1}{3}$

Figure 1

The minimum value of $y_1 + y_2$ is $\frac{5}{17}$ and occurs when $y_1 = \frac{3}{17}$ and $y_2 = \frac{2}{17}$. Further,

$$v = \frac{1}{y_1 + y_2} = \frac{1}{\frac{5}{17}} = \frac{17}{5}$$

$$r_1 = v y_1 = \frac{17}{5} \cdot \frac{3}{17} = \frac{3}{5}$$

$$r_2 = v y_2 = \frac{17}{5} \cdot \frac{2}{17} = \frac{2}{5}.$$

Thus the optimal strategy for R is $\begin{bmatrix} \frac{3}{5} & \frac{2}{5} \end{bmatrix}$.

(b) The expected value against the best counterstrategy is $v = \frac{17}{5}$.   ■

There is a similar linear programming technique for determining the optimal strategy for C.

Optimal Strategy for C   Let the payoff matrix of a game be

$$\begin{bmatrix} a_{11} & a_{12} & \cdots & a_{1n} \\ a_{21} & a_{22} & \cdots & a_{2n} \\ \vdots & \vdots & & \vdots \\ a_{m1} & a_{m2} & \cdots & a_{mn} \end{bmatrix},$$

where all entries of the matrix are positive numbers. Let $z_1, z_2, \ldots, z_n$ be chosen so as to maximize

$$z_1 + z_2 + \cdots + z_n$$

subject to the constraints

$$\begin{cases} z_1 \geq 0, \quad z_2 \geq 0, \quad \ldots, \quad z_n \geq 0 \\ a_{11}z_1 + a_{12}z_2 + \cdots + a_{1n}z_n \leq 1 \\ a_{21}z_1 + a_{22}z_2 + \cdots + a_{2n}z_n \leq 1 \\ \qquad\qquad\qquad \vdots \\ a_{m1}z_1 + a_{m2}z_2 + \cdots + a_{mn}z_m \leq 1. \end{cases}$$

Let $v = 1/(z_1 + z_2 + \cdots + z_n)$. Then an optimal strategy for C is

$$\begin{bmatrix} c_1 \\ c_2 \\ \vdots \\ c_n \end{bmatrix},$$

where $c_1 = vz_1, c_2 = vz_2, \ldots, c_n = vz_n$.

EXAMPLE 3   **Finding the optimal strategy for C**   Determine the optimal strategy for C for the game with payoff matrix

$$\begin{bmatrix} 5 & 3 \\ 1 & 4 \end{bmatrix}.$$

Solution   We must maximize $z_1 + z_2$ subject to the constraints

$$\begin{cases} z_1 \geq 0, \quad z_2 \geq 0 \\ 5z_1 + 3z_2 \leq 1 \\ z_1 + 4z_2 \leq 1. \end{cases}$$

The solution is as follows: The maximum value is $\frac{5}{17}$ and it occurs when $z_1 = \frac{1}{17}$, $z_2 = \frac{4}{17}$. Therefore, $v = \frac{17}{5}$, and the optimal strategy for C is $\begin{bmatrix} c_1 \\ c_2 \end{bmatrix}$, where

$$c_1 = vz_1 = \tfrac{17}{5} \cdot \tfrac{1}{17} = \tfrac{1}{5}$$

$$c_2 = vz_2 = \tfrac{17}{5} \cdot \tfrac{4}{17} = \tfrac{4}{5}. \tag{4}$$

Now Try Exercise 3   ■

Notice that in both Examples 2 and 3 we obtained $v = \frac{17}{5}$. This was not just a coincidence. This phenomenon always occurs, and the number v is called the *value* of the game. An easy computation shows that for the matrix of Examples 2 and

3, when R and C each use their optimal strategies, the expected value is $v = \frac{17}{5}$. That is,

$$\begin{bmatrix} \frac{3}{5} & \frac{2}{5} \end{bmatrix} \begin{bmatrix} 5 & 3 \\ 1 & 4 \end{bmatrix} \begin{bmatrix} \frac{1}{5} \\ \frac{4}{5} \end{bmatrix} = \begin{bmatrix} \frac{17}{5} \end{bmatrix}.$$

Let us briefly reconsider the calculations of optimal strategies for R and C. We begin in every case with the matrix for the game, A. Let us assume that A is an $m \times n$ matrix and that each entry of A is a positive number. To find the optimal strategy for C, we find the matrix Z that maximizes the objective function EZ subject to the constraints $AZ \le B$ and $Z \ge \mathbf{0}$, where

$$E = \underbrace{\begin{bmatrix} 1 & 1 & 1 & \cdots & 1 \end{bmatrix}}_{n \text{ entries}}$$

$$Z = \begin{bmatrix} z_1 \\ z_2 \\ \vdots \\ z_n \end{bmatrix} \quad \text{and} \quad B = \left. \begin{bmatrix} 1 \\ 1 \\ \vdots \\ 1 \end{bmatrix} \right\} m \text{ entries.}$$

The dual of the linear programming problem associated with finding an optimal strategy for C is to find the matrix Y that minimizes the objective function $B^T Y$ subject to the constraints $A^T Y \ge E^T$ and $Y \ge \mathbf{0}$, where

$$Y = \begin{bmatrix} y_1 \\ y_2 \\ \vdots \\ y_m \end{bmatrix}.$$

But this is exactly the problem of finding an optimal strategy for R.

> The problem of finding the optimal strategy for R is a linear programming problem whose dual is the problem of finding an optimal strategy for C, and vice versa.

Our previous work with duality leads us to conclude the following:

1. If there exists an optimal strategy for R, then there exists an optimal strategy for C, and vice versa.

2. The minimum of the objective function $y_1 + y_2 + \cdots + y_m$ and the maximum of the objective function $z_1 + z_2 + \cdots + z_n$ are equal—since the linear programming problems are duals of each other. Hence, the value of the optimal strategy for C is the same as the value of the optimal strategy for R, that is,

$$\frac{1}{z_1 + z_2 + \cdots + z_n} = \frac{1}{y_1 + y_2 + \cdots + y_m}.$$

EXAMPLE 4 **Optimal strategies when matrix entries are not positive** Determine the optimal strategies for R and C for the game with payoff matrix

$$\begin{bmatrix} 1 & -1 \\ -3 & 0 \end{bmatrix}.$$

Solution We cannot apply our technique directly, since only one of the entries of the given matrix is a positive number. However, if we add 4 to each entry, then the new matrix will be

$$\begin{bmatrix} 5 & 3 \\ 1 & 4 \end{bmatrix},$$

which does have all positive entries. These two payoff matrices have the same optimal strategies. The only difference is that the values in the new matrix are 4 more than that of the given matrix. Now, the optimal strategies and the value of the new matrix were found in Examples 2 and 3 to be

$$\begin{bmatrix} \frac{3}{5} & \frac{2}{5} \end{bmatrix}, \qquad \begin{bmatrix} \frac{1}{5} \\ \frac{4}{5} \end{bmatrix}, \qquad \text{and} \qquad \frac{17}{5}.$$

Therefore, the optimal strategies for the given matrix are

$$\begin{bmatrix} \frac{3}{5} & \frac{2}{5} \end{bmatrix} \qquad \text{and} \qquad \begin{bmatrix} \frac{1}{5} \\ \frac{4}{5} \end{bmatrix},$$

and the value is $\frac{17}{5} - 4 = -\frac{3}{5}$.

Now Try Exercise 5

■

EXAMPLE 5 **Finding optimal strategies by the simplex method** Use the simplex method and the resulting tableau to determine the optimal strategies for the game of Example 4.

Solution As in Example 4, add 4 to each entry to get a matrix with positive entries, and set up the tableau for finding the optimal strategy for C. (We choose this linear programming problem because it is a maximization problem.) The transformed matrix A is

$$\begin{bmatrix} 5 & 3 \\ 1 & 4 \end{bmatrix}.$$

To find the optimal strategy for C, we need to find the values of z_1 and z_2 that maximize $z_1 + z_2$ subject to the constraints

$$\begin{cases} 5z_1 + 3z_2 \leq 1 \\ z_1 + 4z_2 \leq 1 \\ z_1 \geq 0, \quad z_2 \geq 0. \end{cases}$$

We set up the tableau using slack variables t and u and display the initial and final tableaux:

	z_1	z_2	t	u	M	
t	5	3	1	0	0	1
u	1	4	0	1	0	1
M	-1	-1	0	0	1	0

	z_1	z_2	t	u	M	
z_1	1	0	$\frac{4}{17}$	$-\frac{3}{17}$	0	$\frac{1}{17}$
z_2	0	1	$-\frac{1}{17}$	$\frac{5}{17}$	0	$\frac{4}{17}$
M	0	0	$\frac{3}{17}$	$\frac{2}{17}$	1	$\frac{5}{17}$

Thus the solution is $z_1 = \frac{1}{17}$, $z_2 = \frac{4}{17}$ with $M = z_1 + z_2 = \frac{5}{17}$. Then $v = \frac{17}{5}$, and the optimal strategy for C is

$$\begin{bmatrix} vz_1 \\ vz_2 \end{bmatrix} = \begin{bmatrix} \frac{1}{5} \\ \frac{4}{5} \end{bmatrix}.$$

This agrees with previous solutions, and the value of the matrix is $\frac{17}{5}$, which is 4 more than the original matrix. Thus the value of the game is $\frac{17}{5} - 4 = -\frac{3}{5}$.

But the optimal strategy for R can be read from the final tableau since y_1 and y_2 are the values of the variables in the dual of the problem we solved. So $y_1 = t = \frac{3}{17}$, $y_2 = u = \frac{2}{17}$, and $M = \frac{5}{17}$. Then $v = \frac{17}{5}$ and the optimal strategy for R is $\begin{bmatrix} vy_1 & vy_2 \end{bmatrix} = \begin{bmatrix} \frac{3}{5} & \frac{2}{5} \end{bmatrix}$.

Now Try Exercise 11 ∎

We actually have a very useful fact.

Fundamental Theorem of Game Theory   Every two-person zero-sum game has a solution.

Verification of the Fundamental Theorem of Game Theory If the given two-person game has a saddle point, then the game is strictly determined, and optimal strategies for R and C are given by the position of the saddle point.

If the game is not strictly determined, then let us assume that the $m \times n$ payoff matrix A has only positive entries. We let B be an $m \times 1$ column matrix in which each entry is 1, and let E be a $1 \times n$ row matrix of 1's. Then

1. There is an optimal feasible solution to the problem:

$$\text{Maximize } M = EZ \text{ subject to } AZ \leq B \text{ and } Z \geq \mathbf{0}. \qquad \text{(P)}$$

2. There is an optimal feasible solution to the problem:

$$\text{Minimize } M = B^T Y \text{ subject to } A^T Y \geq E^T \text{ and } Y \geq \mathbf{0}. \qquad \text{(D)}$$

3. The solutions to (P) and (D) give a solution to the game.

To see that characteristics 1 and 2 hold, we note that there is a feasible solution for the inequalities of the primal problem (P). The $n \times 1$ matrix of zeros, $Z = \mathbf{0}$, satisfies $AZ \leq B$. Also, there is a feasible solution for the inequalities of the dual problem (D). This can be seen by noting that since every element of the matrix A is positive, we can find an $m \times 1$ matrix $Y \geq \mathbf{0}$ with sufficiently large entries to guarantee that $A^T Y \geq E^T$. Since the inequalities of both (P) and (D) have a feasible solution, the fundamental theorem of duality (Chapter 4) tells us that both (P) and (D) have optimal feasible solutions.

Let Z^* and Y^* be optimal feasible solutions of (P) and (D), respectively. Say that

$$Z^* = \begin{bmatrix} z_1^* \\ z_2^* \\ \vdots \\ z_n^* \end{bmatrix} \quad \text{and} \quad Y^* = \begin{bmatrix} y_1^* \\ y_2^* \\ \vdots \\ y_m^* \end{bmatrix}.$$

The maximum for (P),

$$M = z_1^* + z_2^* + \cdots + z_n^*,$$

equals the minimum for (D),

$$M = y_1^* + y_2^* + \cdots + y_m^*,$$

and

$$AZ^* \leq B \quad \text{and} \quad A^T Y^* \geq E^T.$$

Recall that B is an $m \times 1$ matrix of 1's and E^T is an $n \times 1$ matrix of 1's.

M must be strictly greater than zero since at least one of the y_i^* must be >0 in order for $A^T Y \geq E^T$ to hold. Therefore, $1/M$ is defined. We let

$$C = \begin{bmatrix} \dfrac{1}{M} z_1^* \\ \dfrac{1}{M} z_2^* \\ \vdots \\ \dfrac{1}{M} z_n^* \end{bmatrix} = \begin{bmatrix} c_1 \\ c_2 \\ \vdots \\ c_n \end{bmatrix}$$

and

$$R = \begin{bmatrix} \dfrac{1}{M} y_1^* & \dfrac{1}{M} y_2^* & \cdots & \dfrac{1}{M} y_m^* \end{bmatrix} = \begin{bmatrix} r_1 & r_2 & \cdots & r_m \end{bmatrix}.$$

Furthermore, C and R represent optimal strategies for players C and R, respectively. To verify that C and R are legitimate strategies, we note that since $M > 0$, $Z^* \geq \mathbf{0}$, and $Y^* \geq \mathbf{0}$, every entry in C and R is nonnegative. We only need to check that

$$c_1 + c_2 + \cdots + c_n = 1 \quad \text{and} \quad r_1 + r_2 + \cdots + r_m = 1.$$

This follows directly from the definitions of M, C, and R. ∎

PRACTICE PROBLEMS 9.3

1. Determine the optimal strategy for C for the game with payoff matrix

$$\begin{bmatrix} 2 & 14 \\ 6 & 12 \\ 8 & 6 \end{bmatrix}.$$

2. Determine by inspection the optimal strategies for C for the games whose payoff matrices are given.

(a) $\begin{bmatrix} 0 & 12 \\ 4 & 10 \\ 6 & 4 \end{bmatrix}$ (b) $\begin{bmatrix} 6 & 0 \\ 4 & 3 \\ 8 & -1 \end{bmatrix}$

EXERCISES 9.3

1. Suppose that a game has payoff matrix

$$\begin{bmatrix} 1 & 6 \\ 4 & 3 \end{bmatrix}.$$

Reduce the determination of an optimal strategy for R into a linear programming problem. Just set up the problem showing the constraints and the objective function.

2. Suppose a game has a payoff matrix

$$\begin{bmatrix} 10 & 6 \\ 5 & 7 \end{bmatrix}.$$

Reduce the determination of an optimal strategy for R into a linear programming problem. Just set up the problem showing the constraints and the objective function.

In Exercises 3–8, determine optimal strategies for R and for C for the games whose payoff matrices are given.

3. $\begin{bmatrix} 2 & 4 \\ 5 & 3 \end{bmatrix}$ **4.** $\begin{bmatrix} 2 & 3 \\ 3 & 2 \end{bmatrix}$

5. $\begin{bmatrix} 3 & -6 \\ -5 & 4 \end{bmatrix}$ **6.** $\begin{bmatrix} 5 & 2 \\ 7 & 1 \end{bmatrix}$

7. $\begin{bmatrix} 4 & 1 \\ 2 & 4 \end{bmatrix}$ **8.** $\begin{bmatrix} 5 & -8 \\ 3 & 6 \end{bmatrix}$

In Exercises 9 and 10, determine optimal strategies for R for the games whose payoff matrices are given.

9. $\begin{bmatrix} 3 & 5 & -1 \\ 4 & -1 & 6 \end{bmatrix}$ **10.** $\begin{bmatrix} -2 & 1 & 0 \\ 2 & 0 & 1 \end{bmatrix}$

In Exercises 11 and 12, determine optimal strategies for C for the games whose payoff matrices are given.

11. $\begin{bmatrix} -3 & 1 \\ 4 & -1 \\ 1 & 0 \end{bmatrix}$ **12.** $\begin{bmatrix} 0 & 2 \\ 2 & -1 \\ 1 & 0 \end{bmatrix}$

13. Two players, Renée and Carlos, play a game with payoff matrix

$$\begin{bmatrix} 5 & -3 \\ -3 & 1 \end{bmatrix}.$$

Is the game strictly determined? Determine the optimal mixed strategy for each player. What is the value of the game? Explain.

14. Determine the optimal mixed strategy for each player, given the payoff matrix

$$\begin{bmatrix} 4 & -2 \\ -3 & 1 \end{bmatrix}.$$

Give the value of the game.

15. (*Smuggler's Strategy*) A rumrunner attempts to smuggle rum into a country having two ports. Each day the coast guard is able to patrol only one of the ports. If the rumrunner enters via an unpatrolled port, he will be able to sell his rum for a profit of $7000. If he enters the first port and it is patrolled that day, he is certain to be caught and will have his rum (worth $1000) confiscated and be fined $1000. If he enters the second port (which is big and crowded) and it is patrolled that day, he will have time to jettison his cargo and thereby escape a fine.

  (a) What is the optimal strategy for the rumrunner?

  (b) What is the optimal strategy for the coast guard?

  (c) How profitable is rumrunning? That is, what is the value of the game?

16. (*Which Hand?*) Ralph puts a coin in one of his hands and Carl tries to guess which hand holds the coin. If Carl guesses incorrectly, he must pay Ralph $2. If Carl guesses correctly, then Ralph must pay him $3 if the coin was in the left hand and $1 if it was in the right.

  (a) What is the optimal strategy for Ralph?

  (b) What is the optimal strategy for Carl?

  (c) Whom does this game favor?

17. (*Advertising Strategies*) The Carter Company can choose between two advertising strategies (I and II). Its most important competitor, Rosedale Associates, has a choice of three advertising strategies (a, b, c). The estimated payoff to Rosedale Associates away from the Carter Company is given by the payoff matrix

$$\begin{array}{c} \\ a \\ b \\ c \end{array} \begin{array}{cc} \text{I} & \text{II} \\ \begin{bmatrix} -2 & 1 \\ 2 & -3 \\ 1 & -2 \end{bmatrix}, \end{array}$$

where the entries represent thousands of dollars per week. Determine the optimal strategies for each company.

SOLUTIONS TO PRACTICE PROBLEMS 9.3

1. The associated linear programming problem is:
Maximize $z_1 + z_2$ subject to the constraints

$$\begin{cases} z_1 \geq 0, \quad z_2 \geq 0. \\ 2z_1 + 14z_2 \leq 1 \\ 6z_1 + 12z_2 \leq 1 \\ 8z_1 + 6z_2 \leq 1. \end{cases}$$

In Fig. 2 we have sketched the feasible set and evaluated the objective function at each vertex. The maximum value of $z_1 + z_2$ is $\frac{4}{30}$, which is achieved at the vertex $\left(\frac{3}{30}, \frac{1}{30}\right)$. Therefore,

$$v = \frac{1}{z_1 + z_2} = \frac{1}{\frac{4}{30}} = \frac{30}{4}$$

$$c_1 = v \cdot z_1 = \frac{30}{4} \cdot \frac{3}{30} = \frac{3}{4}$$

$$c_2 = v \cdot z_2 = \frac{30}{4} \cdot \frac{1}{30} = \frac{1}{4}.$$

That is, the optimal strategy for C is $\begin{bmatrix} \frac{3}{4} \\ \frac{1}{4} \end{bmatrix}$.

2. (a) $\begin{bmatrix} \frac{3}{4} \\ \frac{1}{4} \end{bmatrix}$. If we add 2 to each entry, we obtain the payoff matrix of Problem 1, so these two games have the same optimal strategies.

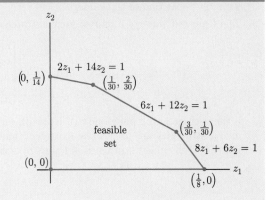

Vertex	$z_1 + z_2$
$(0,0)$	0
$\left(0, \frac{1}{14}\right)$	$\frac{1}{14}$
$\left(\frac{1}{30}, \frac{2}{30}\right)$	$\frac{3}{30}$
$\left(\frac{3}{30}, \frac{1}{30}\right)$	$\frac{4}{30}$
$\left(\frac{1}{8}, 0\right)$	$\frac{1}{8}$

Figure 2

(b) Always play column 2. This game is strictly determined and has the entry 3 as saddle point. It is a good idea always to check for a saddle point before looking for a mixed strategy.

CHAPTER 9 SUMMARY

1. A *zero-sum* game is one in which a payoff to one player results in a loss of the same amount to the other player.

2. The entry in the ith row and jth column of a payoff matrix gives the payoff to the row player (equivalently, the loss to the column player) when the row player chooses row i and the column player chooses column j.

3. A *pure strategy* is one in which the player consistently chooses the same row or column. Strategies involving varied moves are called *mixed strategies.*

4. The *optimal pure strategy* for the row player is to choose the row whose least element is maximal. The optimal pure strategy for the column player is to choose the column whose greatest element is minimal.

5. A *saddle point* is an entry in a payoff matrix that is simultaneously the least element of its row and the greatest element of its column. A game need not have

a saddle point. If a game has more than one saddle point, then the saddle points are equal.

6. A game with a saddle point is called a *strictly determined game.* In a strictly determined game, the optimal pure strategy for each player is to choose a row or column containing a saddle point.

7. In a strictly determined game, if both players use optimal pure strategies, then the saddle point gives the payoff to the row player. The value of the saddle point is called the *value of the game.*

8. If a game is not strictly determined, then the players should use mixed strategies. A *mixed strategy* for the row player is a row matrix whose ith entry is the probability that the row player will choose row i on any repetition of the game. A mixed strategy for the column player is a column matrix whose jth entry is the

probability that the column player will choose column j on any repetition of the game.

9. The expected value of a pair of mixed strategies

$$R = \begin{bmatrix} r_1 & r_2 & \cdots & r_m \end{bmatrix} \quad \text{and} \quad C = \begin{bmatrix} c_1 \\ c_2 \\ \vdots \\ c_n \end{bmatrix}$$

is the average payoff per game to the row player if these mixed strategies are used. The expected value of the pair R, C of mixed strategies is found by computing

the product RAC, where A is the payoff matrix for the game.

10. An *optimal mixed strategy* for the row player is one for which the column player's best counterstrategy results in the greatest possible expected value. Similarly, an optimal mixed strategy for the column player is one for which the row player's best counterstrategy results in the least possible expected value. Optimal mixed strategies for each player are found by solving a pair of dual linear programming problems, as described on pages 471–473.

REVIEW OF FUNDAMENTAL CONCEPTS

1. What do the individual entries of a payoff matrix represent?

2. What is the difference between a pure strategy and a mixed strategy?

3. What is a zero-sum game?

4. Describe the optimal pure strategies for R and for C.

5. When is an entry of a payoff matrix a saddle point?

6. What is a strictly determined game and what is its value?

7. What is the expected value of a pair of mixed strategies, and how is it computed?

8. What is meant by the optimal mixed strategies of R and C, and how are they computed?

KEY FORMULAS

Optimal Pure Strategy for R:

1. For each row of the payoff matrix, determine the least element.

2. Choose the row for which this is as large as possible.

Optimal Pure Strategy for C:

1. For each column of the payoff matrix, determine the largest element.

2. Choose the column for which this is as small as possible.

Value of a Game: If the game is strictly determined, there is a saddle point and the saddle point is $v = $ value of the game.

Let

$$A = \begin{bmatrix} a_{11} & a_{12} & \cdots & a_{1n} \\ a_{21} & a_{22} & \cdots & a_{2n} \\ \vdots & \vdots & & \vdots \\ a_{m1} & a_{m2} & \cdots & a_{mn} \end{bmatrix}$$

be the payoff matrix of a game.

Expected Value of a Game with Strategies

$$R = \begin{bmatrix} r_1 & r_2 & \cdots & r_m \end{bmatrix} \quad \text{and} \quad C = \begin{bmatrix} c_1 \\ c_2 \\ \vdots \\ c_n \end{bmatrix} :$$

$$\begin{bmatrix} e \end{bmatrix} = RAC$$

In the linear programming form of the mixed strategy problem for a game with payoff matrix A, if the solution is $y_1, y_2, \ldots, y_m$ for the minimization problem, then the value

$$v = \frac{1}{y_1 + y_2 + \cdots + y_m}$$

and the optimal strategy for R is $r_1 = vy_1$, $r_2 = vy_2$, $\ldots, r_m = vy_m$.

If the solution is $z_1, z_2, \ldots, z_n$ for the maximization problem, then the optimal strategy for C is $c_1 = vz_1$, $c_2 = vz_2$, $\ldots, c_n = vz_n$.

SUPPLEMENTARY EXERCISES

In Exercises 1–4, state whether or not the games having the given payoff matrices are strictly determined. If so, give the optimal pure strategies and the values of the strategies.

1. $\begin{bmatrix} 5 & -1 & 1 \\ -3 & 5 & 1 \\ 4 & 3 & 2 \end{bmatrix}$ **2.** $\begin{bmatrix} 1 & 2 & 3 \\ 3 & 2 & 1 \end{bmatrix}$

3. $\begin{bmatrix} 0 & 1 \\ 1 & 0 \\ 2 & -1 \end{bmatrix}$ **4.** $\begin{bmatrix} 2 & 1 & 2 \\ -1 & 0 & 3 \\ 4 & 1 & -4 \end{bmatrix}$

In Exercises 5–8, determine the expected value of each pair of mixed strategies for the given payoff matrix.

5. $\begin{bmatrix} \frac{3}{4} & \frac{1}{4} \end{bmatrix}$, $\begin{bmatrix} \frac{1}{3} \\ \frac{2}{3} \end{bmatrix}$; $\begin{bmatrix} 0 & 24 \\ 12 & -36 \end{bmatrix}$

6. $\begin{bmatrix} \frac{1}{2} & \frac{1}{2} \end{bmatrix}$, $\begin{bmatrix} \frac{1}{3} \\ \frac{1}{3} \\ \frac{1}{3} \end{bmatrix}$; $\begin{bmatrix} -6 & 6 & 0 \\ 0 & -12 & 24 \end{bmatrix}$

7. $\begin{bmatrix} .2 & .3 & .5 \end{bmatrix}$, $\begin{bmatrix} .4 \\ .6 \end{bmatrix}$; $\begin{bmatrix} 1 & 0 \\ -3 & 1 \\ 0 & 5 \end{bmatrix}$

8. $\begin{bmatrix} .1 & .1 & .8 \end{bmatrix}$, $\begin{bmatrix} .4 \\ .3 \\ .3 \end{bmatrix}$; $\begin{bmatrix} 0 & 1 & 3 \\ -1 & 0 & 2 \\ -3 & -2 & 0 \end{bmatrix}$

Determine the optimal strategies for R and for C for the games with the payoff matrices of Exercises 9 and 10.

9. $\begin{bmatrix} -3 & 4 \\ 2 & -2 \end{bmatrix}$ **10.** $\begin{bmatrix} 3 & -6 \\ -4 & 4 \end{bmatrix}$

11. Determine the optimal strategy for R for the game with payoff matrix

$$\begin{bmatrix} 5 & -2 & 0 \\ 1 & 4 & 1 \end{bmatrix}.$$

12. Determine the optimal strategy for C for the game with payoff matrix

$$\begin{bmatrix} 1 & 3 \\ 3 & 1 \\ 4 & 2 \end{bmatrix}.$$

13. (*A Card Matching Game*) Ruth and Carol play the following game. Both have two cards, a two and a six. Each puts one of her cards on the table. If both put down the same denomination, Ruth pays Carol $3. Otherwise, Carol pays Ruth as many dollars as the denomination of Carol's card.

(a) Find the optimal strategies for Ruth and Carol.

(b) Whom does this game favor?

14. (*Investment Strategy*) An investor is considering purchasing one of three stocks. Stock A is regarded as conservative, stock B as speculative, and stock C as highly risky. If the economic growth during the coming year is strong, then stock A should increase in value by $3000, stock B by $6000, and stock C by $15,000. If the economic growth during the next year is average, then stock A should increase in value by $2000, stock B by $2000, and stock C by $1000. If the economic growth is weak, then stock A should increase in value by $1000 and stocks B and C decrease in value by $3000 and $10,000, respectively.

(a) Set up the 3×3 payoff matrix showing the investing gains for the possible stock purchases and levels of economic growth.

(b) What is the investor's optimal strategy?

Conceptual Exercises

15. Show that in a two-person game with payoff matrix

$$\begin{bmatrix} a_{11} & a_{12} \\ a_{21} & a_{22} \end{bmatrix}$$

the only games that are *not* strictly determined are those for which either

$$a_{11} > a_{12}, \quad a_{11} > a_{21}, \quad a_{21} > a_{22}, \quad a_{12} > a_{22}$$

or

$$a_{11} < a_{12}, \quad a_{11} < a_{21}, \quad a_{21} > a_{22}, \quad a_{12} > a_{22}$$

16. In Practice Problem 1 in Section 9.3, we determined the optimal strategy for C in a game with payoff matrix

$$\begin{bmatrix} 2 & 14 \\ 6 & 12 \\ 8 & 6 \end{bmatrix}$$

to be

$$\begin{bmatrix} 3/4 \\ 1/4 \end{bmatrix}.$$

Consider the effect of changing the entry in the first row, first column of the payoff matrix to $2 + h$, where h is any positive integer. For what values of h does the optimal strategy for C remain the same?

17. Assume the payoff matrix for a two-person game is given by

$$\begin{bmatrix} a_{11} & a_{12} \\ a_{21} & a_{22} \end{bmatrix}.$$

Then assume that R uses a strategy $\begin{bmatrix} r & 1-r \end{bmatrix}$. If C chooses column 1, then R can expect a return of $a_{11}r + a_{21}(1-r)$. If C chooses column 2, then R can expect a return of $a_{12}r + a_{22}(1-r)$.

(a) Explain why the intersection of the lines

$$y = a_{11}r + a_{21}(1 - r)$$

and

$$y = a_{12}r + a_{22}(1 - r)$$

gives the value of r that is best for R.

(b) How would you find the best strategy for C using a comparable argument?

18. Using the technique of Exercise 17, find a general formula for the best strategy for R in terms of the entries in the payoff matrix.

CHAPTER TEST

1. Consider a game with payoff matrix $\begin{bmatrix} 5 & -1 & 6 \\ -2 & 5 & -8 \\ 8 & -10 & 5 \end{bmatrix}$.

    (a) What is the meaning of the entry 6 in the upper right?

    (b) What is the meaning of the entry -10 in the bottom row?

    (c) Is this game strictly determined? Why or why not?

2. Describe the difference between a pure strategy and a mixed strategy.

3. Each of the following is a payoff matrix for a strictly determined game. Give the value of the game and the optimal pure strategies for each player.

    (a) $\begin{bmatrix} -1 & 0 \\ 2 & -2 \\ 3 & 2 \end{bmatrix}$

    (b) $\begin{bmatrix} 0 & 2 & 4 \\ -4 & 0 & 3 \\ -2 & -3 & 0 \end{bmatrix}$

4. (*A Numbers Game*) Consider the following game. Each player chooses a number from 1 to 5. If the row player's number is higher, then he wins. If the column player's number is higher or if the two numbers match, then

the column player wins. The loser must pay the winner as many dollars as the number chosen by the row player. Construct a payoff matrix for this game. Is the game strictly determined? If so, what is the value of the game? Which player would you rather be?

5. Suppose a game has payoff matrix $\begin{bmatrix} 2 & -2 & 3 \\ -1 & 1 & -3 \end{bmatrix}$.

    (a) Suppose the row player uses the mixed strategy $R = \begin{bmatrix} .4 & .6 \end{bmatrix}$ and the column player uses the mixed strategy $C = \begin{bmatrix} .3 \\ .5 \\ .2 \end{bmatrix}$. Calculate and interpret the expected value for this pair of mixed strategies.

    (b) If the row player uses the mixed strategy from part (a), then is the column player better off using the mixed strategy from part (a) or the mixed strategy $C = \begin{bmatrix} .4 \\ .1 \\ .5 \end{bmatrix}$?

6. Consider a game with payoff matrix $\begin{bmatrix} 0 & -1 \\ -4 & 5 \end{bmatrix}$.

    (a) Determine optimal mixed strategies for each player.

    (b) Whom does this game favor?

CHAPTER 9 PROJECT

Simulating the Outcomes of Mixed-Strategy Games

A mixed strategy requires the use of a device that will randomly select a row (or column) of the payoff matrix subject to a specified probability. For instance, if the strategy for R is $\begin{bmatrix} .6 & .4 \end{bmatrix}$, the device should select the first row 60% of the time and select the second row 40% of the time. The TI-83, TI-83+, and TI-84+ graphing calculators are well-suited to handle this task using the *randInt* function along with the *relational operators*.

Each time **randInt(1,10)** in MATH PRB menu is called, an integer between 0 and 10 is generated. See Fig. 1. A relational operator (found in 2nd [TEST]), often an inequality statement, returns the value 1 when true and the value 0 when false. See Fig. 2. The instructions in Fig. 3 are used to select a row from a matrix with two rows.

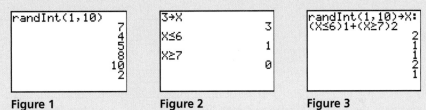

Figure 1	**Figure 2**	**Figure 3**

1. Explain why 60% of the time the instructions in Fig. 3 return a 1 and 40% of the time they return a 2. Type the instructions into your calculator, press the ENTER key twenty times, and count the number of 1's and 2's. Are there approximately twelve 1's and eight 2's?

2. Explain why the instructions in Fig. 3 should not be replaced with

$$\textbf{(randInt(1,10)} \leq \textbf{6)1+(randInt(1,10)} \geq \textbf{7)2}$$

Type this line into your calculator and press the ENTER key several times to convince yourself that it does not always produce 1's and 2's.

In part 1 you pressed the ENTER key repeatedly and manually kept track of the outcomes. The TI-83 can carry out the instructions many times, store the outcomes in a list, and analyze the list. The first line in Fig. 4 generates 100 random numbers between 0 and 1 and places them in the list L_1. The third and fourth lines of Fig. 4 convert each number in the list L_1 to a row number using the strategy $\begin{bmatrix} .6 & .4 \end{bmatrix}$ and place the 100 row choices into the list L_3. Figure 6 uses the window settings shown in Fig. 5 to display the histogram for L_3. We see that the second row was selected 39 times out of 100 times. This is very close to the expected 40%.

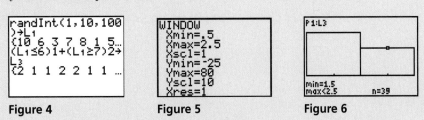

Figure 4	**Figure 5**	**Figure 6**

3. Simulate 100 column selections using the strategy $\begin{bmatrix} .2 \\ .8 \end{bmatrix}$ and create a histogram showing the number of times each column was selected. How many times was the second column selected?

In Section 9.3, the optimal mixed strategies for the payoff matrix $\begin{bmatrix} 5 & 3 \\ 1 & 4 \end{bmatrix}$ were found to be $\begin{bmatrix} .6 & .4 \end{bmatrix}$ for R and $\begin{bmatrix} .2 \\ .8 \end{bmatrix}$ for C and the value of the game was found to be $\frac{17}{5}$. In

Figs. 7 and 8, 100 games are simulated and the payoffs for the games are stored in the list L_3. Fig. 10 shows the histogram for L_3 using the window settings in Fig. 9. For instance, we see that a payoff of 3 occurred in 50 of the 100 games. Fig. 11 shows that the average of the 100 payoffs, 3.35, is very close to the value of the game, 3.4.

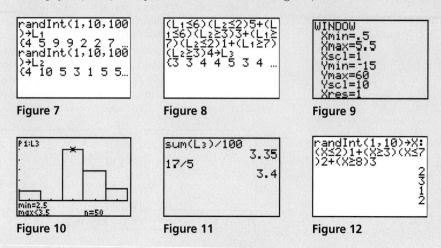

Figure 7   Figure 8   Figure 9

Figure 10   Figure 11   Figure 12

4. Explain in detail what is being calculated in Fig. 8.

5. Create a list of the payoffs for playing 100 games with the payoff matrix $\begin{bmatrix} 5 & 3 \\ 1 & 4 \end{bmatrix}$ using the strategies $\begin{bmatrix} .5 & .5 \end{bmatrix}$ for R and $\begin{bmatrix} .3 \\ .7 \end{bmatrix}$ for C, and compute the average of the payoffs.

6. Suppose a payoff matrix has three rows and that R's strategy is $\begin{bmatrix} .2 & .5 & .3 \end{bmatrix}$. Explain why Fig. 12 simulates the selection of a row by R. Type the instructions into your calculator, press the ENTER key forty times, and count the number of 1's, 2's, and 3's. Are there approximately eight 1's, twenty 2's, and twelve 3's?

7. Consider the 3×2 matrix in the first practice problem of Section 9.3. Suppose the strategies for R and C are $\begin{bmatrix} .2 & .5 & .3 \end{bmatrix}$ and $\begin{bmatrix} .75 \\ .25 \end{bmatrix}$, respectively. Simulate the payoffs for 100 games, display the histogram for the payoffs, and calculate the average value of the payoffs.

The Mathematics of Finance

THIS chapter presents several topics in the mathematics of finance, including compound and simple interest, annuities, and amortization. They are presented in the classical way, using methods that are centuries old but that lend themselves to modern computing techniques.

10.1 Interest

When you deposit money into a savings account, the bank pays you a fee for the use of your money. This fee is called *interest* and is determined by the amount deposited, the duration of the deposit, and the quoted interest rate. The amount deposited is called the *principal*, and the amount to which the principal grows (after the addition of interest) is called the *compound amount* or *balance*.

The entries in a hypothetical bank passbook are shown in Table 1 on the next page. Note the following facts about this passbook:

1. The principal is $100.00. The compound amount after 1 year is $104.06.

2. Interest is being paid four times per year (or, in financial language, *quarterly*).

3. Each quarter, the interest is 1% of the previous balance. That is, $1.00 is 1% of $100.00, $1.01 is 1% of $101.00, and so on. Since $4 \times 1\%$ is 4%, we say that the money is earning 4% *annual interest compounded quarterly*.

TABLE 1				
Date	Deposits	Withdrawals	Interest	Balance
1/1/05	$100.00			$100.00
4/1/05			$1.00	101.00
7/1/05			1.01	102.01
10/1/05			1.02	103.03
1/1/06			1.03	104.06

As in the passbook shown in Table 1, interest rates are usually stated as annual percentage rates, with the interest to be *compounded* (i.e., computed) a certain number of times per year. Some common frequencies for compounding are listed in Table 2. Of special importance is the *interest rate per period*, denoted i, which is calculated by dividing the annual percentage by the number of interest periods per year.

TABLE 2		
Number of interest periods per year	Length of each interest period	Interest compounded
1	One year	Annually
2	Six months	Semiannually
4	Three months	Quarterly
12	One month	Monthly
52	One week	Weekly
365	One day	Daily

For example, in our passbook in Table 1, the annual percentage rate was 4%, the interest was compounded quarterly, and the interest per period was 4%/4 = 1%.

If interest is compounded m times per year and the annual interest rate is r, then the interest rate per period is

$$i = \frac{r}{m}.$$

EXAMPLE 1 **Determining interest rate per period** Determine the interest rate per period for each of the following interest rates.

(a) 5% interest compounded semiannually

(b) 6% interest compounded monthly

Solution (a) The annual percentage rate is 5% and the number of interest periods is 2. Therefore,

$$i = \frac{5\%}{2} = 2\tfrac{1}{2}\%.$$

Note: In decimal form, $i = .025$.

(b) The annual percentage rate is 6% and the number of interest periods is 12. Therefore,

$$i = \frac{6\%}{12} = \frac{1}{2}\%.$$

Note: In decimal form, $i = .005$.

Now Try Exercise 1   ■

Consider a savings account in which the interest rate per period is i. Then the interest earned during a period is i times the previous balance. That is, at the end of an interest period, the new balance, B_{new}, is computed by adding this interest to the previous balance, B_{previous}. Therefore,

$$B_{\text{new}} = B_{\text{previous}} + i \cdot B_{\text{previous}}. \tag{1}$$

EXAMPLE 2   **Computing interest and balances**   Compute the interest and the balance for the first two interest periods for a deposit of $1000 at 4% interest compounded semiannually.

Solution   Here $i = 2\%$, or $.02$. The interest for the first period is 2% of $1000, or $20. Let B_1 be the balance at the end of the first interest period and B_2 the balance after two interest periods. By formula (1),

$$B_1 = 1000 + .02(1000) = 1000 + 20 = \$1020.$$

Similarly, since 2% of $1020 is $20.40, the interest for the second period is $20.40 and the balance is

$$B_2 = B_1 + .02B_1 = 1020 + .02(1020) = 1020 + 20.40 = \$1040.40.$$

Let us compute B_2 using another method:

$$B_2 = B_1 + .02B_1 = 1 \cdot B_1 + .02B_1 = (1 + .02)B_1$$
$$= (1.02)B_1 = (1.02)1020 = \$1040.40.$$

Now Try Exercise 17(a), (b)   ■

The alternative method for computing B_2 just presented can be generalized. Namely, we always have

$$B_{\text{new}} = B_{\text{previous}} + i \cdot B_{\text{previous}} = 1 \cdot B_{\text{previous}} + i \cdot B_{\text{previous}}.$$

Therefore,

$$B_{\text{new}} = (1 + i)B_{\text{previous}}.$$

This last result says that balances for successive time periods are computed by multiplying by $1 + i$.

The formula for the balance after any number of interest periods is now easily derived:

Principal	P
Balance after 1 interest period	$(1 + i)P$
Balance after 2 interest periods	$(1 + i) \cdot (1 + i)P$ or $(1 + i)^2 P$
Balance after 3 interest periods	$(1 + i) \cdot (1 + i)^2 P$ or $(1 + i)^3 P$
Balance after 4 interest periods	$(1 + i)^4 P$
$\vdots$	$\vdots$
Balance after n interest periods	$(1 + i)^n P.$

Denote the compound amount by the letter F (suggestive of "future value"). Then the compound amount after n interest periods is given by the formula

$$F = (1+i)^n P, \tag{2}$$

where i is the interest rate per period and P is the principal.

Values of $(1+i)^n$ for specific values of i and n are easily determined using either a calculator or a table. A brief table of useful values has been included as Table 2 of Appendix A. Financial calculators have a designated key to compute $(1+i)^n$. On an ordinary calculator, the following keystrokes will compute the value of $(1+.005)^{24}$.

$$\boxed{1} \;\; \boxed{+} \;\; \boxed{.005} \;\; \boxed{=} \;\; \boxed{x^y} \;\; \boxed{24} \;\; \boxed{=}$$

On the home screen of a graphing calculator such as the TI-83 or TI-89, use

$$\boxed{(} \;\; \boxed{1} \;\; \boxed{+} \;\; \boxed{.005} \;\; \boxed{)} \;\; \boxed{\wedge} \;\; \boxed{24} \;\; \boxed{\text{ENTER}}$$

EXAMPLE 3 **Computing compound amounts** Apply formula (2) to the savings account passbook discussed at the beginning of this section and calculate the compound amount after 1 year and after 5 years.

Solution The principal P is \$100. Since the interest rate is 4% compounded quarterly, we have $i = 1\%$, or .01. One year consists of four interest periods, so $n = 4$. Therefore, the compound amount after 1 year is

$$F = (1+.01)^4 \cdot 100 = (1.01)^4 \cdot 100 = (1.04060401) \cdot 100 \qquad \text{(from Appendix A)}$$

$$= 104.060401 = \$104.06 \qquad \text{(after rounding to the nearest cent).}$$

Five years consists of $n = 5 \times 4 = 20$ interest periods. Therefore, the compound amount after 5 years is

$$F = (1.01)^{20} \cdot 100 = (1.22019004) \cdot 100 = \$122.02.$$

Now Try Exercise 5 ■

The Excel spreadsheet in Table 3 shows the effects of interest rates (compounded quarterly) on the compound amount. The entries in cells B5 and C5 are **=$B$1*(1+A5/4)^20** and **=$B$1*(1+A5/4)^40**, respectively. The remaining entries in the B and C columns were created by dragging fill handles. The dollar signs in **$B$1** prevent references to this location from being changed during this process.

The next example is a variation of the previous one and introduces a new concept, present value.

EXAMPLE 4 **Computing a present value** How much money must be deposited now in order to have \$1000 after 5 years if interest is paid at a 4% annual rate compounded quarterly?

Solution As in Example 3, we have $i = .01$ and $n = 20$. However, now we are given F and are asked to solve for P.

$$F = (1+i)^n P$$

$$1000 = (1.01)^{20} P$$

$$P = \frac{1000}{(1.01)^{20}}$$

	A	B	C
TABLE 3			
		$100.00	
1	Principal	$100.00	
2			
3		Compound Amount	
4	Interest Rate	5 Years	10 Years
5	3.0%	$116.12	$134.83
6	3.5%	$119.03	$141.69
7	4.0%	$122.02	$148.89
8	4.5%	$125.08	$156.44
9	5.0%	$128.20	$164.36
10	5.5%	$131.41	$172.68
11	6.0%	$134.69	$181.40
12	6.5%	$138.04	$190.56
13	7.0%	$141.48	$200.16
14	7.5%	$144.99	$210.23
15	8.0%	$148.59	$220.80

From Table 2 of Appendix A, $(1.01)^{20} = 1.22019004$. However, the cumbersome arithmetic can be avoided by using Table 3 of Appendix A, which tabulates values of $1 \div (1 + i)^n$ for various values of i and n.

$$P = \frac{1}{(1.01)^{20}} \cdot 1000 = (.81954447) \cdot 1000 = \$819.54$$

We say that $819.54 is the present value of $1000, 5 years from now, at 4% interest compounded quarterly.

Now Try Exercise 11   ■

In general, the *present value* of F dollars at a given interest rate and given length of time is the amount of money P that must be deposited now in order for the compound amount to grow to F dollars in the given length of time. From formula (2), we see that the present value P may be computed from the following formula:

$$P = \frac{1}{(1 + i)^n} \cdot F. \tag{3}$$

EXAMPLE 5 **Computing a present value** Determine the present value of a $10,000 payment to be received on January 1, 2016, if it is now May 1, 2007, and money can be invested at 6% interest compounded monthly.

Solution Here $F = 10,000$, $n = 104$ (the number of months between the two given dates), and $i = \frac{1}{2}\% = .005$. By formula (3),

$$P = \frac{1}{(1 + i)^n} \cdot F = (.59529136)10,000 = \$5952.91.$$

Therefore, $5952.91 invested on May 1, 2007, will grow to $10,000 by January 1, 2016.

Now Try Exercise 7   ■

The interest that we have been discussing so far is the most prevalent type of interest and is known as *compound interest*. There is another type of interest, called *simple interest*, which is used in some financial circumstances. Let us now discuss this type of interest.

Interest rates for simple interest are given as an annual percentage rate r. Interest is earned *only* on the principal P and the interest is rP for each year. Therefore, the interest earned in n years is nrP. So the amount F after n years is the original amount plus the interest earned. That is,

$$F = P + nrP = 1 \cdot P + nrP = (1 + nr)P. \tag{4}$$

EXAMPLE 6 **Computing a balance with simple interest** Calculate the amount after 4 years if $1000 is invested at 5% simple interest.

Solution Apply formula (4) with $P = \$1000$, $n = 4$, and $r = 5\%$ or .05.

$$F = (1 + nr)P = [1 + 4(.05)]1000 = (1.2)1000 = \$1200.$$

Now Try Exercise 27 ∎

In Example 6, had the money been invested at 5% compound interest with annual compounding, then the compound amount would have been $1215.51. Money invested at simple interest is earning interest only on the principal amount. However, with compound interest, after the first interest period, the interest is also earning interest. Thus, if we compare two savings accounts, each earning interest at the same stated annual rate, but with one earning simple interest and the other earning compound interest, the latter will grow at a faster rate.

Effective Rate of Interest The annual rate of interest is also known as the *nominal rate* or the *stated rate*. Its true worth depends on the number of compounding periods. The nominal rate does not help you decide, for instance, whether a savings account paying 3.65% interest compounded quarterly is better than a savings account paying 3.6% interest compounded monthly. A better measure of worth is the *effective rate of interest*. The effective rate is the simple interest rate that yields the same amount after one year as the annual rate of interest. For example, a savings account paying 3.65% interest compounded quarterly has an effective rate of 3.7%, whereas a savings account paying 3.6% compounded monthly has an effective rate of 3.66%. Therefore, the first savings account is better.

Suppose the annual rate of interest r is compounded m times per year. Then with compound interest P dollars will grow to $P(1+i)^m$ in one year, where $i = r/m$. With simple interest r_{eff}, the balance after one year will be $P(1 + r_{\text{eff}})$. Equating the two balances,

$$P(1 + r_{\text{eff}}) = P(1 + i)^m$$
$$1 + r_{\text{eff}} = (1 + i)^m$$
$$r_{\text{eff}} = (1 + i)^m - 1. \tag{5}$$

EXAMPLE 7 **Calculating an effective interest rate** Calculate the effective rate of interest for a savings account paying 3.65% compounded quarterly.

Solution Apply formula (5) with $r = .0365$ and $m = 4$. Then $i = r/m = .0365/4 = .009125$ and

$$r_{\text{eff}} = (1 + .009125)^4 - 1 \approx .037.$$

Therefore, the effective rate is approximately 3.7%.

Now Try Exercise 41   ■

Let us summarize the key formulas developed so far.

Compound Interest

Compound amount: $F = (1 + i)^n P$

Present value: $P = \dfrac{1}{(1 + i)^n} \cdot F$

Effective rate: $r_{\text{eff}} = (1 + i)^m - 1$

where i is the interest rate per period, n is the total number of interest periods, and m is the number of times interest is compounded per year.

Simple Interest

Amount: $F = (1 + nr)P$

where r is the interest rate and n is the number of years.

GC Graphing calculators can easily display successive balances in a savings account. Consider the situation from Table 1, in which \$100 is deposited at 4% interest compounded quarterly. In Figs. 1 and 4 successive balances are displayed in the home screen. In Figs. 2 and 5 successive balances are graphed, and in Figs. 3 and 6 they are displayed in a table.

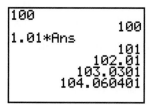
Figure 1

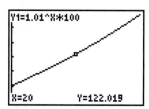
Figure 2

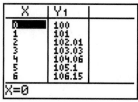
Figure 3

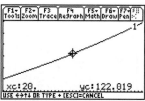
Figure 4

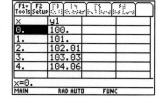
Figure 5

Figure 6

Displaying successive balances on the home screen Successive balances can be determined with the relation $B_{\text{new}} = (1 + i)B_{\text{previous}}$. In Figs. 1 and 4, after the principal (**100**) is entered, the last displayed value (**100**) is assigned to **Ans** (on the TI-83) or **ans(1)** (on the TI-89). The instruction **1.01*Ans** (or **1.01*ans(1)** on the TI-89) generates the next balance (**101**) and assigns it to **Ans** (or **ans(1)**). Each subsequent press of ENTER generates another balance.

Graphs with the TI-83 and TI-89 In order to obtain the graph in Figs. 2 and 5, **Y₁** was set to **1.01^X*100** in the **Y=** editor; that is, $(1 + i)^X \cdot P$. Here X, instead of n, is used to represent the number of interest periods. Then the window was set to $[0, 40]$ *by* $[90, 150]$, with an x-scale of 1 and a y-scale of 10. Then, we pressed GRAPH (on the TI-83) or ◆[GRAPH] (on the TI-89) to display the function, and pressed TRACE (on the TI-83) or F3 (on the TI-89) to obtain the trace cursor. (*Note*: After **Y₁** has been defined, values of **Y₁** can be displayed on the home screen by entering expressions such as **Y₁(20)**.)

Tables with the TI-83 and TI-89 In order to obtain the tables in Figs. 3 and 6, **Y₁** was set to **1.01^X*100** in the **Y=** editor. Then 2nd[TBLSET] (on the TI-83) or ◆ [TBLSET] (on the TI-89) was pressed to bring up the TABLE SETUP screen. **TblStart** was set to 0, and **ΔTbl** was set to 1. Finally, 2nd [TABLE] (on the TI-83) or ◆ [TABLE] (on the TI-89) was pressed to bring up the table.

EXAMPLE 8 **Determining the amount of time for a balance to exceed a specified amount** Consider the situation from Table 1, in which $100 is deposited at 4% interest compounded quarterly. Use a graphing calculator to determine when the balance will exceed $130.

Solution There are three ways to answer this question with a graphing calculator—with the home screen, with a graph, and with a table. (*Note*: The three TI-83 screens shown below have TI-89 analogs that can be created in a similar way.)

1. (Home screen)   Generate successive balances (while keeping track of the number of times the ENTER key has been pressed) until a balance exceeding 130 is reached. See Fig. 7.

2. (Graph)   Set **Y₂ = 130** and find its intersection with the graph of **Y₁ = 1.01^X*100**. See Fig. 8.

3. (Table)   Scroll down the table for **Y₁** until the balance exceeds 130. See Fig. 9.

In each case we see that the balance will exceed $130 after 27 interest periods.

Now Try Exercise 55 ■

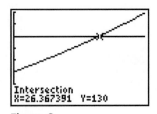

| Figure 7 | Figure 8 |

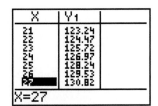

Figure 9

ES The section "Using Excel's Financial Functions" of Appendix C shows how the functions FV, PV, and EFFECT are used to calculate future values (same as compound amounts), present values, and effective rates of interest.

PRACTICE PROBLEMS 10.1

1. (a) In Table 3 of Appendix A, look up the value of $1/(1 + i)^n$ for $i = 6\%$ and $n = 10$.

(b) Calculate the present value of $1000 to be received 10 years in the future at 6% interest compounded annually.

2. Calculate the compound amount after 2 years of $1 at 26% interest compounded weekly.

3. Calculate the future amount of $2000 after 6 months if invested at 6% simple interest.

EXERCISES 10.1

1. Determine i and n for the following situations, where i is the interest rate per period and n is the number of interest periods.

  (a) 12% interest compounded monthly for 2 years

  (b) 8% interest compounded quarterly for 5 years

  (c) 10% interest compounded semiannually for 20 years

2. Determine i and n for the following situations.

  (a) 6% interest compounded annually for 3 years

  (b) 6% interest compounded monthly from January 1, 2001, to July 1, 2005

  (c) 9% interest compounded quarterly from January 1, 2001, to October 1, 2004

3. Determine i, n, P, and F for the following situations.

  (a) $500 invested at 6% interest compounded annually grows to $631.24 in 4 years.

  (b) $800 invested on January 1, 2004, at 6% interest compounded monthly grows to $1455.52 by January 1, 2014.

  (c) $6177.88 is deposited on January 1, 2004. The balance on July 1, 2013, is $9000 and the interest is 4% compounded semiannually.

4. Determine i, n, P, and F for the following situations.

  (a) The amount of money that must be deposited now at 4.5% interest compounded weekly in order to have $7500 in 1 year is $7170.12.

  (b) $3000 deposited at 6% interest compounded monthly will grow to $18,067.73 in 30 years.

  (c) In 1626, Peter Minuit, the first director-general of New Netherlands province, purchased Manhattan Island for trinkets and cloth valued at about $24. Had this money been invested at 8% interest compounded quarterly, it would have amounted to $306,812,469,900,000 by 2007.

5. Calculate the compound amount of $1000 after 2 years if deposited at 6% interest compounded monthly.

6. Calculate the present value of $10,000 payable in 5 years at 4% interest compounded semiannually.

7. Calculate the present value of $100,000 payable in 25 years at 6% interest compounded monthly.

8. Calculate the compound amount of $1000 after 1 year if deposited at 7.3% interest compounded daily.

9. Six thousand dollars is deposited in a savings account at 6% interest compounded monthly. Find the balance after 3 years and the amount of interest earned during that time.

10. Two thousand dollars is deposited in a savings account at 4% interest compounded semiannually. Find the balance after 7 years and the amount of interest earned during that time.

11. (a) How much money would have to be deposited now at 6% interest compounded monthly to accumulate to $4000 in 2 years?

  (b) How much of the $4000 would be interest?

  (c) Prepare a table showing the growth of the account for the first 3 months.

12. (a) How much money would have to be deposited now at 8% interest compounded quarterly to accumulate to $10,000 in 15 years?

  (b) How much of the $10,000 would be interest?

  (c) Prepare a table showing the growth of the account for the first 3 quarters.

13. If you had invested $10,000 on January 1, 2003, at 4% interest compounded quarterly, how much would you have on April 1, 2009?

14. In order to have $10,000 on his 25th birthday, how much would a person just turned 21 have to invest if the money will earn 6% interest compounded monthly?

15. (*Saving for Retirement*) Mr. Smith wishes to purchase a $10,000 sailboat upon his retirement in 3 years. He has just won the state lottery and would like to set aside enough cash in a savings account paying 4% interest compounded quarterly to buy the boat upon retirement. How much should he deposit?

16. (*Savings Account*) Ms. Jones has just invested $100,000 at 6% interest compounded annually. How much money will she have in 20 years?

17. Consider the following savings account passbook:

	Date		
	1/1/06	**2/1/06**	**3/1/06**
Deposit	$1000.00		
Withdrawal			
Interest		$5.00	$5.03
Balance	$1000.00	$1005.00	$1010.03

   (a) What annual interest rate is this bank paying?

   (b) Give the interest and balance on 4/1/06.

   (c) Give the interest and balance on 1/1/08.

18. Consider the following savings account passbook:

	Date		
	1/1/05	**7/1/05**	**1/1/06**
Deposit	$10,000.00		
Withdrawal			
Interest		$200.00	$204.00
Balance	$10,000.00	$10,200.00	$10,404.00

   (a) Give the interest and balance on 7/1/06.

   (b) Give the interest and balance on 1/1/14.

19. Is it more profitable to receive $1000 now or $1700 in 9 years? Assume that money can earn 6% interest compounded annually.

20. Is it more profitable to receive $7000 now or $10,000 in 9 years? Assume that money can earn 4% interest compounded quarterly.

21. (*Comparing Interest Rates*) In 1999 the Nasdaq index grew at a rate of 62.2% compounded weekly. Is this rate better or worse than 85% compounded annually?

22. Would you rather earn 8% interest compounded annually or 7.3% interest compounded daily?

23. On January 1, 2003, a deposit was made into a savings account paying interest compounded quarterly. The balance on January 1, 2006 was $10,000 and the balance on April 1, 2006, was $10,100. How large was the deposit?

24. During the 1990s a deposit was made into a savings account paying 4% interest compounded quarterly. On January 1, 2006, the balance was $2020. What was the balance on October 1, 2005?

Exercises 25–36 concern simple interest.

25. Determine r, n, P, and F for each of the following situations.

   (a) $500 invested at 4% simple interest grows to $510 in 6 months.

   (b) In order to have $550 after 2 years at 5% simple interest, $500 must be deposited.

26. Determine r, n, P, and F for each of the following situations.

   (a) At 6% simple interest, $1000 deposited on January 1, 2006, will be worth $1040 on September 1, 2006.

   (b) At 4% simple interest, in order to have $3600 in 5 years, $3000 must be deposited now.

27. Calculate the amount after 3 years if $1000 is deposited at 5% simple interest.

28. Calculate the amount after 18 months if $2000 is deposited at 6% simple interest.

29. Find the present value of $3000 in 2 years at 10% simple interest.

30. Find the present value of $2000 in 4 years at 7% simple interest.

31. Determine the (simple) interest rate at which $980 grows to $1000 in 6 months.

32. How many years are required for $1000 to grow to $1210 at 7% simple interest?

33. Determine the amount of time required for money to double at 5% simple interest.

34. Derive the formula for the (simple) interest rate at which P dollars grows to F dollars in n years. That is, express r in terms of P, F, and n.

35. Derive the formula for the present value of F dollars in n years at simple interest rate r.

36. Derive the formula for the number of years required for P dollars to grow to F dollars at simple interest rate r.

37. Suppose a principal of $100 is deposited for 5 years in a savings account at 6% interest compounded semiannually. Which of items (a)–(d) can be used to fill in the blank in the following statement? (*Note*: Before computing, use your intuition to guess the correct answer.)

   If the _____ is doubled, then the compound amount will double.

   (a) principal   (b) interest rate

   (c) number of interest periods per year

   (d) number of years

38. Suppose a principal of $100 is deposited for 5 years in a savings account at 6% simple interest. Which of items (a)–(c) can be used to fill in the blank in the following statement? (*Note*: Before computing, use your intuition to guess the correct answer.)

   If the _____ is doubled, then the future amount will double.

   (a) principal   (b) interest rate

   (c) number of years

39. Compute the compound amount after 1 year for $100 invested at 4% interest compounded quarterly. What simple interest rate will yield the same amount in 1 year?

40. Compute the compound amount after 1 year for $100 invested at 6% interest compounded monthly. What simple interest rate will yield the same amount in 1 year?

41. Calculate the effective rate for 4% interest compounded semiannually.

42. Calculate the effective rate for 8.45% interest compounded weekly.

43. Calculate the effective rate for 4.4% interest compounded monthly.

44. Calculate the effective rate for 7.95% interest compounded quarterly.

45. Find the nominal rate corresponding to an effective rate of 4.06% if interest is compounded quarterly.

46. Find the nominal rate corresponding to an effective rate of 6.91% if interest is compounded monthly.

47. If a $1000 investment at compound interest doubles every six years, how long will it take the investment to grow to $8000?

48. If your stock portfolio gained 20% in 2005 and 30% in 2006, then it gained _____ over the two-year period.

 (a) 50%   **(b)** 56%   **(c)** 60%   **(d)** 100%

49. If the value of an investment grows at the rate of 4% compounded annually for 10 years, then it grows about _____ over the ten-year period.

 (a) 25%   **(b)** 40%   **(c)** 44%   **(d)** 48%

50. (*True or False*) An investment growing at the rate of 12% compounded monthly will double (that is increase by 100%) in about 70 months. Therefore, the investment should increase by 50% in about 35 months.

51. (*Comparing Investments*) The same amount of money was invested in each of two different investments on January 1, 2004. Investment A increased by 2.5% in 2004, 3% in 2005, and 8.4% in 2006. Investment B increased by the same amount, r%, in each of the three years and was worth the same as investment A at the end of the three-year period. Determine the value of r.

52. If you increase the interest rate for an investment by 20%, will the future value increase by 20%?

In Exercises 53–56, use a graphing calculator, a spreadsheet, or mathematical software to answer the questions.

53. (*Savings Account*) One thousand dollars is deposited into a savings account at 6.5% interest compounded annually. What are the balances after 1, 2, and 3 years? How many years are required for the balance to reach $1655? After how many years will the balance exceed $2000?

54. (*Savings Account*) Ten thousand dollars is deposited into a savings account at 6% interest compounded monthly. What are the balances after 5, 10, and 15 months? How many months are required for the balance to reach $11,614? After how many months will the balance exceed $13,000?

55. (*Investment*) Tom invests $100,000 at 8% interest compounded annually. When will Tom be a millionaire?

56. (*Savings Account*) How many years are required for $100 to double if deposited at 6% interest compounded quarterly?

SOLUTIONS TO PRACTICE PROBLEMS 10.1

1. **(a)** Look up the entry in Table 3 in Appendix A in the row labeled "10" and the column labeled "6%." That entry is .55839478.

 (b) Here we are given the value in the future $F = $1000 and are asked to find the present value P. Interest compounded annually has just one interest period per year, so $n = 10 \cdot 1 = 10$ and $i = 6\%/1 = 6\%$. By formula (3),

$$P = \frac{1}{(1+i)^n} \cdot F$$

$$= \frac{1}{(1+i)^n} \cdot 1000 \quad \text{(with } i = 6\% \text{ and } n = 10\text{)}$$

$$= (.55839478) \cdot 1000 \quad \text{[from part (a)]}$$

$$= 558.39478$$

$$= \$558.39 \quad \text{(rounding to the nearest cent)}.$$

2. Here we are given the present value, $P = 1$, and are asked to find the value F at a future time. Interest compounded weekly has 52 interest periods each year, so $n = 2 \times 52 = 104$ and $i = 26\%/52 = \frac{1}{2}\%$. By formula (2), we have

$$F = (1+i)^n P$$

$$= (1+i)^n \cdot 1 \quad \text{(with } i = 1/2\% \text{ and } n = 104\text{)}$$

$$= (1.67984969) \cdot 1 \quad \text{(using Table 2)}$$

$$= \$1.68.$$

[*Note:* In general, whenever $P = \$1$, then $F = (1+i)^n \cdot 1 = (1+i)^n$. This explains why $(1+i)^n$ is often referred to as the *compound amount* of 1.]

3. In simple interest problems, time should be expressed in terms of years. Since 6 months is $\frac{1}{2}$ of a year, for- mula (4) gives

$$A = (1 + nr)P = [1 + \tfrac{1}{2}(.06)]2000$$
$$= (1.03)2000$$
$$= \$2060.$$

10.2 Annuities

An *annuity* is a sequence of equal payments made at regular intervals of time. Here are two illustrations.

1. As the proud parent of a newborn daughter, you decide to save for her college education by depositing $100 at the end of each month into a savings account paying 6% interest compounded monthly. Eighteen years from now, after you make the last of 216 payments, the account will contain $38,735.32.

2. Having just won the state lottery, you decide not to work for the next 5 years. You want to deposit enough money into the bank so that you can withdraw $5000 at the end of each month for 60 months. If the bank pays 6% interest compounded monthly, you must deposit $258,627.80.

The payments in the foregoing financial transactions are called *rent*. The amount of a typical rent payment is denoted by the letter R. Thus, in the preceding examples, we have $R = \$100$ and $R = \$5000$, respectively.

In illustration 1, you make equal payments to a bank in order to generate a large sum of money in the future. This sum, namely $38,735.32, is called the *future value of the annuity*. Since the amount of money in the savings account is increased each time a payment is made, we refer to this type of annuity as an *increasing annuity*.

In illustration 2, the bank will make equal payments to you in order to pay back the sum of money that you currently deposit. The value of the current deposit, namely $258,627.80, is called the *present value* of the annuity. Since the amount of money in the savings account is decreased each time a payment is made, we refer to this type of annuity as a *decreasing annuity*.

We now derive formulas for future values and present values of annuities. Suppose that an increasing annuity consists of a sequence of n equal payments, each of R dollars. Suppose that the annuity payments are deposited into an account paying compound interest at the rate of i per interest period. We will further suppose that there is a single annuity payment per interest period and that the payment is made at the end of the interest period. Let us derive a formula for the future value of the annuity: that is, a formula for the balance of the account immediately after the last payment.

Each payment accumulates interest for a different number of interest periods, so let us calculate the balance in the account as the sum of n compound amounts, one corresponding to each payment (Table 1). Denote the future value of the annuity by F. Then F is the sum of the numbers in the right-hand column:

$$F = R + (1+i)R + (1+i)^2 R + (1+i)^3 R + \cdots + (1+i)^{n-1}R.$$

A compact expression for F can be obtained by multiplying both sides of this equation by $(1+i)$ and then subtracting the original equation from the new equation.

TABLE 1

Payment	Amount	Number of interest periods on deposit	Compound amount
1	R	$n-1$	$(1+i)^{n-1}R$
2	R	$n-2$	$(1+i)^{n-2}R$
$\vdots$	$\vdots$	$\vdots$	$\vdots$
$n-2$	R	2	$(1+i)^2 R$
$n-1$	R	1	$(1+i)R$
n	R	0	R

$$(1+i)F = \quad (1+i)R + (1+i)^2 R + (1+i)^3 R + \cdots + (1+i)^{n-1}R + (1+i)^n R$$

$$F = R + (1+i)R + (1+i)^2 R + (1+i)^3 R + \cdots + (1+i)^{n-1}R$$

$$iF = (1+i)^n R - R$$

The last equation can be written $iF = [(1+i)^n - 1] \cdot R$. Dividing both sides by i yields

$$F = \frac{(1+i)^n - 1}{i} \cdot R.$$

The expression $[(1+i)^n - 1]/i$ occurs often in financial analysis and is denoted by the special symbol $s_{\overline{n}|i}$ (read "s sub n angle i"). We may summarize our calculations as follows.

> Suppose that an increasing annuity consists of n payments of R dollars each, deposited at the ends of consecutive interest periods into an account with interest compounded at a rate i per period. Then the future value F of the annuity is given by the formula
>
> $$F = s_{\overline{n}|i}R. \tag{1}$$

Values of $s_{\overline{n}|i}$ may be computed using a calculator or looked up in an appropriate table. We have included a brief table as Table 4 of Appendix A. Financial calculators have a designated key to compute $s_{\overline{n}|i}$. On a typical ordinary calculator, the following keystrokes will compute the value of $s_{\overline{24}|.005}$.

$$\boxed{1.005}\ \boxed{x^y}\ \boxed{24}\ \boxed{=}\ \boxed{-}\ \boxed{1}\ \boxed{=}\ \boxed{\div}\ \boxed{.005}\ \boxed{=}$$

On the home screen of a graphing calculator such as the TI-83 or TI-89, use

$$\boxed{(}\ \boxed{(}\ \boxed{1.005}\ \boxed{\wedge}\ \boxed{24}\ \boxed{)}\ \boxed{-}\ \boxed{1}\ \boxed{)}\ \boxed{\div}\ \boxed{.005}\ \boxed{\text{ENTER}}$$

EXAMPLE 1 **Calculating the future value of an increasing annuity** Calculate the future value of an increasing annuity of $100 per month for 5 years at 6% interest compounded monthly.

Solution Here $R = 100$ and $i = \frac{1}{2}\%$. Since a payment is made at the end of each month for 5 years, there will be $5 \times 12 = 60$ payments. So $n = 60$. Therefore,

$$F = s_{\overline{n}|i}R = s_{\overline{60}|\,1/2\%} \cdot 100 = (69.77003051) \cdot 100 \qquad \text{(by Table 4)}$$
$$= \$6977.00.$$

Now Try Exercise 3 ■

We can easily derive a formula that illustrates exactly how the future value of an increasing annuity changes from period to period. Each new balance can be computed from the previous balance with the formula

$$B_{\text{new}} = (1 + i)B_{\text{previous}} + R.$$

That is, the new balance equals the growth of the previous balance due to interest, plus the amount paid. So, for instance, if B_1 is the balance after the first payment is made, B_2 is the balance after the second payment is made, and so on, then

$$B_1 = R$$
$$B_2 = (1 + i)B_1 + R$$
$$B_3 = (1 + i)B_2 + R$$
$$\vdots$$

Successive unpaid balances are computed by multiplying by $(1 + i)$ and adding R.

If $\$R$ is deposited into an increasing annuity at the end of each interest period with interest rate i per period,

$$B_{\text{new}} = (1 + i)B_{\text{previous}} + R \qquad (2)$$

can be used to calculate each new balance, B_{new}, from the previous balance, B_{previous}.

EXAMPLE 2 **Calculating the future value of an increasing annuity** Consider the annuity of Example 1. Determine the future value after 61 months.

Solution From Example 1, $i = .005$, the balance after 60 months is 6977.00, and $R = 100$. Therefore, by (2)

$$[\text{balance after 61 months}] = (1 + .005)[\text{balance after 60 months}] + 100$$
$$= 1.005(6977) + 100 = 7111.89.$$

Therefore, the annuity will have accumulated to $7111.89 after 61 months. ■

Formula (1) can also be used to determine the rent necessary to achieve a certain future value:

$$F = s_{\overline{n}|i}R$$
$$R = \frac{F}{s_{\overline{n}|i}} = \frac{1}{s_{\overline{n}|i}} \cdot F.$$

Thus we have established the following result.

> Suppose that an increasing annuity of n payments has future value F and has interest compounded at the rate i per period. Then the rent R is given by
>
> $$R = \frac{1}{s_{\overline{n}|\, i}} \cdot F.$$

Table 5 of Appendix A gives values of $1/s_{\overline{n}|\, i}$ for various values of n and i.

EXAMPLE 3 **Determining the rent for an increasing annuity** Ms. Adams would like to buy a \$30,000 airplane when she retires in 8 years. How much should she deposit at the end of each half-year into an account paying 4% interest compounded semiannually so that she will have enough money to purchase the airplane?

Solution For this increasing annuity, $n = 16$, $i = 2\%$, and $F = 30{,}000$. Therefore,

$$R = \frac{1}{s_{\overline{n}|\, i}} \cdot F = \frac{1}{s_{\overline{16}|\, 2\%}} \cdot 30{,}000 = (.05365013) \cdot 30{,}000 \qquad \text{(by Table 5)}$$

$$= \$1609.50.$$

She should deposit \$1609.50 at the end of each half-year period.

Now Try Exercise 9   ■

The *present value* of a decreasing annuity is the amount of money necessary to finance the sequence of annuity payments. More specifically, the present value of the annuity is the amount you would need to deposit in order to provide the desired sequence of annuity payments and leave a balance of zero at the end of the term. Let us now find a formula for the present value of the annuity. There are two ways to proceed. On the one hand, the desired present value could be computed as the sum of the present values of the various annuity payments. This computation would make use of the formula for the sum of the first n terms of a geometric progression. However, there is a much "cleaner" derivation that proceeds indirectly to obtain a formula relating the present value P and the rent R.

As before, let us assume that our annuity consists of n payments made at the ends of interest periods, with interest compounded at a rate i per interest period.

Situation 1: Suppose that the P dollars were just left in the account and that the annuity payments were not withdrawn. At the end of the n interest periods, there would be $(1 + i)^n P$ dollars in the account.

Situation 2: Suppose that the payments are withdrawn but are immediately redeposited into another account having the same rate of interest. At the end of the n interest periods, there would be $s_{\overline{n}|\, i} R$ dollars in the new account.

In both of these situations, P dollars is deposited and it, together with all the interest generated, is earning income at the same interest rate for the same amount of time. Therefore, the final amounts of money in the accounts should be the same. That is,

$$(1 + i)^n P = s_{\overline{n}|\, i} R, \qquad \text{so that} \qquad P = \frac{s_{\overline{n}|\, i}}{(1 + i)^n} \cdot R.$$

Let us now substitute the value of $s_{\overline{n}|\, i}$, namely $[(1 + i)^n - 1]/i$, into the preceding formula. If we denote by $a_{\overline{n}|\, i}$ the expression

$$a_{\overline{n}|\, i} = \frac{(1 + i)^n - 1}{i(1 + i)^n},$$

then the preceding formula may be written in the simple form

$$P = a_{\overline{n}|i}R, \qquad \text{or} \qquad R = \frac{1}{a_{\overline{n}|i}} \cdot P.$$

Tables 6 and 7 of Appendix A give, respectively, the values of $a_{\overline{n}|i}$ and $1/a_{\overline{n}|i}$ for various values of n and i. Let us record the main result of our preceding discussion.

> The present value P and the rent R of a decreasing annuity of n payments with interest compounded at a rate i per interest period are related by the formulas
>
> $$P = a_{\overline{n}|i}R, \qquad R = \frac{1}{a_{\overline{n}|i}} \cdot P.$$

EXAMPLE 4 **Determining the present value for a decreasing annuity** How much money must you deposit now at 6% interest compounded quarterly in order to be able to withdraw $3000 at the end of each quarter-year for 2 years?

Solution For this decreasing annuity, $R = 3000$, $i = 1.5\%$, and $n = 8$. We are asked to calculate the present value of the sequence of payments.

$$P = a_{\overline{n}|i}R = a_{\overline{8}|1.5\%} \cdot 3000 = (7.48592508) \cdot 3000 \qquad \text{(by Table 6)}$$
$$= \$22{,}457.78$$

Now Try Exercise 15 ■

The Excel spreadsheet in Table 2 shows the effect of interest rates (compounded quarterly) on the present value where the duration of the decreasing annuity is two years. [The entry in cell B5 was set to **=PV(A5/4,8,-$B$1)**. The form of the present value function is **PV(**$rate$,$nper$,pmt**)**. The payment was specified as a negative number since money is being withdrawn from the account. The remainder of the B column was created by selecting the cell B5 and dragging its fill handle. The dollar signs in **$B$1** prevent references to this location from being changed.]

TABLE 2

	A	B
1	Rent	$3,000.00
2		
3	Interest	Present
4	rate	value
5	3.0%	$23,209.84
6	3.5%	$23,081.91
7	4.0%	$22,955.03
8	4.5%	$22,829.19
9	5.0%	$22,704.37
10	5.5%	$22,580.57
11	6.0%	$22,457.78
12	6.5%	$22,335.97
13	7.0%	$22,215.16
14	7.5%	$22,095.32
15	8.0%	$21,976.44

EXAMPLE 5 **Determining the rent for a decreasing annuity** If you deposit $10,000 into a fund paying 6% interest compounded monthly, how much can you withdraw at the end of each month for 1 year?

Solution For this decreasing annuity, $P = 10,000$, $i = \frac{1}{2}\%$, and $n = 12$. We are asked to calculate the rent for the sequence of payments.

$$R = \frac{1}{a_{\overline{n}|i}} \cdot P = \frac{1}{a_{\overline{12}|1/2\%}} \cdot 10,000 = (.08606643) \cdot 10,000 \qquad \text{(by Table 7)}$$

$$= \$860.66$$

Now Try Exercise 5   ■

▶ *Remark* In this section we have considered only annuities with payments made at the end of each interest period. Such annuities are called *ordinary annuities*. Annuities that have payments at the beginning of the interest period are called *annuities due*. Annuities whose payment period is different from the interest period are called *general annuities*.   ◀

GC As a time-saving device, the formulas for $s_{\overline{n}|i}$, $\frac{1}{s_{\overline{n}|i}}$, $a_{\overline{n}|i}$, and $\frac{1}{a_{\overline{n}|i}}$ can be assigned to functions in the **Y=** editor. In Fig. 1, the formulas have been stored in **Y₄** through **Y₇** to correspond to the table numbers in Appendix A. Here **X**, instead of n, is used to represent the number of periods. (*Note*: To display **Y₄** with a TI-83, press $\boxed{\text{VARS}}$ ▶ **1 4**.) These functions have been deselected since we want to use them, but not graph them. Figures 2 and 3 show two ways the value of $s_{\overline{60}|1/10\%}$ can be displayed on the home screen.

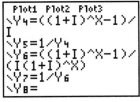

Figure 1

Figure 2

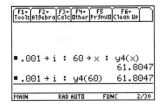

Figure 3

Graphing calculators can easily display successive balances in an annuity.[1] Consider the situation from Example 1 in which $100 is deposited at the end of each month at 6% interest compounded monthly. In Figs. 4 and 7 (see next page) successive balances are displayed in the home screen. In Figs. 5 and 8 successive balances are graphed, and in Figs. 6 and 9 they are displayed in a table.

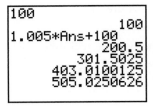

Figure 4

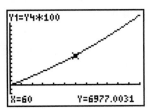

Figure 5

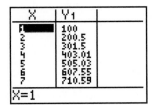

Figure 6

---

[1]We limit our discussion here to increasing annuities. The corresponding analysis for decreasing annuities is similar to the analysis for the amortization of loans presented in detail in Section 10.3.

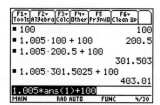

Figure 7

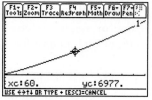

Figure 8

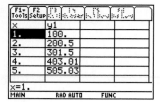

Figure 9

These figures were obtained with the same processes as Figures 1–6 in Section 10.1. Successive balances were determined with the relation $B_{\text{new}} = (1+i)B_{\text{previous}} + R$. The function to be graphed was specified as $\mathbf{Y_1} = \mathbf{Y_4} * \mathbf{100}$ on the TI-83 and as $\mathbf{y1} = \mathbf{y4(x)} * \mathbf{100}$ on the TI-89. The window was set to $[0, 120]$ *by* $[-4000, 17000]$ with an X-scale of 10 and a Y-scale of 1000. Before the function was graphed, the value .005 was stored in I (or i) from the home screen. In the TABLE SETUP menu both **TblStart** and Δ**Tbl** were set to 1.

EXAMPLE 6   **Determining the time for a future value to exceed a specified amount**   Consider the situation from Example 1 in which \$100 is deposited monthly at 6% interest compounded monthly. Use a graphing calculator to determine when the balance will exceed \$10,000.

Solution   There are three ways to answer this question with a graphing calculator—with the home screen, with a graph, and with a table. (*Note*: The three TI-83 screens shown below have TI-89 analogs that can be created in a similar way.)

1. (Home screen)   Generate successive balances (while keeping track of the number of times the ENTER key has been pressed) until a balance exceeding 10,000 is reached. See Fig. 10.

2. (Graph)   Set $\mathbf{Y_2} = \mathbf{10000}$ and find its intersection with the graph of $\mathbf{Y_1} = \mathbf{Y_4 * 100}$. See Fig. 11.

3. (Table)   Scroll down the table for $\mathbf{Y_1}$ until the balance exceeds 10,000. See Fig. 12.

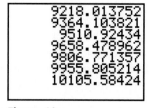

Figure 10

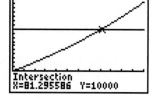

Figure 11

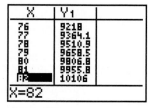

Figure 12

In each case we see that the balance will exceed \$10,000 after 82 months; that is, after 6 years and 10 months.

Now Try Exercise 43   ■

ES   The section "Using Excel's Financial Functions" of Appendix C shows how the functions FV, PV, and PMT are used to calculate future values (F), present values (P), and monthly payments (R) for annuities.

PRACTICE PROBLEMS 10.2

Decide whether or not each of the following annuities is an ordinary annuity, that is, the type of annuity considered in this section. If so, identify n, i, and R and calculate the present value or the future value, whichever is appropriate.

1. You make a deposit at 9% interest compounded monthly into a fund that pays you $1 at the end of each month for 5 years.

2. At the end of each week for 2 years you deposit $10 into a savings account earning 6% interest compounded monthly.

3. At the end of each month for 2 years, you deposit $10 into a savings account earning 6% interest compounded monthly.

EXERCISES 10.2

1. For each of the following increasing annuities, specify i, n, R, and F.

   (a) If at the end of each month, $50 is deposited into a savings account paying 6% interest compounded monthly, the balance after 10 years will be $8193.97.

   (b) Mr. Smith is saving to buy a $65,000 yacht in the year 2007. Since 1997, he has been depositing $2675.19 at the end of each half-year into a fund paying 4% interest compounded semiannually.

2. For each of the following decreasing annuities, specify i, n, R, and P.

   (a) A retiree deposits $72,582 into a bank paying 6% interest compounded monthly and withdraws $520 at the end of each month for 20 years.

   (b) In order to receive $700 at the end of each quarter-year from 2003 until 2008, Ms. Jones deposited $11,446 into a savings account paying 8% interest compounded quarterly.

3. Calculate the future value of an increasing annuity of $1000 per quarter for 5 years at 6% interest compounded quarterly.

4. Calculate the rent of an increasing annuity at 4% interest compounded semiannually if payments are made every half-year and the future value after 7 years is $10,000.

5. Calculate the rent of a decreasing annuity at 8% interest compounded quarterly with payments made every quarter-year for 7 years and present value $100,000.

6. Calculate the present value of a decreasing annuity of $1000 per year for 10 years at 6% interest compounded annually.

7. (*Savings Account*) A person deposits $500 into a savings account at the end of every month for 4 years at 6% interest compounded monthly.

   (a) Find the balance at the end of 4 years.

   (b) How much interest will be earned during the 4 years?

   (c) Prepare a table showing the balance and interest for the first three months.

8. (*Savings Account*) A person deposits $800 into a savings account at the end of every quarter-year for 5 years at 6% interest compounded quarterly.

   (a) Find the balance at the end of 5 years.

   (b) How much interest will be earned during the 5 years?

   (c) Prepare a table showing the balance and interest for the first three quarters.

9. (*Saving for a Car*) How much must Jim save each month in order to have $12,000 to buy a new car in 3 years if the interest rate is 6% compounded monthly? How much of the $12,000 does Jim actually deposit and how much of it is interest?

10. (*Saving to Pay Off a Loan*) Valerie needs $5000 3 years from now in order to pay off a loan. How much must she save each quarter for the next 3 years if interest rates are 8% compounded quarterly? How much of the $5000 will be interest?

11. (*Comparing Bonus Plans*) When Bridget takes a new job, she is offered a $2000 bonus now or the option of an extra $200 each month for the next year. If interest rates are 6% compounded monthly, which choice is better and by how much?

12. (*College Allowance*) When Michael started college, his father gave him $5000 deposited in a savings account earning 6% interest compounded monthly. Michael withdraws $500 a month starting at the beginning of the second month. Prepare a schedule showing the monthly balance in the account for the first four months.

13. (*Saving to Pay Off a Debt*) A city has a debt of $1,000,000 falling due in 15 years. How much money must it deposit at the end of each half-year into a savings fund at 4% interest compounded semiannually in order to raise this amount?

14. (*Saving for a Bicycle*) On January 1, 2005, Tom decided to save for exactly 1 year for a 10-speed bike by depositing $10 at the end of each month into a savings account paying 6% interest compounded monthly. How much did he accumulate?

15. (*College Allowance*) During Jack's first year at college, his father had been sending him $100 per month for incidental expenses. For the sophomore year, his father decided instead to make a deposit into a savings account on August 1 and have his son withdraw $100 on the first of each month from September 1 to May 1. If the bank pays 6% interest compounded monthly, how much should Jack's father deposit?

16. (*Magazine Subscription*) Suppose that a magazine subscription costs $9 per year and that you receive a magazine at the end of each month. At an interest rate of 6% compounded monthly, how much are you actually paying for each issue?

17. Is it more profitable to receive $1000 at the end of each month for 10 years or to receive a lump sum of $163,000 at the end of 10 years? Assume that money can earn 6% interest compounded monthly.

18. Is it more profitable to receive a lump sum of $10,000 at the end of 3 years or to receive $750 at the end of each quarter-year for 3 years? Assume that money can earn 8% interest compounded quarterly.

19. (*Savings Account*) Suppose that you deposited $1000 into a savings account on January 1, 1999, and deposited an additional $100 into the account at the end of each quarter-year. If the bank pays 8% interest compounded quarterly, how much will be in the account on January 1, 2008?

20. (*Savings Account*) Suppose that you opened a savings account on January 1, 2003, and made a deposit of $100. In 2004 you began depositing $10 into the account at the end of each month. If the bank pays 6% interest compounded monthly, how much money will be in the account on January 1, 2008?

21. (*Savings Account*) Ms. Jones deposited $100 at the end of each month for 10 years into a savings account earning 6% interest compounded monthly. However, she deposited an additional $1000 at the end of the seventh year. How much money was in the account at the end of the tenth year?

22. (*Savings Account*) Redo Exercise 21 for the situation where Ms. Jones withdrew $1000 at the end of the seventh year instead of depositing it.

23. Suppose you deposit $600 every six months for 5 years into an annuity at 4% interest compounded semiannually. Which of items (a)–(d) can be used to fill in the blank in the following statement? (*Note:* Before computing, use your intuition to guess the correct answer.)

If the _____ doubles, then the amount accumulated will double.

(a) rent

(b) interest rate

(c) number of interest periods per year

(d) number of years

24. Suppose you deposit $100,000 into an annuity at 4% interest compounded semiannually and withdraw an equal amount at the end of each interest period so that the account is depleted after 10 years. Which of items (a)–(d) can be used to fill in the blank in the following statement? (*Note:* Before computing, use your intuition to guess the correct answer.)

If the _____ doubles, then the total amount withdrawn each year will double.

(a) amount deposited

(b) interest rate

(c) number of interest periods per year

(d) number of years

(*Perpetuity*) *A perpetuity is similar to a decreasing annuity, except that the payments continue forever. Exercises 25 and 26 concern such annuities.*

25. A grateful alumnus decides to donate a permanent scholarship of $1200 per year. How much money should be deposited in the bank at 5% interest compounded annually in order to be able to supply the money for the scholarship at the end of each year?

26. Show that to establish a perpetuity paying R dollars at the end of each interest period, it requires a deposit of R/i dollars, where i is the interest rate per interest period.

(*Deferred Annuity*) *A deferred annuity is a type of decreasing annuity whose term is to start at some future date. Exercises 27–30 concern such annuities.*

27. (*College Fund*) On his tenth birthday, a boy inherits $10,000, which is to be used for his college education. The money is deposited into a trust fund that will pay him R dollars on his 18th, 19th, 20th, and 21st birthdays. Find R if the money earns 6% interest compounded annually.

28. (*College Fund*) Refer to Exercise 27. Find the size of the inheritance that would result in $10,000 per year during the college years (ages 18–21, inclusive).

29. (*Trust Fund*) On December 1, 1998, a philanthropist set up a permanent trust fund to buy Christmas presents for needy children. The fund will provide $6000 each year beginning on December 1, 2008. How much must have been set aside in 1998 if the money earns 6% interest compounded annually?

30. Show that the rent paid by a deferred annuity of n payments that are deferred by m interest periods is given by the formula

$$R = \frac{i(1+i)^{n+m}}{(1+i)^n - 1} \cdot P.$$

31. One dollar is deposited in a savings account with an interest rate of i per interest period. At the end of each interest period, the earned interest is withdrawn

from the account and deposited into a second account earning the same rate of interest.

(a) How much money will be in the first account after n interest periods?

(b) How much money will be in the second account after n interest periods?

(c) Since both the original deposit and the interest are all on deposit and earning interest throughout the entire n interest periods, the amounts in parts (a) and (b) must add up to $(1+i)^n$. Use this fact to derive

$$s_{\overline{n}|i} = \frac{(1+i)^n - 1}{i}.$$

32. Show that $\dfrac{1}{a_{\overline{n}|i}} - \dfrac{1}{s_{\overline{n}|i}} = i$.

33. Show that $s_{\overline{n+1}|i} = (1+i)s_{\overline{n}|i} + 1$.

34. One hundred dollars is deposited in the bank at the end of each year for 20 years. During the first 5 years, the bank paid 6% interest compounded annually and after that paid 7% interest compounded annually. Show that the account balance after 20 years is

$$(1.07)^{15} \cdot 100s_{\overline{5}|6\%} + 100s_{\overline{15}|7\%}.$$

35. (*Municipal Bond*) A municipal bond pays 4% interest compounded semiannually on its face value of $5000. The interest is paid at the end of every half-year period. Fifteen years from now the face value of $5000 will be returned. The current market interest rate for municipal bonds is 3% compounded semiannually. How much should you pay for the bond?

36. (*Second Mortgage*) A businessperson wishes to lend money for a second mortgage on some local real estate. Suppose that the mortgage pays $500 per month for a 5-year period. Suppose that you can invest your money in certificates of deposit paying 6% interest compounded monthly. How much should you offer for the mortgage?

37. (*Business Note*) Suppose that a business note for $50,000 carries an interest rate of 18% compounded monthly. Suppose that the business pays only interest for the first 5 years and then repays the loan amount plus interest in equal monthly installments for the next 5 years.

(a) Calculate the monthly payments during the second 5-year period.

(b) Assume that the current market interest rate for such loans is 12% compounded monthly. How much should you be willing to pay for such a note?

38. (*Lottery Payoff*) A lottery winner is to receive $1000 a month for the next 5 years. How much is this sequence of payments worth today if interest rates are

6% compounded monthly? How is the difference between this amount and the $60,000 paid out beneficial to the agency running the lottery?

39. Formula (2), $B_{new} = (1+i)B_{previous} + R$, applies to an increasing annuity. What is the corresponding formula for a decreasing annuity?

40. (*Lottery Payoff*) A lottery winner is given two options:

Option I: Receive a $20 million lump-sum payment

Option II: Receive 25 equal annual payments totaling $60 million, with the first payment occurring immediately.

If money can earn 6% interest compounded annually, which option is better? *Hint*: With each option, calculate the amount of money earned at the end of 24 years if all the funds are to be deposited into a savings account as soon as they are received.

In Exercises 41–45, use a graphing calculator, a spreadsheet, or mathematical software to answer the questions.

41. A person deposits $1000 at the end of each year into an annuity earning 5% interest compounded annually. What are the balances (that is, future values) after 2, 3, and 4 years? How many years are required for the balance to reach $30,539? After how many years will the balance exceed $50,000?

42. A person deposits $700 at the end of each quarter-year into an annuity earning 8% interest compounded quarterly. What are the balances (that is, future values) after 1, 2, and 3 quarters? How many quarters are required for the balance to reach $5204? After how many quarters will the balance exceed $10,000?

43. (*Bicycle Fund*) Jane has taken a part-time job to save for a $503 racing bike. If she puts $15 each week into a savings account paying 5.2% interest compounded weekly, when will she be able to buy the bike?

44. (*Home Repairs Fund*) Bob needs $3228 to have some repairs done on his house. He decides to deposit $100 at the end of each month into an annuity earning 6% interest compounded monthly. When will he be able to do the repairs?

45. (*Comparing Expense Ratios*) Mutual funds charge an annual fee, called the *expense ratio*, that usually ranges from about .3% to about 1.5% of the amount of money in the account at the end of the year. For instance, if the expense ratio is .5% and you have $10,000 in the account at the end of the year, then you would be charged $.005 \cdot 10000 = \$50$ in fees. Suppose that you put $5000 at the beginning of each year into an IRA and that the funds are invested in a mutual fund earning 6% per year, before expenses. How much more will you have at the end of 10 years if the mutual fund has an expense ratio of .3%, as opposed to 1.5%?

SOLUTIONS TO PRACTICE PROBLEMS 10.2

1. An ordinary decreasing annuity with $n = 60$, $i = \frac{3}{4}\%$, and $R = 1$. You will make a deposit now, in the present, and then withdraw money each month. The amount of this deposit is the present value of the annuity.

$$P = a_{\overline{n}|i}R$$
$$= a_{\overline{60}|.75\%} \cdot 1$$
$$= (48.17337352) \cdot 1 \quad \text{(by Table 6)}$$
$$= \$48.17$$

[*Notes:* (1) When $R = 1$, we have $P = a_{\overline{n}|i}$. This explains why $a_{\overline{n}|i}$ is often called the *present value of an annuity of n payments of $1*. (2) For this transaction, the future value of the annuity needn't be computed. At the end of 5 years, the fund will have a balance of 0.]

2. Not an ordinary annuity since the payment period (1 week) is different from the interest period (1 month).

3. An ordinary increasing annuity with $n = 24$, $i = .5\%$, and $R = \$10$. There is no money in the account now, in the present. However, in 2 years, in the future, money will have accumulated. So for this annuity, only the future value is of concern to you.

$$F = s_{\overline{n}|i}R$$
$$= s_{\overline{24}|.5\%} \cdot 10$$
$$= (25.43195524) \cdot 10 \quad \text{(by Table 4)}$$
$$= \$254.32$$

10.3 Amortization of Loans

In this section we analyze the mathematics of paying off loans. The loans we shall consider will be repaid in a sequence of equal payments at regular time intervals, with the payment intervals coinciding with the interest periods. The process of paying off such a loan is called *amortization*. In order to obtain a feeling for the amortization process, let us consider a particular case, the amortization of a $563 loan to buy a television set. Suppose that this loan charges interest at a 12% rate with interest compounded monthly and that the monthly payments are $116 for 5 months. The repayment process is summarized in Table 1.

TABLE 1

Payment number	Amount	Interest	Applied to principal	Unpaid balance
1	$116	$5.63	$110.37	$452.63
2	116	4.53	111.47	341.16
3	116	3.41	112.59	228.57
4	116	2.29	113.71	114.85
5	116	1.15	114.85	0.00

Note the following facts about the financial transactions.

1. Payments are made at the end of each month. The payments have been carefully calculated to pay off the debt, with interest, in the specified time interval.

2. Since $i = 1\%$, the interest to be paid each month is 1% of the unpaid balance at the end of the previous month. That is, 5.63 is 1% of 563, 4.53 is 1% of 452.63, and so on.

3. Although we write just one check each month for $116, we regard part of the check as being applied to payment of that month's interest. The remainder, namely $116 - $ [interest], is regarded as being applied to repayment of the principal amount.

4. The unpaid balance at the end of each month is the previous unpaid balance minus the portion of the payment applied to the principal. A loan can be paid off early by paying just the current unpaid balance.

The four factors that describe the amortization process just described are as follows:

the principal	$563
the interest rate	12% compounded monthly
the term	5 months
the monthly payment	$116

The important fact to recognize is that the sequence of payments in the preceding amortization constitutes a decreasing annuity, with the person taking out the loan paying the interest. Therefore, the mathematical tools developed in Section 10.2 suffice to analyze the amortization. In particular, we could determine the monthly payment or the principal once the other three factors have been specified.

EXAMPLE 1 **Calculating values associated with a loan** Suppose that a loan has an interest rate of 12% compounded monthly and a term of 5 months.

(a) Given that the principal is $563, calculate the monthly payment.

(b) Given that the monthly payment is $116, calculate the principal.

(c) Given that the monthly payment is $116, calculate the unpaid balance after 3 months.

Solution The sequence of payments constitutes a decreasing annuity with the monthly payments as rent and the principal as the present value. Also, $n = 5$ and $i = 1\%$.

(a) Since $R = (1/a_{\overline{n}|i})P$, we see that

$$R = \frac{1}{a_{\overline{5}|1\%}} \cdot 563 = (.20603980) \cdot 563 \quad \text{(by Table 7)}$$
$$= \$116.$$

(b) Since $P = a_{\overline{n}|i}R$, we see that

$$P = a_{\overline{5}|1\%} \cdot 116 = (4.85343124) \cdot 116 = \$563.00. \quad \text{(by Table 6)}$$

(c) The unpaid balance is most easily calculated by regarding it as the amount necessary to retire the debt. Therefore, it must be sufficient to generate, with interest, the sequence of two remaining payments. That is, the unpaid balance is the present value of a decreasing annuity of two payments of $116. So we see that

$$[\text{unpaid balance after 3 months}] = a_{\overline{2}|1\%} \cdot 116 = (1.97039506) \cdot 116$$
$$= \$228.57.$$

Now Try Exercises 1 and 3 ∎

A *mortgage* is a long-term loan used to purchase real estate. The real estate is used as collateral to guarantee the loan.

EXAMPLE 2 **Calculating values associated with a mortgage** On December 31, 1990, a house was purchased with the buyer taking out a 30-year, $112,475 mortgage at 9% interest, compounded monthly. The mortgage payments are made at the end of each month.

(a) Calculate the amount of the monthly payment.

(b) Calculate the unpaid balance of the loan on December 31, 2016, just after the 312th payment.

(c) How much interest will be paid during the month of January 2017?

(d) How much of the principal will be paid off during the year 2016?

(e) How much interest will be paid during the year 2016?

Solution (a) Denote the monthly payment by R. Then \$112,475 is the present value of an annuity of $n = 360$ payments, with $i = \frac{3}{4}\%$. Therefore,

$$R = \frac{1}{a_{\,\overline{n}|\,i}} \cdot P = \frac{1}{a_{\,\overline{360}|\,.75\%}} \cdot 112{,}475$$

$$= (.00804623) \cdot 112{,}475 \qquad \text{(by Table 7)}$$

$$= \$905.00.$$

(b) The remaining payments constitute a decreasing annuity of 48 payments. Therefore, the unpaid balance is the present value of that annuity.

$$[\text{unpaid balance}] = a_{\,\overline{n}|\,i}R = a_{\,\overline{48}|\,.75\%} \cdot 905$$

$$= (40.18478189) \cdot 905 \qquad \text{(by Table 6)}$$

$$= \$36{,}367.23$$

(c) The interest paid during one month is i, the interest rate per month, times the unpaid balance at the end of the preceding month. Therefore,

$$[\text{interest for January 2017}] = \tfrac{3}{4}\% \text{ of } \$36{,}367.23$$

$$= .0075 \cdot 36{,}367.23$$

$$= \$272.75.$$

(d) Since the portions of the monthly payments applied to repay the principal have the effect of reducing the unpaid balance, this question may be answered by calculating how much the unpaid balance will be reduced during 2016. Reasoning as in part (b), we determine that the unpaid balance on December 31, 2015 (just after the 300th payment), is equal to \$43,596.90. Therefore,

$[\text{amount of principal repaid in 2016}]$

$$= [\text{unpaid balance Dec. 31, 2015}] - [\text{unpaid balance Dec. 31, 2016}]$$

$$= \$43{,}596.90 - \$36{,}367.23 = \$7229.67.$$

(e) During the year 2016, the total amount paid is $12 \times 905 = \$10{,}860$. But by part (d), \$7229.67 is applied to repayment of principal, the remainder being applied to interest.

$$[\text{interest in 2016}] = [\text{total amount paid}] - [\text{principal repaid in 2016}]$$

$$= \$10{,}860 - \$7229.67 = \$3630.33$$

Now Try Exercise 5

■

In the early years of a mortgage, most of each monthly payment is applied to interest. For the mortgage described in Example 2, the interest portion will exceed the principal portion until the twenty-third year.

Each new (unpaid) balance can be computed from the previous balance with the formula

$$B_{\text{new}} = (1 + i)B_{\text{previous}} - R.$$

That is, the new balance equals the growth of the previous balance due to interest, minus the amount paid. Successive balances are computed by multiplying by $(1+i)$ and subtracting R.

> If P is borrowed at interest rate i per period and R is paid back at the end of each interest period, then the formula
>
> $$B_{\text{new}} = (1 + i)B_{\text{previous}} - R \tag{1}$$
>
> can be used to calculate each new balance, B_{new}, from the previous balance, B_{previous}.

The Excel spreadsheet in Table 2 calculates the payment per period and shows the successive balances for the annuity of Example 2. [The entry in cell B4 was set to **=-PMT(B1,B2,B3)**. The form of the payment function is **PMT(** $rate, nper, pv$ **)**. The value of the payment function is a negative number since the money is owed to the bank. B10 was set to B3 and B11 was set to **=(1+$B$1)*B10-$B$4**. The remainder of the B column was created by selecting the cell B11 and dragging its fill handle. C11 was set to **=$B$1*B10** and D11 was set to **=$B$4-C11**. The remainder of the C and D columns was created by dragging fill handles.]

TABLE 2 Amortization table

	A	B	C	D
1	Interest rate	0.75%		
2	Number of periods	360		
3	Principal	$112,475.00		
4	Payment per period	$905.00		
5				
6		Amortization Table		
7				
8		Unpaid		Applied to
9	Payment number	balance	Interest	principal
10	0	$112,475.00		
11	1	$112,413.56	$843.56	$61.44
12	2	$112,351.67	$843.10	$61.90
13	3	$112,289.30	$842.64	$62.36
14	4	$112,226.47	$842.17	$62.83
15	5	$112,163.17	$841.70	$63.30
16	6	$112,099.40	$841.22	$63.78
17	7	$112,035.14	$840.75	$64.25
18	8	$111,970.41	$840.26	$64.74
19	9	$111,905.19	$839.78	$65.22
20	10	$111,839.48	$839.29	$65.71
21	11	$111,773.27	$838.80	$66.20
22	12	$111,706.57	$838.30	$66.70

	A	B	C	D
359	349	$9,521.20	$77.61	$827.38
360	350	$8,687.61	$71.41	$833.59
361	351	$7,847.77	$65.16	$839.84
362	352	$7,001.63	$58.86	$846.14
363	353	$6,149.14	$52.51	$852.49
364	354	$5,290.26	$46.12	$858.88
365	355	$4,424.94	$39.68	$865.32
366	356	$3,553.13	$33.19	$871.81
367	357	$2,674.78	$26.65	$878.35
368	358	$1,789.84	$20.06	$884.94
369	359	$898.26	$13.42	$891.58
370	360	$0.00	$6.74	$898.26
371	Payment number	Unpaid	Interest	Applied to
372		balance		Principal

EXAMPLE 3 **Calculating the unpaid balance on a mortgage** Refer to Example 2. Compute the unpaid balance of the loan on January 31, 2017, just after the 313th payment.

Solution By formula (1),

$$B_{\text{Jan}} = (1 + i)B_{\text{Dec}} - R = (1.0075) \cdot 36{,}367.23 - 905$$
$$= \$35{,}734.98. \qquad \blacksquare$$

Sometimes amortized loans stipulate a *balloon payment* at the end of the term. For instance, you might pay \$200 at the end of each quarter-year for 3 years and an additional \$1000 at the end of the third year. The \$1000 is a balloon payment.

EXAMPLE 4   **Determining how much you can afford to borrow** How much money can you borrow at 8% interest compounded quarterly if you agree to pay \$200 at the end of each quarter-year for 3 years and in addition a balloon payment of \$1000 at the end of the third year?

Solution Here you are borrowing in the present and repaying in the future. The amount of the loan will be the present value of *all* the future payments. The future payments consist of an annuity and a lump-sum payment. Let us calculate the present values of each of these separately. Now, $i = 2\%$ and $n = 12$.

$$[\text{present value of annuity}] = a_{\overline{n}|i}R = a_{\overline{12}|2\%} \cdot 200$$
$$= (10.57534122) \cdot 200 \quad (\text{by Table 6})$$
$$= \$2115.07$$

$$[\text{present value of balloon payment}] = \frac{1}{(1+i)^n} \cdot F$$

$$= \frac{1}{(1+i)^n} \cdot 1000 \quad (\text{for } i = 2\% \text{ and } n = 12)$$

$$= .78849318 \cdot (1000) = \$788.49$$

Therefore, the amount you can borrow is

$$\$2115.07 + \$788.49 = \$2903.56.$$

Now Try Exercise 11 $\blacksquare$

GC As in Section 10.2, we will assume that the formulas for $s_{\overline{n}|i}$, $\frac{1}{s_{\overline{n}|i}}$, $a_{\overline{n}|i}$, and $\frac{1}{a_{\overline{n}|i}}$ have been assigned to the functions $\mathbf{Y_4}$ through $\mathbf{Y_7}$.

Graphing calculators can easily display successive balances for a loan. Consider the situation from Example 2 in which \$112,475 is borrowed at 9% interest compounded monthly and repaid with 360 monthly payments of \$905. In Figs. 1 and 4, successive balances are displayed in the home screen. In Figs. 2 and 5, successive balances are graphed, and in Figs. 3 and 6 they are displayed as part of an amortization table. (The $\mathbf{Y_2}$ column gives the interest portion of the monthly payment.)

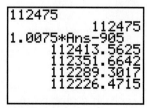

Figure 1

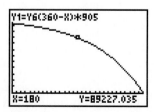

Figure 2

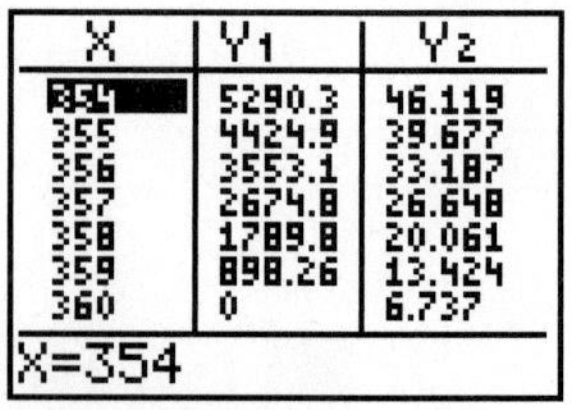

X	Y1	Y2
354	5290.3	46.119
355	4424.9	39.677
356	3553.1	33.187
357	2674.8	26.648
358	1789.8	20.061
359	898.26	13.424
360	0	6.737

X=354

Figure 3

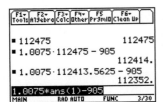

Figure 4

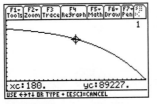

Figure 5

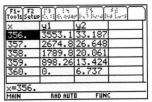

Figure 6

Figures 1–6 were obtained with the same processes as Figures 1–6 in Section 10.1. Successive balances were determined with the relation $B_{\text{new}} = (1+i)B_{\text{previous}} - R$. The function to be graphed was specified as $\mathbf{Y_1 = Y_6(360 - X) * 905}$; that is, $a_{\overline{360-n}|\,i} \cdot R$. The window was set to $[0, 360]$ by $[-15000, 130000]$ with x-scale of 12 and a y-scale of 10000. Before the function was graphed, the value .0075 was stored in I from the home screen. In the TABLE SETUP menu **TblStart** was set to **354** on the TI-83 and to **356** on the TI-89. The second function was defined as $\mathbf{Y_2 = I*Y_1(X-1)}$; that is, the interest rate times the amount of the previous balance. The up-arrow key can be used to generate earlier balances and interest payments.

EXAMPLE 5 **Analyzing the apportionment of mortgage payments** Consider the situation from Example 2 in which \$112,475 is borrowed at 9% interest compounded monthly and repaid with 360 monthly payments of \$905. When will the debt-reduction portion of the payment surpass the interest portion?

Solution The question can be answered with a table or a graph. (*Note*: The three TI-83 screens shown below have TI-89 analogs that can be created in a similar way.)

1. (Table) Consider the table of Fig. 3. Scroll up until an interest greater than half the monthly payment, that is, 452.50 is reached. Set $\mathbf{Y_3 = 900-Y_2}$. Then press ▶ three times to obtain Fig. 7.

2. (Graph) Let $\mathbf{Y_2}$ and $\mathbf{Y_3}$ be the interest and debt-reduction functions as in part 1. Set the window as in Fig. 8, graph the two functions, and find their point of intersection. See Fig. 9.

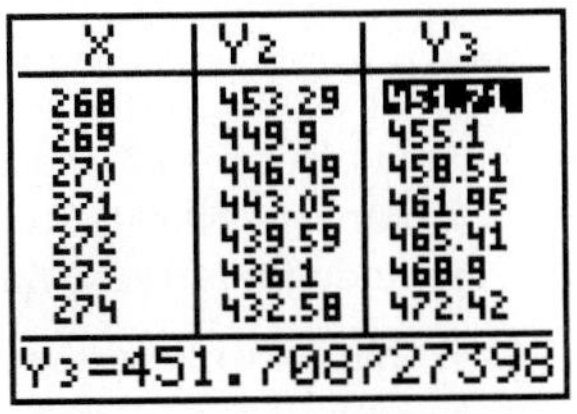

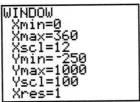

Figure 7

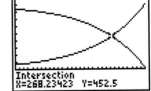

Figure 8

Figure 9

In each case we see that debt reduction will exceed interest after 269 months; that is, after 22 years and 5 months.

Now Try Exercise 33

ES The section "Using Excel's Financial Functions" of Appendix C shows how the functions IPMT and PPMT are used to calculate the interest and principal reduction portions of a specific payment; that is, the values in column C and D of

Table 2. The annuity functions FV, PV, and PMT discussed earlier also apply to mortgages.

1. The word "amortization" comes from the French "à mort," meaning "at the point of death." Justify the word.

2. Explain why only present values and not future values arise in amortization problems.

EXERCISES 10.3

1. A loan of $10,000 is to be repaid with monthly payments for 5 years at 6% interest compounded monthly. Calculate the monthly payment.

2. Find the monthly payment on a $100,000, 25-year mortgage at 12% interest compounded monthly.

3. How much money can you borrow at 12% interest compounded semiannually if the loan is to be repaid at half-year intervals for 10 years and you can afford to pay $1000 per half-year?

4. (*Car Loan*) You buy a car with a down payment of $500 and $100 per month for 3 years. If the interest rate is 9% compounded monthly, how much did the car cost?

5. Consider a $58,331, 30-year mortgage at interest rate 12% compounded monthly with a $600 monthly payment.

  (a) How much interest is paid the first month?

  (b) How much of the first month's payment is applied to paying off the principal?

  (c) What is the unpaid balance after 1 month?

  (d) What is the unpaid balance at the end of 25 years?

  (e) How much of the principal is repaid during the 26th year?

  (f) How much interest is paid during the 301st month?

6. Consider a $21,281.27 loan for 7 years at 8% interest compounded quarterly and a payment of $1000 per quarter-year.

  (a) Compute the unpaid balance after 5 years.

  (b) How much interest is paid during the fifth year?

  (c) How much principal is repaid in the first payment?

  (d) What is the total amount of interest paid on the loan?

7. (*Car Loan*) Susie takes out a car loan for $8000 for a term of 3 years at 12% interest compounded monthly.

  (a) Find her monthly payment.

  (b) Find the total amount she pays for the car.

  (c) Find the total amount of interest she pays.

  (d) Find the amount she still owes after 1 year.

  (e) Find the amount she still owes after 2 years.

  (f) Find the total interest she pays in year 2.

  (g) Prepare an amortization schedule for the first 4 months.

8. (*Home Financing*) James buys a house for $90,000. He puts $10,000 down and then finances the rest at 9% interest compounded monthly for 25 years.

  (a) Find his monthly payment.

  (b) Find the total amount he pays for the house.

  (c) Find the total amount of interest he pays.

  (d) Find the amount he still owes after 23 years.

  (e) Find the amount he still owes after 24 years.

  (f) Find the total amount of interest he pays in year 24.

  (g) Prepare the amortization schedule for the first 3 months.

9. A mortgage at 9% interest compounded monthly with a monthly payment of $1125 has an unpaid balance of $10,000 after 350 months. Find the unpaid balance after 351 months.

10. A loan with a quarterly payment of $1500 has an unpaid balance of $10,000 after 30 quarters and an unpaid balance of $9000 after 31 quarters. If interest is compounded quarterly, find the interest rate.

11. A loan is to be amortized over an 8-year term at 12% interest compounded semiannually, payments of $1000 every 6 months, and a balloon payment of $10,000 at the end of the term. Calculate the amount of the loan.

12. A loan of $105,504.50 is to be amortized over a 5-year term at 12% interest compounded monthly with monthly payments and a $10,000 balloon payment at the end of the term. Calculate the monthly payment.

13. Write out a complete amortization schedule (as in Table 1 at the beginning of this section) for the amortization of a $1000 loan with monthly payments at 12% interest compounded monthly for 4 months.

14. Write out a complete schedule (as in Table 1) for the amortization of a $10,000 loan with payments every 6 months at 12% interest compounded semiannually for 1 year.

15. (*Mortgage Payments*) You purchase a $120,000 house, pay $20,000 down, and take out a 30-year mortgage with monthly payments, at an interest rate of 9% compounded monthly. How much money will you be paying each month?

16. (*Terminating a Mortgage*) In 1995 you purchased a house and took out a 25-year, $50,000 mortgage at 6% interest compounded monthly. In 2005 you sold the house for $150,000. How much money did you have left after you paid the bank the unpaid balance on the mortgage?

17. (*Cash Flow for a Rental Property*) You are considering the purchase of a condominium to use as a rental property. You estimate that you can rent the condominium for $1200 per month and that taxes, insurance, and maintenance costs will run about $200 per month. If interest rates are 9% compounded monthly, how large a 25-year mortgage can you assume and still have the rental income cover the monthly expenses?

18. (*Loan Offers*) A bank makes the following two loan offers to its credit card customers.

Option I: Pay 0% interest for five months and then 9% interest compounded monthly for the remainder of the duration of the loan.

Option II: Pay 6% interest compounded monthly for the duration of the loan.

With either option there will be a minimum payment of $100 due each month. Suppose you would like to borrow $5500 and pay it back in a year and a half. For which option do you pay the least amount to the bank?

19. (*Car Loan*) A car is purchased for $6287.10 with $2000 down and a loan to be repaid at $100 a month for 3 years followed by a balloon payment. If the interest rate is 6% compounded monthly, how large will the balloon payment be?

20. (*Refinancing a Mortgage*) A real estate speculator purchases a tract of land for $1 million and assumes a 25-year mortgage at 12% interest compounded monthly.

(a) What is his monthly payment?

(b) Suppose that at the end of 5 years the mortgage is changed to a 10-year term for the remaining balance. What is the new monthly payment?

(c) Suppose that after 5 more years, the mortgage is required to be repaid in full. How much will then be due?

21. Suppose you borrow $10,000 at 4% interest compounded semiannually and pay off the loan in 7 years. Which of items (a)–(d) can be used to fill in the blank in the following statement? (*Note:* Before computing, use your intuition to guess the correct answer.)

If the _____ is doubled, then the total amount paid will double.

(a) amount of the loan

(b) interest rate

(c) number of interest periods per year

(d) term of the loan

22. Suppose you borrow money at 4% interest compounded semiannually and pay off the loan in payments of $1000 per interest period for 5 years. Which of items (a)–(d) can be used to fill in the blank in the following statement? (*Note:* Before computing, use your intuition to guess the correct answer.)

If the _____ is doubled, then the amount that can be borrowed will double.

(a) amount paid per interest period

(b) interest rate

(c) number of interest periods per year

(d) term of the loan

23. Suppose that you make annual payments of $5000 for 20 years into an annuity paying 6% interest compounded annually.

(a) What is the value of the annuity at the end of the 20th year?

(b) Suppose that you elect to have your annuity repaid to you over a 10-year period in annual installments. What is the annual payment you will receive?

(c) Suppose that after you receive payments for 5 years you elect to have the remainder of your annuity paid to you in a lump sum. How much will you receive?

(*Sinking Fund*) A sinking fund *is a pool of money accumulated by a corporation or government to repay a specific debt at some future date.*

24. A corporation wishes to deposit money into a sinking fund at the end of each half-year in order to repay $50 million in bonds in 10 years. It can expect to receive a 12% (compounded semiannually) return on its deposits to the sinking fund. How much should the deposits be?

25. The Federal National Mortgage Association ("Fannie Mae") puts $30 million at the end of each month into a sinking fund paying 12% interest compounded monthly. The sinking fund is to be used to repay debentures that mature 15 years from the creation of the fund. How large is the face amount of the debentures assuming the sinking fund will exactly pay it off?

26. A corporation borrows $5 million to erect a new headquarters. The financing is arranged using industrial development bonds, to be repaid in 15 years. How much should it deposit into a sinking fund at the end of each quarter if the sinking fund earns 8% interest compounded quarterly?

27. A corporation sets up a sinking fund to replace some aging machinery. It deposits $100,000 into the fund at the end of each month for 10 years. The sinking fund earns 12% interest compounded monthly. The equipment originally cost $13 million. However, the cost of the equipment is rising 6% each year. Will the sinking fund be adequate to replace the equipment? If not, how much additional money is needed?

28. A corporation sets up a sinking fund to replace an aging warehouse. The cost of the warehouse today would be $8 million. However, the corporation plans to replace the warehouse in 5 years. It estimates that the cost of the warehouse will increase by 10% annually. The sinking fund will earn 12% interest compounded monthly. What should be the monthly payment to the sinking fund?

29. (*Comparing Car Loans*) In May 2005, General Motors was offering the choice of a 0% loan for 36 months, or $3500 cash back on the purchase of a $24,000 Buick Century.

   (a) If you take the 0% loan offer, how much will your monthly payment be?

   (b) If you take the $3500 cash-back offer and can borrow money from your local bank at 6% interest compounded monthly for three years, how much will your monthly payment be?

   (c) Which of the two offers is more favorable for you?

30. (*Comparing Car Loans*) In May 2005, General Motors was offering the choice of a 0% loan for 60 months, or $4000 cash back on the purchase of a $26,000 Chevrolet Blazer.

   (a) If you take the 0% loan offer, how much will your monthly payment be?

   (b) If you take the $4000 cash-back offer and can borrow money from your local bank at 9% interest compounded monthly for five years, how much will your monthly payment be?

   (c) Which of the two offers is more favorable for you?

31. Let B_n = balance of a loan after n payments, I_n = the interest portion of the nth payment, and Q_n = the portion of the nth payment applied to the principal. Equation (1) states that $B_n = (1 + i)B_{n-1} - R$.

   (a) Use the fact that $I_n + Q_n = R$ and $I_n = iB_{n-1}$ to show that
   $$B_{n-1} = \frac{R - Q_n}{i}.$$

   (b) Use equation (1) and the results that
   $$B_{n-1} = \frac{R - Q_n}{i} \quad \text{and} \quad B_n = \frac{R - Q_{n+1}}{i}$$
   from part (a) to derive the formula $Q_{n+1} = (1 + i)Q_n$.

   (c) State in your own words the formula obtained in part (b).

   (d) Suppose that for the 10th monthly payment on a loan at 1% interest per month, $100 was applied to the principal. How much of the 11th and 12th payments will be applied to the principal?

32. Let Q_n and I_n be as defined in Exercise 31.

   (a) Use the result in part (b) of Exercise 31 and the facts that $Q_{n+1} = R - I_{n+1}$ and $Q_n = R - I_n$ to show that $I_{n+1} = (1 + i)I_n - iR$.

   (b) State in your own words the formula obtained in part (a).

   (c) Suppose that for a loan with a monthly payment of $400 and an interest rate of 1% per month, $100 of the 9th monthly payment goes toward paying off interest. How much of the 10th and 11th payments will be used to pay off interest?

In Exercises 33–38, use a graphing calculator, a spreadsheet, or mathematical software to answer the questions.

33. (*Financing a Computer*) Bill buys a computer for $2188.91 and pays off the loan (at 9% interest compounded monthly) by paying $100 at the end of each month. Determine the balance after 1, 2, and 3 months. After how many months will the loan be paid off?

34. (*Financing Medical Equipment*) Alice borrows $20,000 to buy some medical equipment and pays off the loan (at 7% interest compounded annually) by paying $4195.92 at the end of each year. Determine the balance after 1, 3, and 5 years. After how many years will the loan be paid off?

35. A loan of $10,000 at 9% interest compounded monthly is repaid in 80 months with monthly payments of $166.68. How much of the loan will have been paid off after 40 months? When will more than half of the loan have been repaid?

36. A loan of $4000 at 6% interest compounded monthly is repaid in 8 years with monthly payments of $52.57. After how many months will the loan be one-quarter paid off? One-half? Three-quarters?

37. A 25-year mortgage of $124,188.57 at 8.5% interest compounded monthly has monthly payments of $1000. After how many months will at least 75% of the monthly payment go toward debt reduction?

38. A 30-year mortgage of $118,135.40 at 8.4% interest compounded monthly has monthly payments of $900. After how many months will the amount applied to debt reduction be more than twice the amount applied to interest?

SOLUTIONS TO PRACTICE PROBLEMS 10.3

1. A portion of each payment is applied to reducing the debt, and by the end of the term the debt is totally annihilated.

2. The debt is formed when the creditor gives you a lump sum of money now, in the present. The lump sum of money is gradually repaid by you, with interest, thereby generating the decreasing annuity. At the end of the term, in the future, the loan is totally paid off, so there is no more debt. That is, the future value is always zero!

▶ **Remark** You are actually functioning like a savings bank, since you are paying the interest. Think of the creditor as depositing the lump sum with you and then making regular withdrawals until the balance is 0.   ◀

10.4 Personal Financial Decisions

You will be faced with several financial decisions shortly after graduating from college. Financial advisers suggest that you open an individual retirement account as soon as you start earning an income. You will probably buy a car and some major appliances with consumer loans. You will consider buying a condominium or a house and will need a mortgage. This section answers the following questions:

- How important is it to open an individual retirement account as soon as possible? What kind of IRA should you open?
- What is the difference between the finance charges on a consumer loan using the common add-on method as opposed to using the method discussed in Section 10.3?
- How do you choose between two mortgages if they have different interest rates and up-front fees?

Individual Retirement Accounts

Money earned in an ordinary savings account is subject to federal, state, and local income taxes. However, a special type of savings account, called an *individual retirement account* (IRA), provides a shelter from these taxes. IRAs are highly touted by financial planners and are available to any individual whose income does not exceed a certain limit. The maximum annual contribution[1] for the years 2006 and 2007 is $4000 (but the amount contributed must not exceed the person's earned income). Typically, money in an IRA cannot be withdrawn without penalty until the person reaches $59\frac{1}{2}$ years of age. However, funds may be withdrawn earlier under certain circumstances.

The two main types of IRAs are known as a traditional IRA and a Roth IRA. Contributions to a traditional IRA are tax deductible, but all withdrawals are taxed. (Interest earned is not taxed until it is withdrawn.) In contrast, contributions to a Roth IRA are not tax deductible, but withdrawals are not taxed. (Therefore, interest earned is never taxed.) With a traditional IRA, withdrawals must begin by age $70\frac{1}{2}$. Deposits for the year 2006 can be made at any time from January 1, 2006 until April 15, 2007.

EXAMPLE 1 **Calculating values associated with a traditional IRA** Suppose that for the year 2006 you deposit $4000 of earned income into a traditional IRA on January 1, you earn an annual interest rate of 6% compounded annually, and you are in a 30%

---

[1]The maximum annual contribution will rise to $5000 in 2008.

marginal tax bracket[2] for the duration of the account. (We recognize that interest rates and tax brackets are subject to change over a long period of time, but some assumptions must be made in order to evaluate the investment.)

(a) How much income tax on earnings will you save for the year 2006?

(b) Assuming no additional deposits are made, how much money will be in the account after 48 years?

(c) Suppose you withdraw all the money in the IRA after 48 years (assuming you are older than $59\frac{1}{2}$). How much will you have left after you pay the taxes on the money?

Solution (a) [income tax saved] = [tax bracket] · [amount]

$$= \quad .30 \quad \cdot \quad 4000$$
$$= 1200$$

Therefore, $1200 in income taxes will be saved.

(b) [balance after 48 years] $= P(1+i)^n$

$$= 4000(1+.06)^{48} = 4000(1.06)^{48}$$
$$= 4000(16.39387173)$$
$$= 65,575.49$$

Therefore, after 48 years the balance in the IRA will be $65,575.49.

(c) Since 30% of the money will be paid in taxes, 70% will be left.

$$[\text{amount after taxes}] = .70(65,575.49)$$
$$= \$45,902.84$$

Now Try Exercise 5 ■

EXAMPLE 2 **Comparing a savings account with a traditional IRA** Rework Example 1 with the money going into an ordinary savings account.

Solution (a) None. There are no tax benefits when money is deposited into an ordinary savings account.

(b) The interest earned each year will be taxed during that year. Therefore, each year 30% of the interest will be paid in taxes and hence only 70% will be kept. Since 70% of 6% is 4.2%, you will effectively earn just 4.2% interest per year.

$$[\text{balance after 48 years}] = P(1+i)^n$$
$$= 4000(1+.042)^{48} = 4000(1.042)^{48}$$
$$= \$28,821.10$$

(c) $28,821.10, since the interest was taxed in the year it was earned. ■

Comparing the results from Examples 1 and 2, we see why financial planners encourage people to set up IRAs. Not only is there a substantial initial tax savings, but the balance upon withdrawal is $17,081.74 greater with the IRA. The amount with the IRA is nearly 60% higher than the amount in an ordinary savings account.

EXAMPLE 3 **Calculating the future value of a Roth IRA** Consider the following variation of Example 1. Suppose you earned $4000, paid your taxes, and then deposited the

---

[2]That is, for each additional dollar you earn, you pay an additional $.30 in taxes. Likewise, for each dollar you can deduct, you save $.30 in taxes.

remainder into a Roth IRA. How much money would you have if you closed out the account after 48 years?

Solution

$$[\text{earnings after income tax}] = [1 - \text{tax bracket}] \cdot [\text{amount}]$$
$$= \quad .70 \quad \cdot \quad 4000$$
$$= \$2800$$

Therefore, $2800 will be deposited into the Roth IRA.

$$[\text{balance after 48 years}] = P(1 + i)^n$$
$$= 2800(1 + .06)^{48} = 2800(1.06)^{48}$$
$$= 2800(16.39387173)$$
$$= \$45,902.84$$

Therefore, after 48 years the balance will be $45,902.84, the same amount found in part (c) of Example 1.

Now Try Exercise 7 ■

The result of Example 3 suggests the following general conjecture that is easily verified.

> **Equivalence of traditional and Roth IRAs**   With the assumption that your tax bracket does not change, the net earnings upon withdrawal from contributing P dollars into a traditional IRA account is the same amount as would result from paying taxes on the P dollars but then contributing the remaining money into a Roth IRA.

Verification of the Equivalence Conjecture: Suppose P dollars is deposited into a traditional IRA at interest rate r compounded annually for n years. Also, suppose that your marginal tax bracket is k. (In Example 1, $k = .30$.) In general, if taxes are paid on A dollars, then the amount remaining after taxes is

$$A - k \cdot A = A \cdot (1 - k).$$

With a traditional IRA, the balance after n years is $[P(1 + r)^n]$ and the amount left after taxes are paid is $[P(1 + r)^n] \cdot (1 - k)$.

With a Roth IRA, taxes are paid initially on the P dollars and so the amount deposited is $[P \cdot (1 - k)]$. The balance after n years is obtained by multiplying the amount deposited by $(1 + r)^n$. That is, the balance is $[P \cdot (1 - k)] \cdot (1 + r)^n$.

Each of the amounts above is the product of the same three numbers; only the order is different. Therefore, the two amounts have the same value. ■

Financial planners not only encourage clients to establish IRAs, but to begin making contributions at an early age. The following example illustrates the advantage of beginning as soon as possible.

EXAMPLE 4   **The value of starting an IRA early**   Earl and Larry each begin full-time jobs in January 2006 and plan to retire in January 2054 after working for 48 years. Assume that any money they deposit into IRAs earns 6% interest compounded annually.

(a) Suppose Earl opens a traditional IRA account immediately and deposits $4000 into his account at the end of each year for twelve years. After that he makes no further deposits and just lets the money earn interest. How much money will Earl have in his account when he retires in January 2054?

(b) Suppose Larry waits 12 years before opening his IRA and then deposits $4000 into the account at the end of each year until he retires. How much money will Larry have in his account when he retires in January 2054?

(c) Who paid the most money into his IRA?

(d) Who had the most money in his account upon retirement?

Solution (a) Earl's contributions consist of an annuity with $R = 4000$, $i = .06$, and $n = 12$. The balance in the account on January 1, 2018 will be

$$F = s_{\overline{n}|i}R = s_{\overline{12}|.06} \cdot 4000$$
$$= (16.86994120) \cdot 4000 = \$67,479.76.$$

This money then earns interest compounded annually for 36 years and grows to

$$67,479.76 \cdot (1.06)^{36} = 67,479.76 \cdot (8.147252) = \$549,774.61.$$

(b) Larry's contributions consist of an annuity with $R = 4000$, $i = .06$, and $n = 36$. The balance in the account on January 1, 2054 will be

$$F = s_{\overline{n}|i}R = s_{\overline{36}|.06} \cdot 4000$$
$$= (119.12086666) \cdot 4000 = \$476,483.47$$

(c) Larry pays in $36 \cdot 4000 = \$144,000$ versus the $12 \cdot 4000 = \$48,000$ paid in by Earl. That is, Larry pays in three times as much money as Earl does.

(d) Earl has $73,291.14 more than Larry.

Now Try Exercise 9

Consumer Loans and the Truth-in-Lending Act

Consumer loans are paid off with a sequence of monthly payments in much the same way as mortgages. However, the most widely used method for determining finance charges on consumer loans is the *add-on method*, which results in an understated interest rate. When an add-on method interest rate is stated as r per year, the term of the loan is t years, and the amount of the loan is P, then the total interest to be paid is $P \cdot r \cdot t$. The monthly payment is then determined by dividing the sum of the principal and the interest by the total number of months. For instance, suppose you take out a two-year loan of $1000 at an annual interest rate of 6%.

$$[\text{monthly payment}] = (1000 + 1000 \cdot .06 \cdot 2)/(12 \cdot 2)$$
$$= 1000(1.12)/24$$
$$= \$46.67$$

The general formula for the monthly payment R is

$$R = \frac{P(1 + rt)}{12t}.$$

This formula can be solved for r to obtain

$$r = \frac{\frac{12Rt - P}{t}}{P}.$$

Since $12Rt - P$ is the amount of interest paid for the duration of the loan, $\frac{12Rt - P}{t}$ is the amount of interest paid per year and $\frac{\frac{12Rt - P}{t}}{P}$ is the annual percent of the principal that consists of the interest, namely r.

For a loan amortized as in Section 10.3, the monthly payment would be

$$R = \frac{1}{a\,\overline{n}|\,i} \cdot P = \frac{1}{a\,\overline{24}|\,1/2\%} \cdot 1000$$

$$= (.04432061) \cdot 1000$$

$$= \$44.32.$$

Therefore, the add-on method results in the borrower being charged an additional $46.67 - 44.32 = 2.35$ dollars per month. The monthly payment of $46.67 actually corresponds to an interest rate of 11.134% when the payment is calculated as in Section 10.3. The rate 11.134% is called the *annual percentage rate* (APR). The problem with the add-on method is that each month the borrower is being charged interest on the entire principal even though after the first month part of the principal has already been repaid.

In 1968, Congress passed the federal Truth-in-Lending Act as a part of the Consumer Protection Act. The law's stated goal is

> to assure a meaningful disclosure of credit terms so that the consumer will be able to compare more readily the various credit terms available . . . .

The Truth-in-Lending Act requires a lender to disclose the APR for any advertised loan.

EXAMPLE 5 **Calculating the monthly payment for a loan with the add-on method** A one-year loan for $5000 is advertised at 8% using the add-on method. What is the monthly payment?

Solution The interest on the loan is $.08 \cdot (5000) = \$400$. The total amount (principal plus interest) of $5400 is to be paid in 12 monthly payments of $5400/12 = \$450$.

Now Try Exercise 11 ■

EXAMPLE 6 **Determining the interest rate on a consumer loan** Suppose a consumer loan of $100,000 for 10 years has an APR of 8% compounded monthly and a monthly payment of $1213.28. What interest rate would be stated with the add-on method?

Solution The total amount paid by the borrower is $120 \cdot 1213.28 = \$145{,}593.60$. Therefore, the interest paid on the loan is $145{,}593.60 - 100{,}000 = \$45{,}593.60$. The interest paid per year is $45{,}593.60/10 = \$4559.36$ which is about 4.56% of $100,000. Therefore, with the add-on method, the interest rate would be given as 4.56%.

Now Try Exercise 15 ■

The problem of determining the APR given the add-on interest rate requires solving an equation of the form $P = a\,\overline{n}|\,i \cdot R$ for i. The computation of the APR is easily accomplished with mathematical software (such as Explorations in Finite Mathematics), graphing calculators, or electronic spreadsheets. For instance, with Excel the APR of 11.134% discussed above is calculated with the RATE function as 12*RATE(24, −46.67, 1000, 0). This same problem can be solved with any graphing calculator as shown in Fig. 1.[3] In Fig. 1(a), the function Y_2 is $a\,\overline{24}|\,x \cdot (46.67)$. The

---

[3]The TI-83 screens presented in this section have TI-89 analogs that can be created in a similar way.

x-coordinate of the intersection of the two functions as shown in Fig. 1(c) is the monthly rate of interest. The annual rate is $12 \cdot (.0092782) = .1113384$.

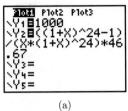

(a)

(b)

(c)

Figure 1

Mortgages with Discount Points

Home mortgages have a stated interest rate, called the *contract rate*. However, some loans also carry *points* or *discount points*. Each point requires that you pay up-front additional interest equal to 1% of the stated loan amount. For instance, for a $200,000 mortgage with 3 discount points you must pay $6000 immediately. This has the effect of reducing the loan to $194,000, yet requiring the same monthly payment as a $200,000 mortgage having no points.

EXAMPLE 7 **Discount points and monthly payments for mortgages** Compute the monthly payment for each of the following 30-year mortgages.

Mortgage A: $200,000 at 6% interest compounded monthly, one point.

Mortgage B: $198,000 at 6.094% compounded monthly, no points

Solution Mortgage A: $\text{[monthly payment]} = \dfrac{1}{a_{\overline{n}|i}} \cdot P = \dfrac{1}{a_{\overline{360}|1/2\%}} \cdot (200{,}000)$

$$= .00599551 \cdot (200{,}000)$$

$$= \$1199.10$$

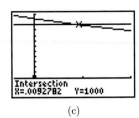

Figure 2

Mortgage B: With an annual interest rate of 6.094%, the monthly interest rate, $.06094/12$, does not appear in Table 7 at the back of the book. Therefore, we must rely on the formula for $a_{\overline{n}|i}$ given in Section 10.2. Figure 2 shows the calculation of $1/a_{\overline{n}|i}$ on a graphing calculator.

$$\text{[monthly payment]} = \frac{1}{a_{\overline{n}|i}} \cdot P$$

$$= \frac{1}{a_{\overline{360}|i}} \cdot (198{,}000), \quad \text{where } i = \frac{.06094}{12} = .00507833.$$

$$= \frac{i(1+i)^{360}}{(1+i)^{360} - 1} \cdot (198{,}000)$$

$$= .0060560728 \cdot (198{,}000)$$

$$= \$1199.10 \qquad\blacksquare$$

The two monthly payments are the same. With mortgage A, the borrower is essentially borrowing $198,000 (since $2000 is paid back to the lender immediately) and then making the same monthly payments as on a $198,000 loan at 6.094%. The Truth-in-Lending Act requires the lender of mortgage A to specify that the loan has an APR of 6.094%.

Discount points are only one type of up-front fee associated with loans. Other types of up-front fees include appraisal, credit report, and document preparation fees. The APR is intended to compare the costs of mortgages having different up-front fees.

The following steps calculate the APR for a mortgage loan having a term of n months.

1. Calculate the monthly payment (call it R) on the stated loan amount.

2. Subtract all up-front fees (including points) from the stated loan amount. Denote the result by P.

3. Find the interest rate that produces the monthly payment R for a loan of P dollars to be repaid in n months.

The interest rate discussed in step 3 can be calculated in Excel with the RATE function as

$$12 * \text{RATE}(n, -R, P, 0).$$

To calculate the interest rate in step 3 with a graphing calculator, solve the equation

$$P = a_{\overline{n}|i} R$$

for i and then multiply the value of i by 12.

EXAMPLE 8 **Calculating the APR for a mortgage with discount points** Use the three steps just given to calculate the APR for mortgage A of Example 7.

Solution **1.** From Example 7, the monthly payment is \$1199.10.

2. $P = 200{,}000 - 2000 = \$198{,}000$ since the up-front fees consist of one discount point.

3. This problem is best solved using technology. The interest rate can be calculated in Excel with the RATE function. The value of 12*RATE(360, −1199.10, 198000, 0) is .06093981.

With a graphing calculator, the interest rate is found by solving $P = a_{\overline{n}|i} R$ for i as shown in Fig. 3. The value of $12 \cdot (.00507832)$ is .06093984.

(a)

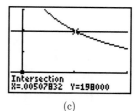
(c)

Figure 3. Solving Example 8 with a graphing calculator.

Therefore, rounded to three decimal places, the APR for mortgage A is 6.094%.

Now Try Exercise 27   ◼

▶ *Note* The answer 6.094% is produced when the problem is solved with the Loan Analysis routine of the Explorations in Finite Mathematics software.   ◀

The APR is of limited use because it is relevant only for loans that are kept for their full terms. The average mortgage is refinanced or terminated after approximately five years; that is, 60 months. Mortgage analysts often rely on the *effective mortgage rate* that takes into account the length of time the loan will be held and the unpaid balance at that time. For mortgage A discussed in Example 7, the monthly payment is $1199.10 and we show in Example 9 that the unpaid balance after 60 months is $186,108.55. With a five-year lifetime, the effective mortgage rate is the interest rate for which a decreasing annuity with a beginning balance of $198,000 and monthly payment of $1199.10 will decline to $186,108.55 after 60 months. A graphing calculator or a spreadsheet can be used to determine that the effective rate of interest for mortgage A is 6.24%. If terminated after 5 years, mortgage A is equivalent to a mortgage of $200,000 at 6.24% interest compounded monthly and having no discount points. That is, with the assumption that mortgage A will be held for only 5 years, the interest rate is effectively 6.24%.

EXAMPLE 9 **Calculating values associated with mortgages** Consider mortgage A of Example 7. The mortgage had an up-front payment of $2000, an interest rate of 6% compounded monthly, and a monthly payment of $1199.10. Let mortgage C be a 360-month loan of $200,000 at 6.24% compounded monthly with no points.

(a) Find the future value of $2000 after 60 months at 6.24% compounded monthly.

(b) Find the future value of an increasing annuity consisting of 60 monthly payments of $1199.10 with an interest rate of 6.24%.

(c) Find the unpaid balance on mortgage A after 60 months.

(d) Find the monthly payment for mortgage C.

(e) Find the future value of an increasing annuity consisting of 60 monthly payments of the amount found in part (d) with an interest rate of 6.24%.

(f) Find the unpaid balance on mortgage C after 60 months.

(g) Compare the sum of the three amounts found in parts (a), (b), and (c) with the sum of the two amounts found in parts (e) and (f).

Solution (a) Since $.0624/12 = .0052$, the future value of $2000 after 60 months is

$$2000(1.0052)^{60} = \$2730.10.$$

(b) $[\text{future value}] = s_{\overline{n}|i}R = s_{\overline{60}|.0052} \cdot (1199.10)$

$$= \frac{(1+i)^{60}-1}{i} \cdot (1199.10), \quad \text{with } i = .0052$$

$$= 70.20206568 \cdot (1199.10) \quad \text{(see Fig. 4)}$$

$$= \$84,179.30$$

```
.0624/12
          .0052
Ans→I
          .0052
((1+I)^60-1)/I
      70.20206568
```

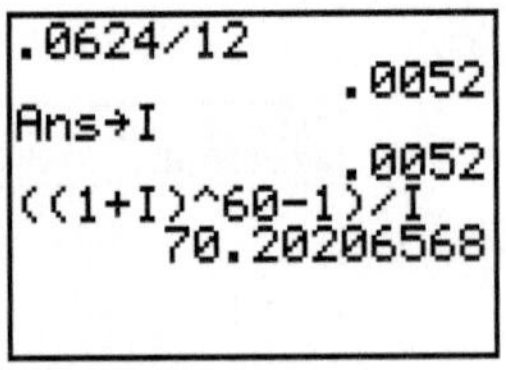

Figure 4

(c) The mortgage has $360 - 60 = 300$ months to go. Therefore,

$$[\text{unpaid balance after 60 months}] = a_{\overline{n}|i}R = a_{\overline{300}|.005} \cdot (1199.10)$$
$$= (155.20686401) \cdot (1199.10)$$
$$= \$186,108.55.$$

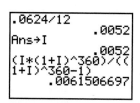

Figure 5

(d) For mortgage C

$$[\text{monthly payment}] = \frac{1}{a_{\,\overline{n}|\,i}} \cdot P$$

$$= \frac{1}{a_{\,\overline{360}|\,i}} \cdot (200{,}000), \quad \text{with } i = \frac{.0624}{12} = .0052$$

$$= \frac{i(1+i)^{360}}{(1+i)^{360} - 1} \cdot (200{,}000)$$

$$= .0061506697 \cdot (200{,}000) \quad (\text{see Fig. 5})$$

$$= \$1230.13$$

(e) $[\text{future value}] = s_{\,\overline{n}|\,i} R = s_{\,\overline{60}|\,.0052} \cdot (1230.13)$

$$= \frac{(1+i)^{60} - 1}{i} \cdot (1230.13), \quad \text{with } i = .0052$$

$$= 70.20206568 \cdot (1230.13) \quad (\text{see Fig. 4})$$

$$= \$86{,}357.67$$

(f) $[\text{unpaid balance}] = a_{\,\overline{n}|\,i} R = a_{\,\overline{300}|\,.0052} \cdot (1230.13)$

$$= \frac{(1+i)^{300} - 1}{i(1+i)^{300}} \cdot (1230.13), \quad \text{with } i = .0052$$

$$= 151.7332387 \cdot (1230.13) \quad (\text{see Fig. 6})$$

$$= \$186{,}651.61$$

Figure 6

(g) $[\text{sum of amounts found for mortgage } A] = 2730.10 + 84{,}179.30 + 186{,}108.55$
$$= \$273{,}017.95$$

$[\text{sum of amounts found for mortgage } C] = 86{,}357.67 + 186{,}651.61$
$$= \$273{,}009.28$$

The two sums are very close. This demonstrates the sense in which the two loans are equivalent after 60 months. (The difference of $8.67 is due to round-off error.) ■

The following steps calculate the effective mortgage rate for a mortgage loan expected to be held for m months.

1. Calculate the monthly payment (call it R) on the stated loan amount.

2. Subtract all up-front fees (including points) from the stated loan amount. Denote the result by P.

3. Determine the unpaid balance on the stated loan amount after m months. Denote the result by B.

4. Find the interest rate that causes a decreasing annuity with a beginning balance of P dollars and monthly payment R to decline to B dollars after m months.

The interest rate discussed in step 4 can be calculated in Excel with the RATE function as

$$12*\text{RATE}(m, -R, P, -B).$$

To calculate the interest rate in step 4 with a graphing calculator, solve the equation[4]

$$R - iB = (R - iP) \cdot (1 + i)^m$$

for i and then multiply the value of i by 12.

EXAMPLE 10 **Calculating effective mortgage rates** Use the four steps just given to calculate the effective mortgage rate for mortgage A of Example 7 assuming the mortgage will be held for 5 years.

Solution
1. The monthly payment $R = \$1199.10$ was calculated in Example 7.
2. $P = 200{,}000 - 2000 = 198{,}000$ since the up-front fee consists only of the one discount point.
3. The balance after 60 months, $B = \$186{,}108.55$, was computed in part (c) of Example 9.
4. This problem is best solved using technology. The interest rate can be calculated in Excel with the RATE function. The value of

$$12*\text{RATE}(60, -1199.10, 198000, -186108.55)$$

is .0624069.

With a graphing calculator, the interest rate is found by solving

$$1199.10 - i(186108.55) = (1199.10 - i(198000)) \cdot (1 + i)^{60}$$

for i as shown in Fig. 7. The value of $12 \cdot (.00520058)$ is .06240696.

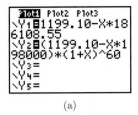   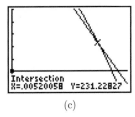

(a)   (b)   (c)

Figure 7. Solving Example 10 with a graphing calculator.

Therefore, rounded to two decimal places, the effective mortgage rate for mortgage A is 6.24%.

Now Try Exercise 31   ■

▶ *Note* Mortgage analysts have a rule of thumb for calculating the effective mortgage rate for a mortgage with points. For mortgages that will be held for 4 to 6 years, each discount point adds about $\frac{1}{4}\%$ to the stated interest rate. For instance, for the mortgage in Example 10, the rule of thumb predicts that the effective mortgage rate should be $\frac{1}{4}\%$ higher than the stated rate for a mortgage having one discount point. This is very close to the actual increase, .24%. Table 1 extends the rule of thumb to other lifetimes. ◀

---

[4]This equation says that the amount of the monthly payment applied to principal in the $(m+1)$st payment is equal to the future value of the amount applied to principal in the first payment. There are several equations that involve m, R, P, and i. We chose this equation since it is one of the simplest.

TABLE 1	Rule of thumb for the significance of discount points
Lifetime	**Difference between effective mortgage rate and stated interest rate, per discount point**
1 year	1 percentage point
2 years	$\frac{1}{2}$ of a percentage point
3 years	$\frac{1}{3}$ of a percentage point
4 to 6 years	$\frac{1}{4}$ of a percentage point
7 to 9 years	$\frac{1}{6}$ of a percentage point
10 to 12 years	$\frac{1}{7}$ of a percentage point
More than 12 years	$\frac{1}{8}$ of a percentage point

Example 11 shows another way to take the expected lifetime of a loan into account when deciding between two mortgages. If you intend to either sell the house or refinance the loan within a few years, then you should avoid a loan carrying excess points since the large amount of interest paid up front will not be recovered. The longer you keep the mortgage, the less expensive points become.

EXAMPLE 11 **Using expected lifetime to compare mortgages** Suppose a lender gives you a choice between the following two 30-year mortgages of $100,000:

Mortgage A: 6.1% interest compounded monthly, two points, monthly payment of $605.99

Mortgage B: 6.6% interest compounded monthly, no points, monthly payment of $638.66

The choice of mortgage depends on the length of time you expect to hold the mortgage. Mortgage B is the better choice if you plan to terminate or refinance the mortgage fairly early. Assuming that you can invest money at 3.6% compounded monthly, determine the length of time you must retain the mortgage in order for mortgage A to be the better choice.

Solution The difference in the monthly payments is $638.66 - 605.99 = \$32.67$. Although mortgage A saves $32.67 per month, it carries two additional points and therefore requires the payment of an additional $2000 up front. With mortgage B the $2000 could be used to generate an income stream of $32.67 per month for a certain number of months, n, by investing it in an annuity at 3.6% compounded monthly. The value of n is best found with technology.

The number of months, n, can be calculated in Excel with the NPER function. The value of NPER(.003, 32.67, −2000) is 67.74. (*Note*: .003 = .036/12.)

With a graphing calculator, the number of months, n, can be found by solving $P = a_{\overline{n}|i}R$ for n. This equation can be shown to be equivalent to

$$\left(\frac{1}{1+i}\right)^n = 1 - \left(\frac{iP}{R}\right).$$

(See Exercise 40 for details.) Set $i = .003$, $P = 2000$, $R = 32.67$, and then find the approximate solution for n, 67.76, as shown in Fig. 8 on the next page. (*Note*: $1/(1 + i) = .99701$ and $1 - (.003 \cdot 2000)/32.67 = .81635$.)

(a)

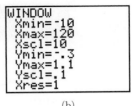
(b)

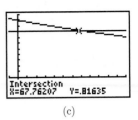
(c)

Figure 8. Solving Example 11 with a graphing calculator.

Therefore, if you plan to hold the loan for 68 months (about $5\frac{2}{3}$ years) or more, you should choose mortgage A.

Now Try Exercise 47   ■

▶ **Note** The rounded answer 68 results when the problem is solved with the Loan Analysis routine of Explorations in Finite Mathematics. ◀

PRACTICE PROBLEMS 10.4

1. Rework Example 11 under the assumption that the $2000 will not earn any interest.

2. Consider the two loans discussed in Example 7. Calculate the balance of each loan after 1 month.

3. Which is the better deal for a 30-year mortgage: (a) 6% interest plus three discount points (6.286% APR), or (b) 6.3% interest plus one discount point (6.396% APR)?

EXERCISES 10.4

In Exercises 1 and 2, fill in the blank with the word "free" or "deferred."

1. Interest earned in a traditional IRA is tax _____.

2. Interest earned in a Roth IRA is tax _____.

3. (*Traditional IRA*) Carlos is 60 years old, is in the 45% marginal tax bracket, and has $300,000 in his traditional IRA. How much money will he have after taxes if he withdraws all the money from the account?

4. (*Roth IRA*) Rework Exercise 3 for a Roth IRA.

5. (*Traditional IRA*) If you are 18 years old, deposit $4000 each year into a traditional IRA for 52 years at 6% interest compounded annually, and retire at age 70, how much money will be in the account upon retirement?

6. (*Comparing IRAs*) Rework Examples 1 and 3 for the case where the marginal tax bracket is 30% when the money is contributed but is only 25% when the balance is withdrawn. Which type of IRA is most advantageous?

7. (*Roth IRA*) Rework Exercise 5 for a Roth IRA.

8. (*Comparing IRAs*) Rework Examples 1 and 3 for the case where the marginal tax bracket is 30% when the money is contributed but rises to 35% when the balance is withdrawn. Which type of IRA is most advantageous?

9. (*Advantages of Prompt Contributions*) Redo Example 4 where Earl and Larry are each in a 40% marginal tax bracket, have Roth IRAs, and contribute the remainder of $4000 after taxes are deducted.

10. (*Value of Starting an IRA Early*) The contribution into an IRA for a particular year can be made any time from January 1 of that year to April 15 of the following year. Suppose Enid and Lucy both set up traditional IRA accounts on January 1 of 2006 and each contributes $4000 into her account for ten years at 6% interest compounded annually. Assume that Enid makes her contributions as soon as possible and Lucy makes her contributions one year later. Calculate the balances in the two accounts at the time Lucy makes her final contribution.

In Exercises 11–14, use the add-on method to determine the monthly payment.

11. $4000 loan at 10% interest for one year

12. $6000 loan at 7% interest for one year

13. $3000 loan at 9% interest for three years

14. $10,000 loan at 8% interest for two years

In Exercises 15–18, give the add-on interest rate.

15. A $2000 loan for 1 year at 5% APR with a monthly payment of $171.21

16. A \$5000 loan for 1 year at 6% APR with a monthly payment of \$430.33

17. A \$20,000 loan for 3 years at 6% APR with a monthly payment of \$608.44

18. A \$10,000 loan for 2 years at 7% APR with a monthly payment of \$447.73

19. (*Discount Method*) Another method used by lenders to determine the monthly payment for a loan is the *discount method*. In this case, the stated interest rate is called the *discount rate* and the formula for determining the total amount paid is

$$[\text{total payment}] = \frac{[\text{loan amount}]}{1 - rt},$$

where r is the discount rate and t is the term of the loan in years. The monthly payment is then $[\text{total payment}]/12t$.

(a) Calculate the monthly payment for a two-year loan of \$880 with a discount rate of 6%.

(b) If the add-on method were used instead, would the monthly payment be greater than or less than the amount found in part (a)?

20. (*True or False*) For any mortgage with discount points, the APR will be greater than the stated interest rate.

21. (*True or False*) For any mortgage that will be terminated before its full term, the effective mortgage rate will be greater than the APR.

22. (*True or False*) If a mortgage is expected to be held for its entire term, the effective mortgage rate will be the same as the APR.

23. (*True or False*) The longer a loan with up-front fees is expected to be held, the greater its effective mortgage rate.

24. (*True or False*) Increasing the up-front fees for a mortgage increases its APR and its effective mortgage rate.

25. (*True or False*) Changing the amount of a loan while keeping the stated interest rate, the term, and the up-front fees the same will have no effect on the APR.

26. (*True or False*) Refer to Example 10. If the stated interest rate were 9% instead of 6%, then the effective mortgage rate would be 9.24%. That is, the interest rate would increase by the same amount as before, .24%.

In multiple-choice Exercises 27–30, assume that the loan amount is \$250,000 and confirm your answer by calculating monthly payments. (Note: The two monthly payments will not be identical due to round-off error.)

27. The APR for a 25-year mortgage at 9% interest compounded monthly and two discount points is

(a) 8.75%   (b) 9.05%   (c) 12%   (d) 9.25%

28. The APR for a 30-year mortgage at 5.9% interest compounded monthly and one discount point is

(a) 9%   (b) 6%   (c) 5.7%

29. The APR for a 20-year mortgage at 5.5% interest compounded monthly and four discount points is

(a) 6%   (b) 5%   (c) 6.5%

30. The APR for a 25-year mortgage at 9% interest compounded monthly and four discount points is

(a) 8.5%   (b) 9.5%   (c) 11%

In multiple-choice Exercises 31–34, assume that the loan amount is \$100,000 and confirm your answer by calculating sums as in Example 9.

31. The effective mortgage rate for a 30-year mortgage at 5.71% interest compounded monthly with 1 discount point that is expected to be kept for 4 years is

(a) 6%   (b) 5.54%   (c) 6.56%   (d) 7%

(*Hint:* For a 5.71%, 30-year mortgage, $R = \$581.03$ and the balance after 4 years is \$94,342.20.)

32. The effective mortgage rate for a 15-year mortgage at 5.481% interest compounded monthly with 3 discount points that is expected to be kept for 10 years is

(a) 5%   (b) 6%   (c) 6.5%   (d) 7%

(*Hint:* For a 5.481%, 15-year mortgage, $R = \$816.08$ and the balance after 10 years is \$42,743.49.)

33. The effective mortgage rate for a 30-year mortgage at 6% interest compounded monthly with 3 discount points that is expected to be kept for 7 years is

(a) 5.5%   (b) 6.2%   (c) 6.56%   (d) 9%

34. The effective mortgage rate for a 30-year mortgage at 9% interest compounded monthly with 2 discount points that is expected to be kept for 7 years is

(a) 9%   (b) 8.5%   (c) 9.4%   (d) 10%

In Exercises 35 and 36, give the two Excel formulas that can be used to obtain the APR and the effective mortgage rate for the mortgage. When calculating the effective mortgage rate, assume the mortgage will be held for 10 years.

35. A 20-year mortgage of \$80,000 carrying a stated interest rate of 6% and three points.

36. A 15-year mortgage of \$180,000 carrying a stated interest rate of 9% and two points.

In Exercises 37 and 38, give the two equations that can be solved for i with a graphing calculator to obtain the monthly interest rates corresponding to the APR and the effective mortgage rate. Assume that the loans will be held for 5 years.

37. A 15-year mortgage of \$120,000 carrying a stated interest rate of 9% and one point.

38. A 20-year mortgage of \$150,000 carrying a stated interest rate of 6% and three points.

39. (*Loan Quandary*) You decide to buy an entertainment system costing $1250 for your apartment. The salesman asks you to pay 20% down and is willing to finance the remaining $1000 with a two-year loan having an APR of 5% and a monthly payment of $43.87. (The total of your payments on the loan will be $1052.88.) Suppose you actually have $1000 in your savings account (after the down payment) and can invest it at 4% compounded monthly. You must decide whether or not to take the loan. The salesman tells you that you will come out ahead if you take the loan since $1000 invested at 4% interest compounded monthly grows to $1083.14, which is greater than the total payments on the loan. This doesn't make sense to you since you would be borrowing money at 5% and only earning 4% on the money you invest. What is wrong with the salesman's reasoning? (*Note:* Ignore taxes.)

40. Consider the equation $P = a_{\overline{n}|i} \cdot R$. Replace $a_{\overline{n}|i}$ with its formula and perform algebraic manipulations on the resulting equation to obtain

$$\left(\frac{1}{1+i}\right)^n = 1 - \left(\frac{iP}{R}\right)$$

Hint: After replacing $a_{\overline{n}|i}$ with its formula, multiply the equation by i/R, on the right side of the equation divide the numerator by the denominator, and solve for $1/(1+i)^n$.

In Exercises 41–48, use a graphing calculator, a spreadsheet, or mathematical software to obtain the solution.

41. Find the APR on a $10,000 loan with an add-on method interest rate of 6% compounded monthly for three years.

42. Find the APR on a $15,000 loan with an add-on method interest rate of 8% compounded monthly for two years.

43. Find the APR on a 25-year mortgage of $300,000 carrying a stated interest rate of 6.5% compounded monthly and three points.

44. Find the APR on a 30-year mortgage of $250,000 carrying a stated interest rate of 6% compounded monthly and two points.

45. Find the effective mortgage rate for a 25-year $140,000 mortgage at 6.5% interest compounded monthly with three discount points. Assume that the mortgage will be held for 7 years.

46. Find the effective mortgage rate for a 30-year $250,000 mortgage at 6% interest compounded monthly with two discount points that is expected to be kept for 6 years.

47. (*Comparing Mortgages*) Suppose a lender gives you a choice between the following two 25-year mortgages of $175,000:

Mortgage A: 6.3% interest compounded monthly, three points, monthly payment of $1159.84

Mortgage B: 6.5% interest compounded monthly, two points, monthly payment of $1181.61

Assuming that you can invest money at 2.4% compounded monthly, determine the length of time you must retain the mortgage in order for mortgage A to be the better choice.

48. (*Comparing Mortgages*) Suppose a lender gives you a choice between the following two 15-year mortgages of $250,000:

Mortgage A: 6.5% interest compounded monthly, one point, monthly payment of $2177.77

Mortgage B: 6.1% interest compounded monthly, two points, monthly payment of $2123.17

Assuming that you can invest money at 3.6% compounded monthly, determine the length of time you must retain the mortgage in order for mortgage B to be the better choice.

SOLUTIONS TO PRACTICE PROBLEMS 10.4

1. In this case, we merely have to solve $n \cdot (32.67) = 2000$.

$$n = \frac{2000}{32.67} = 61.218$$

Therefore, only take the additional points if you expect to hold onto the loan for more than 61 months (that is, a little more than 5 years).

▶ *Note* This computation is much simpler than the one in Example 11 and can be used to get a rough estimate of the proper number of months. The smaller the interest that can be earned on the cost of the points, the better is the approximation. ◀

SOLUTIONS TO PRACTICE PROBLEMS 10.4 (CONTINUED)

2. Successive balances can be calculated with the formula $B_{\text{new}} = (1+i)B_{\text{previous}} - R$. Mortgage A was a \$200,000 loan at 6% interest with a monthly payment of \$1199.10.

$$[\text{balance after 1 month}] = (1.005) \cdot 200,000$$
$$- 1199.10$$
$$= \$199,800.90$$

Mortgage B was a \$198,000 loan at 6.094% interest with a monthly payment of \$1199.10. The monthly interest $i = .06094/12 = .0050783333$.

$$[\text{balance after 1 month}] = (1.0050783333) \cdot 198,000$$
$$- 1199.10$$
$$= \$197,806.41$$

The balance for mortgage A is \$1994.49 higher. The balance will remain higher until the final payment is made.

3. It depends on how long you intend to keep the mortgage. If you expect to hold the mortgage for 30 years, then mortgage A is clearly superior because it has a lower APR. On the other hand, if you expect to terminate the mortgage after two years, then Table 1 on page 525 says that the effective rates are approximately 7.5% and 6.8% respectively, obviously favoring mortgage B.

CHAPTER 10 SUMMARY

1. Money deposited into a *savings account* earns interest at regular time periods. Interest paid on the initial deposit only is called *simple interest*. Interest paid on the current balance (that is, on the initial deposit *and* the accumulated interest) is called *compound interest*.

2. An *increasing* (*decreasing*) *annuity* is a sequence of equal deposits (withdrawals) made at the ends of regular time intervals.

3. A *mortgage* is a type of loan that is paid off in equal payments at the ends of regular time periods.

4. The following letters and symbols are used to describe savings accounts, annuities, and loans.

   P *principal*, the initial amount of money deposited into a savings account *or* the amount of money borrowed in a loan. P also represents the *present value* of a sum of money to be received in the future; that is, the amount of money needed to generate the future money.

   r *annual rate of interest*, interest rate stated by the bank and used to calculate the interest rate per period.

   m *number of* (*compound*) *interest periods per year*, most commonly 1, 4, or 12.

   i *compound interest rate per period*, calculated as $i = \dfrac{r}{m}$.

   n *number of interest periods*.

   F *future value, compound amount,* or *balance*, value in a savings account or an annuity at some point in the future. For a savings account,

   $$F = (1+i)^n P$$

   with compound interest, and $F = (1+nr)P$ with simple interest.

   r_{eff} *effective rate of interest*, the simple interest rate that yields the same amount after one year as the annual rate of interest. $r_{\text{eff}} = (1+i)^m - 1$.

   R *rent*, periodic deposit into or withdrawal from an annuity *or* periodic payment on a loan.

   $s_{\overline{n}|i}$ *s sub n angle i*, future value of an increasing annuity of n \$1 payments at compound interest rate i per period. For an increasing annuity, $F = s_{\overline{n}|i}R$ and $R = \dfrac{1}{s_{\overline{n}|i}} \cdot F$.

   $$s_{\overline{n}|i} = \frac{(1+i)^n - 1}{i}$$

   $a_{\overline{n}|i}$ *a sub n angle i*, present value of a decreasing annuity of n \$1 payments at compound interest rate i per period. For a decreasing annuity or loan, $P = a_{\overline{n}|i}R$ and $R = \dfrac{1}{a_{\overline{n}|i}} \cdot P$.

   $$a_{\overline{n}|i} = \frac{(1+i)^n - 1}{i(1+i)^n}$$

5. At any time, the *balance* of a loan is the amount of money needed to retire (that is, pay off) the loan. It is calculated as the present value of all future payments. A payment used to retire a loan is called a *balloon payment*.

6. Successive balances can be calculated with the following formulas.

$$B_{new} = (1+i)B_{previous} \qquad \text{savings account with compound interest}$$

$$B_{new} = (1+i)B_{previous} + R \quad \text{increasing annuity}$$

$$B_{new} = (1+i)B_{previous} - R \quad \text{loan or decreasing annuity}$$

7. An *individual retirement account* (IRA) is an increasing annuity in which the annual interest earned is either tax free (Roth IRA) or tax-deferred (traditional IRA). Contributions are tax deductible only with a traditional IRA.

8. When finance charges on a consumer loan are calculated with the *add-on method*, the interest paid each month is a fixed percentage of the principal, rather than a fixed percentage of the current balance.

9. For each *discount point* accompanying a mortgage loan, you must pay additional interest up-front equal to 1% of the amount borrowed.

10. The *APR* and the *effective mortgage rate*, which take up-front fees into account, are often more useful than the contract interest rate in appraising a loan. Each of them makes use of the reduced loan amount obtained by subtracting the up-front fees from the stated loan amount. The APR is determined by calculating the interest rate corresponding to the reduced loan amount. The effective mortgage rate also takes into account the expected lifetime of the loan and is the interest rate corresponding to an annuity that decreases the reduced loan amount to the balance after the expected lifetime of the loan.

REVIEW OF FUNDAMENTAL CONCEPTS

1. What is meant by *principal*?

2. What is the difference between compound interest and simple interest?

3. What is meant by the *compound amount* or *balance*?

4. How is interest per period determined from annual interest?

5. Explain how compound interest works.

6. Explain how simple interest works.

7. What is meant by the *present value* of a sum of money to be received in the future?

8. Explain the difference between the *nominal* and *effective* rates for compound interest.

9. What is an annuity?

10. What is meant by the *future value* of an annuity? *present value*? *rent*?

11. Describe the two types of annuities discussed in this chapter. In each case, identify the present and future values.

12. Give the formula for computing a new balance from a previous balance for each type of annuity.

13. Define $s_{\overline{n}|\,i}$ and $a_{\overline{n}|\,i}$.

14. State the four formulas involving $s_{\overline{n}|\,i}$, $a_{\overline{n}|\,i}$, and their reciprocals.

15. What are the components of an amortization table of a loan, and how are they calculated?

16. Give the formula for computing a new balance from a previous balance for a loan.

17. What is a balloon payment?

18. Explain how *traditional* and *Roth IRAs* work.

19. How are finance charges on a consumer loan calculated with the *add-on method*?

20. What are *discount points*?

21. How is the APR of a mortgage calculated?

22. How is the effective mortgage rate of a mortgage calculated?

KEY FORMULAS

Simple Interest: $F = (1+nr)P$, where P is the principal, r is the interest rate, and n is the number of years.

Interest Rate per Period: $i = \dfrac{r}{m}$, where r is the annual interest rate and m is the number of times interest is compounded per year.

Compound Amount: $F = (1+i)^n P$, where P is the principal, n is the number of interest periods, and i is the interest per period.

Present Value: $P = \dfrac{1}{(1+i)^n} \cdot F$, where F is the future value, n is the number of interest periods, and i is the interest per period.

Effective Rate of Interest: $r_{eff} = (1+i)^m - 1$, where m is the number of interest periods, and i is the interest per period.

Future Value of an Increasing Annuity: $F = s_{\overline{n}|\,i}R$, where n is the number of interest periods, i is the interest per period, and R is the payment made at the end of each period.

Periodic Payment into an Increasing Annuity: $R = \dfrac{1}{s_{\overline{n}|i}} \cdot F$, where n is the number of interest periods, i is the interest per period, and F is the future value of the annuity.

Successive Balances in an Increasing Annuity: $B_{\text{new}} = (1+i)B_{\text{previous}} + R$

Present Value of a Decreasing Annuity: $P = a_{\overline{n}|i}R$, where n is the number of interest periods, i is the inter-

est per period, and R is the payment made at the end of each period.

Periodic Payment for a Decreasing Annuity: $R = \dfrac{1}{a_{\overline{n}|i}} \cdot P$, where n is the number of interest periods, i is the interest per period, and P is the present value of the annuity.

Successive Balances in a Mortgage: $B_{\text{new}} = (1+i)B_{\text{previous}} - R.$

SUPPLEMENTARY EXERCISES

1. (PE) If $100 earns 6% interest compounded annually, the compound amount after 10 years is

   (a) $(100.06)^{10}$ (b) $100 + 10(1.06)$ (c) $100(1.6)^{10}$

   (d) $100(1.06)^{10}$ (e) $100 + (0.06)^{10}$

2. (*Saving for Retirement*) Mr. West wishes to purchase a condominium for $80,000 cash upon his retirement 10 years from now. How much should he deposit at the end of each month into an annuity paying 6% interest compounded monthly in order to accumulate the required savings?

3. (*Mortgage Considerations*) The income of a typical family in a certain city is currently $39,200 per year. Family finance experts recommend that mortgage payments not exceed 25% of a family's income. Assuming a current mortgage interest rate of 9% compounded monthly for a 30-year mortgage with monthly payments, how large a mortgage can the typical family in that city afford?

4. Calculate the compound amount of $50 after a year if deposited at 7.3% compounded daily.

5. Which is a better investment: 10% compounded annually or 9% compounded daily?

6. (*Bond Fund*) Ms. Smith deposits $200 per month into a bond fund yielding 6% interest compounded monthly. How much are her holdings worth after 5 years?

7. (*Nonstandard Mortgage*) A real estate investor takes out a $200,000 mortgage subject to the following terms: For the first 5 years, the payments will be the same as the monthly payments on a 15-year mortgage at 12% interest compounded monthly. The balance will then be payable in full.

   (a) What are the monthly payments for the first 5 years?

   (b) What balance will be owed after 5 years?

8. (*College Expenses*) College expenses at a private college currently average $24,000 per year. It is estimated that these expenses are increasing at the rate of $\frac{1}{2}$% per month. What is the estimated cost of a year of college 10 years from now?

9. What is the present value of $50,000 in 10 years at 6% interest compounded monthly?

10. An investment will pay $10,000 in 2 years and then $5000 one year later. If the current market interest rate is 6% compounded monthly, what should a rational person be willing to pay for the investment?

11. (*Car Loan*) A woman purchases a car for $12,000. She pays $3,000 as a down payment and finances the remaining amount at 6% interest compounded monthly for 4 years. What is her monthly car payment?

12. (*Nonstandard Loan*) A businessman buys a $100,000 piece of manufacturing equipment on the following terms: Interest will be charged at a rate of 4% compounded semiannually, but no payments will be made until 2 years after purchase. Starting at that time, equal semiannual payments will be made for 5 years. Determine the semiannual payment.

13. (*Retirement Account*) A retired person has set aside a fund of $105,003.50 for his retirement. This fund is in a bank account paying 6% interest compounded monthly. How much can he draw out of the account at the end of each month so that there is a balance of $30,000 at the end of 15 years? (*Hint*: First compute the present value of the $30,000.)

14. (*Balloon Payment*) A business loan of $509,289.22 is to be paid off in monthly payments for 10 years with a $100,000 balloon payment at the end of the 10th year. The interest rate on the loan is 12% compounded monthly. Calculate the monthly payment.

15. (*Savings Plan*) Ms. Jones saves $100 per month for 30 years at 6% interest compounded monthly. How much are her accumulated savings worth?

16. (*Second Mortgage*) An apartment building is currently generating an income of $2000 per month. Its owners are considering a 10-year second mortgage at 12% interest compounded monthly in order to pay for repairs. How large a second mortgage can the income of the apartment house support?

17. (*Comparing Investments*) Investment A generates $1000 at the end of each year for 10 years. Investment B generates $5000 at the end of the fifth year and $5000 at the end of the tenth year. Assume a market rate of interest of 6% compounded annually. Which is the better investment?

18. (*Bond Value*) A 5-year bond has a face value of $1000 and is currently selling for $800. The bond pays $5 interest at the end of each month and, in addition, will repay the $1000 face value at the end of the fifth year. The market rate of interest is currently 9% compounded monthly. Is the bond a bargain? Why or why not?

19. (*Effective Rate*) Calculate the effective rate for 10% interest compounded semiannually.

20. (*Effective Rate*) Calculate the effective rate for 18% interest compounded monthly.

21. (*Annuities with Extra Deposit*) A person makes an initial deposit of $10,000 into a savings account and then deposits $1000 at the end of each quarter-year for 15 years. If the interest rate is 8% compounded quarterly, how much money will be in the account after 15 years?

22. (*Car Loan*) A $10,000 car loan at 6% interest compounded monthly is to be repaid with 36 equal monthly payments. Write out an amortization schedule for the first 6 months of the loan.

23. (*Savings Plan*) A person pays $200 at the end of each month for 10 years into a fund paying 1% interest per month compounded monthly. At the end of the 10th year, the payments cease, but the balance continues to earn interest. What is the value of the balance at the end of the 20th year?

24. A savings fund currently contains $300,000. It is decided to pay out this amount with 6% interest compounded monthly over a 5-year period. What are the monthly payments?

25. What is the monthly payment on a $150,000, 30-year mortgage at 9% interest?

26. (*Traditional IRA*) Elisa, age 60, is currently in the 30% tax bracket and has $30,000 in a traditional IRA that earns 6% interest compounded annually. She anticipates being in the 35% tax bracket five years from now.
   (a) How much money will Elisa have after paying taxes if she withdraws her money now?
   (b) How much will Elisa have after paying taxes if she waits five years and then withdraws the money from the account?

27. (*Roth IRA*) Rework Exercise 26 for a Roth IRA.

28. (*Consumer Loan*) A consumer loan of $5000 for 2 years has an APR of 9% compounded monthly and a monthly payment of $228.42. What interest rate would be stated with the add-on method?

29. (*Comparing Loans*) Spike is considering purchasing a new computer and has decided to take one of two consumer loans:

Consumer Loan A: 1 year, $3000, 10% APR, monthly payment of $263.75
Consumer Loan B: 1 year, $3000, 6% add-on interest

Which loan should Spike take, and why?

30. Consider a 15-year mortgage of $90,000 at 6% interest compounded monthly with two discount points and a monthly payment of $759.47. The APR for the mortgage is obtained by solving $P = a_{\overline{n}|i}R$ for i and then multiplying by 12. What are the values of P, n, and R?

31. Consider a 20-year mortgage of $100,000 at 6% interest compounded monthly with three discount points and a monthly payment of $716.43. Assume that the loan is expected to be held for six years at which time the unpaid balance is $81,298.44. The effective mortgage is obtained in Excel by evaluating $12*\text{RATE}(m, -R, P, -B)$. What are the values of m, R, P, and B?

32. (*Comparing Mortgages*) Suppose a lender gives you a choice between the following two 25-year mortgages of $200,000:

Mortgage A: 6.5% interest compounded monthly, two points, monthly payment of $1350.41
Mortgage B: 7% interest compounded monthly, one point, monthly payment of $1413.56

Assume that you can invest money at 3.5% compounded monthly. The length of time you must retain the mortgage in order for mortgage A to be the better choice is obtained in Excel by evaluating $\text{NPER}(i, R, -P)$. What are the values of i, R, and P?

33. A 20-year mortgage of $250,000 at 5.75% interest compounded monthly and two discount points has a monthly payment of $1755.21. Show that the APR for this mortgage is 6%.

34. A 30-year mortgage of $100,000 at 5.5% interest compounded monthly and three discount points has a monthly payment of $567.79. Assume that the loan is expected to be terminated after eight years at which time the unpaid balance will be $86,837.98. Show that the effective mortgage rate is 6%.

Conceptual Exercises

35. If interest is compounded semiannually, will the effective rate be higher or lower than the nominal rate?

36. If you decrease the interest rate for an investment by 10%, will the future value decrease by 10%?

37. Consider a decreasing annuity. If the amount withdrawn each month increases by 5%, will the duration decrease by 5%? Give an example to justify your answer.

38. If you double the amount of months in which to pay off a loan, will you halve the monthly payment? Give an example to justify your answer.

39. Give an intuitive explanation why successive payments for a mortgage contribute steadily more towards repayment of the principal.

CHAPTER TEST

1. If $500 is deposited into an account earning 6.5% simple interest, what is the balance in the account after 9 months?

2. Calculate the present value of $5000 payable in 3 years at 4% simple interest.

3. A principal of $1025 is deposited into an account paying 5% interest compounded quarterly for 3 years. What is the compound amount? How much interest is earned?

4. What is the present value of $25,000 payable in 10 years at 8% interest compounded monthly?

5. Victoria is planning for her retirement by depositing $250 at the end of each month into an annuity paying 6% interest compounded monthly. How much money will have accumulated in 15 years?

6. Find the present value of a decreasing annuity that pays $500 per month for 4 years if the money earns 6% interest compounded monthly.

7. (*Increasing Annuity*) Calculate the rent of an increasing annuity at 4% interest compounded semiannually if payments are made every 6 months and the future value after 5 years is $12,000.

8. (*Decreasing Annuity*) Calculate the rent of a decreasing annuity at 9% interest compounded monthly with payments made every month for 10 years and present value $200,000.

9. (*Annuity with Extra Deposit*) Graham opens an account with $300 and then deposits an additional $50 at the end of each month. What is the balance after 4 years if the money earns 6% interest compounded monthly?

10. (*Loan Balance*) After making the fifth semiannual payment of $350 on a loan, Wanda's unpaid balance is $2499.93. If the interest rate is 3% compounded semiannually, what will Wanda's unpaid balance be after she makes her sixth payment?

11. (*Car Loan*) Nigel takes out a 5-year car loan for $8000 at 9% interest compounded monthly.

    (a) What is his monthly payment?

    (b) How much interest is paid the first month?

    (c) What is the unpaid balance after the first month?

    (d) After 4 years, Nigel wishes to retire the loan by making a balloon payment. How much does he owe?

12. (*Nonstandard Mortgage*) If you make a $45,000 down payment on a house and take out a 25-year mortgage with quarterly payments of $3500 at 6% interest compounded quarterly, what was the price of the house?

13. (*Advantage of Starting an IRA Early*) Enid and Lucy each begin full-time jobs in January 2006 and plan to retire in January 2046, after working for 40 years. Assume that any money that they deposit into IRAs earns 6% interest compounded annually.

    (a) Suppose Enid opens a traditional IRA account immediately and deposits $4000 into her account at the end of each year for 11 years. How much money will Enid have in her account when she retires in January 2046?

    (b) Suppose Lucy waits 11 years before opening her traditional IRA and then deposits $4000 into the account at the end of each year until she retires. How much money will Lucy have in her account when she retires in January 2046?

14. A three-year loan for $4000 is advertised at 10% using the add-on method. What is the monthly payment?

15. A 15-year mortgage of $250,000 at 8.829% interest compounded monthly and one discount point has a monthly payment of $2510.31. Show that the APR for this mortgage is 9%.

16. A 30-year mortgage carries two points, is expected to be terminated after 5 years, and has an effective mortgage rate of 6.3%.

    (a) If the expected termination time is increased to 6 years, will the effective mortgage rate increase or decrease?

    (b) If the number of points is increased to three, will the effective mortgage rate increase or decrease?

17. Under what circumstances will the effective mortgage rate and the APR for a mortgage be the same?

Two Items of Interest

Item 1: Successive Interest Computations

1. Suppose a $100,000 investment grows 3% during the first year and 4% during the second year. By what percent will it have grown after the two-year period? (*Note*: The answer is not 7%.)

2. Rework Exercise 1 for the case where the investment earns 4% during the first year and 3% during the second year. Is the answer to Exercise 2 greater than, less than, or equal to the answer to Exercise 1?

3. Consider an annuity in which $100,000 is invested and $10,000 is withdrawn at the end of each year. Suppose the interest rate is 3% during the first year and 4% during the second year. What is the balance at the end of the second year?

4. Rework Exercise 3 where the interest rate is 4% during the first year and 3% during the second year. Is the balance at the end of the second year greater than, less than, or equal to the answer to Exercise 3? Explain why your answer to this question makes sense.

5. Show that if an investment of P dollars declines by 4% during a year, the balance at the end of the year is $P \cdot (1 - .04)$; that is, $P \cdot (.96)$.

6. Show that if an investment of P dollars declines by $r\%$ during a year, the balance at the end of the year is $P \cdot (1 - r/100)$. We say that the investment earned $(-r)\%$.

7. Show that if an investment of P dollars earns $r\%$ one year and $s\%$ the following year, then the balance after the two-year period is $P(1 + r/100)(1 + s/100)$. Conclude that the order of the two numbers r and s does not affect the balance after two years. (*Note*: The numbers r and s can be positive or negative.)

8. Which of the following two statements is true?

   (a) Suppose an investment of P dollars earns 4% one year and loses 4% the next year. Then the value of the investment at the end of the two-year period will be P dollars.

   (b) Suppose investment A earns 4% one year and then loses 3% the next year, and investment B loses 3% one year and then gains 4% the next year. Then the balances of the two investments will be the same after the two-year period.

Item 2: Rule of 72

9. Show that if $1000 is invested at 8% interest compounded annually, then it will double in about 9 years. (*Note*: $9 = 72/8$.)

10. Show that if $1000 is invested for 6 years, then it will approximately double in that time if it appreciates at 12% per year. (*Note*: $12 = 72/6$.)

> **Rule of 72**   If money is invested at $r\%$ interest compounded annually, then it will double in about $72/r$ years. Alternately, if money is invested for n years, then it will double during that time if it appreciates by about $72/n\%$ per year.

11. Conclude from the Rule of 72 that for an interest rate of $r\%$, then $(1 + r/100)^{72/r} \approx 2$. Also, conclude that for a number of years, n, $(1 + .72/n)^n \approx 2$.

We needn't restrict ourselves to doubling. For instance, the *Rule of 114* says that if money is invested at $r\%$ interest, then it will triple in about $114/r$ years. Table 1 gives the numbers associated with several different multiples. Let's denote the number associated with m by N_m. That is, $N_2 = 72$ and $N_3 = 114$.

	TABLE 1
m	Number associated with increasing m-fold
2	72
3	114
4	144
5	167
6	186

12. Notice that $N_6 = N_2 + N_3$. Explain why this makes sense.

13. Explain why $N_{n \cdot p} = N_n + N_p$ for any positive numbers n and p.

14. Find N_{10} and then use that number to estimate the amount of time required for your money to increase tenfold if invested at 8% interest compounded annually. Check your answer by raising (1.08) to that power.

15. Use the values of N_2 and N_3 to find $N_{1.5}$, the number associated with money increasing by one-half. Estimate the amount of time required for $1000 to grow to $1500 when invested at 7% interest compounded annually.

CHAPTER 11

Difference Equations and Mathematical Models

I N this chapter we discuss a number of topics from the mathematics of finance—compound interest, mortgages, and annuities. As we shall see, all such financial transactions can be described by a single type of equation, called a *difference equation*. Furthermore, the same type of difference equation can be used to model many other phenomena, such as the spread of information, radioactive decay, and population growth, to mention a few.

11.1 Introduction to Difference Equations I

To understand what difference equations are and how they arise, consider two examples concerning financial transactions in a savings account.

EXAMPLE 1   **Balance equation**   Suppose that a savings account initially contains $40 and earns 6% interest, compounded annually. Determine a formula that describes how to compute each year's balance from the previous year's balance.

Solution   The balances in the account for the first few years can be given as in the following chart, where y_0 is the initial balance (balance after zero years), y_1 is the balance after one year, and so on.

Year	Balance	Interest for year
0	$y_0 = \$40$	$(.06)40 = \$2.40$
1	$y_1 = \ \ 42.40$	$(.06)42.40 = \$2.54$
2	$y_2 = \ \ 44.94$	$(.06)44.94 = \$2.70$
3	$y_3 = \ \ 47.64$	

Once the balance is known for a particular year, the balance at the end of the next year is computed as follows:

[balance at end of next year] = [balance at end of this year]

$+$ [interest on balance at end of this year].

That is,

$$y_1 = y_0 + .06y_0$$
$$y_2 = y_1 + .06y_1$$
$$y_3 = y_2 + .06y_2.$$

(Notice that since $2 = 3 - 1$, the last equation is $y_3 = y_{3-1} + .06y_{3-1}$.) In general, if y_n is the balance after n years, then y_{n-1} is the balance at the end of the preceding year, so

$$y_n = y_{n-1} + .06y_{n-1} \qquad \text{for} \quad n = 1, 2, 3, \ldots.$$

This equation can be simplified:

$$y_n = y_{n-1} + .06y_{n-1} = 1 \cdot y_{n-1} + .06y_{n-1} = (1 + .06)y_{n-1} = 1.06y_{n-1}. \quad (1)$$

In other words, the balance after n years is 1.06 times the balance after $n - 1$ years. This formula describes how the balance is computed in successive years. For instance, using this formula, we can compute y_4. Indeed, setting $n = 4$ and using the value of y_3 from the preceding chart,

$$y_4 = 1.06y_{4-1} = 1.06y_3 = 1.06(47.64) = \$50.50.$$

In a similar way we can compute all of the year-end balances, one after another, by using formula (1). ■

EXAMPLE 2 **Balance equation** Suppose that a savings account contains $40 and earns 6% interest, compounded annually. At the end of each year a $3 withdrawal is made. Determine a formula that describes how to compute each year's balance from the previous year's balance.

Solution As in Example 1, we compute the first few balances in a straightforward way and organize the data in a chart.

Balance	+ Interest for year	− Withdrawal
$y_0 = \$40$	$(.06)40 = \$2.40$	$\$3$
$y_1 = \ \ 39.40$	$(.06)39.40 = \$2.36$	3
$y_2 = \ \ 38.76$		

Reasoning as in Example 1, let y_n be the balance after n years. Then, by analyzing the calculations of the preceding chart, we see that

$$y_n \quad = \quad y_{n-1} \quad + \quad .06y_{n-1} \quad - \quad 3$$

[new balance] = [old balance] + [interest on old balance] − [withdrawal] or

$$y_n = 1.06y_{n-1} - 3. \tag{2}$$

This formula allows us to compute the values of y_n successively. For example, the preceding chart gives $y_2 = 38.76$. Therefore, from equation (2) with $n = 3$,

$$y_3 = (1.06)y_2 - 3 = (1.06)(38.76) - 3 = 38.09. \qquad \blacksquare$$

Equations of the form (1) and (2) are examples of what are called difference equations. More precisely, a *difference equation* is an equation of the form

$$y_n = ay_{n-1} + b,$$

where a and b are specific numbers. For example, for the difference equation (1) we have $a = 1.06$ and $b = 0$. For the difference equation (2) we have $a = 1.06$ and $b = -3$. A difference equation gives a procedure for calculating the term y_n from the preceding term y_{n-1}, thereby allowing one to compute all the terms—provided, of course, that a place to start is given. For this purpose one is usually given a specific value for y_0. Such a value is called an *initial value*. In both of the previous examples the initial value was 40.

Whenever we are given a difference equation, our goal is to determine as much information as possible about the terms y_0, y_1, y_2, and so on. To this end there are three things we can do:

1. *Generate the first few terms.* This is useful in giving us a feeling for how successive terms are generated.

2. *Graph the terms.* The terms that have been generated can be graphed by plotting the points $(0, y_0)$, $(1, y_1)$, $(2, y_2)$, and so on. Corresponding to the term y_n, we plot the point (n, y_n). The resulting graph (Fig. 1) depicts how the terms increase or decrease as n increases. In Figs. 2 and 3 we have drawn the graphs corresponding to the difference equations of Examples 1 and 2.

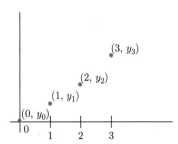

Figure 1

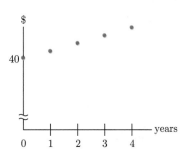

Figure 2

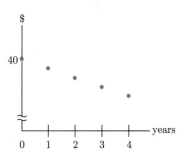

Figure 3

3. *Solve the difference equation.* By a *solution* of a difference equation we mean a general formula from which we can directly calculate any terms without first having to calculate all of the terms preceding it. One can always write down a solution using the values of a, b, and y_0. Assume for now that $a \neq 1$. Then a solution of $y_n = ay_{n-1} + b$ is given by

$$y_n = \frac{b}{1-a} + \left(y_0 - \frac{b}{1-a}\right)a^n, \qquad a \neq 1. \tag{3}$$

(This formula will be derived in Section 11.2.)

EXAMPLE 3   **Solving a difference equation** Solve the difference equation $y_n = 1.06y_{n-1}$, $y_0 = 40$.

Solution Here $a = 1.06$, $b = 0$, $y_0 = 40$. So $b/(1-a) = 0$, and from equation (3),

$$y_n = 0 + (40 - 0)(1.06)^n$$
$$y_n = 40(1.06)^n.$$

EXAMPLE 4   **Solving a difference equation** Solve the difference equation $y_n = 1.06y_{n-1} - 3$, $y_0 = 40$. Determine y_{21}.

Solution Here $a = 1.06$, $b = -3$, $y_0 = 40$. So $b/(1-a) = -3/(1 - 1.06) = -3/-.06 = 50$. Therefore,

$$y_n = 50 + (40 - 50)(1.06)^n$$
$$y_n = 50 - 10(1.06)^n.$$

In particular,

$$y_{21} = 50 - 10(1.06)^{21}.$$

Using a calculator, we find that $(1.06)^{21} \approx 3.4$, so that

$$y_{21} \approx 50 - 10(3.4) = 50 - 34 = 16.$$

So, after 21 years, the bank account of Example 2 will contain about $16. [If we did not have the solution $y_n = 50 - 10(1.06)^n$, we would need to perform 21 successive calculations in order to determine y_{21}.]

In the next two examples we apply our entire three-step procedure to analyze some specific difference equations.

EXAMPLE 5   **Analyzing a difference equation** Apply the three-step procedure to study the difference equation $y_n = .2y_{n-1} + 4.8$, $y_0 = 1$.

Solution First we compute a few terms:

$$y_0 = 1$$
$$y_1 = .2(1) + 4.8 = 5$$
$$y_2 = .2(5) + 4.8 = 5.8$$
$$y_3 = .2(5.8) + 4.8 = 5.96$$
$$y_4 = .2(5.96) + 4.8 = 5.992.$$

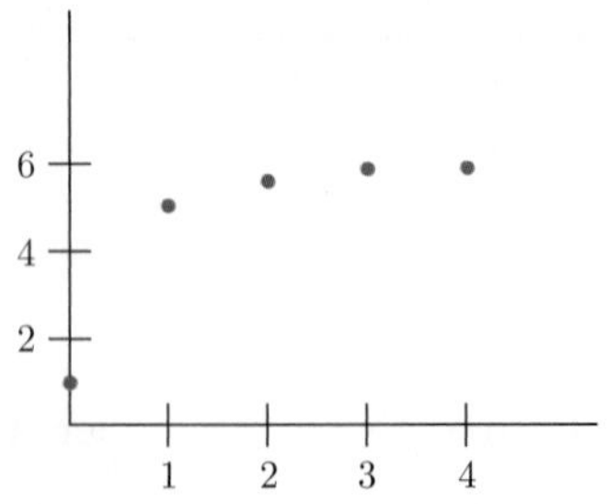

Figure 4

Next, we graph these terms (Fig. 4). That is, we plot $(0, 1)$, $(1, 5)$, $(2, 5.8)$, $(3, 5.96)$, $(4, 5.992)$. Note that the values of y_n increase. Finally, we solve the difference equation. Here $a = .2$, $b = 4.8$, $y_0 = 1$. Thus,

$$\frac{b}{1-a} = \frac{4.8}{1 - .2} = \frac{4.8}{.8} = 6$$

$$y_n = 6 + (1 - 6)(.2)^n = 6 - 5(.2)^n.$$

Now Try Exercise 7

EXAMPLE 6   **Analyzing a difference equation**   Apply the three-step procedure to study the difference equation $y_n = -.8y_{n-1} + 9$, $y_0 = 10$.

Solution   The first few terms are

$$y_0 = 10$$
$$y_1 = -.8(10) + 9 = -8 + 9 = 1$$
$$y_2 = -.8(1) + 9 = -.8 + 9 = 8.2$$
$$y_3 = -.8(8.2) + 9 = 2.44$$
$$y_4 = -.8(2.44) + 9 = 7.048.$$

The graph in Fig. 5 corresponds to these terms. Notice that in this case the points oscillate up and down. To solve the difference equation, note that

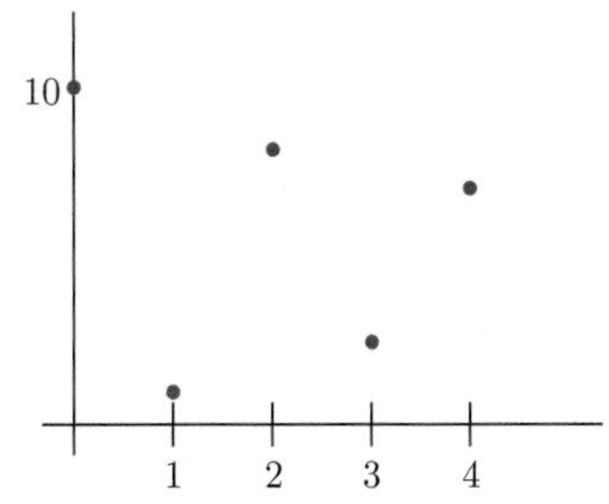

$$\frac{b}{1-a} = \frac{9}{1-(-.8)} = \frac{9}{1.8} = 5.$$

Figure 5

Thus

$$y_n = 5 + (10 - 5)(-.8)^n = 5 + 5(-.8)^n.$$

Now Try Exercise 11   ■

Difference equations can be used to describe many real-life situations. The solution of a particular difference equation in such instances yields a mathematical model of the situation. For example, consider the banking problem of Example 2. We described the activity in the account by a difference equation, where y_n represents the amount of money in the account after n years. Solving the difference equation, we found that

$$y_n = 50 - 10(1.06)^n.$$

And this formula gives a mathematical model of the bank account.

Here is another example of a mathematical model derived using difference equations.

EXAMPLE 7   **Population modeling**   Suppose that the population of a certain country is currently 6 million. The growth of this population attributable to an excess of births over deaths is 2% per year. Further, the country is experiencing immigration at the rate of 40,000 people per year.

(a) Find a mathematical model for the population of the country.

(b) What will be the population of the country after 35 years?

Solution   (a) Let y_n denote the population (in millions) of the country after n years. Then $y_0 = 6$. The growth of the population in year n due to an excess of births over deaths is $.02y_{n-1}$. There are $.04$ (million) immigrants each year. Therefore,

$$y_n = y_{n-1} + .02y_{n-1} + .04 = 1.02y_{n-1} + .04.$$

So the terms satisfy the difference equation

$$y_n = 1.02y_{n-1} + .04, \qquad y_0 = 6.$$

Here

$$\frac{b}{1-a} = \frac{.04}{1-1.02} = \frac{.04}{-.02} = -2$$

$$y_n = -2 + (6 - (-2))(1.02)^n \qquad \text{or}$$

$$y_n = -2 + 8(1.02)^n.$$

This last formula is our desired mathematical model for the population.

(b) To determine the population after 35 years, merely compute y_{35}:

$$y_{35} = -2 + 8(1.02)^{35} \approx -2 + 8(2) = 14,$$

since $(1.02)^{35} \approx 2$. So the population after 35 years will be about 14 million.

Now Try Exercise 19

■

GC Graphing calculators can easily generate successive terms of a difference equation. Consider the difference equation $y_n = .2y_{n-1} + 4.8$, $y_0 = 1$ of Example 5. In Figs. 6 and 9 successive terms are displayed in the home screen. In Figs. 7 and 10 successive terms are graphed, and in Figs. 8 and 11 they are displayed in a table.

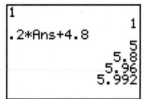

Figure 6

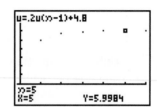

Figure 7.   Trace cursor after being moved several times.

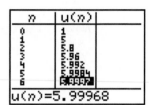

Figure 8

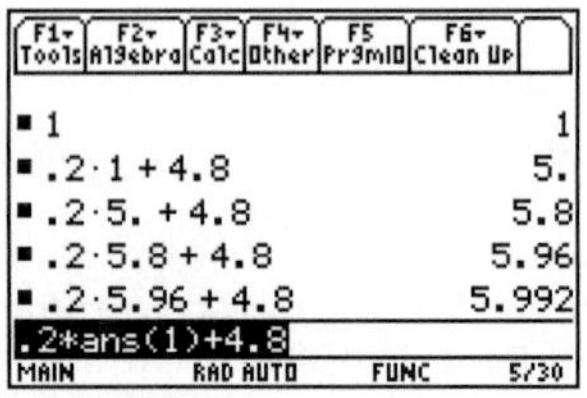

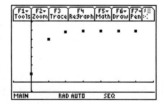

Figure 9

Figure 10

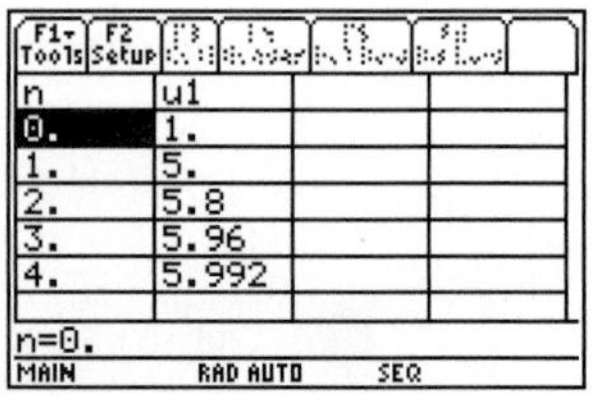

Figure 11

Displaying successive terms on the home screen In Figs. 6 and 9, after the initial value (1) is entered, the last displayed value (1) is assigned to **Ans** (on the TI-83) or **ans(1)** (on the TI-89). The instruction **.2*Ans+4.8** (or **.2*ans(1)+4.8** on the TI-89) generates the next term (5) and assigns it to **Ans** (or **ans(1)**). Each subsequent press of ENTER generates another term.

Graphs with the TI-83 and TI-89 In order to obtain the graphs in Figs. 7 and 10 the calculator must be set to sequence mode. This is accomplished by pressing MODE and selecting **Seq** on the fourth line of the TI-83, or setting **Graph** to **SEQUENCE** on the TI-89. In the **Y=** editor, difference equations are defined with letters such as **u**, **v**, **u1**, and **u2** instead of y. In the **Y=** editor of the TI-83, set $n\text{Min} = 0$, set **u(n)=.2u(n-1)+4.8**, and set **u(nMin)** $= \{1\}$. (**nMin** is the minimum value of n, always 0 for our purposes.) In the **Y=** editor of the TI-89, set

u1 = .2*u1(n-1)+4.8, and set **uil=1**. (**uil** is the initial value of the difference equation.) (*Note*: After u or u1 has been defined, values of u(n) or u1(n) can be displayed on the home screen by entering expressions such as **u(5)** or **u1(5)**.)

The familiar items in the WINDOW screen, **Xmin**, **Xmax**, **Xscl**, **Ymin**, **Ymax**, and **Yscl** have the same meanings as with graphs of functions. Points will be evaluated for values of n from **nMin** to **nMax**, and points will be plotted starting at **PlotStart**. When **PlotStart** is set to 1, graphing begins at y_0. When **PlotStep** is set to 1, no points are skipped. In Figs. 7 and 10, the settings were $[-.1, 6]$ *by* $[-2, 8]$ (on the TI-83) and $[-1, 7]$ *by* $[-1, 7]$ (on the TI-89), **nMin** $= 0$, **nMax** $= 6$, and **PlotStart=PlotStep** $= 1$.

We pressed GRAPH (on the TI-83) or ◆ [GRAPH] (on the TI-89) to display the graphs of the difference equations. You can press TRACE (on the TI-83) or F3 (on the TI-89) to obtain a trace cursor. Each time you press the right-arrow key, the trace cursor will move to the next point and its coordinates will be displayed.

Tables with the TI-83 and TI-89   In order to obtain the tables in Figs. 8 and 11, the setting in the **Y=** editor is the same as for graphs. Then 2nd [TBLSET] (on the TI-83) or ◆ [TblSet] (on the TI-89) was pressed to bring up the TABLE SETUP screen. **TblStart** was set to 0, and **ΔTbl** was set to 1. Finally, 2nd [TABLE] (on the TI-83) or ◆ [TABLE] (on the TI-89) was pressed to bring up the table. You can use the down-arrow key to generate further terms.

ES   The table and graph in Fig. 12 are easily created in an Excel spreadsheet. The following steps create the second column of the table:

1. To enter the numbers 0 through 6 in the first column, enter **0** into cell A2, enter **1** into cell A3, select the two cells, and drag the fill handle down to A8.
2. Type **1** into cell B2 and press Enter.
3. Type **=.2*B2+4.8** into cell B3 and press Enter.
4. Click on cell B3 and drag its fill handle down to B8.

To create the graph, carry out the following steps:

1. Select the range A2:B8.
2. Click on the Chart Wizard icon in the standard toolbar.
3. Click on XY (Scatter) in the "Chart type" list.
4. Click the Finish button in the lower-right corner of the window.
5. Optionally, right-click on the legend "Series 1" and click on Clear.

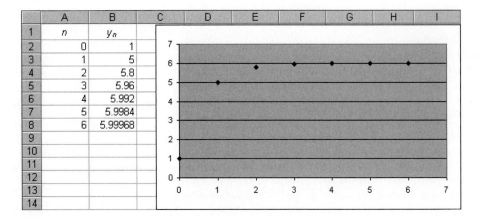

Figure 12.   Excel table and graph for $y_n = .2y_{n-1} + 4.8$.

PRACTICE PROBLEMS 11.1

1. Consider the difference equation $y_n = -2y_{n-1} + 21$, $y_0 = 7.5$.

   (a) Generate $y_0, y_1, y_2, y_3,$ and y_4 from the difference equation.

   (b) Graph these first few terms.

   (c) Solve the difference equation.

2. Use the solution in Problem 1(c) to obtain the terms $y_0, y_1,$ and y_2.

EXERCISES 11.1

For each of the difference equations in Exercises 1–6, identify a and b and compute $b/(1-a)$.

1. $y_n = 4y_{n-1} - 6$ 2. $y_n = -3y_{n-1} + 16$

3. $y_n = -\frac{1}{2}y_{n-1}$ 4. $y_n = \frac{1}{3}y_{n-1} + 4$

5. $y_n = -\frac{2}{3}y_{n-1} + 15$ 6. $y_n = .5y_{n-1} - 4$

In Exercises 7–14, (a) generate y_0, y_1, y_2, y_3, y_4 from the difference equation. (b) Graph these first few terms. (c) Solve the difference equation.

7. $y_n = \frac{1}{2}y_{n-1} - 1$, $y_0 = 10$

8. $y_n = .5y_{n-1} + 5$, $y_0 = 2$

9. $y_n = 2y_{n-1} - 3$, $y_0 = 3.5$

10. $y_n = 5y_{n-1} - 32$, $y_0 = 8$

11. $y_n = -.4y_{n-1} + 7$, $y_0 = 17.5$

12. $y_n = -2y_{n-1}$, $y_0 = \frac{1}{2}$

13. $y_n = 2y_{n-1} - 16$, $y_0 = 15$

14. $y_n = 2y_{n-1} + 3$, $y_0 = -2$

15. The solution to $y_n = .2y_{n-1} + 4.8$, $y_0 = 1$ is $y_n = 6 - 5(.2)^n$. Use the solution to compute y_0, y_1, y_2, y_3, y_4.

16. The solution to $y_n = -.8y_{n-1} + 9$, $y_0 = 10$ is $y_n = 5 + 5(-.8)^n$. Use the solution to compute y_0, y_1, y_2.

17. (*Account Balance*) One thousand dollars is deposited into a savings account paying 5% interest compounded annually. Let y_n be the amount after n years. What is the difference equation showing how to compute y_n from y_{n-1}?

18. (*Population Decline*) The population of a certain country is currently 70 million but is declining at the rate of 1% each year. Let y_n be the population after n years. Find a difference equation showing how to compute y_n from y_{n-1}.

19. (*Population Decline*) The population of a certain country is currently 70 million but is declining at the rate of 1% each year due to an excess of deaths over births. In addition, the country is losing one million people each year due to emigration. Let y_n be the population after n years. Find a difference equation satisfied by y_n.

20. (*Account Balance*) One thousand dollars is deposited into an account paying 5% interest compounded annually. At the end of each year $100 is added to the account. Let y_n be the amount in the account after n years. Find a difference equation satisfied by y_n.

21. Consider the difference equation $y_n = y_{n-1} + 2$, $y_0 = 1$.

   (a) Generate y_0, y_1, y_2, y_3, y_4.

   (b) Sketch the graph.

   (c) Why cannot formula (3) be used to obtain the solution?

22. Rework Exercise 21 for $y_n = y_{n-1} - 2$, $y_0 = 10$.

23. (*Loan Balance*) Suppose that you take a consumer loan for $55 at 20% annual interest and pay off $36 at the end of each year for two years. Compute the balance on the loan immediately after you make the first payment (i.e., the amount you would pay if you wanted to settle the account at the beginning of the second year).

24. (*Loan Balance*) Refer to Exercise 23. Set up the difference equation for y_n, the balance after n years.

25. (*Savings Plan*) Ferdinand knows that he will need to replace his home office equipment someday. He deposits $800 into a savings account paying 4% interest compounded annually. At the end of each year he deposits $250 into the account. Let y_n be the amount in the account after n years.

   (a) Find a difference equation satisfied by y_n.

   (b) Solve the difference equation.

   (c) How much money will he have in the account after 7 years?

26. (*Loan Balance*) Anne takes out a $317 loan at 10% annual interest. She pays off $100 at the end of each year for 4 years. Compute the balance on the loan immediately after she makes the first payment. Then compute the balance after the second payment.

27. (*Automobile Depreciation*) A rule of thumb states that cars in personal use depreciate by 15% each year. Suppose a new car is purchased for $20,000. Let y_n be the value of the car after n years.

   (a) Find a difference equation satisfied by y_n.

   (b) Solve the difference equation.

   (c) How much will the car be worth after 5 years?

Exercises 28–35 require the use of a graphing calculator, a spreadsheet, or mathematical software.

In these exercises,

(a) Generate a table displaying the values of y_0 to y_6.

(b) Graph the first 24 terms of the difference equation.

(c) Use the table or graph to answer the questions.

28. $y_n = 1.2y_{n-1} - 2$, $y_0 = 12.5$. What is y_{11}? For what value of n does y_n first exceed 100?

29. $y_n = 1.05y_{n-1} - 7.8$, $y_0 = 102.67$. What is y_{14} (rounded to two decimal places)? For what value of n does $y_n = 0$ (rounded to two decimal places)?

30. $y_n = .8y_{n-1} + .6$, $y_0 = 10$. What is y_{12}? For what values of n is y_n within .1 units of 3?

31. $y_n = -.85y_{n-1} + 11.1$, $y_0 = 4$. What is y_{12}? For what values of n is y_n within .1 units of 6?

32. $y_n = -1.3y_{n-1} + 1.5$, $y_0 = 1$. What is y_{14}? For what value of n is y_n approximately 39.77?

33. $y_n = y_{n-1} + .2$, $y_0 = 1$. What is y_9? For what value of n does $y_n = 4.6$?

34. $y_n = -y_{n-1} + .2$, $y_0 = 1$. What special behavior do the terms exhibit?

35. $y_n = .75y_{n-1} + 1.25$, $y_0 = 2$. What is y_{10}? For what values of n is y_n within .01 units of 5?

SOLUTIONS TO PRACTICE PROBLEMS 11.1

1. (a) $y_0 = 7.5$.

$y_1 = -2(7.5) + 21 = -15 + 21 = 6$.

$y_2 = -2(6) + 21 = -12 + 21 = 9$.

$y_3 = -2(9) + 21 = -18 + 21 = 3$.

$y_4 = -2(3) + 21 = -6 + 21 = 15$.

(b) We graph the points $(0, 7.5)$, $(1, 6)$, $(2, 9)$, $(3, 3)$, $(4, 15)$ in Fig. 13. (In order to accommodate the points, we use a different scale for each axis.)

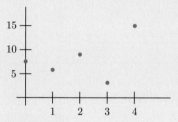

Figure 13

(c) $y_n = \dfrac{b}{1-a} + \left(y_0 - \dfrac{b}{1-a}\right)a^n$.

Since $a = -2$ and $b = 21$,

$$\frac{b}{1-a} = \frac{21}{1-(-2)} = \frac{21}{3} = 7.$$

$$y_n = 7 + (7.5 - 7)(-2)^n = 7 + \tfrac{1}{2}(-2)^n.$$

2. $y_n = 7 + \tfrac{1}{2}(-2)^n$.

$y_0 = 7 + \tfrac{1}{2}(-2)^0 = 7 + \tfrac{1}{2} = 7.5$.

(Here we have used the fact that any number to the zeroth power is 1.)

$y_1 = 7 + \tfrac{1}{2}(-2)^1 = 7 + \tfrac{1}{2}(-2) = 7 - 1 = 6$.

$y_2 = 7 + \tfrac{1}{2}(-2)^2 = 7 + \tfrac{1}{2}(4) = 7 + 2 = 9$.

11.2 Introduction to Difference Equations II

The main result of Section 11.1 is the following:

The difference equation $y_n = ay_{n-1} + b$, with $a \neq 1$, has solution

$$y_n = \frac{b}{1-a} + \left(y_0 - \frac{b}{1-a}\right)a^n. \tag{1}$$

When $a = 1$, the difference equation $y_n = a \cdot y_{n-1} + b$ is $y_n = 1 \cdot y_{n-1} + b$ or just $y_n = y_{n-1} + b$. In this case the solution is given by a formula different from (1):

The difference equation $y_n = y_{n-1} + b$ has the solution

$$y_n = y_0 + bn. \tag{2}$$

EXAMPLE 1 **Solving a difference equation of the form $y_n = y_{n-1} + b$**
(a) Solve the difference equation $y_n = y_{n-1} + 2$, $y_0 = 3$.
(b) Find y_{100}.

Solution (a) Since $a = 1$, the solution is given by (2). Here $b = 2$, $y_0 = 3$. Therefore,

$$y_n = 3 + 2n.$$

(b) $y_{100} = 3 + 2(100) = 203$.
  Note: Without (2), determining y_{100} would require 100 computations.

Now Try Exercise 1 ■

Both (1) and (2) are derived at the end of this section. Before deriving them, however, let us see what they tell us about simple interest, compound interest, and consumer loans.

Simple Interest Suppose that a certain amount of money is deposited in a savings account. If interest is paid only on the initial deposit (and not on accumulated interest), then the interest is called *simple*. For example, if $40 is deposited at 6% simple interest, then each year the account earns .06(40), or $2.40. So the bank balance accumulates as follows:

Year	Amount	Interest
0	$40.00	$2.40
1	42.40	2.40
2	44.80	2.40
3	47.20	

EXAMPLE 2 **Modeling simple interest**
(a) Find a formula for y_n, the amount in the aforementioned account at the end of n years.
(b) Find the amount at the end of 10 years.

Solution (a) Let y_n = the amount at the end of n years. So $y_0 = 40$. Moreover,

[amount at end of n years] = [amount at end of $n - 1$ years] + [interest].
  y_n = y_{n-1} +   2.40.

This difference equation has $a = 1$, $b = 2.40$, so from formula (2),

$$y_n = y_0 + bn = 40 + (2.40)n.$$

This is the desired formula.
(b) $y_{10} = 40 + (2.40)10 = 40 + 24.00 = \$64$.

Now Try Exercise 7 ■

Compound Interest When interest is calculated on the current amount in the account (instead of on the amount initially deposited), the interest is called *compound*. The interest discussed in Section 11.1 was compound, being computed on the balance at the end of each year. Such interest is called *annual* compound interest. Often interest is compounded more than once a year. For example, an interest rate might be stated as 6% compounded *semiannually*. This means that interest is computed every 6 months, with 3% given for each 6-month period. At the end of each such period the interest is added to the balance, which is then used to compute the interest for the next 6-month period. Similarly, 6% *interest compounded quarterly* means .06/4 or .015 interest four times a year. And 6% interest compounded six times a year means .06/6 or .01 interest each interest period of 2 months. Or, in general, if 6% interest is compounded k times a year, then .06/k interest is earned k times a year. This illustrates the following general principle:

> If interest is at a yearly rate r and is compounded k times per year, then the interest rate per period (denoted i) is $i = r/k$.

EXAMPLE 3 **Modeling compound interest** Suppose that the interest rate is 6% compounded monthly. Find a formula for the amount after n months.

Solution Here $r = .06$, $k = 12$, so that the monthly interest rate is $i = .005$. Let y_n denote the amount after n months. Then, reasoning as in Section 11.1,

$$y_n = y_{n-1} + .005y_{n-1} = 1.005y_{n-1}.$$

The solution of this difference equation is obtained as follows:

$$\frac{b}{1-a} = 0 \qquad y_n = 0 + (y_0 - 0)(1.005)^n = y_0(1.005)^n.$$

Now Try Exercise 3 ■

EXAMPLE 4 **Modeling simple and compound interest** Suppose that interest is computed at the rate i per interest period. Find a general formula for the balance y_n after n interest periods under (a) simple interest, (b) compound interest.

Solution (a) With simple interest, the interest each period is just i times the initial amount— that is, $i \cdot y_0$. Therefore,

$$y_n = y_{n-1} + iy_0.$$

Apply (2) with $b = iy_0$. Then

$$y_n = y_0 + (iy_0)n.$$

(b) With compound interest, the interest each period is i times the current balance, or $i \cdot y_n$. Therefore,

$$y_n = y_{n-1} + iy_{n-1} = (1 + i)y_{n-1}.$$

Here $b/(1 - a) = 0$, so that

$$y_n = 0 + (y_0 - 0)(1 + i)^n = y_0(1 + i)^n.$$ ■

Summarizing,

> If y_0 dollars is deposited at interest rate i per period, then the amount after n interest periods is
>
> Simple interest: $\qquad y_n = y_0 + (iy_0)n$ $\qquad$ (3)
>
> Compound interest: $\qquad y_n = y_0(1 + i)^n.$ $\qquad$ (4)

EXAMPLE 5   **Account balance**   How much money will you have after seven years if you deposit $40 at 8% interest compounded quarterly?

Solution   Apply formula (4). Here a period is a quarter, or 3 months. In 7 years there are $7 \cdot 4 = 28$ periods. The interest rate per period is

$$i = \frac{.08}{4} = .02.$$

The amount after 28 periods is

$$y_{28} = 40(1.02)^{28} = \$69.64.$$

[To compute $(1.02)^{28}$, we used a calculator. The answer could have been left in the form $40(1.02)^{28}$.]

Now Try Exercise 9   ■

Consumer Loans   It is common for people to buy cars or appliances "on time." Basically, they are borrowing money from the dealer and repaying it (with interest) with several equal payments until the loan is paid off. Each time period, part of the payment goes toward paying off the interest and part goes toward reducing the balance of the loan. A consumer loan used to purchase a house is called a *mortgage*.

EXAMPLE 6   **Loan balance**   Suppose that a consumer loan of $2400 carries an interest rate of 12% compounded annually and a yearly payment of $1000.

(a) Write down a difference equation for y_n, the balance owed after n years.

(b) Compute the balances after 1, 2, and 3 years.

Solution   (a) At the end of each year the new balance is computed as follows:

$$[\text{new balance}] = [\text{previous balance}] + [\text{interest}] - [\text{payment}].$$

Since the interest is compound, it is computed on the previous balance:

$$y_n = y_{n-1} + .12y_{n-1} - 1000.$$

Thus

$$y_n = 1.12y_{n-1} - 1000, \qquad y_0 = 2400$$

is the desired difference equation.

(b) $y_1 = 1.12y_0 - 1000 = (1.12)(2400) - 1000 = \$1688.$

$y_2 = 1.12y_1 - 1000 = (1.12)(1688) - 1000 = 890.56.$

$y_3 = 1.12y_2 - 1000 = (1.12)(891) - 1000 \approx 0.$

Now Try Exercise 17   ■

Although savings accounts using simple interest are practically unheard of, consumer loans are often computed using simple interest. Under simple interest, interest is paid on the entire amount of the initial loan, not just on the outstanding balance.

EXAMPLE 7   **Loan balance**   Rework Example 6, except assume that the interest is now simple.

Solution (a) As before,

$$[\text{new balance}] = [\text{previous balance}] + [\text{interest}] - [\text{payment}].$$

Now, however, since the interest is simple, it is computed on the original loan y_0. So

$$y_n = y_{n-1} + .12y_0 - 1000, \qquad y_0 = 2400.$$

(b) $y_1 = y_0 + .12y_0 - 1000 = \$1688.$
$y_2 = y_1 + .12y_0 - 1000 = 976.$
$y_3 = y_2 + .12y_0 - 1000 = 264.$ ■

▶ *Remark* Note that after three years the loan of Example 7 is not yet paid off, whereas the loan of Example 6 is. Therefore, the loan at 12% simple interest is more expensive than the one at 12% compound interest. Actually, a 12% simple interest loan might be comparable to a 16% compound interest loan. At one time it was a common practice to advertise loans in terms of simple interest to make the interest rate seem cheaper. But the Federal Truth-in-Lending Act now requires that all loans be stated in terms of their equivalent compound interest rate. ◀

Verification of Formula (1)   From the difference equation $y_n = ay_{n-1} + b$ we get

$$y_1 = ay_0 + b$$
$$y_2 = ay_1 + b = a(ay_0 + b) + b$$
$$\quad = a^2 y_0 + ab + b$$
$$y_3 = ay_2 + b = a(a^2 y_0 + ab + b) + b$$
$$\quad = a^3 y_0 + a^2 b + ab + b$$
$$y_4 = ay_3 + b = a(a^3 y_0 + a^2 b + ab + b) + b$$
$$\quad = a^4 y_0 + a^3 b + a^2 b + ab + b.$$

The pattern that clearly develops is

$$y_n = a^n y_0 + a^{n-1} b + a^{n-2} b + \cdots + a^2 b + ab + b. \tag{5}$$

Multiply both sides of (5) by a and then subtract the new equation from (5). Notice that many terms drop out.

$$y_n = \quad a^n y_0 + \qquad a^{n-1} b + a^{n-2} b + \cdots + a^2 b + ab + b$$
$$ay_n = a^{n+1} y_0 + a^n b + a^{n-1} b + a^{n-2} b + \cdots + a^2 b + ab$$
$$\overline{y_n - ay_n = a^n y_0 - a^{n+1} y_0 - a^n b + b}$$

The last equation can be written

$$(1-a)y_n = (1-a)a^n y_0 - ba^n + b.$$

Now, divide both sides of the equation by $(1-a)$. [*Note:* Since $a \neq 1$, $1 - a \neq 0$.]

$$y_n = y_0 a^n - \frac{b}{1-a} \cdot a^n + \frac{b}{1-a}$$
$$\quad = \frac{b}{1-a} + \left(y_0 - \frac{b}{1-a} \right) a^n,$$

which is formula (1). ■

Verification of Formula (2)   Formula (1) gives the solution of the difference equation $y_n = ay_{n-1} + b$ in the case $a \neq 1$. The preceding reasoning also gives the solution in case $a = 1$—namely equation (5) holds for any value of a. In particular, for $a = 1$ equation (5) reads

$$y_n = a^n y_0 + a^{n-1}b + a^{n-2}b + \cdots + a^2 b + ab + b$$
$$= 1^n y_0 + 1^{n-1}b + 1^{n-2}b + \cdots + 1^2 b + 1b + b$$
$$= y_0 + b + b + \cdots + b + b + b$$
$$= y_0 + bn.$$

Thus we have derived formula (2). ∎

PRACTICE PROBLEMS 11.2

1. Solve the following difference equations.
   (a) $y_n = -.2y_{n-1} + 24$, $y_0 = 25$
   (b) $y_n = y_{n-1} - 3$, $y_0 = 7$
2. (*Land Value*) In 1626, Peter Minuit, the first director-general of New Netherlands Province, purchased Manhattan Island for trinkets and cloth valued at about \$24. Suppose that this money had been invested at 7% interest compounded quarterly. How much would it have been worth by the U.S. bicentennial year, 1976?

EXERCISES 11.2

1. Solve the difference equation $y_n = y_{n-1} + 5$, $y_0 = 1$.
2. Solve the difference equation $y_n = y_{n-1} - 2$, $y_0 = 50$.

In Exercises 3–8, find an expression for the amount of money in the bank after five years when the initial deposit is \$80 and the annual interest rate is as given.

3. 9% compounded monthly
4. 4% simple interest
5. 100% compounded daily
6. 8% compounded quarterly
7. 7% simple interest
8. 12% compounded semiannually
9. Determine the amount of money accumulated after one year when \$1 is deposited at 40% interest compounded
   (a) annually   (b) semiannually   (c) quarterly.
10. Find the general formula for the amount of money accumulated after t years when A dollars is invested at annual interest rate r compounded k times per year.
11. For the difference equation $y_n = 2y_{n-1} - 10$, generate y_0, y_1, y_2, y_3 and draw the graph corresponding to the initial condition:
    (a) $y_0 = 10$ (b) $y_0 = 11$ (c) $y_0 = 9$
12. For the difference equation $y_n = \frac{1}{2}y_{n-1} + 5$, generate y_0, y_1, y_2, y_3 and draw the graph corresponding to the initial condition:
    (a) $y_0 = 10$ (b) $y_0 = 18$ (c) $y_0 = 2$

In Exercises 13–16, solve the difference equation and, by inspection, determine the long-run behavior of the terms (i.e., the behavior of the terms as n gets large).

13. $y_n = .4y_{n-1} + 3$, $y_0 = 7$
14. $y_n = 3y_{n-1} - 12$, $y_0 = 10$
15. $y_n = -5y_{n-1}$, $y_0 = 2$
16. $y_n = -.7y_{n-1} + 3.4$, $y_0 = 3$
17. (*Mortgage Balance*) A bank loan of \$38,900 at 9% interest compounded monthly is made in order to buy a house and is paid off at the rate of \$350 per month for 20 years. (Such a loan is called a *mortgage.*) The balance at any time is the amount still owed on the loan, that is, the amount that would have to be paid out to repay the loan all at once at that time. Find the difference equation for y_n, the balance after n months.
18. (*Mortgage Balance*) Refer to Exercise 17. Express in mathematical notation the fact that the loan is paid off after 20 years.
19. (*Depreciation Modeling*) A house is purchased for \$50,000 and depreciated over a 25-year period. Let y_n be the (undepreciated) value of the house after n years. Determine and solve the difference equation for y_n, assuming straight-line depreciation (i.e., each year the house depreciates by $\frac{1}{25}$ of its original value).
20. (*Depreciation Modeling*) Refer to Exercise 19. Determine and solve the difference equation for y_n, assuming the double-declining balance method of depreciation (i.e., each year the house depreciates by $\frac{2}{25}$ of its value at the beginning of that year).

21. (*Population Growth*) (PE) If a population of 500 bacteria doubles in size every 5 minutes, approximately how many minutes will it take for the population to grow to 500,000?
(a) 45 (b) 50 (c) 55 (d) 60 (e) 65

22. (*Savings Account*) (PE) The balance in a savings account doubles every 12 years. Approximately how many years will it take for the balance to grow from $100 to $25,000?
(a) 60 (b) 72 (c) 84 (d) 96 (e) 108

23. (*Investment Value*) (PE) By how much does the value of an investment increase in 2 weeks if it grows at the rate of $50 per day?
(a) $100 (b) $640 (c) $700
(d) $1000 (e) $1400

24. (*Population Growth*) (PE) A city's population grows at the rate of one person every 20 minutes. By how many people does the population grow in 6 hours?
(a) 18 (b) 20 (c) 60 (d) 120 (e) 180

25. (*Population Growth*) (PE) A population doubles every year for 5 years. If the size of the population after the 5 years is 100,000, what was the population at the end of the second year?
(a) 7500 (b) 10,000 (c) 12,500
(d) 25,000 (e) 50,000

26. (*Investment Value*) (PE) The value of an investment doubles every month for 8 months. After the eighth month, the investment was worth $400,000. What was the investment worth after the fourth month?
(a) $25,000 (b) $50,000 (c) $75,000
(d) $100,000 (e) $200,000

27. (*Bacteria Growth*) (PE) If a culture of 10^5 bacteria doubles in size every 15 minutes, how many bacteria are present after 1 hour?
(a) $10^4(10^5)$ (b) $(10^5)^4$ (c) $2^4(10^5)$
(d) $2(10^5)$ (e) $4(10^5)$

28. (*Population Growth*) (PE) A population of 10^7 doubles in size every 6 months. How big is the population after 8 years?
(a) $(10^7)^{16}$ (b) $10^{16}(10^7)$ (c) $16(10^7)$
(d) $2(10^7)$ (e) $2^{16}(10^7)$

Exercises 29–35 require the use of a graphing calculator, a spreadsheet, or mathematical software.

In these exercises, answer the questions by generating a table or graph of an appropriate difference equation.

29. One thousand dollars is deposited into a savings account at 6% interest compounded quarterly. Determine the balance after 3 years. When will the balance reach $1659? How many quarters are required for the initial deposit to double?

30. Redo Exercise 29 for simple interest of 1.5% per quarter-year.

31. (*Account Balance*) Two hundred dollars is deposited into a savings account at simple interest of 4.5% annually. Determine the balance after 5 years. When will the balance reach or exceed $308? How many years are required for the initial deposit to double?

32. (*Account Balance*) Redo Exercise 31 for compound interest of 4.5% compounded annually.

33. (*Credit Card Debt*) Roberto has a $2000 debt on a credit card charging 18% interest compounded monthly. If he makes a $105 payment at the end of each month (and has no new charges), how many months will it take for his debt to be less than $105?

34. Consider the difference equation $y_n = -.9y_{n-1} + 19$, $y_0 = 5$. What value do the terms approach as n gets large?

35. Consider the difference equation $y_n = .85y_{n-1} + 9$, $y_0 = 5$. What value do the terms approach as n gets large?

SOLUTIONS TO PRACTICE PROBLEMS 11.2

1. (a) Since $a = -.2 \neq 1$, use formula (1).

$$\frac{b}{1-a} = \frac{24}{1-(-.2)} = \frac{24}{1.2} = \frac{240}{12} = 20$$

$$y_n = 20 + (25-20)(-.2)^n = 20 + 5(-.2)^n$$

(b) Since $a = 1$, use formula (2).

$$y_n = 7 + (-3)n = 7 - 3n$$

2. Since interest is compounded quarterly, the interest per period is $.07/4 = .0175$. Three hundred and fifty years consists of $4(350) = 1400$ interest periods. Therefore, by (4), the amount accumulated is

$$y_{1400} = 24(1.0175)^{1400}.$$

(This amount is approximately $850 billion, which is more than Manhattan Island was worth in 1976.)

11.3 Graphing Difference Equations

In this section we introduce a method for sketching the graph of the difference equation $y_n = ay_{n-1} + b$ (with initial value y_0) directly from the three numbers a, b, and y_0. As we shall see, the graphs arising from difference equations can be described completely by two characteristics—vertical direction and long-run behavior. To begin we introduce some vocabulary to describe graphs.

The *vertical direction* of a graph refers to the up-and-down motion of successive terms. A graph *increases* if it rises when read from left to right—that is, if the terms get successively larger. A graph *decreases* if it falls when read from left to right—that is, if the terms get successively smaller. Figure 1 shows the graphs of two difference equations. Both graphs increase. Figure 2 shows two examples of decreasing graphs. A graph that is either increasing or decreasing is called *monotonic*. That is, a graph is monotonic if it always heads in one direction—up or down.

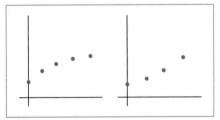

Figure 1. Increasing graphs.

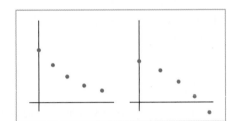

Figure 2. Decreasing graphs.

The extreme opposite of a monotonic graph is one that changes its direction with every term. Such a graph is called *oscillating*. Figure 3 shows two examples of oscillating graphs. A difference equation having an oscillating graph has terms y_n that alternately increase and decrease.

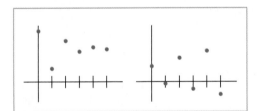

Figure 3. Oscillating graphs.

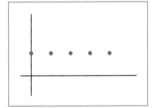

Figure 4. Constant graph.

In addition to the monotonic and oscillating graphs, there are the *constant* graphs, which always remain at the same height. That is, all the terms y_n are the same. A constant graph is illustrated in Fig. 4.

One of the main results of this section is that the graph of a difference equation $y_n = ay_{n-1} + b$ is always either monotonic, oscillating, or constant. This threefold classification gives us all the possibilities for the vertical direction of the graph.

Long-run behavior refers to the eventual behavior of the graph. Most graphs of difference equations exhibit one of two types of long-run behavior. Some approach a horizontal line and are said to be *asymptotic* to the line. Some go indefinitely high or indefinitely low and are said to be *unbounded*. These phenomena are illustrated in Figs. 5 and 6.

The dashed horizontal lines in Figs. 5 and 6 each have the equation $y = b/(1-a)$. Note that in Fig. 5 the terms move steadily closer to the dashed line and in Fig. 6 they move steadily farther away from the dashed line. In the first case we say that

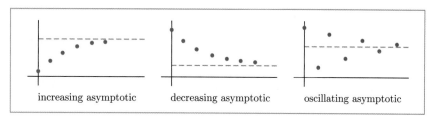

Figure 5. Asymptotic graphs.

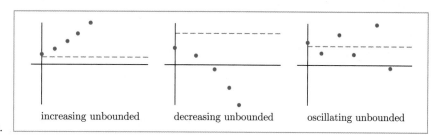

Figure 6. Unbounded graphs.

the terms are *attracted* to the dashed line and in the second case we say that they are *repelled* by it.

Constant Graphs   The general formula for the nth term of a difference equation where $a \neq 1$ is

$$y_n = \frac{b}{1-a} + \left(y_0 - \frac{b}{1-a}\right)a^n.$$

It is clear that as n varies so does a^n, and this makes y_n vary with n—unless, of course, the coefficient of a^n is 0, in which case the graph is constant. Thus we have the following result:

> The graph of $y_n = ay_{n-1} + b$ $(a \neq 1)$ is constant if $y_0 - b/(1-a) = 0$; that is, if $y_0 = b/(1-a)$.

Thus, when the graph starts out on the line $y = b/(1-a)$, it stays on the line.

EXAMPLE 1   **Graphing a difference equation**   Sketch the graph of the difference equation $y_n = 2y_{n-1} - 1$, $y_0 = 1$.

Solution   First compute $b/(1-a)$:

$$\frac{b}{1-a} = \frac{-1}{1-2} = 1.$$

But $y_0 = 1$. So $b/(1-a)$ and y_0 are the same and the graph is constant, always equal to 1 (Fig. 7).

Now Try Exercise 17   ∎

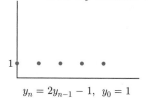

$y_n = 2y_{n-1} - 1, \ y_0 = 1$

Figure 7

Throughout this section assume that $a \neq \pm 1$. Then the graph of a given difference equation either is constant or is one of the types shown in Figs. 5 and 6. We can determine the nature of the graph by looking at the coefficients a and b and the initial value y_0. We have seen how constant graphs are handled. Thus, assume that we are dealing with a nonconstant graph—that is, the graph of a difference equation for which a^n actually affects the formula for y_n. Monotonic graphs may be differentiated from oscillating graphs by the following test.

> ***Test 1***   If $a > 0$, then the graph of $y_n = ay_{n-1} + b$ is monotonic.
>
> If $a < 0$, then the graph is oscillating.

The next two examples provide a convincing argument for Test 1.

EXAMPLE 2   **Vertical direction of a difference equation**   Discuss the vertical direction of the graph of $y_n = -.8y_{n-1} + 9$, $y_0 = 50$.

Solution   The formula for y_n yields

$$y_n = 5 + 45(-.8)^n.$$

Note that the term $(-.8)^n$ is alternately positive and negative, since any negative number to an even power is positive and any negative number to an odd power is negative. Therefore, the expression $45(-.8)^n$ is alternately positive and negative. So $y_n = 5 + 45(-.8)^n$ is computed by alternately adding and subtracting something from 5. Thus y_n oscillates around 5. In this example $a = -.8$ (a negative number), so that the behavior just observed is consistent with that predicted by Test 1. The graph is sketched in Fig. 8. Note that in this case 5 is just $b/(1-a)$, so the graph oscillates about the line $y = b/(1-a)$. ∎

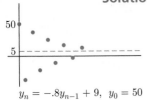

$y_n = -.8y_{n-1} + 9$, $y_0 = 50$

Figure 8

The reasoning of Example 2 works whenever $a < 0$. The oscillation of a^n from positive to negative and back forces the term

$$\left(y_0 - \frac{b}{1-a} \right) a^n$$

to swing back and forth from positive to negative. So the value of

$$y_n = \frac{b}{1-a} + \left(y_0 - \frac{b}{1-a} \right) a^n$$

swings above and below $b/(1-a)$. In other words, the graph oscillates about the line $y = b/(1-a)$.

EXAMPLE 3   **Vertical direction of a difference equation**   Discuss the vertical direction of the graph of $y_n = .8y_{n-1} + 9$, $y_0 = 50$.

Solution   In this case $b/(1-a) = 45$, and the formula for y_n gives

$$y_n = 45 + 5(.8)^n.$$

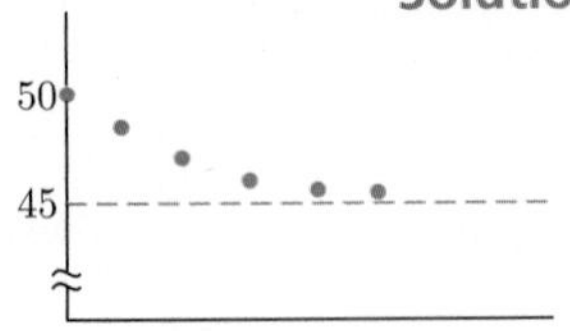

The expression $5(.8)^n$ is always positive. As n gets larger, $5(.8)^n$ gets smaller, so that y_n decreases to 45. That is, the graph is steadily decreasing to $y = 45$ (Fig. 9). Here $a = .8 > 0$, so Test 1 correctly predicts that the graph is monotonic. ∎

$y_n = .8y_{n-1} + 9$, $y_0 = 50$

Figure 9

Using the following result, it is possible to determine whether a graph is asymptotic or unbounded.

> ***Test 2***   If $|a| < 1$, then the graph of $y_n = ay_{n-1} + b$ is asymptotic to the line $y = b/(1-a)$. If $|a| > 1$, then the graph is unbounded and moves away from the line $y = b/(1-a)$.

Let us examine some difference equations in light of Test 2.

EXAMPLE 4 **Vertical direction and long-run behavior of a difference equation** Discuss the graphs of the difference equation $y_n = .2y_{n-1} + 4.8$, with $y_0 = 1$ and with $y_0 = 11$.

Solution The graphs are shown in Fig. 10. Since $a = .2$, which has absolute value less than 1, Test 2 correctly predicts that each graph is asymptotic to the line $y = b/(1-a) = 4.8/(1-.2) = 6$. When the initial value is less than 6, the graph increases and moves toward the line $y = 6$. When the initial value is greater than 6, the graph decreases toward the line $y = 6$. In each case the graph is *attracted* to the line $y = 6$. ■

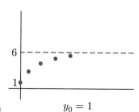
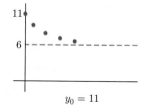

Figure 10

$y_0 = 1$ $\qquad\qquad$ $y_0 = 11$

EXAMPLE 5 **Applying Test 2** Apply Test 2 to the difference equation $y_n = -.8y_{n-1} + 9$, $y_0 = 50$.

Solution $|a| = |-.8| = .8 < 1$. So by Test 2 the graph is asymptotic to the line $y = b/(1-a) = 5$. This agrees with the graph as drawn in Fig. 8. ■

EXAMPLE 6 **Vertical direction and long-run behavior of a difference equation** Discuss the graphs of $y_n = 1.4y_{n-1} - 8$, with $y_0 > 20$ and with $y_0 < 20$.

Solution The graphs are shown in Fig. 11. Since $a = 1.4 > 0$, Test 1 predicts that the graphs are monotonic. Since $|a| = |1.4| = 1.4 > 1$, Test 2 predicts that the graphs are unbounded. Here

$$\frac{b}{1-a} = \frac{-8}{1-(1.4)} = \frac{-8}{-.4} = 20.$$

Both graphs move away from the line $y = 20$ as if being *repelled* by a force. ■

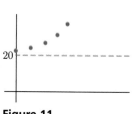
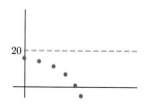
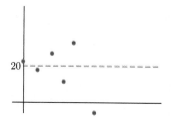

Figure 11 $\qquad\qquad\qquad\qquad\qquad\qquad\qquad\qquad\qquad$ **Figure 12**

EXAMPLE 7 **Vertical direction and long-run behavior of a difference equation** Discuss the graph of the difference equation $y_n = -2y_{n-1} + 60$, where $y_0 > 20$.

Solution A graph is drawn in Fig. 12. Since $a = -2 < 0$, Test 1 predicts that the graph is oscillating. Since $|a| = |-2| = 2 > 1$, Test 2 predicts that the graph is unbounded.

Here

$$\frac{b}{1-a} = \frac{60}{1-(-2)} = \frac{60}{3} = 20.$$

Notice that successive points move farther away from $y = 20$; that is, they are repelled by the line. ∎

Verification of Test 2 The key to verifying Test 2 is to make the following two observations:

1. If $|a| < 1$, then the powers a, a^2, a^3, a^4, ... become successively smaller and approach 0. For example, if $a = .4$, then this sequence of powers is

$$.4, .16, .064, .0256, \ldots .$$

2. If $|a| > 1$, then the powers $a, a^2, a^3, a^4, \ldots$ become unbounded; that is, they become arbitrarily large. For example, if $a = 3$, then this sequence is $3, 9, 27, 81, \ldots$.

To verify Test 2, look at the formula for y_n:

$$y_n = \frac{b}{1-a} + \boxed{\left(y_0 - \frac{b}{1-a}\right)a^n}.$$

If $|a| < 1$, then the powers of a get smaller and smaller; so the boxed term approaches 0. That is, y_n approaches $b/(1-a)$ and the graph is asymptotic to $y = b/(1-a)$. If $|a| > 1$, the powers of a are unbounded, so that the boxed term becomes arbitrarily large in magnitude. Thus y_n is unbounded, and so is the graph. ∎

In the case where $|a| < 1$, the graphs (whether oscillating or monotonic) are *attracted* steadily to the line $y = b/(1-a)$. In the case where $|a| > 1$, the graphs are *repelled* by the line $y = b/(1-a)$. Thinking of graphs as being attracted or repelled helps us in making a rough sketch.

Sign of a:	$\begin{cases}\text{Positive} \\ \text{Negative}\end{cases}$	$\begin{array}{l}\text{Monotonic} \\ \text{Oscillating}\end{array}$
Size of a:	$\begin{cases}|a| < 1 \\ |a| > 1\end{cases}$	$\begin{array}{l}\text{Attract} \\ \text{Repel}\end{array}$

Figure 13 shows some of the graphs already examined with the appropriate descriptive words labeling the line $y = b/(1-a)$.

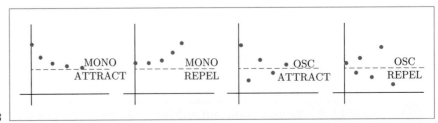

Figure 13

Based on Tests 1 and 2 and the discussion of constant graphs, we can state a procedure for making a rough sketch of a graph without solving or generating terms.

> **To sketch the graph of $y_n = a\,y_{n-1} + b$, $a \neq 0, \pm 1$:**
>
> **1.** Draw the line $y = b/(1-a)$ as a dashed line.
>
> **2.** Plot y_0. If y_0 is on the line $y = b/(1-a)$, the graph is constant and the procedure terminates.
>
> **3.** If a is positive, write MONO, since the graph is then monotonic. If a is negative, write OSC, since the graph is then oscillating.
>
> **4.** If $|a| < 1$, write ATTRACT, since the graph is attracted to the line $y = b/(1-a)$. If $|a| > 1$, write REPEL, since the graph is repelled from the line.
>
> **5.** Use all the information to sketch the graph.

EXAMPLE 8   **Graphing a difference equation**   Sketch the graph of $y_n = .6y_{n-1} + 8$, $y_0 = 50$.

Solution **1.** $\dfrac{b}{1-a} = \dfrac{8}{1-(.6)} = \dfrac{8}{.4} = 20$. So draw the line $y = 20$.

2. $y_0 = 50$, which is not on the line $y = 20$. So the graph is not constant.

3. $a = .6$, which is positive. Write MONO above the dashed line.

4. $|a| = |.6| = .6 < 1$. Write ATTRACT below the dashed line.

5. The preceding information tells us to start at 50 and move monotonically toward the line $y = 20$.

Now Try Exercise 17   ■

EXAMPLE 9   **Graphing a difference equation**   Sketch the graph of $y_n = -1.5y_{n-1} + 5$, $y_0 = 2.6$.

Solution **1.** $\dfrac{b}{1-a} = \dfrac{5}{1-(-1.5)} = \dfrac{5}{2.5} = 2$. So draw the line $y = 2$.

2. $y_0 = 2.6$, which is not on the line $y = 2$. So the graph is not constant.

3. $a = -1.5$, which is negative. Write OSC above the dashed line.

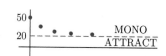

4. $|a| = |-1.5| = 1.5 > 1$. Write REPEL below the dashed line.

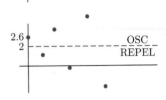

5. The preceding information says that the graph begins at 2.6, oscillates about, and is steadily repelled by the line $y = 2$.

Now Try Exercise 19

The preceding procedure does not give an exact graph, but it shows the nature of the graph. This is exactly what is needed in many applications.

EXAMPLE 10 **Using a mortgage balance equation** Suppose that the yearly interest rate on a mortgage is 9% compounded monthly and that you can afford to make payments of $300 per month. How much can you afford to borrow?

Solution Let i = the monthly interest rate, R = the monthly payment, and y_n = the balance after n months. Recall that y_0 = the initial amount of the loan. Then the balance after n months equals the balance after $n - 1$ months plus the interest on that balance minus the monthly payment. That is,

$$y_n = y_{n-1} + iy_{n-1} - R = (1 + i)y_{n-1} - R.$$

In this particular example, $i = .09/12 = .0075$ and $R = 300$, so the difference equation reads

$$y_n = 1.0075y_{n-1} - 300.$$

Apply our graph-sketching technique to this difference equation. Here

$$\frac{b}{1-a} = \frac{-300}{1 - (1.0075)} = \frac{-300}{-.0075} = 40{,}000.$$

Since $a = 1.0075$ is positive and $|a| = 1.0075 > 1$, the words MONO and REPEL describe the graphs.

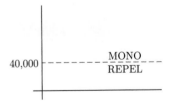

Let us see what happens for various initial values. If $y_0 > 40{,}000$, the balance increases indefinitely.

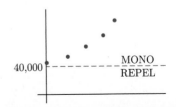

If $y_0 = 40{,}000$, the balance will always be 40,000.

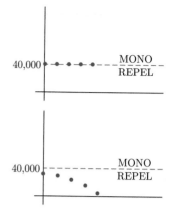

If $y_0 < 40{,}000$, the balance decreases steadily and eventually reaches 0, at which time the mortgage is paid off.

Therefore, the loan must be less than $40,000 in order for it to be paid off eventually.

Now Try Exercise 25

PRACTICE PROBLEMS 11.3

1. (*Parachuting*) A parachutist opens her parachute after reaching a speed of 100 feet per second. Suppose that y_n, the speed n seconds after opening the parachute, satisfies $y_n = .1y_{n-1} + 14.4$. Sketch the graph of y_n.

2. Rework Problem 1 if the parachute opens when the speed is 10 feet per second.

3. (*Retirement Fund*) Upon retirement, a person de-posits a certain amount of money into the bank at 6% interest compounded monthly and withdraws $100 at the end of each month.

   (a) Set up the difference equation for y_n, the amount in the bank after n months.

   (b) How large must the initial deposit be so that the money will never run out?

EXERCISES 11.3

Each of the graphs in Fig. 14 on the next page comes from a difference equation of the form $y_n = ay_{n-1} + b$. In Exercises 1–8, list all graphs that have the stated property.

1. Monotonic
2. Increasing
3. Unbounded
4. Constant
5. Repelled from $y = b/(1-a)$
6. Decreasing
7. $|a| < 1$
8. $a < 0$

In Exercises 9–14, sketch a graph having the given characteristics.

9. $y_0 = 4$, monotonic, repelled from $y = 2$
10. $y_0 = 5$, monotonic, attracted to $y = -3$
11. $y_0 = 2$, oscillating, attracted to $y = 6$
12. $y_0 = 7$, monotonic, repelled from $y = 10$
13. $y_0 = -2$, monotonic, attracted to $y = 5$
14. $y_0 = 1$, oscillating, repelled from $y = 2$

In Exercises 15–20, make a rough sketch of the graph of the difference equation without generating terms or solving the difference equation.

15. $y_n = 3y_{n-1} + 4$, $y_0 = 0$
16. $y_n = .3y_{n-1} + 3.5$, $y_0 = 2$
17. $y_n = .5y_{n-1} + 3$, $y_0 = 6$
18. $y_n = 4y_{n-1} - 18$, $y_0 = 5$
19. $y_n = -2y_{n-1} + 12$, $y_0 = 5$
20. $y_n = -.6y_{n-1} + 6.4$, $y_0 = 1$

21. (*Spread of Information*) A particular news item was broadcast regularly on radio and TV. Let y_n be the number of people who had heard the news within n hours after broadcasting began. Sketch the graph of y_n, assuming that it satisfies the difference equation $y_n = .7y_{n-1} + 3000$, $y_0 = 0$.

22. (*Radioactive Decay*) The radioactive element strontium 90 emits particles and slowly decays. Let y_n be the amount left after n years. Then y_n satisfies the difference equation $y_n = .98y_{n-1}$. Sketch the graph of y_n if initially there are 10 milligrams of strontium 90.

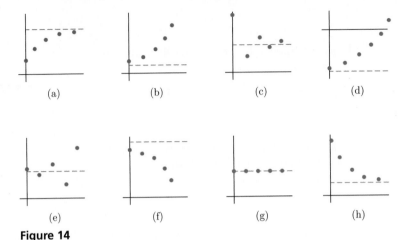

Figure 14

23. (*Price Equation*) Laws of supply and demand cause the price of oats to fluctuate from year to year. Suppose that the current price is $1.72 per bushel and that the price n years from now, p_n, satisfies the difference equation $p_n = -.6p_{n-1} + 1.6$. (Prices are assumed to have been adjusted for inflation.) Sketch the graph of p_n.

24. (*Population Growth*) Under ideal conditions a bacteria population satisfies the difference equation $y_n = 1.4y_{n-1}$, $y_0 = 1$, where y_n is the size of the population (in millions) after n hours. Sketch a graph that shows the growth of the population.

25. (*Mortgage Planning*) Suppose that the interest rate on a mortgage is 9% compounded monthly. If you can afford to pay $450 per month, how much money can you borrow?

26. (*Mortgage Planning*) A municipal government can take out a long-term construction loan at 8% interest compounded quarterly. Assuming that it can pay back $100,000 per quarter, how much money can it borrow?

27. (*Savings Account Withdrawals*) A person makes an initial deposit into a savings account paying 6% interest compounded annually. He plans to withdraw $120 at the end of each year.

   (a) Find the difference equation for y_n, the amount after n years.

   (b) How large must y_0 be such that the money will not run out?

28. (*Savings Account Withdrawals*) Casimir makes an initial deposit into a savings fund paying 4% compounded quarterly for his teenage daughter, Julie. She plans to withdraw $200 at the end of each quarter to buy some new clothes.

   (a) Find the difference equation for y_n, the amount in the account after n quarters.

   (b) How large must the initial deposit be so that Julie will not run out of money?

29. (*Mortgage Planning*) Gabriella is looking to take out a loan at 10% interest compounded semiannually for her greenhouse and flower business. If she can make a $1500 payment at the end of each half-year, how much can she afford to borrow?

30. (*Loan Payments*) Suppose that a loan of $10,000 is to be repaid at $120 per month and that the annual interest rate is 12% compounded monthly. Then the interest for the first month is $.01 \cdot (10,000)$, or $100. The $120 paid at the end of the first month can be thought of as paying the $100 interest and paying $20 toward the reduction of the loan. Therefore, the balance after 1 month is $10,000 - $20 = $9980. How much of the $120 paid at the end of the second month goes for interest and how much is used to reduce the loan? What is the balance after 2 months?

Exercises 31–36 require the use of a graphing calculator, a spreadsheet, or mathematical software.

In these exercises, find a difference equation with the given properties. Then confirm your answer by graphing the difference equation.

31. The terms begin at 1 and monotonically approach 8.

32. The terms begin at 9 and monotonically approach 4.

33. The terms begin at 1 and monotonically increase without bound.

34. The terms begin at 12, oscillate, and approach 6.

35. The terms begin at 5, oscillate, and are unbounded.

36. The terms begin at 50 and monotonically decrease without bound.

SOLUTIONS TO PRACTICE PROBLEMS 11.3

1. $\frac{b}{1-a} = \frac{14.4}{1-(.1)} = \frac{14.4}{.9} = \frac{144}{9} = 16.$ $y_0 = 100,$
which is greater than $b/(1-a)$. Since $a = .1$ is positive, the terms y_n are monotonic. Since $|a| = .1 < 1$, the terms are attracted to the line $y = 16$. Now, plot $y_0 = 100$, draw the line $y = 16$, and write in the words MONO and ATTRACT [Fig. 15(a)]. Since the terms are attracted to the line monotonically they must move downward and asymptotically approach the line [Fig. 15(b)]. (Notice that the terminal speed, 16 feet per second, does not depend on the speed of the parachutist when the parachute is opened.)

2. Everything is the same as before except that now $y_0 < 16$. The speed increases to a terminal speed of 16 feet per second.

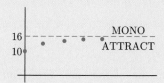

3. **(a)** $i = .06/12 = .005$

$$y_n = y_{n-1} + (\text{interest}) - (\text{withdrawal})$$
$$= y_{n-1} + .005y_{n-1} - 100$$
$$= (1.005)y_{n-1} - 100$$

(b) $\frac{b}{1-a} = \frac{-100}{1-(1.005)} = \frac{-100}{-.005} = 20,000.$ Since $a = 1.005 > 0$, the graph is monotonic. Since $|a| = 1.005 > 1$, the graph is repelled from the line $y = 20,000$.

If $y_0 < 20,000$, the amount of money in the account decreases and eventually runs out.

If $y_0 = 20,000$, the amount of money in the account stays constant at 20,000.

If $y_0 > 20,000$, the amount of money in the account grows steadily and is unbounded.

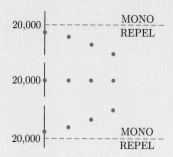

Answer: $y_0 \geq 20,000.$

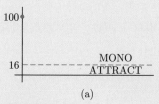 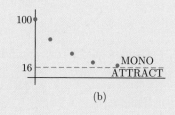

Figure 15

11.4 Mathematics of Personal Finance

In this section we apply the theory of difference equations to the study of mortgages and annuities.

Mortgages Most families take out bank loans to pay for a new house. Such a loan, called a mortgage, is used to purchase the house and then is repaid with interest in monthly installments over a number of years, usually 25 or 30. The monthly payments are computed so that after exactly the correct length of time the unpaid balance[1] is 0 and the loan is thereby paid off.

Mortgages can be described by difference equations, as follows. Let y_n be the unpaid balance on the mortgage after n months. In particular, y_0 is the initial

---

[1]The balance after n months is the amount that would have to be paid at that time in order to retire the debt.

amount borrowed. Let i denote the monthly interest rate and R the monthly mortgage payment. Then

$$\begin{bmatrix} \text{balance after} \\ n \text{ months} \end{bmatrix} = \begin{bmatrix} \text{balance after} \\ n-1 \text{ months} \end{bmatrix} + \begin{bmatrix} \text{interest for} \\ \text{month} \end{bmatrix} - \begin{bmatrix} \text{payment} \end{bmatrix}$$

$$y_n = y_{n-1} + iy_{n-1} - R$$

$$y_n = (1+i)y_{n-1} - R. \tag{1}$$

EXAMPLE 1 **Finding the amount of a mortgage** Suppose that you can afford to pay \$300 per month and the yearly interest rate is 9% compounded monthly. Exactly how much can you borrow if the mortgage is to be paid off in 30 years?

Solution The monthly interest rate is $.09/12 = .0075$, so equation (1) becomes in this case

$$y_n = 1.0075y_{n-1} - 300.$$

Further, we are given that the mortgage runs for 30 years, or 360 months. Thus

$$y_{360} = 0. \tag{2}$$

Our problem is to determine y_0, the amount of the loan. From our general theory we first compute $b/(1-a)$:

$$\frac{b}{1-a} = \frac{-300}{1-(1.0075)} = 40{,}000.$$

The formula for y_{360} is then given by

$$y_{360} = \frac{b}{1-a} + \left(y_0 - \frac{b}{1-a}\right)a^{360}$$

$$= 40{,}000 + (y_0 - 40{,}000)(1.0075)^{360}.$$

Using a calculator, we find that

$$(1.0075)^{360} = 14.73057612.$$

Therefore,

$$y_{360} = 40{,}000 + (y_0 - 40{,}000)(14.73057612).$$

However, by equation (2), $y_{360} = 0$, so that

$$0 = 40{,}000 + (y_0 - 40{,}000)(14.73057612) = 14.73057612y_0 - 549{,}223.0449$$

$$y_0 = \frac{549{,}223.0449}{14.73057612} = 37{,}284.56.$$

Thus, the initial amount, that is, the amount that can be borrowed, is \$37,284.56.

Now Try Exercise 5 ■

Table 1 shows the status of the mortgage after selected payments. After the first payment of \$300 is made, the balance of the loan is $y_1 = \$37{,}264.19$. This payment consists of \$279.63 interest and \$20.37 repayment of the loan. In the early months almost all of the monthly payment goes toward interest. With each passing month, however, the amount of interest declines and the amount applied to reducing the debt increases.

TABLE 1

Payment Number	1	2	120	240	348	360
Balance on loan	$37,264.19	$37,243.68	$33,343.49	$23,682.51	$3430.48	$0
Interest	279.63	279.48	250.45	178.53	27.77	2.23
Reduction of debt	20.37	20.52	49.55	121.47	272.23	297.77

EXAMPLE 2   **Determining mortgage payments**   Suppose that we have a 30-year mortgage for $140,000 at 7.5% interest, compounded monthly. Find the monthly payment.

Solution   Here the monthly interest rate i is just $.075/12 = .00625$ and $y_0 = 140{,}000$. Since the mortgage is for 30 years, we have

$$y_{360} = 0.$$

The difference equation for the mortgage is

$$y_n = 1.00625 y_{n-1} - R \qquad y_0 = 140{,}000.$$

Now

$$\frac{b}{1-a} = -\frac{R}{1-1.00625} = \frac{R}{.00625} = 160R.$$

The formula for y_n is

$$y_n = 160R + (140{,}000 - 160R)(1.00625)^n$$

Thus

$$0 = y_{360} = 160R + (140{,}000 - 160R)(1.00625)^{360}.$$

To complete the problem, solve this equation for R. Using a calculator, $(1.00625)^{360} = 9.421533905$. Thus

$$160R + (140{,}000 - 160R)(9.421533905) = 0$$
$$1347.445425R = 1{,}319{,}014.747$$
$$R = \$978.90.$$

Now Try Exercise 11   ■

Annuities   The term "annuity" has several meanings. For our purposes an annuity is a bank account into which equal sums are deposited at regular intervals, either weekly, monthly, quarterly, or annually. The money draws interest and accumulates for a certain number of years, after which it becomes available to the investor. Annuities are often used to save for a child's college education or to generate funds for retirement.

The growth of money in an annuity can be described by a difference equation. Let $y_n =$ [the amount of money in the annuity after n time periods], $i =$ [the interest rate per time period], $D =$ [the deposit per time period]. Then $y_0 =$ [the amount after 0 time periods] $= 0$.[2] Moreover,

$$y_n = [\text{previous amount}] + [\text{interest}] + [\text{deposit}]$$
$$= \quad y_{n-1} \quad + \quad i y_{n-1} \quad + \quad D$$
$$= (1+i)y_{n-1} + D.$$

This last equation is the difference equation of the annuity.

---

[2]One model for the situation $y_0 = 0$ is a payroll savings plan, where a person signs up at time zero and has the first deduction made at the end of the next pay period.

EXAMPLE 3 **Finding the future balance of an annuity** Suppose that $20 is deposited into an annuity at the end of every quarter-year and that interest is earned at the annual rate of 8%, compounded quarterly. How much money is in the annuity after 10 years?

Solution Since 10 years = 40 quarters, the problem is to compute y_{40}. Now $D = 20$ and $i = .08/4 = .02$. So the difference equation reads

$$y_n = 1.02y_{n-1} + 20.$$

In this case

$$\frac{b}{1-a} = \frac{20}{1-1.02} = -1000.$$

Thus

$$y_n = -1000 + [0 - (-1000)](1.02)^n = -1000 + 1000(1.02)^n.$$

In particular, setting $n = 40$,

$$y_{40} = -1000 + 1000(1.02)^{40}.$$

Using a calculator, $(1.02)^{40} = 2.208039664$. Thus

$$y_{40} = -1000 + 1000(2.208039664) = -1000 + 2208.039664 = 1208.039664.$$

Thus after 10 years the account contains $1208.04.

Now Try Exercise 7   ■

EXAMPLE 4 **Determining the periodic deposit with an annuity** How much money must be deposited at the end of each quarter into an annuity at 8% interest compounded quarterly in order to have $10,000 after 15 years?

Solution Here D is unknown. But $i = .02$. Also, 15 years = 60 quarters, so that $y_{60} = 10,000$. The difference equation in this case reads

$$y_n = 1.02y_{n-1} + D, \qquad y_0 = 0.$$

Then

$$\frac{b}{1-a} = \frac{D}{1-1.02} = \frac{D}{-.02} = -50D$$

and

$$y_n = -50D + (0 - (-50D))(1.02)^n$$
$$= -50D + 50D(1.02)^n.$$

Set $n = 60$. Then

$$y_{60} = -50D + 50D(1.02)^{60}.$$

However, from the statement of the problem, $y_{60} = 10,000$. Thus we have the equation

$$10,000 = -50D + 50D(1.02)^{60}$$

to be solved for D. But $(1.02)^{60} = 3.281030788$, so that

$$-50D + 50(3.281030788)D = 10,000$$
$$-50D + 164.0515394D = 10,000$$
$$114.0515394D = 10,000$$
$$D = \frac{10,000}{114.0515394} = \$87.68.$$

Now Try Exercise 9 ■

GC A complete table similar to Table 1 is called an *amortization table* and is easily created with a graphing calculator. Consider the mortgage of Example 1 with difference equation $y_n = 1.0075y_{n-1} - 300$, $y_0 = 37{,}284.56$. Let v_n be the interest portion of the nth payment. Then $v_n = .0075y_{n-1}$ for $n = 1, 2, 3, \ldots$. These two difference equations are entered into the **Y=** editor as follows.

TI-83	TI-89
`nMin = 0`	`u1=1.0075*u1(n-1)-300`
`u(n)=1.0075u(n-1)-300`	`ui1=37284.56`
`u(nMin)=37284.56`	`u2=.0075*u1(n-1)`
`v(n)=.0075u(n-1)`	`ui2=0`
`v(nMin)=0`	

To create the tables in Figs. 1 and 2, press [2nd] [TBLSET] (on the TI-83) or [♦] [TblSet] (on the TI-89) to bring up the TABLE SETUP screen. **TblStart** and **ΔTbl** should be set to 1. Then, pressing [2nd] [TABLE] (on the TI-83) or [♦] [TABLE] (on the TI-89) brings up the table. You can use the down-arrow key to generate further balances and interest payments.

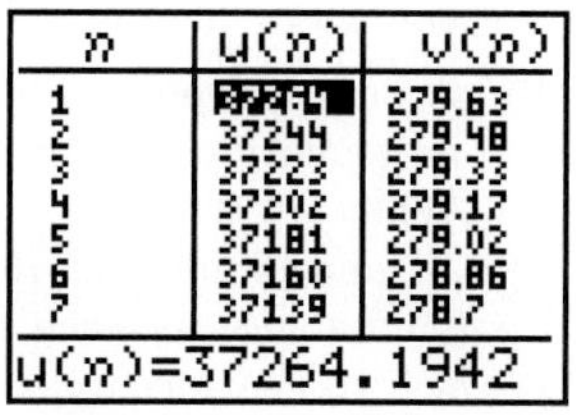

Figure 1

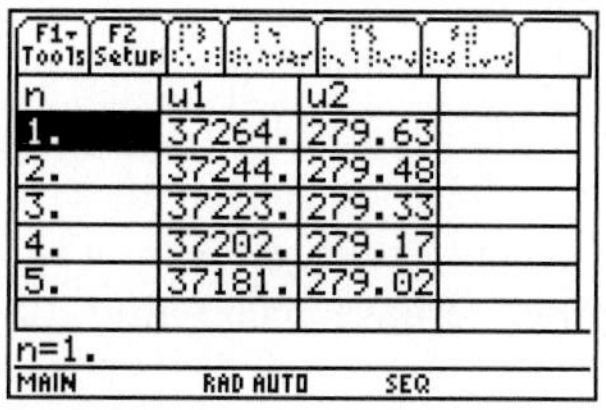

Figure 2

	A	B	C
1		Balance	Interest
2	0	37284.56	
3	1	37264.19	279.63
4	2	37243.68	279.48
5	3	37223.00	279.33
6	4	37202.18	279.17
7	5	37181.19	279.02
8	6	37160.05	278.86
9	7	37138.75	278.70

Figure 3

ES Figure 3 contains an amortization table similar to the ones in Figs. 1 and 2 but shown in an Excel spreadsheet. To obtain the numbers in the B column, **37284.56** was entered into B2, **=1.0075*B2−300** was entered into B3, and B3 was selected and its fill handle dragged down to cell B9. To obtain the numbers in the C column, **=0.0075*B2** was entered into C3, and C3 was selected and its fill handle dragged down to cell C9. *Note*: The cells in the B and C columns were formatted to show two decimal places.

PRACTICE PROBLEMS 11.4

1. (*Savings Account*) Suppose that you deposit \$650,000 into a bank account paying 5% interest compounded annually and you withdraw \$50,000 at the end of each year. Find a difference equation for y_n, the amount in the account after n years.
2. Refer to Problem 1. How much money will be in the account after 20 years?
3. Refer to Problem 1. Assume that the money is tax-free and that you could earn 5% interest compounded annually. Would you rather have \$650,000 now or \$50,000 a year for 20 years?

EXERCISES 11.4

In Exercises 1–4, give the difference equation for y_n, the amount (or balance) after n interest periods.

1. A mortgage loan of $32,500 at 9% interest compounded monthly and having monthly payments of $261.50.

2. A bank deposit of $1000 at 6% interest compounded semiannually.

3. An annuity for which $4000 is deposited into an account at 6% interest compounded quarterly and $200 is added to the account at the end of each quarter.

4. A bank account into which $20,000 is deposited at 6% interest compounded monthly and $100 is withdrawn at the end of each month.

5. How much money can you borrow at 12% interest compounded monthly if the loan is to be paid off in monthly installments for 10 years and you can afford to pay $660 per month?

6. Find the monthly payment on a $38,000, 25-year mortgage at 12% interest compounded monthly.

7. Find the amount accumulated after 20 years if, at the end of each year, $300 is deposited into an account paying 6% interest compounded annually.

8. How much money would you have to put into an account initially at 8% interest compounded quarterly in order to have $6000 after 14 years?

9. How much money would you have to deposit at the end of each month into an annuity paying 6% interest compounded monthly in order to have $6000 after 12 years?

10. How much money would you have to put into a bank account paying 6% interest compounded monthly in order to be able to withdraw $150 each month for 30 years?

11. (*Car Loan*) In order to buy a used car, a person borrows $4000 from the bank at 12% interest compounded monthly. The loan is to be paid off in 3 years with equal monthly payments. What will the monthly payments be?

12. How much money would you have to deposit at the end of each month into an annuity paying 6% interest compounded monthly in order to have $1620 after 4 years?

Exercises 13–16 require the use of a graphing calculator or a computer.

In these exercises, use a table or graph of an appropriate difference equation to answer the questions.

13. (*Corporate Loan*) The ABC Corporation borrows $45 million (at 8% interest compounded annually) to buy a shopping center. The loan is repaid with annual payments of $4 million per year. Determine the balance after 10 years. After how many years will the loan be paid off?

14. (*Merchandise Financing*) John buys a stereo system for $605.54 and pays for it with monthly payments of $90. The interest rate is 12% compounded monthly. Determine the amount still owed after 4 months. After how many months will the loan be repaid?

15. A person deposits $100 at the end of each month into an annuity earning 6% interest compounded monthly. What are the balances (that is, future values) after 5, 10, and 15 months? How many months are required for the balance to reach $3228? After how many months will the balance exceed $4000?

16. (*Savings Plan*) Alice embarks on a savings routine. She decides to deposit $75 at the end of every three months into a savings account paying 6% interest compounded quarterly. How much money will she have accumulated after 5 years? When will she have accumulated $2474? After how many quarters will her savings exceed $4000?

SOLUTIONS TO PRACTICE PROBLEMS 11.4

1. $\begin{bmatrix} \text{amount after} \\ n \text{ years} \end{bmatrix} = \begin{bmatrix} \text{amount after} \\ n-1 \text{ years} \end{bmatrix} + \begin{bmatrix} \text{interest for} \\ n\text{th year} \end{bmatrix}$

   $\qquad - \begin{bmatrix} \text{withdrawal at} \\ \text{end of year} \end{bmatrix}$

   $y_n = y_{n-1} + .05y_{n-1} - 50,000 = (1.05)y_{n-1} - 50,000$

   $y_0 = 650,000$

2. Solve the difference equation and set $n = 20$.

   $\dfrac{b}{1-a} = \dfrac{-50,000}{1-(1.05)} = \dfrac{-50,000}{-.05} = 1,000,000$

   $y_n = 1,000,000 + (650,000 - 1,000,000)(1.05)^n$

   $\qquad = 1,000,000 - 350,000(1.05)^n$

   $y_{20} = 1,000,000 - 350,000(2.653297705)$

   $\qquad = 71,345.80$

3. $650,000. According to Problem 2, this money could be deposited into the bank. You could take out $50,000 per year and still have $71,345.80 left over after 20 years.

11.5 Modeling with Difference Equations

In this section we show how difference equations may be used to build mathematical models of a number of phenomena. Since the difference equation describing a situation contains as much data as the formula for y_n, we shall regard the difference equation itself as a mathematical model. We demonstrate in this section that the difference equation and a sketch of its graph can often be of more use than an explicit solution.

The concept of proportionality will be needed to develop some of these models.

Proportionality To say that quantities are proportional is the same as saying that one quantity is equal to a constant times the other quantity. For instance, in a state having a 4% sales tax, the sales tax on an item is proportional to the price of the item, since

$$[\text{sales tax}] = .04[\text{price}].$$

Here, the constant of proportionality is .04. In general, for two proportional quantities we have

$$[\text{first quantity}] = k[\text{second quantity}],$$

where k is some fixed constant of proportionality. Note that if both quantities are positive and if the first quantity is known always to be smaller than the second, then k is a positive number less than 1; that is, $0 < k < 1$.

EXAMPLE 1 **Radioactive decay** Certain forms of natural elements, such as uranium 238, strontium 90, and carbon 14, are radioactive. That is, they decay, or dissipate, over a period of time. Physicists have found that this process of decay obeys the following law: Each year the amount that decays is proportional to the amount present at the start of the year. Construct a mathematical model describing radioactive decay.

Solution Let y_n = the amount left after n years. The physical law states that

$$[\text{amount that decays in year } n] = k \cdot y_{n-1},$$

where $k \cdot y_{n-1}$ represents a constant of proportionality times the amount present at the start of that year. Therefore,

$$y_n \quad = \quad y_{n-1} \quad - \quad k \cdot y_{n-1}$$

$$\begin{bmatrix} \text{amount after} \\ \text{year } n \end{bmatrix} = \begin{bmatrix} \text{amount after} \\ \text{year } n-1 \end{bmatrix} - \begin{bmatrix} \text{amount that decays} \\ \text{in year } n \end{bmatrix}$$

or

$$y_n = (1-k)y_{n-1},$$

where k is a constant between 0 and 1. This difference equation is the mathematical model for radioactive decay. ■

EXAMPLE 2 **Radioactive decay** Experiment shows that for cobalt 60 (a radioactive form of cobalt used in cancer therapy) the decay constant k is given by $k = .12$.

(a) Write the difference equation for y_n in this case.

(b) Sketch the graph of the difference equation.

Solution (a) Setting $k = .12$ in the result of Example 1, we get

$$y_n = (1 - .12)y_{n-1} \quad \text{or} \quad y_n = .88y_{n-1}.$$

(b) Since .88 is positive and less than 1, the graph is monotonic and attracted to the line $y = b/(1-a) = 0/(1-.88) = 0$. The graph is sketched in Fig. 1.

Figure 1

MONO
ATTRACT

Now Try Exercise 3

EXAMPLE 3 **Growth of a bacteria culture** A bacteria culture grows in such a way that each hour the increase in the number of bacteria in the culture is proportional to the total number present at the beginning of the hour. Sketch a graph depicting the growth of the culture.

Solution Let y_n = the number of bacteria present after n hours. Then the increase from the previous hour is ky_{n-1}, where k is a positive constant of proportionality. Therefore,

$$y_n = y_{n-1} + ky_{n-1}$$

$$\begin{bmatrix} \text{number after} \\ n \text{ hours} \end{bmatrix} = \begin{bmatrix} \text{number after} \\ n-1 \text{ hours} \end{bmatrix} + \begin{bmatrix} \text{increase during} \\ n\text{th hour} \end{bmatrix}$$

or

$$y_n = (1+k)y_{n-1}.$$

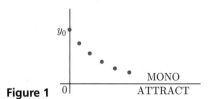

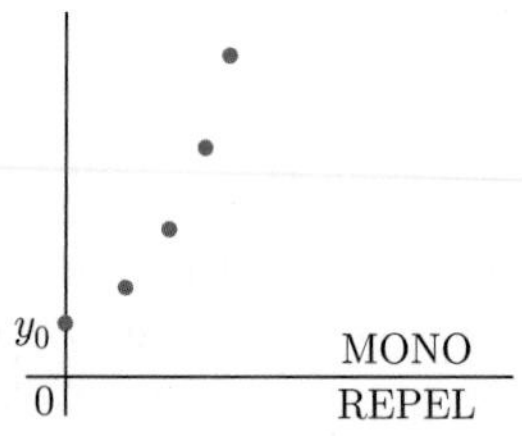

Figure 2

MONO
REPEL

This last equation gives a mathematical model of the growth of the culture. Since $a = 1+k > 0$ and $|a| > 1$, the graph is monotonic and repelled from $b/(1-a) = 0$. Thus the graph is as drawn in Fig. 2.

EXAMPLE 4 **Spread of information** Suppose that at 8 A.M. on a Saturday the local radio and TV stations in a town start broadcasting a certain piece of news. The number of people learning the news each hour is proportional to the number who had not yet heard it by the end of the preceding hour.

(a) Write a difference equation describing the spread of the news through the population of the town.

(b) Sketch the graph of this difference equation for the case where the constant of proportionality is .3 and the population of the town is 50,000.

Solution The example states that two quantities are proportional. The first quantity is the number of people learning the news each hour and the second is the number who have not yet heard it by the end of the preceding hour.

(a) Let y_n = the number of people who have heard the news after n hours. The terms $y_0, y_1, y_2, \ldots$ are increasing.

Let P = the total population of the town. Then the number of people who have not yet heard the news after $n-1$ hours is $P - y_{n-1}$. Thus the assumption can be stated in the mathematical form

$$y_n = y_{n-1} + k(P - y_{n-1})$$

$$\begin{bmatrix} \text{number who know} \\ \text{after } n \text{ hours} \end{bmatrix} = \begin{bmatrix} \text{number who know} \\ \text{after } n-1 \text{ hours} \end{bmatrix} + \begin{bmatrix} \text{number who learn} \\ \text{during } n\text{th hour} \end{bmatrix},$$

where k is a constant of proportionality. The constant k tells how fast the news is traveling. It measures the percentage of the uninformed population that hears the news each hour. It is clear that $0 < k < 1$.

(b) For part (b) we assume that $k = .3$ and that the town has a population of 50,000. Further, we measure the number of people in thousands. Then $P = 50$ and the mathematical model of the spread of the news is

$$y_n = y_{n-1} + .3(50 - y_{n-1}) = y_{n-1} + 15 - .3y_{n-1} = .7y_{n-1} + 15.$$

Suppose that initially no one had heard the news, so that

$$y_0 = 0.$$

Now we may sketch the graph:

$$\frac{b}{1-a} = \frac{15}{1-.7} = \frac{15}{.3} = 50.$$

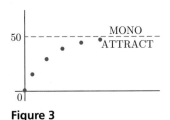

Figure 3

Since $a = .7$, the graph is monotonic and attracted to the line $y = 50$ (Fig. 3). Note that the number of people who hear the news increases and approaches the population of the entire town. The increases between consecutive terms are large at first and then become smaller. This coincides with the intuitive impression that the news spreads rapidly at first and then progressively slower.

Now Try Exercise 5 ∎

EXAMPLE 5 **Supply and demand** This year's level of production and price for most agricultural products greatly affect the level of production and price next year. Suppose that the current crop of soybeans in a certain country is 80 million bushels. Let q_n denote the quantity[3] of soybeans grown n years from now, and let p_n denote the market price in n years. Suppose experience has shown that q_n and p_n are related by the following equations:

$$p_n = 20 - .1q_n \qquad q_n = 5p_{n-1} - 10.$$

Draw a graph depicting the changes in production from year to year.

Solution What we seek is the graph of a difference equation for q_n.

$$q_n = 5p_{n-1} - 10 = 5(20 - .1q_{n-1}) - 10 = 100 - .5q_{n-1} - 10 = -.5q_{n-1} + 90$$

This is a difference equation for q_n with

$$a = -.5 \qquad b = 90 \qquad \frac{b}{1-a} = \frac{90}{1-(-.5)} = 60.$$

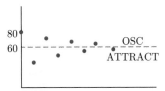

Figure 4

Since a is negative, the graph is oscillating. Since $|a| = .5 < 1$, the graph is attracted to the line $y = b/(1-a) = 60$. The initial condition is $q_0 = 80$. So the graph is as drawn in Fig. 4. Note that the graph oscillates and approaches 60. That is, the crop size fluctuates from year to year and eventually gets close to 60 million bushels.

Now Try Exercise 13 ∎

---

[3]q_n in terms of millions of bushels, p_n in terms of dollars per bushel.

PRACTICE PROBLEMS 11.5

1. (*Glucose Infusion*) Glucose is being given to a patient intravenously at the rate of 100 milligrams per minute. Let y_n be the amount of glucose in the blood after n minutes, measured in milligrams. Suppose that each minute the body takes from the blood 2% of the amount of glucose present at the beginning of that minute. Find a difference equation for y_n and sketch its graph. (*Hint:* Each minute the amount of glucose in the blood is increased by the intravenous infusion and decreased by the absorption into the body.)

2. (*Light at Ocean Depths*) Sunlight is absorbed by water and so, as one descends into the ocean, the intensity of light diminishes. Suppose that at each depth, going down one more meter causes a 20% decrease in the intensity of the sunlight. Find a difference equation for y_n, the intensity of the light at a depth of n meters, and sketch its graph.

EXERCISES 11.5

1. (*Population Dynamics*) In a certain country with current population 100 million, each year the number of births is 3% and the number of deaths 1% of the population at the beginning of the year. Find the difference equation for y_n, the population after n years. Sketch its graph.

2. (*Population Dynamics*) A small city with current population 50,000 is experiencing an emigration of 600 people each year. Assuming that each year the increase in population due to natural causes is 1% of the population at the start of that year, find the difference equation for y_n, the population after n years. Sketch its graph.

3. (*Drug Absorption*) After a certain drug is injected, each hour the amount removed from the bloodstream by the body is 25% of the amount in the bloodstream at the beginning of the hour. Find the difference equation for y_n, the amount in the bloodstream after n hours, and sketch the graph.

4. (*Elevation and Atmospheric Pressure*) The atmospheric pressure at sea level is 14.7 pounds per square inch. Suppose that at any elevation an increase of 1 mile results in a decrease of 20% of the atmospheric pressure at that elevation. Find the difference equation for y_n, the atmospheric pressure at elevation n miles, and sketch its graph.

5. (*Spread of Information*) A sociological study[4] was made to examine the process by which doctors decide to adopt a new drug. Certain doctors who had little interaction with other physicians were called "isolated." Out of 100 isolated doctors, each month the number who adopted the new drug that month was 8% of those who had not yet adopted the drug at the beginning of the month. Find a difference equation for y_n, the number of isolated physicians using the drug after n months, and sketch its graph.

6. (*Solution Concentration*) A cell is put into a fluid containing an 8 milligram/liter concentration of a solute. (This concentration stays constant throughout.) Initially, the concentration of the solute in the cell is 3 milligrams/liter. The solute passes through the cell membrane at such a rate that each minute the increase in concentration in the cell is 40% of the difference between the outside concentration and the inside concentration. Find the difference equation for y_n, the concentration of the solute in the cell after n minutes, and sketch its graph.

7. (*Learning Curve*) Psychologists have found that in certain learning situations in which there is a maximum amount that can be learned, the additional amount learned each minute is proportional to the amount yet to be learned at the beginning of that minute. Let 12 units of information be the maximum amount that can be learned and let the constant of proportionality be 30%. Find a difference equation for y_n, the amount learned after n minutes, and sketch its graph.

8. (*Genetics*) Consider two genes A and a in a population, where A is a dominant gene and a is a recessive gene controlling the same genetic trait. (That is, A and a belong to the same locus.) Suppose that initially 80% of the genes are A and 20% are a. Suppose that in each generation .003% of genes A mutate to gene a. Find a difference equation for y_n, the percentage of genes a after n generations, and sketch its graph. [*Note:* The percentage of genes $A = 1 -$ (the percentage of genes a).]

9. (*Account Balance*) Thirty thousand dollars is deposited in a savings account paying 5% interest compounded annually, and $1000 is withdrawn from the account at the end of each year. Find the difference equation for y_n, the amount in the account after n years, and sketch its graph.

10. (*Account Balance*) Rework Exercise 9 where $15,000 is deposited initially.

[4] James S. Coleman, Elihu Katz, and Herbert Menzel, "The Diffusion of an Innovation Among Physicians," *Sociometry*, 20, 1957, pp. 253–270.

11. (*Thermodynamics*) When a cold object is placed in a warm room, each minute its increase in temperature is proportional to the difference between the room temperature and the temperature of the object at the beginning of the minute. Suppose that the room temperature is 70°F, the initial temperature of the object is 40°F, and the constant of proportionality is 20%. Find the difference equation for y_n, the temperature of the object after n minutes, and sketch its graph.

12. (*Electricity Usage*) Suppose that the annual amount of electricity used in the United States will increase at a rate of 1.8% each year and that this year 3.2 trillion kilowatt-hours are being used. Find a difference equation for y_n, the number of kilowatt-hours to be used during the year that is n years from now, and sketch its graph.

13. (*Supply and Demand*) Suppose that in Example 5 the current price of soybeans is $4.54 per bushel. Find the difference equation for p_n and sketch its graph.

14. (*Drug Absorption*) The popular sleep-aid drug Zaleplon has a half-life of one hour. That is, the amount of the drug present in the body is halved each hour. If 10 mg of Zaleplon is taken at bedtime, what is the difference equation for the amount of the drug present in the body after n hours? How much will be eliminated after 8 hours?

15. (*Drug Absorption*) The anti-inflammatory drug Advil® has a half-life of 2 hours. That is, the amount of the drug present in the body is halved every two hours. After 4 hours _____ of the drug will remain in the body.

    (a) 0%   **(b)** 25%   **(c)** 66%   **(d)** 33%

Exercises 16–18 require the use of a graphing calculator, a spreadsheet, or mathematical software.
In these exercises, answer the questions by generating successive terms of the appropriate difference equation.

16. (*Thermodynamics*) A steel rod of temperature 600° is immersed in a large vat of water at temperature 75°. Each minute its decrease in temperature is proportional to the difference between the water temperature

and the temperature of the rod at the beginning of the minute, where the constant of proportionality is .3. What is the temperature after 2 seconds? When will the temperature drop below 80°?

17. (*Population Dynamics*) The birth rate in a certain city is 3.5% per year and the death rate is 2% per year. Also, there is a net movement of population out of the city at a steady rate of 300 people per year. The population in 2005 was 5 million. Estimate the 2010 population. When will the population exceed 6 million? When will the population have doubled to 10 million?

18. (*Spread of Information*) Consider the sociological study of Exercise 5. Out of a group of *nonisolated* doctors, let y_n be the percent who adopted the new drug after n months. Then y_n satisfies the difference equation

$$y_n = .0025y_{n-1}(500 - y_{n-1}), \qquad y_0 = 3.$$

That is, initially 3% were using the drug. What percent were using the drug after 1 year? When were over half of the doctors using the drug? When were over 99% using the drug?

19. (*Caffeine Absorption*) After caffeine is absorbed into the body, 13% is eliminated from the body each hour. Assume a person drinks an 8-oz cup of brewed coffee containing 130 mg of caffeine, and the caffeine is absorbed immediately into the body. Find the difference equation for y_n, the amount of caffeine in the body after n hours. Solve the difference equation. After how many hours will 65 mg (one-half the original amount) remain in the body? How much caffeine will be in the body 24 hours after the person drank the coffee?

20. (*Caffeine Absorption*) Refer to Exercise 19. Suppose the person drinks a cup of coffee at 7 A.M. and then drinks a cup of coffee at the end of each hour until 7 A.M. the next day. Find the difference equation for y_n, the amount of caffeine in the body after n hours. Solve the difference equation. How much caffeine will be in the body at the end of the 24 hours?

SOLUTIONS TO PRACTICE PROBLEMS 11.5

1. The amount of glucose in the blood is affected by two factors. It is being increased by the steady infusion of glucose; each minute the amount is increased by 100 milligrams. On the other hand, the amount is being decreased each minute by $.02y_{n-1}$. Therefore,

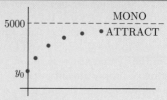

Figure 5

$$
\underset{\substack{\begin{bmatrix}\text{amount after}\\ n \text{ minutes}\end{bmatrix}}}{y_n} = \underset{\substack{\begin{bmatrix}\text{amount after}\\ n-1 \text{ minutes}\end{bmatrix}}}{y_{n-1}}
$$

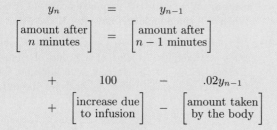

$$
+ \underset{\substack{\begin{bmatrix}\text{increase due}\\ \text{to infusion}\end{bmatrix}}}{100} - \underset{\substack{\begin{bmatrix}\text{amount taken}\\ \text{by the body}\end{bmatrix}}}{.02y_{n-1}}
$$

or

$$
y_n = .98y_{n-1} + 100.
$$

To sketch the graph, first compute $b/(1-a)$.

$$
\frac{b}{1-a} = \frac{100}{1-.98} = \frac{100}{.02} = 5000.
$$

The amount of glucose in the blood rises and approaches 5000 milligrams (Fig. 5).

2. The change in intensity in going from depth $n-1$ meters to n meters is 20% of the intensity at $n-1$ meters. Therefore,

$$
\underset{\substack{\begin{bmatrix}\text{intensity at}\\ n \text{ meters}\end{bmatrix}}}{y_n} = \underset{\substack{\begin{bmatrix}\text{intensity at}\\ n-1 \text{ meters}\end{bmatrix}}}{y_{n-1}} - \underset{\substack{\begin{bmatrix}\text{change in}\\ \text{intensity}\end{bmatrix}}}{.20y_{n-1}}
$$

$$
y_n = .8y_{n-1}.
$$

Thus $b/(1-a) = 0$, and the graph is as shown in Fig. 6.

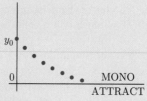

Figure 6

CHAPTER 11 SUMMARY

1. A *difference equation* of the form $y_n = ay_{n-1} + b$, where a, b, and y_0 are specified, determines a sequence of numbers in which each of the numbers (that is, $y_1, y_2, \ldots$) is obtained from the preceding number by multiplying the preceding number by a and adding b. The first number in the sequence, y_0, is called the *initial value*.

2. The graph of a difference equation is obtained by graphing the points $(0, y_0), (1, y_1), (2, y_2), \ldots$.

3. The value of the nth term of a difference equation $y_n = ay_{n-1} + b$ (y_0 given) can be obtained directly (that is, without generating the preceding terms) with one of the following formulas:

$$
y_n = \frac{b}{1-a} + \left(y_0 - \frac{b}{1-a}\right)a^n \quad (a \neq 1)
$$
$$
y_n = y_0 + bn \qquad\qquad\qquad (a = 1).
$$

4. Money deposited in a savings account is called the *principal*. The amount of money earned by the deposit is determined by the type of interest (*simple* or *compound*), the yearly rate of interest, and, for compound interest, the number of times interest is compounded annually. Suppose P dollars is deposited at yearly interest rate r. With simple interest, the balance after n years, y_n, satisfies the difference equation

$$
y_n = y_{n-1} + rP, \quad y_0 = P.
$$

With interest compounded k times per year, the *interest rate per period* is $i = r/k$ and the balance after n years, y_n, satisfies the difference equation

$$
y_n = (1+i)y_{n-1}, \quad y_0 = P.
$$

5. Terminology has been developed to describe the behavior of graphs of difference equations.

increasing   rises when read from left to right
decreasing   falls when read from left to right
monotonic   either increasing or decreasing
oscillating   changes its direction from rising to falling with every term
constant   always has the same value, y_0
attracted   moves closer and closer to the line $y = b/(1-a)$
repelled   moves farther and farther away from the line $y = b/(1-a)$ without bound

6. The five-step procedure for sketching the graph of a difference equation appears on page 557.

7. A *mortgage* is a loan that is paid off in equal payments. The sequence of successive balances (that is, amounts still owed) satisfies a difference equation of the form

$$y_n = (1+i)y_{n-1} - R,$$

where i is the interest rate per period and R is the payment per period.

8. An *annuity* is a sequence of equal deposits into a savings account. The sequence of successive balances satisfies a difference equation of the form

$$y_n = (1+i)y_{n-1} + D,$$

where i is the interest rate per period and D is the payment per period.

REVIEW OF FUNDAMENTAL CONCEPTS

1. Explain how a sequence of numbers is generated by a difference equation of the form $y_n = ay_{n-1} + b$.

2. What is meant by an initial value for the difference equation in Question 1?

3. How do you obtain the graph of the difference equation in Question 1?

4. Give the formula that is the solution of the difference equation $y_n = ay_{n-1} + b$ with $a \neq 1$ and then with $a = 1$.

5. Explain how both simple interest and compound interest work.

6. What does it mean for a graph to be increasing? decreasing? monotonic? oscillating? constant?

7. Describe the possible long-run behaviors of the graph of the difference equation $y_n = ay_{n-1} + b$, $a \neq 1$.

8. Give the five-step procedure for sketching the graph of $y_n = ay_{n-1} + b$, $a \neq 1$.

9. What is a mortgage, and how are its successive balances calculated?

10. What is an annuity, and how are its successive balances calculated?

KEY FORMULAS

The solution of the difference equation $y_n = ay_{n-1} + b$ with y_0 given is

$$y_n = \frac{b}{1-a} + \left(y_0 - \frac{b}{1-a}\right)a^n, \quad \text{if } a \neq 1$$

$$y_n = y_0 + bn, \quad \text{if } a = 1.$$

With P dollars invested at simple interest rate r, the balance after n years, y_n, satisfies the difference equation $y_n = y_{n-1} + rP$, $y_0 = P$.

With P dollars invested at compound interest with periodic interest rate i, the balance after n interest periods, y_n, satisfies the difference equation $y_n = (1+i)y_{n-1}$, $y_0 = P$.

The balance y_n after the nth payment on a mortgage at periodic interest rate i and periodic payment R satisfies the difference equation $y_n = (1+i)y_{n-1} - R$.

The balance y_n after the nth deposit of D dollars into an annuity earning periodic interest rate i satisfies the difference equation $y_n = (1+i)y_{n-1} + D$.

SUPPLEMENTARY EXERCISES

1. Consider the difference equation $y_n = -3y_{n-1} + 8$, $y_0 = 1$.

(a) Generate y_1, y_2, y_3 from the difference equation.

(b) Solve the difference equation.

(c) Use the solution in part (b) to obtain y_4.

2. Consider the difference equation $y_n = y_{n-1} - \frac{3}{2}$, $y_0 = 10$.

(a) Generate y_1, y_2, y_3 from the difference equation.

(b) Solve the difference equation.

(c) Use the solution in part (b) to obtain y_6.

3. (*Account Balance*) How much money would you have to put into a savings account initially at 8% interest compounded quarterly in order to have \$6600 after 10 years?

4. (*Account Balance*) How much money would you have in the bank after 2 years if you deposited \$1000 at 5.2% interest compounded weekly?

5. Make a rough sketch of the graph of the difference equation $y_n = -\frac{1}{3}y_{n-1} + 8$, $y_0 = 10$, without generating terms.

6. Make a rough sketch of the graph of the difference equation $y_n = 1.5y_{n-1} - 2$, $y_0 = 5$, without generating terms.

7. (*Population Dynamics*) The population of a certain city is currently 120,000. The growth rate of the population due to more births than deaths is 3% per year. Furthermore, each year 600 people move out of the city. Let y_n be the population after n years.
   (a) Find the difference equation for the population growth.
   (b) What will be the population after 20 years?

8. (*Mortgage*) The monthly payment on a \$35,000, 30-year mortgage at 12% interest compounded monthly is \$360. Let y_n be the unpaid balance of the mortgage after n months.
   (a) Give the difference equation expressing y_n in terms of y_{n-1}. Also, give y_0.
   (b) What is the unpaid balance after 7 years?

9. (*Annuity*) How much money must be deposited at the end of each week into an annuity at 5.2% (= .052) interest compounded weekly in order to have \$40,000 after 21 years?

10. (*Mortgage Payments*) Find the monthly payment on a \$33,100, 20-year mortgage at 6% interest compounded monthly.

11. (*Loan*) How much money can you borrow at 8% interest compounded annually if the loan is to be paid off in yearly installments for 18 years and you can afford to pay \$2400 per year?

12. (*Alumnus Gift*) A college alumnus pledges to give his alma mater \$50 at the end of each month for 4 years. If the college puts the money in a savings account earning 6% interest compounded monthly, how much money will be in the account at the end of 4 years?

13. (*Political Campaign*) An unknown candidate for governor in a state having 1,000,000 voters mounts an extensive media campaign. Each day, 10% of the voters who do not yet know about him become aware of his candidacy. Let y_n be the number of voters who are aware of his candidacy after n days. Find a difference equation for y_n, and sketch its graph.

14. (*State Legislature*) Suppose that 100 people were just elected to a certain state legislature and that after each term 8% of those still remaining from this original group will either retire or not be reelected. Let y_n be the number of legislators from the original group of 100 who are still serving after n terms. Find a difference equation for y_n, and sketch its graph.

CHAPTER TEST

1. Consider the difference equation $y_n = \frac{1}{2}y_{n-1} + 3$, $y_0 = -2$.
   (a) Determine y_1, y_2, y_3.
   (b) Solve the difference equation.
   (c) Use the solution in part (b) to determine y_4.

2. Consider the difference equation $y_n = y_{n-1} + 50$, $y_0 = 60$.
   (a) Determine y_1, y_2, y_3.
   (b) Solve the difference equation.
   (c) Use the solution in part (b) to determine y_4.

3. Make a rough sketch of the graph of each difference equation without generating terms or solving the difference equation.
   (a) $y_n = .6y_{n-1} + 20$, $y_0 = 10$
   (b) $y_n = -1.25y_{n-1} + 9$, $y_0 = 2$

4. (*Chlorine Concentration*) Ideally, the concentration of chlorine in a swimming pool should be between 1 and 2 parts per million (ppm). Due to the sun and bacteria, about 20% of the chlorine dissipates each day.

Suppose that initially a swimming pool has a concentration of 1 ppm. How much chlorine, in ppm, should be added to the pool at the end of each day so that the concentration will approach 1.5 ppm as the days progress?

5. (*Bank Loan*) Brittany takes out a loan from a bank and repays it in quarterly payments. The balance after n quarters, y_n, satisfies the difference equation $y_n = 1.02y_{n-1} - 3000$, $y_0 = 31,726$.
   (a) How much money did Brittany borrow?
   (b) What was the yearly rate of interest?
   (c) How much did Brittany pay each quarter?
   (d) What will the balance of the loan be after the first quarterly payment?

6. (*Loan Status*) The following table shows partial data for the initial status of a 2-year loan with monthly payments.

Payment Number	1	2	3
Balance	A	$8,888.01	B
Interest	$100	$96.31	$92.58
Reduction of debt	$354.15	$357.84	C

(a) What is the monthly payment?

(b) What is the yearly rate of interest?

(c) Find A, B, and C.

7. (*Mortgage*) I plan to take out a 25-year mortgage at 9% interest compounded monthly. If I am able to pay $1000 per month, how much can I borrow?

8. (*Annuity*) Suppose that $500 is deposited into an annuity at the end of each year and that interest is 6% compounded annually. Let y_n be the balance at the end of n years.

(a) Give the difference equation for y_n. (Include the initial value.)

(b) How much money will be in the annuity after 17 years?

9. (*Annuity*) How much money would you have to deposit at the end of each month into an annuity paying 6% interest compounded monthly in order to have $20,000 after 10 years?

10. (*Drug Absorption*) Patients with certain heart problems are often treated with digitoxin, a derivative of the digitalis plant. The rate at which a person's body eliminates digitoxin is proportional to the amount of digitoxin present. Each day about 10% of any given amount of the drug will be eliminated. Suppose that a dose of .05 milligram is given daily to a patient. Find the difference equation for y_n, the amount in the bloodstream after n doses, and sketch the graph.

CHAPTER 11 PROJECT

Connections to Markov Processes

This chapter project uses difference equations to find the stable distribution for a Markov process and gain insights into the rate with which distribution matrices approach the stable distribution.

1. Show that any 2×2 stochastic matrix can be written in the form

$$A = \begin{bmatrix} 1 - s & r \\ s & 1 - r \end{bmatrix},$$

where r and s are numbers between 0 and 1.

In this discussion, we will make the assumption that r and s are not both zero and are not both one. Denote the successive distribution matrices associated with matrix A as

$$\begin{bmatrix} x_0 \\ y_0 \end{bmatrix}, \quad \begin{bmatrix} x_1 \\ y_1 \end{bmatrix}, \quad \begin{bmatrix} x_2 \\ y_2 \end{bmatrix}, \ldots,$$

where

$$\begin{bmatrix} x_n \\ y_n \end{bmatrix} = A \cdot \begin{bmatrix} x_{n-1} \\ y_{n-1} \end{bmatrix} \quad \text{for } n = 1, 2, 3, \ldots.$$

2. Show that $x_n = [1 - (r + s)]x_{n-1} + r$ and $y_n = [1 - (r + s)]y_{n-1} + s$. (*Hint*: Use the fact that $x_{n-1} + y_{n-1} = 1$.)

3. Show that the values of x_n approach $\dfrac{r}{r + s}$ as n gets large.

4. Show that the values of y_n approach $\dfrac{s}{r + s}$ as n gets large.

5. Express the stable distribution for A in terms of r and s.

6. Apply the result in part 5 to the 2×2 matrix in Example 1 of Section 8.2 to obtain the stable distribution for the working women Markov process.

7. Use the result in part 2 to solve for $\begin{bmatrix} x_n \\ y_n \end{bmatrix}$. That is, obtain an explicit formula for each entry that does not depend directly on A or on the preceding terms in the sequence other than the initial terms x_0 and y_0.

8. Explain why successive distribution matrices approach the stable distribution rapidly when $r + s$ is close to 1.

9. When will successive distributions oscillate about the stable distribution as they approach it?

CHAPTER 12

Logic

IN this chapter we introduce logic: the foundation for mathematics, coherent argument, and computing. The idea of logical argument was introduced in writing by Aristotle. For centuries his principles have formed the basis for systematic thought, communication, debate, law, mathematics, and science. More recently, the concept of a computing machine and the essence of programming are applications of the ancient ideas we present here. Also, new developments in mathematical logic have helped to promote significant advances in artificial intelligence.

12.1 Introduction to Logic

The building blocks of logic are statements, connectives, and the rules for calculating the truth or falsity of compound statements. We begin with statements.

A *statement* is a declarative sentence that is either true or false.
Here are some examples of statements:

1. George Washington was the first president of the United States.

2. The New York Knicks won the NBA basketball championship in 1989.

3. The number of atoms in the universe is 10^{75}.

4. Ronald Reagan knew what Oliver North was doing.

5. If x is a positive real number and $x^2 = 9$, then $x = 3$.

We know that statement 1 is true. Statement 2 is false. Statements 3 and 4 are either true or false, but we do not know which. Neither is both true *and* false. Statement 5 is true.

To better understand what a statement is, we list some nonstatements.

6. He is a real nice guy.

7. Do your homework!

8. If $x^2 = 9$, then $x = 3$.

9. Is this course fun?

Items 6 and 8 are not statements because the person "he" and the variable x are not specified. There are circumstances under which we would claim that item 6 is true and others under which we would claim that item 6 is false; clearly, it depends on who "he" is. Similarly, if x were a positive real number, item 8 would be true. If x were -3, item 8 would be false. Items 7 and 9 are not declarative sentences.

The *truth value* of a statement is either *true* or *false*. The problem of deciding the truth value of a declarative sentence will be dealt with in a simple manner, although it is far from simple. The assignment of a truth value to a statement may be obvious, as in the statement "George Washington was the first president of the United States." The truth value of the statement "The number of atoms in the universe is 10^{75}," on the other hand, is much more problematic.

Logic forms the basis for analyzing legal briefs, political rhetoric, and family discussions. It allows us to understand another's point of view as well as to expose weaknesses in an argument that lead to unfounded conclusions. We combine statements naturally when we speak, using connective words like "and" and "or." We state implications with the words "if" and "then." Frequently, we negate statements with the word "not." A *compound statement* is formed by combining statements using the words "and," "or," "not," or "if, then." A *simple statement* is a statement that is not a compound statement. A compound statement is analyzed as a combination of simple statements. Logic cannot be used to determine the truth value of a simple statement. However, if there is agreement on the truth value of simple statements that are components of a compound statement, the rules of mathematical logic determine the truth value of the compound statement. Its truth value is determined according to mathematical rules given in the next sections.

EXAMPLE 1 **Simple statements** Give the simple statements in each of the following compound statements.

(a) The number 6 is even and the number 5 is odd.

(b) Tom Jones does a term paper or takes the final exam.

(c) If England is in the Common Market, then the British eat Spanish oranges and Italian melons.

Solution (a) The simple statements are "The number 6 is even" and "The number 5 is odd."

(b) The simple statements are "Tom Jones does a term paper" and "Tom Jones takes the final exam."

(c) The simple statements are "England is in the Common Market," "The British eat Spanish oranges," and "The British eat Italian melons."

It is sometimes helpful to add English words to clarify the implied meaning in simple statements.

Now Try Exercise 11   ■

To be able to develop the rules of logic and logical argument, we need to deal with any logical statement, rather than specific examples. We use the letters p, q, r, and so on to represent simple statements. They are not really the statements themselves, but variables for which a statement may be substituted. For example, we can let p represent a statement in general. Then, if we wish, we can specify that for the moment p will represent the statement "The number 5 is odd." In this particular case, p has the truth value *true*. We might decide to let p represent the statement "There are 13 states in the United States today." Then the truth value of the statement p is *false*.

The use of these logical variables is similar to the use of variables x, y, z, and so on to represent unspecified numbers in algebra. In algebra, we manipulate the symbols (such as x, y, $+$, and the implied multiplication and exponentiation) in expressions according to specified rules to establish identities such as

$$(x + y)^2 = x^2 + 2xy + y^2,$$

which are true no matter which numbers are substituted for x and y. In logic, we will manipulate the symbols in compound statements (p, q, r, "and," "or," "not," and "if, then") to look for expressions that are true no matter which statements are substituted for p, q, r, and so on.

It is useful to be able to write a compound statement in terms of its component parts and to use symbols to represent the connectives. We use $\wedge$ to represent the word "and" and $\vee$ to represent the word "or."

EXAMPLE 2 **Symbolic "and"** Write the compound statement "Ina likes popcorn and Fred likes peanuts" in symbolic form.

Solution We first let p represent the statement "Ina likes popcorn" and we let q represent the statement "Fred likes peanuts." We use the symbol $\wedge$ to represent the word "and." Thus, we represent our compound statement symbolically as $p \wedge q$.   ■

EXAMPLE 3 **Symbolic "or"** Write the compound statement "The United States trades with Japan or Germany trades with Japan" in symbolic form.

Solution We use the letter p to represent the statement "The United States trades with Japan" and the letter q to represent the statement "Germany trades with Japan." We use the symbol $\vee$ to represent the word "or." Then the compound sentence is of the form $p \vee q$.   ■

The symbol $\sim$ represents "not" so that, with p as defined in the previous example, we would use $\sim p$ to represent the statement "The United States does not trade with Japan."

EXAMPLE 4 **Symbolic form of compound statements** Let p represent "Fred likes Cindy" and let q represent "Cindy likes Fred." Use the connectives $\wedge$, $\vee$, and $\sim$ to represent the following compound sentences:

(a) Fred and Cindy like each other.

    (b) Fred likes Cindy but Cindy does not like Fred.

    (c) Fred and Cindy dislike each other.

    (d) Fred likes Cindy or Cindy likes Fred.

Solution (a) "Fred and Cindy like each other" should be rewritten as "Fred likes Cindy and Cindy likes Fred." This is expressed symbolically as $p \wedge q$. Statement (b) can be written $p \wedge \sim q$. Note that the word "but" here means "and." Statement (c) can be rewritten as "Fred dislikes Cindy and Cindy dislikes Fred." This is written symbolically as $\sim p \wedge \sim q$. The last statement (d) is simply $p \vee q$.

Now Try Exercise 17 ■

EXAMPLE 5 **Translating from symbols to English** Let p represent the statement "The interest rate is 10%" and let q be the statement "The Dow Jones average is over 10,000." Write the English statements corresponding to each of the following.

    (a) $p \vee q$ (b) $p \wedge q$ (c) $p \wedge \sim q$ (d) $\sim p \vee \sim q$

Solution (a) The statement $p \vee q$ can be written, "The interest rate is 10% or the Dow Jones average is over 10,000."

    (b) We write $p \wedge q$ as "The interest rate is 10% and the Dow Jones average is over 10,000."

    (c) The statement $p \wedge \sim q$ becomes "The interest rate is 10% and the Dow Jones average is less than or equal to 10,000."

    (d) The statement $\sim p \vee \sim q$ can be written as "The interest rate is not 10% or the Dow Jones average is less than or equal to 10,000."

Now Try Exercise 23 ■

The symbol $\rightarrow$ represents an implication. The statement $p \rightarrow q$ is read "p implies q" or "if p, then q." Thus, using the representations of Example 5, the English sentence "If the interest rate is 10%, then the Dow Jones average is over 10,000" can be written as $p \rightarrow q$.

EXAMPLE 6 **Translating from English to symbols** Let p denote the statement "The train stops in Washington" and let q denote the statement "The train stops in New York." Write the following statements in symbolic form.

    (a) The train stops in New York and Washington.

    (b) The train stops in Washington but not in New York.

    (c) The train does not stop in New York.

    (d) The train stops in New York or Washington.

    (e) The train stops in New York or Washington but not in both.

    (f) If the train stops in New York, then it does not stop in Washington.

Solution (a) $p \wedge q$. (b) $p \wedge \sim q$. (c) Negate q to get $\sim q$. (d) $p \vee q$. (e) $(p \vee q) \wedge \sim(p \wedge q)$. The parentheses make the statement clear. We discuss the need for them in Section 12.3. (f) $q \rightarrow \sim p$.

Now Try Exercise 25 ■

How are truth values assigned to compound statements? The assignment depends on the truth values of the simple statements and the connectives in the compound form. The rules are discussed in the next section.

PRACTICE PROBLEMS 12.1

1. Determine which of the following sentences are statements.

   (a) The earnings of IBM went up from 1993 to 1994.

   (b) The national debt of the United States is $6 trillion.

   (c) What an exam that was!

   (d) Abraham Lincoln was the sixteenth president of the United States.

   (e) Lexington is the capital of Kentucky or Albany is the capital of the United States.

   (f) When was the Civil War?

2. Let p denote the statement "Sally is the class president" and let q denote "Sally is an accounting major." Translate the symbolic statements into proper English.

   (a) $p \wedge q$ (b) $\sim p$ (c) $p \vee q$

   (d) $\sim p \vee q$ (e) $p \wedge \sim q$ (f) $\sim p \vee \sim q$

EXERCISES 12.1

In Exercises 1–15, determine which sentences are statements.

1. The number 3 is even.
2. The 1939 World's Fair was held in Miami.
3. The price of coffee depends on the rainfall in Brazil.
4. The Nile River flows through Asia.
5. What a way to go!
6. If snow falls on the Rockies, people are skiing in Aspen.
7. Why is the sky blue?
8. Moisture in the atmosphere determines the type of cloud formation.
9. No aircraft carrier is assigned to the Indian Ocean.
10. The number of stars in the universe is 10^{60}.
11. $x + 3 \geq 0$.
12. He is a brave fellow.
13. Let us pray.
14. The Louvre and the Metropolitan Museum of Art contain paintings by Leonardo da Vinci.
15. If a United States coin is fair, the chance of getting a head is $\frac{1}{3}$.

Find the simple statements in each of the compound statements in Exercises 16–19 and write the compound statement symbolically.

16. China is in Asia and Chicago is in North America.
17. The Phelps Library is either in New York or in Dallas.
18. Randy studies German on either Tuesday or Friday.
19. The Smithsonian Museum of Natural History has displays of rocks and bugs.

In Exercises 20 and 21, write the compound statements in symbolic form, using p and q for each of the simple statements.

20. The number 7 is odd and the number 14 is even.

21. No Amtrak trains go to Chicago or Cincinnati.

22. Let p denote the statement "Arizona has the largest U.S. Indian population" and let q denote the statement "Arizona is the site of the O.K. Corral." Write out the following statements in proper English sentences.

    (a) $\sim p$ (b) $\sim p \vee q$ (c) $\sim q \wedge p$

    (d) $p \vee q$ (e) $\sim p \wedge \sim q$ (f) $\sim(p \vee q)$

23. Let p denote the statement "Ozone is opaque to ultraviolet light" and let q denote the statement "Life on earth requires ozone." Write out the following statements in proper English.

    (a) $p \wedge q$ (b) $\sim p \vee q$ (c) $\sim p \vee \sim q$ (d) $\sim(\sim q)$

24. Let p denote the statement "Papyrus is the earliest form of paper" and let q denote "The papyrus reed is found in Africa." Put the following statements into symbolic form.

    (a) Papyrus is not the earliest form of paper.

    (b) The papyrus reed is not found in Africa or papyrus is not the earliest form of paper.

    (c) Papyrus is the earliest form of paper and the papyrus reed is not found in Africa.

25. Let p denote the statement "Florida borders Alabama" and let q denote the statement "Florida borders Mississippi." Put the following into symbolic form.

    (a) Florida borders Alabama or Mississippi.

    (b) Florida borders Alabama but not Mississippi.

    (c) Florida borders Mississippi but not Alabama.

    (d) Florida borders neither Alabama nor Mississippi.

SOLUTIONS TO PRACTICE PROBLEMS 12.1

1. Statements appear in (a), (b), (d), and (e). Both (c) and (f) are not statements.

2. (a) "Sally is the class president and Sally is an accounting major."

   (b) "Sally is not the class president."

   (c) "Sally is the class president or Sally is an accounting major."

   (d) "Sally is not the class president or Sally is an accounting major."

   (e) "Sally is the class president and Sally is not an accounting major."

   (f) "Sally is not the class president or Sally is not an accounting major."

12.2 Truth Tables

In this section we discuss how the truth values of the statements $p \wedge q$, $p \vee q$, and $\sim p$ depend on the truth values of p and q.

The simple statements will be denoted p, q, r, and so on. A *statement form* is an expression formed from simple statements and connectives according to the following rules.

A simple statement is a statement form.

If p is a statement form, $\sim p$ is a statement form.

If p and q are statement forms, then so are $p \wedge q$, $p \vee q$, and $p \rightarrow q$.

EXAMPLE 1   **Defining statement forms** Show that each of the following is a statement form according to the definition. Assume that p, q, and r are simple statements.

(a) $(p \wedge \sim q) \rightarrow r$ (b) $\sim(p \rightarrow (q \vee \sim r))$

Solution (a) Since p, q, and r are simple statements, they are statement forms. Since q is a statement form, so is $\sim q$. Then $p \wedge \sim q$ is a statement form. Since r is a statement form, $(p \wedge \sim q) \rightarrow r$ is a statement form.

(b) Since r is a statement form, so is $\sim r$. Since both q and $\sim r$ are statement forms, $q \vee \sim r$ is a statement form. However, p is also a statement form; thus $p \rightarrow (q \vee \sim r)$ is a statement form. The negation of a statement form is a statement form, so $\sim(p \rightarrow (q \vee \sim r))$ is also.

Now Try Exercise 1   ◼

The *propositional calculus* is the manipulation, verification, and simplification of logical statement forms. It allows us to see for example whether two different statement forms are logically equivalent in the sense that they must always have the same truth values.

The main mechanism for determining the truth values of statement forms is known as a *truth table*. It is also possible to use tree diagrams (see Chapter 6) to determine the truth values of statement forms. We begin with truth tables.

Consider any simple statement p. Then p has one of the two truth values T (TRUE) or F (FALSE). We can list the possible values of p in a table:

p
T
F

Clearly, $\sim p$ (the negation of p) is a form derived from p and has two possible values also. When p has the truth value T, $\sim p$ has the truth value F, and vice versa. We represent the truth value of $\sim p$ in a truth table.

p	$\sim p$
T	F
F	T

The statement form $p \wedge q$ (p and q) is made up of two statements represented by p and q and the connective $\wedge$. The statement p can have one of the truth values T or F. Similarly, q can have one of the truth values T or F. Hence, by the multiplication principle, there are four possible pairs of truth values for p and q.

Their *conjunction*, $p \wedge q$, is true if and only if both p is true *and* q is true. The truth table for conjunction is

p	q	$p \wedge q$
T	T	T
T	F	F
F	T	F
F	F	F

The *disjunction* of p and q, $p \vee q$ (p or q), is true if either p is true *or* q is true or both p and q are true. In English, the word "or" is ambiguous; we have to distinguish between the exclusive and inclusive "or." For example, in the sentence "Ira will go to either Princeton or Stanford," the word "or" is assumed to be exclusive since Ira will choose one of the schools, but not both. On the other hand, in the sentence, "Diana is smart or she is rich," the "or" probably is inclusive since either Diana is smart, Diana is rich, or Diana is both smart and rich. The mathematical statement form $p \vee q$ uses the inclusive "or" and is unambiguous, as the truth table makes clear.

p	q	$p \vee q$
T	T	T
T	F	T
F	T	T
F	F	F

Note that T stands for TRUE and F stands for FALSE. Each line of the truth table corresponds to a combination of truth values for the components p and q. There are only two lines in the table for $\sim p$, while for the others there are four.

What we have described so far is a system of calculating the truth value of a statement form based on the truth of its component statement forms. The beauty of truth tables is that they can be used to determine the truth values of more elaborate statement forms by reapplying the basic rules.

EXAMPLE 2 **Constructing a truth table** Construct a truth table for the statement form

$$\sim(p \vee q).$$

Solution We write all the components of the statement form so that they can be evaluated for the four possible pairs of values for p and q. Note that we enter the truth values for p and q in the same order as in the previous tables. It is a good idea to use this order all the time; that is,

p	q
T	T
T	F
F	T
F	F

The truth table for the form $\sim(p \vee q)$ contains a column for each calculation.

p	q	$p \vee q$	$\sim(p \vee q)$
T	T	T	F
T	F	T	F
F	T	T	F
F	F	F	T

The third column represents the truth values of the disjunction of the first two columns, and the fourth column represents the negation of the third column. The statement form $\sim(p \vee q)$ is TRUE in one case: when both p and q are FALSE. Otherwise, $\sim(p \vee q)$ is FALSE.

Now Try Exercise 5 ■

EXAMPLE 3 **Tautology** Construct a truth table for the statement form

$$\sim(p \wedge \sim q) \vee p.$$

Solution

p	q	$\sim q$	$p \wedge \sim q$	$\sim(p \wedge \sim q)$	$\sim(p \wedge \sim q) \vee p$
T	T	F	F	T	T
T	F	T	T	F	T
F	T	F	F	T	T
F	F	T	F	T	T
(1)	(2)	(3)	(4)	(5)	(6)

We fill in each column by using the rules already established. Column 3 is the negation of the truth values in column 2. Column 4 represents the conjunction of columns 1 and 3. Column 5 is the negation of the statement form whose values appear in column 4. Finally, column 6 is the disjunction of columns 5 and 1. We see that this statement form has truth value TRUE no matter what the truth values of the statements p and q are. Such a statement is called a tautology.

Now Try Exercise 7 ■

A statement form that has truth value TRUE regardless of the truth values of the individual statement variables it contains is called a *tautology*.

A statement form that has truth value FALSE regardless of the truth values of the individual statement variables it contains is called a *contradiction*.

There is a more efficient way to prepare the truth table of Example 3. Use the same order for entering possible truth values of p and q. Put the statement form at the top of the table. Fill in columns under each operation as you need them, working from the inside out. We label the columns in the order in which we fill them in.

p	q	$\sim$	$(p$	$\wedge$	$\sim q)$	$\vee$	p
T	T	T		F	F		T
T	F	F		T	T		T
F	T	T		F	F		T
F	F	T		F	T		T
(1)	(2)	(5)		(4)	(3)		(6)

We entered the values of p and q in columns 1 and 2 and then entered the values for $\sim q$ in column 3. We used the conjunction of columns 1 and 3 to fill in column 4, applied the negation to column 4 to get column 5, and the disjunction of columns 5 and 1 to get the values in column 6.

EXAMPLE 4 **Constructing truth tables** Construct the truth table for the statement form

$$(\sim p \vee \sim q) \wedge (p \wedge \sim q).$$

Solution Again, we will use the short form of the table, putting the statement form at the top of the table, and labeling each column in the order in which it was completed.

p	q	$(\sim p$	$\vee$	$\sim q)$	$\wedge$	$(p$	$\wedge$	$\sim q)$
T	T	F	F	F	F	T	F	F
T	F	F	T	T	T	T	T	T
F	T	T	T	F	F	F	F	F
F	F	T	T	T	F	F	F	T
(1)	(2)	(3)	(5)	(4)	(7)		(6)	

There are some unlabeled columns. These were just recopied for convenience and clarity. It is not necessary to include them. Column 7, the conjunction of columns 5 and 6, tells us that the statement form is TRUE if and only if p is TRUE and q is FALSE.

Now Try Exercise 9 ■

EXAMPLE 5 **Compressing truth tables** Find the truth table for the statement form

$$(p \vee q) \wedge \sim (p \wedge q).$$

Solution

p	q	$(p \vee q)$	$\wedge$	$\sim$	$(p \wedge q)$
T	T	T	F	F	T
T	F	T	T	T	F
F	T	T	T	T	F
F	F	F	F	T	F
(1)	(2)	(3)	(6)	(5)	(4)

Column 6, the conjunction of columns 3 and 5, gives the truth values of the statement form. ■

We note that the statement form in Example 5 can be read as "p or q but not both p and q." This is the *exclusive* p "or" q. We will find it a useful connective and denote it by $\oplus$. The statement form $p \oplus q$ is true exactly when p is true or q is true but not both p and q are true. The truth table of $p \oplus q$ is

p	q	$p \oplus q$
T	T	F
T	F	T
F	T	T
F	F	F

If there are three simple statements in a statement form and we denote them p, q, and r, then each could take on the truth values TRUE or FALSE. Hence, by the multiplication principle, there are $2 \times 2 \times 2 = 8$ different assignments to the three variables together. There are eight lines in a truth table for such a statement form. Again, in the truth table we will list all the T's and then all the F's for the first variable. Then we alternate TT followed by FF in the second column. Finally, for the third variable we alternate T with F. This ensures that we have listed all the possibilities and facilitates comparison of the final results.

EXAMPLE 6 **Constructing truth tables for three simple statements** Construct a truth table for $(p \vee q) \wedge [(p \vee r) \wedge \sim r]$.

Solution Note the order in which the T's and F's are listed in the first three columns. The columns are numbered in the order in which they were filled.

p	q	r	$(p \vee q)$	$\wedge$	$[(p \vee r)$	$\wedge$	$\sim r]$
T	T	T	T	F	T	F	F
T	T	F	T	T	T	T	T
T	F	T	T	F	T	F	F
T	F	F	T	T	T	T	T
F	T	T	T	F	T	F	F
F	T	F	T	F	F	F	T
F	F	T	F	F	T	F	F
F	F	F	F	F	F	F	T
(1)	(2)	(3)	(4)	(8)	(5)	(7)	(6)

The statement form has truth value TRUE if and only if p is TRUE and r is FALSE (regardless of the truth value of q).

Now Try Exercise 25 ■

The formal evaluation of the truth table of a statement form can help in deciding the truth value when particular sentences are substituted for the statement variables.

EXAMPLE 7 **Logically true or false** Let p denote the statement "London is the capital of England" and let q denote the statement "Venice is the capital of Italy." Determine the truth value of each of the following statements.

(a) $p \land q$ (b) $p \lor q$ (c) $\sim p \land q$

(d) $\sim p \lor \sim q$ (e) $\sim(p \land q)$ (f) $p \oplus q$

Solution First, we note that p has truth value TRUE and q has truth value FALSE. (Rome is the capital of Italy.) Thus the conjunction in (a) is FALSE. The disjunction in (b) is TRUE because p is TRUE. In (c), both $\sim p$ and q are false, so their conjunction is FALSE. Since $\sim q$ is TRUE, statement (d) is TRUE. Since (a) is FALSE, and the statement in (e) is its negation, (e) is TRUE. Since p is TRUE and q is FALSE, $p \oplus q$ is TRUE.

Now Try Exercise 37   ■

Logic and Computer Languages In many computer languages, such as BASIC, the logical connectives are incorporated into programs that depend on logical decision making. The symbols are replaced by their original English words in the BASIC language; for example,

AND	$\land$
OR	$\lor$
NOT	$\sim$
XOR	$\oplus$ (XOR stands for "exclusive or").

If the statements p and q are assigned truth values T and F, then the truth tables for the connectives AND, OR, XOR, and NOT are as shown in Fig. 1.

p	q	p AND q	p OR q	p XOR q		p	NOT p
T	T	T	T	F		T	F
T	F	F	T	T		F	T
F	T	F	T	T			
F	F	F	F	F			

Figure 1

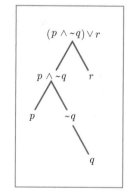

Figure 2

Use of a Tree to Represent a Statement Form Another method of describing a statement form is by using a mathematical structure called a *tree*. We have already seen tree diagrams in Chapter 6. The statement form $(p \land \sim q) \lor r$ is diagrammed in Fig. 2. We begin by writing the statement form at the top of the tree. Two branches are then required, one for each of the operands: $(p \land \sim q)$ and r. We continue until we get to forms involving only a simple statement alone. From $(p \land \sim q)$, we require two branches, one for each of the operands p and $\sim q$. Finally, one branch is needed to perform the negation of q.

If we have specific truth values for the statement variables, the tree can be used to determine the truth value of the statement form by entering the values for all the variables and filling in truth values for the forms, working from the bottom to the top of the tree. Suppose that we are interested in the value of the statement form $(p \land \sim q) \lor r$ when p is TRUE, q is FALSE, and r is FALSE. Figure 3 shows how we fill in values at the nodes and proceed up the tree to its top for the truth value of the statement form.

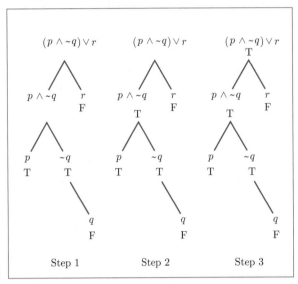

Figure 3

GC Graphing calculators can evaluate logical expressions involving the following four logical operators: AND, OR, XOR, NOT. See Fig. 4, which was obtained on the TI-83 by pressing 2nd [TEST] ▶. TRUE and FALSE are represented by the numbers 1 and 0, respectively. See Figs. 5 and 6. With the TI-89, the four operators along with the words *true* and *false*, can be either typed into the entry line, or by pressing 2nd [MATH] **8**, or can be selected from the CATALOG menu.

Figure 4

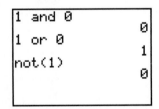

Figure 5

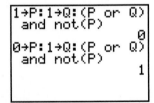

Figure 6

With a TI-83, an entire truth table can be displayed by creating lists of zeros and ones for the variables and attaching a formula containing logical operators to a list name (see Fig. 7). (The lists **P** and **Q** can be created from the home screen as in Fig. 8. A formula is attached to **L1** by entering a logical expression surrounded by quotation marks. The list names **LP** and **LQ** are displayed from the LIST menu.)

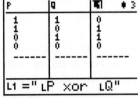

Figure 7

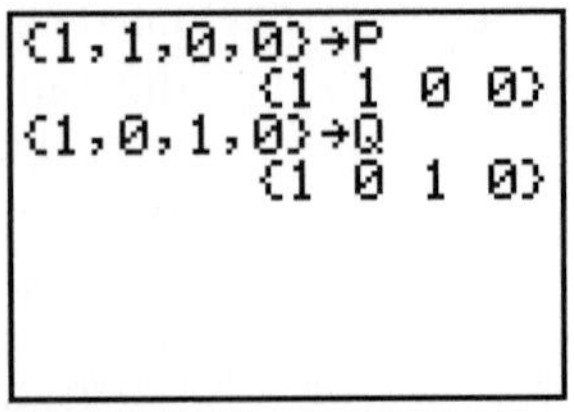

Figure 8

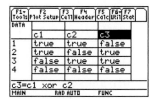

Figure 9

With the TI-89, an entire truth table can be created as shown in Fig. 9. Begin by creating a Data variable from the Data/Matrix editor found in the APPS menu. In the truth table of Fig. 9, the entries in the **c1** and **c2** columns were entered manually. The entries in the c3 column appeared when **c3** was set to **c1 xor c2**.

ES An Excel spreadsheet can display an entire truth table for logical expressions involving AND, OR, and NOT. If x, y, ..., z are addresses of cells containing the words TRUE or FALSE, then the truth values of the three logical functions are given in Table 1.

TABLE 1	
Function	Truth Value
AND(x,y,...,z)	(value of x)$\wedge$(value of y)$\wedge$...$\wedge$(value of z)
OR(x,y,...,z)	(value of x)$\vee$(value of y)$\vee$...$\vee$(value of z)
NOT(x)	$\sim$(value of x)

Figure 10 shows the truth table for the logical expression in Example 5. After the formula was entered into cell C1, the cell was selected and its fill handle was dragged down to C4.

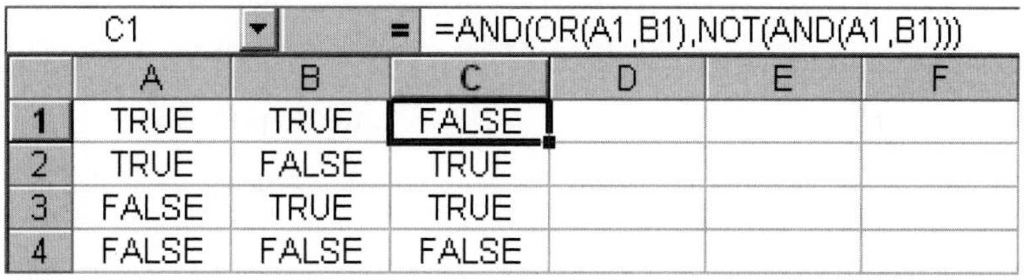

	C1		=	=AND(OR(A1,B1),NOT(AND(A1,B1)))		
	A	B	C	D	E	F
1	TRUE	TRUE	FALSE			
2	TRUE	FALSE	TRUE			
3	FALSE	TRUE	TRUE			
4	FALSE	FALSE	FALSE			

Figure 10. Truth table for Example 5.

PRACTICE PROBLEMS 12.2

1. Construct the truth table for $(p \vee \sim r) \wedge q$.

2. Construct the truth table for $p \oplus \sim q$.

3. Let p denote "May follows April" and let q denote "June follows May." Determine the truth values of

the following.

(a) $p \wedge \sim q$ **(b)** $\sim(p \vee \sim q)$

(c) $p \oplus q$ **(d)** $\sim[(p \wedge q) \oplus \sim q]$

EXERCISES 12.2

*Show that the expressions in Exercises 1–4 are **statement forms** according to the definition. Assume p, q, and r are simple statements.*

1. $\sim(p \wedge \sim r) \vee q$ **2.** $(p \rightarrow \sim q) \wedge r$

3. $(\sim p \vee r) \rightarrow (q \wedge r)$ **4.** $\sim((p \wedge r) \rightarrow q)$

In Exercises 5–28, construct truth tables for the given statement forms.

5. $p \wedge \sim q$ **6.** $\sim(p \vee \sim q) \wedge (p \wedge \sim q)$

7. $(p \vee \sim q) \vee \sim p$ **8.** $(p \wedge q) \vee (p \wedge \sim q)$

9. $\sim((p \vee q) \wedge (p \wedge q))$ **10.** $(p \vee \sim q) \oplus (\sim p)$

11. $p \oplus (\sim p \vee q)$ **12.** $(p \oplus q) \wedge r$

13. $(p \wedge \sim r) \vee q$ **14.** $\sim(p \wedge r) \vee q$

15. $\sim[(p \wedge r) \vee q]$ **16.** $p \wedge \sim p$

17. $p \vee \sim p$ **18.** $(p \vee q) \wedge \sim r$

19. $p \oplus (q \vee r)$ **20.** $p \vee (q \wedge r)$

21. $(p \vee q) \wedge (p \vee r)$ **22.** $(p \vee q) \wedge (p \vee \sim r)$

23. $(p \vee q) \wedge \sim(p \vee q)$ **24.** $(\sim p \vee q) \wedge r$

25. $\sim(p \vee q) \wedge r$ **26.** $\sim[(p \vee q) \wedge r]$

27. $\sim p \vee (q \wedge r)$ **28.** $(p \oplus q) \oplus \sim(p \oplus q)$

29. Compare the truth tables of $\sim p \vee \sim q$ and $\sim(p \wedge q)$.

30. Compare the truth tables of $p \wedge (q \vee r)$ and $(p \wedge q) \vee (p \wedge r)$.

31. Compare the truth tables of $\sim(p \oplus q)$ and $(p \wedge q) \vee \sim(p \vee q)$.

32. How many possible truth tables can you construct for statement forms involving two variables, p and q? (*Hint*: Consider the number of ways to complete the last column of a truth table for each of the four possible pairs of values for p and q.)

33. Compare the truth tables of $(p \wedge q) \vee r$ and $p \wedge (q \vee r)$.

34. (*The Symbol $\ominus$ (NXOR)*) Define the connective $\ominus$ by the truth table

p	q	$p \ominus q$
T	T	T
T	F	F
F	T	F
F	F	T

Construct the truth tables of the following.

(a) $p \ominus \sim q$ **(b)** $(p \ominus q) \ominus r$

(c) $p \ominus (q \ominus r)$ **(d)** $\sim (p \ominus q) \wedge (p \oplus q)$

35. (*The Sheffer Stroke (NAND)*) Define $p|q$ by the truth table

p	q	$p|q$
T	T	F
T	F	T
F	T	T
F	F	T

This connective is called the *Sheffer stroke* (or NAND). Construct truth tables for the following.

(a) $p|p$ **(b)** $(p|p)|(q|q)$

(c) $(p|q)|(p|q)$ **(d)** $p|((p|q)|q)$

36. Compare the truth tables of the connectives $\sim$, $\vee$, and $\wedge$ with those constructed in Exercise 35. Do you see any similarities?

37. Let p denote "John Lennon was a member of the Beatles" and let q be the statement "The Beatles came from Spain." Determine the truth values of the following.

(a) $p \vee \sim q$ **(b)** $\sim p \wedge q$

(c) $p \oplus q$ **(d)** $\sim p \oplus q$

(e) $\sim (p \oplus q)$ **(f)** $(p \vee q) \oplus \sim q$

38. Let p be the statement "There were 14 original American colonies" and let q denote "Utah was one of the original colonies." Determine the truth value of each of the following.

(a) $p \vee q$ **(b)** $\sim p \vee q$

(c) $\sim p \wedge \sim q$ **(d)** $p \oplus \sim q$

39. Let p be the statement "A rectangle has three sides" and q be the statement "A right angle has 90 degrees." Determine the truth value of each of the following.

(a) $p \wedge \sim q$ **(b)** $\sim (p \oplus q)$

(c) $p \wedge q$ **(d)** $\sim p \wedge \sim q$

40. (*Using a Tree Diagram*) Draw a tree for the statement form given in Exercise 5. Assuming p has truth value T and q has value F, use the tree to find the truth value of the statement form.

41. (*Using a Tree Diagram*) Draw a tree for the statement form given in Exercise 12. Assuming that p has truth value T, q has truth value F, and r has truth value F, use the tree to find the truth value of the statement form.

42. Repeat Exercise 40 for the statement in Exercise 9.

43. Repeat Exercise 41 for the statement in Exercise 13.

In Exercises 44 and 45, what value will be displayed when the Enter key is pressed?

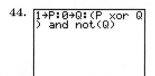

44. `1→P:0→Q:(P xor Q) and not(Q)`

45. `1→P:0→Q:(P and Q) or not(P)`

Exercises 46–51 require the use of a graphing calculator.

In Exercises 46–49, use a graphing calculator to determine the value of the logical expression when (a) p is TRUE, *q is* TRUE, *and (b) p is* TRUE, *q is* FALSE.

46. $(p \oplus q) \wedge \sim q$ **47.** $(p \vee q) \oplus \sim p$

48. $\sim p \vee (p \wedge q)$ **49.** $(p \wedge \sim q) \oplus p$

In Exercises 50 and 51, if your calculator can display a truth table, generate a truth table for the given expression.

50. $(p \vee \sim q) \oplus p$ **51.** $\sim (p \oplus q) \vee (p \wedge q)$

SOLUTIONS TO PRACTICE PROBLEMS 12.2

1.

p	q	r	$(p \lor \sim r) \land q$
T	T	T	T F T
T	T	F	T T T
T	F	T	T F F
T	F	F	T T F
F	T	T	F F F
F	T	F	T T T
F	F	T	F F F
F	F	F	T T F
(1)	(2)	(3)	(5) (4) (6)

2.

p	q	$p \oplus \sim q$
T	T	T F
T	F	F T
F	T	F F
F	F	T T
(1)	(2)	(4) (3)

3. Both p and q are TRUE. Thus (a) is FALSE. Item (b) is the negation of the statement $(p \lor \sim q)$, which is TRUE since p is TRUE. Hence, the statement in (b) is FALSE. Since both p and q are TRUE, $p \oplus q$ is FALSE. To analyze (d), we consider that $(p \land q)$ is TRUE and $\sim q$ is FALSE, so $[(p \land q) \oplus \sim q]$ is TRUE. Its negation is FALSE.

12.3   Implication

We are quite familiar with implications. For example, "If Jay is caught smoking in the restroom, he is suspended from school" is an implication. Note that, although Jay may avoid being caught smoking in the restroom, he still might be suspended for some other infraction. Symbolically, we can represent the statement "Jay is caught smoking in the restroom" by p, while we represent "Jay is suspended from school" by q. The implication is represented by the *conditional* connective $\rightarrow$ and we write $p \rightarrow q$. The truth table for the statement form using the conditional is as follows:

p	q	$p \rightarrow q$
T	T	T
T	F	F
F	T	T
F	F	T

The *conditional* statement form $p \rightarrow q$ has truth value FALSE only when p has truth value TRUE and q has truth value FALSE.

In the preceding case, if p is TRUE, then "Jay is caught smoking in the restroom" is TRUE, while if q is FALSE, then "Jay is not suspended" is TRUE. The implication $p \rightarrow q$ is FALSE in this case.

We call p the *hypothesis* and q the *conclusion*. Note that if the hypothesis is FALSE, then the implication $p \rightarrow q$ is TRUE. There are several ways to read $p \rightarrow q$ in English:

1. p implies q.
2. If p, then q.
3. p only if q.
4. q, if p.
5. p is sufficient for q.
6. q is a necessary condition for p.

Note that $p \rightarrow q$ means that p is sufficient for q, so if you have a true p, then you get a true q. On the other hand, $q \rightarrow p$ means that p is necessary for q, so if you want a true q, then you must have a true p.

EXAMPLE 1 **Hypothesis and conclusion** In each of the following statements, determine the hypothesis and the conclusion.

(a) Bill goes to the party only if Greta goes to the party.

(b) Sue goes to the party if Craig goes to the party.

(c) For 6 to be even, it is sufficient that its square, 36, be even.

Solution (a) The statement is in the form of p only if q. Thus the hypothesis is "Bill goes to the party" and the conclusion is "Greta goes to the party." We could rewrite the statement as "If Bill goes to the party, then Greta goes to the party." The statement "If Bill goes to the party, then Greta goes to the party" seems to mean that Greta is following Bill around, whereas the statement "Bill goes to the party only if Greta goes to the party" makes the romance seem quite the opposite. However, this last statement should be interpreted as follows: If "Bill goes to the party" is true, then that means Greta must also have gone— for that was the ONLY reason he would go. The unemotional interpretation of the logical form shows the equivalence of the two statements.

(b) The statement is of the form q, if p. It has hypothesis "Craig goes to the party" and conclusion "Sue goes to the party."

(c) This is of the form p is sufficient for q, with the hypothesis p of the form "The square of the integer 6 is even" and the conclusion q of the form "The integer 6 is even."

Now Try Exercise 43 ■

EXAMPLE 2 **Truth value of an implication** Determine whether each of the following statements is true or false.

(a) If Paris is in France, then the Louvre is in Paris.

(b) If the Louvre is in Paris, then $2 + 3 = 7$.

(c) If $2 + 3 = 7$, then the Louvre is in Paris.

(d) If $2 + 3 = 7$, then Paris is in Spain.

(e) Paris is in Spain only if $2 + 3 = 7$.

Solution The statement (a) is TRUE because both the hypothesis and the conclusion are TRUE. The statement (b) is FALSE because the hypothesis is TRUE but the conclusion is FALSE. The last three statements are TRUE because the hypothesis in each is FALSE. Note that the last statement has hypothesis "Paris is in Spain" and conclusion "$2 + 3 = 7$."

Now Try Exercise 41 ■

EXAMPLE 3 **Using a truth table for an implication** Construct the truth table of the statement "If the president dies or becomes incapacitated, then the vice president becomes president."

Solution We let p be the statement "The president dies"; we let q be the statement "The president becomes incapacitated." We let r be the statement "The vice president becomes president." The statement form is $(p \vee q) \rightarrow r$. A truth table for the statement form is as follows:

p	q	r	$(p \vee q)$	$\rightarrow r$
T	T	T	T	T
T	T	F	T	F
T	F	T	T	T
T	F	F	T	F
F	T	T	T	T
F	T	F	T	F
F	F	T	F	T
F	F	F	F	T
(1)	(2)	(3)	(4)	(5)

The statement is FALSE only in the cases where the president dies or becomes incapacitated but the vice president does not become president.

Now Try Exercise 11   ■

It is important to note that $p \rightarrow q$ is not the same as $q \rightarrow p$. We should not confuse the hypothesis and the conclusion in an implication. An example will demonstrate the difference between $p \rightarrow q$ and $q \rightarrow p$. Consider the implication "If the truck carries ice cream, then it is refrigerated." If the implication is in the form $p \rightarrow q$, then $q \rightarrow p$ is the implication "If the truck is refrigerated, then it carries ice cream." These implications need not have the same truth values. The first implication is probably true; the second need not be. The truth tables of $p \rightarrow q$ and $q \rightarrow p$ show the differences.

p	q	$p \rightarrow q$	$q \rightarrow p$
T	T	T	T
T	F	F	T
F	T	T	F
F	F	T	T

The implication $q \rightarrow p$ is called the *converse* of the statement $p \rightarrow q$. Thus, given the statement "If the Los Angeles Dodgers won the pennant, some games of the World Series are in California," its converse is "If some games of the World Series are in California, then the Los Angeles Dodgers won the pennant." While the original statement is true, the converse is not. There are situations in which the hypothesis of the converse is true, but the conclusion is false. Some games of the World Series may be in California because the San Francisco Giants, the Oakland Athletics, the San Diego Padres, or the Anaheim Angels, rather than the Los Angeles Dodgers, won the pennant.

EXAMPLE 4   **The conjunction of a statement and its converse**   Construct the truth table for the conjunction of $p \rightarrow q$ and its converse. That is, find the truth table of the statement form $(p \rightarrow q) \wedge (q \rightarrow p)$.

Solution

p	q	$(p \to q)$	$\land$	$(q \to p)$
T	T	T	T	T
T	F	F	F	T
F	T	T	F	F
F	F	T	T	T
(1)	(2)	(3)	(5)	(4)

We note that the statement form has truth value TRUE whenever p and q have the same truth value: either both TRUE or both FALSE.

Now Try Exercise 45   ■

The statement form $(p \to q) \land (q \to p)$ is referred to as the *biconditional*, which we write as $p \leftrightarrow q$. The statement $p \leftrightarrow q$ is read as "p, if and only if q" or "p is necessary and sufficient for q." The truth table for the biconditional is as follows:

p	q	$p \leftrightarrow q$
T	T	T
T	F	F
F	T	F
F	F	T

The biconditional is a statement form simply because it can be expressed as $(p \to q) \land (q \to p)$. In some sense, which we will discuss further in the next section, the biconditional expresses a kind of equivalence of p and q. That is, the biconditional statement form is TRUE whenever p and q are both TRUE or both FALSE. The biconditional is FALSE whenever p and q have different truth values.

Using connectives, we have seen how to string together several simple statements into compound and fairly complex statement forms. Such forms should not be confusing: What is the meaning of $\sim p \lor q$? Do we mean $\sim(p \lor q)$ or do we mean $(\sim p) \lor q$? To clarify matters and to avoid the use of too many parentheses, we define an order of *precedence* for the connectives. This dictates which of the connectives should be applied first.

The order of precedence for logical connectives is

$$\sim, \quad \land, \quad \lor, \quad \to, \quad \leftrightarrow.$$

One applies $\sim$ first, then $\land$, and so on. If there is any doubt, insert parentheses to clarify the statement form. Using the order of precedence, we see that $\sim p \lor q$ is $(\sim p) \lor q$.

EXAMPLE 5   **Order of precedence**   Compare the truth tables of $(\sim p) \lor q$ and $\sim(p \lor q)$.

Solution

p	q	$(\sim p) \vee q$		$\sim(p \vee q)$	
T	T	F	T	F	T
T	F	F	F	F	T
F	T	T	T	F	T
F	F	T	T	T	F
(1)	(2)	(3)	(4)	(6)	(5)

Note that columns 4 and 6 are different.

Now Try Exercise 17 ■

EXAMPLE 6 **Order of precedence** Insert parentheses in the statement to show the proper order for the application of the connectives.

$$p \wedge q \vee r \rightarrow \sim s \wedge r$$

Solution Reading from the left, we apply the $\sim$ first, so $\sim s \wedge r$ becomes $(\sim s) \wedge r$. We then scan for the connective $\wedge$, since that is next in the precedence list. So $p \wedge q \vee r$ becomes $(p \wedge q) \vee r$. We apply the $\rightarrow$ last, and the statement can be written

$$[(p \wedge q) \vee r] \rightarrow [(\sim s) \wedge r].$$

Now Try Exercise 19 ■

Parentheses have highest priority in the precedence. Clearly,

$$p \wedge (q \vee (r \rightarrow \sim s)) \wedge r$$

has the same symbols as the statement of Example 6, but the parentheses have defined a different statement form.

That the statement form $p \vee q \vee r$ may be written either as $(p \vee q) \vee r$ or as $p \vee (q \vee r)$ is shown in another section. The statement form $p \wedge q \wedge r$ can be written as $(p \wedge q) \wedge r$ or as $p \wedge (q \wedge r)$, as is shown later.

Implications and Computer Languages The logical connective $\rightarrow$ is used in writing computer programs, although in some programming languages it is expressed with the words IF, THEN. Thus we might see an instruction of the form "IF $a = 3$ THEN LET $s = 0$." Strictly speaking, this is not a statement form, because "LET $s = 0$" is not a statement. However, that command means "the computer will set $s = 0$." Thus, if the value of a is 3, the computer sets s to 0. If $a \neq 3$, the program just continues. The instruction

"IF ... THEN ... ELSE"

allows for a branch in the program. For example,

"IF $a = 3$ THEN LET $s = 0$ ELSE LET $s = 1$"

assigns value 0 to s if $a = 3$ and assigns value 1 to s if $a \neq 3$.

EXAMPLE 7 **Implication and computing** For the given input values of A and B, use the program to determine the value of C. The asterisk ($*$) denotes multiplication.

IF $(A * B) + 6 \geq 10$

THEN LET $C = A * B$

ELSE LET $C = 10$

(a) $A = -2$, $B = -7$ (b) $A = -2$, $B = 3$ (c) $A = 2$, $B = 2$

Solution (a) We determine that $(A * B) + 6 = 14 + 6 = 20$. Hence, we set $C = 14$.

(b) In this case, $(A * B) + 6 = 0$. So we set $C = 10$.

(c) $(A * B) + 6 = 10$. Hence, $C = 4$.

Now Try Exercise 51

■

Implication and Common Language The mathematical uses of the conditional and the biconditional are very precise. However, colloquial speech is rarely as precise. For example, consider the statement "Peter gets dessert only if he eats his broccoli." Technically, this means that if Peter gets dessert, he eats his broccoli. However, Peter and his parents probably interpret the statement to mean that Peter gets dessert if and only if he eats his broccoli. This imprecision in language is sometimes confusing; we make every effort to avoid confusion in mathematics by adhering to strict rules for using the conditional and biconditional.

PRACTICE PROBLEMS 12.3

1. Using

p: "A square is a rectangle"

and

q: "A rectangle has four sides,"

write out the statements in symbolic form. Name the hypothesis and the conclusion in each statement.

(a) A square is a rectangle if a rectangle has four sides.

(b) A rectangle has four sides if a square is a rectangle.

(c) A rectangle has four sides only if a square is a rectangle.

(d) A square is a rectangle is sufficient to show that a rectangle has four sides.

(e) For a rectangle to have four sides, it is necessary for a square to be a rectangle.

2. Let p denote the statement "There are 48 states in the United States," let q denote the statement "The

American flag is red, white, and blue," and let r be the statement "Maine is on the East Coast." Determine the truth value of each of the following statement forms.

(a) $p \wedge q \to r$ (b) $p \vee q \to r$

(c) $\sim p \wedge q \leftrightarrow r$ (d) $p \wedge \sim q \to \sim r$

3. Use the program to assign a value to D in each of the cases listed.

LET $C = A + B$

IF $(C > 0)$ OR $(A > 0)$

THEN LET $D = 1$

ELSE LET $D = -1$

(a) $A = 4$, $B = 6$ (b) $A = 4$, $B = -6$

(c) $A = 4$, $B = -4$ (d) $A = -2$, $B = 4$

(e) $A = -6$, $B = 4$ (f) $A = -2$, $B = -2$

EXERCISES 12.3

Construct a truth table for each of the statement forms in Exercises 1–15.

1. $p \to \sim q$

2. $p \vee (q \to \sim r)$

3. $(p \oplus q) \to q$

4. $(p \oplus q) \to r$

5. $(\sim p \wedge q) \to r$

6. $\sim (p \to q)$

7. $(p \to q) \leftrightarrow (\sim p \vee q)$

8. $p \oplus (q \to r)$

9. $(p \to q) \to r$

10. $p \to (q \to r)$

11. $\sim (p \vee q) \to (\sim p \wedge r)$

12. $p \to (p \oplus q)$

13. $(p \vee q) \leftrightarrow (p \wedge q)$

14. $[(\sim p \wedge q) \oplus q] \to p$

15. $[p \wedge (q \vee r)] \leftrightarrow [(p \wedge q) \vee (p \wedge r)]$

For each of the Exercises 16–19, insert parentheses in accordance with the precedence of logical operations.

16. $p \wedge q \to p \vee \sim q$

17. $\sim p \wedge \sim q \to \sim p \wedge q$

18. $p \vee \sim q \wedge r \to p \vee r$

19. $\sim p \wedge \sim q \vee r \to \sim q \wedge r$

Let p be the TRUE statement "Abraham Lincoln was the sixteenth president of the United States" and let q be the FALSE statement "The battle of Gettysburg took place at the O.K. Corral." Determine the truth value of each of the statement forms in Exercises 20–29.

20. $p \to q$

21. $\sim p \to q$

22. $p \to \sim q$

23. $q \to p$

24. $p \leftrightarrow q$

25. $(p \oplus q) \to p$

26. $(p \vee q) \to q$

27. $(p \wedge \sim q) \to (\sim p \oplus q)$

28. $[p \wedge (\sim q \to \sim p)] \oplus q$

29. $p \to [p \wedge (p \oplus q)]$

In Exercises 30–37, write the statement forms in symbols using the conditional ($\to$) or the biconditional ($\leftrightarrow$) connective. Name the hypothesis and the conclusion in each conditional form. Let p be the statement "Sally studied" and q be the statement "Sally passes."

30. If Sally studied, then Sally passes.

31. Sally studied if and only if Sally passes.

32. If Sally passes, then Sally studied.

33. Sally passes only if Sally studied.

34. That Sally studied is sufficient for Sally to pass.

35. Sally's studying is necessary for Sally to pass.

36. Sally passes implies that Sally studied.

37. If Sally did not study, then Sally does not pass.

In Exercises 38–42, let p denote the TRUE statement "The die is fair" and let q be the TRUE statement "The probability of a 2 is $\frac{1}{6}$." Write each of the statement forms in symbols, name the hypothesis and the conclusion in each, and determine whether the statement is TRUE or FALSE.

38. If the die is fair, the probability of a 2 is $\frac{1}{6}$.

39. If the die is not fair, the probability of a 2 is not $\frac{1}{6}$.

40. The probability of a 2 is $\frac{1}{6}$ only if the die is fair.

41. The die is not fair if the probability of a 2 is not $\frac{1}{6}$.

42. The die is fair implies that the probability of a 2 is $\frac{1}{6}$.

43. Give the hypothesis and the conclusion in each statement.

(a) Healthy people live a long life.

(b) The train stops at the station only if a passenger requests it.

(c) For the azalea plant to grow, it is necessary that it be exposed to sunlight.

(d) Only if Jane goes to the store, I will go to the store.

44. State the hypothesis and conclusion in each statement form.

(a) Copa Beach is crowded if the weather is hot and sunny.

(b) Our team wins a game only if I carry a rabbit's foot.

(c) If I carry a rabbit's foot, our team wins a game.

(d) Ivy is green is sufficient for it to be healthy.

45. State the converse of each of the following statements.

(a) City Sanitation collects the garbage if the mayor calls.

(b) The price of beans goes down only if there is no drought.

(c) If goldfish swim in Lake Erie, Lake Erie is fresh water.

(d) If tap water is not salted, then it boils slowly.

46. State the converse of each of the following statements.

(a) If Jane runs 20 miles, Jane is tired.

(b) Cindy loves Fred only if Fred loves Cindy.

(c) Jon cashes a check if the bank is open.

(d) Errors are clear only if the documentation is complete.

(e) Sally's eating the vegetables is a necessary condition for Sally's getting dessert.

(f) Sally's eating the vegetables is sufficient for Sally's getting dessert.

47. Determine the output value of A in the program for the given input values of X and Y.

$$\text{LET } Z = X + Y$$
$$\text{IF } (Z \neq 0) \text{ AND } (X > 0)$$
$$\text{THEN LET } A = 6$$
$$\text{ELSE LET } A = 4$$

(a) X = 0, Y = 0   (b) X = 8, Y = −8

(c) X = −3, Y = 3   (d) X = −3, Y = 8

(e) X = 8, Y = −3   (f) X = 3, Y = −8

48. For the given input values of A and B, use the program to find the output values of X.

$$\text{LET } C = A + B$$
$$\text{IF } ((A > 0) \text{ OR } (B > 0)) \text{ AND } (C > 0)$$
$$\text{THEN LET } X = 100$$
$$\text{ELSE LET } X = -100$$

(a) A = 2, B = 2   (b) A = 2, B = −2

(c) A = 2, B = −5   (d) A = −2, B = −5

(e) A = −5, B = 3   (f) A = −5, B = 8

49. For the input values of A and B, use the program to determine the value of Y.

$$\text{LET } C = A * B$$
$$\text{IF } ((C \geq 10) \text{ AND } (A < 0)) \text{ OR } (B < 0)$$
$$\text{THEN LET } Y = 7$$
$$\text{ELSE LET } Y = 0$$

(a) $A = -2$, $B = -6$ (b) $A = -1$, $B = -6$

(c) $A = -2$, $B = 6$ (d) $A = 6$, $B = -1$

(e) $A = 4$, $B = 3$ (f) $A = 3$, $B = 1$

50. For the given input values of X and Y, use the program to determine the output value of Z.

$$\text{IF } Y \neq 0$$
$$\text{THEN LET } Z = X/Y$$
$$\text{ELSE LET } Z = -1{,}000{,}000$$

(a) $X = 6$, $Y = 2$ (b) $X = 10$, $Y = 5$

(c) $X = 5$, $Y = 10$ (d) $X = 0$, $Y = 10$

(e) $X = 10$, $Y = 0$ (f) $X = -1{,}000{,}000$, $Y = 1$

51. For the given input values of A and B, use the program to determine the value of C.

$$\text{IF } ((A < 0) \text{ AND } (B < 0)) \text{ OR } (B \geq 6)$$
$$\text{THEN LET } C = (A * B) + 4$$
$$\text{ELSE LET } C = 0$$

(a) $A = -1$, $B = -2$ (b) $A = -2$, $B = 8$

(c) $A = -2$, $B = 3$ (d) $A = 3$, $B = -2$

(e) $A = 3$, $B = 8$ (f) $A = 3$, $B = -3$

52. For the given input values for A and B, use the program to determine the output value of C.

$$\text{IF } ((A \geq 5) \text{ OR } (B \geq 5)) \text{ AND } (B \leq 10)$$
$$\text{THEN LET } C = A - B$$
$$\text{ELSE LET } C = -30$$

(a) $A = 7$, $B = 7$ (b) $A = 7$, $B = 3$

(c) $A = 7$, $B = 12$ (d) $A = 4$, $B = 7$

(e) $A = 4$, $B = 3$ (f) $A = 4$, $B = 12$

53. For the given input values of A and B, find the value of X in the program.

$$\text{LET } C = A - B$$
$$\text{IF } (C < 0) \text{ OR } (B < 0)$$
$$\text{THEN LET } D = 5 * C$$
$$\text{ELSE LET } D = 0$$
$$\text{LET } X = D + 3$$

(a) $A = 0$, $B = 0$ (b) $A = 6$, $B = 3$

(c) $A = -5$, $B = 3$ (d) $A = 3$, $B = 5$

(e) $A = 5$, $B = -3$ (f) $A = -5$, $B = -3$

54. We let S be the weekly salary for a consultant who works H hours at a rate of R dollars per hour. We take a deduction of 4% on salaries up to $1000 per week and deduct the maximum of $40 from salaries above $1000. Use the following program to find the pay P for each of the consultants given the rate and the number of hours worked in a week.

$$\text{LET } S = R * H$$
$$\text{IF } S < 1000$$
$$\text{THEN LET } D = .04 * S$$
$$\text{ELSE LET } D = 40$$
$$\text{LET } P = S - D$$

(a) $R = 10$, $H = 40$ (b) $R = 20$, $H = 40$

(c) $R = 25$, $H = 20$ (d) $R = 25$, $H = 40$

(e) $R = 30$, $H = 40$ (f) $R = 35$, $H = 40$

SOLUTIONS TO PRACTICE PROBLEMS 12.3

1. (a) This is of the form $q \to p$ with hypothesis q and conclusion p. Statement (b) is of the form $p \to q$, where p is the hypothesis and q is the conclusion. Statement (c) is of the form $q \to p$, since it states that "If a rectangle has four sides, then a square is a rectangle." Note that "only if" signals the clause that is the conclusion. In (c), q is the hypothesis and p is the conclusion. Statement (d) is of the form $p \to q$ with hypothesis p and conclusion q. Statement (e) is the converse of (d) and is of the form $q \to p$ with hypothesis q and conclusion p.

SOLUTIONS TO PRACTICE PROBLEMS 12.3 (CONTINUED)

2. First, we note that p, q, and r have truth values FALSE, TRUE, and TRUE, respectively. Use the order of precedence to decide which connectives are applied first.

(a) TRUE.   Since $p \wedge q$ is FALSE, the implication $(p \wedge q) \to r$ is TRUE.

(b) TRUE.   Here $p \vee q$ is TRUE and r is TRUE, so $(p \vee q) \to r$ is TRUE.

(c) TRUE. Since p is FALSE and q is TRUE, $(\sim p) \wedge q$ is TRUE. Since r is also TRUE, $(\sim p \wedge q) \leftrightarrow r$ is TRUE.

(d) TRUE.   Since p is FALSE and $\sim q$ is also FALSE, $p \wedge (\sim q)$ is FALSE. Thus, $(p \wedge \sim q) \to r$ is TRUE.

	A	B	C	D
(a)	4	6	10	1
(b)	4	−6	−2	1
(c)	4	−4	0	1
(d)	−2	4	2	1
(e)	−6	4	−2	−1
(f)	−2	−2	−4	−1

For example, in (b) we check first to see if (C > 0) is TRUE. It is not. We check to see if (A > 0) is TRUE. It is. So we let D = 1. In (e), however, both (C > 0) and (A > 0) are FALSE, so we let D = −1.

12.4 Logical Implication and Equivalence

Different statement forms may have the same truth tables. Suppose, for example, that p and q are simple statements and we construct the truth tables of both $p \to q$ and $\sim p \vee q$.

p	q	$p \to q$	$\sim p \vee q$	
T	T	T	F	T
T	F	F	F	F
F	T	T	T	T
F	F	T	T	T
(1)	(2)	(3)	(4)	(5)

We may compare columns 3 and 5 to see that, for any given pair of truth values assigned to p and q, the statement forms $p \to q$ and $\sim p \vee q$ have the same truth values.

Two statement forms that have the same truth tables are called *logically equivalent*.

Recall that we defined a *tautology* to be a statement form that has truth value TRUE regardless of the truth values of its component statements. We denote a tautology with the letter t. A statement form that always has truth value FALSE regardless of the truth values of its component statements is called a *contradiction*. We use the letter c to denote a contradiction.

EXAMPLE 1 **Tautology and contradiction** Show that $p \vee \sim p$ is a tautology and that $p \wedge \sim p$ is a contradiction.

Solution We construct the truth tables for the statement forms.

p	$\sim p$	$p \vee \sim p$	$p \wedge \sim p$
T	F	T	F
F	T	T	F

No matter what the value of p, $p \vee \sim p$ is TRUE and $p \wedge \sim p$ is FALSE.

Now Try Exercise 1   ■

EXAMPLE 2   **Logical equivalence** Show that $\sim(p \wedge q)$ is logically equivalent to $\sim p \vee \sim q$.

Solution We construct a truth table to demonstrate that $\sim(p \wedge q)$ is logically equivalent to $\sim p \vee \sim q$.

p	q	$\sim(p \wedge q)$		$\sim p \vee \sim q$		
T	T	F	T	F	F	F
T	F	T	F	F	T	T
F	T	T	F	T	T	F
F	F	T	F	T	T	T
(1)	(2)	(4)	(3)	(5)	(7)	(6)

Since columns 4 and 7 are the same, $\sim(p \wedge q)$ and $\sim p \vee \sim q$ have the same truth tables and therefore are logically equivalent. We write $[\sim(p \wedge q)] \Leftrightarrow [\sim p \vee \sim q]$.

Now Try Exercise 3   ■

For convenience, we denote compound statement forms by capital letters. We will use P, Q, and R, and so on, to denote statement forms such as $p \rightarrow q$, $p \wedge (q \rightarrow \sim r)$, $r \vee (p \wedge \sim(r \vee q))$, and so on. As we saw in the last section, $P \leftrightarrow Q$ is TRUE whenever P and Q have the same truth values (either both TRUE or both FALSE). And $P \leftrightarrow Q$ is FALSE if the truth values of P and Q differ. Thus P is logically equivalent to Q if and only if $P \leftrightarrow Q$ is a tautology. We write $P \Leftrightarrow Q$ when P and Q are logically equivalent. In other words,

> P and Q are logically equivalent, and we write $P \Leftrightarrow Q$, whenever $P \leftrightarrow Q$ is a tautology.

We note that in comparing columns 4 and 7 in Example 2 we find that $\sim(p \wedge q)$ has the same truth table as $\sim p \vee \sim q$. This is reflected in the entries in column 8 of the following table; $\sim(p \wedge q) \leftrightarrow \sim p \vee \sim q$ is a tautology.

p	q	$\sim(p \wedge q)$		$\leftrightarrow$	$\sim p$	$\vee$	$\sim q$
T	T	F	T	T	F	F	F
T	F	T	F	T	F	T	T
F	T	T	F	T	T	T	F
F	F	T	F	T	T	T	T
(1)	(2)	(4)	(3)	(8)	(5)	(7)	(6)

This particular logical equivalence is important and is one of *De Morgan's laws*. We summarize some important logical equivalences in Table 1.

The logical equivalences are important for understanding common English as well as for more formal analysis of mathematical statements and applications to computing. For example, the *contrapositive rule* is used frequently in speech. It states the equivalence of $p \rightarrow q$ with $\sim q \rightarrow \sim p$.

The *contrapositive* of the statement form $p \rightarrow q$ is the statement form $\sim q \rightarrow \sim p$.

TABLE 1	Logical equivalences	
1.	$\sim\sim p \Leftrightarrow p$	Double negation
2a.	$(p \vee q) \Leftrightarrow (q \vee p)$	Commutative laws
b.	$(p \wedge q) \Leftrightarrow (q \wedge p)$	
c.	$(p \leftrightarrow q) \Leftrightarrow (q \leftrightarrow p)$	
3a.	$[(p \vee q) \vee r] \Leftrightarrow [p \vee (q \vee r)]$	Associative laws
b.	$[(p \wedge q) \wedge r] \Leftrightarrow [p \wedge (q \wedge r)]$	
4a.	$[p \vee (q \wedge r)] \Leftrightarrow [(p \vee q) \wedge (p \vee r)]$	Distributive laws
b.	$[p \wedge (q \vee r)] \Leftrightarrow [(p \wedge q) \vee (p \wedge r)]$	
5a.	$(p \vee p) \Leftrightarrow p$	Idempotent laws
b.	$(p \wedge p) \Leftrightarrow p$	
6a.	$(p \vee c) \Leftrightarrow p$	Identity laws
b.	$(p \vee t) \Leftrightarrow t$	
c.	$(p \wedge c) \Leftrightarrow c$	
d.	$(p \wedge t) \Leftrightarrow p$	
7a.	$(p \vee \sim p) \Leftrightarrow t$	
b.	$(p \wedge \sim p) \Leftrightarrow c$	
8a.	$\sim(p \vee q) \Leftrightarrow (\sim p \wedge \sim q)$	De Morgan's laws
b.	$\sim(p \wedge q) \Leftrightarrow (\sim p \vee \sim q)$	
9.	$(p \rightarrow q) \Leftrightarrow (\sim q \rightarrow \sim p)$	Contrapositive
10a.	$(p \rightarrow q) \Leftrightarrow (\sim p \vee q)$	Implication
b.	$(p \rightarrow q) \Leftrightarrow \sim(p \wedge \sim q)$	
11.	$(p \leftrightarrow q) \Leftrightarrow [(p \rightarrow q) \wedge (q \rightarrow p)]$	Equivalence
12.	$(p \rightarrow q) \Leftrightarrow [(p \wedge \sim q) \rightarrow c]$	Reductio ad absurdum

EXAMPLE 3   **Stating the contrapositive of a statement**   State the contrapositive of "If Jay does not play Lotto, Jay will not win the Lotto jackpot."

Solution   We let p be the statement "Jay does not play Lotto" and q be the statement "Jay will not win the Lotto jackpot." We use the *double negation* rule from Table 1 to see that the negation of p is the statement "Jay plays Lotto," while the negation of q is the statement "Jay will win the Lotto jackpot." The contrapositive is the statement "If Jay will win the Lotto jackpot, then Jay plays Lotto." It is logically equivalent to the original statement form (see rule 9).

Now Try Exercise 27   ∎

To prove the *contrapositive* rule, we use the truth table to show that $(p \rightarrow q) \Leftrightarrow (\sim q \rightarrow \sim p)$.

p	q	$(p \rightarrow q)$	$\leftrightarrow$	$(\sim q$	$\rightarrow$	$\sim p)$
T	T	T	T	F	T	F
T	F	F	T	T	F	F
F	T	T	T	F	T	T
F	F	T	T	T	T	T
(1)	(2)	(3)	(7)	(4)	(6)	(5)

A comparison of columns 3 and 6 shows that $p \rightarrow q$ and $\sim q \rightarrow \sim p$ have the same truth tables. Column 7 shows that the biconditional is a tautology. Hence

$$p \rightarrow q$$

is logically equivalent to its contrapositive

$$\sim q \rightarrow \sim p.$$

EXAMPLE 4 **Stating the contrapositive of a statement** A recent Associated Press article quoted a prominent official as having said, "If Mr. Jones is innocent of a crime, then he is not a suspect." State the contrapositive. Do you think the statement is TRUE or FALSE?

Solution The contrapositive of the statement is "If Mr. Jones is a suspect, then he is guilty of a crime." That seems FALSE, so the original statement is also FALSE.

Now Try Exercise 29

De Morgan's laws allow us to negate statements having the disjunctive or conjunctive connectives. It is important to recognize that when the negation is brought inside the parentheses it changes the connective from $\wedge$ to $\vee$, or vice versa.

EXAMPLE 5 **Using DeMorgan's laws** Negate the statement "The earth's orbit is round and a year has 365 days."

Solution We are asked to negate a statement of the form $p \wedge q$. By rule 8b, the negation can be written as "The earth's orbit is not round or a year does not have 365 days." This is TRUE whenever the original statement is FALSE.

Now Try Exercise 17

EXAMPLE 6 **Using DeMorgan's laws** Rewrite the statement "It is false that Jack and Jill went up the hill."

Solution We use De Morgan's laws to negate a statement of the form $(p \wedge q)$ to get "Jack did not go up the hill or Jill did not go up the hill."

EXAMPLE 7 **Using DeMorgan's laws** Use De Morgan's laws to negate the statement form $(p \wedge q) \wedge r$.

Solution $\sim ((p \wedge q) \wedge r) \Leftrightarrow (\sim(p \wedge q) \vee (\sim r)) \Leftrightarrow ((\sim p \vee \sim q) \vee \sim r)$. Note that each time the negation is applied to the conjunction or the disjunction, the connective changes.

Now Try Exercise 19

The rules in Table 1 allow us to simplify statement forms in a variety of ways. In place of any of the simple statements p, q, and r, we may substitute any compound statement. Provided we substitute consistently throughout, the equivalence holds. The rules are similar to the rules of algebra in that the need and the purpose for simplification depend on the situation. Frequently, however, we can simplify statement forms containing $\rightarrow$ by eliminating the implication. As the *implication* rule

$$(p \rightarrow q) \Leftrightarrow (\sim p \vee q)$$

shows, we can reduce a statement involving $\rightarrow$ to one involving other connectives. In fact, as the next examples demonstrate, we may express statements involving any of the connectives using only $\sim$ and $\vee$ or only $\sim$ and $\wedge$.

EXAMPLE 8   **Eliminating the $\rightarrow$ symbol**   Eliminate the connective $\rightarrow$ in the statement form and simplify as much as possible.

$$(\sim p \wedge \sim q) \rightarrow (q \rightarrow r).$$

Solution   Eliminate both implications and simplify by first replacing p by $(\sim p \wedge \sim q)$ and q by $(q \rightarrow r)$ in rule 10a. Then reapply 10a to the implication $(q \rightarrow r)$. We have

$$(\sim p \wedge \sim q) \rightarrow (q \rightarrow r)$$

$\Leftrightarrow \sim(\sim p \wedge \sim q) \vee (q \rightarrow r)$	Implication
$\Leftrightarrow \sim(\sim p \wedge \sim q) \vee (\sim q \vee r)$	Implication
$\Leftrightarrow (\sim\sim p \vee \sim\sim q) \vee (\sim q \vee r)$	De Morgan's law
$\Leftrightarrow (p \vee q) \vee (\sim q \vee r)$	Double negation
$\Leftrightarrow [p \vee (q \vee \sim q) \vee r]$	Associative law
$\Leftrightarrow [(q \vee \sim q) \vee p \vee r]$	Commutative law
$\Leftrightarrow [t \vee (p \vee r)]$	Rule 7a
$\Leftrightarrow t$	Identity rule

Thus the original statement is a tautology.

Now Try Exercise 13   ■

    We used the principle of *substitution* in the solution to the previous example. This allowed us to substitute statement forms for the simple statements in the rules of Table 1. We also substituted equivalent statement forms in interpreting the rules. The *substitution principles* are formally stated as follows:

Substitution principles:

Suppose P and R are statement forms and $P \Leftrightarrow R$. If R is substituted for P in a statement form Q, the resulting statement form Q' is logically equivalent to Q.

Suppose that P and Q are statement forms containing the simple statement p. Assume that $P \Leftrightarrow Q$. If R is any statement form and is substituted for every p in P to yield P' and for every p in Q to yield Q', then $P' \Leftrightarrow Q'$.

EXAMPLE 9   **Eliminating the $\rightarrow$ symbol**   Rewrite the statement form

$$(p \rightarrow q) \wedge (q \vee r)$$

(a) using only the connectives $\sim$ and $\vee$.

(b) using only the connectives $\sim$ and $\wedge$.

Solution (a) $(p \to q) \wedge (q \vee r)$

$$\Leftrightarrow (\sim p \vee q) \wedge (q \vee r) \qquad \text{Implication}$$
$$\Leftrightarrow (q \vee \sim p) \wedge (q \vee r) \qquad \text{Commutative law}$$
$$\Leftrightarrow q \vee (\sim p \wedge r) \qquad \text{Distributive law}$$
$$\Leftrightarrow q \vee (\sim p \wedge \sim \sim r) \qquad \text{Double negation}$$
$$\Leftrightarrow q \vee \sim (p \vee \sim r) \qquad \text{De Morgan's law}$$

(b) $(p \to q) \wedge (q \vee r)$

$$\Leftrightarrow (\sim p \vee q) \wedge (q \vee r) \qquad \text{Implication}$$
$$\Leftrightarrow (\sim p \vee \sim \sim q) \wedge (q \vee r) \qquad \text{Double negation}$$
$$\Leftrightarrow \sim (p \wedge \sim q) \wedge (q \vee r) \qquad \text{De Morgan's law}$$
$$\Leftrightarrow \sim (p \wedge \sim q) \wedge \sim \sim (q \vee r) \qquad \text{Double negation}$$
$$\Leftrightarrow \sim (p \wedge \sim q) \wedge \sim (\sim q \wedge \sim r) \qquad \text{De Morgan's law} \qquad \blacksquare$$

EXAMPLE 10 **Using the implication rule** Use the implication rule to show that

$$[(p \wedge \sim r) \to (q \to r)] \Leftrightarrow [\sim (p \wedge \sim r) \vee (q \to r)].$$

Solution The implication rule states that $p \to q \Leftrightarrow \sim p \vee q$. Substitute the statement form $p \wedge \sim r$ for each p on both sides of the logical equivalence. Substitute $q \to r$ for each q in the logical equivalence. $\blacksquare$

Recall that we have already seen De Morgan's laws in another form in Chapter 5. In that case, S and T were sets and

$$(S \cup T)' = S' \cap T' \qquad \text{and} \qquad (S \cap T)' = S' \cup T'.$$

De Morgan's laws for unions and intersections of sets are closely related to De Morgan's laws for the connectives $\vee$ and $\wedge$. In Section 12.6, we explore this further.

We have seen in this section how statement forms that look quite different can be logically equivalent. Statement forms sometimes can be simplified by using the logical equivalences of Table 1. Logical argument is based not only on logical equivalences, but also on those statements that logically imply others.

Given statement forms P and Q, we say that P *logically implies* Q whenever $P \to Q$ is a tautology. We write this as $P \Rightarrow Q$. $P \Rightarrow Q$ means that $P \to Q$ is a tautology.

This means that $P \Rightarrow Q$ if, whenever P is TRUE, then Q is TRUE. To determine whether a statement form P logically implies a statement form Q, we only need to consider those rows of the truth table for which P is TRUE or those for which Q is FALSE.

Table 2 summarizes some useful logical implications.

EXAMPLE 11 **Logical implication: Disjunctive syllogism** Verify Table 2, 5: $[(p \vee q) \wedge \sim q] \Rightarrow p$.

Solution Let statement form $P = (p \vee q) \wedge \sim q$ and statement form $Q = p$. We are interested in verifying that $P \to Q$ is a tautology. In a straightforward approach, we can

	TABLE 2 Logical implications	
1.	$p \Rightarrow (p \vee q)$	Addition
2.	$(p \wedge q) \Rightarrow p$	Simplification
3.	$[(p \wedge (p \rightarrow q)] \Rightarrow q$	Modus ponens
4.	$[(p \rightarrow q) \wedge \sim q] \Rightarrow \sim p$	Modus tollens
5.	$[(p \vee q) \wedge \sim q] \Rightarrow p$	Disjunctive syllogism
6.	$[(p \rightarrow q) \wedge (q \rightarrow r)] \Rightarrow (p \rightarrow r)$	Hypothetical syllogism
7a.	$[(p \rightarrow q) \wedge (r \rightarrow s)] \Rightarrow [(p \vee r) \rightarrow (q \vee s)]$	Constructive dilemmas
7b.	$[(p \rightarrow q) \wedge (r \rightarrow s)] \Rightarrow [(p \wedge r) \rightarrow (q \wedge s)]$	

construct the entire truth table.

p	q	$[(p \vee q)$	$\wedge$	$\sim q]$	$\rightarrow$	p
T	T	T	F	F	T	
T	F	T	T	T	T	
F	T	T	F	F	T	
F	F	F	F	T	T	
(1)	(2)	(3)	(5)	(4)	(6)	

Column 6 shows that the implication is a tautology. We could have shortened our work. The only situation in which an implication $P \rightarrow Q$ is FALSE is when P is TRUE and Q is FALSE. We check out these possibilities by first letting P be TRUE. This requires that both $(p \vee q)$ and $\sim q$ be TRUE. This occurs when q is FALSE and p is TRUE, and this is the only possibility in which P is TRUE. The only row of the truth table we must consider is

p	q	$[(p \vee q)$	$\wedge$	$\sim q]$	$\rightarrow$	p
T	F	T	T	T	T.	
(1)	(2)	(3)	(5)	(4)	(6)	

In this example it was simple to construct the entire truth table, but, in general, we can save a lot of work by considering only the relevant cases.

Now Try Exercise 9   ■

EXAMPLE 12 **Verification of modus ponens**   Verify modus ponens (logical implication 3).

Solution We must show that

$$[p \wedge (p \rightarrow q)] \rightarrow q$$

is a tautology. The first row of the truth table would have sufficed.

p	q	$[(p \wedge$	$(p \rightarrow q)]$	$\rightarrow$	q
T	T	T	T	T	
T	F	F	F	T	
F	T	F	T	T	
F	F	F	T	T	
(1)	(2)	(4)	(3)	(5)	

Looking at the last column, we see that the implication is a tautology.

PRACTICE PROBLEMS 12.4

1. Prove $(p \to q) \Leftrightarrow (\sim p \lor q)$.

2. Rewrite the statement

$$(p \to \sim q) \to (r \land p)$$

eliminating the connective $\to$ and simplifying if possible.

3. Assume that the statement "Jim goes to the ballgame only if Ted gets tickets" is TRUE.

  (a) Write the contrapositive and give its truth value.

  (b) Write the converse and give its truth value.

  (c) Write the negation and give its truth value.

EXERCISES 12.4

1. Show that $[(p \to q) \land q] \to p$ is not a tautology.

2. Show that the distributive laws hold:

  (a) $[p \lor (q \land r)] \Leftrightarrow [(p \lor q) \land (p \lor r)]$

  (b) $[p \land (q \lor r)] \Leftrightarrow [(p \land q) \lor (p \land r)]$.

3. Show that the second law of implication holds:

$$(p \to q) \Leftrightarrow \sim(p \land \sim q).$$

4. Write a statement equivalent to

$$\sim(p \land \sim q) \to (r \land p)$$

using only $\sim$ and $\land$.

5. The Sheffer stroke is defined by the truth table

p	q	$p|q$
T	T	F
T	F	T
F	T	T
F	F	T

  (a) Show that $\sim p \Leftrightarrow p|p$.

  (b) Show that $p \lor q \Leftrightarrow (p|p)|(q|q)$.

  (c) Show that $p \land q \Leftrightarrow (p|q)|(p|q)$.

  (d) Write $p \to q$ using only the Sheffer stroke.

  (e) Write $p|q$ using only $\sim$ and $\land$.

6. Without using truth tables, show that

$$[(p \lor \sim q) \land r] \to p \Leftrightarrow (p \lor q) \lor \sim r.$$

7. Show that

$$(p \to q) \Leftrightarrow [(p \land \sim q) \to c]$$

using the fact that c denotes a contradiction; that is, c is always FALSE. Read the statement aloud.

8. (a) Prove that

$$p \Rightarrow [q \to (p \land q)].$$

  (b) True or false?

$$p \Leftrightarrow [q \to (p \land q)]$$

9. True or false?

$$(p \lor q) \Rightarrow [q \to (p \land q)]$$

10. True or false?

$$[p \lor (p \land q)] \Leftrightarrow p$$

11. Write an equivalent form of $p \oplus q$ using only $\sim$ and $\lor$.

12. Write each statement using only the connectives $\sim$ and $\lor$.

  (a) $p \land \sim q \to p$ **(b)** $(p \to r) \land (q \lor r)$

  (c) $p \to [r \land (p \lor q)]$ **(d)** $(p \land q) \to (\sim q \lor r)$

13. Write an equivalent form of the statement below without using $\to$.

$$(p \lor q) \to (q \land \sim r)$$

14. Write an equivalent form of the statement below without using $\to$.

$$(q \to \sim r) \to (p \land q)$$

15. Write an equivalent form of the statement below without using $\to$.

$$\sim(p \land \sim q) \to (p \lor \sim r)$$

16. Write an equivalent form of the statement below without using $\to$.

$$\sim(\sim(p \to q) \to r)$$

17. Negate the following statements.

  (a) Arizona borders California and Arizona borders Nevada.

  (b) There are tickets available or the agency can get tickets.

  (c) The killer's hat was either white or gray.

18. Negate the following statements.

    (a) Montreal is a province in Canada and Ottawa is a province in Canada.

    (b) The salesman goes to the customer or the customer calls the salesman.

    (c) The hospital does not admit psychiatric patients or orthopedic patients.

19. Use DeMorgan's laws to negate the statement $p \lor \sim q \land r$ without using parentheses.

20. Use DeMorgan's laws to negate the statement $(p \lor \sim q) \lor (\sim r)$ without using parentheses.

21. Negate each of the statements.

    (a) Either Jeremy takes 12 credits or Jeremy takes 15 credits this semester.

    (b) Sandra received a gift from Sally and a gift from Sacha.

22. Negate the following statements.

    (a) The plane from California to Maine was on time and every seat on it was taken.

    (b) Kenneth was not on time nor was he dressed properly.

23. Write each statement as an OR or as an AND statement and then write its negation.

    (a) "The Old Man and the Sea" was written by Ernest Hemingway or by Jack London.

    (b) "H. M. S. Pinafore" was written by Gilbert and Sullivan.

24. Negate each of the following statements.

    (a) Issac Newton and Henry Ford invented the calculus.

    (b) James Galway or Paul Simon plays the piano.

25. Negate the following statements.

    (a) If I have a ticket to the theater, I spent a lot of money.

    (b) Basketball is played on an indoor court only if the players wear sneakers.

    (c) The stock market is going up implies that the interest rates are going down.

    (d) For humans to stay healthy, it is sufficient that humans have enough water.

26. For each of the following statements, give the contrapositive and determine its truth value.

    (a) If the lists in the book are not in alphabetical order, it is not a telephone directory. '

    (b) The water is salty if it comes from the Atlantic Ocean.

27. For each of the following statements, give the contrapositive and determine its truth value.

    (a) If the sum of odd numbers is odd, then the number of items in the sum is odd.

    (b) If the computer keyboard is standard, then "K" is not next to "W."

28. For each statement, give the contrapositive, the converse, and the negation. Determine the truth value in each case.

    (a) If a rectangle has equal sides, it is a square.

    (b) An airplane flies faster than the speed of sound only if it is a Concorde.

    (c) If the intersection of two sets is not empty, then the union of the two sets is not empty.

    (d) If a coin is fair, the probability of a head is $\frac{1}{2}$.

29. Give the contrapositive and the converse of each statement and then give the truth value of each.

    (a) If a bird is small, it is a hummingbird.

    (b) If two nonvertical lines have the same slope, they are parallel.

    (c) If we are in Paris, we must be in France.

    (d) If a road is one-way, you cannot legally make a U-turn.

30. Bill, Sue, and Alice are lined up facing forward with Bill first, then Sue, then Alice. From a collection that they know contains three blue and two red hats, hats are placed on their heads while they are blindfolded. Blindfolds are removed, but they continue to face forward and see only those hats in front of them. Bill sees none, Sue sees Bill's, and Alice sees both Sue's and Bill's. Alice claims she does not know what color her hat is. Sue also claims that she does not know what color her hat is. Now Bill knows what color his hat is. What is it?

31. A prisoner is given one chance for freedom. He may ask one yes-or-no question of either of two guards. Each guard allows access to one of two unmarked doors. One is the door to freedom, the other the door to death. One guard always tells the truth, the other always lies. What question should the prisoner ask?

32. Verify modus tollens (logical implication 4 in Table 2).

33. Verify hypothetical syllogism (logical implication 6 in Table 2).

SOLUTIONS TO PRACTICE PROBLEMS 12.4

1. We show that $(p \to q) \leftrightarrow (\sim p \vee q)$ is a tautology.

p	q	$(p \to q)$	$\leftrightarrow$	$(\sim p$	$\vee$	$q)$
T	T	T	T	F		T
T	F	F	T	F		F
F	T	T	T	T		T
F	F	T	T	T		T
(1)	(2)	(3)	(6)	(4)		(5)

2. $(p \to \sim q) \to (r \wedge p)$

$\Leftrightarrow (\sim p \vee \sim q) \to (r \wedge p)$ Implication

$\Leftrightarrow [\sim(\sim p \vee \sim q)] \vee (r \wedge p)$ Implication

$\Leftrightarrow (p \wedge q) \vee (r \wedge p)$ De Morgan's law; double negation

$\Leftrightarrow (p \wedge q) \vee (p \wedge r)$ Commutative law

$\Leftrightarrow p \wedge (q \vee r)$ Distributive law

3. (a) Contrapositive: "If Ted does not get tickets, Jim does not go to the ballgame." TRUE.

(b) Converse: "If Ted gets tickets, Jim goes to the ballgame." Truth value is unknown.

(c) We negate a statement of the form $p \to q$. We use the implication law $(p \to q) \Leftrightarrow (\sim p \vee q)$, form the negative, and apply De Morgan's law to get $[\sim(p \to q)] \Leftrightarrow (p \wedge \sim q)$. Negation: "Jim goes to the ballgame and Ted does not get tickets." The negation of the original true statement is FALSE.

12.5   Valid Argument

In this section we study methods of valid argument. We rely heavily on Table 2, the table of logical implications, presented in Section 12.4. Let us begin with an example.

EXAMPLE 1   **Valid argument**   Analyze the argument, given that the first two statements are true.

If Marvin studies mathematics, then he is smart.

Marvin is not smart.

Therefore, Marvin does not study mathematics.

Solution   Let p denote "Marvin studies mathematics" and let q denote "Marvin is smart." The argument is of the form

$$\text{If } [(p \to q) \wedge \sim q], \text{ then } \sim p.$$

Note that

$$[(p \to q) \wedge \sim q] \Rightarrow \sim p$$

is the rule of modus tollens of Table 2 of the previous section. Thus the implication

$$[(p \to q) \wedge \sim q] \to \sim p$$

is a tautology and, therefore, always TRUE. The argument presented is valid.

Now Try Exercise 1   ■

The technique used in the example can be extended to more complex arguments. An *argument* or a *proof* is a set of statements

$$H_1, H_2, \ldots, H_n$$

each of which is assumed to be true and a statement C that is claimed to have been deduced from them. The statements

$$H_1, H_2, \ldots, H_n$$

are called *hypotheses* and the statement C is called the *conclusion*. We say that the argument is *valid* if, and only if,

$$H_1 \wedge H_2 \wedge \cdots \wedge H_n \Rightarrow C.$$

The statement $H_1 \wedge H_2 \wedge \cdots \wedge H_n \to C$ is a tautology provided it is never false; if each hypothesis is true, the conclusion is also true.

In the example just given, the argument is valid because, for the hypotheses

$$H_1 = (p \to q) \quad \text{and} \quad H_2 = {\sim}q \quad \text{and the conclusion} \quad C = {\sim}p,$$

we have the logical implication $H_1 \wedge H_2 \Rightarrow C$.

There are several important points to make here. First, although the conclusion may be a true statement, the argument presented may or may not be valid. Also, if one or more of the premises is false, it is possible for a valid argument to result in a conclusion that is false. The logical implications stated earlier in this chapter can be restated in the form of *rules of inference*, as given in Table 1.

TABLE 1   Rules of inference

	From:	Conclude:	
1.	P	$P \vee Q$	Addition
2.	$P \wedge Q$	P	Subtraction
3.	$P \wedge (P \to Q)$	Q	Modus ponens
4.	$(P \to Q) \wedge {\sim}Q$	${\sim}P$	Modus tollens
5.	$(P \vee Q) \wedge {\sim}P$	Q	Disjunctive syllogism
6.	$(P \to Q) \wedge (Q \to R)$	$P \to R$	Hypothetical syllogism
7a.	$(P \to Q) \wedge (R \to S)$	$(P \vee R) \to (Q \vee S)$	Constructive dilemmas
b.	$(P \to Q) \wedge (R \to S)$	$(P \wedge R) \to (Q \wedge S)$	

EXAMPLE 2   **Multiple-step argument**   Suppose that the following statements are true.

Pat is going to the office or to dinner with Jan.

If Pat is going to the office, then it is not sunny.

It is sunny.

Prove: Pat is going to dinner with Jan.

Solution   Let s, p, and d denote the statements

$$s = \text{“It is sunny.”}$$
$$p = \text{“Pat is going to the office.”}$$
$$d = \text{“Pat is going to dinner with Jan.”}$$

We write the argument step by step. Numbers in parentheses refer to the previous steps used in the argument

1. $p \vee d$ Hypothesis

2. $p \to {\sim}s$ Hypothesis

3. s Hypothesis
4. $\sim p$ Modus tollens (2, 3)
5. d Disjunctive syllogism (1, 4)

Step 4 results from the implication $[(p \rightarrow \sim s) \wedge s] \Rightarrow \sim p$. And step 5 results from the implication $[(p \vee d) \wedge \sim p] \Rightarrow d$. We have made convenient substitutions in the rules in Table 1.

Now Try Exercise 3   ■

EXAMPLE 3   **Verify validity**   Show that the following argument is valid.

I study either mathematics or economics.
If I have to take English, then I do not study economics.
I do not study mathematics.
Therefore, I do not have to take English.

Solution   Let p, q, and r represent the statements

$$p = \text{``I study mathematics.''}$$
$$q = \text{``I study economics.''}$$
$$r = \text{``I have to take English.''}$$

The hypotheses are

$$p \vee q, \quad r \rightarrow \sim q, \quad \text{and} \quad \sim p.$$

We write the argument step by step. Numbers in parentheses refer to the previous steps used in the argument.

1. $p \vee q$ Hypothesis
2. $r \rightarrow \sim q$ Hypothesis
3. $\sim p$ Hypothesis
4. q, since $[(p \vee q) \wedge \sim p] \Rightarrow q$ Disjunctive syllogism (1, 3)
5. $\sim r$, since $[(r \rightarrow \sim q) \wedge q] \Rightarrow \sim r$ Modus tollens (2, 4)

Thus we have shown that r can be proven false by a valid argument from the given hypotheses.

Now Try Exercise 7   ■

EXAMPLE 4   **Verify validity**   Verify that the following argument is valid.

If it is raining, then my car stalls or will not start.
If my car stalls, then I arrive late for school.
If my car does not start, I do not go to school.
Therefore, if it is raining, either I arrive late for school or I do not go to school.

Solution   Let p, q, r, s, and u be the following statements.

$$p = \text{``It is raining.''}$$
$$q = \text{``My car stalls.''}$$
$$r = \text{``My car starts.''}$$
$$s = \text{``I arrive late for school.''}$$
$$u = \text{``I go to school.''}$$

The argument proceeds as follows.

1. $p \rightarrow (q \vee \sim r)$ Hypothesis
2. $q \rightarrow s$ Hypothesis
3. $\sim r \rightarrow \sim u$ Hypothesis
4. $(q \vee \sim r) \rightarrow (s \vee \sim u)$ Constructive dilemma (2, 3)
5. $p \rightarrow (s \vee \sim u)$ Hypothetical syllogism (1, 4)

Therefore, a valid conclusion is "If it rains, either I arrive late for school or I do not go to school." ∎

In trying to deduce that

$$H_1 \wedge H_2 \wedge \cdots \wedge H_n \Rightarrow C,$$

we sometimes find it easier to prove that the contrapositive of

$$H_1 \wedge H_2 \wedge \cdots \wedge H_n \rightarrow C$$

is a tautology. Thus we would need to show that

$$\sim C \rightarrow \sim(H_1 \wedge H_2 \wedge \cdots \wedge H_n)$$

is a tautology. If we assume that $\sim C$ is TRUE, then, of course, the premise is that C is FALSE. The only case we need to consider to prove the tautology is the case where $\sim C$ is TRUE. In that case, we must show that $\sim(H_1 \wedge H_2 \wedge \cdots \wedge H_n)$ is TRUE also. This requires that the statement

$$(\sim H_1) \vee (\sim H_2) \vee \cdots \vee (\sim H_n)$$

be TRUE (use De Morgan's law). The disjunction is TRUE if and only if at least one of its components is TRUE. So we must show that $(\sim H_i)$ is TRUE for some subscript i. Hence, we are required to show that H_i is FALSE for at least one i. This is called an *indirect proof*.

We summarize the idea.

Indirect Proof   To prove

$$(H_1 \wedge H_2 \wedge \cdots \wedge H_n) \Rightarrow C,$$

we assume the conclusion C is FALSE and then prove that at least one of the hypotheses H_i must be FALSE.

EXAMPLE 5 **Using indirect proof**   Use an indirect proof to show that the following argument is valid.

If I am happy, then I do not eat too much.

I eat too much or I spend money.

I do not spend money.

Therefore, I am not happy.

Solution   We will begin symbolically. Let

$$p = \text{"I am happy."}$$
$$q = \text{"I eat too much."}$$
$$r = \text{"I spend money."}$$

We wish to show that $[(p \rightarrow \sim q) \wedge (q \vee r) \wedge \sim r] \Rightarrow \sim p$. Assume, by way of indirect proof, that the conclusion is FALSE.

1. p Negation of conclusion

2. $p \rightarrow \sim q$ Hypothesis

3. $\sim q$ Modus ponens (1, 2)

4. $q \vee r$ Hypothesis

5. r Disjunctive syllogism (3, 4)

Therefore, the given hypothesis $\sim r$ is FALSE. We express the proof in words. Let us assume the negative of the conclusion; that is, we assume "I am happy." Then "I do not eat too much" (modus ponens). We were to assume "I eat too much or I spend money"; hence (disjunctive syllogism), it is true that "I spend money." This means that the hypothesis "I do not spend money" cannot be true. Assumption of the negation of the conclusion leads us to claim one of the hypotheses is false. Thus we have a valid indirect proof.

Now Try Exercise 21   ■

PRACTICE PROBLEMS 12.5

1. Show that the argument is valid.

If goldenrod is yellow, then violets are blue.

Either pine trees are not green or goldenrod is yellow.

Pine trees are green.

Therefore, violets are blue.

2. Show by indirect proof that the argument is valid.

If I go to the beach, I cannot study.

Either I study or I work as a waiter.

I go to the beach.

Therefore, I work as a waiter.

EXERCISES 12.5

In Exercises 1–10, show that the argument is valid.

1. Either Sue goes to the movies or she reads. Sue does not go to the movies. Therefore, Sue reads.

2. If the class votes for an oral final, the teacher is glad. If the exam is not scheduled for a Monday, the teacher is sad. The exam is scheduled for a Friday. Therefore, the class doesn't vote for an oral final.

3. If my allowance comes this week and I pay the rent, then my bank account will be in the black. If I do not pay the rent, I will be evicted. I am not evicted and my allowance comes. Therefore, my bank account is in the black.

4. Either Jane is in sixth grade implies that Jane understands fractions or Jane is in sixth grade implies that Jane is in a remedial math class. Jane is in sixth grade. Therefore, either Jane understands fractions or she is in a remedial math class.

5. If the price of oil increases, the OPEC countries are in agreement. If there is no U.N. debate, the price of oil increases. The OPEC countries are in disagreement. Therefore, there is a U.N. debate.

6. If Jill wins, then Jack loses. If Peter wins, then Paul loses. Either Jill wins or Peter wins. Therefore, either Jack loses or Paul loses.

7. If the germ is present, then the rash and the fever are present. The fever is present. The rash is not present. Therefore, the germ is not present.

8. If Hal is a politician, he is a liar or a fraud. Hal is not a liar. He is not a fraud. Therefore, Hal is not a politician.

9. If the material is cotton or rayon, it can be made into a dress. The material cannot be made into a dress. Therefore, it is not rayon.

10. If there is money in my account and I have a check, then I will pay the rent. If I do not have a check, then I am evicted. Therefore, if I am not evicted and if I do not pay the rent, then there is no money in my account.

In Exercises 11–20, test the validity of the arguments.

11. If the salaries go up, then more people apply. Either more people apply or the salaries go up. Therefore, the salaries go up.

12. If Rita studies, she gets good grades. Rita gets bad grades. Therefore, Rita does not study.

13. Either the balloon is yellow or the ribbon is pink. If the balloon is filled with helium, then the balloon is a green one. The balloon is filled with helium. Therefore, the ribbon is pink.

14. If the job offer is for at least $30,000 or has 5 weeks vacation, I will accept the position. If the offer is for less than $30,000, then I will not accept the job and I will owe rent money. I will not accept the job. Therefore, I will owe rent.

15. If the papa bear sits, the mama bear stands. If the mama bear stands, the baby bear crawls on the floor. The baby bear is standing. Therefore, the papa bear is not sitting.

16. If it is snowing, I wear my boots. It is not snowing. Therefore, I am not wearing boots.

17. If wheat prices are steady, exports will increase or the GNP will be steady. Wheat prices are steady and the GNP is steady. Therefore, exports will increase.

18. If we eat out, either Mom or Dad will treat. I pay for dinner. Therefore, we do not eat out.

19. If Tim is industrious, then Tim is in line for a promotion. Either Tim is in line for a promotion or he is thinking of leaving. Therefore, if Tim is thinking of leaving, Tim is industrious.

20. If I pass history, then I do not go to summer school. If I go to summer school, I will take a course in French. Therefore, if I go to summer school, then either I do not pass history or I take a course in French.

In Exercises 21–24, use indirect proof to show the argument is valid.

21. Sam goes to the store only if he needs milk. Sam does not need milk. Therefore, Sam does not go to the store.

22. If it rains hard, there will be no picnic. If Dave brings the Frisbee, the kids will be happy. The kids are not happy and there is a picnic. Therefore, it does not rain hard and Dave did not bring the Frisbee.

23. If the newspaper and television both report a crime, then it is a serious crime. If a person was killed, then the newspaper reports the crime. A person is killed. Television reports the crime. Therefore, the crime is serious.

24. If Linda feels ill, she takes aspirin. If she runs a fever, she does not take a bath. If Linda does not feel ill, she takes a bath. Linda runs a fever. Therefore, she takes aspirin.

Show that each of the arguments in Exercises 25 and 26 is valid by first using a direct proof and then by using an indirect proof.

25. If Jimmy does not find the keys, he will do his homework. Jimmy does not do his homework. Show that Jimmy found his keys.

26. If spring vacation includes Easter, Simone goes to France. If George goes to France, Simone does not go. Spring vacation includes Easter. Therefore, George does not go to France.

Show that each of the arguments in Exercises 27 and 28 is valid by using a direct proof or by using an indirect proof.

27. If Marissa does not go to the movies, then she is not idle. If Marissa is not in a knitting class, she is idle. Marissa is not in a knitting class. Therefore, Marissa goes to the movies.

28. Sydney collects a penalty payment if Norman leaves a mess in the apartment. Either Rachel cleans the apartment or Norman leaves a mess in the apartment. Rachel does not clean the apartment. Prove that Sydney collects a penalty payment.

SOLUTIONS TO PRACTICE PROBLEMS 12.5

1. Let

$$g = \text{"Goldenrod is yellow."}$$
$$v = \text{"Violets are blue."}$$
$$p = \text{"Pine trees are green."}$$

The following steps show the argument is valid.

1. $g \rightarrow v$	Hypothesis
2. $\sim p \vee g$	Hypothesis
3. p	Hypothesis
4. g	Disjunctive syllogism (2, 3)
5. v	Modus ponens (1, 4)

2. Let

$$p = \text{"I go to the beach."}$$
$$q = \text{"I study."}$$
$$r = \text{"I work as a waiter."}$$

Begin by assuming the negation of the conclusion.

1. $\sim r$	Negation of conclusion
2. $p \rightarrow \sim q$	Hypothesis
3. $q \vee r$	Hypothesis
4. q	Disjunctive syllogism (1, 3)
5. $\sim p$	Modus tollens (2, 4)

Since one of the hypotheses has been shown to be false, the argument is valid.

12.6 Predicate Calculus

In a previous discussion, we noted that a sentence of the form

"The number x is even"

is not a statement. Although the sentence is a declarative one, we cannot determine whether the statement is TRUE or FALSE because we do not know to what x refers. For example, if $x = 3$, then the statement is FALSE. If $x = 6$, then the statement is TRUE.

We define an *open sentence* $p(x)$ to be a declarative sentence that becomes a statement when x is given a particular value chosen from a universe of values. An open sentence is also known as a *predicate*.

Consider the open sentence $p(x) =$ "If x is even, $x - 6 > 8$." Let us consider as possible values for x all integers greater than zero, so the universe $U = \{1, 2, 3, 4, \dots\}$. Then the open sentence $p(x)$ represents many statements, one for each value of x chosen from U. Let us state $p(x)$ and record its truth value for several possible values of x. We note that the open sentence $p(x)$ becomes the statement $p(1)$ when we substitute the specific value 1 for the indeterminate letter x. Recall that an implication $p \rightarrow q$ is false if and only if p is true and q is false.

$p(1) =$ "If 1 is even, $-5 > 8$" is TRUE (since the hypothesis is FALSE).

$p(2) =$ "If 2 is even, $-4 > 8$" is FALSE.

$p(3) =$ "If 3 is even, $-3 > 8$" is TRUE.

$p(4) =$ "If 4 is even, $-2 > 8$" is FALSE.

$p(20) =$ "If 20 is even, $14 > 8$" is TRUE.

$p(21) =$ "If 21 is even, $15 > 8$" is TRUE.

In fact, $p(x)$ is TRUE for any odd integer x. And $p(x)$ is TRUE for all values of x such that x is even and $x > 14$. The only values of x for which $p(x)$ is FALSE are x even and between 2 and 14, inclusive. That is, $p(2)$, $p(4)$, $p(6)$, $p(8)$, $p(10)$, $p(12)$, and $p(14)$ are FALSE. All other values of x from U make $p(x)$ a TRUE statement.

EXAMPLE 1   **Analyzing predicates**   Let

$$p(x) = \text{"If } x > 4, \text{ then } x + 10 > 14\text{"}$$

be an open sentence. Let $x \in U$ (that is, x is an element of U), where $U = \{1, 2, 3, 4, \dots\}$. Find the truth value of each statement formed when these values are substituted for x in $p(x)$.

Solution $p(1)$ is TRUE because if $x = 1$, the hypothesis is FALSE.

$p(2)$ is TRUE because if $x = 2$, the hypothesis is FALSE.

$p(3)$ is TRUE because if $x = 3$, the hypothesis is FALSE.

$p(4)$ is TRUE because if $x = 4$, the hypothesis is FALSE.

$p(5)$ is TRUE because if $x = 5$, then the hypothesis is TRUE and $x + 10 > 14$.

In fact, $p(x)$ is TRUE for all values of $x \in U$. We say that "for all $x \in U$, if $x > 4$, then $x + 10 > 14$."

Now Try Exercise 1   ■

The statement

For all $x \in U$, $p(x)$

is symbolized by

$$\forall x \in U \ p(x).$$

The statement $\forall x \in U \ p(x)$ is TRUE if, and only if, $p(x)$ is TRUE for every $x \in U$.

We call the symbol $\forall$ the *universal quantifier* and we read it as "for all," "for every," or "for each." The universal set must be known in order to decide if $\forall x \in U \ p(x)$ is TRUE or FALSE. We emphasize that, whereas $p(x)$ is an open sentence with no assignable truth value, $\forall x \in U \ p(x)$ is a legitimate logical statement. At times, the notation will be abbreviated to $\forall x \, p(x)$ when the universe U is clear, or to $\forall x \, [p(x)]$ for further clarity.

From the example, we can see that the statement "For all $x \in U$, if $x > 4$, then $x + 10 > 14$" is a TRUE statement for $U = \{1, 2, 3, \dots\}$.

EXAMPLE 2 **Truth of a predicate in a fixed universe** Let $U = \{1, 2, 3, 4, 5, 6\}$. Determine the truth value of the statement

$$\forall x \, [(x - 4)(x - 8) > 0].$$

Solution We let $p(x)$ be the open statement "$(x - 4)(x - 8) > 0$." We consider the truth values of $p(1)$, $p(2)$, $p(3)$, $p(4)$, $p(5)$, and $p(6)$. We note that $p(1)$ is TRUE because $(1 - 4)(1 - 8) = (-3)(-7) = 21$ and $21 > 0$. We find that $p(2)$ and $p(3)$ are also TRUE. However, $p(4)$ is FALSE because $(4 - 4)(4 - 8) = 0$. We need not check any other values from U. Already, we know that

$$[\forall x \in U \ p(x)] \text{ is FALSE.}$$

We note that there are values of x in U for which $p(x)$ is TRUE.

Now Try Exercise 9(b) ■

The statement

There exists an x in U such that $p(x)$

is symbolized by

$$\exists x \in U \ p(x).$$

The statement $\exists x \in U \ p(x)$ is TRUE if, and only if, there is at least one element $x \in U$ such that $p(x)$ is TRUE.

The symbol $\exists$ is called the *existential quantifier*. We read the existential quantifier as "there exists x such that $p(x)$," "for some x, $p(x)$," or "there is some x for which $p(x)$." We may write $\exists x \, p(x)$ when the universal set is clear. When the universal set does not appear explicitly in the statement, U must be made clear in order for $\exists x \, p(x)$ to be a statement and to have a truth value.

EXAMPLE 3 **Universal and existential quantifiers** Determine the truth value of the following statements where $U = \{1, 2, 3, 4, 5, 6, 7, 8\}$.

(a) $\forall x \, (x + 3 < 15)$ (b) $\exists x \, (x > 5)$

(c) $\exists x \, (x^2 = 0)$ (d) $\forall x \, (0 < x < 10)$

(e) $\forall x \, [(x + 2 > 5) \vee (x \leq 3)]$ (f) $\exists x \, [(x - 1)(x + 2) = 0]$

Solution   (a) is TRUE because for every $x \in U$, $x + 3$ is less than 15. The statement in (b) is TRUE because, for example, $7 \in U$ and $7 > 5$. Statement (c) is FALSE because there is no $x \in U$ that satisfies the equation $x^2 = 0$. Statement (d) is TRUE. Statement (e) is TRUE because for each $x \in U$ either $x + 2 > 5$ or $x \leq 3$. Statement (f) is TRUE because $1 \in U$ and $(1 - 1)(1 + 2) = 0$.

Now Try Exercise 11   ■

Sometimes it is convenient to let U be a very large set—the real numbers, for example—and to restrict U within the statement itself. Consider the statement

For every positive integer x, either x is even or $x + 1$ is even.

One way to write this is to let U be the set of positive integers and to write

$$\forall x \in U \, [(x \text{ is even}) \vee (x + 1 \text{ is even})].$$

An equivalent statement that more explicitly shows the universal set would be

$$\forall x \in U \, [(x \text{ is a positive integer}) \rightarrow ((x \text{ is even}) \vee (x + 1 \text{ is even}))].$$

The universal set can then be the set of all integers. The universe is restricted by using an implication with the hypothesis that x is a positive integer in the open sentence $p(x)$. We can allow U to be any set containing the positive integers and restrict U by using such an implication.

A statement with the existential quantifier can be rephrased to include the universe in the statement. For example, let U be the set of positive integers, and consider the statement

$$\exists x \in U \, [(x - 3)(x + 6) = 0].$$

We can rewrite the statement as

$$\exists x \in U \, [(x \text{ is a positive integer}) \wedge ((x - 3)(x + 6) = 0)].$$

We could let U be the set of all integers or the reals, if we wish. Any set containing the positive integers would be an acceptable choice for U. The existential operator and the connective $\wedge$ can be used to restrict the universe to the positive integers.

We frequently have reason to negate a quantified statement. Of course, the negation changes the truth value from TRUE to FALSE or from FALSE to TRUE. The negations should be properly worded.

EXAMPLE 4   **Negation of a universal statement**   Negate the statement

"For every positive integer x, either x is even or $x + 1$ is even."

Solution   We could say, "It is not the case that for every positive integer x, either x is even or $x + 1$ is even." That is not entirely satisfactory, however. What we mean is that there exists a positive integer for which it is not the case that x is even or $x + 1$ is even. Assuming that U is the set of positive integers, this is written symbolically as

$$\sim[\forall x \, ((x \text{ is even}) \vee (x + 1 \text{ is even}))] \Leftrightarrow \exists x \, [\sim((x \text{ is even}) \vee (x + 1 \text{ is even}))]$$
$$\Leftrightarrow \exists x \, [(x \text{ is odd}) \wedge (x + 1 \text{ is odd})].$$

The last equivalence follows from De Morgan's laws, which allow us to rewrite a statement of the form $\sim(p \lor q)$ as $(\sim p \land \sim q)$. The negation of the original statement then is "There exists a positive integer x such that x is odd and $x + 1$ is odd." ■

EXAMPLE 5   **Negation of an existential statement**   Negate the statement "There exists a positive integer x such that $x^2 - x - 2 = 0$."

Solution   One way to negate the statement is to say, "It is not the case that there exists a positive integer x such that $x^2 - x - 2 = 0$." A better way is to say, "For all positive integers x, $x^2 - x - 2 \neq 0$." Note that

$$\sim[\exists x \ (x^2 - x - 2 = 0)] \Leftrightarrow [\forall x \ (x^2 - x - 2 \neq 0)].$$

Now Try Exercise 13

■

> We summarize the rules for the negation of quantified statements. The negation of a quantified statement can be rewritten by bringing the negation inside the open statement and replacing $\exists$ by $\forall$, or vice versa.
>
> $$\sim [\exists x \ p(x)] \Leftrightarrow [\forall x \ \sim p(x)]$$
> $$\sim [\forall x \ p(x)] \Leftrightarrow [\exists x \ \sim p(x)]$$
>
> These rules are also called De Morgan's laws.

EXAMPLE 6   **Negation of statements with quantifiers**   Write the negation of each of the following statements.

(a) All university students like football.

(b) There is a mathematics textbook that is both short and clear.

(c) For every positive integer x, if x is even, then $x + 1$ is odd.

Solution   (a) Here the universe is "university students" and $p(x)$ is the statement "Student x likes football." We negate $\forall x \ p(x)$ to get $\exists x \ [\sim p(x)]$, which in words is "There exists a university student who does not like football."

(b) The universe is the set of all mathematics textbooks. We have a statement of the form $\exists x \ [p(x) \land q(x)]$, where $p(x)$ denotes "The mathematics textbook is short" and $q(x)$ denotes "The mathematics textbook is clear." The negation of the statement is of the form $\sim[\exists x \ (p(x) \land q(x))] \Leftrightarrow [\forall x \ \sim(p(x) \land q(x))] \Leftrightarrow [\forall x \ (\sim p(x) \lor \sim q(x))]$. This translates into the English sentence "Every mathematics textbook is either not short or not clear."

(c) The universe is the set of positive integers. The statement is of the form

$$\forall x \ (p(x) \to q(x)),$$

where $p(x)$ is the statement "x is even" and $q(x)$ is "$x + 1$ is odd." We write the negation

$$\sim[\forall x \ (p(x) \to q(x))] \Leftrightarrow [\exists x \ \sim(p(x) \to q(x))] \Leftrightarrow [\exists x \ \sim(\sim p(x) \lor q(x))]$$
$$\Leftrightarrow [\exists x \ (p(x) \land \sim q(x))].$$

In English, this is "There exists a positive integer x such that x is even and $x + 1$ is also even." ■

We sometimes have occasion to use sentences that have two quantified variables. For example, the sentence "For every child there exists an adult who cares for him or her" has two variables: the child and the adult. We can write the statement symbolically using two variables x and y. We let the universe for the x variable be the set of all children and the universe for the y variable be the set of all adults. Then we let $p(x, y) =$ "y cares for x." The statement can be written symbolically as

$$\forall x \, \exists y \, p(x, y).$$

Binding each of the variables with the universal or existential quantifier and a universal set gives a legitimate logical statement.

EXAMPLE 7   **Predicates with two variables**   Let the universe for each of the variables be the set of all Americans. Let

$$p(x, y) = x \text{ is taller than } y$$
$$q(x, y) = x \text{ is heavier than } y.$$

Write the English sentences from the symbolic statements.

(a) $\forall x \, \forall y \, [p(x, y) \rightarrow q(x, y)]$

(b) $\forall x \, \exists y \, [p(x, y) \wedge q(x, y)]$

(c) $\forall y \, \exists x \, [p(x, y) \vee q(x, y)]$

(d) $\exists x \, \exists y \, [p(x, y) \wedge q(x, y)]$

(e) $\exists x \, \exists y \, [p(x, y) \wedge q(y, x)]$

Solution   (a) For all Americans x and y, if x is taller than y, then x is heavier than y.

(b) For every American x, there is an American y such that x is taller and heavier than y.

(c) For every American y, there is an American x such that x is taller than y or x is heavier than y.

(d) There are Americans x and y such that x is taller than y and x is heavier than y.

(e) There are Americans x and y such that x is taller than y and y is heavier than x.   ■

The rules for the negation of statements with more than one variable are just repeated applications of De Morgan's laws mentioned for single-variable statements. We have

$$\sim [\forall x \, \forall y \, p(x, y)] \Leftrightarrow \exists x \, [\sim \forall y \, p(x, y)] \Leftrightarrow \exists x \, \exists y \, [\sim p(x, y)]$$

and

$$\sim [\exists x \, \exists y \, p(x, y)] \Leftrightarrow \forall x \, [\sim \exists y \, p(x, y)] \Leftrightarrow \forall x \, \forall y \, [\sim p(x, y)].$$

We summarize these rules and several others in Table 1.

Notice that rule 7 is a logical implication and not a logical equivalence. Any other exchanges of $\exists$ and $\forall$ should be handled very carefully. They are unlikely to give equivalent statements even in specific cases.

TABLE 1

1. $\sim [\forall x \; \forall y \; p(x,y)] \Leftrightarrow \exists x \; \exists y \; [\sim p(x,y)]$
2. $\sim [\forall x \; \exists y \; p(x,y)] \Leftrightarrow \exists x \; \forall y \; [\sim p(x,y)]$
3. $\sim [\exists x \; \forall y \; p(x,y)] \Leftrightarrow \forall x \; \exists y \; [\sim p(x,y)]$
4. $\sim [\exists x \; \exists y \; p(x,y)] \Leftrightarrow \forall x \; \forall y \; [\sim p(x,y)]$
   $\left.\vphantom{\begin{array}{c}1\\2\\3\\4\end{array}}\right\}$ De Morgan's laws
5. $\forall x \; \forall y \; p(x,y) \Leftrightarrow \forall y \; \forall x \; p(x,y)$
6. $\exists x \; \exists y \; p(x,y) \Leftrightarrow \exists y \; \exists x \; p(x,y)$
7. $\exists x \; \forall y \; p(x,y) \Rightarrow \forall y \; \exists x \; p(x,y)$

EXAMPLE 8 **English translation and truth value for predicates with two variables** Let the universe for both variables be the nonnegative integers $0, 1, 2, 3, 4, \ldots$. Write out the statements in words and decide on the truth value of each.

(a) $\forall x \; \forall y \; [2x = y]$ (b) $\forall x \; \exists y \; [2x = y]$

(c) $\exists x \; \forall y \; [2x = y]$ (d) $\exists x \; \forall y \; [2y = x]$

(e) $\exists y \; \forall x \; [2x = y]$ (f) $\exists x \; \exists y \; [2x = y]$

(g) $\forall y \; \exists x \; [2x = y]$

Solution (a) For every pair of nonnegative integers x and y, $2x = y$. This is FALSE because, for example, it is FALSE in the case $x = 3$, $y = 10$.

(b) For every nonnegative integer x, there is a nonnegative integer y such that $2x = y$. This is TRUE because, once having chosen any nonnegative number x, we can let y be the double of x.

(c) There exists a nonnegative integer x such that, for every nonnegative integer y, $2x = y$. This statement is FALSE because, no matter what x we choose, $2x = y$ is TRUE only for y equal to twice x and for no other nonnegative number.

(d) There is a nonnegative number x such that, for all nonnegative numbers y, $2y = x$. This is FALSE for similar reasons as in (c).

(e) There exists a nonnegative integer y such that, for all nonnegative integers x, $2x = y$. The statement is FALSE. For it to be TRUE, we would need to find a specific value of y that can be fixed and for which, no matter what nonnegative integer x we choose, $2x = y$. This is the same as (d).

(f) There exist nonnegative integers x and y such that $2x = y$. This is TRUE. One choice that makes the statement $2x = y$ TRUE is $x = 4$ and $y = 8$.

(g) For every nonnegative integer y, there exists a nonnegative integer x such that $2x = y$. This is FALSE because if $y = 5$ there exists no nonnegative integer x such that $2x = 5$.

Now Try Exercise 17 ■

A statement of the form $\forall x \; p(x)$ is FALSE if $\exists x \; [\sim p(x)]$ is TRUE. If we want to show that the statement $\forall x \; p(x)$ is FALSE, we need only find a value for x for which $p(x)$ is FALSE. Such an example is called a *counterexample*. We used such a counterexample in parts (a) and (g) of Example 8.

Analogy between Sets and Statements Actually, the symbols and the truth values in the tables are suggestive of set theory definitions introduced in Chapter 5. Recall that we use U to denote the universe and the letters S and T to denote sets in that universe. Then the union of the sets S and T is defined as

$$S \cup T = \{x \in U : x \in S \text{ or } x \in T\}.$$

If we allow $p(x)$ to represent "$x \in S$" and $q(x)$ to represent "$x \in T$," then it is TRUE that $x \in S \cup T$ if, and only if, the statement $p(x) \vee q(x)$ is TRUE. Thus

$$\forall x \ [x \in S \cup T \leftrightarrow p(x) \vee q(x)].$$

Similarly, we recall the definition of the intersection of two sets:

$$S \cap T = \{x \in U : x \in S \text{ and } x \in T\}.$$

It is TRUE that $x \in S \cap T$ if, and only if, the statement $p(x) \wedge q(x)$ is TRUE, where $p(x)$ and $q(x)$ are as above. Thus

$$\forall x \ [x \in S \cap T \leftrightarrow p(x) \wedge q(x)].$$

Using the symbol $\notin$ for "is not an element of," in addition, we have

$$S' = \{x \in U : x \notin S\}.$$

Thus $x \in S'$ if and only if $\sim p(x)$ is TRUE. Hence,

$$\forall x \ [x \in S' \leftrightarrow \sim p(x)].$$

The exclusive "or" corresponds to the *symmetric difference of S and T* defined by

$$S \oplus T = \{x \in U : x \in S \cup T \quad \text{but} \quad x \notin S \cap T\}.$$

That is, it is TRUE that $x \in S \oplus T$ if and only if the statement $p(x) \oplus q(x)$ is TRUE or

$$\forall x \ [x \in S \oplus T \leftrightarrow p(x) \oplus q(x)].$$

As with the connectives $\vee$ and $\wedge$, the connectives $\rightarrow$ and $\leftrightarrow$ have counterparts in set theory. We will assume that the universe is U and that S and T are subsets of U. We let $p(x)$ be the statement "$x \in S$" and let $q(x)$ be the statement "$x \in T$." Then we have seen that S is a subset of T if, and only if, every element of S is also an element of T. We say that

S is a subset of T if and only if $(x \in S) \rightarrow (x \in T)$ is TRUE for all x,

or, equivalently,

S is a subset of T if and only if $\forall x \ [p(x) \rightarrow q(x)]$ is TRUE.

Two sets are equal if they have the same elements. We claim that

$S = T$ if and only if $(x \in S) \leftrightarrow (x \in T)$ is TRUE for all x,

that is,

$S = T$ if and only if $\forall x \ [p(x) \leftrightarrow q(x)]$ is TRUE.

EXAMPLE 9 **Sets and statements** Let $U = \{1, 2, 3, 4, 5, 6\}$, let $S = \{x \in U : x \leq 3\}$, and let $T = \{x \in U : x \text{ divides } 6\}$. Show $S \subseteq T$.

SOLUTIONS TO PRACTICE PROBLEMS 12.6 (CONTINUED)

3. (a) For all nonnegative integers x and y, $x < y$. FALSE.

(b) For every nonnegative integer x, there exists a nonnegative integer y such that $x < y$. TRUE. For any nonnegative integer x, we can let $y = x + 1$.

(c) For every nonnegative integer y, there exists a nonnegative integer x such that $x < y$. FALSE. A counterexample is found by letting $y = 0$. Then there is no nonnegative integer x such that $x < y$.

(d) There exists a nonnegative integer x such that, for all nonnegative integers y, $x < y$. FALSE. No matter what x is selected, letting $y = 0$ gives a counterexample.

(e) There exist nonnegative integers x and y such that $x < y$. TRUE. Just let $x = 3$ and $y = 79$.

(f) For every pair of nonnegative integers x and y, either $x < y$ or $y < x$. FALSE. Consider $x = y$; say, $x = 4$ and $y = 4$.

12.7 Logic Circuits

Logic circuits are ubiquitous in the devices that we use every day. We find them in computers, telephones, digital clocks, and television sets. These are electrical circuits that are designed to control current flow depending on the input activity. Current flow passes through gates to an output line. The gates operate according to strict rules that we will discuss. There are several simple types of circuits from which more complex circuits can be developed.

The simplest form of circuit has one input line with current input that is either ON (1) or OFF (0). This gate performs the function of negating the input. The output is the inverse of the input, so that if the input is ON (1), the output is OFF (0) and vice versa. In the language of logic, we call this a *NOT gate*. In the language of logic circuits, the gate is called an *inverter*.

There are other basic gates, each with two inputs. The first is the *AND gate*, which behaves exactly as the logical connective AND ($\wedge$). Current flows through the gate from two inputs if, and only if, both input currents are ON. Thus, the output is ON (1) if, and only if, both inputs are ON.

The *OR gate* behaves exactly as the logical connective OR ($\vee$). Current flows through the OR gate if, and only if, at least one of the input currents is ON. Thus, the output of the OR gate is ON if, and only if, at least one of the inputs is ON.

Engineers and designers find it helpful to draw diagrams of the circuits to reflect the connective with an image of a gate-type device. These are pictured below in Fig. 1.

Figure 1

The output from the various inputs is reflected by Tables 1 through 3.

TABLE 1	The NOT Gate
Input	**Output**
p	$\sim p$
0	1
1	0

TABLE 2		The AND Gate
Input		**Output**
p	q	$p \wedge q$
0	0	0
0	1	0
1	0	0
1	1	1

TABLE 3		The OR Gate
Input		**Output**
p	q	$p \vee q$
0	0	0
0	1	1
1	0	1
1	1	1

Solution We must show that $\forall x \, [(x \in S) \rightarrow (x \in T)]$ is TRUE. An implication is FALSE only in the case where the hypothesis is TRUE and the conclusion is FALSE. Let us suppose then that the hypothesis is TRUE and show that the conclusion must also be TRUE in that case. We assume that $x \in S$. By the definition of the set S, this means that x is 1 or 2 or 3. Then $x \in T$, since in each of the cases x divides 6 and is therefore an element of T. Hence, $(x \in S) \rightarrow (x \in T)$ cannot be FALSE and must be TRUE for all $x \in U$.

Now Try Exercise 21 ∎

Recall that we have already seen De Morgan's laws in another form in Chapter 5. In that case, we assumed that S and T were sets and showed that

$$(S \cup T)' = S' \cap T' \quad \text{and} \quad (S \cap T)' = S' \cup T'.$$

We know that $x \in S \cup T$ if, and only if, $(x \in S) \vee (x \in T)$ is TRUE. So $x \notin S \cup T$ if, and only if, the negation of the statement is TRUE. Thus

$$x \in (S \cup T)' \text{ if, and only if, } \sim[(x \in S) \vee (x \in T)] \text{ is TRUE.}$$

By De Morgan's law in Table 1 of Section 12.4, we know that

$$x \in (S \cup T)' \text{ if, and only if, } \sim(x \in S) \wedge \sim(x \in T) \text{ is TRUE.}$$

Thus

$$x \in (S \cup T)' \text{ if, and only if, } (x \notin S) \wedge (x \notin T) \text{ is TRUE.}$$

This means that $x \in S' \cap T'$. The De Morgan laws for unions and intersections of sets are closely related to the De Morgan laws for the connectives $\vee$ and $\wedge$.

Colloquial Usage Again, we point out that colloquial usage is frequently not precise. For example, the statement "All students in finite mathematics courses do not fail" is strictly interpreted to mean that $\forall x \in U \, [x \text{ does not fail}]$ with U = the set of students in finite mathematics courses. This means, of course, that no one fails. On the other hand, some might loosely interpret this to mean that not every student in finite mathematics courses fails. This is the statement $\exists x \in U \, [x \text{ does not fail}]$. We require precision of language in mathematics and do *not* accept the second interpretation as correct.

PRACTICE PROBLEMS 12.6

1. Write the following statement symbolically and write out its negation in English: "There exists a flower that can grow in sand and is not subject to mold."

2. For the universe $U = \{0, 1, 2, 3, 4, 5, 6, 7, 8\}$, determine the truth values of the following statements.

(a) $\forall x \, (x > 2)$ **(b)** $\exists x \, (x > 2)$

(c) $\forall x \, (x^2 < 100)$

(d) $\exists x \, [(x - 1 = 4) \wedge (3x + 5 = 20)]$

3. Consider the universe for both variables x and y to be the set of nonnegative integers $\{0, 1, 2, 3, 4, \ldots\}$. Write out the statements in English and determine the truth value of each.

(a) $\forall x \, \forall y \, [x < y]$ **(b)** $\forall x \, \exists y \, [x < y]$

(c) $\forall y \, \exists x \, [x < y]$ **(d)** $\exists x \, \forall y \, [x < y]$

(e) $\exists x \, \exists y \, [x < y]$ **(f)** $\forall x \, \forall y \, [(x < y) \vee (y < x)]$

EXERCISES 12.6

1. Let $U = \{1, 2, 3, 4, 5, 6\}$. Determine the truth value of

$$p(x) = [(x \text{ is even}) \text{ or } (x \text{ is divisible by } 3)]$$

for the given values of x.

(a) $x = 1$ (b) $x = 4$ (c) $x = 3$

(d) $x = 6$ (e) $x = 5$

2. Determine the truth value of $p(x)$ for the values of x chosen from the universe of all letters of the alphabet where

$$p(x) = [(x \text{ is a vowel}) \text{ and } (x \text{ is in the word ABLE})].$$

(a) $x = a$ (b) $x = d$

(c) $x = b$ (d) $x = i$

3. Consider the universe of all college students. Let $p(x)$ denote the open statement "x takes a writing course."

(a) Write the statement "Every college student takes a writing course" in symbols.

(b) Write the statement "Not all college students take a writing course" symbolically.

(c) Write the statement "Every college student does not take a writing course" symbolically.

(d) Do any of the statements in (a), (b), or (c) imply one another? Explain.

4. Recently, as the Amtrak train pulled into the Baltimore station, the conductor announced, "All doors do not open." Since passengers were permitted to get off at the station, what do you think the conductor meant to say?

5. An alert California teacher chided "Dear Abby" (*Baltimore Sun*, March 1, 1989) for her statement "Confidential to Eunice: All men do not cheat on their wives." Let $p(x)$ be the statement "x cheats on his wife" in the universe of all men. Write out Abby's statement symbolically. Rewrite the statement using the existential quantifier. Do you think the statement is TRUE or FALSE? What do you think Abby really meant to say?

6. Consider the universe of all orange juice. Write the following symbolically, using $p(x) = $ "x comes from Florida."

(a) All orange juice does not come from Florida.

(b) Some orange juice comes from Florida.

(c) Not all orange juice comes from Florida.

(d) Some orange juice does not come from Florida.

(e) No orange juice comes from Florida.

(f) Are any of the statements (a) through (e) equivalent to one another? Explain.

7. Let the universe be all university professors. Let $p(x)$ be the open statement "x likes poetry." Write the following statements symbolically.

(a) All university professors like poetry.

(b) Some university professors do not like poetry.

(c) Some university professors like poetry.

(d) Not all university professors like poetry.

(e) All university professors do not like poetry.

(f) No university professors like poetry.

(g) Are any of the statements in (a) through (f) equivalent? Explain.

8. Let $U = \{0, 1, 2, 3, 4\}$. Let

$$p(x) = [x^2 > 9].$$

Find the truth value of

(a) $\exists x \, p(x)$ (b) $\forall x \, p(x)$

9. Let $U = \{1, 2, 3, 4, 5, 6, 7, 8, 9\}$. Let

$$p(x) = \left[(x \text{ is prime}) \rightarrow (x^2 + 1 \text{ is even})\right].$$

Find the truth value of

(a) $\exists x \, p(x)$ (b) $\forall x \, p(x)$

10. Let $U = \{3, 4, 5, 6, \ldots\}$. Let

$$p(x) = \left[(x \text{ is prime}) \rightarrow (x^2 + 1 \text{ is even})\right].$$

Find the truth value of

(a) $\exists x \, p(x)$ (b) $\forall x \, p(x)$

11. Let the universe consist of all nonnegative integers. Let $p(x)$ be the statement "x is even." Let $q(x)$ be the statement "x is odd." Determine the truth value of

(a) $\forall x \, [p(x) \vee q(x)]$ (b) $[\forall x \, p(x)] \vee [\forall x \, q(x)]$

(c) $\exists x \, [p(x) \vee q(x)]$ (d) $\exists x \, [p(x) \wedge q(x)]$

(e) $\forall x \, [p(x) \wedge q(x)]$ (f) $[\exists x \, p(x)] \wedge [\exists x \, q(x)]$

(g) $\forall x \, [p(x) \rightarrow q(x)]$ (h) $[\forall x \, p(x)] \rightarrow [\forall x \, q(x)]$

12. Let the universe consist of all real numbers. Let

$$p(x) = (x \text{ is positive})$$

and

$$q(x) = (x \text{ is a perfect square}).$$

Determine the truth value of

(a) $\forall x \, p(x)$ (b) $\forall x \, q(x)$

(c) $\exists x \, p(x)$ (d) $\exists x \, q(x)$

(e) $\exists x \, [p(x) \rightarrow q(x)]$ (f) $\exists x \, [q(x) \rightarrow p(x)]$

(g) $\forall x \, [p(x) \rightarrow q(x)]$ (h) $\forall x \, [q(x) \rightarrow p(x)]$

(i) $\forall x \, p(x) \rightarrow \exists x \, q(x)$ (j) $[\forall x \, p(x)] \rightarrow [\forall x \, q(x)]$

13. Negate the following statements.

(a) Every dog has its day.

(b) Some men fight wars.

(c) All mothers are married.

(d) For every pot, there is a cover.

(e) Not all children have pets.

(f) Not every month has 30 days.

14. Negate each statement by changing existential quantifiers to universal quantifiers, or vice versa.

(a) Every stitch saves time.

(b) All books have hard covers.

(c) Some children are afraid of snakes.

(d) Not all computers have a hard disk.

(e) Some chairs do not have arms.

15. Consider the universe of nonnegative integers $= \{0, 1, 2, 3, 4, \ldots\}$. Write the English sentence for each symbolic statement. Determine the truth value of each statement. If the statement is FALSE, give a counterexample and write its negation out in words.

(a) $\forall x \forall y \, [x + y > 12]$ (b) $\forall x \exists y \, [x + y > 12]$

(c) $\exists x \forall y \, [x + y > 12]$ (d) $\exists x \exists y \, [x + y > 12]$

16. Consider the universe of all subsets of the set $A = \{a, b, c\}$. Let the variables x and y denote subsets of A. Find the truth value of each of the following statements and explain your answer. If the statement is FALSE, give a counterexample.

(a) $\forall x \forall y \, [x \subseteq y]$

(b) $\forall x \forall y \, [(x \subseteq y) \vee (y \subseteq x)]$

(c) $\exists x \forall y \, [x \subseteq y]$

(d) $\forall x \exists y \, [x \subseteq y]$

17. Let the universe for both variables x and y be the set $\{1, 2, 3, 4, 5, 6\}$. Let $p(x, y) = $ "x divides y." Give the truth values of each of the following statements; explain your answer, and give a counterexample in the case that the statement is FALSE.

(a) $\forall x \forall y \, p(x, y)$ (b) $\forall x \exists y \, p(x, y)$

(c) $\exists x \forall y \, p(x, y)$ (d) $\exists y \forall x \, p(x, y)$

(e) $\forall y \exists x \, p(x, y)$ (f) $\forall x \, p(x, x)$

18. If $p(x)$ denotes "$x \in S$" and $q(x)$ denotes "$x \in T$," describe the following using logical statement forms with $p(x)$ and $q(x)$.

(a) $x \in S' \cup T$ (b) $x \in S \oplus T'$

(c) $x \in S' \cap T'$ (d) $x \in (S \cup T)'$

19. Let the universal set be

$$U = \{0, 1, 2, 3, 4, 5, 6, 7, 8, 9, 10, 11, 12, 13\},$$

let $S = \{x \in U : x \geq 8\}$, and let $T = \{x \in U : x \leq 10\}$.

(a) What implication must be TRUE if, and only if, S is a subset of T?

(b) Is S a subset of T? Explain.

20. Is S a subset of T? Let

$$U = \{1, 2, 3, 4, 5, 6, 7, 8, 9, 10, 11, 12\}$$
$$S = \{x \in U : x \text{ divides } 12 \text{ evenly}\}$$
$$T = \{x \in U : x \text{ is a multiple of } 2\}.$$

21. Prove that S is a subset of T, where

$$U = \{1, 2, 3, 4, 5, 6, 7, 8, 9\}$$
$$S = \{x \in U : x \text{ is a multiple of } 2\}$$
$$T = \{x \in U : x \text{ divides } 24 \text{ evenly}\}.$$

22. Prove that S is a subset of T, where

$$U = \{a, b, c, d, e, f, g, h\}$$
$$S = \{x \in U : x \text{ is a letter in } bad\}$$
$$T = \{x \in U : x \text{ is a letter in } badge\}.$$

23. Prove that $S = T$, where

$$U = \{1, 2, 3, 4, 5, 6\}$$
$$S = \{x \in U : x \text{ is a solution to } (x - 8)(x - 3) = 0\}$$
$$T = \{x \in U : x \text{ is a solution to } x^2 = 9\}.$$

SOLUTIONS TO PRACTICE PROBLEMS 12.6

1. Let the universe be the collection of all flowers. Let $p(x)$ be the open sentence "x can grow in sand" and let $q(x)$ be "x is subject to mold." The statement is then

$$\exists x \, [p(x) \wedge \sim q(x)].$$

The negation is

$$\forall x \, [\sim p(x) \vee q(x)].$$

This can be stated in English as

"Every flower cannot grow in sand or is subject to mold."

2. (a) FALSE. As a counterexample, consider $x = 2$.

(b) TRUE. Consider $x = 3$.

(c) TRUE. In fact, for every $x \in U$, $x^2 \leq 64$.

(d) TRUE. Consider $x = 5$.

Solution We must show that $\forall x \ [(x \in S) \to (x \in T)]$ is TRUE. An implication is FALSE only in the case where the hypothesis is TRUE and the conclusion is FALSE. Let us suppose then that the hypothesis is TRUE and show that the conclusion must also be TRUE in that case. We assume that $x \in S$. By the definition of the set S, this means that x is 1 or 2 or 3. Then $x \in T$, since in each of the cases x divides 6 and is therefore an element of T. Hence, $(x \in S) \to (x \in T)$ cannot be FALSE and must be TRUE for all $x \in U$.

Now Try Exercise 21   ■

Recall that we have already seen De Morgan's laws in another form in Chapter 5. In that case, we assumed that S and T were sets and showed that

$$(S \cup T)' = S' \cap T' \quad \text{and} \quad (S \cap T)' = S' \cup T'.$$

We know that $x \in S \cup T$ if, and only if, $(x \in S) \vee (x \in T)$ is TRUE. So $x \notin S \cup T$ if, and only if, the negation of the statement is TRUE. Thus

$$x \in (S \cup T)' \text{ if, and only if, } \sim[(x \in S) \vee (x \in T)] \text{ is TRUE.}$$

By De Morgan's law in Table 1 of Section 12.4, we know that

$$x \in (S \cup T)' \text{ if, and only if, } \sim(x \in S) \wedge \sim(x \in T) \text{ is TRUE.}$$

Thus

$$x \in (S \cup T)' \text{ if, and only if, } (x \notin S) \wedge (x \notin T) \text{ is TRUE.}$$

This means that $x \in S' \cap T'$. The De Morgan laws for unions and intersections of sets are closely related to the De Morgan laws for the connectives $\vee$ and $\wedge$.

Colloquial Usage Again, we point out that colloquial usage is frequently not precise. For example, the statement "All students in finite mathematics courses do not fail" is strictly interpreted to mean that $\forall x \in U$ [x does not fail] with $U =$ the set of students in finite mathematics courses. This means, of course, that no one fails. On the other hand, some might loosely interpret this to mean that not every student in finite mathematics courses fails. This is the statement $\exists x \in U$ [x does not fail]. We require precision of language in mathematics and do *not* accept the second interpretation as correct.

PRACTICE PROBLEMS 12.6

1. Write the following statement symbolically and write out its negation in English: "There exists a flower that can grow in sand and is not subject to mold."

2. For the universe $U = \{0, 1, 2, 3, 4, 5, 6, 7, 8\}$, determine the truth values of the following statements.

   (a) $\forall x \ (x > 2)$ (b) $\exists x \ (x > 2)$

   (c) $\forall x \ (x^2 < 100)$

   (d) $\exists x \ [(x - 1 = 4) \wedge (3x + 5 = 20)]$

3. Consider the universe for both variables x and y to be the set of nonnegative integers $\{0, 1, 2, 3, 4, \ldots\}$. Write out the statements in English and determine the truth value of each.

   (a) $\forall x \ \forall y \ [x < y]$ (b) $\forall x \ \exists y \ [x < y]$

   (c) $\forall y \ \exists x \ [x < y]$ (d) $\exists x \ \forall y \ [x < y]$

   (e) $\exists x \ \exists y \ [x < y]$ (f) $\forall x \ \forall y \ [(x < y) \vee (y < x)]$

EXERCISES 12.6

1. Let $U = \{1, 2, 3, 4, 5, 6\}$. Determine the truth value of

$$p(x) = [(x \text{ is even}) \text{ or } (x \text{ is divisible by } 3)]$$

for the given values of x.

(a) $x = 1$ (b) $x = 4$ (c) $x = 3$

(d) $x = 6$ (e) $x = 5$

2. Determine the truth value of $p(x)$ for the values of x chosen from the universe of all letters of the alphabet where

$$p(x) = [(x \text{ is a vowel}) \text{ and } (x \text{ is in the word ABLE})].$$

(a) $x = a$ (b) $x = d$

(c) $x = b$ (d) $x = i$

3. Consider the universe of all college students. Let $p(x)$ denote the open statement "x takes a writing course."

(a) Write the statement "Every college student takes a writing course" in symbols.

(b) Write the statement "Not all college students take a writing course" symbolically.

(c) Write the statement "Every college student does not take a writing course" symbolically.

(d) Do any of the statements in (a), (b), or (c) imply one another? Explain.

4. Recently, as the Amtrak train pulled into the Baltimore station, the conductor announced, "All doors do not open." Since passengers were permitted to get off at the station, what do you think the conductor meant to say?

5. An alert California teacher chided "Dear Abby" (*Baltimore Sun*, March 1, 1989) for her statement "Confidential to Eunice: All men do not cheat on their wives." Let $p(x)$ be the statement "x cheats on his wife" in the universe of all men. Write out Abby's statement symbolically. Rewrite the statement using the existential quantifier. Do you think the statement is TRUE or FALSE? What do you think Abby really meant to say?

6. Consider the universe of all orange juice. Write the following symbolically, using $p(x) = $ "x comes from Florida."

(a) All orange juice does not come from Florida.

(b) Some orange juice comes from Florida.

(c) Not all orange juice comes from Florida.

(d) Some orange juice does not come from Florida.

(e) No orange juice comes from Florida.

(f) Are any of the statements (a) through (e) equivalent to one another? Explain.

7. Let the universe be all university professors. Let $p(x)$ be the open statement "x likes poetry." Write the following statements symbolically.

(a) All university professors like poetry.

(b) Some university professors do not like poetry.

(c) Some university professors like poetry.

(d) Not all university professors like poetry.

(e) All university professors do not like poetry.

(f) No university professors like poetry.

(g) Are any of the statements in (a) through (f) equivalent? Explain.

8. Let $U = \{0, 1, 2, 3, 4\}$. Let

$$p(x) = [x^2 > 9].$$

Find the truth value of

(a) $\exists x\, p(x)$ (b) $\forall x\, p(x)$

9. Let $U = \{1, 2, 3, 4, 5, 6, 7, 8, 9\}$. Let

$$p(x) = \left[(x \text{ is prime}) \rightarrow (x^2 + 1 \text{ is even})\right].$$

Find the truth value of

(a) $\exists x\, p(x)$ (b) $\forall x\, p(x)$

10. Let $U = \{3, 4, 5, 6, \dots\}$. Let

$$p(x) = \left[(x \text{ is prime}) \rightarrow (x^2 + 1 \text{ is even})\right].$$

Find the truth value of

(a) $\exists x\, p(x)$ (b) $\forall x\, p(x)$

11. Let the universe consist of all nonnegative integers. Let $p(x)$ be the statement "x is even." Let $q(x)$ be the statement "x is odd." Determine the truth value of

(a) $\forall x\, [p(x) \vee q(x)]$ (b) $[\forall x\, p(x)] \vee [\forall x\, q(x)]$

(c) $\exists x\, [p(x) \vee q(x)]$ (d) $\exists x\, [p(x) \wedge q(x)]$

(e) $\forall x\, [p(x) \wedge q(x)]$ (f) $[\exists x\, p(x)] \wedge [\exists x\, q(x)]$

(g) $\forall x\, [p(x) \rightarrow q(x)]$ (h) $[\forall x\, p(x)] \rightarrow [\forall x\, q(x)]$

12. Let the universe consist of all real numbers. Let

$$p(x) = (x \text{ is positive})$$

and

$$q(x) = (x \text{ is a perfect square}).$$

Determine the truth value of

(a) $\forall x\, p(x)$ (b) $\forall x\, q(x)$

(c) $\exists x\, p(x)$ (d) $\exists x\, q(x)$

(e) $\exists x\, [p(x) \rightarrow q(x)]$ (f) $\exists x\, [q(x) \rightarrow p(x)]$

(g) $\forall x\, [p(x) \rightarrow q(x)]$ (h) $\forall x\, [q(x) \rightarrow p(x)]$

(i) $\forall x\, p(x) \rightarrow \exists x\, q(x)$ (j) $[\forall x\, p(x)] \rightarrow [\forall x\, q(x)]$

13. Negate the following statements.

(a) Every dog has its day.

(b) Some men fight wars.

(c) All mothers are married.

(d) For every pot, there is a cover.

(e) Not all children have pets.

(f) Not every month has 30 days.

14. Negate each statement by changing existential quantifiers to universal quantifiers, or vice versa.

(a) Every stitch saves time.

(b) All books have hard covers.

(c) Some children are afraid of snakes.

(d) Not all computers have a hard disk.

(e) Some chairs do not have arms.

15. Consider the universe of nonnegative integers $= \{0, 1, 2, 3, 4, \ldots\}$. Write the English sentence for each symbolic statement. Determine the truth value of each statement. If the statement is FALSE, give a counterexample and write its negation out in words.

(a) $\forall x \, \forall y \, [x + y > 12]$ **(b)** $\forall x \, \exists y \, [x + y > 12]$

(c) $\exists x \, \forall y \, [x + y > 12]$ **(d)** $\exists x \, \exists y \, [x + y > 12]$

16. Consider the universe of all subsets of the set $A = \{a, b, c\}$. Let the variables x and y denote subsets of A. Find the truth value of each of the following statements and explain your answer. If the statement is FALSE, give a counterexample.

(a) $\forall x \, \forall y \, [x \subseteq y]$

(b) $\forall x \, \forall y \, [(x \subseteq y) \vee (y \subseteq x)]$

(c) $\exists x \, \forall y \, [x \subseteq y]$

(d) $\forall x \, \exists y \, [x \subseteq y]$

17. Let the universe for both variables x and y be the set $\{1, 2, 3, 4, 5, 6\}$. Let $p(x, y) =$ "x divides y." Give the truth values of each of the following statements; explain your answer, and give a counterexample in the case that the statement is FALSE.

(a) $\forall x \, \forall y \, p(x, y)$ **(b)** $\forall x \, \exists y \, p(x, y)$

(c) $\exists x \, \forall y \, p(x, y)$ **(d)** $\exists y \, \forall x \, p(x, y)$

(e) $\forall y \, \exists x \, p(x, y)$ **(f)** $\forall x \, p(x, x)$

18. If $p(x)$ denotes "$x \in S$" and $q(x)$ denotes "$x \in T$," describe the following using logical statement forms with $p(x)$ and $q(x)$.

(a) $x \in S' \cup T$ **(b)** $x \in S \oplus T'$

(c) $x \in S' \cap T'$ **(d)** $x \in (S \cup T)'$

19. Let the universal set be

$$U = \{0, 1, 2, 3, 4, 5, 6, 7, 8, 9, 10, 11, 12, 13\},$$

let $S = \{x \in U : x \geq 8\}$, and let $T = \{x \in U : x \leq 10\}$.

(a) What implication must be TRUE if, and only if, S is a subset of T?

(b) Is S a subset of T? Explain.

20. Is S a subset of T? Let

$$U = \{1, 2, 3, 4, 5, 6, 7, 8, 9, 10, 11, 12\}$$
$$S = \{x \in U : x \text{ divides } 12 \text{ evenly}\}$$
$$T = \{x \in U : x \text{ is a multiple of } 2\}.$$

21. Prove that S is a subset of T, where

$$U = \{1, 2, 3, 4, 5, 6, 7, 8, 9\}$$
$$S = \{x \in U : x \text{ is a multiple of } 2\}$$
$$T = \{x \in U : x \text{ divides } 24 \text{ evenly}\}.$$

22. Prove that S is a subset of T, where

$$U = \{a, b, c, d, e, f, g, h\}$$
$$S = \{x \in U : x \text{ is a letter in } bad\}$$
$$T = \{x \in U : x \text{ is a letter in } badge\}.$$

23. Prove that $S = T$, where

$$U = \{1, 2, 3, 4, 5, 6\}$$
$$S = \{x \in U : x \text{ is a solution to } (x - 8)(x - 3) = 0\}$$
$$T = \{x \in U : x \text{ is a solution to } x^2 = 9\}.$$

SOLUTIONS TO PRACTICE PROBLEMS 12.6

1. Let the universe be the collection of all flowers. Let $p(x)$ be the open sentence "x can grow in sand" and let $q(x)$ be "x is subject to mold." The statement is then

$$\exists x \, [p(x) \wedge {\sim} q(x)].$$

The negation is

$$\forall x \, [{\sim} p(x) \vee q(x)].$$

This can be stated in English as

"Every flower cannot grow in sand or is subject to mold."

2. (a) FALSE. As a counterexample, consider $x = 2$.

(b) TRUE. Consider $x = 3$.

(c) TRUE. In fact, for every $x \in U$, $x^2 \leq 64$.

(d) TRUE. Consider $x = 5$.

SOLUTIONS TO PRACTICE PROBLEMS 12.6 (CONTINUED)

3. (a) For all nonnegative integers x and y, $x < y$. FALSE.

(b) For every nonnegative integer x, there exists a nonnegative integer y such that $x < y$. TRUE. For any nonnegative integer x, we can let $y = x + 1$.

(c) For every nonnegative integer y, there exists a nonnegative integer x such that $x < y$. FALSE. A counterexample is found by letting $y = 0$. Then there is no nonnegative integer x such that $x < y$.

(d) There exists a nonnegative integer x such that, for all nonnegative integers y, $x < y$. FALSE. No matter what x is selected, letting $y = 0$ gives a counterexample.

(e) There exist nonnegative integers x and y such that $x < y$. TRUE. Just let $x = 3$ and $y = 79$.

(f) For every pair of nonnegative integers x and y, either $x < y$ or $y < x$. FALSE. Consider $x = y$; say, $x = 4$ and $y = 4$.

12.7 Logic Circuits

Logic circuits are ubiquitous in the devices that we use every day. We find them in computers, telephones, digital clocks, and television sets. These are electrical circuits that are designed to control current flow depending on the input activity. Current flow passes through gates to an output line. The gates operate according to strict rules that we will discuss. There are several simple types of circuits from which more complex circuits can be developed.

The simplest form of circuit has one input line with current input that is either ON (1) or OFF (0). This gate performs the function of negating the input. The output is the inverse of the input, so that if the input is ON (1), the output is OFF (0) and vice versa. In the language of logic, we call this a *NOT gate*. In the language of logic circuits, the gate is called an *inverter*.

There are other basic gates, each with two inputs. The first is the *AND gate*, which behaves exactly as the logical connective AND ($\wedge$). Current flows through the gate from two inputs if, and only if, both input currents are ON. Thus, the output is ON (1) if, and only if, both inputs are ON.

The *OR gate* behaves exactly as the logical connective OR ($\vee$). Current flows through the OR gate if, and only if, at least one of the input currents is ON. Thus, the output of the OR gate is ON if, and only if, at least one of the inputs is ON.

Engineers and designers find it helpful to draw diagrams of the circuits to reflect the connective with an image of a gate-type device. These are pictured below in Fig. 1.

Figure 1

$$p \;-\!\!\!\triangleright\!\!\!\text{NOT} \square\!\!-\; \sim p \qquad \begin{array}{l} p \\ q \end{array} \!-\!\boxed{\text{AND}}\!- p \wedge q \qquad \begin{array}{l} p \\ q \end{array} \!-\!\boxed{\text{OR}}\!- p \vee q$$

The output from the various inputs is reflected by Tables 1 through 3.

TABLE 1 The NOT Gate	
Input	**Output**
p	$\sim p$
0	1
1	0

TABLE 2 The AND Gate		
Input		**Output**
p	q	$p \wedge q$
0	0	0
0	1	0
1	0	0
1	1	1

TABLE 3 The OR Gate		
Input		**Output**
p	q	$p \vee q$
0	0	0
0	1	1
1	0	1
1	1	1

We can construct more complex circuits by combining these basic building blocks.

EXAMPLE 1   **Drawing a circuit from a logical statement** Draw a logic circuit for three inputs, p, q, and r and output $(\sim p) \wedge (q \vee r)$. What is the output when the inputs are $p = 1$, $q = 1$, and $r = 0$?

Solution   We begin at the end with the AND gate and prepare the diagram for input from the two components, $\sim p$ and $q \vee r$. See Figure 2.

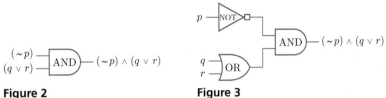

Figure 2   **Figure 3**

Then draw the circuits for each of the components and complete the diagram as in Figure 3.

To see the output for the inputs $p = 1$, $q = 1$, and $r = 0$, we note that the inverter changes the input at the first gate to the output of 0. The OR gate converts the inputs $q = 1$, and $r = 0$ to output value 1. The final AND gate has inputs 0 and 1, and therefore the output from the AND gate is 0.

Now Try Exercise 5   ■

The logic we have studied to this point helps to simplify complicated circuits. This is especially useful in the efficient design of complicated circuitry for the many devices dependent on electrical current. We repeatedly use the equivalences in Table 1 of Section 12.4.

EXAMPLE 2   **Finding the logical statement from a circuit** Write the logical statement represented by the circuit drawn in Fig. 4.

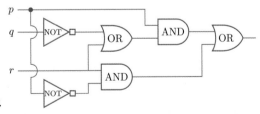

Figure 4

Solution $(p \wedge (\sim q \vee r)) \vee (\sim p \wedge r)$

Now Try Exercise 1   ■

EXAMPLE 3   **Simplifying a circuit** Simplify the circuit in Example 2.

Solution The expression $(p \wedge (\sim q \vee r)) \vee (\sim p \wedge r)$ can be simplified by using Table 1 of Section 12.4.

$$(p \wedge (\sim q \vee r)) \vee (\sim p \wedge r) \Leftrightarrow (p \wedge \sim q) \vee (p \wedge r) \vee (\sim p \wedge r)$$
$$\Leftrightarrow (p \wedge \sim q) \vee (r \wedge (p \vee \sim p))$$
$$\Leftrightarrow (p \wedge \sim q) \vee (r \wedge t)$$
$$\Leftrightarrow (p \wedge \sim q) \vee r$$

We draw the circuit in its simplified form in Figure 5.

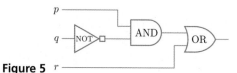

Figure 5

Now Try Exercise 9

We say that two logic circuits are *equivalent* if they perform the same funtion. That is, for every possible input, the output produced in the first circuit is the same as the output produced in the second circuit. We can show this equivalency by using truth tables or by relying on the Table of Logical Equivalences (Table 1, Section 12.4).

Circuits with NAND, NOR, and XOR

We can design circuits that behave according to the logical rules of the connectives NAND, NOR, and XOR. The symbols for these gates are given in Tables 4–6. The input and output values are also displayed. Figures 6–8 illustrate the elementary circuits for NAND, NOR, and XOR.

TABLE 4	The NAND Gate	
Input		**Output**
p	q	p NAND q
0	0	1
0	1	1
1	0	1
1	1	0

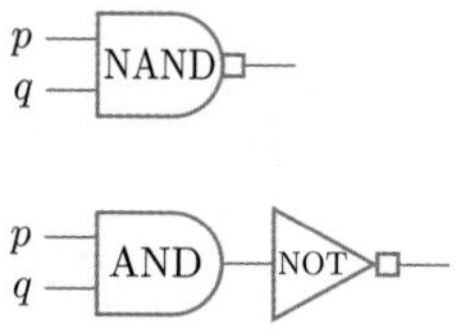

Figure 6

TABLE 5	The NOR Gate	
Input		**Output**
p	q	p NOR q
0	0	1
0	1	0
1	0	0
1	1	0

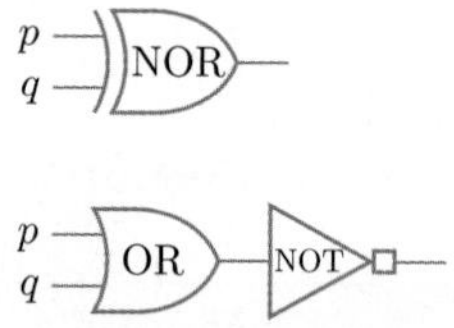

Figure 7

TABLE 6	The XOR Gate
Input	**Output**
p q	p XOR q
0 0	0
0 1	1
1 0	1
1 1	0

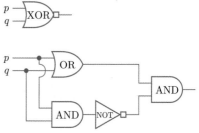

Figure 8

PRACTICE PROBLEMS 12.7

1. (a) Simplify the circuit shown in Figure 9 by using the Table of Logical Equivalences (Table 1, Section 12.4).

(b) Check your result with a truth table.

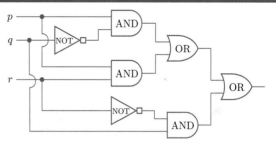

Figure 9

EXERCISES 12.7

1. Write the logic statement represented by Figure 10.

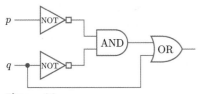

Figure 10

2. Write the logic statement represented by Figure 11.

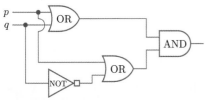

Figure 11

3. Write the logic statement represented by Figure 12.

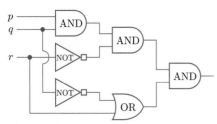

Figure 12

4. Write the logic statement represented by Figure 13.

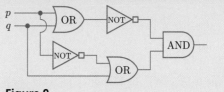

Figure 13

Draw the logic circuit that represents each of the logical statements in Exercises 5–8. Determine the output when the variables have the values given.

5. $p \wedge (\sim q \vee r)$; for $p = 1$, $q = 0$, $r = 1$

6. $(p \wedge q) \vee (p \vee r)$; for $p = 1$, $q = 0$, $r = 1$

7. $(\sim p \vee \sim q) \wedge (\sim p \wedge r)$; for $p = 1$, $q = 0$, $r = 1$

8. $\sim((p \wedge \sim q) \vee (p \wedge r))$; $p = 0$, $q = 0$, $r = 1$

9. Simplify the logic circuit in Figure 14 as much as possible.

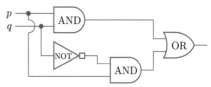

Figure 14

10. Simplify the logic circuit in Figure 15 as much as possible.

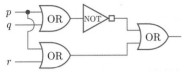

Figure 15

(11.) Simplify the logic circuit in Figure 16 as much as possible.

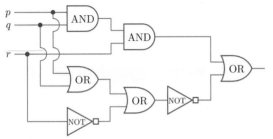

Figure 16

12. Simplify the logic circuit in Figure 17 as much as possible.

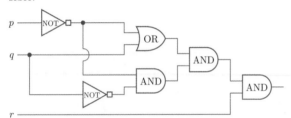

Figure 17

13. Design a logic circuit with two inputs (use p and q) that always has output 1.

14. Design a logic circuit with two inputs (use p and q) that always has output 0.

(15.) Design a circuit that acts as an OR gate for inputs p and q using three NOR gates.

16. Design a circuit that acts as an AND gate for inputs p and q using three NAND gates.

17. Design a circuit that acts as an XOR gate for inputs p and q using three gates other than XOR or OR.

18. Design a circuit for (p NAND q) NAND (p NAND q) and show the outputs for all values of p and q on a truth table.

19. (*The Game of Match*) In the game of "match," each of two players, A and B, has control of an input switch. At the signal, each player allows the current to flow (ON) or stops the current (OFF). If both players choose the same input (both ON or both OFF), player A wins. If their inputs differ, player B wins. Design a circuit so that the output is ON only when player A wins.

20. (*Switch Design for a Lecture Hall*) In designing a large four-walled lecture hall, the architect has placed a door on each wall, and next to each door there is a light switch. Design a circuit so that the light can be turned on or off from any switch. (*Hint*: The light should go on if an even number of switches are on, and the light should go off if an odd number of switches are on. Why?)

SOLUTIONS TO PRACTICE PROBLEMS 12.7

1. **(a)** The circuit can be translated into logical symbols:
$$\sim(p \vee q) \wedge (\sim p \vee q).$$
Using the Table of Logical Equivalences (Table 1, Section 12.4), we have

$$\sim(p \vee q) \wedge (\sim p \vee q) \Leftrightarrow (\sim p \wedge \sim q) \wedge (\sim p \vee q)$$
$$\Leftrightarrow ((\sim p \wedge \sim q) \wedge \sim p) \vee ((\sim p \wedge \sim q) \wedge q)$$
$$\Leftrightarrow (\sim p \wedge \sim q) \vee (\sim p \wedge (\sim q \wedge q))$$
$$\Leftrightarrow (\sim p \wedge \sim q) \vee (\sim p \wedge c)$$

$$\Leftrightarrow (\sim p \wedge \sim q) \vee c$$
$$\Leftrightarrow \sim p \wedge \sim q.$$

The circuit equivalent is displayed in Fig. 18.

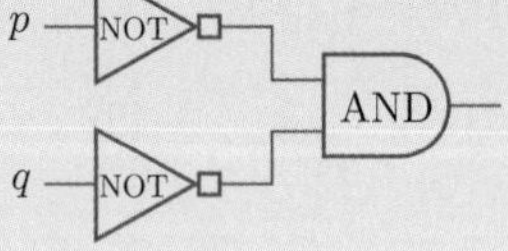

p —[NOT]—
q —[NOT]— →[AND]—

Figure 18

(b) To check the calculation, we use truth tables.

p	q	$p \vee q$	$\sim(p \vee q)$	$\wedge$	$\sim p$	$\vee$	q	$\sim p$	$\wedge$	$\sim q$
T	T	T	F	F	F	T	T	F	F	F
T	F	T	F	F	F	F	F	F	F	T
F	T	T	F	F	T	T	T	T	F	F
F	F	F	T	T	T	T	F	T	T	T
(1)	(2)	(3)	(4)	(7)	(5)	(6)		(8)		

Compare columns (7) and (8) to see that the statements are equivalent and that the circuits are equivalent.

CHAPTER 12 SUMMARY

1. A logical statement (proposition) is a declarative sentence that is either TRUE or FALSE.

2. Logical statements frequently have connectives such as *and* $\wedge$, *or* $\vee$, *not* $\sim$, *implies* $\rightarrow$, *if and only if* $\leftrightarrow$. The order of precedence for the connectives is $\sim$, $\wedge$, $\vee$, $\rightarrow$, $\leftrightarrow$. A statement without any connectives is called a *simple statement*. The truth value of a statement depends only on the truth values of the simple statements it contains, as summarized in the following truth table.

p	q	$\sim p$	$p \vee q$	$p \wedge q$	$p \rightarrow q$	$p \leftrightarrow q$
T	T	F	T	T	T	T
T	F	F	T	F	F	F
F	T	T	T	F	T	F
F	F	T	F	F	T	T

3. A statement that is TRUE no matter what truth values are assigned to the simple statements of which it is constructed is called a *tautology*. A *contradiction* is a statement that is always FALSE.

4. The statement $p \rightarrow q$ is the implication "if p, then q (or p implies q)." It is FALSE only when p is TRUE and q is FALSE. The two-way implication $p \leftrightarrow q$ is TRUE whenever p and q have the same truth value and FALSE otherwise.

5. We say two statements p and q are *logically equivalent* and write $p \Leftrightarrow q$ if they have the same truth table, that is, if $p \leftrightarrow q$ is a tautology. We say that p *logically implies* q and write $p \Rightarrow q$ if $p \rightarrow q$ is a tautology.

6. There are several statements related to $p \rightarrow q$: The converse is $q \rightarrow p$ and the contrapositive is $\sim q \rightarrow \sim p$. If the statement is TRUE, the converse may be TRUE or FALSE. However, every statement is logically equivalent to its contrapositive. That is, $(p \rightarrow q) \leftrightarrow (\sim q \rightarrow \sim p)$ is a tautology.

7. We define the connective $\oplus$ in terms of the other connectives as follows: $p \oplus q \Leftrightarrow (p \wedge \sim q) \vee (\sim p \wedge q)$. Therefore, the statement $p \oplus q$ is TRUE if and only if p and q have opposite truth values (one is TRUE and the other is FALSE).

8. The logical connectives used in computer languages AND, OR, NOT, IF...THEN, and XOR correspond to the logical symbols $\wedge$, $\vee$, $\sim$, $\rightarrow$, $\oplus$, respectively.

9. Rules for simplifying statements containing connectives follow algebraic principles so that there is a "calculus" of propositions. To simplify statements, we usually use either the truth table or the propositional calculus. The most common laws of propositional calculus are the commutative, associative, and distributive laws:

$$p \wedge q \Leftrightarrow q \wedge p \qquad\qquad p \vee q \Leftrightarrow q \vee p$$
$$p \wedge (q \wedge r) \Leftrightarrow (p \wedge q) \wedge r \qquad p \vee (q \vee r) \Leftrightarrow (p \vee q) \vee r$$
$$p \wedge (q \vee r) \Leftrightarrow (p \wedge q) \vee (p \wedge r)$$
$$p \vee (q \wedge r) \Leftrightarrow (p \vee q) \wedge (p \vee r)$$

10. De Morgan's laws are

$$\sim(p \vee q) \Leftrightarrow \sim p \wedge \sim q$$
$$\sim(p \wedge q) \Leftrightarrow \sim p \vee \sim q.$$

11. The most common rules of inference are given in Table 1 on page 609.

12. Statements containing variables are called *predicates* and can be made into logical statements with *quantifiers*. The quantifiers are the symbols $\forall$ ("for all") and $\exists$ ("there exists"). These symbols refer to the particular universal set for the variables in the predicate.

13. Important rules of predicate calculus include the following laws:

$$\left.\begin{array}{l} \sim[\forall x\, p(x)] \Leftrightarrow [\exists x \sim p(x)] \\ \sim[\exists x\, p(x)] \Leftrightarrow [\forall x \sim p(x)] \end{array}\right\} \text{ De Morgan's laws}$$

$$\forall x\, \forall y\, p(x, y) \Leftrightarrow \forall y\, \forall x\, p(x, y)$$
$$\exists x\, \exists y\, p(x, y) \Leftrightarrow \exists y\, \exists x\, p(x, y)$$
$$\exists x\, \forall y\, p(x, y) \Rightarrow \forall y\, \exists x\, p(x, y).$$

14. To prove the statement

$$(H_1 \wedge H_2 \wedge \cdots \wedge H_n) \Rightarrow C$$

by the method of indirect proof, we assume that the conclusion C is FALSE and prove that at least one of the hypotheses H_i must be FALSE.

15. Logic circuits frequently can be simplified using the Table of Logical Equivalences.

REVIEW OF FUNDAMENTAL CONCEPTS

1. What is a logical statement? How do you decide if a collection of words is or is not a logical statement?

2. Write down the truth tables of the simple logical connectives $\wedge$, $\vee$, $\sim$, $\rightarrow$, $\leftrightarrow$.

3. When is $p \rightarrow q$ a TRUE statement?

4. What do we mean by "logical equivalence"? Explain

how you might use a truth table to establish logical equivalence.

5. Explain how you might use the "algebraic" nature of the rules of Table 1, Section 12.4, to establish logical equivalence.

6. State De Morgan's laws. When should you use them?

7. Given the implication $p \rightarrow q$, what is the hypothesis? What is the conclusion?

8. Given the implication $p \rightarrow q$, write down the contrapositive and the converse. If the implication is TRUE, what can we say about the truth of the contrapositive and that of the converse?

9. Give an example (in words) of an implication that is TRUE and write its contrapositive and converse, making sure you know the difference.

10. Write the negation of $p \rightarrow q$ without using the arrow symbol.

11. Associate each of the logical connectives AND, OR, NOT, IF...THEN, and XOR with the logical symbols and an appropriate truth table.

12. Draw a logic tree for each of the symbols $\wedge$, $\vee$, $\rightarrow$, and $\leftrightarrow$ when the input has a TRUE p and a FALSE q.

13. What is a tautology? Describe how you would prove that a statement is a tautology.

14. Demonstrate each rule of inference in Table 1 of Section 12.5 with an English statement.

15. Write an English statement corresponding to $\forall x \, p(x)$. Write its negation.

16. Write an English statement corresponding to $\exists x \, [p(x) \rightarrow q(x)]$. Write its negation.

17. State De Morgan's laws for quantified statements.

18. Write an English statement of the form $\forall x \, \exists y \, p(x, y)$. Write the negation of the statement.

19. Draw the logic circuits that correspond to logical equivalences 1–8 in Table 1, Section 12.4.

KEY FORMULAS

We refer the student to Section 12.4, Tables 1 and 2, and Section 12.6, Table 1.

SUPPLEMENTARY EXERCISES

1. Determine which of the following are statements.
   (a) The universe is 1 billion years old.
   (b) What a beautiful morning!
   (c) Mathematics is an important part of our culture.
   (d) He is a gentleman and a scholar.
   (e) All poets are men.

2. Write each of the following statements in "if ..., then ..." format.
   (a) The lines are perpendicular implies that their slopes are negative reciprocals of each other.
   (b) Goldfish can live in a fish bowl only if the water is aerated.
   (c) Jane uses her umbrella if it rains.
   (d) Only if Morris eats all his food does Sally give him a treat.

3. Write the contrapositive and the converse of each of the statements.
   (a) The Yankees are playing in Yankee Stadium if they are in New York City.
   (b) If the Richter scale indicates the earthquake is a 7, then the quake is considered major.
   (c) If a coat is made of fur, it is warm.

   (d) If Jane is in Russia, she is in Moscow.

4. Negate the statements.
   (a) If two triangles are similar, their sides are equal.
   (b) There exists a real number x such that $x^2 = 5$.
   (c) For every positive integer n, if n is even, then n^2 is even also.
   (d) There exists a real number x such that $x^2 + 4 = 0$.

5. Determine which of the statements are tautologies.
   (a) $p \vee \sim p$
   (b) $(p \rightarrow q) \leftrightarrow (\sim p \vee q)$
   (c) $(p \wedge \sim q) \leftrightarrow \sim(\sim p \wedge q)$
   (d) $[p \rightarrow (q \rightarrow r)] \leftrightarrow [(p \rightarrow q) \rightarrow r]$

6. Construct a truth table for each of the following statements.
   (a) $p \rightarrow (\sim q \vee r)$ (b) $p \wedge (q \leftrightarrow (r \wedge p))$

7. Which of the logical implications are true?
   (a) $[p \wedge (\sim p \vee q)] \Rightarrow q$ (b) $[(p \rightarrow q) \wedge q] \Rightarrow p$

8. True or false?
   (a) $(\sim q \rightarrow \sim p) \Leftrightarrow (p \rightarrow q)$
   (b) $[(p \rightarrow q) \wedge (r \rightarrow p)] \Leftrightarrow q$

9. True or false?

   (a) $[(p \to q) \land \sim p] \Rightarrow \sim q$ (b) $[(p \to q) \land \sim q] \Rightarrow \sim p$

10. For the given input for A and B, determine the output for Z.

    LET $C = 3 * A + B$

    IF $((C > 0)$ AND $(B > 3))$

    THEN LET $Z = C$

    ELSE LET $Z = 100$

    (a) $A = 4, B = 5$ (b) $A = 10, B = 2$

    (c) $A = -4, B = 5$ (d) $A = -10, B = -2$

11. For the given inputs, determine the output for Z.

    IF $((X > 0)$ AND $(Y > 0))$ OR $(C \geq 10)$

    THEN LET $Z = X * Y$

    ELSE LET $Z = X + Y$

    (a) $X = 5, Y = 10, C = 10$

    (b) $X = 5, Y = -5, C = 10$

    (c) $X = -10, Y = -5, C = 2$

    (d) $X = 2, Y = 5, C = 4$

12. The following statement is TRUE.

    If the deduction is allowed,
    then the tax law has been revised.

    Give the truth value of each of the following statements if it can be determined directly from the truth of the original.

    (a) If the tax law has been revised, the deduction is allowed.

    (b) If the tax law has not been revised, the deduction is not allowed.

    (c) The deduction is allowed only if the tax law has been revised.

    (d) The deduction is allowed if the tax law has been revised.

    (e) The deduction is allowed or the tax law has been revised.

13. Assume the following statement is TRUE.
    If the voter is over the age of 21 and has a driver's license, the voter is eligible for free driver education. Determine the truth value of each of the following directly from the original, if possible.

    (a) If the voter is eligible for free driver education, the voter is over 21 and has a driver's license.

    (b) If the voter is not eligible for free driver education, the voter is not over 21 and does not have a driver's license.

    (c) If the voter is not eligible for free driver education, the voter either is not over 21 or does not hold a driver's license.

14. Assume the statement "All mathematicians like rap music" is TRUE. Determine the truth value of the following from this assumption, if possible.

    (a) Some mathematicians like rap music.

    (b) Some mathematicians do not like rap music.

    (c) There exists no mathematician who does not like rap music.

15. Assume the statement "Some apples are not rotten" is TRUE. Determine the truth value of each of the following using only that fact, if possible.

    (a) Some apples are rotten.

    (b) All apples are not rotten.

    (c) Not all apples are rotten.

16. Show that the argument is valid: If taxes go up, I sell the house and move to India. I do not move to India. Therefore, taxes do not go up.

17. Show that the argument is valid: I study mathematics and I study business. If I study business, I cannot write poetry or I cannot study mathematics. Therefore, I cannot write poetry.

18. Show that the argument is valid: If I shop for a dress, I wear high heels. If I have a sore foot, I do not wear high heels. I shop for a dress. Therefore, I do not have a sore foot.

19. Show that the argument is valid: Asters or dahlias grow in the garden. If it is spring, asters do not grow in the garden. It is spring. Therefore, dahlias grow in the garden.

20. Use indirect proof to show the argument is valid: If the professor gives a test, Nancy studies hard. If Nancy has a date, she takes a shower. If the professor does not give a test, she does not take a shower. Nancy has a date. Therefore, Nancy studies hard.

21. Draw the logic circuit corresponding to the expression $(\sim p \lor q) \land (q \land r)$ and determine the output when the input values are $p = 0$, $q = 0$, $r = 1$.

22. Simplify the logic circuit in Fig. 19 by using the Table of Logical Equivalences to calculate its equivalent.

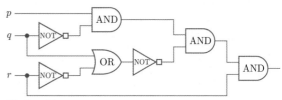

Figure 19

Conceptual Exercises

23. Construct a statement equivalent to p XOR q using only NOR and AND.

24. Design a logic circuit with three inputs for which the output (current) is always OFF (0) regardless of the input.

25. (*Designing a Voting Machine*) Design a voting machine for a committee of three people. A motion is presented to the committee for a vote. Each person can vote "yes" by pushing a button (on) and may vote "no" by not pushing the button. The majority rules. The circuit you design should light up when the motion passes. Design a similar machine for a committee of five people. An abstention is considered a "no" vote.

CHAPTER TEST

1. Derive the truth table for the statement

$$(p \wedge \sim q) \to q.$$

2. Which of the following expressions are statements?
   (a) Books belong in libraries.
   (b) Are mathematicians poets?
   (c) All trains from New York to Washington stop at Philadelphia.
   (d) There is a teacher at each middle school in Los Angeles who tells bad jokes.
   (e) What a terrific poem!

3. Write the contrapositive of

   If Bob hits a triple or a home run, then the coach buys him ice cream.

4. Assume the universe is the set of all integers. Write the negation of

   There exists an integer that is neither even nor greater than 8.

5. Let $U = \{1, 2, 3, 4, 5, 6, 7, 8, 9\}$. Let $p(x)$ be the open statement

$$x^2 \text{ is even} \to x^3 + 1 \text{ is prime}.$$

   Determine the truth values of the following:

$$\forall x \, [p(x)]$$
$$\exists x \, [\sim p(x)].$$

6. Classify each statement form as a tautology, a contradiction, or neither.
   (a) $(p \wedge \sim q) \vee (q \to \sim p)$
   (b) $(p \oplus \sim q) \leftrightarrow (p \leftrightarrow q)$
   (c) $\sim(p \to q) \wedge q$

7. Tell whether the following statements are TRUE or FALSE and give a reason for your answer.
   (a) If Chicago is in Illinois, then Toronto is in Canada.
   (b) If Toronto is in Canada, then Chicago is in Louisiana.
   (c) If Chicago is in Louisiana, then $7 + 4 = 15$.
   (d) If $7 + 4 = 15$ or Chicago is in Illinois, then Louisiana is in Canada.

8. Assume that the following statements are TRUE.

   If you win the lottery or have a good job, you will have a lot of money.

   If you get robbed, you will not have a lot of money.

   You get robbed.

   Can you conclude that you do not have a good job?

9. Write the negation of the following:
   (a) There is an English dictionary that does not contain the word "Internet."
   (b) Every student at the university listens to jazz.
   (c) The elevator stops at all floors.

10. Let p be the statement "Spanish textbooks are expensive." Let q be the statement "Publishing costs are high." Write the following in symbolic "if, then" form:
   (a) Spanish books are expensive is sufficient for publishing costs to be high.
   (b) Publishing costs are high is a sufficient condition for Spanish books to be expensive.
   (c) Spanish books are expensive only if publishing costs are high.
   (d) Only if publishing costs are high are Spanish books expensive.
   (e) Spanish books are expensive if publishing costs are high.

11. Use the following program to assign a value to the variable G for each of the inputs given.

$$\text{LET } D = A - (B * C)$$
$$\text{IF } D = 5 \text{ OR } C = 0$$
$$\text{THEN } G = 0$$
$$\text{ELSE } G = D + A$$

   (a) $A = 4$, $B = 9$, $C = 3$
   (b) $A = 185$, $B = 30$, $C = 6$
   (c) $A = 0$, $B = 30$, $C = -6$
   (d) $A = 10$, $B = 9$, $C = 0$

12. True or false:
   (a) $p \wedge (q \vee r) \Leftrightarrow (p \wedge q) \vee r$
   (b) $p \wedge (q \vee r) \Rightarrow (\sim q \to r)$

13. Simplify the logic circuit in Fig. 20 by using the Table of Logical Equivalences to calculate its equivalent.

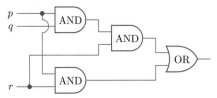

Figure 20

14. Draw the logic circuit corresponding to the expression

$$(\sim p \lor (q \land \sim r)) \land (\sim q \lor r)$$

and determine the output when the input values are $p = 1$, $q = 0$, $r = 1$.

A Logic Puzzle

Denise, Miriam, Sally, Nelson, and Bob are students at the same university. Each is in a different mathematics course and each is in a different year at the university. From the clues given, determine what mathematics course each is enrolled in and the year (freshman, sophomore, junior, senior, graduate student) each is in.

A. Sally is a sophomore.

B. Miriam (who is neither a junior nor a freshman) is taking a statistics course.

C. Neither the person who is taking calculus (who is not Bob) nor the one taking finite math is the person who is a freshman.

D. The student who is a senior is enrolled in algebra.

E. Neither Nelson (who is not a junior) nor Bob is taking precalculus.

Solve the puzzle by filling in the chart below with O to signify "yes" and X to signify "no." For example, clue A indicates that Sally is a sophomore, so put an O at the intersection of the "Sally" column and the "Sophomore" row. Notice that you can conclude that the other four students are not sophomores and that Sally is not a freshman, junior, senior, or graduate student, so that you can put eight X's in the chart to represent these conclusions. Solve the problem by working with both the clues and the chart.

	Bob	Denise	Miriam	Nelson	Sally	Freshman	Sophomore	Junior	Senior	Graduate
Algebra										
Finite Math										
Statistics										
Precalculus										
Calculus										
Freshman					
Sophomore					
Junior					
Senior					
Graduate					

Appendices

Appendix A
Tables

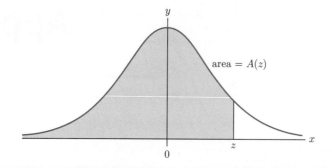

area = $A(z)$

TABLE 1	Areas under the standard normal curve								
z	$A(z)$	z	$A(z)$	z	$A(z)$	z	$A(z)$	z	$A(z)$
−3.50	.0002	−2.00	.0228	−.50	.3085	1.00	.8413	2.45	.9929
−3.45	.0003	−1.95	.0256	−.45	.3264	1.05	.8531	2.50	.9938
−3.40	.0003	−1.90	.0287	−.40	.3446	1.10	.8643	2.55	.9946
−3.35	.0004	−1.85	.0322	−.35	.3632	1.15	.8749	2.60	.9953
−3.30	.0005	−1.80	.0359	−.30	.3821	1.20	.8849	2.65	.9960
−3.25	.0006	−1.75	.0401	−.25	.4013	1.25	.8944	2.70	.9965
−3.20	.0007	−1.70	.0446	−.20	.4207	1.2813	.9000	2.75	.9970
−3.15	.0008	−1.65	.0495	−.15	.4404	1.30	.9032	2.80	.9974
−3.10	.0010	−1.60	.0548	−.10	.4602	1.35	.9115	2.85	.9978
−3.05	.0011	−1.55	.0606	−.05	.4801	1.40	.9192	2.90	.9981
−3.00	.0013	−1.50	.0668	.00	.5000	1.45	.9265	2.95	.9984
−2.95	.0016	−1.45	.0735	.05	.5199	1.50	.9332	3.00	.9987
−2.90	.0019	−1.40	.0808	.10	.5398	1.55	.9394	3.05	.9989
−2.85	.0022	−1.35	.0885	.15	.5596	1.60	.9452	3.10	.9990
−2.80	.0026	−1.30	.0968	.20	.5793	1.65	.9505	3.15	.9992
−2.75	.0030	−1.25	.1056	.25	.5987	1.70	.9554	3.20	.9993
−2.70	.0035	−1.20	.1151	.30	.6179	1.75	.9599	3.25	.9994
−2.65	.0040	−1.15	.1251	.35	.6368	1.80	.9641	3.30	.9995
−2.60	.0047	−1.10	.1357	.40	.6554	1.85	.9678	3.35	.9996
−2.55	.0054	−1.05	.1469	.45	.6736	1.90	.9713	3.40	.9997
−2.50	.0062	−1.00	.1587	.50	.6915	1.95	.9744	3.45	.9997
−2.45	.0071	−.95	.1711	.55	.7088	2.00	.9772	3.50	.9998
−2.40	.0082	−.90	.1841	.60	.7257	2.05	.9798		
−2.35	.0094	−.85	.1977	.65	.7422	2.10	.9821		
−2.30	.0107	−.80	.2119	.70	.7580	2.15	.9842		
−2.25	.0122	−.75	.2266	.75	.7734	2.20	.9861		
−2.20	.0139	−.70	.2420	.80	.7881	2.25	.9878		
−2.15	.0158	−.65	.2578	.85	.8023	2.30	.9893		
−2.10	.0179	−.60	.2743	.90	.8159	2.35	.9906		
−2.05	.0202	−.55	.2912	.95	.8289	2.40	.9918		

TABLE 2 $(1 + i)^n$ Compound amount of $1 invested for n interest periods at interest rate i per period

n	1/50 %	1/2 %	1 %	2 %	6 %
1	1.000200000	1.00500000	1.01000000	1.02000000	1.06000000
2	1.000400040	1.01002500	1.02010000	1.04040000	1.12360000
3	1.000600120	1.01507513	1.03030100	1.06120800	1.19101600
4	1.000800240	1.02015050	1.04060401	1.08243216	1.26247696
5	1.001000400	1.02525125	1.05101005	1.10408080	1.33822558
6	1.001200600	1.03037751	1.06152015	1.12616242	1.41851911
7	1.001400840	1.03552940	1.07213535	1.14868567	1.50363026
8	1.001601120	1.04070704	1.08285671	1.17165938	1.59384807
9	1.001601447	1.04591058	1.09368527	1.19509257	1.68947896
10	1.002001806	1.05114013	1.10462213	1.21899442	1.79084770
11	1.002202201	1.05639583	1.11566835	1.24337431	1.89829856
12	1.002402642	1.06167781	1.12682503	1.26824179	2.01219647
13	1.002603122	1.06698620	1.13809328	1.29360663	2.13292826
14	1.002803643	1.07232113	1.14947421	1.31947876	2.26090396
15	1.003004204	1.07768274	1.16096896	1.34586834	2.39655819
16	1.003204804	1.08307115	1.17257864	1.37278571	2.54035168
17	1.003405445	1.08848651	1.18430443	1.40024142	2.69277279
18	1.003505100	1.09392894	1.19614748	1.42824625	2.85433915
19	1.003806848	1.09939858	1.20810895	1.45681117	3.02559950
20	1.004007609	1.10489558	1.22019004	1.48594740	3.20713547
21	1.004208411	1.11042006	1.23239194	1.51566634	3.39956360
22	1.004409252	1.11597216	1.24471586	1.54597967	3.60353742
23	1.004610134	1.12155202	1.25716302	1.57689926	3.81974966
24	1.004811056	1.12715978	1.26973465	1.60843725	4.04893464
25	1.005012018	1.13279558	1.28243200	1.64060599	4.29187072
26	1.005213022	1.13845955	1.29525631	1.67341811	4.54938296
27	1.005414063	1.14415185	1.30820888	1.70688648	4.82234594
28	1.005615146	1.14987261	1.32129097	1.74102421	5.11168670
29	1.005816269	1.15562197	1.33450388	1.77584469	5.41838790
30	1.006017433	1.16140008	1.34784892	1.81136158	5.74349117
36	1.007225257	1.19668052	1.43076878	2.03988734	8.14725200
48	1.009645259	1.27048916	1.61222608	2.58707039	16.39387173
52	1.010453217	1.29609015	1.67768892	2.80032819	20.69688534
60	1.012071075	1.34885015	1.81669670	3.28103079	32.98769085
104	1.021015784	1.67984969	2.81464012	7.84183795	428.36106292
120	1.024287860	1.81939673	3.30038689	10.76516303	1,088.18774784
180	1.036652115	2.45409356	5.99580198	35.32083136	35,896.80101597
240	1.049165620	3.31020448	10.89255365	115.88873515	
300	1.061830136	4.46496981	19.78846626	380.23450806	
360	1.074647608	6.02257521	35.94964133	1,247.56112775	
365	1.075722685	6.17465278	37.78343433	1,377.40829197	

TABLE 3 $1/(1 + i)^n$ Present value of $1. Principal that will accumulate to $1 in n interest periods at a compound rate of i per period

n	1/50 %	1/2 %	1 %	2 %	6 %
1	.99980004	.99502488	.99009901	.98039216	.94339623
2	.99960012	.99007450	.98029605	.96116878	.88999644
3	.99940024	.98514876	.97059015	.94232233	.83961928
4	.99920040	.98024752	.96098034	.92384543	.79209366
5	.99900060	.97537067	.95146569	.90573081	.74725817
6	.99880084	.97051808	.94204524	.88797138	.70496054
7	.99860112	.96568963	.93271805	.87056018	.66505711
8	.99840144	.96088520	.92348322	.85349037	.62741237
9	.99820180	.95610468	.91433982	.83675527	.59189846
10	.99800220	.95134794	.90528695	.82034830	.55839478
11	.99780264	.94661487	.89632372	.80426304	.52678753
12	.99760312	.94190534	.88744923	.78849318	.49696936
13	.99740364	.93721924	.87866260	.77303253	.46883902
14	.99720420	.93255646	.86996297	.75787502	.44230096
15	.99700479	.92791688	.86134947	.74301473	.41726506
16	.99680543	.92330037	.85282126	.72844581	.39364628
17	.99660611	.91870684	.84437749	.71416256	.37136442
18	.99640683	.91413616	.83601731	.70015937	.35034379
19	.99620759	.90958822	.82773992	.68643076	.33051301
20	.99600839	.90506290	.81954447	.67297133	.31180473
21	.99580923	.90056010	.81143017	.65977582	.29415540
22	.99561010	.89607971	.80339621	.64683904	.27750510
23	.99541102	.89162160	.79544179	.63415592	.26179726
24	.99521198	.88718567	.78756613	.62172149	.24697855
25	.99501298	.88277181	.77976844	.60953087	.23299863
26	.99481401	.87837991	.77204796	.59757928	.21981003
27	.99461509	.87400986	.76440392	.58586204	.20736795
28	.99441621	.86966155	.75683557	.57437455	.19563014
29	.99421736	.86533488	.74934215	.56311231	.18455674
30	.99401856	.86102973	.74192292	.55207089	.17411013
36	.99282657	.83564492	.69892495	.49022315	.12274077
48	.99044688	.78709841	.62026041	.38653761	.06099840
52	.98965492	.77155127	.59605806	.35710100	.04831645
60	.98807290	.74137220	.55044962	.30478227	.03031434
104	.97941686	.59529136	.35528521	.12752113	.00233448
120	.97628805	.54963273	.30299478	.09289223	.00091896
180	.96464377	.40748243	.16678336	.02831190	.00002786
240	.95313836	.30209614	.09180584	.00862897	.00000084
300	.94177018	.22396568	.05053449	.00262996	.00000003
360	.93053759	.16604193	.02781669	.00080156	.00000000
365	.92960762	.16195243	.02646663	.00072600	.00000000

TABLE 4 $s_{\overline{n}|i}$ Future value of an ordinary annuity of n \$1 payments each, immediately after the last payment at compound interest rate of i per period

				i		
n	1/2 %	3/4 %	1 %	1.5 %	2 %	6 %
1	1.00000000	1.00000000	1.00000000	1.00000000	1.00000000	1.00000000
2	2.00500000	2.00750000	2.01000000	2.01500000	2.02000000	2.06000000
3	3.01502500	3.02255625	3.03010000	3.04522500	3.06040000	3.18360000
4	4.03010013	4.04522542	4.06040100	4.09090337	4.12160800	4.37461600
5	5.05025063	5.07556461	5.10100501	5.15226693	5.20404016	5.63709296
6	6.07550188	6.11363135	6.15201506	6.22955093	6.30812096	6.97531854
7	7.10587939	7.15948358	7.21353521	7.32299419	7.43428338	8.39383765
8	8.14140879	8.21317971	8.28567056	8.43283911	8.58296905	9.89746791
9	9.18211583	9.27477856	9.36852727	9.55933169	9.75462843	11.49131598
10	10.22802641	10.34433940	10.46221254	10.70272167	10.94972100	13.18079494
11	11.27916654	11.42192194	11.56683467	11.86326249	12.16871542	14.97164264
12	12.33556237	12.50758636	12.68250301	13.04121143	13.41208973	16.86994120
13	13.39724018	13.60139325	13.80932804	14.23682960	14.68033152	18.88213767
14	14.46422639	14.70340370	14.94742132	15.45038205	15.97393815	21.01506593
15	15.53654752	15.81367923	16.09689554	16.68213778	17.29341692	23.27596988
16	16.61423026	16.93228183	17.25786449	17.93236984	18.63928525	25.67252808
17	17.69730141	18.05927394	18.43044314	19.20135539	20.01207096	28.21287976
18	18.78578791	19.19471849	19.61474757	20.48937572	21.41231238	30.90565255
19	19.87971685	20.33867888	20.81089504	21.79671636	22.84055863	33.75999170
20	20.97911544	21.49121897	22.01900399	23.12366710	24.29736980	36.78559120
21	22.08401101	22.65240312	23.23919403	24.47052211	25.78331719	39.99272668
22	23.19443107	23.82229614	24.47158598	25.83757994	27.29898354	43.39229028
23	24.31040322	25.00096336	25.71630183	27.22514364	28.84496321	46.99582769
24	25.43195524	26.18847059	26.97346485	28.63352080	30.42186247	50.81557735
25	26.55911502	27.38488412	28.24319950	30.06302361	32.03029972	54.86451200
26	27.69191059	28.59027075	29.52563150	31.51396896	33.67090572	59.15638272
27	28.83037015	29.80469778	30.82088781	32.98667850	35.34432383	63.70576568
28	29.97452200	31.02823301	32.12909669	34.48147867	37.05121031	68.52811162
29	31.12439461	32.26094476	33.45038766	35.99870085	38.79223451	73.63979832
30	32.28001658	33.50290184	34.78489153	37.53868137	40.56807921	79.05818622
36	39.33610496	41.15271612	43.07687836	47.27596921	51.99436719	119.12086666
48	54.09783222	57.52071111	61.22260777	69.56521929	79.35351927	256.56452882
52	59.21803075	63.31106835	67.76889215	77.92489152	90.01640927	328.28142239
60	69.77003051	75.42413693	81.66966986	96.21465171	114.05153942	533.12818089
104	135.96993732	156.68432202	181.46401172	246.93411381	342.09189731	7,122.68438195
120	163.87934681	193.51427708	230.03868946	331.28819149	488.25815171	18,119.79579725
180	290.81871245	378.40576900	499.58019754	905.62451261	1,716.04156785	
240	462.04089516	667.88686993	989.25536539	2,308.85437027	5,744.43675765	
300	692.99396243	1,121.12193732	1,878.84662619	5,737.25330834	18,961.72540308	
360	1,004.51504245	1,830.74348307	3,494.96413277	14,113.58539279	62,328.05638744	
365	1,034.93055669	1,905.50947396	3,678.34343329	15,209.49204803	68,820.41459830	

TABLE 5 $1/s_{\overline{n}|i}$ Rent per period for an ordinary annuity of n payments, with compound interest rate i per period, and future value $1

n	1/2 %	3/4 %	1 %	1.5 %	2 %	6 %
1	1.00000000	1.00000000	1.00000000	1.00000000	1.00000000	1.00000000
2	.49875312	.49813200	.49751244	.49627792	.49504950	.48543689
3	.33167221	.33084579	.33002211	.32838296	.32675467	.31410981
4	.24813279	.24720501	.24628109	.24444479	.24262375	.22859149
5	.19800997	.19702242	.19603980	.19408932	.19215839	.17739640
6	.16459546	.16356891	.16254837	.16052521	.15852581	.14336263
7	.14072854	.13967488	.13862828	.13655616	.13451196	.11913502
8	.12282886	.12175552	.12069029	.11858402	.11650980	.10103594
9	.10890736	.10781929	.10674036	.10460982	.10251544	.08702224
10	.09777057	.09667123	.09558208	.09343418	.09132653	.07586796
11	.08865903	.08755094	.08645408	.08429384	.08217794	.06679294
12	.08106643	.07995148	.07884879	.07667999	.07455960	.05927703
13	.07464224	.07352188	.07241482	.07024036	.06811835	.05296011
14	.06913609	.06801146	.06690117	.06472332	.06260197	.04758491
15	.06436436	.06323639	.06212378	.05994436	.05782547	.04296276
16	.06018937	.05905879	.05794460	.05576508	.05365013	.03895214
17	.05650579	.05537321	.05425806	.05207966	.04996984	.03544480
18	.05323173	.05209766	.05098205	.04880578	.04670210	.03235654
19	.05030253	.04916740	.04805175	.04587847	.04378177	.02962086
20	.04766645	.04653063	.04541531	.04324574	.04115672	.02718456
21	.04528163	.04414543	.04303075	.04086550	.03878477	.02500455
22	.04311380	.04197748	.04086372	.03870332	.03663140	.02304557
23	.04113465	.03999846	.03888584	.03673075	.03466810	.02127848
24	.03932061	.03818474	.03707347	.03492410	.03287110	.01967900
25	.03765186	.03651650	.03540675	.03326345	.03122044	.01822672
26	.03611163	.03497693	.03386888	.03173196	.02969923	.01690435
27	.03468565	.03355176	.03244553	.03031527	.02829309	.01569717
28	.03336167	.03222871	.03112444	.02900108	.02698967	.01459255
29	.03212914	.03099723	.02989502	.02777878	.02577836	.01357961
30	.03097892	.02984816	.02874811	.02663919	.02464992	.01264891
36	.02542194	.02429973	.02321431	.02115240	.01923285	.00839483
48	.01848503	.01738504	.01633384	.01437500	.01260184	.00389765
52	.01688675	.01579503	.01475603	.01283287	.01110909	.00304617
60	.01433280	.01325836	.01224445	.01039343	.00876797	.00187572
104	.00735457	.00638226	.00551073	.00404966	.00292319	.00014040
120	00610205	.00516758	.00434709	.00301852	.00204810	.00005519
180	.00343857	.00264267	.00200168	.00110421	.00058274	.00000167
240	.00216431	.00149726	.00101086	.00043312	.00017408	.00000005
300	.00144301	.00089196	.00053224	.00017430	.00005274	.00000000
360	.00099551	.00054623	.00028613	.00007085	.00001604	.00000000
365	.00096625	.00052479	.00027186	.00006575	.00001453	.00000000

TABLE 6 $a_{\overline{n}|i}$ Present value of an ordinary annuity of n payments of $1, one period before the first payment, with interest compounded at i per period

n	1/2%	3/4%	1%	1.5%	2%	6%
1	.99502488	.99255583	.99009901	.98522167	.98039216	.94339623
2	1.98509938	1.97772291	1.97039506	1.95588342	1.94156094	1.83339267
3	2.97024814	2.95555624	2.94098521	2.91220042	2.88388327	2.67301195
4	3.95049566	3.92611041	3.90196555	3.85438465	3.80772870	3.46510561
5	4.92586633	4.88943961	4.85343124	4.78264497	4.71345951	4.21236379
6	5.89638441	5.84559763	5.79547647	5.69718717	5.60143089	4.91732433
7	6.86207404	6.79463785	6.72819453	6.59821396	6.47199107	5.58238144
8	7.82295924	7.73661325	7.65167775	7.48592508	7.32548144	6.20979381
9	8.77906392	8.67157642	8.56601758	8.36051732	8.16223671	6.80169227
10	9.73041186	9.59957958	9.47130453	9.22218455	8.98258501	7.36008705
11	10.67702673	10.52067452	10.36762825	10.07111779	9.78684805	7.88687458
12	11.61893207	11.43491267	11.25507747	10.90750521	10.57534122	8.38384394
13	12.55615131	12.34234508	12.13374007	11.73153222	11.34837375	8.85268296
14	13.48870777	13.24302242	13.00370304	12.54338150	12.10624877	9.29498393
15	14.41662465	14.13699495	13.86505252	13.34323301	12.84926350	9.71224899
16	15.33992502	15.02431261	14.71787378	14.13126405	13.57770931	10.10589527
17	16.25863186	15.90502492	15.56225127	14.90764931	14.29187188	10.47725969
18	17.17276802	16.77918107	16.39826858	15.67256089	14.99203125	10.82760348
19	18.08235624	17.64682984	17.22600850	16.42616837	15.67846201	11.15811649
20	18.98741915	18.50801969	18.04555297	17.16863879	16.35143334	11.46992122
21	19.88797925	19.36279870	18.85698313	17.90013673	17.01120916	11.76407662
22	20.78405896	20.21121459	19.66037934	18.62082437	17.65804820	12.04158172
23	21.67568055	21.05331473	20.45582113	19.33086145	18.29220412	12.30337898
24	22.56286622	21.88914614	21.24338726	20.03040537	18.91392560	12.55035753
25	23.44563803	22.71875547	22.02315570	20.71961120	19.52345647	12.78335616
26	24.32401794	23.54218905	22.79520366	21.39863172	20.12103576	13.00316619
27	25.19802780	24.35949286	23.55960759	22.06761746	20.70689780	13.21053414
28	26.06768936	25.17071251	24.31644316	22.72671671	21.28127236	13.40616428
29	26.93302423	25.97589331	25.06578530	23.37607558	21.84438466	13.59072102
30	27.79405397	26.77508021	25.80770822	24.01583801	22.39645555	13.76483115
36	32.87101624	31.44680525	30.10750504	27.66068431	25.48884248	14.62098713
48	42.58031778	40.18478189	37.97395949	34.04255365	30.67311957	15.65002661
52	45.68974664	42.92761812	40.39419423	35.92874185	32.14494992	15.86139252
60	51.72556075	48.17337352	44.95503841	39.38026889	34.76088668	16.16142771
104	80.94172854	72.03438325	64.47147918	52.49436634	43.62394373	16.62775868
120	90.07345333	78.94169267	69.70052203	55.49845411	45.35538850	16.65135068
180	118.50351467	98.59340884	83.32166399	62.09556231	48.58440478	16.66620237
240	139.58077168	111.14495403	90.81941635	64.79573209	49.56855168	16.66665259
300	155.20686401	119.16162216	94.94655125	65.90090069	49.86850220	16.66666624
360	166.79161439	124.28186568	97.21833108	66.35324174	49.95992180	16.66666665
365	167.60951473	124.61379021	97.35333747	66.37572674	49.96369994	16.66666666

TABLE 7 $1/a_{\overline{n}|i}$ Rent per period for an ordinary annuity of n payments whose present value is $1, with interest compounded at i per period

n	1/2 %	3/4 %	1 %	1.5 %	2 %	6 %
1	1.00500000	1.00750000	1.01000000	1.01500000	1.02000000	1.06000000
2	.50375312	.50563200	.50751244	.51127792	.51504950	.54543689
3	.33667221	.33834579	.34002211	.34338296	.34675467	.37410981
4	.25313279	.25470501	.25628109	.25944479	.26262375	.28859149
5	.20300997	.20452242	.20603980	.20908932	.21215839	.23739640
6	.16959546	.17106891	.17254837	.17552521	.17852581	.20336263
7	.14572854	.14717488	.14862828	.15155616	.15451196	.17913502
8	.12782886	.12925552	.13069029	.13358402	.13650980	.16103594
9	.11390736	.11531929	.11674036	.11960982	.12251544	.14702224
10	.10277057	.10417123	.10558208	.10843418	.11132653	.13586796
11	.09365903	.09505094	.09645408	.09929384	.10217794	.12679294
12	.08606643	.08745148	.08884879	.09167999	.09455960	.11927703
13	.07964224	.08102188	.08241482	.08524036	.08811835	.11296011
14	.07413609	.07551146	.07690117	.07972332	.08260197	.10758491
15	.06936436	.07073639	.07212378	.07494436	.07782547	.10296276
16	.06518937	.06655879	.06794460	.07076508	.07365013	.09895214
17	.06150579	.06287321	.06425806	.06707966	.06996984	.09544480
18	.05823173	.05959766	.06098205	.06380578	.06670210	.09235654
19	.05530253	.05666740	.05805175	.06087847	.06378177	.08962086
20	.05266645	.05403063	.05541531	.05824574	.06115672	.08718456
21	.05028163	.05164543	.05303075	.05586550	.05878477	.08500455
22	.04811380	.04947748	.05086372	.05370332	.05663140	.08304557
23	.04613465	.04749846	.04888584	.05173075	.05466810	.08127848
24	.04432061	.04568474	.04707347	.04992410	.05287110	.07967900
25	.04265186	.04401650	.04540675	.04826345	.05122044	.07822672
26	.04111163	.04247693	.04386888	.04673196	.04969923	.07690435
27	.03968565	.04105176	.04244553	.04531527	.04829309	.07569717
28	.03836167	.03972871	.04112444	.04400108	.04698967	.07459255
29	.03712914	.03849723	.03989502	.04277878	.04577836	.07357961
30	.03597892	.03734816	.03874811	.04163919	.04464992	.07264891
36	.03042194	.03179973	.03321431	.03615240	.03923285	.06839483
48	.02348503	.02488504	.02633384	.02937500	.03260184	.06389765
52	.02188675	.02329503	.02475603	.02783287	.03110909	.06304617
60	.01933280	.02075836	.02224445	.02539343	.02876797	.06187572
104	.01235457	.01388226	.01551073	.01904966	.02292319	.06014040
120	.01110205	.01266758	.01434709	.01801852	.02204810	.06005519
180	.00843857	.01014267	.01200168	.01610421	.02058274	.06000167
240	.00716431	.00899726	.01101086	.01543312	.02017408	.06000005
300	.00644301	.00839196	.01053224	.01517430	.02005274	.06000000
360	.00599551	.00804623	.01028613	.01507085	.02001604	.06000000
365	.00596625	.00802479	.01027186	.01506575	.02001453	.06000000

Appendix B
Using the TI-83/TI-83+/TI-84+ Graphing Calculators

Functions

Functions are graphed in a rectangular window like the one shown in Fig. 1. The numbers on the x-axis range from **Xmin** to **Xmax**, and the numbers on the y-axis range from **Ymin** to **Ymax**. The distances between tick marks are **Xscl** and **Yscl** on the x- and y-axes, respectively. To specify these quantities, press $\boxed{\text{WINDOW}}$ and type in the values of the six variables. Figure 2 gives the settings associated with the window in Fig. 4. The axis ranges corresponding to this setting are often denoted by $[-4, 4]$ by $[-5, 8]$. (*Note*: To enter a negative number, use the $\boxed{(-)}$ key on the bottom row of the calculator.)

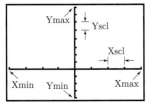

Figure 1. Typical window

Figure 2. Settings for Fig. 4

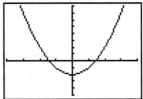

Figure 3. Function declarations

To specify functions, press the $\boxed{\text{Y=}}$ key and type expressions next to the function names $\mathbf{Y_1}, \mathbf{Y_2}, \dots$. (*Note*: To erase an expression, use the arrow keys to move the cursor anywhere on the expression, and press $\boxed{\text{CLEAR}}$.) In the screen of Fig. 3, called the "**Y=** editor," several functions have been specified. The expressions were produced with the following keystrokes:

$\mathbf{Y_1}$: $\boxed{\text{X,T,}\theta\text{,}n}$ $\boxed{x^2}$ $\boxed{-}$ 2

$\mathbf{Y_2}$: 2 $\boxed{\text{X,T,}\theta\text{,}n}$ (Notice that a multiplication sign is not needed.)

$\mathbf{Y_3}$: $\boxed{\text{2nd}}$ $\boxed{\sqrt{}}$ 1 $\boxed{+}$ $\boxed{\text{X,T,}\theta\text{,}n}$ $\boxed{)}$

$\mathbf{Y_4}$: $\boxed{(-)}$ 1 $\boxed{\div}$ $\boxed{(}$ 1 $\boxed{-}$ $\boxed{\text{X,T,}\theta\text{,}n}$ $\boxed{)}$

$\mathbf{Y_5}$: 2 $\boxed{\wedge}$ $\boxed{\text{X,T,}\theta\text{,}n}$ ($\boxed{\wedge}$ is the symbol for exponentiation.)

$\mathbf{Y_6}$: 5 $\boxed{\text{VARS}}$ $\boxed{\blacktriangleright}$ 1 3   (The standard method for entering a function.)

Notice that in Fig. 3 the equal sign in $\mathbf{Y_1}$ is highlighted, whereas the other equal signs are not

highlighted. This highlighting can be toggled by moving the cursor to an equal sign and pressing $\boxed{\text{ENTER}}$. Functions with highlighted equal signs are said to be *selected*. Pressing the $\boxed{\text{GRAPH}}$ key instructs the calculator to graph all selected functions. (*Note*: The words **Plot1 Plot2 Plot3** on the first row are used for statistical plots. The backslash symbol preceding each function is used to specify one of seven possible styles for the graph of the function.)

Press $\boxed{\text{GRAPH}}$ to obtain Fig. 4. Then press $\boxed{\text{TRACE}}$ and press $\boxed{\blacktriangleright}$, the right-arrow, 29 times to obtain Fig. 5. Each time a right- or left-arrow key is pressed, the trace cursor moves along the curve and the coordinates of the trace cursor are displayed.

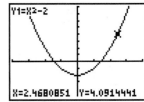

Figure 4   **Figure 5**

To approximate the function value for a specific value of x, move the trace cursor as close as possible to the value of x and read the y-coordinate of the point. For a more precise function value, press $\boxed{\text{2nd}}$ [CALC] **1**, type in a value for **X** (such as 2.5), and press $\boxed{\text{ENTER}}$. See Fig. 6. Alternately, you can just enter 2.5 without first pressing $\boxed{\text{2nd}}$ [CALC] **1**.

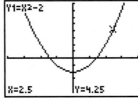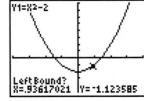

Figure 6   **Figure 7**

A point where the graph crosses the x-axis is called an x-*intercept*. The coordinates of an x-intercept can be approximated by tracing. The x-coordinate of an x-intercept of the function $\mathbf{Y_1}$ is called a *zero* of $\mathbf{Y_1}$ or a *root* of the equation $\mathbf{Y_1} = \mathbf{0}$. Figure 4 shows that $\mathbf{Y_1}$ has a zero between 1 and 2 and the value of the zero is about 1.5. For a precise value of the zero, press $\boxed{\text{2nd}}$ [CALC] **2** and answer the questions. See Fig. 7. Reply to **"Left Bound?"** by moving the trace cursor to a point whose x-coordinate is less than the root and pressing $\boxed{\text{ENTER}}$. Reply to **"Right Bound?"** by moving the trace cursor to a point whose x-coordinate is greater than the root and pressing ENTER. See Fig. 8. Reply to **"Guess?"** by moving the trace cursor near the x-intercept and pressing ENTER. Figure 9 shows the resulting display. An alternate way to find a zero of a function is to reply to the questions by entering appropriate numbers. For instance, after pressing 2nd [CALC] **2**, respond to **"Left Bound?"** by typing in the number 1 and pressing ENTER, respond to **"Right Bound?"** by typing in the number 2 and pressing ENTER, and respond to **"Guess?"** by typing in the number 1.5 and pressing ENTER. The final screen will be identical to Fig. 9.

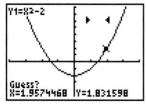

Figure 8

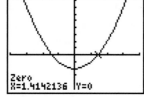

Figure 9

So far, all operations were carried out while looking at the graph of a function. These same operations also can be carried out in the home screen, which is invoked by pressing 2nd [QUIT]. For instance, the function Y_1 can be evaluated at 4 by entering $Y_1(4)$ and pressing ENTER.

Return to the **Y=** editor by pressing Y= and then select Y_2 so that now both Y_1 and Y_2 are selected. Press GRAPH to obtain the graphs of the two functions. Press TRACE and then press the right-arrow key several times. Now press the down-arrow key several times. Each time the down-arrow key is pressed, the trace cursor moves from one curve to the other. The identity of the function containing the cursor is given in the upper left part of the screen. Move the cursor as close as possible to the point of intersection of the two curves to approximate the coordinates of the intersection point. For more precise values, press 2nd [CALC] **5** and answer the questions. Reply to **"First curve?"** by moving the trace cursor to one of the curves and pressing ENTER. Reply to **"Second curve?"** by moving the trace cursor to the other curve and pressing ENTER. Reply to **"Guess?"** by either moving the trace cursor near the point of intersection or typing in a number close to the x-coordinate of the point and pressing ENTER. Figure 10 shows the resulting display.

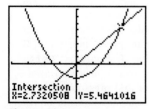

Figure 10

Tables

When you press 2nd [TABLE], a table of function values is displayed with a column for each selected function. (If more than two functions are selected, you must press the right-arrow key to see the third and later functions.) Prior to invoking a table, you should press 2nd [TBLSET] to specify certain properties of the table. See Fig. 11. The values for **X** will begin with the setting of **TblStart** and increase by the setting of **ΔTbl**. For our purposes, the settings for **Indpnt** and **Depend** should always be **Auto**. After you press 2nd [TABLE] to display the table (see Fig. 12), you can use the up- and down-arrow keys to generate further values.

Figure 11

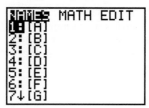

Figure 12

Matrices

The TI-83 can store up to 10 matrices. The matrices are referred to as [**A**], [**B**], [**C**], ... With a TI-83 Plus or TI-84 Plus, pressing 2nd [MATRX] produces the screen in Fig. 13 with the three menus **NAMES**, **MATH**, and **EDIT**. (With an ordinary TI-83, press MATRX.) The **NAMES** menu is used to display the name of a matrix on the home screen, the **MATH** menu is used to perform certain operations on a single matrix, and the **EDIT** menu is used to define a new matrix or alter an existing matrix.

```
NAMES MATH EDIT
1▪[A]
2: [B]
3: [C]
4: [D]
5: [E]
6: [F]
7↓[G]
```

Figure 13

To create a new matrix   Press 2nd [MATRX] ◄ or MATRX ◄ to call up the **MATRIX EDIT** menu, and then press a number corresponding to one of the unused matrix names to obtain a matrix-entry screen. Type in the number of rows, press ENTER, type in the number of columns, and press ENTER to specify the size of the matrix. Then type in the first entry of the matrix, press ENTER, type in the next entry of the matrix, press ENTER, and so on until all entries have been entered. See Fig. 14.

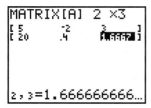

Figure 14

To alter an existing matrix   Press 2nd [MATRX] ◄ or MATRX ◄ to call up the matrix **EDIT** menu, and then press a number corresponding to one of the matrix names to be altered. Move the cursor to any entry you want to change, type in the new number, and press ENTER. You can change as many entries as you like, even the number of rows and columns.

To delete a matrix   Press 2nd [MEM] **2 5** to obtain a list of all matrices that have been created. Use the down-arrow key to select the desired matrix, and then press ENTER DEL **2**.

To display the name of a matrix on the home screen   Press 2nd [MATRX] or MATRX to obtain a list of all matrix names. Use the down-arrow key to select the desired matrix, and then press ENTER. Alternately, from the matrix **NAMES** menu, type the number associated with the matrix.

Lists

A list can be thought of as a sequence or an ordered set of up to 999 numbers. Although lists can have custom names, we will use the built-in names $L_1, L_2, \ldots, L_6$ that are found above the numeric keys. To display L_1 on the home screen, press 2nd [L_1]. The elements of list L_1 can be referred to as $L_1(1), L_1(2), L_1(3), \ldots$. A set of numbers can be placed into a list by enclosing them with set braces and storing them in the list. However, often the stat list editor provides the best way to place numbers into lists and to view lists.

To invoke the stat list editor, press STAT **1**. Initially, the stat list editor has columns labeled L_1, L_2, and L_3. See Fig. 15. Three more columns for lists are off the screen.

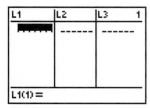

Figure 15

To place a number into a list in the stat list editor   Move the cursor to the dashed line, type in the number, and press either the down-arrow key or the ENTER key.

To remove a number from a list in the stat list editor   Move the cursor to the number and press DEL.

To delete all numbers in a list from the stat list editor   Move the cursor to the list name at the top of the screen and press CLEAR ENTER.

To create a scatterplot of a set of points whose x-coordinates are in L₁ and whose y-coordinates are in L₂   Press 2nd [STAT PLOT] **1**, create the screen in Fig. 16, and press GRAPH. (*Note:* Check that the window settings are adequate for displaying the points.)

Histograms

The following steps display a frequency histogram with rectangles centered above the integers $a, a+1, \ldots, b$.

- Press STAT **1** to invoke the list editing screen shown in Fig. 15.
- Enter each number from a to b having nonzero frequency in the L_1 column and its associated frequency to its right in the L_2 column. Or, ignore the L_2 column and place each number (repeated according to its frequency) in the L_1 column.
- Press 2nd [STAT PLOT] **1** to invoke a screen similar to Fig. 16.

Figure 16

- Select "On" in the second line, the histogram icon in the Type line, and L_1 in the Xlist line. If you ignored the L_2 column in the second step, place 1 in the Freq line; otherwise, place L_2 in the Freq line.
- Press WINDOW to invoke the window screen in Fig. 2.
- Set Xmin = $a - .5$, Xmax = $b + .5$, Xscl = 1, Ymin $\approx -.3*$(greatest frequency), Ymax $\approx 1.2*$(greatest frequency), Yscl $\approx .1*$(greatest frequency). *Note:* These Y settings allow ample space for the display of values while tracing.
- Press GRAPH to view the histogram.
- To view the height of a rectangle, press TRACE and use the arrow keys to move the cursor to the top center of the rectangle. The values of *min*, *max*, and *n* will be the x-coordinate of the left side of the rectangle, the x-coordinate of the right side of the rectangle, and the height of the rectangle, respectively.

A relative frequency histogram can be displayed by placing the relative frequencies in the L_2 column of the list editor. If so, the *greatest relative frequency* should be used in place of the *greatest frequency* when setting the values of Ymin, Ymax, and Yscl.

Appendix C
Spreadsheet Fundamentals

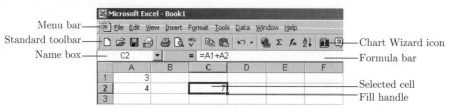

Figure 1.   Upper-left corner of a Microsoft Excel spreadsheet.

Cell

A *cell* is a small rectangle located at the intersection of a column and a row. Its *address* or *reference* (also known as its *relative address* or *relative reference*) consists of its column letter followed by its row number. When you click on a cell, its border thickens and its address is displayed in the Name box. Such a cell is said to be *selected* or *active*. After you type text, a number, or a formula into an active cell and press the Enter key, the text, number, or the value of the formula is displayed in the cell. The exact information you typed (known as the *content* of the cell) is displayed in the Formula bar whenever you reselect the cell.

Range

A *range of cells* or *range* is a rectangular array of more than one cell. The *address* or *reference* of a range consists of the address of the cell in the upper-left corner of the range, a colon (:), and then the address of the cell in the lower-right corner. To *select* a range, hold down the left mouse button while you drag the cursor from the upper-left cell to the lower-right cell. You can fill a range by entering information into each cell individually. However, often you fill an entire range with a single formula. If so, you should press Ctrl+Shift+Enter (instead of just Enter) after typing in the formula. Such a formula, called a *range formula* or an *array formula*, is automatically displayed in the Formula bar surrounded by braces.

Moving a Cell or Range

You can move a selected cell or range by dragging its border, or by pressing Ctrl+X, clicking on the new location, and pressing Ctrl+V. If the content of the original cell or range uses relative addresses, these addresses automatically change to reflect the new position of the cell or range. (You can prevent an address from changing by preceding its column letter and row number with dollar signs. Such an address is called an *absolute address*.)

Naming a Cell or Range

As an alternative to identifying cells or ranges with addresses, you can give a *name* to a cell or range. To specify a name, select the cell or range, click on the Name box, type in the name, and press the Enter key. (*Note*: The name will not be registered if you forget to press the Enter key.) Names can contain up to 255 characters (no spaces), must begin with a letter or underscore character, and cannot look like addresses. Deleting the contents of a cell or range does not delete its name. To delete a name, click on Name in the Insert menu, click on Define, click on the name in the list of names that appears, click the Delete button, and click the OK button.

Entering a Formula

Formulas are what makes a spreadsheet a spreadsheet. To specify that the entry typed into a selected cell or range is a formula, you first type an equal sign (=). The equal sign is followed by a expression containing numbers, addresses, names, functions, and operators such as +, −, *, /, and ^ (exponentiation).

Copying a Formula

Select the cell or range whose content is the original formula. There are two methods to copy the formula. (*Note*: With both methods, all relative addresses will be altered appropriately.)

First method: Press Ctrl+C, select the location(s) to hold the copy or copies, and press Ctrl+V.

Second method: Hover the mouse pointer over the cell's fill handle until the pointer becomes a crosshair (+). Drag the pointer vertically or horizontally along the cells that are to hold the copy or copies.

Installing Add-Ins

Two Excel tools that we use in this book are called *Solver* and *Data Analysis*. Solver is used to find solutions to systems of linear equations and to linear programming problems. Data Analysis is used to analyze data. To see if these tools have been installed on your computer, click the

Tools button on the menu bar and see if the tools are listed. If not, click on Add-Ins in the Tools menu, check the boxes next to Analysis Toolpack and Solver, and then click on the OK button. In some cases, they will not become available until you exit Excel and reinvoke it. Also, you might be instructed to insert your Microsoft Office CD and run the setup routine to install them.

Entering a Formula by Pointing

Instead of typing the address of a cell into a formula manually, just click on the cell. A dotted rectangle will surround the cell, the cell's name or address will appear in the formula, and the word *Point* will appear in the status bar below the spreadsheet. To enter the address of a range, click on the upper-left cell of the range and drag the mouse cursor to the lower-right cell of the range. A dotted line will surround the range. When you release the mouse, the range's name or address will appear in the formula.

Converting Formulas to Values

Sometimes the formulas that were used to fill a cell or range are no longer needed. To replace the contents of the cells with their values, select the cell or range, press Ctrl+C, click on *Paste Special* in the Edit menu, click on the Values circle, and click on the OK button.

Summing a Column or Row of Numbers

Click on the top number in the column or row, drag the mouse pointer to one or more cells beyond the column or row of numbers, and click on the *Autosum button* (Σ) in the Standard toolbar. The contents of the last cell will be a formula containing the SUM function and the value of the cell will be the sum of the numbers in the column or row.

Creating a Histogram

Excel's *Chart Wizard* can create a histogram from a frequency or probability distribution with a few simple steps. In this discussion, we refer to the first column of numbers in the distribution table as the "category range" and the second column of numbers as the "values range." A histogram can be created as follows:

- Select the values range of the table.
- Click on the Chart Wizard icon in the Standard toolbar. A window titled "Chart Wizard-Step 1 of 4-Chart Type" will appear.
- Click the Next button. A window titled "Chart Wizard-Step 2 of 4-Chart Source Data" will appear.
- Click the Series tab at the top of the window.
- Click the box labeled "Category (X) axis labels:".
- Use pointing to fill the box with the address of the category range. That is, click on the top cell of the category range, and drag the mouse cursor down to the

lowest cell of the range. A dotted line will surround the range. When you release the mouse, the range's name or address will appear in the Category (X) axis labels box.
- Click the Finish button to create the histogram.

Notes

1. You can get rid of the Series legend to the right of the histogram by right-clicking on it and selecting Clear.
2. You can determine the height of one of the rectangles by hovering the mouse pointer on the rectangle.
3. You can resize the histogram by clicking on it and then dragging one of the eight small black squares found at the corners and the centers of the sides.
4. You can remove the spaces between the rectangles by right-clicking on one of the rectangles, clicking on Format Data Series, clicking the Options tab, and setting the Gap width to zero.
5. You can change the font size (and other characteristics) of the numbers on an axis by right-clicking on the axis, clicking on Format Axis, clicking on the Font tab, making the changes, and clicking the OK button.

Simulating a Random Sample from a Probability Distribution

The following steps use Excel's Data Analysis tool to create a random sample from a discrete probability distribution. The sample numbers will be displayed in a range of m rows and n columns.

- Place the probability table in the first two columns of the spreadsheet. That is, column A should contain the possible outcomes (which must be numeric) and column B should contain the associated probabilities.
- Click on Data Analysis in the Tools menu.
- Double-click on Random Number Generation in the Data Analysis dropdown list.
- Enter the number of columns (n) of random numbers into the "Number of Variables" box.
- Enter the number of random numbers (m) that you would like to generate in each column into the Number of Random Numbers box.
- Select "Discrete" as the Distribution.
- Place the address of the range containing the probability distribution in the "Value and Probability Input Range" box. This can be accomplished by placing the cursor in the box, clearing the box if necessary, dragging the mouse pointer from cell A1 to the lower-right corner of the probability table, and releasing the mouse button.
- Click on the Output Range circle, and key in the address of an $m \times n$ array of cells.

- Click on the OK button to generate the desired random sample.

Creating a Frequency Distribution Table and Histogram for the Random Sample Generated Above

The following steps use Excel's Data Analysis tool to create a frequency distribution and optionally a histogram for a range of cells containing the random numbers generated above.

- Click on Data Analysis in the Tools menu.
- Double-click on Histogram in the Data Analysis list.
- Place into the Input Range box the address of the range of cells containing the random sample.
- Place into the Bin Range box the address of the range of possible outcomes; that is, the first column of the probability table.
- Click the Output Range circle and key in the address of any cell to the right of the cells already filled.
- If you would like also to display a histogram, click the small rectangle to the left of Chart Output.
- Click the OK button to see the frequency distribution table and possibly the histogram. (*Note*: The word More and the number 0 appear in the last row of the table and in the histogram. To delete them, select the row, click the right mouse button, click Delete, select "Shift cells up", and click OK.)

Using Goal Seek to Solve an Equation Having One Unknown

Suppose the contents of cell B1 contains a formula whose value depends on the value of A1, and you know the value you would like B1 to have. Then the *Goal Seek* tool can be used to determine the value of A1 that will produce the sought-after value in B1. For example, you can use Goal Seek to find the interest rate (compounded annually) for which $100 will grow to $146 after 11 years. Specify the percentage format[1] for cell A1, enter

$$=100*(1 + A1)\char94 11$$

into cell B1, click on the Tools menu, and click on the Goal Seek tool. A Goal Seek window will appear. Fill in the window as shown in Fig. 2, and then click the OK button. The solution, .0035 (that is, 3.50%), will appear in cell A1 of the spreadsheet. *Note*: Using the Goal Seek tool is sometimes referred to as *backsolving*.

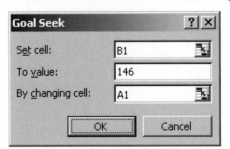

Figure 2

Using the Insert Function Dialog Box

To invoke the Insert Function dialog box with Excel 2002, click on the Insert Function button located next to the Formula bar. With earlier versions of Excel, click on the Paste Function button located on the Standard toolbar. The dialog box allows you to display the more than 300 available functions and descriptions of their arguments and what they do. When you select a function and click on OK (or double-click on the function), a Function Arguments dialog box appears to help you specify the arguments of the function. You can fill in each argument box by typing or pointing. To assist you when using the pointing method, you can click on the icon at the right side of the argument box to temporarily collapse the dialog box and thereby make more of the spreadsheet visible. After you have filled the boxes for all the arguments, press OK (or the Enter key) to fill a single cell with the value of the function, or press Ctrl+Shift+Enter to fill a range of cells.

---

[1]To specify the percentage format for a selected cell or range of cells, click on Format, click on Cells, select the Number tab, and click on Percentage.

Using Excel's Financial Functions

The five most important financial functions for our purposes are shown in Table 1.

The values of these functions and their parameters are signed according to their flows. Money received by you is given a positive value, and money paid by you is given a negative value. For instance, the values of savings account deposits and loan payments are negative, and the values of savings account withdrawals and loan amounts are positive. Table 2 shows how to use these functions to solve some examples from the text.

Table 3 contains some other useful financial functions. Example 5 of Section 10.3 can be solved by keying

$$=\textbf{IPMT(0.75\%,A1,360,112475,0)}$$
$$-\textbf{PPMT(0.75\%,A1,360,112475,0)}$$

into B1 and using Goal Seek to set cell B1 to the value 0 by changing cell A1.

TABLE 1		
Function	**Used to calculate**	**Function and required arguments**
FV	future value	FV(rate, nper, pmt, pv)
PV	present value	PV(rate, nper, pmt, fv)
PMT	periodic payment	PMT(rate, nper, pv, fv)
NPER	number of interest periods	NPER(rate, pmt, pv, fv)
EFFECT	effective rate of interest	EFFECT(nominal_rate, nper)

TABLE 2		
Section	**Example**	**Function and required arguments**
10.1	3	FV(1%, 4, 0, −100)
10.1	5	PV(0.5%, 104, 0, 10000)
10.1	7	EFFECT(3.65%, 4)
10.2	1	FV(0.5%, 60, −100, 0)
10.2	3	PMT(2%, 16, 0, 30000)
10.2	4	PV(1.5%, 8, 3000, 0)
10.2	5	PMT(0.5%, 12, −10000, 0)
10.2	6	NPER(0.5%, −100, 0, 10000)
10.3	2(a)	PMT(0.75%, 360, 112475, 0)
10.3	4	PV(2%, 12, 200, 1000)

TABLE 3		
Function	**Used to calculate**	**Function and required arguments**
RATE	interest rate per period	RATE(nper, pmt, pv, fv)
IPMT	interest portion of loan payment number *per*	IPMT(rate, per, nper, pv, fv)
PPMT	principal reduction portion of loan payment number *per*	PPMT(rate, per, nper, pv, fv)
NOMINAL	nominal rate of interest	NOMINAL(effective_rate, nper)

Appendix D
Using the TI-89 Graphing Calculator

Functions

Functions are graphed in rectangular windows like the one shown in Fig. 1. The numbers on the x-axis range from **xmin** to **xmax**, and the numbers on the y-axis range from **ymin** to **ymax**. The distance between the tick marks are **xscl** and **yscl** on the x- and y-axes, respectively. To specify these quantities, press ◆ [WINDOW] and type in the values of these variables. Figure 2 shows the settings associated with the windows in Figs. 1 and 4. The axis ranges associated with this setting are often written $[-4, 4]$ by $[-5, 8]$. (*Note*: To enter a negative number you must use the $\boxed{(-)}$ key on the bottom row of the calculator.)

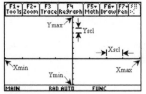

Figure 1. Typical window

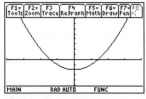

Figure 2. Settings for Figs. 1 and 4

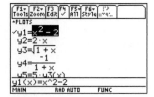

Figure 3. Function declarations

To input functions, press ◆ [Y=] and type expressions next to the function names **y1**, **y2**, ... (*Note*: To erase an expression, use the arrow keys to move to that function and press $\boxed{\text{CLEAR}}$ or $\boxed{\leftarrow}$. To edit an expression once it has been entered, use the arrow keys to highlight the expression and press $\boxed{\text{F3}}$.) In the screen of Fig. 3, called the "**Y=** editor", several functions have been specified. The expressions were input with the following keystrokes:

y1: $\boxed{\text{X}}$ $\boxed{\frown}$ 2 $\boxed{-}$ 2 ($\frown$ is the symbol for exponentiation.)

y2: 2 $\boxed{\text{X}}$ (Notice that a multiplication symbol is not necessary.)

y3: $\boxed{\text{2nd}}$ $\boxed{\sqrt{\ }}$ 1 $\boxed{+}$ $\boxed{\text{X}}$ $\boxed{)}$

y4: $\boxed{(-)}$ 1 $\boxed{\div}$ $\boxed{((}$ 1 $\boxed{-}$ $\boxed{\text{X}}$ $\boxed{)}$

y5: 5 $\boxed{\text{Y}}$ 3 $\boxed{((}$ $\boxed{\text{X}}$ $\boxed{)}$

Notice that in Fig. 3 there is a check mark in front of **y1** while the other functions do not have check marks in front of them. The check mark can be toggled with the $\boxed{\text{F4}}$ button. Functions with a check mark are said to be

selected. Pressing ◆ [GRAPH] instructs the calculator to graph all selected functions. *Note*: Arrowing up from **y1** shows **Plot1**, **Plot2**, (up to **Plot6**) which are used for statistical plots. There are eight possible styles for the graph of a function found in the STYLE menu accessed by $\boxed{\text{2nd}}$ [F6].

Press ◆ [GRAPH] to obtain Fig. 4. Then press $\boxed{\text{F3}}$ to trace and press $\boxed{\blacktriangleright}$, the right arrow, 49 times to obtain Fig. 5. Each time a right- or left-arrow key is pressed, the trace cursor moves along the curve and the coordinates of the trace cursor are displayed along the bottom.

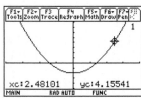

Figure 4

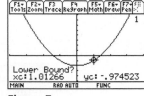

Figure 5

To approximate the function value for a specific value of x, move the cursor as close as possible to the value of x and read the y-coordinate of the point. For a more precise function value, there are two options. While using the trace cursor, input the value of x and press $\boxed{\text{ENTER}}$. Alternatively, press $\boxed{\text{F5}}$, for the Math menu, then 1, then type in the value of x (such as 2.5), and press $\boxed{\text{ENTER}}$. See Fig. 6.

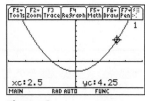

Figure 6

Figure 7

A point where the graph crosses the x-axis is called an *x-intercept*. Tracing can approximate the coordinates of an x-intercept. The x-coordinate of an x-intercept of the function **y1** is called a *zero* of **y1** or a *root* of the equation **y1=0**. Figure 4 shows that **y1** has a zero between 1 and 2 and the value is about 1.5. For a more precise value of the zero, press $\boxed{\text{F5}}$ **2** and answer the questions. See Fig. 7. Reply to "**Lower Bound?**" by either moving the trace cursor to a point whose x-coordinate is less than the root and pressing $\boxed{\text{ENTER}}$ or by simply typing in a value for the x-coordinate which is less than the root (say 1) and pressing $\boxed{\text{ENTER}}$. See Fig. 8. Reply to "**Upper Bound?**" by either moving the trace cursor to a point whose x-coordinate is greater than the root and pressing $\boxed{\text{ENTER}}$ or by simply typing in a value for the x-coordinate which is greater than the root (say 2) and pressing $\boxed{\text{ENTER}}$. Figure 9 shows the resulting display.

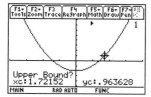

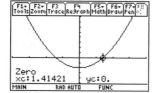

Figure 8   **Figure 9**

Many operations can be carried out using the home screen, which is invoked by pressing $\boxed{\text{HOME}}$. For instance, the function **y1** can be evaluated at -2 by entering **y1(-2)** and pressing $\boxed{\text{ENTER}}$.

Return to the **Y=** editor by pressing $\boxed{\blacklozenge}$[Y=] and then select **y2** so that both **y1** and **y2** are selected. Press $\boxed{\blacklozenge}$ [GRAPH] to obtain the graphs of both functions. Press $\boxed{\text{F3}}$ to trace and then press the right-arrow key several times. Now press the down-arrow key several times. Notice that each time the down-arrow key is pressed, the trace cursor moves from one curve to the other. The identity of the function containing the cursor is indicated in the upper right hand part of the screen. Move the cursor as close as possible to the point of intersection to approximate the coordinates of the point of intersection. For more precise values, press $\boxed{\text{F5}}$ **5** and answer the questions. Reply to "**1st Curve?**" by moving the trace cursor to one of the curves and pressing $\boxed{\text{ENTER}}$. Reply to "**2nd Curve?**" by moving the trace cursor to the other curve and pressing $\boxed{\text{ENTER}}$. Reply to "**Lower Bound?**" by either moving the trace cursor to the left of the point of intersection or typing in a number less than the x-coordinate of the point and pressing $\boxed{\text{ENTER}}$. Reply to "**Upper Bound?**" by either moving the trace cursor to the right of the point of intersection or typing in a number greater than the x-coordinate of the point and pressing $\boxed{\text{ENTER}}$. Figure 10 shows the resulting display.

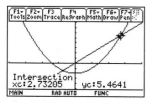

Figure 10

Tables

When $\boxed{\blacklozenge}$ [TABLE] is pressed, a table of function values is displayed showing a column for each selected function. (If more than three functions are selected, use the right-arrow key to see the fourth and later functions.) Prior to invoking a table, press $\boxed{\blacklozenge}$ [TBLSET] to specify certain properties of the table. See Fig. 11. The values for **x** will

begin with the setting of **tblStart** and increase by the setting of Δ**tbl**. For our purposes, the setting for **Graph $\leftrightarrow$ Table** should be **OFF** and the setting for **Independent** should be **AUTO**. After the table is displayed by pressing $\boxed{\blacklozenge}$ [TABLE] (see Fig. 12), use the up- and down-arrow keys to generate further values.

Figure 11   **Figure 12**

Matrices

Although matrices can be created directly on the home screen, it is usually better to use the Data/Matrix Editor

To create a new matrix   Press $\boxed{\text{APPS}}$, move the cursor to Data/Matrix Editor, and press $\boxed{\text{ENTER}}$ followed by **3** to open a **New** matrix. Press $\boxed{\blacktriangleright}$ **2** to set the **Type** to **Matrix**. Suppose we want to input a 2×3 matrix. Arrow down to **Variable** and type in a name for the matrix (say **a**). Next input the **Row dimension** (2) and the **Col dimension** (3). See Fig. 13. Then press $\boxed{\text{ENTER}}$ twice to obtain the matrix-entry screen. Finally, type in the first entry of the matrix, press $\boxed{\text{ENTER}}$, type in the next entry of the matrix, press $\boxed{\text{ENTER}}$, and so on until all entries have been entered. See Fig. 14.

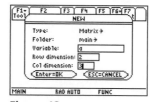

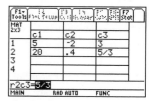

Figure 13   **Figure 14**

To alter an existing matrix   Press $\boxed{\text{APPS}}$, move the cursor to Data/Matrix Editor, and press $\boxed{\text{ENTER}}$ followed by **2** to **Open** an existing matrix. Press $\boxed{\blacktriangleright}$ **2** to set the **Type** to **Matrix**. Arrow down to **Variable** and press $\boxed{\blacktriangleright}$ to view a list of all saved matrices. Use the down-arrow key to select the matrix you wish to change and press $\boxed{\text{ENTER}}$. Press $\boxed{\text{ENTER}}$ again to obtain the matrix-entry screen. Use the arrow keys to highlight the entry to be altered, type the new value, and press $\boxed{\text{ENTER}}$. Change as many entries as needed. If an extra row or column needs to be added, press $\boxed{\text{2nd}}$ [F6] **1**, and then press **2** to insert a new row above the current row or press **3** to insert a new column to the left of the current column. If a row or column needs to be deleted, press $\boxed{\text{2nd}}$ [F6] **2**, and then press **2** to remove the current row or press **3** to remove the current column. The matrix can also be directly resized by pressing 2nd [F6] **6** and then inputting the new row and column dimensions.

To delete a matrix   Press 2nd [Var-Link] to view a list of all variables that have been saved including matrices. Use the down-arrow key to highlight the matrix to be deleted, and then press ←. A screen appears asking for confirmation; press ENTER to delete. If there are multiple matrices to be deleted, highlight each matrix in turn and press F4 each time, then press ←. The confirmation screen will appear; press ENTER to delete.

To display the name of a matrix on the home screen   To display a matrix on the home screen, simply type its name on the entry line and press ENTER. Alternately, press 2nd [Var-Link] to view a list of all variables that have been saved including matrices. Use the down-arrow key to highlight the matrix to be displayed, press ENTER to place the matrix name on the entry line, and press ENTER again to see the matrix displayed on the home screen.

Lists

A list can be thought of as a sequence or an ordered set of up to 999 numbers. There are two basic ways to create a list. In the home screen, a set of numbers can be placed into a list by enclosing the numbers with set braces and storing in a variable. See Fig. 15. Lists can also be created in the Data/Matrix Editor by pressing APPS, selecting the Data/Matrix editor, and pressing ENTER followed by **3** to open a new **List**. Press ▶ **3** to set the **Type** to **List**. Then type in a name for the **Variable**. Press ENTER twice to open the list editor. Type the numbers into the first column that is labeled **c1**. (*Note*: There are 99 columns available.) See Fig. 16.

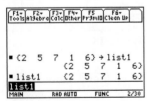

Figure 15

Figure 16

To place a number into a list in the list editor
Type in the number, and press either an arrow key or the ENTER key.

To remove a number from a list in the list editor
Move the cursor to the number and press ←.

To remove all numbers in a list in the list editor
Move the cursor to the column, then press 2nd [F6] **5** to clear the entire column.

To create a scatter plot of a set of points whose *x*-coordinates are in c1 and whose *y*-coordinates are in c2   First note that when a second column is added, the variable type automatically changes from **List** to **Data**. From the list/data editor press F2 F1 to see Fig. 17. Change the **Plot Type** to **Scatter**, and change the **Mark** to **Dot** (or **Box** or **Cross** or **Plus** or **Square**). Set *x* to **c1** and **y** to **c2**. Leave **Use Freq and Categories?** set to **NO**. See Fig. 18. Press ENTER to save the plot definitions. To view the scatter plot, first set the window to the appropriate ranges and then press ♦ [GRAPH] or type F2 **9** (ZoomData) to have the calculator automatically set the ranges to the data values and graph the scatter plot. (*Note*: Make sure all functions have been deselected before you graph the scatter plot.)

 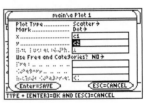

Figure 17   **Figure 18**

Histograms

The following steps display a frequency histogram with rectangles centered above the integers $a, a+1, \ldots, b$.

- Press APPS, move the cursor to Data/Matrix Editor, and press ENTER followed by **3** to open a **New** data screen. Then type in a name for the variable. Press ENTER twice to open a data editing screen similar to that in Fig. 16.

- Enter each number from a to b having nonzero frequency in the **c1** column and its associated frequency to its right in the **c2** column. Or, ignore the **c2** column and place each number (repeated according to its frequency) in the **c1** column.

- Press F2 F1 to invoke a screen similar to Fig. 17. Or, press ♦ [Y=], select **Plot1**, and press F3 to invoke a screen similar to Fig. 17. (*Note*: Make sure to uncheck any other functions defined in the **Y=** editor.)

- Press ▶ **4** to set the **Plot Type** to **Histogram**. Set **x** to **c1**, set **Hist. Bucket Width** to 1. If the frequencies are in **c2**, set **Use Freq and Categories?** to **YES** by pressing ▶ **2** and then type **c2** in the **Freq** box. If only **c1** is used, leave **Use Freq and Categories?** set to **NO**. Press ENTER to save the plot definitions.

- Press ♦ [WINDOW] to invoke a window screen similar to Fig. 2.

- Set **xmin** $= a - .5$, **xmax** $= b + .5$, **xscl** $= 1$, **ymin** $\approx -.3*$(greatest frequency), **ymax** $\approx 1.2*$(greatest frequency), **yscl** $\approx .1*$(greatest frequency). (*Note*:

These y settings will allow ample space for the display of values while tracing.)

- Press ◆ [GRAPH] to display the histogram.
- To view the height of each rectangle, press F3 and use the arrow keys to move the cursor to the top center of the rectangles. The value of **min** is the x-coordinate of the left edge of the rectangle, the value of **max** is the x-coordinate of the right edge of the rectangle, and the value of n is the height of the rectangle.

A relative frequency histogram can be displayed by placing the relative frequencies in the **c2** column of the data editor. If so, the *greatest relative frequency* should be used in place of the *greatest frequency* when setting **ymin**, **ymax**, and **yscl**.

Answers

CHAPTER 1

EXERCISES 1.1

1., 3., 5.

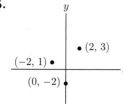

7.

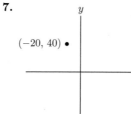

9. e   **11.** Yes   **13.** No   **15.** $m = 5, b = 8$
17. $m = 0, b = 3$ **19.** $y = -2x + 3$
21. $x = \frac{5}{3}$ **23.** $(2, 0), (0, 8)$ **25.** $(7, 0)$, none

27.

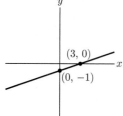

29.

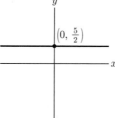

31.

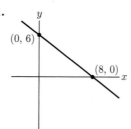

33.

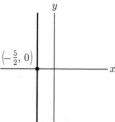

35. a, b, c, e   **37. (a)** L_3 **(b)** L_1 **(c)** L_2 **39. (a)** $4\frac{2}{3}$ minutes, or 4 minutes, 40 seconds **(b)** 72° water was placed in the kettle. **(c)** No

41.(a)

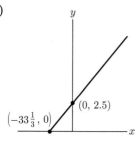

(b) In 1960, 2.5 trillion cigarettes were sold.
(c) 1980
(d) 7 trillion

43. (a)

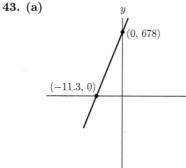

(b) In 1997 the yearly insurance rate for a small car was $678.
(c) $858
(d) The year 2007

45. $y = 0$ **47.** $y = b$ **49.** $2x - y = -3$ **51.** $x + 0 \cdot y = -3$ **53.** $2x + 3y = -15$

57. (a)

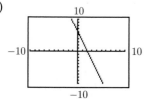

59. (a)

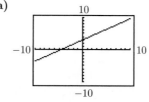

61. $[-10, 110]$ *by* $[-10, 60]$

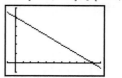

(b) 0   **(c)** $(2, 0), (0, 6)$

(b) $\frac{13}{3}$ **(c)** $(-4.5, 0), (0, 3)$

EXERCISES 1.2

1. False   **3.** True   **5.** $x \geq 4$ **7.** $x \geq 3$ **9.** $y \leq -2x + 5$ **11.** $y \geq 15x - 18$ **13.** $x \geq -\frac{3}{4}$ **15.** Yes   **17.** No

19. Yes   **21.** Yes   **23.**

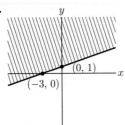

25.

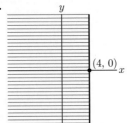

27.

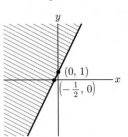

29.

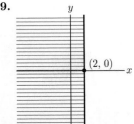

31.

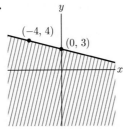

33.

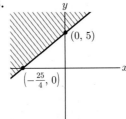

35.

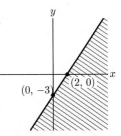

37.

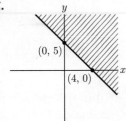

39.

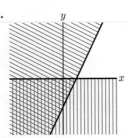

41.

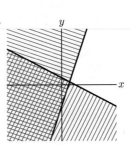

43.

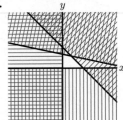

45. Yes   **47.** No   **49.** Below   **51.** Above   **53.** $\begin{cases} y \geq 2x - 1 \\ y \leq 2x \end{cases}$ **55.** d   **57.** e   **59.** (a) $(6, 2.5)$ (b) Above

61.

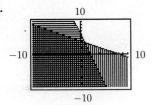

EXERCISES 1.3

1. $(2, 3)$ **3.** $(2, 1)$ **5.** $(12, 3)$ **7.** Yes
9. $x = \frac{10}{3}, y = \frac{1}{3}$ **11.** $x = -\frac{7}{9}, y = -\frac{22}{9}$
13. $A = (3, 4), B = (6, 2)$
15. $A = (0, 0), B = (2, 4), C = \left(5, \frac{11}{2}\right), D = (5, 0)$

17.

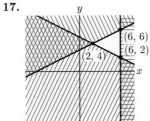

19.

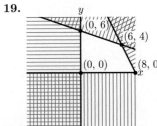

21.

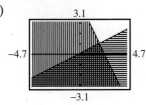

(0, 8)
(1, 4)
(3, 2)
(9, 0)

23. (a) $2.00 **(b)** $.05 or less **25.** 29,500 units; $3.00 **27.** 180; $13.60

29. (18, 63); Method A is best for calls of more than 18 minutes duration.

31. d **33.** (1.69, 3.38) **35.** (1, 1)

37. (a)

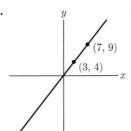

(b) (2, 1) **(c)**

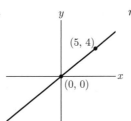

(d) Yes

EXERCISES 1.4

1. $\frac{2}{3}$ **3.** 5 **5.**

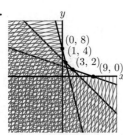

$m = \frac{5}{4}$ **7.**

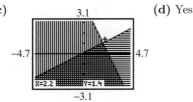

(5, 4)
(0, 0)

$m = \frac{4}{5}$ **9.** Undefined

(7, 9)
(3, 4)

11.

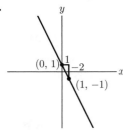

(0, 1)
−2
(1, −1)

13.

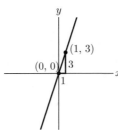

(0, 0)
(1, 3)
3
1

15. $y = -2x + 7$ **17.** $y = -2x + 4$

19. $y = \frac{1}{4}x + \frac{3}{2}$ **21.** $y = -x$

23. $y = 3$ **25.** (0, 3)

27. Each unit sold yields a commission of $5. In addition, she receives $60 per week base pay.

29. (a) (0, 1200); at $1200 no one will buy the item.

(b) (400, 0); even if the item is given away, only 400 will be taken.

(c) −3; to sell an additional item, the price must be reduced by $3.

(d) $150

(e) 300 items

(f)

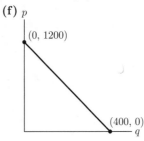

(0, 1200)
(400, 0)

31. (a) $y = 90x + 5000$ **(d)**

(b) $5000

(c) $90

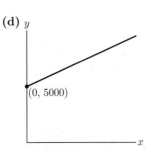

(0, 5000)

33. (a) $30,000

(b) 60 coats

(c) (0, 0); if no coats are sold, there is no revenue.

(d) 100; each additional coat yields an additional $100 in revenue.

35. (a)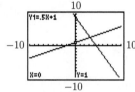

(b) 17,600 gallons

(c) 12,000 gallons

(d) $(0, 30{,}000)$; the tank holds 30,000 gallons.

(e) $(75, 0)$; the tank is empty after 75 days.

37. (a) $y = .1x + 160$ **(b)** \$260 **(c)** \$3400   **39.** $y = 3x - 1$ **41.** $y = x + 1$ **43.** $y = -7x + 35$ **45.** $y = 4$

47. $y = \frac{1}{2}x$ **49.** $y = -2x$ **51.** 5; 1; −1   **53.** $-\frac{5}{4}; -\frac{3}{2}; -\frac{3}{4}$ **55. (a)** (C)   **(b)** (B)   **(c)** (D)   **(d)** (A)

57. $F = \frac{9}{5}C + 32$ **59.** $y = \frac{3097}{14}x + 2035$; \$2919.86   **61.** $y = -\frac{1}{2}x + 25$; 21 mpg   **63.** $y = (96{,}463/22)x + 184{,}867$;

$\approx 303{,}253$ **65.** Counterclockwise   **67.** $y \geq 4x + 3$ **69.** $\begin{cases} y \leq -\frac{1}{2}x + 4 \\ y \leq -x + 5 \\ y \geq x - 3 \\ x \geq 0, y \geq 0 \end{cases}$ **71.** $k = 9$ **73.** $a = -.05$ **77.** c

79. e   **81.** c   **83.** d   **85.**

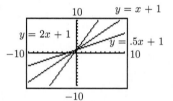

87.

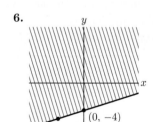

EXERCISES 1.5

1. 4   **3.** 6.70   **5.** $m = -1.4; b = 8.5$ **7.** $y = 4.5x - 3$ **9.** $y = -2x + 11.5$ **11. (a)** $y = .338x + 21.6$ **(b)** About 393

13. (a) $y = .448x + 14.4$ **(b)** About 22.5%   **(c)** 2005   **15. (a)** $y = .155x + 72.4$ **(b)** 77 **(c)** 80 **(d)** 86 This is an example of a fit that is not capable of extrapolating beyond the given data.   **17. (a)** $y = .046x + 2.669$ **(b)** \$3.22 **(c)** 2008

CHAPTER 1: SUPPLEMENTARY EXERCISES

1. $x = 0$ **2.**

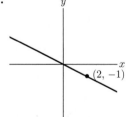

3. $\left(2, -\frac{4}{5}\right)$

4. $\frac{3}{4}$

5. $y = -\frac{1}{2}x + 5$

6.

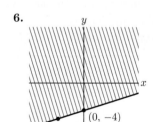

7. Yes

8. (3, 5)

9. $y = \frac{1}{5}x + 13$

10. 10

11. (5, 0)

12.

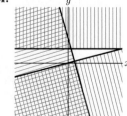

13. (7, 10)   **14.**

15. (0, 7)

16. The rate is \$35 per hour plus a flat fee of \$20.

17. No   **18.** $y = \frac{2}{3}x - 2$ **19.** d

20. $y \leq \frac{2}{3}x + \frac{3}{2}$

21. $y \geq 2.4x - 5.8$

22. $x = -.3, y = .4$

23. $y = -\frac{2}{5}x + \frac{7}{5}$

24. $x \geq \frac{6}{5}$ **25.** $m = -2$;
y-intercept: $(0, 8)$;
x-intercept: $(4, 0)$;

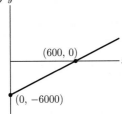

26. No

28. (a) (C) **(b)** (A) **(c)** (B) **(d)** (D)

29. (a) L_3 **(b)** L_1 **(c)** L_2

30. $\begin{cases} y \leq -\frac{7}{8}x + 5 \\ y \geq \frac{8}{7}x - \frac{43}{14} \\ x \geq 0, y \geq 0 \end{cases}$; $\left(\frac{43}{16}, 0\right)$

31. 300 units; \$2

32. $(0, 0), (10, 0), (9, 5), (6, 7), (0, 4)$

33. $y = x + \frac{5}{2}$

34. (a) $y = 10x - 6000$
(b) x-intercept: $(600, 0)$,
y-intercept: $(0, -6000)$

(c)

35. (a) A: $y = .1x + 50$;
B: $y = .2x + 40$
(b) B **(c)** A
(d) 100 miles

36. (a) $y = .019x + .51$
(b) 2000

37.

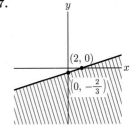

38. \$5000 **39.** $\begin{cases} y \leq \frac{5}{4}x + 5 \\ y \leq -\frac{2}{5}x + 2 \\ y \geq \frac{3}{5}x - 3 \\ y \geq -\frac{5}{2}x - 5 \end{cases}$ **40.** $\begin{cases} y \leq -\frac{2}{3}x + 2 \\ y \geq -3 \\ x \geq -2 \\ x \leq 4 \end{cases}$ **41.** $y = \frac{53,500}{3}x + 365,000$; 507,667

42. 102,145 **43.** 19.2% **44. (a)** $y = 1.089x - .264$ **(b)** 84.22 **(c)** 73.73 **45. (a)** $y = .47x + 1.98$ **(b)** 4.8% **(c)** 2007
46. (a) $y = .152x - 3.063$ **(b)** About 21 **(c)** About 165 grams

CHAPTER 1: CHAPTER TEST

1.

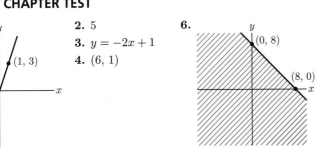

2. 5
3. $y = -2x + 1$
4. $(6, 1)$

6.

7. $y = 2x - \frac{71}{11}$
8. $(16, 0); (25, 0); (13, 36); (2, 14)$
9. Less than \$2500; greater than \$2500
10. (a) $y = .019x + .771$
(b) \$0.87 **(c)** 2010 (after 14.7 yrs)

CHAPTER 2

EXERCISES 2.1

1. $\xrightarrow{2[1]} \begin{cases} x - 6y = 4 \\ 5x + 4y = 1 \end{cases}$ **3.** $\xrightarrow{[2]+5[1]} \begin{cases} x + 2y = 3 \\ 14y = 16 \end{cases}$ **5.** $\xrightarrow{[3]+(-4)[1]} \begin{cases} x - 2y + z = 0 \\ y - 2z = 4 \\ 9y - z = 5 \end{cases}$ **7.** $\xrightarrow{[1]+\frac{1}{2}[2]} \begin{bmatrix} 1 & 0 & | & 5 \\ 0 & 1 & | & 4 \end{bmatrix}$

9. Multiply the second row of the matrix by $\frac{1}{3}$. **11.** Change the first row of the matrix by adding to it 3 times the second row. **13.** $\begin{bmatrix} 1 & 2 \\ 0 & 10 \end{bmatrix}$ **15.** $\begin{bmatrix} 1 & 2 \\ 3 & -2 \end{bmatrix}$ **17.** $[2] + 2[1]$ **19.** $[1] + (-2)[2]$ **21.** Interchange rows 1 and 2 or rows 1 and 3 **23.** $[1] + (-3)[3]$ **25.** $x = -1, y = 1$ **27.** $x = -\frac{8}{7}, y = -\frac{9}{7}, z = -\frac{3}{7}$ **29.** $x = -1, y = 1$
31. $x = 1, y = 2, z = -1$ **33.** $x = -2.5, y = 15$ **35.** $x = 1, y = -6, z = 2$ **37.** $x = -1, y = -2, z = 5$ **39.** b
41. d **43.** 150 short sleeve, 200 long sleeve **45.** 262 adults, 88 children **47.** $x = \$25,000, y = \$50,000, z = \$25,000$
49. $23\frac{1}{3}$ pounds of first type, 85 pounds of second type, $201\frac{2}{3}$ pounds of third type
55. $x = 1, y = -6, z = 2$ **57.** $x = 3, y = -3, z = 2, w = -4$

EXERCISES 2.2

1. $\begin{bmatrix} 1 & -2 & 3 \\ 0 & 13 & -8 \end{bmatrix}$ **3.** $\begin{bmatrix} 9 & -1 & 0 & -7 \\ -\frac{1}{2} & \frac{1}{2} & 1 & 3 \\ 5 & -1 & 0 & -3 \end{bmatrix}$ **5.** $\begin{bmatrix} 1 & \frac{3}{2} \\ 0 & -9 \\ 0 & \frac{7}{2} \end{bmatrix}$ **7.** $\begin{bmatrix} 4 & 3 & 0 \\ 1 & 1 & 0 \\ \frac{1}{6} & \frac{1}{2} & 1 \end{bmatrix}$

9. $y =$ any value, $x = 3 + 2y$ **11.** $x = 1, y = 2$ **13.** No solution **15.** $z =$ any value, $x = -6 - z, y = 5$

17. No solution **19.** $z =$ any value, $w =$ any value, $x = 2z + w, y = 5 - 3w$ **21.** $x = 5, y = 7$

23. Possible answers: $z = 0, x = -13, y = 9; z = 1, x = -8, y = 6; z = 2, x = -3, y = 3$

25. Possible answers: $y = 0, x = 23, z = 5; y = 1, x = 16, z = 5; y = 2, x = 9, z = 5$ **27.** $x =$ food 1, $y =$ food 2, $z =$ food 3; $z =$ any amount, $x = 300 - z, y = 100 - z$ $(0 \le z \le 100)$ **29.** 6 floral squares, the other 90 any mix of solid green and solid blue. **31.** $x = \pm 1, y = \pm 2, z = \pm 3$ **33.** No solution if $k \ne -12$. Infinitely many if $k = -12$.

35. One; $x = 7, y = 3$ **37.** No solution **39.** No solution **41.** $z =$ any value, $x = -21z - 19, y = -9z - 7$

EXERCISES 2.3

1. 2×3 **3.** 1×3, row matrix **5.** 2×2, square matrix **7.** $-4; 0$ **9.** $i = 1, j = 3$ **11.** $\begin{bmatrix} 9 & 3 \\ 7 & -1 \end{bmatrix}$

13. $\begin{bmatrix} 1 & 3 \\ 1 & 2 \\ 4 & -2 \end{bmatrix}$ **15.** $[11]$ **17.** $[10]$ **19.** Yes; 3×5 **21.** No **23.** Yes; 3×1 **25.** $\begin{bmatrix} 6 & 17 \\ 6 & 10 \end{bmatrix}$ **27.** $\begin{bmatrix} 21 \\ -4 \\ 8 \end{bmatrix}$

29. $\begin{bmatrix} 5 & 6 \\ 7 & 8 \end{bmatrix}$ **31.** $\begin{bmatrix} .48 & .39 \\ .52 & .61 \end{bmatrix}$ **33.** $\begin{bmatrix} 25 & 17 & 2 \\ 3 & -1 & 2 \\ 1 & 1 & 4 \end{bmatrix}$ **35.** $\begin{cases} 2x + 3y = 6 \\ 4x + 5y = 7 \end{cases}$ **37.** $\begin{cases} x + 2y + 3z = 10 \\ 4x + 5y + 6z = 11 \\ 7x + 8y + 9z = 12 \end{cases}$

39. $\begin{bmatrix} 3 & 2 \\ 7 & -1 \end{bmatrix}\begin{bmatrix} x \\ y \end{bmatrix} = \begin{bmatrix} -1 \\ 2 \end{bmatrix}$ **41.** $\begin{bmatrix} 1 & -2 & 3 \\ 0 & 1 & 1 \\ 0 & 0 & 1 \end{bmatrix}\begin{bmatrix} x \\ y \\ z \end{bmatrix} = \begin{bmatrix} 5 \\ 6 \\ 2 \end{bmatrix}$

47. (a) $\begin{bmatrix} 340 \\ 265 \end{bmatrix}$ **(b)** Mike's clothes cost \$340; Don's clothes cost \$265.

49. (a) $[1130 \quad 1885 \quad 1571.25]$, total cost for the plain, chocolate-covered, and yogurt-covered items

(b) $\begin{bmatrix} 2962.50 \\ 2181.25 \\ 7918.75 \end{bmatrix}$, total revenue from peanuts, raisins, and coffee beans **51. (a)** I: 2.75, II: 2, III: 1.3 **(b)** A: 74, B: 112, C: 128, D: 64, F: 22 **53.** 10,100 voting Democratic, 7900 voting Republican **55.** Carpenters: \$1000, bricklayers: \$1050, plumbers: \$600 **57. (a)** $[162 \quad 150 \quad 143]$, number of units of each nutrient consumed at breakfast **(b)** $[186 \quad 200 \quad 239]$, number of units of each nutrient consumed at lunch **(c)** $[288 \quad 300 \quad 344]$, number of units of each nutrient consumed at dinner **(d)** $[5 \quad 8]$, total number of ounces of each food that Mikey eats during a day **(e)** $[636 \quad 650 \quad 726]$, number of units of each nutrient consumed per day **59. (a)** $\begin{bmatrix} 720 \\ 646 \end{bmatrix}$ **(b)** \$720

63. $\begin{bmatrix} 3 & -2 & 1 \\ -5 & 6 & 7 \end{bmatrix}$ **65.** 4×4 **67.** $[7698 \quad 55{,}382 \quad 99{,}919]\begin{bmatrix} 15.1 \\ 15.7 \\ 11.8 \end{bmatrix} = [2{,}164{,}781]$ **69.** $\begin{bmatrix} 27.9 & 130.6 & -69.88 \\ 106.75 & -149.44 & 26.1 \\ -47.5 & 336.2 & -18.7 \end{bmatrix}$

71. $\begin{bmatrix} -69.14 & 147.9 & -43.26 \\ 158.05 & -3.69 & 33.46 \\ -176.1 & 259.5 & 59.3 \end{bmatrix}$ **73.** $\begin{bmatrix} 160.16 & -26.7 & 4 \\ 2.7 & 150.85 & -53 \\ 187.4 & -35.5 & 48.6 \end{bmatrix}$ **75.** They match.

EXERCISES 2.4

1. $x = 2, y = 0$ **3.** $\begin{bmatrix} 1 & -2 \\ -3 & 7 \end{bmatrix}$ **5.** $\begin{bmatrix} 1 & -1 \\ -\frac{5}{2} & 3 \end{bmatrix}$ **7.** $\begin{bmatrix} 1.6 & -.4 \\ -.6 & 1.4 \end{bmatrix}$ **9.** $\left[\frac{1}{3}\right]$

11. $x = 4, y = -\frac{1}{2}$ **13.** $x = 32, y = -6$ **15. (a)** $\begin{bmatrix} .8 & .3 \\ .2 & .7 \end{bmatrix} \begin{bmatrix} x \\ y \end{bmatrix} = \begin{bmatrix} m \\ s \end{bmatrix}$ **(b)** $\begin{bmatrix} x \\ y \end{bmatrix} = \begin{bmatrix} 1.4 & -.6 \\ -.4 & 1.6 \end{bmatrix} \begin{bmatrix} m \\ s \end{bmatrix}$

(c) 110,000 married; 40,000 single **(d)** 130,000 married; 20,000 single **17. (a)** $\begin{bmatrix} .7 & .1 \\ .3 & .9 \end{bmatrix} \begin{bmatrix} x \\ y \end{bmatrix} = \begin{bmatrix} u \\ v \end{bmatrix}$

(b) $\begin{bmatrix} x \\ y \end{bmatrix} = \begin{bmatrix} \frac{3}{2} & -\frac{1}{6} \\ -\frac{1}{2} & \frac{7}{6} \end{bmatrix} \begin{bmatrix} u \\ v \end{bmatrix}$ **(c)** 8500; 4500 **19.** $x = 9, y = -2, z = -2$ **21.** $x = 1, y = 5, z = -4, w = 9$

25. (a) $\begin{bmatrix} 1 & 2 \\ .9 & 0 \end{bmatrix} \begin{bmatrix} x \\ y \end{bmatrix} = \begin{bmatrix} a \\ b \end{bmatrix}$ **(b)** After 1 year: 1,170,000 in group I and 405,000 in group II. After 2 years: 1,980,000 in group I and 1,053,000 in group II. **(c)** 700,000 in group I and 55,000 in group II.

29. One possible answer is $\begin{bmatrix} 1 & 1 \\ 1 & 1 \end{bmatrix} \begin{bmatrix} x \\ y \end{bmatrix} = \begin{bmatrix} 2 \\ 3 \end{bmatrix}$. **31.** $\begin{bmatrix} -\frac{10}{73} & \frac{75}{292} \\ \frac{25}{73} & -\frac{5}{292} \end{bmatrix}$ **33.** $\begin{bmatrix} \frac{1020}{8887} & \frac{2910}{8887} & -\frac{500}{8887} \\ \frac{3050}{8887} & \frac{860}{8887} & \frac{1990}{8887} \\ \frac{125}{8887} & \frac{618}{8887} & \frac{810}{8887} \end{bmatrix}$

35. $x = -\frac{4}{5}, y = \frac{28}{5}, z = 5$ **37.** $x = 0, y = 2, z = 0, w = 2$ **39.** Displays ERR:SINGULAR MAT.

EXERCISES 2.5

1. $\begin{bmatrix} -2 & 3 \\ 5 & -7 \end{bmatrix}$ **3.** $\begin{bmatrix} \frac{1}{19} & \frac{3}{19} \\ \frac{3}{76} & -\frac{5}{38} \end{bmatrix}$ **5.** No inverse **7.** $\begin{bmatrix} -1 & 2 & -4 \\ 1 & -1 & 3 \\ 0 & 0 & 1 \end{bmatrix}$ **9.** No inverse **11.** $\begin{bmatrix} -5 & 6 & 0 & 0 \\ 1 & -1 & 0 & 0 \\ 0 & 0 & -\frac{1}{46} & \frac{1}{46} \\ 0 & 0 & \frac{25}{46} & -\frac{1}{23} \end{bmatrix}$

13. $x = 2, y = -3, z = 2$ **15.** $x = 4, y = -4, z = 3, w = -1$ **17.** $\begin{bmatrix} \frac{3}{13} & \frac{7}{13} \\ \frac{1}{13} & -\frac{2}{13} \end{bmatrix}$ **19.** $\begin{bmatrix} -3 & 5 \\ 10 & -16 \end{bmatrix}$ **21.** $\begin{bmatrix} 2 & -1 \\ -7 & 4 \end{bmatrix}$

23. $\begin{bmatrix} -2.4 & 2.2 & 1.8 \\ .8 & -.4 & -.6 \\ .6 & -.8 & -.2 \end{bmatrix}$

EXERCISES 2.6

1. 20 cents **3.** Energy sector **5.** Coal: $8.84 billion, steel: $3.725 billion, electricity: $9.895 billion

7. Computers: $354 million, semiconductors: $172 million **11. (a)** $A = \begin{bmatrix} .25 & .30 \\ .20 & .15 \end{bmatrix} \begin{matrix} T \\ E \end{matrix}$ where columns are T and E **(b)** $\begin{bmatrix} 1.47 & .52 \\ .35 & 1.30 \end{bmatrix}$

(c) Transportation: $8.91 billion, energy: $5.65 billion. **13.** Plastics: $955,000, industrial equipment: $590,000

15. Manufacturing: $398 million, transportation: $313 million, agriculture: $452 million **19.** $\begin{bmatrix} 10.25 \\ 13.82 \\ 8.65 \end{bmatrix}$

CHAPTER 2: SUPPLEMENTARY EXERCISES

1. $\begin{bmatrix} 1 & -2 & \frac{1}{3} \\ 0 & 8 & \frac{16}{3} \end{bmatrix}$ **2.** $\begin{bmatrix} 1 & 0 & 1 \\ 2 & 1 & 0 \\ -12 & 0 & 7 \end{bmatrix}$ **3.** $x = 4, y = 5$ **4.** $x = 50, y = 2, z = -12$ **5.** $x = -1, y = \frac{2}{3}, z = \frac{1}{3}$

6. No solution **7.** $z = $ any value, $x = 1 - 3z, y = 4z, w = 5$ **8.** $x = 7, y = 3$ **9.** $\begin{bmatrix} 5 \\ 3 \\ 7 \end{bmatrix}$ **10.** $\begin{bmatrix} 6 & 17 \\ 12 & 26 \end{bmatrix}$

11. $x = -2, y = 3$ **12. (a)** $x = 13, y = 23, z = 19$ **(b)** $x = -4, y = 13, z = 14$ **13.** $\begin{bmatrix} -1 & 3 \\ \frac{1}{2} & -1 \end{bmatrix}$

14. $\begin{bmatrix} 5 & -1 & -1 \\ -3 & 1 & 0 \\ -1 & 0 & 1 \end{bmatrix}$ **15.** Corn: 500 acres; Wheat: 0 acres; Soybeans: 500 acres

16. (a) $\begin{bmatrix} 1860 \\ 1650 \end{bmatrix}$; total month's costs for each store **(b)** $\begin{bmatrix} 4405 \\ 3945 \end{bmatrix}$; total month's revenue for each store

(c) $\begin{bmatrix} 75 \\ 45 \\ 65 \end{bmatrix}$; profit for each piece of equipment   **(d)** $\begin{bmatrix} 2545 \\ 2295 \end{bmatrix}$; total month's profit for each store

17. 4 apples, 9 bananas, 5 oranges   **18. (a)** A: 9400, 8980; B: 7300, 7510   **(b)** A: 10,857, 12,082; B: 6571, 5969
19. Industry I: 20; industry II: 20   **20.** a   **21. (a)** True **(b)** False **(c)** True   **23.** No

CHAPTER 2: CHAPTER TEST

1. $x = 2, y = 4, z = -5$ **2. (a)** $x = 4, y = -3, z = 6$ **(b)** $x = 2, y = 3, z = 5$ **(c)** No solution
(d) $z = $ any value; $x = z + 2, y = -2z + 2$ **(e)** $y = $ any value; $x = -2y, z = 0$ **3.** $w = 1, z = 2, x = 3, y = 5$
4. (1) $z = 0, y = 6, x = 9$; (2) $z = 1, y = 2, x = 15$; (3) $z = -1, y = 10, x = 3$
5. Not defined; $\begin{bmatrix} 2 & 2 & 2 \\ 2 & 3 & 2 \end{bmatrix}$; $\begin{bmatrix} -2 & 3 \\ -2 & 4 \end{bmatrix}$; not defined; $\begin{bmatrix} -1 & 0 & -1 \\ -1 & 1 & 1 \\ 1 & 0 & 1 \end{bmatrix}$

6. (a) $.5x + y + z = m$; $3x + 2.5y + 2z = v$; $4x + 3y + 2z = p$ **(b)** $\begin{bmatrix} .5 & 1 & 1 \\ 3 & 2.5 & 2 \\ 4 & 3 & 2 \end{bmatrix} \cdot \begin{bmatrix} x \\ y \\ z \end{bmatrix} = \begin{bmatrix} m \\ v \\ p \end{bmatrix}$

(c) 550 hours molding time, 1450 hours oven time, 1700 hours painting time   **7.** $\begin{bmatrix} \frac{1}{2} & -\frac{1}{2} & \frac{1}{2} \\ \frac{1}{2} & -\frac{1}{2} & -\frac{1}{2} \\ -\frac{1}{2} & \frac{3}{2} & \frac{1}{2} \end{bmatrix}$

8. 400 students   **9.** Wood: $1.98, steel: $7.39, coal: $3.87

CHAPTER 3

EXERCISES 3.1

1. Yes   **3.** No   **5. (a)**

	A	**B**	**Truck capacity**
Volume	4 cubic feet	3 cubic feet	300 cubic feet
Weight	100 pounds	200 pounds	10,000 pounds
Earnings	$13	$9	

(b) $4x + 3y \le 300$; $100x + 200y \le 10,000$ **(e)**
(c) $y \le 2x, x \ge 0, y \ge 0$
(d) $13x + 9y$

[graph: $y = 2x$; $y = -\frac{4}{3}x + 100$; $y = -\frac{1}{2}x + 50$; feasible set; axes marked 100, 50, 75, 100]

7. (a)

	Essay questions	Short-answer questions	Available
Time to answer	10 minutes	2 minutes	90 minutes
Quantity	10	50	
Required	3	10	
Worth	20 points	5 points	

(b) $10x + 2y \le 90$
(c) $x \ge 3, x \le 10, y \ge 10, y \le 50$
(d) $20x + 5y$

(e)

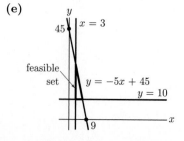

9. (a)

	Alfalfa	Corn	Requirements
Protein	.13 pound	.065 pound	4550 pounds
TDN	.48 pound	.96 pound	26,880 pounds
Vitamin A	2.16 IUs	0	43,200 IUs
Cost/lb	$0.01	$0.016	

(b) $.13x + .065y \geq 4550;$
$.48x + .96y \geq 26,880;$
$2.16x \geq 43,200; \ y \geq 0$

(c)

(d) $.01x + .016y$

EXERCISES 3.2

1. $(20, 0)$ **3.** $(6, 0)$ **5.** $(0, 5)$ **7.** $(3, 3)$ **9.** $(0, 7)$ **11.** $(2, 1)$ **13.** Ship 75 crates of cargo A and no crates of cargo B.   **15.** Answer 3 essay questions and 30 short-answer questions.   **17.** The minimum cost of $504 is achieved by buying 28,000 pounds of alfalfa and 14,000 pounds of corn.   **19.** Make 16 chairs and no sofas.   **21.** The minimum value is 30 and occurs at $(2, 6)$.   **23.** The maximum value is 49 and occurs at $(2, 9)$.   **25.** The maximum value is 6600 and occurs at $(12, 36)$.   **27.** The minimum value is 40 and occurs at $(4, 3)$.   **29.** Produce 9 hockey games and 8 soccer games each day.   **31.** Supply 6 tubes of food A and 6 tubes of food B.   **33.** Make 400 cans of Fruit Delight and 500 cans of Heavenly Punch.   **35.** The farmer should plant $83\frac{1}{3}$ acres of oats and $16\frac{2}{3}$ acres of corn to make a profit of $6933.33.   **37. (b)** The farmer should plant 78 acres of oats and 22 acres of corn to make a profit of $7040.   **39.** The feasible set contains no points.   **41.** The maximum value is 84 and occurs at $(12, 0)$.

EXERCISES 3.3

1. (a) $y = -\frac{3}{2}x + \frac{c}{14}$ **(b)** Up **(c)** B **3.** Possible answer: $5x + y$ **5.** Possible answer: $2x + y$ **7.** Possible answer: $x + 5y$ **9.** Possible answer: $2x + 3y$ **11.** C **13.** D **15.** D **17.** C **19.** $\frac{1}{4} \leq k \leq 3$ **21.** Feed 1 can of brand A and 3 cans of brand B.   **23.** Mr. Jones should invest $2000 in low-risk stocks, $3000 in medium-risk stocks, and $4000 in high-risk stocks.   **25.** Let (a, b) correspond to a cars shipped from Baltimore to Philadelphia, b cars shipped from Baltimore to Trenton, $4 - a$ cars shipped from NY to Philadelphia, and $7 - b$ cars shipped from NY to Trenton. Then the minimum cost of $990 is achieved at $(0, 5)$, $(4, 1)$, or anywhere on the line segment connecting these two points.   **27.** Produce 90,000 gallons of gasoline, 5000 gallons of jet fuel, and 5000 gallons of diesel fuel.   **29.** Buy 9 high-capacity trucks and 21 low-capacity trucks.   **31.** Ship 400 pounds of coffee from Seattle to Salt Lake City and 350 pounds from San Jose to Reno.   **33.** Create 22 Basic I kits, 10 Basic II kits, and 12 Deluxe kits.   **35.** rice: $[9.33, 42]$; soybeans: $[7, 31.5]$

CHAPTER 3: SUPPLEMENTARY EXERCISES

1. Use 10 type A planes and 3 type B planes.   **2.** Use 2 ounces of wheat germ and 1 ounce of enriched oat flour.   **3.** Produce 9 hardtops and 16 sports cars.   **4.** Make 500 boxes of mixture A and 200 boxes of mixture B.   **5.** Publish 60 elementary books, 8 intermediate books, and 4 advanced books.   **6.** Transport 80 computers from Rochester and 45 computers from Queens.   **7.** Transport no computers from warehouse A to outlet I, 200 computers from warehouse A to outlet II, 200 computers from warehouse B to outlet I, and 100 computers from warehouse B to outlet II.   **8.** Invest $2000 in the CD, $0 in mutual funds, and $8000 in stocks.   **9.** Yes; No   **10.** Yes; No

CHAPTER 3: CHAPTER TEST

1. Step 1: Translate the problem into mathematical language. Identify variables and write the inequalities and objective function.
Step 2: Graph the feasible set.
Step 3: Determine the vertices of the feasible set.

Step 4: Evaluate the objective function at each vertex. Determine the optimal point.

2. (a) Minimize $70x + 90y$ subject to $\begin{cases} 20x + 30y \geq 300 \\ 10x + 20y \geq 200 \\ x \geq 0, y \geq 0 \end{cases}$

    (b) Maximize $.07x + .06y + .045z$ subject to $\begin{cases} y \geq 150,000 \\ z \leq 200,000 \\ x + y + z \leq 500,000 \\ x \geq .5y \\ x \geq 0, y \geq 0, z \geq 0 \end{cases}$

3. (a)

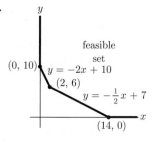

    (b) No

4. $(0, 10)$; $(7, 5)$; $(9, 3)$; $(9, 0)$

5. (a) $x = 10, y = 0$ **(b)** $x = 3, y = 6$

6. (a) $y = -\frac{8}{5}x + 640$ **(b)** $y = -\frac{8}{5}x + \frac{c}{5}$ **(c)** down   **(d)** E

7. C **8.** 8 bears and 6 wreaths

FEASIBLE SETS FOR CHAPTER 3

EXERCISES 3.2

21.

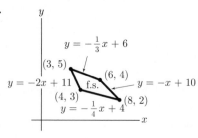

23.

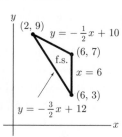

25.

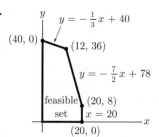

27.

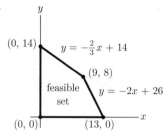

29.

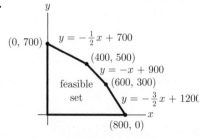

31.

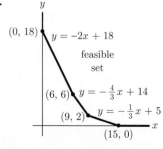

33.

35.

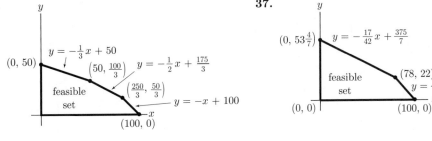

37.

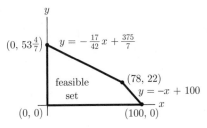

EXERCISES 3.3

21.

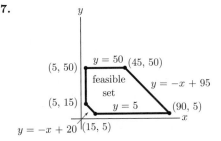

23.

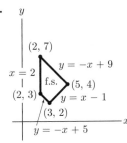

25.

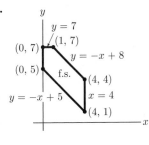

27.

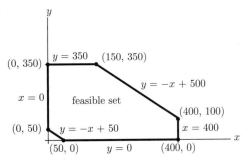

29.

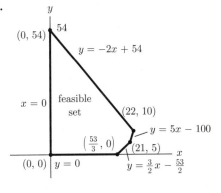

31.

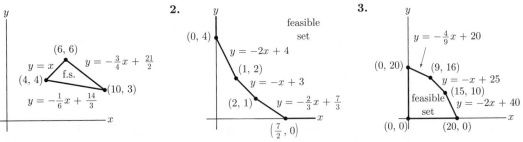

33.

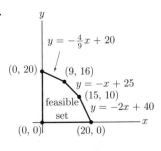

CHAPTER 3: SUPPLEMENTARY EXERCISES

1.

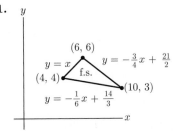

2.

3.

4.

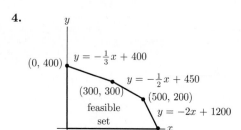

5.

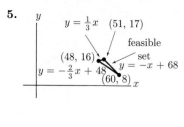

6.

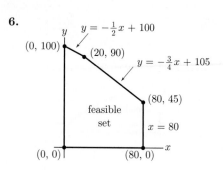

7.

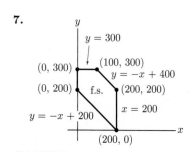

8.

y

$(2000, 5000)$　　　$y = 5000$

$(10,000, 5000)$

$x = 2000$

feasible set

$x = 10,000$

$(2000, 0)$　$y = 0$

$(10,000, 0)$

CHAPTER 4

EXERCISES 4.1

1. $\begin{cases} 20x + 30y + u & = 3500 \\ 50x + 10y \quad + v & = 5000 \\ -8x - 13y \qquad + M & = 0 \end{cases}$ Maximize M given $x \geq 0, y \geq 0, u \geq 0, v \geq 0$.

3. $\begin{cases} x + y + z + u & = 100 \\ 3x \quad + z \quad + v & = 200 \\ 5x + 10y \qquad\qquad + w & = 100 \\ -x - 2y + 3z \qquad + M & = 0 \end{cases}$ Maximize M given $x \geq 0, y \geq 0,$ $z \geq 0, u \geq 0, v \geq 0, w \geq 0$.

5. $\begin{cases} 4x + 6y - 7z + u & = 16 \\ 3x + 2y \quad + v & = 11 \\ 9y + 3z \quad + w & = 21 \\ -3x - 5y - 12z \quad + M & = 0 \end{cases}$ Maximize M given $x \geq 0, \ y \geq 0, \ z \geq 0, \ u \geq 0, \ v \geq 0, \ w \geq 0$.

7. (a)
$$\begin{array}{ccccccc c}
x & y & t & u & v & w & M & \\
\left[\begin{array}{ccccccc|c}
3 & 2 & 1 & 0 & 0 & 0 & 0 & 10 \\
1 & 0 & 0 & 1 & 0 & 0 & 0 & 15 \\
0 & 1 & 0 & 0 & 1 & 0 & 0 & 3 \\
1 & 1 & 0 & 0 & 0 & 1 & 0 & 5 \\
-1 & -15 & 0 & 0 & 0 & 0 & 1 & 0
\end{array}\right]
\end{array}$$
(b) $x = 0, y = 0, t = 10, u = 15, v = 3, w = 5, M = 0$

9. (a)
$$\begin{array}{cccccc c}
x & y & u & v & w & M & \\
\left[\begin{array}{cccccc|c}
1 & 3 & 1 & 0 & 0 & 0 & 24 \\
0 & 1 & 0 & 1 & 0 & 0 & 5 \\
1 & 7 & 0 & 0 & 1 & 0 & 10 \\
-2 & -1 & 0 & 0 & 0 & 1 & 50
\end{array}\right]
\end{array}$$
(b) $x = 0, y = 0, u = 24, v = 5, w = 10, M = 50$

11. $x = 15, y = 0, u = 10, v = 0, M = 20$　　**13.** $x = 10, y = 0, z = 15, u = 23, v = 0, w = 0, M = -11$

15. (a)

$$
\begin{array}{ccccc}
x & y & u & v & M
\end{array}
$$
$$
\left[
\begin{array}{ccccc|c}
1 & \frac{3}{2} & \frac{1}{2} & 0 & 0 & 6 \\
0 & -\frac{1}{2} & -\frac{1}{2} & 1 & 0 & 4 \\
0 & -5 & 5 & 0 & 1 & 60
\end{array}
\right]
$$
$x = 6, y = 0, u = 0, v = 4, M = 60$

(b)

$$
\begin{array}{ccccc}
x & y & u & v & M
\end{array}
$$
$$
\left[
\begin{array}{ccccc|c}
\frac{2}{3} & 1 & \frac{1}{3} & 0 & 0 & 4 \\
\frac{1}{3} & 0 & -\frac{1}{3} & 1 & 0 & 6 \\
\frac{10}{3} & 0 & \frac{20}{3} & 0 & 1 & 80
\end{array}
\right]
$$
$x = 0, y = 4, u = 0, v = 6, M = 80$

(c)

$$
\begin{array}{ccccc}
x & y & u & v & M
\end{array}
$$
$$
\left[
\begin{array}{ccccc|c}
0 & 1 & 1 & -2 & 0 & -8 \\
1 & 1 & 0 & 1 & 0 & 10 \\
0 & -10 & 0 & 10 & 1 & 100
\end{array}
\right]
$$
$x = 10, y = 0, u = -8, v = 0, M = 100$

(d)

$$
\begin{array}{ccccc}
x & y & u & v & M
\end{array}
$$
$$
\left[
\begin{array}{ccccc|c}
-1 & 0 & 1 & -3 & 0 & -18 \\
1 & 1 & 0 & 1 & 0 & 10 \\
10 & 0 & 0 & 20 & 1 & 200
\end{array}
\right]
$$
$x = 0, y = 10, u = -18, v = 0, M = 200$

17. After operation (d).   **19. (a)** Group I: x, y; Group II: u, v **(b)** (i) feasible; I: y, u; II: x, v (ii) feasible; I: x, u; II: y, v (iii) not feasible; I: y, v; II: x, u (iv) not feasible; I; x, y; II: u, v **(c)** Operation (i).

EXERCISES 4.2

1. (a) 3 **(b)**
$$
\begin{array}{c}
u \\ y \\ M
\end{array}
\left[
\begin{array}{ccccc|c}
\frac{16}{3} & 0 & 1 & -\frac{2}{3} & 0 & 6 \\
\frac{1}{3} & 1 & 0 & \frac{1}{3} & 0 & 2 \\
\hline
0 & 0 & 0 & 4 & 1 & 24
\end{array}
\right]
$$
(c) $x = 0, y = 2, u = 6, v = 0; M = 24$

3. (a) 10 **(b)**
$$
\begin{array}{c}
u \\ y \\ M
\end{array}
\left[
\begin{array}{ccccc|c}
-13 & 0 & 1 & -\frac{6}{5} & 0 & 6 \\
\frac{3}{2} & 1 & 0 & \frac{1}{10} & 0 & \frac{1}{2} \\
\hline
7 & 0 & 0 & \frac{1}{5} & 1 & 1
\end{array}
\right]
$$
(c) $x = 0, y = \frac{1}{2}, u = 6, v = 0, M = 1$

5. $x = 0, y = 5; M = 15$ **7.** $x = 12, y = 20; M = 88$ **9.** $x = 0, y = \frac{19}{3}, z = 5; M = 44$ **11.** $x = 0, y = 30; M = 90$
13. $x = 50, y = 100; M = 1300$ **15.** 24 basketballs, 224 footballs, for profit of $508.   **17.** 98 chairs, 4 sofas, 21 tables
19. 11 hours bicycling, 4 hours swimming, 15 hours jogging; 3 pounds   **21.** 65 type A restaurants and 5 type C
restaurants   **23.** 100 bags of mix B and 25 bags of mix C   **25.** $x = 100, y = 50; M = 45,000$ **27.** $x = \frac{8}{5}, y = \frac{3}{5}$;
$M = 10$ **29.** $x = 5, y = 0, z = 0; M = 80$

EXERCISES 4.3

1. $x = \frac{3}{5}, y = \frac{22}{5}; 156$ **3.** $x = \frac{5}{2}, y = \frac{1}{2}; 8$ **5.** $x = 3, y = 5; 59$ **7.** $x = 5, y = 4; 38$ **9.** 1 serving of food A, 3
servings of food B   **11.** Stock 100 of brand A, 50 of brand B, and 450 of brand C.   **13.** Supply all computers from
Chicago at a cost of $8800.   **15.** $x = 1, y = 1; M = -1$

EXERCISES 4.4

1. $x = 4, y = 22$; profit $= 122 **3.** $x = 25, y = 25$; cost $= 250 **5.** $-400 \le h \le \frac{400}{3}$ **7.** $\begin{bmatrix} 9 & 1 & 1 \\ 4 & 8 & -3 \end{bmatrix}$ **9.** $\begin{bmatrix} 7 \\ 6 \\ 5 \\ 1 \end{bmatrix}$

11. Yes

13. Minimize $\begin{bmatrix} 7 & 5 & 4 \end{bmatrix} \begin{bmatrix} x \\ y \\ z \end{bmatrix}$ subject to the constraints $\begin{bmatrix} 3 & 8 & 9 \\ 1 & 2 & 5 \\ 4 & 1 & 7 \end{bmatrix} \begin{bmatrix} x \\ y \\ z \end{bmatrix} \ge \begin{bmatrix} 75 \\ 80 \\ 67 \end{bmatrix}$ and $\begin{bmatrix} x \\ y \\ z \end{bmatrix} \ge \begin{bmatrix} 0 \\ 0 \\ 0 \end{bmatrix}$.

15. Maximize $\begin{bmatrix} 3 & 5 \end{bmatrix} \begin{bmatrix} x \\ y \end{bmatrix}$ subject to the constraints $\begin{bmatrix} 3 & 6 \\ 7 & 5 \\ 4 & 3 \end{bmatrix} \begin{bmatrix} x \\ y \end{bmatrix} \le \begin{bmatrix} 90 \\ 138 \\ 120 \end{bmatrix}$ and $\begin{bmatrix} x \\ y \end{bmatrix} \ge \begin{bmatrix} 0 \\ 0 \end{bmatrix}$.

17. Minimize $2x + 3y$ subject to the constraints $\begin{cases} 7x + 4y \ge 33 \\ 5x + 8y \ge 44 \\ x + 3y \ge 55 \\ x \ge 0, y \ge 0. \end{cases}$ **19.** $-2; [25, 55]$

EXERCISES 4.5

1. Minimize $80u + 76v$ subject to the constraints $\begin{cases} 5u + 3v \geq 4 \\ u + 2v \geq 2 \\ u \geq 0, v \geq 0. \end{cases}$

3. Maximize $u + 2v + w$ subject to the constraints $\begin{cases} u - v + 2w \leq 10 \\ 2u + v + 3w \leq 12 \\ u \geq 0, v \geq 0, w \geq 0. \end{cases}$

5. Maximize $-7u + 10v$ subject to the constraints $\begin{cases} -2u + 8v \leq 3 \\ 4u + v \leq 5 \\ 6u + 9v \leq 1 \\ u \geq 0, v \geq 0. \end{cases}$

7. $x = 12$, $y = 20$, $M = 88$; $u = \frac{2}{7}$, $v = \frac{6}{7}$, $M = 88$ **9.** $x = 0, y = 2, M = 24$; $u = 0, v = 12, w = 0, M = 24$

11. Maximize $3u + 5v$ subject to the constraints $\begin{cases} u + 2v \leq 3 \\ u \quad\;\; \leq 1 \\ u \geq 0, v \geq 0. \end{cases}$

$x = \frac{5}{2}$, $y = \frac{1}{2}$, minimum $= 8$; $u = 1$, $v = 1$, maximum $= 8$

13. Minimize $6u + 9v + 12w$ subject to the constraints $\begin{cases} u + 3v \quad\quad \geq 10 \\ -2u \quad\;\; + w \geq 12 \\ v + 3w \geq 10 \\ u \geq 0, v \geq 0, w \geq 0. \end{cases}$

$x = 3$, $y = 12$, $z = 0$, maximum $= 174$; $u = 0$, $v = \frac{10}{3}$, $w = 12$, minimum $= 174$

15. Suppose we can hire workers out at a profit of u dollars per hour, sell the steel at a profit of v dollars per unit, and sell the wood at a profit of w dollars per unit. To find the minimum profit at which that should be done, minimize

$90u + 138v + 120w$ subject to the constraints $\begin{cases} 3u + 7v + 4w \geq 3 \\ 6u + 5v + 3w \geq 5 \\ u \geq 0, v \geq 0, w \geq 0. \end{cases}$

17. Suppose we can buy anthracite at u dollars per ton, ordinary coal at v dollars per ton, and bituminous coal at w dollars per ton. To find the maximum cost at which this should be done, maximize $80u + 60v + 75w$ subject to the

constraints $\begin{cases} 4u + 4v + 7w \leq 150 \\ 10u + 5v + 5w \leq 200 \\ u \geq 0, v \geq 0, w \geq 0. \end{cases}$ **19.** \$3.63 **21.** $x = 2$, $y = 1$, maximum $= 74$

CHAPTER 4: SUPPLEMENTARY EXERCISES

1. $x = 2, y = 3$, max. $= 18$ **2.** $x = 0, y = 7$, max. $= 35$ **3.** $x = 4, y = 5$, max. $= 23$ **4.** $x = 2, y = 4$, max. $= 34$
5. $x = 5, y = 1$, min. $= 6$ **6.** $x = 0, y = 6$, min. $= 12$ **7.** $x = 4, y = 1$, min. $= 110$ **8.** $x = 4, y = 3$, min. $= 41$
9. $x = 1, y = 6, z = 8$, max. $= 884$ **10.** $x = 60, y = 8, z = 20, w = 0$, max. $= 312$

11. Minimize $14u + 9v + 24w$ subject to the constraints $\begin{cases} u + v + 3w \geq 2 \\ 2u + v + 2w \geq 3 \\ u \geq 0, v \geq 0, w \geq 0. \end{cases}$

12. Maximize $8u + 5v + 7w$ subject to the constraints $\begin{cases} u + v + 2w \leq 20 \\ 4u + v + w \leq 30 \\ u \geq 0, v \geq 0, w \geq 0. \end{cases}$

13. Primal: $x = 4, y = 5$, max. $= 23$; Dual: $u = 1, v = 1, w = 0$, min. $= 23$
14. Primal: $x = 4, y = 1$, min. $= 110$; Dual: $u = \frac{10}{3}$, $v = \frac{50}{3}$, $w = 0$, max. $= 110$

15. $A = \begin{bmatrix} 1 & 2 \\ 1 & 1 \\ 3 & 2 \end{bmatrix}$, $B = \begin{bmatrix} 14 \\ 9 \\ 24 \end{bmatrix}$, $C = \begin{bmatrix} 2 & 3 \end{bmatrix}$, $X = \begin{bmatrix} x \\ y \end{bmatrix}$ Primal: Maximize CX subject to $AX \leq B, X \geq \mathbf{0}$.

Dual: $U = \begin{bmatrix} u \\ v \\ w \end{bmatrix}$ Minimize $B^T U$ subject to $A^T U \geq C^T, U \geq \mathbf{0}$.

16. $A = \begin{bmatrix} 1 & 4 \\ 1 & 1 \\ 2 & 1 \end{bmatrix}$, $B = \begin{bmatrix} 8 \\ 5 \\ 7 \end{bmatrix}$, $C = \begin{bmatrix} 20 & 30 \end{bmatrix}$, $X = \begin{bmatrix} x \\ y \end{bmatrix}$ Primal: Minimize CX subject to $AX \geq B, X \geq \mathbf{0}$.

Dual: $U = \begin{bmatrix} u \\ v \\ w \end{bmatrix}$ Maximize $B^T U$ subject to $A^T U \le C^T, U \ge \mathbf{0}$.

17. **(a)** 30 type A sticks, 40 type B sticks **(b)** $11   **18.** $210

CHAPTER 4: CHAPTER TEST

1.

x	y	z	u	v	w	M	
1	1	−2	1	0	0	0	10
2	−1	3	0	1	0	0	18
1	3	1	0	0	1	0	21
−2	−1	3	0	0	0	1	0

2. $x = 35$, $y = 0$, $z = 30$, $u = 0$, $v = 0$, $w = 42$, $M = 560$

Dual: $x = 0$, $y = 14$, $z = 0$, $u = 0$, $v = \frac{7}{2}$, $w = 0$, $M = 560$

3. Maximize $3x - 4y$ subject to $\begin{cases} 6x + 7y \le 120 \\ 15x + 5y \le 195 \\ x \ge 0, y \ge 0. \end{cases}$; $x = 13$, $y = 0$, $u = 42$, $v = 0$, $M = 39$

4. $x = 2$, $y = 4$, $u = 0$, $v = 4$, $w = 0$, $M = 44$.   **5.** Maximize $6u + 3v$ subject to $\begin{cases} u + 2v \le 3 \\ u - v \le 2 \\ u \ge 0, v \ge 0. \end{cases}$

6. **(a)** Maximize $\begin{bmatrix} .50 & .35 \end{bmatrix} \begin{bmatrix} x \\ y \end{bmatrix}$ subject to the constraints $\begin{bmatrix} 2 & 1 \\ 2 & 3 \end{bmatrix} \begin{bmatrix} x \\ y \end{bmatrix} \le \begin{bmatrix} 6000 \\ 9600 \end{bmatrix}$ and $\begin{bmatrix} x \\ y \end{bmatrix} \ge \begin{bmatrix} 0 \\ 0 \end{bmatrix}$.

(b) Minimize $\begin{bmatrix} 6000 & 9600 \end{bmatrix} \begin{bmatrix} u \\ v \end{bmatrix}$ subject to the constraints $\begin{bmatrix} 2 & 2 \\ 1 & 3 \end{bmatrix} \begin{bmatrix} u \\ v \end{bmatrix} \ge \begin{bmatrix} .50 \\ .35 \end{bmatrix}$ and $\begin{bmatrix} u \\ v \end{bmatrix} \ge \begin{bmatrix} 0 \\ 0 \end{bmatrix}$.

(c) Minimize $6000u + 9600v$ subject to the constraints $\begin{cases} 2u + 2v \ge .50 \\ u + 3v \ge .35 \\ u \ge 0, v \ge 0. \end{cases}$

u is a measure of the value of a pound of paper

v is a measure of the value of a minute of labor

(d) The dual gives the minimum acceptable profit that can be achieved by selling the paper and hiring out the workers.

CHAPTER 5

EXERCISES 5.1

1. **(a)** $\{5, 6, 7\}$ **(b)** $\{1, 2, 3, 4, 5, 7\}$ **(c)** $\{1, 3\}$ **(d)** $\{5, 7\}$ **3.** **(a)** $\{a, b, c, d, e, f\}$ **(b)** $\{c\}$ **(c)** $\emptyset$ **5.** $\emptyset, \{1\}, \{2\}, \{1, 2\}$
7. **(a)** {all male college students who like football} **(b)** {all female college students}
(c) {all female college students who don't like football}
(d) {all male college students or all college students who like football}
9. **(a)** $S = \{1979, 1983, 1984, 1987, 1999, 2003\}$
(b) $T = \{1980, 1983, 1985, 1989, 1991, 1995, 1996, 1997, 1998, 1999, 2003\}$
(c) $S \cap T = \{1983, 1999, 2003\}$
(d) $S \cup T = \{1979, 1980, 1983, 1984, 1985, 1987, 1989, 1991, 1995, 1996, 1997, 1998, 1999, 2003\}$
(e) $S' \cap T = \{1980, 1985, 1989, 1991, 1995, 1996, 1997, 1998\}$ **(f)** $S \cap T' = \{1979, 1984, 1987\}$
11. From 1974 to 2001, during only three years did the Standard and Poor's Index increase by 2% or more during the first five days and not increase by 16% or more for that year.   **13.** **(a)** $\{d, f\}$ **(b)** $\{a, b, c, e, f\}$ **(c)** $\emptyset$ **(d)** $\{a, c\}$ **(e)** $\{e\}$
(f) $\{a, c, e, f\}$ **(g)** $\{a, b, c, e\}$ **(h)** $\{a, c\}$ **(i)** $\{d\}$ **15.** S **17.** U **19.** $\emptyset$ **21.** $L \cup T$ **23.** $L \cap P$ **25.** $P \cap L \cap T$
27. S' **29.** $S \cup A \cup D$ **31.** $(A \cap S)' \cap D$ **33.** {male students at Mount College}   **35.** {people who are both teachers and students at Mount College}   **37.** {males or students at Mount College}   **39.** {females at Mount College}
41. S' **43.** $(V \cup C) \cap S'$ **45.** $(V \cup C)'$ **47.** **(a)** $\{B, C, D, E\}$ **(b)** $\{C, D, E, F\}$ **(c)** $\{A, D, E, F\}$ **(d)** $\{A, C, D, E, F\}$
(e) $\{A, F\}$ **(f)** $\{D, E\}$ **49.** Possible answer: $\{2\}$ **51.** S is a subset of T.

EXERCISES 5.2

1. 7 **3.** 0 **5.** 11 **7.** S is a subset of T. **9.** 17 million **11.** 10 **13.** 452 **15.**

17.  **19.** 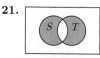 **21.** **23.** 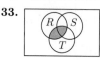 **25.**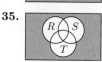

27. 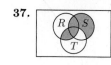 **29.** 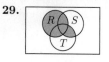 **31.** 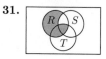 **33.**  **35.**

37. 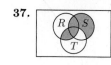 **39.** $S' \cup T'$ **41.** $S \cap T'$ **43.** U **45.** S' **47.** $R \cap T$ **49.** $R' \cap S \cap T$
51. $T \cup (R \cap S')$ **53.** $(R \cap S \cap T) \cup (R' \cap S' \cap T')$ **55.** People who are not illegal aliens or everyone over the age of 18 who is employed **57.** Everyone over the age of 18 who is unemployed **59.** Noncitizens who are unemployed

EXERCISES 5.3

1. 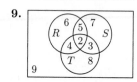 **3.** 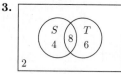 **5.** 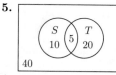 **7.**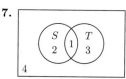

9. 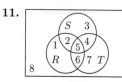 **11.** 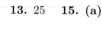 **13.** 25 **15. (a)** 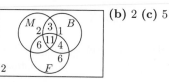 **(b)** 2 **(c)** 5

17. 30 **19.** 4 **21.** 4, 2, 1 **23.** 28 **25.** 12 **27.** 2000 **29.** 200 **31.** 1600 **33.** 90 **35.** 6 **37.** 140
39. 30 **41.** 190 **43.** 180 **45.** 210 **47.** 35 **49.** d **51.** 450 **53.** 3500

EXERCISES 5.4

1. 15 **3.** 676 **5.** 380 **7.** 20 **9.** 120 **11.** 64 **13.** 6840 **15.** 870 **17.** 24 **19.** 32 **21.** 360,000

23. (a) 2401 **(b)** 840 **(c)** 343 **(d)** 240 **25. (a)** 362,880 **(b)** 40,320 **(c)** 720 **27.** 256 **29.** 729 **31.** 168
33. 175,760,000 **35.** 160 **37.** 145 **39.** 972 **41.** 1,048,576 **43.** 16 **45.** d **47.** 24 **49.** 8250 **51.** 33
53. 324 **55.** 1512 **57.** 999,999,999 **59.** 16 **61.** 48 **63.** 48; 1152

EXERCISES 5.5

1. 12 **3.** 120 **5.** 120 **7.** 5 **9.** 5 **11.** n **13.** 1 **15.** $\dfrac{n(n-1)}{2}$ **17.** 720 **19.** 72 **21.** 24 **23.** 36
25. 2730 **27.** 210 **29.** 210 **31.** 120 **33.** 28 **35.** 30,240 **37.** 252 **39.** 38,955,840 **41.** 161,700; 35
43. 15,600 **45.** 75,287,520 **47.** 2,598,960 **49.** 1287 **51.** 1140 **53.** 120 **55.** 70 **57.** Yes; Moe: 36; Joe: 35
59. 479,001,600 **61.** 14,400 **63.** 840 **65.** 10,080 **67.** 120 **69.** 379,236 **71.** 96,875 **73.** 10 **75.** 680
77. semester system **79. (a)** 1,221,759 **(b)** 3,921,225 **(c)** first lottery **(d)** second lottery

EXERCISES 5.6

1. (a) 64 **(b)** 20 **(c)** 22 **(d)** 57 **3.** 126 **5. (a)** 120 **(b)** 56 **(c)** 64 **7.** $C(100, 25) \cdot C(75, 40)$ **9.** 60 **11.** 67,800,320
13. 80 **15.** 3,527,160 **17.** 3528 **19.** 225,225 **21.** 165,765,600 **23.** 210 **25.** 30,240 **27.** 870,912,000
29. 24 **31.** 24 **33.** 3744 **35.** 3050 **37.** 369 **39.** 1500 **41.** 3,628,800 **43.** 64 **45.** 16 **47.** .264%
49. four aces

EXERCISES 5.7

1. 15   **3.** 8   **5.** 153   **7.** 1   **9.** 1   **11.** n **13.** 1   **15.** $n!$ **17.** 64   **19.** $x^{10}, 10x^9 y, 45x^8 y^2$
21. $105x^2 y^{13}, 15xy^{14}, y^{15}$ **23.** $184,756x^{10}y^{10}$ **25.** 330   **27.** 64   **29.** 16   **31.** 32   **33.** 255   **35.** 196,608
37. 120   **39.** 16   **41.** 247   **43.** 4096   **47.** 3696   **49.** $2835x^3$ **51.** 32,767

EXERCISES 5.8

1. 20   **3.** 180   **5.** 210   **7.** 34,650   **9.** 166,320   **11.** 1,401,400   **13.** 2,858,856   **15.** 99,768,240
17. 2,546,168,625   **19.** 488,864,376   **21.** 12   **23.** 135,135   **25.** 126   **27.** 115,166,175,166,136,334,240

CHAPTER 5: SUPPLEMENTARY EXERCISES

1. $\emptyset, \{a\}, \{b\}, \{a, b\}$ **2.**   **3.** 120   **4.** 240   **5.**   **6.** $x^{12} + 12x^{11}y + 66x^{10}y^2$

7. 840   **8.** 15   **9.** 35   **10.** 13,860   **11.** 0   **12.** 136   **13.** 6   **14.** 49   **15.** 47   **16.** 11   **17.** 46   **18.** 58
19. 22   **20.** 23   **21.** 126   **22.** 2^{40} **23.** 550   **24.** 480   **25.** 20   **26.** 450,000   **27.** 8,100,000,000
28. 146,813,779,479,510   **29.** 390,625; 325,089   **30.** 792   **31.** C(100, 14)   **32.** 27,500   **33.** 59,049   **34.** C(60, 10)
35. 9,765,625   **36.** 85,766,121   **37.** 210   **38.** 66,512,160   **39.** 14!   **40.** $\dfrac{1}{5!} \cdot \dfrac{100!}{(20!)^5}$ **41.** $\dfrac{80}{99} \cdot \dfrac{1}{5!} \cdot \dfrac{100!}{(20!)^5}$ **42.** 100
43. 100   **44.** 1000   **45.** 34,560   **46.** 95,040   **47.** 5148   **48.** 648   **49.** 243   **50.** 4512   **51.** 12,696   **52.** 6
53. 72   **54.** 12   **55.** 45; 35   **56.** Second teacher   **57.** 4200   **58.** 5   **59.** 5040   **60. (a)** 14,520
(b) 25,344   **61.** 29   **62.** 70   **63.** 36,504   **64.** 12   **65.** 16,360,143,800   **66. (a)** 224,640,000
(b) There are 20 times as many plates in (a).

CHAPTER 5: CHAPTER TEST

1. (a) 24   **(b)** 210   **(c)** 153   **(d)** 1   **(e)** 30   **2. (a)** True   **(b)** True   **(c)** False   **3. (a)** {a, e}   **(b)** $\emptyset$
4. $C \cap E$ represents the set of certified public accountants who are self-employed. $C \cup E'$ represents the set of people who
are either certified public accountants or are not self-employed.   **5.**   **6.** 10   **7.** 4   **8.** 60

9. (a) 2,209,413,024   **(b)** 210   **10.** 210   **11.** 144   **12.** 136,800   **13.** 792   **14.** 128   **15.** 2520

CHAPTER 6

EXERCISES 6.2

1. (a) {RS, RT, RU, RV, ST, SU, SV, TU, TV, UV}   **(b)** {RS, RT, RU, RV}   **(c)** {TU, TV, UV}
3. (a) {HH, HT, TH, TT}   **(b)** {HH, HT}   **5. (a)** {(I, red), (I, white), (II, red), (II, white)}
(b) {(I, red), (I, white)}   **7. (a)** $S = \{$All positive numbers of minutes$\}$ **(b)** "More than 5 minutes but less than 8
minutes," $\emptyset$, "5 minutes or less," "8 minutes or more," "5 minutes or less," "Less than 4 minutes," S
9. (a) {(Fr, Lib), (Fr, Con), (So, Lib), (So, Con), (Jr, Lib), (Jr, Con), (Sr, Lib), (Sr, Con)}
   (b) {(Fr, Con), (So, Con), (Jr, Con), (Sr, Con)}
   (c) {(Jr, Lib)}
   (d) {(So, Lib), (Jr, Lib), (Sr, Lib)}
11. (a) No   **(b)** Yes   **13.** $\emptyset, \{a\}, \{b\}, \{c\}, \{a, b\}, \{a, c\}, \{b, c\}, S$
15. Yes   **17. (a)** {0, 1, 2, 3, 4, 5, 6, 7, 8, 9, 10}   **(b)** {6, 7, 8, 9, 10}
19. (a) No   **(b)** Yes   **(c)** Yes   **21.** The set of nonnegative integers   **23.** The set of nonnegative numbers

25. $\{(000,000,000),(000,000,001),\ldots,(999,999,999)\}$; $E' =$ the event that at least one number is odd; $E \cap F =$ the event that all numbers are even and are more than 699

27. (a) 324 **(b)** "The murder occurred in the library with a gun." **(c)** "Either the murder occurred in the library or it was done with a gun."

EXERCISES 6.3

1. (a) $\frac{46,277}{774,746}$ **(b)** $\frac{48,132}{774,746}$ **(c)** $\frac{726,614}{774,746}$ **3. (a)** $\frac{5}{36}$ **(b)** $\frac{1}{6}$ **5.** $\frac{1}{19}$ **7. (a)** $\frac{1}{6}$ **(b)** 1 to 5 **9. (a)** .7 **(b)** .7
11. (a) $\frac{10}{11}$ **(b)** $\frac{1}{3}$ **(c)** $\frac{4}{9}$ **13.** 9 to 91 **15.** $\frac{11}{18}; \frac{7}{18}$ **17. (a)** .7 **(b)** .2 **19.** .9 **21.** .6 **23. (a)** .7 **(b)** 7000
25.

Number of Colleges Applied to	Probability
1	.19
2	.13
3	.15
4	.17
5 to 20	.36

27. (a) .15, .55, .20, .10 **(b)** .30

29. (a) Some categories are left out—people who use a computer for both school and work, for example. **(b)** 83%
31. (a) $\frac{3}{8}, \frac{1}{4}, \frac{2}{5}, \frac{1}{5}$ **(b)** $\frac{49}{40}$, which is greater than 1 **(c)** Bookies have to make a living. The payoffs are a little lower than they should be; thus allowing the bookie to make a profit.

EXERCISES 6.4

1. (a) $\frac{1}{9}$ **(b)** $\frac{2}{9}$ **3. (a)** $\frac{7}{13}$ **(b)** $\frac{6}{13}$ **(c)** $\frac{4}{13}$ **(d)** $\frac{9}{13}$ **5. (a)** $\frac{2}{429}$ **(b)** $\frac{7}{429}$ **(c)** $\frac{427}{429}$ **7.** $\frac{11}{12}$ **9.** $\frac{5}{6}$
11. $\frac{16}{17}$ **13.** $\frac{5}{7106}$ **15.** $\frac{47}{250}$ **17.** $\frac{13}{49}$ **19. (a)** .25 **(b)** .75 **(c)** .8 **21.** 0 **23.** $\frac{1}{11}$ **25.** $\frac{2}{5}$ **27. (a)** $\frac{15}{28}$
(b) $\frac{15}{56}$ **(c)** $\frac{9}{56}$ **(d)** $\frac{9}{14}$ **29.** $\frac{5}{6}$ **31.** $\frac{5}{16}$ **33.** $\frac{1}{1,919,190}$ **35.** .90055 **37.** .066 **39. (a)** .119 **(b)** .152
41. .625 **43. (a)** 1:80,089,127 **(b)** $\frac{1}{80,089,128}$ **45.** $\frac{5}{7}$ **47.** $\frac{8815}{499422} \approx .01765$ **49.** $\frac{12}{25}$ **51.** 13 **53. (a)** 4 **(b)** .2139
55. 281

EXERCISES 6.5

1. $\frac{1}{3}, \frac{1}{5}$ **3.** $\frac{4}{7}$ **5.** No **7. (a)** .36 **(b)** .81 **9.** .7967 **11.** $\frac{3}{4}, \frac{1}{2}$ **13. (a)** .4 **(b)** .6 **(c)** .75 **15.** $\frac{1}{4}$ **17.** .94
19. No **21.** .009975 **23.** .2401 **27.** .4, .24, .36 **31. (a)** $\frac{1}{10}$ **(b)** 100 per 1000 **(c)** $\frac{1}{20}$
33. (a) .28 **(b)** .07 **(c)** .6065 **35. (a)** .16 **(b)** .64 **(c)** .36 **(d)** .42 **(e)** .64 **(f)** .31
37. (a) .40 **(b)** .56 **(c)** .18 **(d)** .70 **(e)** .45 **(f)** .67 **39. (a)** .7599 **(b)** .0642 **(c)** .7348
43. $\frac{10}{143}; \frac{25}{286}$ **45.** $\frac{1}{6}; \frac{1}{3}$ **47.** .0986 **49.** No; Pr(woman has 2 boys) $= \frac{1}{3}$; Pr(man has 2 boys) $= \frac{1}{2}$ **51.** Yes **53.** $\frac{1}{16}$
55. (a) $1000p$ **(b)** $1000p - 499,500p^2$ **57.** Not independent **59. (a)** 3.69×10^{-6} **(b)** 6.16×10^{-7} **(c)** a **61.** 26

EXERCISES 6.6

1.

3.

5. .08 **7.** .295 **9.** $\frac{7}{12}$

11.

$\frac{1201}{5525} \approx .22$ **13.** .262 **15.** $\frac{25}{27}$ **17.** .86 **19.** $\frac{4}{7}$ **21. (a)** .60 **(b)** .75
23. $1 - (.9999)^n$ **25.** Same shape; probability of winning card game is greater. **27. (a)** $\frac{1}{4}; \frac{3}{4}$ **(b)** .7 **29.** $\frac{6}{11}$
31. .999996 **33.** $\frac{25}{26}$ **35.** $\frac{3}{8}$ **37.** .66

EXERCISES 6.7

1. $\frac{8}{53}$ **3.** $\frac{3}{7}$ **5.** .075   **7.** $\frac{8}{9}$ **9. (a)** .1325   **(b)** $\frac{12}{53} \approx .23$ **11.** $\frac{5}{103} \approx .049$
13. (a) $\frac{1}{4}$ **(b)** $\frac{13}{17} \approx .765$ **(c)** .130   **15. (a)** $\frac{5}{9}$ **(b)** 11% **17. (a)** .01   **(b)** $\frac{33}{34} \approx .971$ **19.** $\frac{31}{37} \approx .838$

EXERCISES 6.8

1. Theoretical probabilities: $\frac{1}{6}$ for each face.

CHAPTER 6: SUPPLEMENTARY EXERCISES

1. $\frac{31}{32}$ **2.** No   **3.** $\frac{4}{9}$ **4.** $\frac{1}{12}$ **5.** $\frac{2}{5}$ **6.** $\frac{2}{11}$ **7.** $\frac{1}{5}$ **8.** $\frac{1}{5}$ **9.** .4667   **10.** .3333   **11.** $\frac{7}{15}$
12. (a) $\frac{1}{10}$ **(b)** $\frac{2}{5}$ **13. (a)** $\frac{2}{15}$ **(b)** $\frac{1}{3}$ **14. (a)** $\left(\frac{1}{36}\right)^3$ **(b)** $\left(\frac{1}{10}\right)^4$ **(c)** $\frac{167}{500}$ **(d)** $\left(\frac{13}{18}\right)^3 \left(\frac{1}{2}\right)^4$
15. (a) $\frac{1}{12}$ **(b)** $\frac{1}{2}$ **16.** $\frac{7}{12}$ **17.** $\frac{1}{21}$ **18.** $\frac{2}{3}$ **19.** No   **20.** $\frac{2}{3}$ **21.** $\frac{1}{3}$ **22.** $\frac{19}{49}$ **23.** $\frac{1}{6}$ **24.** $\frac{1}{21}$ **25.** $\frac{2}{3}$
26. Switch   **27.** $\frac{13}{25}$ **28.** 13 to 37   **29.** $\frac{1}{120,960}$ **30.** $\frac{1}{6}$ **31.** $\frac{5}{16}$ **32.** $\frac{1}{3}$ **33.** $\frac{4}{25}$ **34.** $\frac{1}{8}$ **35.** $\frac{138}{301}$ **36.** e
37. a   **38.** $\frac{6}{7}$ **39.** $\frac{5}{324}$ **40.** $\frac{95}{1827}$ **41. (a)** $\frac{1}{2197}$ **(b)** $\frac{469}{2197}$ **42.** 35   **43.** 421   **45.** Theoretical probability: $\frac{5}{32}$

CHAPTER 6: CHAPTER TEST

1. (a) $S = \{$PN, PD, PQ, PH, ND, NQ, NH, DQ, DH, QH$\}$ **(b)** $E = \{$PN, PQ, NQ, DH$\}$
2. (a) $\frac{3}{8}$ **(b)** $\frac{1,000,000}{1,000,001}$ **3. (a)** 2 to 3   **(b)** 7 to 3   **4. (a)** A male junior is elected.   **(b)** A female junior is not
elected.   **(c)** A male or a junior is elected.   **5.** $\frac{15}{16}$ **6.** .9612   **7. (a)** $\frac{5}{8}$ **(b)** $\frac{3}{4}$ **(c)** $\frac{2}{3}$ **(d)** No   **(e)** No
8. .6513; .3874   **9. (a)** $\frac{1}{3}$ **(b)** $\frac{7}{13}$ **(c)** No; No   **(d)** $\frac{1}{3}$ **10. (a)** .50   **(b)** .50   **11. (a)** $\frac{18}{25}$ **(b)** $\frac{7}{13}$

CHAPTER 7

EXERCISES 7.1

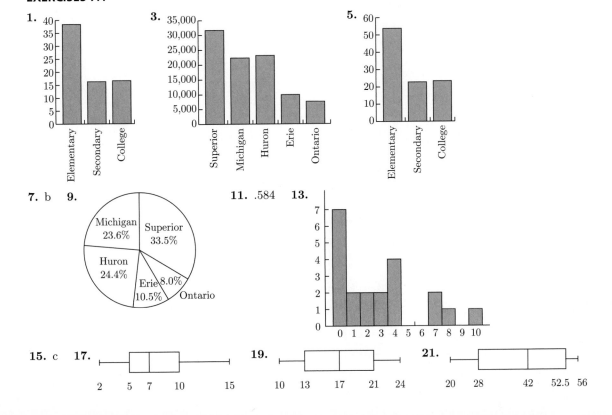

7. b   **11.** .584   **15.** c

23. (A)–(c), (B)–(d), (C)–(a), (D)–(b) **25.** (a) min $= 200, Q_1 = 400, Q_2 = 600, Q_3 = 700, \text{max} = 800$

(b) 25% (c) 25% (d) 50% (e) 75% **27.**

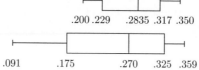

EXERCISES 7.2

1.

Grade	Relative frequency
0	.08
1	.12
2	.40
3	.24
4	.16

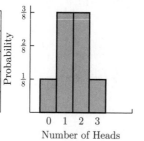

3.

Number of calls during minute	Relative frequency
20	.05
21	.05
22	0
23	.10
24	.30
25	.20
26	0
27	.15
28	.10
29	.05

5.

Number of heads	Probability
0	$\frac{1}{8}$
1	$\frac{3}{8}$
2	$\frac{3}{8}$
3	$\frac{1}{8}$

7.

Number of red balls	Probability
0	$\frac{4}{35}$
1	$\frac{18}{35}$
2	$\frac{12}{35}$
3	$\frac{1}{35}$

9.

Number of red balls	Player's earnings	Probability
2	$5	$\frac{1}{15}$
1	$1	$\frac{8}{15}$
0	−$1	$\frac{6}{15}$

11. .6 **13.**

k	$\Pr(X^2 = k)$
0	.1
1	.2
4	.3
9	.2
16	.2

15.

k	$\Pr(X - 1 = k)$
-1	.1
0	.2
1	.3
2	.2
3	.2

17.

k	$\Pr(\frac{1}{15}Y = k)$
1	.3
2	.4
3	.1
4	.1
5	.1

19.

k	$\Pr((X + 1)^2 = k)$
1	.1
4	.2
9	.3
16	.2
25	.2

21.

	Relative frequency	
Grade	9 AM class	10 AM class
F	.17	.16
D	.25	.23
C	.33	.15
B	.17	.21
A	.08	.25

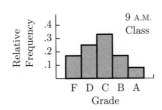

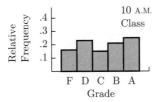

The 9 A.M. class has the distribution centered on the C grade with relatively few A's. The 10 A.M. class has a large percentage of A's and D's with fewer C's.

23. 80%   **25. (a)** 25%   **(b)** 60%   **(c)**   **(d)** ≈ 25

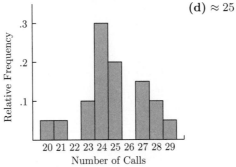

27. (a) 59   **(b)** 5%   **(c)** 54   **(d)** 35%   **(e)** ≈ 54

29. (a) $\Pr(U = 4) = \frac{1}{15}$ **(b)** $\frac{2}{3}$ **(c)** $\frac{14}{15}$ **(d)** $\frac{1}{3}$ **(e)**

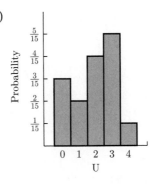

EXERCISES 7.3

1. $\frac{25}{216}$ **3.** $\frac{3}{64}$ **5.** .3174   **7.** .1488   **9.** .8507   **11.** .2007   **13.** .2618   **15.**

k	$\Pr(X = k)$
0	.0168
1	.0896
2	.2090
3	.2787
4	.2322
5	.1239
6	.0413
7	.0079
8	.0007

17. d   **19.** .5583   **21.** .0820   **23.** $\frac{20}{27}$ **25.** 9   **31.** (a) .0290 (b)   (c) .0047
27. .20736   **29.** (a) .0003976 (b) .02624

k	$\Pr(X = k)$
0	.0105
1	.0524
2	.1258
3	.1929
4	.2122
5	.1782
6	.1188

EXERCISES 7.4

1. 2.35   **3.** (a) 2.9   (b)   (c) 2.9   **5.** $\overline{x}_A = 1.5, \overline{x}_B = 1.6$

Grade	Relative Frequency
4	.3
3	.4
2	.2
1	.1

7. $E(X) \approx -.0526$ **9.** $E(X) \approx .1667$

Earnings	Probability
−$1	$\frac{37}{38}$
$35	$\frac{1}{38}$

Earnings	Probability
−50¢	$\frac{1}{3}$
0¢	$\frac{4}{15}$
50¢	$\frac{1}{5}$
$1	$\frac{2}{15}$
$1.50	$\frac{1}{15}$

11. $1000   **13.** 4.47   **15.** 10   **17.** Three free throws   **19.** d   **21.** b   **23.** c   **25.** d   **27.** b   **29.** a   **31.** a

EXERCISES 7.5

1. 1.4   **3.** B **5.** (a) $\mu_A = 15, \mu_B = 13, \sigma_A^2 = 160, \sigma_B^2 = 141$ (b) A (c) B
7. (a) $\mu_A = 103, \sigma_A^2 = 4.6, \mu_B = 104, \sigma_B^2 = 3.4$ (b) B (c) B **9.** (a) $\geq .75$ (b) $\geq .89$ (c) $\geq .31$ **11.** ≥ 4688
13. 8   **15.** (a) $\mu = 7, \sigma^2 = \frac{35}{6}$ (b) $\frac{5}{6}$ (c) $\geq \frac{19}{54}$ **17.** .64   **21.** $\mu = 46,021, \sigma \approx 3441.3$ **23.** $\mu = 4.72, \sigma \approx 1.40$

EXERCISES 7.6

1. .8944   **3.** .4013   **5.** .2417   **7.** .6170   **9.** 1.75   **11.** .75   **13.** ≈ 1.28 **15.** $\mu = 6, \sigma = 2$ **17.** $\mu = 9, \sigma = 1$
19. $-\frac{8}{3}$ **21.** a   **23.** .9772   **25.** .6247   **27.** .0002   **29.** .9876   **31.** .0122   **33.** (a) 616   (b) Between 396 and
644 (c) 674   **35.** 19,750 miles   **37.** (a) 5.35 ounces   (b) 4.76 ounces   **39.** The normal curve is translated to the
right.   **41.** .2358

EXERCISES 7.7

1. (a) .1974   (b) .7888   (c) .9878   **3.** .0062   **5.** .2743   **7.** .0013   **9.** .6368   **11.** .1056   **13.** 2, .1469
15. Exact: .2356; normal approximation: .2358   **17.** Exact: .0812; normal approximation: .0813

CHAPTER 7: SUPPLEMENTARY EXERCISES

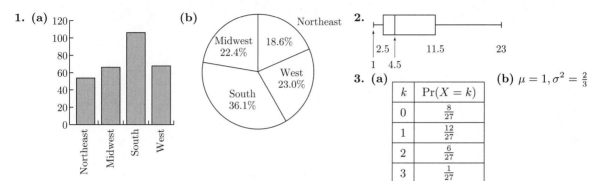

1. (a)   (b)   **2.**   **3.** (a)   (b) $\mu = 1, \sigma^2 = \frac{2}{3}$

k	$\Pr(X = k)$
0	$\frac{8}{27}$
1	$\frac{12}{27}$
2	$\frac{6}{27}$
3	$\frac{1}{27}$

4. .2266   **5.** .2857   **6.** .2646   **7.** $\geq \frac{8}{9}$ **8.** $\mu = 4.8, \sigma^2 = 19.76$ **9.** 10.56%

10.

k	$\Pr(X = k)$
0	$\frac{1}{70}$
1	$\frac{16}{70}$
2	$\frac{36}{70}$
3	$\frac{16}{70}$
4	$\frac{1}{70}$

$\mu = 2, \sigma^2 = \frac{4}{7}$ **11.** .0122   **12.** .84   **13.** 22.5   **14.** −.75   **15.** $\frac{1}{8}; \frac{1}{4}; \frac{5}{16}; \frac{5}{16}$

16. 52; choose true for all the questions.

CHAPTER 7: CHAPTER TEST

1. .58   **2.**

interquartile range = 11

3. (a)

Number of Heads, k	$\Pr(X = k)$
0	.25
1	.50
2	.25

(b)

k	$\Pr(2X + 5 = k)$
5	.25
7	.50
9	.25

4. (a) .296   (b) .619   (c) 2   **5.** .4, 3.24, 1.8   **6.** Lucy; $\approx$ 11 cents per roll   **7.** (a) .8413   (b) .9878   (c) .7162
8. (a) .1587   (b) .7745   (c) 99 mg/100 ml, 113 mg/100 ml   **9.** (a) .3983   (b) .0118; .0092; difference is .0026
(c) .9962; .9878; difference is .0084

CHAPTER 8

EXERCISES 8.1

1. Yes   **3.** No   **5.** Yes   **7.** 53.5%, 56.75%   **9.** 40% will be in zone I, 40% will be in zone II, and 20% will be in

zone III.   **11.** (a) $\begin{array}{c} L \\ R \end{array} \begin{bmatrix} .9 & .7 \\ .1 & .3 \end{bmatrix}$ (b) $\begin{bmatrix} .88 & .84 \\ .12 & .16 \end{bmatrix}$ (c) $\begin{bmatrix} .8 \\ .2 \end{bmatrix}_1, \begin{bmatrix} .86 \\ .14 \end{bmatrix}_2$ (d) 87.5%   **13.** (a) $\begin{array}{c} D \\ R \end{array} \begin{bmatrix} .7 & .4 \\ .3 & .6 \end{bmatrix}$

(b) $\begin{bmatrix} .61 & .52 \\ .39 & .48 \end{bmatrix}; \begin{bmatrix} .583 & .556 \\ .417 & .444 \end{bmatrix}$ (c) 58.3%

$$\begin{array}{c} & U & S & R \\ U \\ \textbf{15. (a)} \; S \\ R \end{array} \begin{bmatrix} .86 & .05 & .03 \\ .08 & .86 & .05 \\ .06 & .09 & .92 \end{bmatrix}$$ **(b)** 11.4%   **17.** $\begin{bmatrix} .44 \\ .56 \end{bmatrix}_3$, $\begin{bmatrix} .44 \\ .56 \end{bmatrix}_4$ **19.** All powers are $\begin{bmatrix} \frac{1}{3} & \frac{1}{3} \\ \frac{2}{3} & \frac{2}{3} \end{bmatrix}$.

21. $\begin{bmatrix} .1 & .3 \\ .9 & .7 \end{bmatrix}$; $\begin{bmatrix} .28 & .24 \\ .72 & .76 \end{bmatrix}$; $\begin{bmatrix} .24 & .25 \\ .76 & .75 \end{bmatrix}$; $\begin{bmatrix} .25 & .25 \\ .75 & .75 \end{bmatrix}$; $\begin{bmatrix} .25 & .25 \\ .75 & .75 \end{bmatrix}$ **23.** No

25. (a) $\begin{bmatrix} .35 \\ .65 \end{bmatrix}$; $\begin{bmatrix} .425 \\ .575 \end{bmatrix}$; $\begin{bmatrix} .3875 \\ .6125 \end{bmatrix}$; $\begin{bmatrix} .40625 \\ .59375 \end{bmatrix}$ **(b)** $\begin{bmatrix} .40625 \\ .59375 \end{bmatrix}$ **27.** $\begin{bmatrix} .4 \\ .6 \end{bmatrix}$ **29.** $\begin{bmatrix} .4 & .4 \\ .6 & .6 \end{bmatrix}$

EXERCISES 8.2

1. Yes   **3.** Yes   **5.** Yes   **7.** $\begin{bmatrix} \frac{1}{6} \\ \frac{5}{6} \end{bmatrix}$ **9.** $\begin{bmatrix} .6 \\ .4 \end{bmatrix}$ **11.** $\begin{bmatrix} \frac{5}{14} \\ \frac{3}{7} \\ \frac{3}{14} \end{bmatrix}$ **13.** 87.5%   **15.** 25%   **17.** 40%   **19.** $\begin{bmatrix} .5 \\ .5 \end{bmatrix}$ is

a stable distribution for the matrix $A = \begin{bmatrix} 0 & 1 \\ 1 & 0 \end{bmatrix}$ because $.5 + .5 = 1$ and $\begin{bmatrix} 0 & 1 \\ 1 & 0 \end{bmatrix}\begin{bmatrix} .5 \\ .5 \end{bmatrix} = \begin{bmatrix} .5 \\ .5 \end{bmatrix}$. However, given an

arbitrary initial distribution $\begin{bmatrix} \\ \end{bmatrix}_0 \neq \begin{bmatrix} .5 \\ .5 \end{bmatrix}$, $A^n \begin{bmatrix} \\ \end{bmatrix}_0$ will not approach $\begin{bmatrix} .5 \\ .5 \end{bmatrix}$ as n gets large, so the existence of a

stable distribution for A does not contradict the main premise of this section.   **21.** $\begin{bmatrix} .7 & .7 \\ .3 & .3 \end{bmatrix}$; $\begin{bmatrix} .7 \\ .3 \end{bmatrix}$

23. $\begin{bmatrix} \frac{8}{35} & \frac{8}{35} & \frac{8}{35} \\ \frac{3}{7} & \frac{3}{7} & \frac{3}{7} \\ \frac{12}{35} & \frac{12}{35} & \frac{12}{35} \end{bmatrix}$; $\begin{bmatrix} \frac{8}{35} \\ \frac{3}{7} \\ \frac{12}{35} \end{bmatrix}$

EXERCISES 8.3

1. No   **3.** Yes   **5.** $R = [.5]$; $S = \begin{bmatrix} .3 \\ .2 \end{bmatrix}$; $F = [2]$; $\begin{bmatrix} 1 & 0 & .6 \\ 0 & 1 & .4 \\ \hline 0 & 0 & 0 \end{bmatrix}$

7. $R = \begin{bmatrix} .3 & .6 \\ .1 & .2 \end{bmatrix}$; $S = \begin{bmatrix} .1 & 0 \\ .5 & .2 \end{bmatrix}$; $F = \begin{bmatrix} 1.6 & 1.2 \\ .2 & 1.4 \end{bmatrix}$; $\begin{bmatrix} 1 & 0 & .16 & .12 \\ 0 & 1 & .84 & .88 \\ \hline 0 & 0 & 0 & 0 \\ 0 & 0 & 0 & 0 \end{bmatrix}$

9. $R = \begin{bmatrix} .5 & 0 \\ .1 & .6 \end{bmatrix}$; $S = \begin{bmatrix} .1 & .2 \\ .3 & 0 \\ 0 & .2 \end{bmatrix}$; $F = \begin{bmatrix} 2 & 0 \\ .5 & 2.5 \end{bmatrix}$; $\begin{bmatrix} 1 & 0 & 0 & .3 & .5 \\ 0 & 1 & 0 & .6 & 0 \\ 0 & 0 & 1 & .1 & .5 \\ \hline 0 & 0 & 0 & 0 & 0 \\ 0 & 0 & 0 & 0 & 0 \end{bmatrix}$

11. If the gambler begins with $2, he should have $1 for an expected number of .79 plays.

13. (a) $\begin{array}{c} & D & G & F & S \\ D \\ G \\ F \\ S \end{array} \begin{bmatrix} 1 & 0 & .2 & .1 \\ 0 & 1 & 0 & .9 \\ \hline 0 & 0 & 0 & 0 \\ 0 & 0 & .8 & 0 \end{bmatrix}$ **(b)** $\begin{array}{c} & D & G & F & S \\ D \\ G \\ F \\ S \end{array} \begin{bmatrix} 1 & 0 & .28 & .1 \\ 0 & 1 & .72 & .9 \\ \hline 0 & 0 & 0 & 0 \\ 0 & 0 & 0 & 0 \end{bmatrix}$ **(c)** .72   **(d)** 1.8 years

15. (a) $\frac{13}{14}$; $\frac{11}{14}$ **(b)** $\frac{20}{7}$ months   **(c)** 5786; $1214   **17. (a)** $\frac{3}{4}$; $\frac{1}{2}$; $\frac{1}{4}$ **(b)** 4   **19.** $\frac{50}{81}$; $5\frac{1}{2}$

CHAPTER 8: SUPPLEMENTARY EXERCISES

1. Stochastic, neither   **2.** Stochastic, regular   **3.** Stochastic, regular   **4.** Stochastic, absorbing

5. Not stochastic   **6.** Stochastic, absorbing   **7.** $\begin{bmatrix} \frac{5}{9} \\ \frac{4}{9} \end{bmatrix}$ **8.** $\left[\begin{array}{ccc|cc} 1 & 0 & 0 & \frac{1}{2} & \frac{1}{2} \\ 0 & 1 & 0 & \frac{1}{4} & \frac{1}{8} \\ 0 & 0 & 1 & \frac{1}{4} & \frac{3}{8} \\ \hline 0 & 0 & 0 & 0 & 0 \\ 0 & 0 & 0 & 0 & 0 \end{array}\right]$

9. (a)
$$\begin{array}{c} \\ H \\ L \\ M \end{array} \begin{array}{ccc} H & L & M \\ \left[\begin{array}{ccc} .5 & .4 & .3 \\ .4 & .3 & .5 \\ .1 & .3 & .2 \end{array}\right] \end{array}$$
; **(b)** 38%   **(c)** $\frac{19}{97}$ **10. (a)**
$$\begin{array}{c} \\ P \\ N \end{array} \begin{array}{cc} P & N \\ \left[\begin{array}{cc} .8 & .3 \\ .2 & .7 \end{array}\right] \end{array}$$
(b) 30%   **(c)** 60%

11. $\left[\begin{array}{cc|ccc} 1 & 0 & \frac{11}{12} & 1 & \frac{1}{2} \\ 0 & 1 & \frac{1}{12} & 0 & \frac{1}{2} \\ \hline 0 & 0 & 0 & 0 & 0 \\ 0 & 0 & 0 & 0 & 0 \\ 0 & 0 & 0 & 0 & 0 \end{array}\right]$ **12. (a)**
$$\begin{array}{c} \\ I \\ II \\ III \\ IV \end{array} \begin{array}{cccc} I & II & III & IV \\ \left[\begin{array}{cc|cc} 1 & 0 & 0 & \frac{1}{4} \\ 0 & 1 & \frac{1}{3} & \frac{1}{4} \\ \hline 0 & 0 & 0 & \frac{1}{2} \\ 0 & 0 & \frac{2}{3} & 0 \end{array}\right] \end{array}$$
(b) $\frac{1}{6}$ **(c)** $\frac{5}{8}$ **(d)** $2\frac{1}{2}$ minutes   **13.** c   **14.** 58.5%

15. (a) If the traffic is moderate on a particular day, then for the next day the probability of Light is .2, the probability of Moderate is .75, and the probability of Heavy is .05. **(b)** About 37.0%, Light; 47.8%, Moderate; 15.2%, Heavy **(c)** About 3 days   **16.** State G: ≈ 1.48 months; state S: ≈ 1.83 months.

17. (a) $\begin{bmatrix} \frac{122}{1683} \\ \frac{23}{99} \\ \frac{4}{9} \\ \frac{422}{1683} \end{bmatrix}$; In the long run, the probability of having 1, 2, 3, or 4 units of water in the reservoir at any given time will be $\frac{122}{1683}, \frac{23}{99}, \frac{4}{9}$, or $\frac{422}{1683}$, respectively. **(b)** About $6881

CHAPTER 8: CHAPTER TEST

1. (a) stochastic   **(b)** not stochastic   **(c)** not stochastic   **(d)** stochastic
2. (a) not regular   **(b)** regular   **(c)** regular   **3.** b   **4.** $\begin{bmatrix} \frac{3}{7} \\ \frac{4}{7} \end{bmatrix}$; $\begin{bmatrix} \frac{3}{7} & \frac{3}{7} \\ \frac{4}{7} & \frac{4}{7} \end{bmatrix}$

5. (a) $\begin{bmatrix} .70 & .60 \\ .30 & .40 \end{bmatrix}$ **(b)** $\begin{bmatrix} .50 \\ .50 \end{bmatrix}$ **(c)** 66.5%   **(d)** $\begin{bmatrix} .70 & .60 \\ .30 & .40 \end{bmatrix}\begin{bmatrix} \frac{2}{3} \\ \frac{1}{3} \end{bmatrix} = \begin{bmatrix} \frac{2}{3} \\ \frac{1}{3} \end{bmatrix}$

6. (a) absorbing   **(b)** not absorbing   **(c)** absorbing

7. (a)
$$\begin{array}{ccc} IBM & App & None \\ \left[\begin{array}{ccc} 1 & 0 & .3 \\ 0 & 1 & .1 \\ 0 & 0 & .6 \end{array}\right] \end{array}$$
(b) $\left[\begin{array}{cc|c} 1 & 0 & \frac{3}{4} \\ 0 & 1 & \frac{1}{4} \\ \hline 0 & 0 & 0 \end{array}\right]$ **(c)** 25%   **(d)** 2.5 years

CHAPTER 9

EXERCISES 9.1

1. R: row 2; C: column 1   **3.** R: row 2; C: column 1   **5.** R: row 1; C: column 1
7. (a) row 1, column 2   **(b)** 0   **9.**
$$\begin{array}{c} \\ H \\ T \end{array} \begin{array}{cc} H & T \\ \left[\begin{array}{cc} 2 & -1 \\ -1 & -4 \end{array}\right] \end{array}$$
; strictly determined; R shows heads, C shows tails

11.
$$\begin{array}{c} \\ F \\ A \\ N \end{array} \begin{array}{ccc} F & A & N \\ \left[\begin{array}{ccc} 8000 & -1000 & 1000 \\ -7000 & 4000 & -2000 \\ 3000 & 3000 & 2000 \end{array}\right] \end{array}$$
; strictly determined; both should be neutral   **13.**
$$\begin{array}{c} \\ 5 \\ 10 \end{array} \begin{array}{ccc} 6 & 7 & 8 \\ \left[\begin{array}{ccc} 1 & -5 & -5 \\ -6 & 3 & 2 \end{array}\right] \end{array}$$
; not strictly determined

EXERCISES 9.2

1. (a) 0 **(b)** 1 **(c)** -1.12 **(d)** .5 [b is most advantageous to R] **3.** \$29,200 **5.** 0

7. No, no saddle point, yes, $\begin{bmatrix} \frac{11}{16} \\ \frac{5}{16} \end{bmatrix}$ **9.** No, no saddle point. Yes, $\begin{bmatrix} \frac{1}{30} \\ \frac{29}{30} \end{bmatrix}$ **11.** Strictly determined, Renée row 1, Carlos column 1; value $= 1$; not fair

EXERCISES 9.3

1. Minimize $y_1 + y_2$ subject to $\begin{cases} y_1 + 4y_2 \geq 1 \\ 6y_1 + 3y_2 \geq 1 \\ y_1 \geq 0, \ y_2 \geq 0 \end{cases}$ **3.** $\begin{bmatrix} \frac{1}{2} & \frac{1}{2} \end{bmatrix}$; $\begin{bmatrix} \frac{1}{4} \\ \frac{3}{4} \end{bmatrix}$ **5.** $\begin{bmatrix} \frac{1}{2} & \frac{1}{2} \end{bmatrix}$; $\begin{bmatrix} \frac{5}{9} \\ \frac{4}{9} \end{bmatrix}$

7. $\begin{bmatrix} \frac{2}{5} & \frac{3}{5} \end{bmatrix}$; $\begin{bmatrix} \frac{3}{5} \\ \frac{2}{5} \end{bmatrix}$ **9.** $\begin{bmatrix} \frac{7}{13} & \frac{6}{13} \end{bmatrix}$ **11.** $\begin{bmatrix} \frac{1}{5} \\ \frac{4}{5} \end{bmatrix}$ **13.** No, C: $\begin{bmatrix} \frac{1}{3} \\ \frac{2}{3} \end{bmatrix}$, R: $\begin{bmatrix} \frac{1}{3} \\ \frac{2}{3} \end{bmatrix}$, $v = -\frac{1}{3}$

15. (a) $\begin{bmatrix} \frac{8}{17} & \frac{9}{17} \end{bmatrix}$ **(b)** $\begin{bmatrix} \frac{8}{17} \\ \frac{9}{17} \end{bmatrix}$ **(c)** about \$2765 **17.** $\begin{bmatrix} \frac{5}{8} & \frac{3}{8} & 0 \end{bmatrix}$; $\begin{bmatrix} \frac{1}{2} \\ \frac{1}{2} \end{bmatrix}$

CHAPTER 9: SUPPLEMENTARY EXERCISES

1. Strictly determined; R 3, C 3; 2 **2.** Not strictly determined **3.** Not strictly determined

4. Strictly determined; R 1, C 2; 1 **5.** 7 **6.** 2 **7.** 1.4 **8.** -1.3 **9.** $\begin{bmatrix} \frac{4}{11} & \frac{7}{11} \end{bmatrix}$; $\begin{bmatrix} \frac{6}{11} \\ \frac{5}{11} \end{bmatrix}$

10. $\begin{bmatrix} \frac{8}{17} & \frac{9}{17} \end{bmatrix}$; $\begin{bmatrix} \frac{10}{17} \\ \frac{7}{17} \end{bmatrix}$ **11.** $\begin{bmatrix} 0 & 1 \end{bmatrix}$ **12.** $\begin{bmatrix} \frac{1}{4} \\ \frac{3}{4} \end{bmatrix}$ **13. (a)** Ruth should play the two $\frac{9}{14}$ of the time and the six $\frac{5}{14}$ of the time. Carol should play the two $\frac{5}{14}$ of the time and the six $\frac{9}{14}$ of the time. **(b)** Ruth

14. (a)

	S	A	W
A	3000	2000	1000
B	6000	2000	-3000
C	15,000	1000	$-10,000$

(b) Buy stock A

15. Consider the optimal strategies using the maxima/minima technique. **16.** $h < 4$

CHAPTER 9: CHAPTER TEST

1. (a) A move R_1 by R and C_3 by C results in a payoff of 6 to R. **(b)** A move R_3 by R and C_2 by C results in a payoff of 10 to C. **(c)** No **3. (a)** row 3, column 2, value $= 2$ **(b)** row 1, column 1, value $= 0$

4.

	1	2	3	4	5
1	-1	-1	-1	-1	-1
2	2	-2	-2	-2	-2
3	3	3	-3	-3	-3
4	4	4	4	-4	-4
5	5	5	5	5	-5

; Yes; -1; C

5. (a) On average, C gains .16 every time the game is played. **(b)** $C = \begin{bmatrix} .4 \\ .1 \\ .5 \end{bmatrix}$

6. (a) $C = \begin{bmatrix} \frac{3}{5} \\ \frac{2}{5} \end{bmatrix}$; $R = \begin{bmatrix} \frac{9}{10} & \frac{1}{10} \end{bmatrix}$ **(b)** C

CHAPTER 10

EXERCISES 10.1

1. (a) $i = .01$, $n = 24$ **(b)** $i = .02$, $n = 20$ **(c)** $i = .05$, $n = 40$ **3. (a)** $i = .06$, $n = 4$, $P = \$500$, $F = \$631.24$
(b) $i = .005$, $n = 120$, $P = \$800$, $F = \$1455.52$ **(c)** $i = .02$, $n = 19$, $P = \$6177.88$, $F = \$9000$ **5.** \$1127.16

7. $22,396.57   **9.** $7180.08; $1180.08   **11. (a)** $3548.74   **(b)** $451.26   **(c)**

Month	Interest	Balance
0		$3548.74
1	$17.74	$3566.48
2	$17.83	$3584.31
3	$17.92	$3602.23

13. $12,824.32   **15.** $8874.49   **17. (a)** 6%   **(b)** $5.05; $1015.08   **(c)** $5.61; $1127.16   **19.** $1700 in 9 years

21. better   **23.** $8874.49   **25. (a)** $r = .04$, $n = \frac{1}{2}$, $P = \$500$, $F = \$510$ **(b)** $r = .05$, $n = 2$, $P = \$500$, $F = \$550$

27. $1150   **29.** $2500   **31.** 4.08%   **33.** 20 years   **35.** $P = \dfrac{F}{1 + nr}$ **37.** a   **39.** $104.06; 4.06%   **41.** 4.04%

43. 4.49%   **45.** 4%   **47.** 18 yrs.   **49.** d   **51.** 4.6   **53.** $1065; $1134.23; $1207.95; 8; 12   **55.** 30 years

EXERCISES 10.2

1. (a) $i = .005$, $n = 120$, $R = \$50$, $F = \$8193.97$ **(b)** $i = .02$, $n = 20$, $R = \$2675.19$, $F = \$65,000$ **3.** $23,123.67

5. $4698.97   **7. (a)** $27,048.92   **(b)** $3048.92   **(c)**

Month	Interest	Balance
1		500
2	2.50	1002.50
3	5.01	1507.51

9. $305.06; $10,982.16, $1017.84   **11.** $200 each month; $343.75   **13.** $24,649.92   **15.** $877.91

17. $1000 at the end of each month   **19.** $7239.32   **21.** $17,584.62   **23.** a   **25.** $24,000   **27.** $4339.35

29. $59,189.85   **31. (a)** $1   **(b)** $s_{\overline{m}\,i} \cdot i$ **35.** $5600.40   **37. (a)** $1269.67   **(b)** $65,134.88

39. $B_{\text{new}} = (1 + i)B_{\text{previous}} - R$ **41.** $2050, $3152.50, $4310.13; 19; 26   **43.** After 33 weeks   **45.** $4734.59

EXERCISES 10.3

1. $193.33   **3.** $11,469.92   **5. (a)** $583.31   **(b)** $16.69   **(c)** $58,314.31   **(d)** $26,973.02   **(e)** $4188.65

(f) $269.73   **7. (a)** $265.71   **(b)** $9565.56   **(c)** $1565.56   **(d)** $5644.58   **(e)** $2990.59   **(f)** $534.53

(g)

Payment number	Amount	Interest	Applied to principal	Unpaid balance
1	$265.71	$80.00	$185.71	$7814.29
2	265.71	78.14	187.57	7626.72
3	265.71	76.27	189.44	7437.28
4	265.71	74.37	191.34	7245.94

9. $8950   **11.** $14,042.36

13.

Payment number	Amount	Interest	Applied to principal	Unpaid balance
1	$256.28	$10.00	$246.28	$753.72
2	256.28	7.54	248.74	504.98
3	256.28	5.05	251.23	253.74
4	256.28	2.54	253.74	0.00

15. $804.62   **17.** $119,161.62   **19.** $1196.68

21. a   **23. (a)** $183,927.96   **(b)** $24,989.92   **(c)** $105,266.63   **25.** $2.5 billion   **27.** No; $277,151.11

29. (a) $666.67 **(b)** $623.65 **(c)** b   **31. (d)** $101, $102.01   **33.** $2105.33, $2021.12, $1936.28, 24 months

35. $4258.21; after 46 months   **37.** 261

EXERCISES 10.4

1. deferred   **3.** $165,000   **5.** $1,313,125.69   **7.** $1,313,125.69   **9. (a)** $329,864.79 **(b)** $285,890.08 **(c)** Larry
(d) Earl   **11.** $366.67   **13.** $105.83   **15.** 2.73%   **17.** 3.17%   **19. (a)**$41.67   **(b)** $41.07; the monthly payment is
less.   **21.** False   **23.** False   **25.** False   **27.** d   **29.** a   **31.** a   **33.** c
35. 12*RATE(240, -573.14, 77600, 0); 12*RATE(120, -573.14, 77600, -51624.70)
37. $(((1+i)^{180} - 1)/(i(1+i)^{180})) \cdot 1217.12 = 118800;\ 1217.12 - i \cdot 96081.51 = (1217.12 - i \cdot 118800)(1+i)^{60}$
39. The salesman should compare the present value of the loan payments with the $1000.   **41.** 11.08%   **43.** 6.83%
45. 7.08%   **47.** 87.72 months

CHAPTER 10: SUPPLEMENTARY EXERCISES

1. (d)   **2.** $488.16   **3.** $101,497   **4.** $53.79   **5.** 10% compounded annually   **6.** $13,954.01
7. (a) $2400.34   **(b)** $167,304.68   **8.** $43,665.52   **9.** $27,481.64   **10.** $13,050.08   **11.** $211.37   **12.** $12,050.34
13. $782.92   **14.** $6872.11   **15.** $100,451.50   **16.** $139,401.04   **17.** Investment A
18. Yes, it is a bargain since the present value is $879.57.   **19.** 10.25%   **20.** 19.56%   **21.** $146,861.85

Payment number	Amount	Interest	Applied to principal	Unpaid balance
1	$304.22	$50.00	$254.22	$9745.78
2	304.22	48.73	255.49	9490.29
3	304.22	47.45	256.77	9233.52
4	304.22	46.17	258.05	8975.47
5	304.22	44.88	259.34	8716.13
6	304.22	43.58	260.64	8455.49

22. (table above)
23. $151,843.34   **24.** $5799.84   **25.** $1206.93
26. (a) $21,000   **(b)** $26,095.40
27. (a) $30,000   **(b)** $40,146.77   **28.** 4.82%
29. Consumer Loan A
30. $P = 88{,}200,\ n = 180,\ R = 759.47$
31. $P = 97{,}000,\ m = 72,\ R = 716.43;\ B = 81{,}298.32$
32. $i = .035/12,\ R = 63.15,\ P = 2000$
35. Higher   **36.** No   **37.** No   **38.** No

CHAPTER 10: CHAPTER TEST

1. $524.38   **2.** $4464.29   **3.** $1189.77; $164.77   **4.** $11,263.09   **5.** $72,704.68   **6.** $21,290.16   **7.** $1095.92
8. $2533.52   **9.** $3086.03   **10.** $2187.43   **11. (a)** $166.07   **(b)** $60   **(c)** $7893.93 **(d)** $1899.00
12. $225,686.46   **13. (a)** $324,488.67   **(b)** $294,559.19   **14.** $144.44   **16. (a)** It will decrease because the cost of the
points will be spread over a longer period of time.   **(b)** Increasing the number of points will increase the effective cost
of the loan, therefore, increasing its interest rate.   **17.** If there are no points, or the mortgage is not terminated early.

CHAPTER 11

EXERCISES 11.1

1. $4, -6; 2$ **3.** $-\frac{1}{2}, 0; 0$ **5.** $-\frac{2}{3}, 15; 9$
7. (a) $10, 4, 1, -\frac{1}{2}, -\frac{5}{4}$ **(b)**   **(c)** $y_n = -2 + 12\left(\frac{1}{2}\right)^n$

9. (a) 3.5, 4, 5, 7, 11 **(b)** 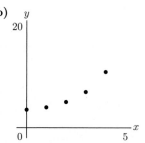 **(c)** $y_n = 3 + (.5)2^n$

11. (a) 17.5, 0, 7, 4.2, 5.32 **(b)** 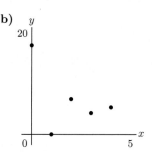 **(c)** $y_n = 5 + 12.5(-.4)^n$

13. (a) 15, 14, 12, 8, 0 **(b)** 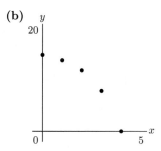 **(c)** $y_n = 16 - 2^n$

15. 1, 5, 5.8, 5.96, 5.992 **17.** $y_n = 1.05y_{n-1}, y_0 = 1000$ **19.** $y_n = .99y_{n-1} - 1{,}000{,}000, y_0 = 70{,}000{,}000$

21. (a) 1, 3, 5, 7, 9 **(b)** 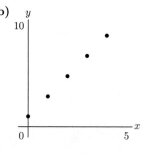 **(c)** $a = 1$, so the denominator of $\frac{b}{1-a}$ is zero. **23.** \$30

25. (a) $y_n = 1.04y_{n-1} + 250$; $y_0 = 800$ **(b)** $y_n = -6250 + 7050(1.04)^n$ **(c)** about \$3027.32

27. (a) $y_n = .85y_{n-1}, y_0 = 20{,}000$ **(b)** $y_n = 20{,}000(.85)^n$ **(c)** \$8,874.11

29. (a) (b)

n	y_n
0	102.67
1	100.0035
2	97.20368
3	94.26386
4	91.17705
5	87.9359
6	84.5327
7	80.95933
8	77.2073
9	73.26767
10	69.13105
11	64.7876
12	60.22698
13	55.43833
14	50.41025
15	45.13076
16	39.5873
17	33.76666
18	27.655
19	21.23775
20	14.49963
21	7.424615
22	−0.004154
23	−7.804362
24	−15.99458

(c) 50.41; 22

31. (a) (b)

n	y_n
0	4
1	7.7
2	4.555
3	7.22825
4	4.955988
5	6.887411
6	5.245701
7	6.641154
8	5.455019
9	6.463234
10	5.606251
11	6.334686
12	5.715516
13	6.241811
14	5.794461
15	6.174708
16	5.851498
17	6.126227
18	5.892707
19	6.091199
20	5.922481
21	6.065891
22	5.943992
23	6.047606
24	5.959535

(c) 5.715516; $n \geq 19$

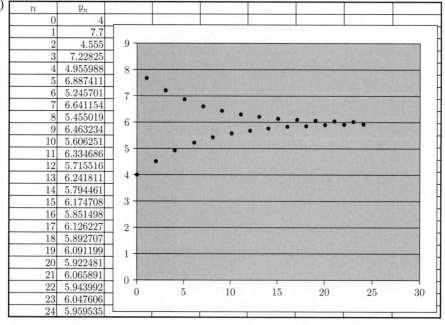

33. (a) (b)

n	y_n
0	1
1	1.2
2	1.4
3	1.6
4	1.8
5	2
6	2.2
7	2.4
8	2.6
9	2.8
10	3
11	3.2
12	3.4
13	3.6
14	3.8
15	4
16	4.2
17	4.4
18	4.6
19	4.8
20	5
21	5.2
22	5.4
23	5.6
24	5.8

(c) 2.8; 18

35. (a) (b)

n	y_n
0	2
1	2.75
2	3.3125
3	3.734375
4	4.050781
5	4.288086
6	4.466064
7	4.599548
8	4.699661
9	4.774746
10	4.831059
11	4.873295
12	4.904971
13	4.928728
14	4.946546
15	4.95991
16	4.969932
17	4.977449
18	4.983087
19	4.987315
20	4.990486
21	4.992865
22	4.994649
23	4.995986
24	4.99699

(c) 4.831059; $n \geq 20$

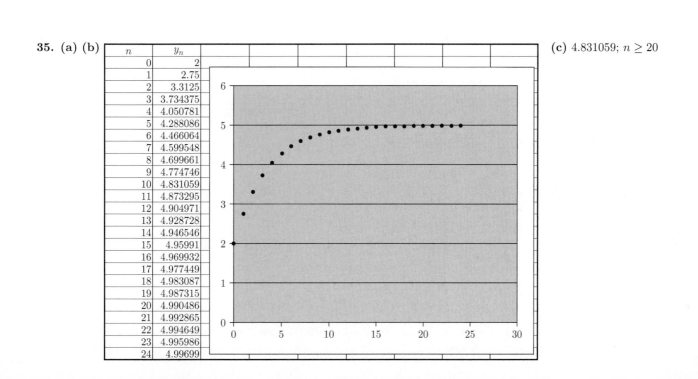

EXERCISES 11.2

1. $y_n = 1 + 5n$ **3.** $80(1.0075)^{60}$ **5.** $80\left(1 + \frac{1}{365}\right)^{1825}$ **7.** 108   **9. (a)** \$1.40 **(b)** \$1.44 **(c)** \$1.46
11. (a) 10; 10; 10; 10   **(b)** 11; 12; 14; 18   **(c)** 9; 8; 6; 2

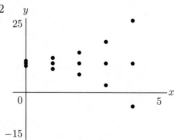

13. $y_n = 5 + 2(.4)^n$; y_n approaches 5   **15.** $y_n = 2(-5)^n$; y_n gets arbitrarily large, alternating between being positive and negative.   **17.** $y_n = 1.0075y_{n-1} - 350$, $y_0 = 38,900$ **19.** $y_n = y_{n-1} - 2000$, $y_0 = 50,000$; $y_n = 50,000 - 2000n$
21. (b)   **23. (c)**   **25. (c)**   **27. (c)**   **29.** \$1195.62; $8\frac{1}{2}$ years; 47   **31.** \$245; 12 years; 23 years   **33.** 22   **35.** 60

EXERCISES 11.3

1. (a), (b), (d), (f), (h)   **3. (b), (d), (e), (f)**   **5. (b), (d), (e), (f)**   **7. (a), (c), (h)**, possibly **(g)**

9. Possible answer: **11.** Possible answer: **13.** Possible answer:

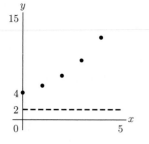   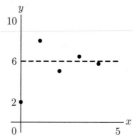   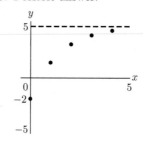

15.   **17.**   **19.**

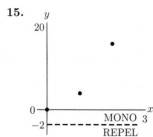   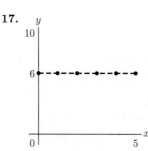   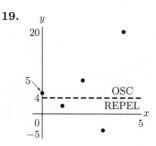

21.   **23.**

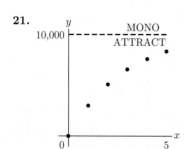   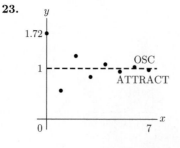

25. less than \$60,000   **27. (a)** $y_n = 1.06y_{n-1} - 120$ **(b)** at least \$2000   **29.** less than \$30,000   **31.** Possible answer: $y_n = .5y_{n-1} + 4$, $y_0 = 1$ **33.** Possible answer: $y_n = 2y_{n-1}$, $y_0 = 1$ **35.** Possible answer: $y_n = -2y_{n-1}$, $y_0 = 5$

EXERCISES 11.4

1. $y_n = 1.0075y_{n-1} - 261.50$, $y_0 = 32,500$ **3.** $y_n = 1.015y_{n-1} + 200$, $y_0 = 4000$ **5.** \$46,002.34 **7.** \$11,035.68
9. \$28.55 **11.** \$132.86 **13.** \$39.205 million; 30 **15.** \$505.03; \$1022.80; \$1553.65; 30; 37

EXERCISES 11.5

1. $y_n = 1.02y_{n-1}$, $y_0 = 100$ million

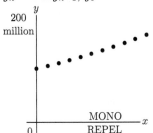

3. $y_n = .75y_{n-1}$

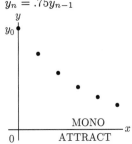

5. $y_n = .92y_{n-1} + 8$, $y_0 = 0$

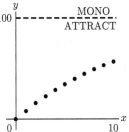

7. $y_n = .7y_{n-1} + 3.6$, $y_0 = 0$

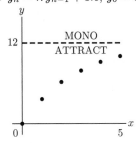

9. $y_n = 1.05y_{n-1} - 1000$,
$y_0 = 30,000$

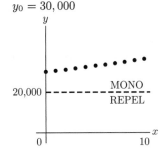

11. $y_n = .8y_{n-1} + 14$, $y_0 = 40$

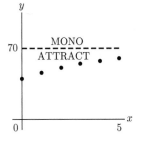

13. $p_n = -.5p_{n-1} + 21$, $p_0 = 4.54$

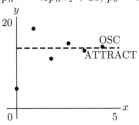

15. b **17.** 5.38 million; 2015; 2049
19. $y_n = .87y_{n-1}$, $y_0 = 130$; $y_n = 130(.87)^n$; 5 hours; 4.6 mg

CHAPTER 11: SUPPLEMENTARY EXERCISES

1. (a) 5; −7; 29 **(b)** $y_n = 2 - (-3)^n$ **(c)** −79 **2.** (a) $\frac{17}{2}$; 7; $\frac{11}{2}$ **(b)** $y_n = 10 - \frac{3}{2}n$ **(c)** 1 **3.** \$2989.08
4. \$1109.54 **5.**

6.

7. (a) $y_n = 1.03y_{n-1} - 600$, $y_0 = 120,000$ **(b)** 200,611 **8.** (a) $y_n = 1.01y_{n-1} - 360$, $y_0 = 35,000$ **(b)** \$33,693.28
9. \$20.22 **10.** \$237.14 **11.** \$22,492.53 **12.** \$2,704.89

13. $y_n = .9y_{n-1} + 100{,}000$, $y_0 = 0$ **14.** $y_n = .92y_{n-1}$, $y_0 = 100$

CHAPTER 11: CHAPTER TEST

1. (a) 2, 4, 5   **(b)** $y_n = 6 - 8\left(\tfrac{1}{2}\right)^n$ **(c)** 5.5   **2. (a)** 110, 160, 210   **(b)** $y_n = 60 + 50n$ **(c)** 260

3. (a)   **(b)**

4. .3 ppm

5. (a) \$31.726   **(b)** 8%   **(c)** \$3000
    (d) \$29,360.52

6. (a) \$454.15   **(b)** 12.5%
    (c) \$9,245.85; \$8,526.44; \$361.57

7. \$119,161.62

8. (a) $y_n = 1.06y_{n-1} + 500$; $y_0 = 0$ **(b)** \$14,106.44   **9.** \$122.04   **10.** $y_n = .9y_{n-1} + .05$; $y_0 = 0$

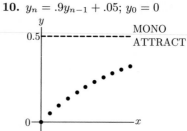

CHAPTER 12

EXERCISES 12.1

1. Statement   **3.** Statement   **5.** Not a statement   **7.** Not a statement   **9.** Statement

11. Not a statement   **13.** Not a statement   **15.** Statement

17. p: The Phelps Library is in New York.
    q: The Phelps Library is in Dallas.
    $p \vee q$

19. p: The Smithsonian Museum of Natural History has displays of rocks.
    q: The Smithsonian Museum of Natural History has displays of bugs.
    $p \wedge q$

21. p: Amtrack trains go to Chicago.
    q: Amtrack trains go to Cincinnati.
    $\sim p \wedge \sim q$ or $\sim(p \vee q)$

23. (a) Ozone is opaque to ultraviolet light, and life on earth requires ozone.   **(b)** Ozone is not opaque to ultraviolet light, or life on earth requires ozone.   **(c)** Ozone is not opaque to ultraviolet light, or life on earth does not require ozone. **(d)** It is not the case that life on earth does not require ozone. **25. (a)** $p \vee q$ **(b)** $p \wedge \sim q$ **(c)** $q \wedge \sim p$ **(d)** $\sim p \wedge \sim q$

EXERCISES 12.2

1. $\sim r$ is a statement, $p \wedge \sim r$ is a statement, $\sim(p \wedge \sim r)$ is a statement, $\sim(p \wedge \sim r) \vee q$ is a statement.   **3.** $\sim p$ is a statement, $\sim p \vee r$ is a statement, $q \wedge r$ is a statement, $(\sim p \vee r) \rightarrow (q \wedge r)$ is a statement.

5.

p	q	p∧~q
T	T	F
T	F	T
F	T	F
F	F	F

7.

p	q	(p∨~q)	∨	~p
T	T	T F	**T**	F
T	F	T T	**T**	F
F	T	F F	**T**	T
F	F	T T	**T**	T

9.

p	q	~	((p∨q)	∧	(p∧q))
T	T	**F**	T	T	T
T	F	**T**	T	F	F
F	T	**T**	T	F	F
F	F	**T**	F	F	F

11.

p	q	p⊕(~p∨q)
T	T	T **F** F T T
T	F	T **T** F F F
F	T	F **T** T T T
F	F	F **T** T T F

13.

p	q	r	(p∧~r)∨q
T	T	T	T
T	T	F	T
T	F	T	F
T	F	F	T
F	T	T	T
F	T	F	T
F	F	T	F
F	F	F	F

15.

p	q	r	~[(p∧r)∨q]
T	T	T	F
T	T	F	F
T	F	T	F
T	F	F	T
F	T	T	F
F	T	F	F
F	F	T	T
F	F	F	T

17.

p	p∨~p
T	T
F	T

19.

p	q	r	p⊕(q∨r)
T	T	T	F
T	T	F	F
T	F	T	F
T	F	F	T
F	T	T	T
F	T	F	T
F	F	T	T
F	F	F	F

21.

p	q	r	(p∨q)∧(p∨r)
T	T	T	T
T	T	F	T
T	F	T	T
T	F	F	T
F	T	T	T
F	T	F	F
F	F	T	F
F	F	F	F

23.

p	q	(p∨q)∧~(p∨q)
T	T	F
T	F	F
F	T	F
F	F	F

25.

p	q	r	~(p∨q)∧r
T	T	T	F
T	T	F	F
T	F	T	F
T	F	F	F
F	T	T	F
F	T	F	F
F	F	T	T
F	F	F	F

27.

p	q	r	~p∨(q∧r)
T	T	T	T
T	T	F	F
T	F	T	F
T	F	F	F
F	T	T	T
F	T	F	T
F	F	T	T
F	F	F	T

29. They are identical.   **31.** They are identical.

33. $(p∧q)∨r$ is T and $p∧(q∨r)$ is F when p is F and r is T. Otherwise, the tables are identical.

35. (a)

p	p|p
T	F
F	T

(b)

p	q	(p|p)|(q|q)
T	T	T
T	F	T
F	T	T
F	F	F

(c)

p	q	(p|q)|(p|q)
T	T	T
T	F	F
F	T	F
F	F	F

(d)

p	q	p|((p|q)|q)
T	T	F
T	F	F
F	T	T
F	F	T

37. (a) T **(b)** F **(c)** T **(d)** F **(e)** F **(f)** F **39. (a)** F **(b)** F **(c)** F **(d)** F

41. $(p⊕q)∧r$

```
            (p⊕q) ∧ r
                F
              /   \
         p ⊕ q     r
            T       F
          /   \
         p     q
         T     F
```

43. $(p∨q)∧(p∨~r)$

```
          (p∨q) ∧ (p∨~r)
                T
             /     \
         p ∨ q       p ∨ ~r
           T           T
          / \         /   \
         p   q       p    ~r
         T   F       T     T
                           |
                           r
                           F
```

45. 0 **47. (a)** T **(b)** T

49. (a) T   (b) F   **51.**

```
 p      q      ▓    ◆ 3
 1      1      1
 1      0      0
 0      1      0
 0      0      1
------  ------ ------
L1 ="not( LP  xor  L
```

EXERCISES 12.3

1.

p	q	$p \to \sim q$
T	T	F
T	F	T
F	T	T
F	F	T

3.

p	q	$(p \oplus q) \to q$
T	T	T
T	F	F
F	T	T
F	F	T

5.

p	q	r	$(\sim p \wedge q) \to r$
T	T	T	T
T	T	F	T
T	F	T	T
T	F	F	T
F	T	T	T
F	T	F	F
F	F	T	T
F	F	F	T

7.

p	q	$(p \to q) \leftrightarrow (\sim p \vee q)$
T	T	T
T	F	T
F	T	T
F	F	T

9.

p	q	r	$(p \to q) \to r$
T	T	T	T
T	T	F	F
T	F	T	T
T	F	F	T
F	T	T	T
F	T	F	F
F	F	T	T
F	F	F	F

11.

p	q	r	$\sim (p \vee q) \to (\sim p \wedge r)$
T	T	T	T
T	T	F	T
T	F	T	T
T	F	F	T
F	T	T	T
F	T	F	T
F	F	T	T
F	F	F	F

13.

p	q	$(p \vee q) \leftrightarrow (p \wedge q)$
T	T	T
T	F	F
F	T	F
F	F	T

15.

p	q	r	$[p \wedge (q \vee r)] \leftrightarrow [(p \wedge q) \vee (p \wedge r)]$
T	T	T	T
T	T	F	T
T	F	T	T
T	F	F	T
F	T	T	T
F	T	F	T
F	F	T	T
F	F	F	T

17. $((\sim p) \wedge (\sim q)) \to ((\sim p) \wedge q)$

19. $(((\sim p) \wedge (\sim q)) \vee r) \to ((\sim q) \wedge r)$ **21.** T   **23.** T   **25.** T   **27.** F   **29.** T   **31.** $p \leftrightarrow q$

33. $q \to p$; hyp: q; con: p **35.** $q \to p$; hyp: q; con: p

37. $\sim p \to \sim q$; hyp: $\sim p$; con: $\sim q$ **39.** $\sim p \to \sim q$; hyp: $\sim p$; con: $\sim q$; TRUE

41. $\sim q \to \sim p$; hyp: $\sim q$; con: $\sim p$; TRUE   **43. (a)** hyp: A person is healthy. con: The person lives a long life.
(b) hyp: The train stops at the station. con: A passenger requests the stop.   **(c)** hyp: The azalea grows. con: The azalea is exposed to sunlight.   **(d)** hyp: I will go to the store. con: Jane goes to the store.   **45. (a)** If City Sanitation collects the garbage, then the mayor calls.   **(b)** The price of beans goes down if there is no drought.   **(c)** Goldfish swim in Lake Erie if Lake Erie is fresh water.   **(d)** Tap water is not salted if it boils slowly.

47. (a) 4   **(b)** 4   **(c)** 4   **(d)** 4   **(e)** 6   **(f)** 6   **49. (a)** 7   **(b)** 7   **(c)** 0   **(d)** 7   **(e)** 0   **(f)** 0
51. (a) 6   **(b)** -12 **(c)** 0   **(d)** 0   **(e)** 28   **(f)** 0   **53. (a)** 3   **(b)** 3   **(c)** -37 **(d)** -7 **(e)** 43   **(f)** -7

EXERCISES 12.4

1. When p is FALSE and q is TRUE, the statement is FALSE.   **5. (d)** $p|(q|q)$ **(e)** $\sim (p \wedge q)$

7.

p	q	c	$(p \to q) \leftrightarrow [(p \wedge \sim q) \to c]$
T	T	F	T
T	F	F	T
F	T	F	T
F	F	F	T

   9. False   **11.** $\sim [\sim (p \vee q) \vee \sim (\sim p \vee \sim q)]$

13. $\sim (p \vee q) \vee (q \wedge \sim r)$ **15.** $(p \wedge \sim q) \vee (p \vee \sim r)$

17. (a) Arizona does not border California or Arizona does not border Nevada. **(b)** There are no tickets available, and the agency cannot get tickets. **(c)** The killer's hat was neither white nor gray. **19.** $\sim p \wedge q \vee \sim r$ **21. (a)** Jeremy does not take 12 credits and Jeremy does not take 15 credits this semester. **(b)** Sandra did not receive a gift from Sally or Sandra did not receive a gift from Sacha. **23. (a)** "The Old Man and the Sea" was written by Ernest Hemingway or "The Old Man and the Sea" was written by Jack London. "The Old Man and the Sea" was not written by Ernest Hemingway and not written by Jack London. **(b)** "H. M. S. Pinafore" was written by Gilbert and "H. M. S. Pinafore" was written by Sullivan. "H. M. S. Pinafore" was not written by Gilbert or not written by Sullivan. **25. (a)** I have a ticket to the theater, and I did not spend a lot of money. **(b)** Basketball is played on an indoor court, and the players do not wear sneakers. **(c)** The stock market is going up, and interest rates are not going down. **(d)** Humans have enough water, and humans are not staying healthy. **27. (a)** If the number of odd numbers in a sum is even, then the sum is even. (T) **(b)** If K is next to W, then the computer keyboard is not standard. (T) **29. (a)** Contrapositive: If a bird is not a hummingbird, then it is not small. (F) Converse: If a bird is a hummingbird, then it is small. (T) **(b)** Contrapositive: If two nonvertical lines are not parallel, they do not have the same slope. (T) Converse: If two nonvertical lines are parallel, they have the same slope. (T) **(c)** Contrapositive: If we are not in France, then we are not in Paris. (T) Converse: If we are in France, we must be in Paris. (F) **(d)** Contrapositive: If you can legally make a U-turn, then the road is not one-way. (T) Converse: If you cannot legally make a U-turn, then the road is one-way. (F)

31. Ask either guard, "If I asked you whether your door was the door to freedom, would you say yes?"

33. $[(p \to q) \wedge (q \to r)] \Longrightarrow p \to r$

p	q	r	$(p \to q)$	$\wedge$	$(q \to r)$	$\Longrightarrow$	$(p \to r)$
T	T	T	T	T	T	**T**	T
T	T	F	T	F	F	**T**	F
T	F	T	F	F	T	**T**	T
T	F	F	F	F	T	**T**	F
F	T	T	T	T	T	**T**	T
F	T	F	T	F	F	**T**	T
F	F	T	T	T	T	**T**	T
F	F	F	T	T	T	**T**	T

EXERCISES 12.5

1. m = "Sue goes to the movies."
r = "Sue reads."

1.	$m \vee r$	hyp.
2.	$\sim m$	hyp.
3.	r	disj. syll. (1, 2)

3. a = "My allowance comes this week."
p = "I pay the rent."
b = "My bank account will be in the black."
e = "I will be evicted."

1.	$(a \wedge p) \to b$	hyp.
2.	$\sim p \to e$	hyp.
3.	$\sim e \wedge a$	hyp.
4.	$\sim e$	subtr. (3)
5.	p	mod. tollens (2, 4)
6.	a	subtr. (3)
7.	b	mod. ponens (5, 6, 1)

5. p = "The price of oil increases."
a = "The OPEC countries are in agreement."
d = "There is a U.N. debate."

1.	$p \to a$	hyp.
2.	$\sim d \to p$	hyp.
3.	$\sim a$	hyp.
4.	$\sim p$	mod. tollens (1, 3)
5.	d	mod. tollens (2, 4)

7. g = "The germ is present."
r = "The rash is present."
f = "The fever is present."

1.	$g \to (r \wedge f)$	hyp.
2.	f	hyp.
3.	$\sim r$	hyp.
4.	$\sim r \vee \sim f$	addition (3)
5.	$\sim (r \wedge f)$	DeMorgan (4)
6.	$\sim g$	mod. tollens (1, 5)

9. $c =$ "The material is cotton."
$r =$ "The material is rayon."
$d =$ "The material can be made into a dress."

1. $(c \vee r) \to d$ hyp.
2. $\sim d$ hyp.
3. $\sim (c \vee r)$ mod. tollens (1, 2)
4. $\sim c \wedge \sim r$ DeMorgan (3)
5. $\sim r$ subtraction (4)

11. Invalid **13.** Valid **15.** Valid **17.** Invalid **19.** Invalid

21. $s =$ "Sam goes to the store."
$m =$ "Sam needs milk."
$H_1 = s \to m$
$H_2 = \sim m$
$C = \sim s$

1. s $\sim C$
2. $s \to m$ H_1
3. m $\sim H_2$; mod. ponens (1, 2)

23. $n =$ "The newspaper reports the crime."
$t =$ "Television reports the crime."
$s =$ "The crime is serious."
$k =$ "A person is killed."
$H_1 = (n \wedge t) \to s$
$H_2 = k \to n$
$H_3 = k$
$H_4 = t$
$C = s$

1. $\sim s$ $\sim C$
2. $(n \wedge t) \to s$ H_1
3. $\sim (n \wedge t)$ mod. tollens (1, 2)
4. $\sim n \vee \sim t$ DeMorgan (3)
5. t H_4
6. $\sim n$ disj. syllogism (4, 5)
7. $k \to n$ H_2
8. $\sim k$ $\sim H_3$; mod. tollens (6, 7)

25. $j =$ "Jimmy finds his keys."
$h =$ "He does his homework."

Direct proof
1. $\sim j \to h$ H_1
2. $\sim h$ H_2
3. j mod. tollens (1, 2)

Indirect proof
1. $\sim j$ $\sim C$
2. $\sim j \to h$ H_1
3. h mod. ponens (1, 2) } contradiction
4. $\sim h$ H_2

27. $m =$ "Marissa goes to the movies."
$k =$ "Marissa is in a knitting class."
$i =$ "Marissa is idle."

1. $\sim m \to \sim i$ H_1
2. $\sim k \to i$ H_2
3. $\sim k$ H_3
4. i mod. ponens (2, 3)
5. m mod. tollens (1, 4)

EXERCISES 12.6

1. (a) F (b) T (c) T (d) T (e) F **3.** (a) $\forall x p(x)$ (b) $\sim [\forall x p(x)]$ (c) $\forall x \sim p(x)$ (d) (c) implies (b)
5. $\forall x \sim p(x)$, or $\sim [\exists x p(x)]$. This is FALSE. Abby meant to say, "Not all men cheat on their wives."
7. (a) $\forall x p(x)$ (b) $\exists x \sim p(x)$ (c) $\exists x p(x)$ (d) $\sim [\forall x p(x)]$ (e) $\forall x \sim p(x)$ (f) $\sim [\exists x p(x)]$
  (g) (b) and (d); (e) and (f)
9. (a) T (b) F **11.** (a) T (b) F (c) T (d) F (e) F (f) T (g) F (h) T
13. (a) Not every dog has his day. (b) No men fight wars. (c) Some mothers are unmarried. (d) There exists a pot
without a cover. (e) All children have pets. (f) Every month has 30 days.
15. (a) "The sum of any two nonnegative integers is greater than 12." FALSE: let $x = 1$, $y = 2$. "There exist two
nonnegative integers whose sum is not greater than 12." (b) "For any nonnegative integer, there is another nonnegative
integer that, added to the first, makes a sum greater than 12." TRUE. (c) "There is a nonnegative integer that, added to
any other nonnegative integer, makes a sum greater than 12." TRUE. (d) "There are two nonnegative integers the sum
of which is greater than 12." TRUE.
17. (a) FALSE: let $x = 2, y = 3$. (b) TRUE: for any x, let $y = x$. (c) TRUE: let $x = 1$. (d) FALSE: no y is divisible by
every x. (e) TRUE: for any y, let $x = y$. (f) TRUE: any x divides itself. **19.** (a) $\forall x[x \geq 8 \to x \leq 10]$ (b) No;
consider $x = 11$. **21.** $S = \{2, 4, 6, 8\}, T = \{1, 2, 3, 4, 6, 8\}$. So $\forall x[x \in S \to x \in T]$. **23.** The solutions to
$(x - 8)(x - 3) = 0$ are 8 and 3. The solutions to $x^2 = 9$ are -3 and 3. So, $\forall x(x \in S \leftrightarrow x \in T)$. Therefore, $S = T = \{3\}$.

EXERCISES 12.7

1. $(p \lor q) \land (p \lor \sim q)$ **3.** $((p \land q) \land \sim r) \land (\sim q \lor r)$

5. 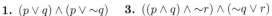 ; 1   **7.** ; 0

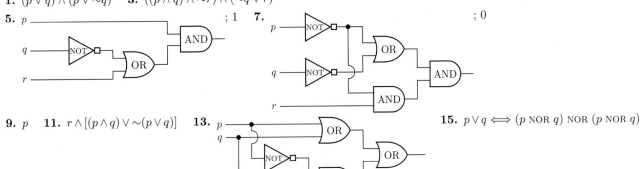

9. p **11.** $r \land [(p \land q) \lor \sim(p \lor q)]$ **13.**   **15.** $p \lor q \Longleftrightarrow (p \text{ NOR } q) \text{ NOR } (p \text{ NOR } q)$

17. $p \text{ XOR } q \Longleftrightarrow (p \text{ NOR } q) \text{ NOR } (p \land q)$ **19.**

CHAPTER 12: SUPPLEMENTARY EXERCISES

1. (a) Statement **(b)** Not a statement **(c)** Statement **(d)** Not a statement **(e)** Statement

2. (a) If two lines are perpendicular, then their slopes are negative reciprocals of each other. **(b)** If goldfish can live in a fishbowl, then the water is aerated. **(c)** If it rains, then Jane uses her umbrella. **(d)** If Sally gives Morris a treat, then he ate all his food.

3. (a) Contrapositive: If the Yankees are not playing in Yankee Stadium, then they are not in New York City; Converse: If the Yankees are playing in Yankee Stadium, then they are in New York City. **(b)** Contrapositive: If the earthquake is not considered major, then the Richter scale does not indicate the quake is a 7; Converse: If the earthquake is considered major, then the Richter scale indicates the quake is a 7. **(c)** Contrapositive: If the coat is not warm, then it is not made of fur; Converse: If the coat is warm, then it is made of fur. **(d)** Contrapositive: If Jane is not in Moscow, then she is not in Russia; Converse: If Jane is in Moscow, then she is in Russia.

4. (a) Two triangles are similar but their sides are unequal. **(b)** For every real number $x, x^2 \neq 5$. **(c)** There exists a positive integer n such that n is even and n^2 is not even. **(d)** For every real number $x, x^2 + 4 \neq 0$.

5. (a) Tautology **(b)** Tautology **(c)** Not a tautology **(d)** Not a tautology

6. (a)

p	q	r	$p \to (\sim q \lor r)$
T	T	T	T
T	T	F	F
T	F	T	T
T	F	F	T
F	T	T	T
F	T	F	T
F	F	T	T
F	F	F	T

(b)

p	q	r	$p \land (q \leftrightarrow (r \land p))$
T	T	T	T
T	T	F	F
T	F	T	F
T	F	F	T
F	T	T	F
F	T	F	F
F	F	T	F
F	F	F	F

7. (a) True **(b)** False   **8. (a)** True **(b)** False   **9. (a)** False **(b)** True   **10. (a)** 17 **(b)** 100 **(c)** 100 **(d)** 100

11. (a) 50 **(b)** -25 **(c)** -15 **(d)** 10   **12. (a)** Cannot be determined **(b)** TRUE **(c)** TRUE **(d)** Cannot be determined **(e)** Cannot be determined   **13. (a)** Cannot be determined **(b)** Cannot be determined **(c)** TRUE

14. (a) TRUE **(b)** FALSE **(c)** TRUE   **15. (a)** Cannot be determined **(b)** Cannot be determined **(c)** TRUE

16. $t =$ "Taxes go up."
$s =$ "I sell the house."
$m =$ "I move to India."

1. $t \to (s \land m)$	hyp.
2. $\sim m$	hyp.
3. $\sim s \lor \sim m$	addition (2)
4. $\sim (s \land m)$	DeMorgan (3)
5. $\sim t$	mod. tollens (1, 4)

17. m = "I study mathematics."
b = "I study business."
p = "I can write poetry."

1. $m \wedge b$	hyp.
2. $b \rightarrow (\sim p \vee \sim m)$	hyp.
3. b	subtraction (1)
4. $\sim p \vee \sim m$	mod. ponens (2, 3)
5. m	subtraction (1)
6. $\sim p$	disj. syllogism (4, 5)

18. d = "I shop for a dress."
h = "I wear high heels."
s = "I have a sore foot."

1. $d \rightarrow h$	hyp.
2. $s \rightarrow \sim h$	hyp.
3. d	hyp.
4. h	mod. ponens (1, 3)
5. $\sim s$	mod. tollens (2, 4)

19. a = "Asters grow in the garden."
d = "Dahlias grow in the garden."
s = "It is spring."

1. $a \vee d$	hyp.
2. $s \rightarrow \sim a$	hyp.
3. s	hyp.
4. $\sim a$	mod. ponens (2, 3)
5. d	disj. syllogism (1, 4)

20. t = "The professor gives a test."
h = "Nancy studies hard."
d = "Nancy has a date."
s = "Nancy takes a shower."
$H_1 = t \rightarrow h$
$H_2 = d \rightarrow s$
$H_3 = \sim t \rightarrow \sim s$
$H_4 = d$
$C = h$

1. $\sim h$	$\sim C$
2. $t \rightarrow h$	H_1
3. $\sim t$	mod. tollens (1, 2)
4. $\sim t \rightarrow \sim s$	H_3
5. $\sim s$	mod. ponens (3, 4)
6. $d \rightarrow s$	H_2
7. $\sim d$	$\sim H_4$; mod tollens (5, 6)

**21.

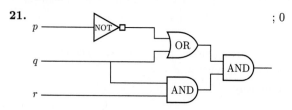

; 0

22. $p \wedge \sim q \wedge r$ **23.** $(p$ NOR $q)$ NOR $(p$ AND $q)$

**24.

**25.

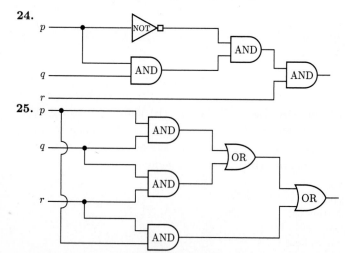

CHAPTER 12: CHAPTER TEST

1.

p	q	$(p \wedge \sim q) \to q$
T	T	T
T	F	F
F	T	T
F	F	T

2. (a) Statement   **(b)** Not a Statement   **(c)** Statement   **(d)** Statement   **(e)** Not a Statement

3. If the coach does not buy Bob ice cream, then Bob did not hit a triple or a homerun.   **4.** Every integer is either even or greater than 8.   **5.** False; True   **6. (a)** Neither   **(b)** Tautology   **(c)** Contradiction   **7. (a)** True   **(b)** False   **(c)** True   **(d)** False   **8.** The argument is valid.   **9. (a)** Every English dictionary contains the word "Internet." **(b)** There is a student at the university who does not listen to jazz.   **(c)** There is a floor at which the elevator does not stop.   **10. (a)** $p \to q$ **(b)** $q \to p$ **(c)** $p \to q$ **(d)** $p \to q$ **(e)** $q \to p$ **11. (a)** -19 **(b)** 0 **(c)** 180 **(d)** 0

12. (a) False   **(b)** True

13.

14. ; 0

Solutions to Circled Exercises

CHAPTER 1

EXERCISES 1.1

19. $14x + 7y = 21$
$7y = -14x + 21$
$y = -2x + 3$

31. $3x + 4y = 24$. To find y-intercept: set $x = 0$

$$3(0) + 4y = 24$$
$$4y = 24$$
$$y = 6$$

To find x-intercept: set $y = 0$

$$3x + 4(0) = 24$$
$$3x = 24$$
$$x = 8$$

Plot the points $(8,0)$ and $(0,6)$ and draw the line through these two points:

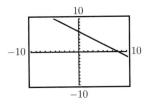

EXERCISES 1.3

31. Let x = number of $13''$ TVs and y = number of $19''$ TVs sold. Then

$$y = x + 5$$
$$90x + 120y = 4800$$

The second equation has standard form $y = -\frac{3}{4}x + 40$. Equate:

$$x + 5 = -\frac{3}{4}x + 40$$
$$\frac{7}{4}x = 35$$
$$x = 20$$

So, $y = 20 + 5 = 25$. Thus, 20 $13''$ TVs and 25 $19''$ TVs were sold, and the total number is 45.

EXERCISES 1.4

75. The vertical line shown in the figure is $x = 1$.

$$l_1 : y_1 = m_1 x$$
$$l_2 : y_2 = m_2 x$$

Let a and b be the legs of a right triangle formed by the vertical line $x = 1$ and the two lines graphed. Then

$$a^2 + b^2 = (m_1 - m_2)^2$$
$$a^2 + b^2 = m_1^2 - 2m_1 m_2 + m_2^2$$

because m_2 is negative. But we know that

$$a^2 = 1 + m_1^2$$
$$b^2 = 1 + m_2^2$$

Using substitution, we have $2 + m_1^2 + m_2^2 = m_1^2 - 2m_1 m_2 + m_2^2$. Thus, $2 = -2m_1 m_2$ or $m_2 = -\frac{1}{m_1}$. We have shown that if two lines are perpendicular then one slope is the negative reciprocal of the other. Reversing the steps in order, the converse of this statement can be shown. That is, if one slope is the negative reciprocal of the other, then the two lines are perpendicular.

79. Let $x =$ number of units. Then $100x =$ variable cost, $1{,}000{,}000 =$ fixed cost, $R(x) = 130x$, and Profit $= \$2{,}000{,}000$.

$$\text{Profit} = \text{Revenue} - \text{Costs}$$
$$2{,}000{,}000 = 130x - (100x + 1{,}000{,}000) = 30x - 1{,}000{,}000$$
$$3{,}000{,}000 = 30x$$
$$100{,}000 = x$$

EXERCISES 1.5

9. Ordered pairs are $(1, 9)$, $(2, 8)$, $(3, 6)$ and $(4, 3)$

$$\sum x = 10; \quad \sum y = 26$$
$$\sum xy = 9 + 16 + 18 + 12 = 55$$
$$\sum x^2 = 1 + 4 + 9 + 16 = 30$$
$$m = \frac{4 \cdot 55 - 10(26)}{4(30) - 10^2} = -2$$
$$b = \frac{26 - (-2)(10)}{4} = 11.5$$
$$y = -2x + 11.5$$

15. (a) $\sum x = 200; \quad \sum y = 393; \quad \sum x^2 = 12{,}000; \quad \sum xy = 16{,}340$

$$m = \frac{5(16{,}340) - 200(393)}{5(12{,}000) - 200^2} = .155$$
$$b = \frac{393 - .155(200)}{5} = 72.4$$
$$y = .155x + 72.4$$

(b) $y = .155(30) + 72.4 = 77.05$ **(c)** $y = .155(50) + 72.4 = 80.15$ **(d)** $y = .155(90) + 72.4 = 86.35$ (This is an example of a fit for which we cannot extrapolate beyond the given data.)

CHAPTER 2

EXERCISES 2.1

39. Let x = number of grams of cheddar cheese, and y = number of grams of baked potato.

$$\begin{cases} x + y = 180 \\ .25x + .02y = 10.5 \end{cases} \xrightarrow{[2] + (-.25)[1]} \begin{cases} x + y = 180 \\ -.23y = -34.5 \end{cases} \xrightarrow{-\frac{1}{.23}[2]} \begin{cases} x + y = 180 \\ y = 150 \end{cases} \xrightarrow{[1] + (-1)[2]} \begin{cases} x = 30 \\ y = 150 \end{cases}$$

Answer: (b)

41. Let x = cost of golf balls, and y = cost of golf glove. Then $x + y = 20$. Using **Statement I**: $y = 3x$. Then

$$\begin{cases} x + y = 20 \\ -3x + y = 0 \end{cases} \xrightarrow{[2] + 3[1]} \begin{cases} x + y = 20 \\ 4y = 60 \end{cases} \xrightarrow{\frac{1}{4}[2]} \begin{cases} x + y = 20 \\ y = 15 \end{cases} \xrightarrow{[1] + (-1)[2]} \begin{cases} x = 5 \\ y = 15 \end{cases}$$

Using **Statement II**: $y = 15$. Then

$$\begin{cases} x + y = 20 \\ y = 15 \end{cases} \xrightarrow{[1] + (-1)[2]} \begin{cases} x = 5 \\ y = 15 \end{cases}$$

The box of balls costs \$5. Either statement is sufficient, so the answer is (d).

45. Let x = number of adults and y = number of children. Then

$$\begin{cases} x + y = 350 \\ 8.5x + 5.5y = 2711 \end{cases}$$

In matrix form:

$$\begin{bmatrix} 1 & 1 & | & 350 \\ 8.5 & 5.5 & | & 2711 \end{bmatrix} \xrightarrow{[2] + (-8.5)[1]} \begin{bmatrix} 1 & 1 & | & 350 \\ 0 & -3 & | & -264 \end{bmatrix} \xrightarrow{-\frac{1}{3}[2]} \begin{bmatrix} 1 & 1 & | & 350 \\ 0 & 1 & | & 88 \end{bmatrix} \xrightarrow{[1] + (-1)[2]} \begin{bmatrix} 1 & 0 & | & 262 \\ 0 & 1 & | & 88 \end{bmatrix}$$

Therefore, $x = 262$ and $y = 88$. That is, 262 adults and 88 children.

EXERCISES 2.2

25. $\begin{cases} x + 7y - 3z = 8 \\ z = 5 \end{cases}$ $\begin{array}{l} x + 7y - 3(5) = 8 \\ x + 7y - 15 = 8 \\ x = 23 - 7y \end{array}$

Let $y = 2$: then $x = 23 - 14 = 9$; so $(9, 2, 5)$ is a solution.
Let $y = 0$; then $x = 23$; so $(23, 0, 5)$ is a solution.
Let $y = 1$; then $x = 23 - 7 = 16$; so $(16, 1, 5)$ is a solution.

27. Let x = food 1, y = food 2, and z = food 3.

$$\begin{cases} 2x + 4y + 6z = 1000 \\ 3x + 7y + 10z = 1600 \\ 5x + 9y + 14z = 2400 \end{cases}$$

$$\begin{bmatrix} ②& 4 & 6 & | & 1000 \\ 3 & 7 & 10 & | & 1600 \\ 5 & 9 & 14 & | & 2400 \end{bmatrix} \rightarrow \begin{bmatrix} 1 & 2 & 3 & | & 500 \\ 0 & ① & 1 & | & 100 \\ 0 & -1 & -1 & | & -100 \end{bmatrix} \rightarrow \begin{bmatrix} 1 & 0 & 1 & | & 300 \\ 0 & 1 & 1 & | & 100 \\ 0 & 0 & 0 & | & 0 \end{bmatrix} \qquad \begin{cases} x + z = 300 \\ y + z = 100 \end{cases}$$

z = any amount, $x = 300 - z$, $y = 100 - z$ ($0 \le z \le 100$)

33. $\begin{cases} 2x - 3y = 4 \\ -6x + 9y = k \end{cases}$ $\begin{bmatrix} ② & -3 & | & 4 \\ -6 & 9 & | & k \end{bmatrix} \rightarrow \begin{bmatrix} 1 & -1.5 & | & 2 \\ 0 & 0 & | & 12 + k \end{bmatrix}$

No solution: any value of k not equal to -12. Infinitely many solutions if $k = -12$.

EXERCISES 2.3

21. $A = 3 \times 2$, $B = 3 \times 2$. Since the number of columns of A (2) is not equal to the number of rows of B (3), matrix multiplication is not possible.

63. $([A] + [B]) - [A] = \begin{bmatrix} 3 & -2 & 1 \\ -5 & 6 & 7 \end{bmatrix}$

EXERCISES 2.4

25. (a) $x + 2y = a$; $.9x = b$. $\begin{bmatrix} 1 & 2 \\ .9 & 0 \end{bmatrix} \begin{bmatrix} x \\ y \end{bmatrix} = \begin{bmatrix} a \\ b \end{bmatrix}$ **(b)** $\begin{bmatrix} 1 & 2 \\ .9 & 0 \end{bmatrix} \begin{bmatrix} 450{,}000 \\ 360{,}000 \end{bmatrix} = \begin{bmatrix} 1{,}170{,}000 \\ 405{,}000 \end{bmatrix}$ after one year

$\begin{bmatrix} 1 & 2 \\ .9 & 0 \end{bmatrix} \begin{bmatrix} 1{,}170{,}000 \\ 405{,}000 \end{bmatrix} = \begin{bmatrix} 1{,}980{,}000 \\ 1{,}053{,}000 \end{bmatrix}$ after the second year.

(c) $\begin{bmatrix} x \\ y \end{bmatrix} = \begin{bmatrix} 1 & 2 \\ .9 & 0 \end{bmatrix}^{-1} \begin{bmatrix} 810{,}000 \\ 630{,}000 \end{bmatrix} = \begin{bmatrix} 0 & \frac{10}{9} \\ \frac{1}{2} & -\frac{5}{9} \end{bmatrix} \begin{bmatrix} 810{,}000 \\ 630{,}000 \end{bmatrix} = \begin{bmatrix} 700{,}000 \\ 55{,}000 \end{bmatrix}$ one year earlier.

EXERCISES 2.5

13. $\begin{bmatrix} 1 & 1 & 2 \\ 3 & 2 & 2 \\ 1 & 1 & 3 \end{bmatrix} \begin{bmatrix} x \\ y \\ z \end{bmatrix} = \begin{bmatrix} 3 \\ 4 \\ 5 \end{bmatrix}$. Use the Gauss–Jordan method to find the inverse of the 3×3 matrix.

$$\begin{bmatrix} \textcircled{1} & 1 & 2 & | & 1 & 0 & 0 \\ 3 & 2 & 2 & | & 0 & 1 & 0 \\ 1 & 1 & 3 & | & 0 & 0 & 1 \end{bmatrix} \longrightarrow \begin{bmatrix} 1 & 1 & 2 & | & 1 & 0 & 0 \\ 0 & \textcircled{-1} & -4 & | & -3 & 1 & 0 \\ 0 & 0 & 1 & | & -1 & 0 & 1 \end{bmatrix} \longrightarrow \begin{bmatrix} 1 & 0 & -2 & | & -2 & 1 & 0 \\ 0 & 1 & 4 & | & 3 & -1 & 0 \\ 0 & 0 & \textcircled{1} & | & -1 & 0 & 1 \end{bmatrix}$$

$$\longrightarrow \begin{bmatrix} 1 & 0 & 0 & | & -4 & 1 & 2 \\ 0 & 1 & 0 & | & 7 & -1 & -4 \\ 0 & 0 & 1 & | & -1 & 0 & 1 \end{bmatrix}$$

$$\begin{bmatrix} x \\ y \\ z \end{bmatrix} = \begin{bmatrix} -4 & 1 & 2 \\ 7 & -1 & -4 \\ -1 & 0 & 1 \end{bmatrix} \begin{bmatrix} 3 \\ 4 \\ 5 \end{bmatrix} = \begin{bmatrix} 2 \\ -3 \\ 2 \end{bmatrix}$$

Therefore, $x = 2$, $y = -3$, $z = 2$.

17. Since $A \cdot A^{-1} = A^{-1} \cdot A = I_3$, A is the inverse of A^{-1}. To find the inverse of A^{-1}, let $\Delta = 2 \cdot (-3) - 7 \cdot 1 = -13$. Then

$$A = \begin{bmatrix} \frac{-3}{-13} & \frac{-7}{-13} \\ \frac{-1}{-13} & \frac{2}{-13} \end{bmatrix} = \begin{bmatrix} \frac{3}{13} & \frac{7}{13} \\ \frac{1}{13} & -\frac{2}{13} \end{bmatrix}.$$

CHAPTER 3

EXERCISES 3.2

35. Since the farmer can spend \$2400 for labor at \$8 per hour, the available labor is $2400/8 = 300$ hours.

	Oats	Corn	Available
Capital	\$18	\$36	\$2100
Labor	2 hours	6 hours	300 hours
Land	1 acre	1 acre	100 acres
Revenue	\$55	\$125	

Let x be the number of acres of oats, and let y be the number of acres of corn.

$$\text{Capital}: 18x + 36y \leq 2100$$
$$\text{Labor}: 2x + 6y \leq 300$$
$$\text{Land}: x + y \leq 100$$

In standard form the inequalities are:

$$\begin{cases} y \leq -\dfrac{1}{2}x + \dfrac{175}{3} \\ y \leq -\dfrac{1}{3}x + 50 \\ y \leq -x + 100 \\ x \geq 0,\ y \geq 0 \end{cases}$$

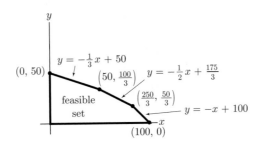

Find the vertices. Lower left corner: $(0,0)$; y-intercept of $y = -\frac{1}{3}x + 50$: $(0, 50)$

Intersection of $y = -\frac{1}{3}x + 50$ and $y = -\frac{1}{2}x + \frac{175}{3}$:

$$-\frac{1}{3}x + 50 = -\frac{1}{2}x + \frac{175}{3}$$
$$\frac{1}{6}x = \frac{25}{3}$$
$$x = 50$$
$$y = -\frac{1}{3}x + 50 = -\frac{1}{3}(50) + 50 = \frac{100}{3}$$

$\left(50, \frac{100}{3}\right)$

Intersection of $y = -\frac{1}{2}x + \frac{175}{3}$ and $y = -x + 100$:

$$-\frac{1}{2}x + \frac{175}{3} = -x + 100$$
$$\frac{1}{2}x = \frac{125}{3}$$
$$x = \frac{250}{3}$$
$$y = -x + 100 = -\frac{250}{3} + 100 = \frac{50}{3}$$

$\left(\frac{250}{3}, \frac{50}{3}\right)$

x-intercept of $y = -x + 100$: $(100, 0)$

Find the objective function for the profit.

Revenue: $55x + 125y$

Leftover capital: $2100 - 18x - 36y$

Leftover labor cash reserve: $2400 - 8(2x + 6y) = 2400 - 16x - 48y$

The profit is the sum of the above: $21x + 41y + 4500$

Vertex	Profit $= 21x + 41y + 4500$
$(0,0)$	$21(0) + 41(0) + 4500 = 4500$
$(0,50)$	$21(0) + 41(50) + 4500 = 6550$
$\left(50, \frac{100}{3}\right)$	$21(50) + 41\left(\frac{100}{3}\right) + 4500 \approx 6916.67$
$\left(\frac{250}{3}, \frac{50}{3}\right)$	$21\left(\frac{250}{3}\right) + 41\left(\frac{50}{3}\right) + 4500 \approx 6933.33$
$(100, 0)$	$21(100) + 41(0) + 4500 = 6600$

The maximum profit is achieved at $\left(\frac{250}{3}, \frac{50}{3}\right)$, or $\left(83\frac{1}{3}, 16\frac{2}{3}\right)$. The farmer should plant $83\frac{1}{3}$ acres of oats and $16\frac{2}{3}$ acres of corn to make a profit of \$6933.33

37. (a) Since the conditions are less restrictive than they were in Exercise 35, the optimal solution found in Exercise 35 will still be in the feasible set.

(b) The total cost to plant an acre of oats is $18 capital plus $16 for labor, or $34. The total cost to plant an acre of corn is $36 capital plus $48 for labor, or $84. Note that the objective function for the profit is the same as in Exercise 35.

	Oats	Corn	Available
Cost	$34	$84	$4500
Land	1 acre	1 acre	100 acres
Revenue	$55	$125	

Cost: $34x + 84y \leq 4500$

Land: $x + y \leq 100$

In standard form the inequalities are:

$$\begin{cases} y \leq -\dfrac{17}{42}x + \dfrac{375}{7} \\ y \leq -x + 100 \\ x \geq 0, \ y \geq 0 \end{cases}$$

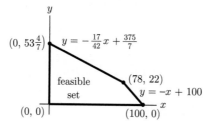

Find the vertices. Lower left corner: $(0,0)$

y-intercept of $y = -\frac{17}{42}x + \frac{375}{7}$: $\left(0, \frac{375}{7}\right)$ or $\left(0, 53\frac{4}{7}\right)$

Intersection of $y = -\frac{17}{42}x + \frac{375}{7}$ and $y = -x + 100$:

$$-\frac{17}{42}x + \frac{375}{7} = -x + 100$$
$$\frac{25}{42}x = \frac{325}{7}$$
$$x = 78$$
$$y = -x + 100 = -78 + 100 = 22 \quad (78, 22)$$

x-intercept of $y = -x + 100$: $(100, 0)$

Vertex	Profit $= 21x + 41y + 4500$
$(0,0)$	$21(0) + 41(0) + 4500 = 4500$
$\left(0, \frac{375}{7}\right)$	$21(0) + 41\left(\frac{375}{7}\right) + 4500 \approx 6696.43$
$(78, 22)$	$21(78) + 41(22) + 4500 = 7040$
$(100, 0)$	$21(100) + 41(0) + 4500 = 6600$

The maximum profit is achieved at $(78, 22)$. Plant 78 acres of oats and 22 acres of corn. Yes, the profit of $7040 is more than that achieved in Exercise 35.

EXERCISES 3.3

27. Let the variables represent the number of thousands of gallons.

	Gasoline	Jet fuel	Diesel fuel
Profit	$.15	$.12	$.10
Variables	x	y	$100 - x - y$

The required inequalities are:

$$\begin{cases} x \geq 5, \ y \geq 5 \\ 100 - x - y \geq 5 \\ x + y \geq 20 \\ x + (100 - x - y) \geq 50 \end{cases} \quad \text{or} \quad \begin{cases} x \geq 5, \ y \geq 5 \\ y \leq -x + 95 \\ y \geq -x + 20 \\ y \leq 50 \end{cases}$$

(Note that the inequalities above have ignored the somewhat subtle issue of whether there will be enough gasoline for *both* the airlines and the trucking firm at the same time. This is only a potential issue if both suppliers require gasoline—that is, if $y \leq 20$ and $100 - x - y \leq 50$. In this case, we require that the gasoline requirements of each firm be less than the amount of gasoline actually produced—that is, $(20 - y) + (50 - (100 - x - y)) \leq x$. This inequalitity is equivalent to $-30 \leq 0$, so it is always satisfied.)

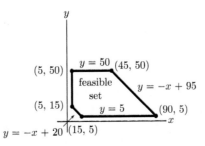

The objective function for the profit is $.15x + .12y + .1(100 - x - y)$, or $10 + .05x + .02y$.

Vertex	Profit $= 10 + .05x + .02y$
$(5, 50)$	11.25
$(45, 50)$	13.25
$(90, 5)$	14.60
$(15, 5)$	10.85
$(5, 15)$	10.55

The maximum profit of $14,600 is achieved at $(90, 5)$. Then $100 - x - y = 100 - 90 - 5 = 5$. Produce 90,000 gallons of gasoline, 5000 gallons of jet fuel, and 5000 gallons of diesel fuel.

CHAPTER 4

EXERCISES 4.2

13. Maximize $M = 6x + 7y + 300$. Rewrite the objective function as $-6x - 7y + M = 300$. The constraints are written as

$$2x + 3y + u = 400, \quad x + y + v = 150, \quad x \geq 0, \quad y \geq 0, \quad u \geq 0, \quad v \geq 0.$$

The initial tableau is

$$
\begin{array}{ccccc}
x & y & u & v & M \\
\end{array}
$$
$$
\left[
\begin{array}{ccccc|c}
2 & ③ & 1 & 0 & 0 & 400 \\
1 & 1 & 0 & 1 & 0 & 150 \\
\hline
-6 & -7 & 0 & 0 & 1 & 300 \\
\end{array}
\right]
$$

and the second column is the pivot column since -7 is most negative number in the last row. Compare ratios in rows 1 and 2 using the rightmost columns and the second column to find the smallest. Since $\frac{400}{3} < \frac{150}{1}$, the pivot element is in the first row of column 2. Pivot on that element. The resulting tableau is

$$
\begin{array}{ccccc}
x & y & u & v & M \\
\end{array}
$$
$$
\left[
\begin{array}{ccccc|c}
\frac{2}{3} & 1 & \frac{1}{3} & 0 & 0 & \frac{400}{3} \\
①\frac{1}{3} & 0 & -\frac{1}{3} & 1 & 0 & \frac{50}{3} \\
\hline
-\frac{4}{3} & 0 & \frac{7}{3} & 0 & 1 & \frac{3700}{3} \\
\end{array}
\right].
$$

Clearly the first column is the pivot column. The ratios are $\frac{400}{3} \cdot \frac{3}{2} = 200$ *and* $\frac{50}{3} \cdot \frac{3}{1} = 50$. The smaller is 50, so the pivot element is row 2 in column 1. The final tableau is given by

$$
\begin{array}{ccccc}
x & y & u & v & M \\
\end{array}
$$
$$
\left[
\begin{array}{ccccc|c}
0 & 1 & 1 & -2 & 0 & 100 \\
1 & 0 & -1 & 3 & 0 & 50 \\
\hline
0 & 0 & 1 & 4 & 1 & 1300 \\
\end{array}
\right].
$$

The solution is $x = 50$, $y = 100$, $u = 0$, $v = 0$, $M = 1300$.

19. Let $x =$ number of hours of bicycling, $y =$ number of hours of jogging, $z =$ number of hours of swimming. We are to maximize $M = 200x + 475y + 275z$ subject to:

$$\begin{cases} x + y + z \le 30 \\ x \le 4 \\ y \le x + z \\ x \ge 0, \quad y \ge 0, \quad z \ge 0. \end{cases}$$

The initial tableau is

$$\begin{array}{ccccccc}
x & y & z & u & v & w & M \\
\end{array}$$
$$\left[\begin{array}{ccccccc|c}
1 & 1 & 1 & 1 & 0 & 0 & 0 & 30 \\
0 & 0 & 1 & 0 & 1 & 0 & 0 & 4 \\
-1 & 1 & -1 & 0 & 0 & 1 & 0 & 0 \\
\hline
-200 & -475 & -275 & 0 & 0 & 0 & 1 & 0
\end{array}\right]$$

and the solution is 11 hours should be spent bicycling, 15 hours at jogging, and 4 hours of swimming for a total of 10,425 calories expended. He will lose 3 pounds ($3 = 10425/3500$).

23. Let a be the number of bags of mix A, b the number of bags of mix B, and c the number of bags of mix C. Maximize $3a + 5b + 6c$ subject to the constraints

$$\begin{cases} 12a + 10b + 8c \le 1200 \\ 5a + 6b + 8c \le 800 \\ 3a + 4b + 4c \le 600 \\ a \ge 0, \quad b \ge 0, \quad c \ge 0. \end{cases}$$

$$\begin{array}{c}
\begin{array}{ccccccc}
a & b & c & u & v & w & M
\end{array} \\
\begin{array}{c} u \\ v \\ w \\ M \end{array}
\left[\begin{array}{ccccccc|c}
12 & 10 & 8 & 1 & 0 & 0 & 0 & 1200 \\
5 & 6 & \boxed{8} & 0 & 1 & 0 & 0 & 800 \\
3 & 4 & 4 & 0 & 0 & 1 & 0 & 600 \\
\hline
-3 & -5 & -6 & 0 & 0 & 0 & 1 & 0
\end{array}\right]
\end{array}$$

$$\begin{array}{c}
\begin{array}{ccccccc}
a & b & c & u & v & w & M
\end{array} \\
\begin{array}{c} u \\ c \\ w \\ M \end{array}
\left[\begin{array}{ccccccc|c}
7 & \boxed{4} & 0 & 1 & -1 & 0 & 0 & 400 \\
\frac{5}{8} & \frac{3}{4} & 1 & 0 & \frac{1}{8} & 0 & 0 & 100 \\
\frac{1}{2} & 1 & 0 & 0 & -\frac{1}{2} & 1 & 0 & 200 \\
\hline
\frac{3}{4} & -\frac{1}{2} & 0 & 0 & \frac{3}{4} & 0 & 1 & 600
\end{array}\right]
\end{array}$$

$$\begin{array}{c}
\begin{array}{ccccccc}
a & b & c & u & v & w & M
\end{array} \\
\begin{array}{c} b \\ c \\ u \\ M \end{array}
\left[\begin{array}{ccccccc|c}
\frac{7}{4} & 1 & 0 & \frac{1}{4} & -\frac{1}{4} & 0 & 0 & 100 \\
-\frac{11}{16} & 0 & 1 & -\frac{3}{16} & \frac{5}{16} & 0 & 0 & 25 \\
-\frac{5}{4} & 0 & 0 & -\frac{1}{4} & -\frac{1}{4} & 1 & 0 & 100 \\
\hline
\frac{13}{8} & 0 & 0 & \frac{1}{8} & \frac{5}{8} & 0 & 1 & 650
\end{array}\right]
\end{array}$$

No bags of Mix A, 100 bags of Mix B, and 25 bags of Mix C.

EXERCISES 4.3

9. Let $a =$ number of servings of food A, $b =$ number of servings of food B. Minimize $3a + 1.5b$ subject to the constraints

$$\begin{cases} 30a + 10b \ge 60 \\ 10a + 10b \ge 40 \\ 20a + 60b \ge 120 \\ a \ge 0, \quad b \ge 0. \end{cases}$$

	a	b	u	v	w	M	
u	$\left(-30\right)$	-10	1	0	0	0	-60
v	-10	-10	0	1	0	0	-40
w	-20	-60	0	0	1	0	-120
M	3	$\frac{3}{2}$	0	0	0	1	0

	a	b	u	v	w	M	
a	1	$\frac{1}{3}$	$-\frac{1}{30}$	0	0	0	2
v	0	$-\frac{20}{3}$	$-\frac{1}{3}$	1	0	0	-20
w	0	$\left(-\frac{160}{3}\right)$	$-\frac{2}{3}$	0	1	0	-80
M	3	$\frac{1}{2}$	$\frac{1}{10}$	0	0	1	-6

	a	b	u	v	w	M	
a	1	0	$-\frac{3}{80}$	0	$\frac{1}{160}$	0	$\frac{3}{2}$
v	0	0	$-\frac{1}{4}$	1	$\left(-\frac{1}{8}\right)$	0	-10
b	0	1	$\frac{1}{80}$	0	$-\frac{3}{160}$	0	$\frac{3}{2}$
M	0	0	$\frac{3}{32}$	0	$\frac{3}{320}$	1	$-\frac{27}{4}$

	a	b	u	v	w	M	
a	1	0	$-\frac{1}{20}$	$\frac{1}{20}$	0	0	1
w	0	0	2	-8	1	0	80
b	0	1	$\frac{1}{20}$	$-\frac{3}{20}$	0	0	3
M	0	0	$\frac{3}{40}$	$\frac{3}{40}$	0	1	$-\frac{15}{2}$

1 serving of food A, 3 servings of food B.

EXERCISES 4.4

5. Refer to the solution of circled Exercise 23 in Section 4.2. If we change the amount of peat by h units, the last column of the initial tableau would be $\begin{bmatrix} 1200 + h \\ 800 \\ 600 \\ 0 \end{bmatrix}$. Consider the u column of the final tableau and the last column in that tableau. The adjusted final column of the tableau would be

$$\begin{bmatrix} 100 + \frac{1}{4}h \\ 25 - \frac{3}{16}h \\ 100 - \frac{1}{4}h \\ 650 + \frac{1}{8}h \end{bmatrix}.$$

All of these entries must be ≥ 0 in order to have a feasible solution. Hence we require

$$100 + \frac{1}{4}h \geq 0$$

$$25 - \frac{3}{16}h \geq 0$$

$$100 - \frac{1}{4}h \geq 0$$

$$650 + \frac{1}{8}h \geq 0.$$

The solution is $-400 \leq h \leq \frac{400}{3}$.

CHAPTER 5

EXERCISES 5.2

11. Let $U = \{$all letters of the alphabet$\}$, let $V = \{$letters with vertical symmetry$\}$, and $H = \{$letters with horizontal symmetry$\}$. Then $V \cup H = \{$letters with vertical or horizontal symmetry$\}$ and $V \cap H = \{$letters with both vertical and horizontal symmetry$\}$.

$$n(U) = 26,\ n(V) = 11,\ n(H) = 9,\ n(V \cap H) = 4$$
$$n(V \cup H) = n(V) + n(H) - n(V \cap H)$$
$$n(V \cup H) = 11 + 9 - 4 = 16$$
$$n((V \cup H)') = n(U) - n(V \cup H) = 26 - 16 = 10$$

There are 10 letters with no symmetry.

39. $S' \cup (S \cap T)' = S' \cup (S' \cup T') = (S' \cup S') \cup T' = S' \cup T'$

EXERCISES 5.3

19. Let $U = \{$students$\}$, $D = \{$students who passed the diagnostic test$\}$, $C = \{$students who passed the course$\}$. Draw the Venn diagram showing the following data.

$$n(U) = 30;\ n(D) = 21;\ n(C) = 23; n(D \cap C') = 2$$
$$n(D \cap C) = n(D) - n(D \cap C') = 21 - 2 = 19$$
$$n(D' \cap C) = n(C) - n(D \cap C) = 23 - 19 = 4$$

Four students passed the course even though they failed the diagnostic test.

21. Let $U = \{$lines$\}$, $V = \{$lines with verbs$\}$, $A = \{$lines with adjectives$\}$

$$n(U) = 14;\ n(V) = 11;\ n(A) = 9;$$
$$n(V \cap A) = 7$$
$$n(V \cap A') = n(V) - n(V \cap A) = 11 - 7 = 4$$

Four lines have a verb with no adjective.

$$n(V' \cap A) = n(A) - n(V \cap A) = 9 - 7 = 2$$

Two lines have an adjective but no verb.

$$n(V \cup A) = n(V) + n(A) - n(V \cap A) = 11 + 9 - 7 = 13$$
$$n((V \cup A)') = n(U) - n(V \cup A) = 14 - 13 = 1$$

One line has neither an adjective nor a verb.

EXERCISES 5.4

37. The number of handshakes between opposing players is n (select a player from the winning team) times n (select a player from the losing team) $= 10 \cdot 10 = 100$. At first glance, the number of handshakes among two members of the winning team is n (select first player)$\cdot n$ (select second player) $= 10 \cdot 9 = 90$. However, we have counted each handshake twice, since the handshake between player A and player B is the same as the handshake between player B and player A. Hence, the number of handshakes among teammates of the winning team is $90/2 = 45$. Therefore, the total number of handshakes is $100 + 45 = 145$.

57. There are a total of 10^9 different social security numbers that could be selected since there are 10 digits (0 through 9) and each social security number consists of 9 digits. So we have $10 \cdot 10 \cdot 10 \cdot 10 \cdot 10 \cdot 10 \cdot 10 \cdot 10 \cdot 10 = 10^9$ different social security numbers. If we eliminate the one selection of 000-00-0000 then we have $10^9 - 1$ different social security numbers or 999,999,999.

EXERCISES 5.5

57. Moe has

$$C(9,2) = \frac{P(9,2)}{2!} = \frac{9 \cdot 8}{2 \cdot 1} = 36 \text{ choices.}$$

Joe has

$$C(7,3) = \frac{P(7,3)}{3!} = \frac{7 \cdot 6 \cdot 5}{3 \cdot 2 \cdot 1} = 35 \text{ choices.}$$

Thus Joe is correct.

59. This is the same as ordering 12 different pictures. Then there are $P(12, 12) = 12! = 479{,}001{,}600$ arrangements.

EXERCISES 5.6

25. There are 10 letters in the word *ABSTEMIOUS*, so each arrangement is formed by filling 10 boxes with appropriate letters. A two-step procedure consists of first selecting the 5 boxes for the vowels (in alphabetical order) and then placing the five consonants in the five remaining boxes. The first step can be accomplished in $C(10, 5)$ ways and, after the first step is carried out, the second step can be accomplished in 5! ways. Since $C(10, 5) = 252$, and $5! = 120$, there are $252 \cdot 120$, or 30,240, arrangements with the vowels in alphabetical order.

EXERCISES 5.7

17. $\binom{6}{0} + \binom{6}{1} + \binom{6}{2} + \binom{6}{3} + \binom{6}{4} + \binom{6}{5} + \binom{6}{6} =$ number of subsets of a set of 6 elements $= 2^6 = 64$.

49. $(x - 3y)^7 = [x + (-3y)]^7$ The term with y^4 is

$$\binom{7}{4} x^3 (-3y)^4 = 35x^3 \cdot 81y^4 = 2835x^3 y^4.$$

The coefficient of y^4 is $2835x^3$.

51. The number of members of a Scrabble club with at least one person can be found by determining the number of subsets of size one or larger that can be obtained from a 15 element set. The total number of subsets of a set consisting of 15 elements is 2^{15}, but this number includes the empty set. Thus, the number of subsets of size one or larger is $2^{15} - 1 = 32{,}767$.

EXERCISES 5.8

23. Use the formula $\frac{1}{m!} \cdot \frac{n!}{(r!)^m}$. In this example, $n = 14$ (total number of children), $m = 7$ (number of groups), $r = 2$ (the number of children in each group). Substituting, we have

$$\frac{1}{7!} \cdot \frac{14!}{(2!)^7} = 135{,}135.$$

25. Use the formula $\frac{1}{m!} \cdot \frac{n!}{(r!)^m}$. In this example, $n = 10$ (total number of students), $m = 2$ (the number of groups; two teams), $r = 5$ (the number of students on each team). Substituting, we have $\frac{1}{2!} \cdot \frac{10!}{(5!)^2} = 126$.

CHAPTER 6

EXERCISES 6.2

5. (a) {(I, red), (I, white), (II, red), (II, white)}   **(b)** All combinations with I: {(I, red), (I, white)}

EXERCISES 6.3

7. (a) $1 - \left(\frac{1}{3} + \frac{1}{2} \right) = \frac{1}{6}$

(b) The probability is $\frac{1}{6}$.

$$\frac{\Pr(E)}{1 - \Pr(E)} = \frac{\frac{1}{6}}{1 - \frac{1}{6}} = \frac{\frac{1}{6}}{\frac{5}{6}} = \frac{1}{5}.$$

The odds are 1 to 5.

EXERCISES 6.4

23. After the first sock is picked, only one of the remaining 11 socks will be its mate. Therefore,

$$\Pr\left(\text{second sock matches first}\right) = \frac{1}{11}.$$

25. There are $5! = 120$ ways to arrange the family. In order to determine the number of arrangements with the parents standing together, treat the parents as a single unit with the man on the right of the woman. Then, there are $4! = 24$ ways to arrange the three children and the parental unit. There are another 24 ways with the man on the left of the woman, or 48 total ways with the parents standing together. Therefore, the probability of the parents standing together is $\frac{48}{120} = \frac{2}{5}$.

35. The total number of ways to select 5 senators is

$$C(100, 5) = \frac{100 \cdot 99 \cdot 98 \cdot 97 \cdot 96}{5 \cdot 4 \cdot 3 \cdot 2 \cdot 1} = 75{,}287{,}520.$$

The total number of ways to select senators from different states is

$$\frac{100 \cdot 98 \cdot 96 \cdot 94 \cdot 92}{5!} = 67{,}800{,}320.$$

$$\Pr\left(\text{no two members from same state}\right) = \frac{67{,}800{,}320}{75{,}287{,}520} \approx .90055.$$

EXERCISES 6.5

25. $\Pr\left(E \cap F\right) = \Pr\left(E\right) \cdot \Pr\left(F\right)$ Definition of "independent"
$\Pr\left(E\right) + \Pr\left(F\right) - \Pr\left(E \cup F\right) = \Pr\left(E\right) \cdot \Pr\left(F\right)$ Substitution
$[1 - \Pr(E')] + [1 - \Pr(F')] - [1 - \Pr(E' \cap F')] = [1 - \Pr\left(E'\right)] \cdot [1 - \Pr\left(F'\right)]$ Substitution
$\Pr\left(E' \cap F'\right) = \Pr\left(E'\right) \cdot \Pr\left(F'\right)$ Simplification

27. 0 points : $1 - .6 = .4$
1 point : $.6 \times .4 = .24$
2 points : $.6 \times .6 = .36$

39. The probability of a hit $= .3$

(a) Probability of at least one hit $= 1 - \Pr\left(\text{no hits}\right) = 1 - (.7)^4 = .76$

(b) $\Pr\left(\text{at least one hit in each of the first 10 games}\right) = (.76)^{10} = .064$

(c) $\Pr\left(\text{he does not get at least one hit in each of the first ten games}\right) = 1 - .064 = .936$. $\Pr$ (20 players with .300 averages do not get a hit in the first ten games) $= (.936)^{20} = .2664$. So, the $\Pr$ (at least one of them gets at least one hit in the first 10 games) $= 1 - \Pr$ (none of them get a hit in the first 10 games) $= 1 - .2664 = .7336$.

53. $\Pr\left(\text{couple has at least five children}\right) = \Pr\left(BBBBG\right) + \Pr\left(BBBBB\right) = \frac{1}{32} + \frac{1}{32} = \frac{1}{16}$.

EXERCISES 6.6

9. $\Pr\left(\text{white, then red}\right) + \Pr\left(\text{red, then red}\right) = \left(\frac{2}{3} \times \frac{1}{2}\right) + \left(\frac{1}{3} \times \frac{3}{4}\right) = \frac{7}{12}$

19.

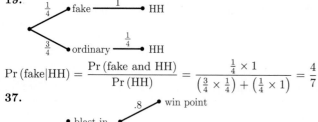

$\Pr\left(\text{fake}|\text{HH}\right) = \dfrac{\Pr\left(\text{fake and HH}\right)}{\Pr\left(\text{HH}\right)} = \dfrac{\frac{1}{4} \times 1}{\left(\frac{3}{4} \times \frac{1}{4}\right) + \left(\frac{1}{4} \times 1\right)} = \dfrac{4}{7}$

37.

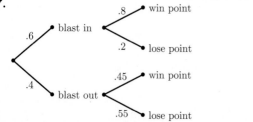

Pr (wins the point) = Pr (wins and gets first serve in) + Pr (wins and gets second serve in) = Pr (first serve) · Pr (wins given first serve) + Pr (second serve)· Pr (wins given second serve) = $(.6)(.8) + (.4)(.45) = .66$.

EXERCISES 6.7

7. $\text{Pr (passed exam | A)} = \dfrac{\text{Pr (passed exam)} \cdot \text{Pr (A|passed exam)}}{\text{Pr (passed exam)} \cdot \text{Pr (A|passed exam)} + \text{Pr (failed exam)} \cdot \text{Pr (A|failed exam)}}$

$$= \frac{.80 \times .40}{(.80 \times .40) + (.20 \times .20)} = \frac{.32}{.36} = \frac{8}{9} \approx .89$$

EXERCISES 6.8

3. (TI-83) `randInt(1,100,10)`
(TI-89) `seq(rand(100),x,1,10)`
where the numbers 1–83 represent a successful free-throw and 84–100 represent a miss.

CHAPTER 7

EXERCISES 7.1

15. The median of $\{2, 3, 5, 20\}$ is $\dfrac{3+5}{2} = 4$.
If $n < 4$, then the median of $\{n, 1, 3, 5, 15\}$ is the larger of 3 and n.
If $n > 4$, then the median of $\{n, 1, 3, 5, 15\}$ is the smaller of 5 and n.
If $n = 4$, then the median of $\{n, 1, 3, 5, 15\}$ is 4.
25. (a) min = 200, $Q_1 = 400$, $Q_2 = 600$, $Q_3 = 700$, max = 800.
(b) Since $400 = Q_1$, the first quartile, approximately 25% of the scores lie between min and Q_1.
(c) Since $400 = Q_1$ and $600 = Q_2$, approximately 25% of the scores lie between 400 and 600.
(d) Since $600 = Q_2$, the median, approximately one-half, (or 50%) of the scores are higher than 600.
(e) Since $200 = $ min and $700 = Q_3$ approximately three-quarters (or 75%) of the scores lie between 200 and 700.

EXERCISES 7.2

9.

No. Red Balls	Player's Earnings	Probability
2	$5	$\dfrac{\binom{2}{2}\binom{4}{0}}{\binom{6}{2}} = \dfrac{1}{15}$
1	$1	$\dfrac{\binom{2}{1}\binom{4}{1}}{\binom{6}{2}} = \dfrac{8}{15}$
0	−$1	$\dfrac{\binom{2}{0}\binom{4}{2}}{\binom{6}{2}} = \dfrac{6}{15}$

EXERCISES 7.3

19. Let "success" be "a guilty vote." Then $p = .80$, $q = .20$, $n = 12$.

$$\text{Pr}(X \geq 10) = \text{Pr}(X = 10) + \text{Pr}(X = 11) + \text{Pr}(X = 12)$$

$$= \binom{12}{10}(.8)^{10}(.2)^2 + \binom{12}{11}(.8)^{11}(.2)^1 + \binom{12}{12}(.8)^{12}(.2)^0$$

$$\approx .5583$$

21. The first nine tosses must contain exactly three heads, and the tenth toss must be a head. Therefore the probability is

$$\text{Pr (3 heads in nine tosses)} \cdot \text{Pr (10th toss is a head)} = \binom{9}{3}\left(\frac{1}{2}\right)^9 \cdot \left(\frac{1}{2}\right) = 84\left(\frac{1}{2}\right)^{10} \approx .0820.$$

27. Let A be the server and B be the other player. In order for A to win after 6 points, A must win the sixth point and 3 of the 5 earlier points. Therefore, the probability is

$$\text{Pr}\,(A \text{ wins 3 out of the first five points}) \cdot \text{Pr}\,(A \text{ wins the sixth point}) = \binom{5}{3}(.6)^3(.4)^2 \cdot (.6) = 10(.6)^4(.4)^2 = .20736.$$

EXERCISES 7.4

25. Let d = size of Dick's card collection. Let t = size of Tom's card collection.

$$\frac{\frac{3}{2}d + d + \frac{1}{2}d}{3} = 120$$
$$d = 120$$
$$t = \frac{3}{2}(120) = 180$$

Answer (d) is correct.

31. Let x = chance of rain. Expected gain from insurance = $40{,}000x$. We must solve

$$40{,}000x = 8000$$
$$x = .20 \quad \text{or} \quad 20\%$$

Answer (a) is correct. (Attendance revenue is irrelevant, since it is not affected by the decision whether or not to purchase insurance.)

EXERCISES 7.5

13. Probability $= 1 - \dfrac{6^2}{c^2} = \dfrac{7}{16}$

$$16c^2 - 576 = 7c^2$$
$$9c^2 = 576$$
$$c = \sqrt{64} = 8$$

EXERCISES 7.6

35. $\mu = 30{,}000$, $\sigma = 5000$

$\text{Pr}\,(Z \le z) = .02 \Rightarrow z = z_{02} \approx -2.05$

$$x_{02} = \mu + z_{02} \cdot \sigma$$
$$x_{02} = 30{,}000 + (-2.05) \cdot 5000 = 19{,}750$$

The warranty should expire after 19,750 miles.

EXERCISES 7.7

5. $n = 90$, $p = \frac{9}{19}$, $\mu = 90\left(\frac{9}{19}\right) = \frac{810}{19}$, $\sigma = \sqrt{90\left(\frac{9}{19}\right)\left(\frac{10}{19}\right)} = \frac{90}{19}$

Let X be the number of wins in 90 trials.

$$\text{Pr}\,(X > 45) \approx \text{Pr}\left(Z \ge \frac{45.5 - \frac{810}{19}}{\frac{90}{19}}\right) \approx \text{Pr}\,(Z \ge .61) \approx \text{Pr}\,(Z \ge .60) = 1 - .7257 = .2743$$

13. probability of failure = $(.01)(.02)(.01) = .000002$

$n = 1{,}000{,}000$, $E(X) = \mu = 1{,}000{,}000(.000002) = 2$

$\sigma = \sqrt{1{,}000{,}000(.000002)(.999998)} \approx 1.414$

$\text{Pr}\,(X > 3) \approx \text{Pr}\left(Z \ge \frac{3.5-2}{1.414}\right) \approx \text{Pr}\,(Z \ge 1.06) \approx \text{Pr}\,(Z \ge 1.05) = 1 - .8531 = .1469$

CHAPTER 8

EXERCISES 8.1

$$
\begin{array}{c}
 \quad U \quad\ S \quad\ R \\
\begin{array}{c} U \\ \textbf{15. (a)}\ S \\ R \end{array}
\begin{bmatrix} .86 & .05 & .03 \\ .08 & .86 & .05 \\ .06 & .09 & .92 \end{bmatrix}
\end{array}
$$

(b) Since we are concerned only with the people who live in urban areas in 2006, we use

$$
\begin{bmatrix} \ \ \\ \ \ \\ \ \ \end{bmatrix}_0 = \begin{bmatrix} 1 \\ 0 \\ 0 \end{bmatrix}_0 .
$$

$$
\begin{bmatrix} \ \ \\ \ \ \\ \ \ \end{bmatrix}_0 = A^2 \begin{bmatrix} \ \ \\ \ \ \\ \ \ \end{bmatrix}_0 = \begin{bmatrix} .86 & .05 & .03 \\ .08 & .86 & .05 \\ .06 & .09 & .92 \end{bmatrix}\begin{bmatrix} .86 & .05 & .03 \\ .08 & .86 & .05 \\ .06 & .09 & .92 \end{bmatrix}\begin{bmatrix} 1 \\ 0 \\ 0 \end{bmatrix}_0 = \begin{bmatrix} .7454 \\ .1406 \\ .114 \end{bmatrix}_2
$$

11.4% of people who live in urban areas in 2006 will live in rural areas in 2008.

CHAPTER 9

EXERCISES 9.1

3. $\begin{bmatrix} -2 & 4 & 1 \\ -1 & 3 & 5 \\ -3 & 5 & 2 \end{bmatrix}$

For R: we underline the minima for the rows. The maximum of these is in row 2.

$\begin{bmatrix} -2 & 4 & 1 \\ \underline{-1} & 3 & 5 \\ -3 & \underline{5} & 2 \end{bmatrix}$

For C: we underline the maxima for the columns. The minimum of these is in column 1.

$$
\begin{array}{c}
\quad F \qquad\ A \qquad\ N \\
\begin{array}{c} F \\ \textbf{11.}\ A \\ N \end{array}
\begin{bmatrix} 8000 & -1000 & 1000 \\ -7000 & 4000 & -2000 \\ 3000 & 3000 & 2000 \end{bmatrix}
\end{array}
$$

Row minima: $\begin{bmatrix} 8000 & -1000 & 1000 \\ -7000 & 4000 & -2000 \\ 3000 & 3000 & \underline{2000} \end{bmatrix}$

Column maxima: $\begin{bmatrix} \underline{8000} & -1000 & 1000 \\ -7000 & \underline{4000} & -2000 \\ 3000 & 3000 & \underline{2000} \end{bmatrix}$

Row 3, column 3 is a saddle point, so the game is strictly determined. Both candidates should be neutral.

EXERCISES 9.2

1. (a) $\begin{bmatrix} .5 & .5 \end{bmatrix}\begin{bmatrix} 3 & -1 \\ -7 & 5 \end{bmatrix}\begin{bmatrix} .5 \\ .5 \end{bmatrix} = \begin{bmatrix} 0 \end{bmatrix}$ **(b)** $\begin{bmatrix} 1 & 0 \end{bmatrix}\begin{bmatrix} 3 & -1 \\ -7 & 5 \end{bmatrix}\begin{bmatrix} .5 \\ .5 \end{bmatrix} = \begin{bmatrix} 1 \end{bmatrix}$

(c) $\begin{bmatrix} .3 & .7 \end{bmatrix}\begin{bmatrix} 3 & -1 \\ -7 & 5 \end{bmatrix}\begin{bmatrix} .6 \\ .4 \end{bmatrix} = \begin{bmatrix} -1.12 \end{bmatrix}$ **(d)** $\begin{bmatrix} .75 & .25 \end{bmatrix}\begin{bmatrix} 3 & -1 \\ -7 & 5 \end{bmatrix}\begin{bmatrix} .2 \\ .8 \end{bmatrix} = \begin{bmatrix} .5 \end{bmatrix}$

7. $\begin{bmatrix} 3 & -1 \\ -2 & 2 \end{bmatrix}$ R: underline minima in rows. The maximum is in row 1.

$\begin{bmatrix} 3 & \underline{-1} \\ -2 & \underline{2} \end{bmatrix}$ C: underline maxima in columns. The minimum is in column 2.

But there is no saddle point.

Robert's strategy is $\begin{bmatrix} .3 & .7 \end{bmatrix}$. Carol's strategy is $\begin{bmatrix} c \\ 1-c \end{bmatrix}$ and the game should be fair, so the value $= 0$. Hence

$$
\begin{bmatrix} .3 & .7 \end{bmatrix}\begin{bmatrix} 3 & -1 \\ -2 & 2 \end{bmatrix}\begin{bmatrix} c \\ 1-c \end{bmatrix} = \begin{bmatrix} 0 \end{bmatrix} .
$$

Then

$$[-.5 \quad 1.1]\begin{bmatrix} c \\ 1-c \end{bmatrix} = [0] \quad \text{and} \quad -.5c + 1.1(1-c) = 0.$$

From this, we find $c = 11/16$ and $1 - c = 5/16$ and Carol's strategy is $\begin{bmatrix} 11/16 \\ 5/16 \end{bmatrix}$.

EXERCISES 9.3

9. Add 2 to each entry to make all entries of the payoff matrix positive. We get

$$\begin{bmatrix} 5 & 7 & 1 \\ 6 & 1 & 8 \end{bmatrix}.$$

Set the tableaux up to find C's optimal strategy, then read the dual's solution off the bottom row.

	z_1	z_2	z_3	t	u	M	
t	5	⑦	1	1	0	0	1
u	6	1	8	0	1	0	1
M	-1	-1	-1	0	0	1	0

	z_1	z_2	z_3	t	u	M	
z_2	$\frac{5}{7}$	1	$\frac{1}{7}$	$\frac{1}{7}$	0	0	$\frac{1}{7}$
u	$\frac{37}{7}$	0	$\frac{55}{7}$	$-\frac{1}{7}$	1	0	$\frac{6}{7}$
M	$-\frac{2}{7}$	0	$-\frac{6}{7}$	$\frac{1}{7}$	0	1	$\frac{1}{7}$

	z_1	z_2	z_3	t	u	M	
z_2	$\frac{34}{55}$	1	0	$\frac{8}{55}$	$-\frac{1}{55}$	0	$\frac{7}{55}$
z_3	$\frac{37}{55}$	0	1	$-\frac{1}{55}$	$\frac{7}{55}$	0	$\frac{6}{55}$
M	$\frac{16}{55}$	0	0	$\frac{7}{55}$	$\frac{6}{55}$	1	$\frac{13}{55}$

$t = \frac{7}{55}$, $u = \frac{6}{55}$, $M = \frac{13}{55}$, $v = \frac{1}{M} = \frac{55}{13}$, and R's optimal strategy is $[vt \quad vu] = [\frac{7}{13} \quad \frac{6}{13}]$.

15. The payoff matrix, with entries in thousands of dollars is

$$\begin{array}{cc} & \begin{array}{cc} 1 & 2 \end{array} \\ \begin{array}{c} 1 \\ 2 \end{array} & \begin{bmatrix} -2 & 7 \\ 7 & -1 \end{bmatrix}. \end{array}$$

Add 3 to each entry to make all the entries positive. We get

$$\begin{bmatrix} 1 & 10 \\ 10 & 2 \end{bmatrix}.$$

Then maximize $M = z_1 + z_2$ subject to

$$\begin{cases} z_1 + 10z_2 \le 1 \\ 10z_1 + 2z_2 \le 1 \\ z_1 \ge 0,\ z_2 \ge 0 \end{cases}.$$

$$\begin{array}{c} \begin{array}{ccccc} z_1 & z_2 & t & u & M \end{array} \\ \begin{array}{c} t \\ u \\ M \end{array} \left[\begin{array}{ccccc|c} 1 & \boxed{10} & 1 & 0 & 0 & 1 \\ 10 & 2 & 0 & 1 & 0 & 1 \\ -1 & -1 & 0 & 0 & 1 & 0 \end{array} \right] \end{array}$$

$$\begin{array}{c} \begin{array}{ccccc} z_1 & z_2 & t & u & M \end{array} \\ \begin{array}{c} t \\ u \\ M \end{array} \left[\begin{array}{ccccc|c} \frac{1}{10} & 1 & \frac{1}{10} & 0 & 0 & \frac{1}{10} \\ \boxed{\frac{49}{5}} & 0 & -\frac{1}{5} & 1 & 0 & \frac{4}{5} \\ -\frac{9}{10} & 0 & \frac{1}{10} & 0 & 1 & \frac{1}{10} \end{array} \right] \end{array}$$

$$\begin{array}{c} \begin{array}{ccccc} z_1 & z_2 & t & u & M \end{array} \\ \begin{array}{c} t \\ u \\ M \end{array} \left[\begin{array}{ccccc|c} 0 & 1 & \frac{5}{49} & -\frac{1}{98} & 0 & \frac{9}{98} \\ 1 & 0 & -\frac{1}{49} & \frac{5}{49} & 0 & \frac{4}{49} \\ 0 & 0 & \frac{4}{49} & \frac{9}{98} & 1 & \frac{17}{98} \end{array} \right] \end{array}$$

$v = \frac{1}{M} = \frac{98}{17}$

(a) R's optimal strategy is given by the bottom entries under t and u: $\begin{bmatrix} vt & vu \end{bmatrix} = \begin{bmatrix} \frac{8}{17} & \frac{9}{17} \end{bmatrix}$.

(b) C's optimal strategy is $\begin{bmatrix} vz_1 \\ vz_2 \end{bmatrix} = \begin{bmatrix} \frac{8}{17} \\ \frac{9}{17} \end{bmatrix}$.

(c) $v - 3 = \frac{1}{M} - 3 = \frac{98}{17} - 3 \approx 2.765$ thousand dollars.

CHAPTER 10

EXERCISES 10.1

47. Since we start with $1000 and this amount doubles every six years, we have $2000 at the end of six years, $4000 at the end of twelve years and $8000 at the end of eighteen years. So, it will take 18 years for the investment to grow to $8000.

51. Since the amount A of the investment is arbitrary, let's use $A = 100$. For the first year, we have $100 + 0.025(100) = 102.50$. Then at the end of year two, the amount is $102.50(1.03) = 105.575$. At the end of year three, the amount is $105.575(1.084) = 114.44$. Since investment B increased at the same amount each year $(r\%)$, this is equivalent to compounding $100 at the interest rate of $r\%$ for three years. The equation is $114.44 = 100 \left(1 + \frac{r}{100}\right)^3$. This means that

$$r = 100 \left(\sqrt[3]{\frac{114.44}{100}} - 1 \right) \approx 4.6.$$

EXERCISES 10.2

37. (a) $R = \dfrac{1}{a\,\overline{_{60|}}\,1.5\%}(\$50{,}000) = .02539343 \cdot \$50{,}000 = \$1269.67$

(b) Treat as two annuities. The first has a present value of

$$P = a\,\overline{_{60|}}\,1\%(.015)(\$50{,}000) = 44.95503841(.015)(\$50{,}000) = \$33{,}716.28.$$

The second has a present value of

$$P = a\,\overline{_{60|}}\,1\%(\$1269.67) \div (1.01)^{60} = 44.95503841(1269.67)/1.81669670 = \$31{,}418.60.$$

The total present value = $65,134.88.

EXERCISES 10.3

27. Future value needs to be $(1.06)^{10}(\$13 \text{ million}) = \$23.281$ million.

$F = s \overline{_{120|}} \, _{1\%}(\$100,000) = 230.03868946 \cdot (\$100,000) = \$23.004$ million

The sinking fund will not be adequate. Another $277,000 will be needed to cover the shortfall.

29. (a) Monthly payment for a 0% loan for 36 months is $24,000/36 = \$666.67$.

(b) For $3500 cash-back offer, the amount of loan is $20,500. To amortize that loan over 36 months, use the formula:

$$R = \frac{1}{a \, \overline{_{36|}} \, _{1/2\%}}(20,500) = .03042194 \cdot (20,500) = 623.65.$$

(c) b is the better value.

EXERCISES 10.4

19. (a) $[\text{total repayment}] = \dfrac{[\text{loan amount}]}{1 - rt} = \dfrac{880}{1 - .06(2)} = 1000$

$[\text{monthly payment}] = \dfrac{[\text{total payment}]}{12t} = \dfrac{1000}{12(2)} = \$41.67$

(b) $R = \dfrac{P(1 + rt)}{12t} = \dfrac{880(1 + .06 \cdot 2)}{12(2)} = \$41.07.$ The monthly payment is less.

CHAPTER 11

EXERCISES 11.1

19. $y_n = y_{n-1} - .01y_{n-1} - 1{,}000{,}000; \; y_0 = 70{,}000{,}000$

$y_n = .99y_{n-1} - 1{,}000{,}000, \; y_0 = 70{,}000{,}000$

25. (a) $y_n = 1.04y_{n-1} + 250, \; y_0 = 800$

(b) $y_n = \dfrac{250}{1 - 1.04} + \left(800 - \dfrac{250}{1 - 1.04}\right)(1.04)^n$

$y_n = -6250 + 7050(1.04)^n$

(c) $y_7 = -6250 + 7050(1.04)^7 \approx 3027.32.$ About $3027.32.

EXERCISES 11.2

9. $y_0 = 1$

(a) $i = .40, \; n = 1, \; y_n = 1(1.40)^1 = \$1.40$

(b) $i = \dfrac{.40}{2} = .20, \; n = 2 \times 1 = 2, \; y_n = 1(1.20)^2 = \$1.44$

(c) $i = \dfrac{.40}{4} = .10, \; n = 4 \times 1 = 4, \; y_n = 1(1.10)^4 \approx \$1.46$

13. $a = .4, \, b = 3; \; y_n = \dfrac{3}{1 - .4} + \left(7 - \dfrac{3}{1 - .4}\right)(.4)^n = 5 + 2(.4)^n;$ as n gets large, $(.4)^n$ approaches 0, and y_n approaches 5.

15. $a = -5, \, b = 0; \; y_n = \dfrac{0}{1 + 5} + \left(2 - \dfrac{0}{1 + 5}\right)(-5)^n = 2(-5)^n;$ as n gets large, y_n gets arbitrarily large, alternating between being positive and negative.

EXERCISES 11.3

27. (a) $y_n = 1.06y_{n-1} - 120$

(b) $a = 1.06, \, b = -120.$ The deposit, y_0, must be at least

$$\frac{b}{1 - a} = \frac{-120}{1 - 1.06} = \frac{-120}{-.06} = \$2000.$$

EXERCISES 11.5

7. $y_n = y_{n-1} + .30(12 - y_{n-1}) = .7y_{n-1} + 3.6, \; y_0 = 0$

$a = .7, \, b = 3.6, \; \dfrac{b}{1 - a} = 12.$

$a > 0;$ monotonic; $|a| < 1$: attracted to $y = 12$.

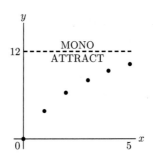

11. $y_n = y_{n-1} + .20(70 - y_{n-1}) = .8y_{n-1} + 14,\ y_0 = 40$

$a = .8,\ b = 14,\ \dfrac{b}{1-a} = 70$

$a > 0$: monotonic; $|a| < 1$: attracted to $y = 70$

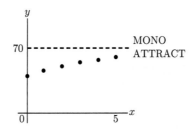

CHAPTER 12

EXERCISES 12.1

13. Not a statement because it is not a declarative sentence

25. (a) $p \vee q$ (b) $p \wedge \sim q$ (c) $q \wedge \sim p$ (d) $\sim p \wedge \sim q$

EXERCISES 12.2

3. $\sim p$ is a statement, $(\sim p \vee r)$ is a statement, $(q \wedge r)$ is a statement and therefore $(\sim p \vee r) \to (q \wedge r)$ is a statement.

19. Recall the truth table for $p \oplus q$ is

p	q	$p \oplus q$
T	T	F
T	F	T
F	T	T
F	F	F

p	q	r	$[p$	$\oplus$	$(q \vee r)]$
T	T	T	F	T	
T	T	F	F	T	
T	F	T	F	T	
T	F	F	T	F	
F	T	T	T	T	
F	T	F	T	T	
F	F	T	T	T	
F	F	F	F	F	
(1)	(2)	(3)	(5)	(4)	

39. p is FALSE, q is TRUE

(a) $p \wedge \sim q$ is FALSE because p is FALSE.

(b) $p \oplus q$ is TRUE in this case, hence $\sim(p \oplus q)$ is FALSE.

(c) $p \wedge q$ is FALSE because p is FALSE.

(d) $\sim p$ is TRUE, but $\sim q$ is FALSE. Hence $\sim p \wedge \sim q$ is FALSE.

EXERCISES 12.3

7.

p	q	$(p \to q)$	$\leftrightarrow$	$(\sim p$	$\vee$	$q)$
T	T	T	T	F	T	T
T	F	F	T	F	F	F
F	T	T	T	T	T	T
F	F	T	T	T	T	F
(1)	(2)	(3)	(6)	(4)		(5)

25. $(p \oplus q) \to p$ is TRUE. Because p is TRUE and q is FALSE, $p \oplus q$ is TRUE. Since $(p \oplus q)$ is TRUE and p is TRUE, the statement is TRUE.

35. The statement is "p is necessary for q." Hence $q \to p$.

39. The statement is $\sim p \to \sim q$. Since p is TRUE, $\sim p$ is FALSE. Any implication with a FALSE hypothesis ($\sim p$) implies any conclusion ($\sim q$) is TRUE.

47. (a) $Z = X + Y = 0$. Since $Z = 0$, $A = 4$. **(b)** $Z = X + Y = 0$. Since $Z = 0$, $A = 4$.

(c) $Z = X + Y = 0$. Since $Z = 0$, $A = 4$. **(d)** $Z = 5$ and $X < 0$. Since $Z \neq 0$ but $X < 0$, $A = 4$.

(e) $Z = 5$ and $X > 0$. Since $Z \neq 0$ and $X > 0$, $A = 6$. **(f)** $Z = -5$ and $X > 0$. Since $Z \neq 0$ and $X > 0$, $A = 6$.

EXERCISES 12.4

11. The statement $p \oplus q$ is true when exactly one of p or q is true. Therefore, $p \oplus q$ is false when both p and q are true or both p and q are false. Hence $p \oplus q$ is false when $p \wedge q$ is true or $\sim p \wedge \sim q$ is true. This means $p \oplus q$ is true when we negate $(p \wedge q) \vee (\sim p \wedge \sim q)$. Hence $p \oplus q \Longleftrightarrow \sim[(p \wedge q) \vee (\sim p \wedge \sim q)]$. We cannot use the connective $\wedge$, so we use DeMorgan's laws to rewrite this expression using only $\sim$ and $\vee$. $p \oplus q \Longleftrightarrow \sim[\sim(\sim p \vee \sim q) \vee \sim(p \vee q)]$.

15. Note $p \to q \Longleftrightarrow \sim p \vee q$. Hence $\sim(p \wedge \sim q) \to (p \vee \sim r) \Longleftrightarrow \sim(\sim(p \wedge \sim q)) \vee (p \vee \sim r) \Longleftrightarrow (p \wedge \sim q) \vee (p \vee \sim r)$.

EXERCISES 12.7

11. We simplify $[(p \wedge q) \wedge r] \vee \sim[(p \vee q) \vee \sim r]$. Note: $\sim[(p \vee q) \vee \sim r] \Longleftrightarrow [\sim(p \vee q) \wedge \sim(\sim r)] \Longleftrightarrow [\sim(p \vee q) \wedge r]$. Using Table 1 of Section 12.4, parts 2 and 4,

$[(p \wedge q) \wedge r] \vee [\sim(p \vee q) \wedge r] \Longleftrightarrow [r \wedge (p \wedge q)] \vee [r \wedge \sim(p \vee q)] \Longleftrightarrow r \wedge [(p \wedge q) \vee \sim(p \vee q)]$. The diagram follows.

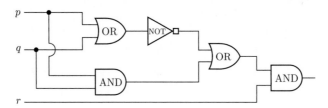

15.

p	q	$p \vee q$	$(p$ NOR $q)$	NOR	$(p$ NOR $q)$
T	T	T	F	T	F
T	F	T	F	T	F
F	T	T	F	T	F
F	F	F	T	F	T
(1)	(2)	(3)	(4)	(6)	(5)

Note that columns (3) and (6) are identical. Hence $p \vee q \Longleftrightarrow (p$ NOR $q)$ NOR $(p$ NOR $q)$. Draw the circuit corresponding to this expression.

Index